THEORETICAL AND COMPUTATIONAL DYNAMICS OF A COMPRESSIBLE FLOW

THEORETICAL AND COMPUTATIONAL DYNAMICS OF A COMPRESSIBLE FLOW

Shih-I Pai, Ph. D. & Dr. Tech. hc

Professor Emeritus
Institute for Physical Science and Technology
University of Maryland,
Honorary Professor
Northwestern Polytechnical University

Shijun Luo, Ph. D.

Professor of Aerodynamics
Northwestern Polytechnical University

CRC Press
Taylor & Francis Group
Boca Raton London New York

CRC Press is an imprint of the
Taylor & Francis Group, an **informa** business

First published 1991 by Van Nostrand Reinhold and Science press

Published 2019 by CRC Press
Taylor & Francis Group
6000 Broken Sound Parkway NW, Suite 300
Boca Raton, FL 33487-2742

First issued in paperback 2019

ISBN-13: 978-0-367-44831-8 (pbk)
ISBN-13: 978-0-442-30310-5 (hbk)

Library of Congress Catalog Card Number 90-12602

Library of Congress Cataloging-in-Publication Data
Pai, Shih-I .
Theoretical and computational dynamics of a compressible flow/Shih-I Pai, Shijun Luo.
p. 699 cm. 160 × 240
Includes bibliographical references and indexes.
ISBN 0-442-30310-6
1. Fluid dynamics. 2. Compressibility.
I. Lo. Shih-chǔn. II. Title.
QA913. P29 1991 532'. 0535 –dc 20 90-12602 CIP

To
Mrs. Alice Yen-Lan Wang Pai
and
Mrs. Guiying Sheng Luo

PREFACE

The purpose of this book is to give an introduction to the theoretical and computational fluid dynamics of a compressible fluid. Emphasis is laid on the basic assumptions and the formulation of the theory of compressible flow as well as on the methods of solving problems. This book is intended for the students of fluid dynamics who are interested in the essential results and the useful techniques in the theoretical analysis and numerical methods of compressible flow. The authors also hope that the book may serve as a useful reference to research workers in this field.

After the introduction, the thermodynamical and physical properties of gases are briefly reviewed. These serve as a foundation for the fluid dynamics of compressible fluid. In order to limit the size of this book, we consider mainly the flow of an ideal compressible fluid in which the effects of transport phenomena are neglected (Chapter I to XVI). However, the transport phenomena do have much influence on the flow of a compressible fluid. Hence in Chapter XVII, some basic concepts of transport phenomena are discussed, which prepare the students for further study of some important aspects of a compressible fluid flow.

It is almost thirty years since the publication of the book: "Introduction to the Theory of Compressible Flow" by the senior author of this book. Over thirty years, considerable advances in the theory of compressible flow have been made; and of course, some of these advances are included in this book. Among these last advances, two aspects should be specially mentioned: one is the inclusion of new physical phenomena in the theory of compressible flow and the other is the maturity of computational fluid dynamics. Let us briefly state some of the advances as follows:

(I) New physical phenomena in compressible flow

(i) Thermal radiation

It is well known that there is a very close relation between compressible flow and heat transfer. Among the three modes of heat transfer— heat convection, heat conduction and thermal radiation, the thermal radiation has been completely neglected in ordinary discussions of compressible fluid flow. Because the velocity and the temperature are not high enough, it can be shown that the thermal radiation effects are negligible in ordinary flow problems of a compressible flow. However, during flying at a very high speed in the space of very low density, the thermal radiation may become a very important factor in the flow problems. Hence we should consider some main effects of thermal radiation on the compressible flow.

(ii) Electromagnetofluid dynamics

At very high temperature, gas will be ionized. Ionized gas is also known as plasma which is the fourth state of matter and the most abundant state in the universe. In general, both the electric field and the magnetic field have great influence on the flow of a plasma. Hence in Chapter XV of this book, we shall discuss both the effects of electric field and magnetic field and title this chapter *electromagnetogasdynamics*.

(iii) Chemical reaction and phase changes

For high temperature flow, both the chemical reaction and the phase changes become important, particularly in the multi-phase flows. The authors write a chapter on *Multi-phase flows* (Chapter XVI) in this book in which some essential features of multi-phase flows will be discussed.

(II) Computational fluid dynamics

During the last thirty years, because of the advance of high speed computers, a new branch of fluid dynamics, the computational fluid dynamics, has been developed. For a modern fluid mechanics' book, the computational fluid dynamics has to be included, particularly some basic concepts and algorithms. The senoir author is very glad that Professor Shijun Luo, a leading authority of computational gasdynamics, joins him as co-author for this book. Most of the text on computational gasdynamics is contributed by Professor Luo, though they have discussed every aspect of this book together.

The authors wish to express their sincere appreciation to the authorities of Northwestern Polytechnical University for providing a nice surrounding so that the authors could work together for the final draft of this manuscript.

The junior author would like to thank the Institute of Physical Science and Technology, University of Maryland for the support of starting the work under Exchange Visitor Program P-I-0793, the National Aeronautical and Space Administration for the support of finishing the work under Interchange NCA 2-201 and Dr. T. L. Holst of the NASA-Ames Research Center for reading the computational part and making suggestions.

In conclusion, the authors would express their sincere gratitude to their wives: Mrs. Alice Yen-Lan Wang Pai and Mrs. Guiying Sheng Luo for their constant encouragement and patience during the preparation of the manuscript. The authors wish to offer their appreciation to Messrs. Guo-Bao Guo, Ping Liu, Jin-Sheng Cai and Chao Gao of the Northwestern Polytechnical University for the careful preparation of the final typescript.

Shih-I Pai
and
Shijun Luo

Xi'an, China
December, 1987

CONTENTS

Chapter I

INTRODUCTION

1. Microscopic and macroscopic points of view

Even though a fluid is composed of a large number of particles, ordinarily the fluid is so dense that we may consider it as continuous media in the treatment of fluid mechanics. However, if the fluid is very rarefied, one would expect that the coarse structure of molecules would affect the flow phenomena, and in the extremely rarefied gas, the gas should behave like individual particles. There are two different approaches in the theoretical investigations of the flow of a fluid: the microscopic point of view and the macroscopic point of view. The microscopic treatment is the more accurate way of analysis in which the motion of individual particles and their interactions are analyzed. Such an analysis is known as the kinetic theory of fluids. Because of many physical and mathematical difficulties, it is not possible at present to treat the flow problem of fluids exactly by molecular theory. Many simplified assumptions about molecular forces and collision phenomena have to be made in the formation of the theory[1, 2, 3] and the resultant equations can only be solved approximately. The most successful molecular theory for fluids is the kinetic theory of gases. Recently considerable efforts have been made to develop the kinetic theory of plasma[2]. The basic equation of the kinetic theory of gases is known as the Boltzmann equation which is a nonlinear partial differentio-integral equation. At the present time it is not possible to solve the Boltzmann equation even for rather simple practical flow problems. We would not expect to use the molecular theory of fluids to analyze flow problems in the near future. But, the Boltzmann equation serves two important aspects in the study of gasdynamics. In the first place, the fundamental equations of gasdynamics may be derived from the Boltzmann equation as the first approximation. Thus we may have some guides about the validity of the fundamental equations for a macroscopic description of the fluid flow from the analysis of Boltzmann equation. In the second place, the Boltzmann equation may give valuable information on the transport coefficients such as the coefficient of viscosity and heat conductivity. In the macroscopic analysis, these transport coefficients are simply introduced as known functions of physical quantities of gasdynamics such as temperature and pressure. We shall discuss these points in detail in Chapter XVII.

For many practical problems, the description of the motion of the fluid by molecular theory is too detailed to be useful. We do not care about the motion of individual particles in the fluid and are interested in the resultant effects due to the motion of a large number of particles. In other words, we are interested in the macroscopic quantities only, such as pressure, density, temperature, mean flow velocities and electric current density. It is possible to postulate the fundamental equations for the mechanics of fluids based on the conservation laws of mass, momentum, energy and electric charge in terms of these macroscopic quantities only without considering the fluids as composed of individual particles. These fundamental equations should, of course, be consistent with those derived from the microscopic description. In this book, we treat the flow problem from the macroscopic point of view only, yet a few remarks of microscopic point of view may be added.

The fundamental equations of fluids from the macroscopic point of view are known as the Navier-Stokes equations which are useful in the analysis of ordinarily dense fluid. For rarefied gas, these equations should be modified and sometimes, they should not be used at all. Hence in the study of fluid mechanics, we should have a criterion to determine the degree of rarefication of the fluid. For liquid, the Navier-Stokes equations can always be used to analyze the flow problems. For a gas or a plasma, the degree of rarefication may be expressed by the non-dimensional parameter known as the Knudsen number K_f which is defined as:

$$K_f = \frac{L_f}{L} \tag{1.1}$$

where L_f is the mean free path of the gas or plasma and L is the characteristic length of the flow field.

From the kinetic theory of gases[3], the mean free path of a natural gas is defined approximately by

$$\nu = \frac{1}{2} L_f \bar{c}_a \tag{1.2}$$

where ν is the coefficient of kinematic viscosity and $\bar{c}_a$ is the mean molecular velocity which is related to the sound speed "a" of the gas by

$$a = \bar{c}_a (\gamma \pi / 8)^{1/2} \tag{1.3}$$

where γ is the ratio of specific heat at constant pressure c_p to that at constant volume c_v and sometimes known as isentropic exponent. From Eqs. (1.1) to (1.3), we have

$$K_f = \frac{L_f}{L} = 1.255\sqrt{\gamma}\, M/Re \tag{1.4}$$

where $M=U/a$ is the Mach number of the flow field, U is the typical velocity of the flow field and $Re=UL/\nu$ is the Reynolds number of the flow field. When the Knudsen number is negligibly small, the fluid may be considered as a continuum. If the Knudsen number is not negligibly small, the flow field will be different from that of a continuum as we shall discuss in §2. The choice of the characteristic length L depends on the problem considered. Hence we may choose the typical dimension of the body as L as we study the forces on this body in a gas flow. We may use the boundary layer thickness on the body as L when we are interested in the skin friction and heat transfer through the boundary layer. When we investigate the transition region in a shock wave, the thickness of the shock may be used as L. Because of the various choices of the characteristic length L, whether a gas flow should be considered as rarefied or not depends on the particular problems considered. Once the value of L is chosen, the Knudsen number tells us the degree of rarefication. From Eq. (1.4), we see that when the Mach number is large and the Reynolds number is small, the Knudsen number becomes important and the effect of rarefication will be important. Since the modern trend for the flow problems is towards high speed, i.e., high Mach number and high altitude, i.e., low Re, the rarefied gasdynamics becomes important, particularly in space flight. But we shall not discuss these flow problems in this book. Special treatise should be referred to for rarefied gasdynamics[3].

2. General discussions of compressible fluid-flow regimes

There are four states of matter: solid, liquid, gas and plasma (ionized gas). Except for the solid state, matter may be deformed without applying any force, provided the change of shape that takes place over a sufficiently long time. The term *fluid* has been used as a general name for the three states of matter: liquid, gas and plasma. There are many similar properties of all fluids which may be treated by the method known as Fluid Mechanics.

The molecules of a liquid in the three states are quite close together. One of the properties of the liquid is its great resistance to the change of volume. Ordinarily, the liquid is considered as an incompressible fluid. However, at very high temperature range, the liquid may be gradually changed into vapor or gas and its density changes. Under such a condition, the liquid should be considered as a compressible fluid. We shall discuss such cases in Chapter XVI.

On the other hand, the volume of a gas or a plasma can easily be changed. Ordinarily, the flow of a gas or a plasma is considered as the main topics of compressible flow, particularly when the velocity and/or the temperature are high. In a major portion of this book, we shall study mainly the flow of a gas or a plasma as the compressible fluid. In the study of the flow of a compressible fluid we have to study thermodynamics simultaneously with the mechanics of a continuous medium. While by Professor Theodore von Kármán's suggestion the name *aerothermodynamics*[4] may be used for the fluid dynamics of a compressible fluid, some authors use the term gasdynamics, because the subject deals essentially with gas. We prefer the term of *dynamics of a compressible fluid* because in general we may discuss all the three states of a fluid.

In this book, we consider mainly the macroscopic approach of the fluid mechanics. It is advisable to divide the flow field into various regimes according to the value of Knudsen number. Since in ordinary treatment of fluid mechanics, the Mach number and the Reynolds number are more familiar than the Knudsen number, we classify the flow regimes according to their values of Mach number M and Reynolds number Re as follows[5]:

Regimes	Range of M and Re
Free molecule flow	$M/Re \geqslant 3$
Transition region	$M/Re \leqslant 3$ & $M/(Re)^{1/2} \geqslant 0.1$
Slip flow	$0.1 \geqslant M/Re^{1/2} \geqslant 0.01$
Continuum flow	$M/(Re)^{1/2} \leqslant 0.01$

In the slip flow regime, the fluid still behaves as a continuous medium but there is a slip of velocity on the solid boundary and a jump in temperature on the solid boundary due to the rarefied effects. In the transition regime, the flow greatly depends on the discrete character of the fluid. Finally when the Knudsen number is much larger than unity, the collision between molecules is negligible and we have the free molecule flow. In this book, we consider only the continuum flow regimes with no slip boundary conditions. Special treatises should be referred to for the study of the other three regimes[1–3].

For the continuum flow regime, the macroscopic properties such as pressure, temperature and flow velocity are the mean values of a large number of molecules. In most of the flow problems of practical importance, such a condition can be fulfilled. For instance, air at normal temperature and pressure contains 2.7×10^{19} molecules per cubic centimeter. If we consider a "point" in space for an ordinary aeronautical problem as a cube with sides of 1/1000 millimeter, this volume will contain 2.7×10^{7} molecules, which is a number generally quite adequate for taking a mean value. Hence the assumption of a continuum is very good for most gasdynamics problem.

Even though we discuss only the macroscopic properties of gases in this book, our results should be consistent with the more refined theory of gas motion such as the kinetic theory of gases.

In the main part of this book, we shall consider the compressible fluid as a perfect gas with constant values of specific heats. However, at very high temperature, the heat capacities of perfect gas are greatly influenced by molecular structure. Included here are the variations of heat capacities or specific heat with temperature, dissociation, ionization and thermal radiation. We shall point out how we can take these effects into account in our analysis.

In a majority of flow problems, the gas may be considered as an inviscid and non-heat-conducting fluid, i. e., fluid without the effect of transport phenomena. These include the calculation of the pressure distributions in the flow field not too close to a solid wall or other transition regions. We shall discuss the flow of an inviscid and non-heat-conducting fluid in Chapters III to XIV. The effects of transport phenomena will be discussed in Chapter XVII.

Another important group of problems involves the flow of a gas in which chemical reaction occurs and the flow of a fluid in which phase changes may occur. We shall discuss these problems in Chapter XVI and Chapter XVII.

At very high temperature, above 10,000 K, the gas may be ionized. The interaction of electromagnetic force and the fluid dynamic force should be considered. We call it *electromagnetofluid dynamics*, which will be discussed in Chapter XV.

A new chapter of multi-phase flow (Chapter XVI) is introduced in this second edition because many new practical applications of multiphase flows, particularly the two-phase flow, have been developed recently.

3. Fundamental equations of a continuous compressible fluid

In the study of fluid flow, we wish to find the velocity distributions as well as the states of the fluid over all the space for all times. If the fluid is a single gas, ordinarily a knowledge of the three velocity components (u, v, w), the temperature T, the pressure p, and the density ρ of the fluid, which are functions of spatial coordinates and time, is required. We must find six relations connecting these six unknowns. These are:

(a) Equation of state which connects the temperature, the pressure, and the density of the fluid. The existence of such a relation is an empirical fact and may be proved in the statistical mechanics.

(b) Equation of continuity which expresses the conservation of mass in the fluid.

(c) Equations of motion which are generally three in number and express the relations of conservation of momentum in the fluid.

(d) Equation of energy which expresses the conservation of energy in the fluid.

We shall derive and discuss these fundamental equations in the following chapters. In Chapter V, the fundamental equations for the flow of a compressible, inviscid and non-heat-conducting fluid will be derived; and in Chapter XVII, those for a viscous, heat-conducting compressible fluid will be presented.

In the fluid of a mixture of gases, it is necessary to include the conservation relations for each of the constituents of the mixture. We shall discuss those in Chapter XVII. Similarly, for multi-phase flow, the conservation relations for each phase should be considered. We shall discuss these relations in Chapter XVI. Between the conservation relations of each species in the mixture in multi-phase flow, new additional interaction factors should be considered.

In electromagnetofluid dynamics (Chapter XV), the gasdynamic equations and the electromagnetic equations are combined. We have special coupling terms between the electromagnetic and fluid dynamic phenomena.

Unless otherwise specified (in part of Chapters XV to XVII) the treatment in this book will be restricted to the problems of ordinary gasdynamics and only those equations listed in (a) to (d) will be discussed.

4. Boundary and initial conditions

For every particular problem of fluid dynamics, we have given certain initial and boundary conditions. Our problem is to find solutions of the fundamental equations which also satisfy these initial and boundary conditions.

By initial conditions we mean the velocity distributions and the states of the fluid at certain initial time $t=0$. Customarily in fluid dynamics problems we do not give the spatial distribution of these initial values, but we only require that the initial conditions be consistent with the boundary conditions for $t=0$ and the fundamental equations.

In the case of a dynamical system with a finite number of degrees of freedom, the motion is determined by initial position and velocity. For a continuous medium, which has an infinite number of degrees of freedom, the motion is determined not only by the initial conditions but also by boundary conditions, such as conditions on the velocity on the boundary of the domain considered at all the time $t \geqslant 0$. For ordinary fluid dynamics problems, the boundary conditions are usually assumed as follows:

Across a surface separating a body and a fluid or two fluids, the velocity components, the stresses, and the temperature are all continuous.

These boundary conditions are based on the experimental fact that no slip occurs at the surface of a solid body for ordinary gases or fluids. For ordinary gases, the relative velocity at the surface of solid wall is zero. Such a fluid is usually referred to as a Newtonian fluid. It is this type of fluid in which we are interested. In the case of very rarefied gases, the mean relative velocity of the gas at the wall is not zero[3-6]. This is the so-called slip flow in which the boundary conditions should be changed accordingly. We shall not discuss slip flow in this book.

The boundary conditions of no-slip can be satisfied by the fundamental equations of a viscous fluid. As we shall show in Chapter XVII, the viscous effect for the flow at large Reynolds number, which is the case for most engineering problems, is confined to a narrow region near the boundary known as the boundary layer[7]. Outside the boundary layer region, the flow behaves like an inviscid fluid flow. Furthermore the boundary layer is so thin that the pressure across the boundary is constant. Thus if we are interested in the flow outside the boundary layer or the pressure on the solid body, the solution for an inviscid fluid gives fairly good results. Since the equations of an inviscid fluid are much simpler than those of a viscous fluid, it is easier to solve the inviscid equations and thus obtain much important knowledge of the fluid flow. For inviscid fluid, the tangential stresses are zero, hence the no-slip condition cannot be satisfied. In the inviscid fluid, we allow the existence of surface of discontinuity and relax some of the boundary conditions for the real fluid. For instance, in the inviscid fluid on the solid surface, we impose only the condition of zero normal velocity to the surface and allow a discontinuity in tangential velocity component. Such a surface of discontinuity may be regarded as a first approximation for the boundary layer over the surface. There are two types of surface of discontinuity in the inviscid fluid: one, known as the vortex sheet, is a surface of discontinuity in the tangential velocity; the other, a surface discontinuity in the normal velocity component, is known as a shock wave. Shock waves can exist only in adiabatic supersonic flows, but vortex sheets may exist in both subsonic and supersonic flows. We shall discuss their properties in detail in the following chapters.

For electromagnetofluid dynamics, additional boundary conditions for electromagnetic variables are needed as we shall discuss them in Chapter XV. Similarly, for multi-phase flow and for the flow of a mixture of gases, additional boundary conditions for each species are needed. We shall discuss them in Chapters XVI and XVII.

5. Kinematics of fluid flow

Even though the fluid is considered as a continuum medium, it is still convenient to use the concept of fluid particle or fluid element in fluid mechanics. The fluid particle or fluid element is a group of molecules of a fluid in a small volume which may be considered as a point from a practical point of view.

Such a group of molecules may be observed in the actual flow of fluid. For instance, when we put a small amount of dye in a flow of liquid, we may see clearly the motion of colored liquid spots which may be regarded as fluid elements. Various techniques of visualization of flow pattern verify the existence of fluid particles or fluid elements. One of the objects of fluid mechanics is to predict the motion of fluid elements under various conditions. The first question is: how do we describe the motion of such fluid elements? There are two methods of describing the fluid motion: the Lagrangian method and the Eulerian method, although both methods are in reality due to Euler[8].

(1) Lagrangian method. In this method, we are concerned with the history of the individual fluid elements. We want to know their velocities, accelerations, and other properties at various places and different times. If at any given time $t=t_0$, a fluid element has Cartesian coordinates (x_0, y_0, z_0), then at time $t=t$, it will move to the position (x, y, z). It is evident that the coordinates (x, y, z) of the fluid element are functions of the initial coordinates (x_0, y_0, z_0) and the time t. We may write

$$\left.\begin{aligned} x&=F_1(x_0, y_0, z_0, t) \\ y&=F_2(x_0, y_0, z_0, t) \\ z&=F_3(x_0, y_0, z_0, t) \end{aligned}\right\} \tag{1.5}$$

Without loss of generality, the three initial coordinates (x_0, y_0, z_0) representing the name of the fluid element can also be considered as Cartesian coordinates. Eqs. (1.5) give the position of all fluid particles at different times. The first partial derivatives of Eqs. (1.5) with respect to time give the velocity components of the fluid elements and the second partial derivatives with respect to time of Eqs. (1.5) give the corresponding accelerations.

(2) Eulerian method. In this method, we are concerned with what is happening at a given time t at a point (x, y, z) in the space filled with the fluid. The velocity components of the fluid at any time and at any point in space may be written as

$$\left.\begin{aligned} u&=f_1(x, y, z, t) \\ v&=f_2(x, y, z, t) \\ w&=f_3(x, y, z, t) \end{aligned}\right\} \tag{1.6}$$

Here the coordinates (x, y, z) represent a fixed point in space and not the location of a given fluid element.

The acceleration of the fluid element used in the Eulerian method may be derived in the following manner:

Consider a fluid element at the location (x, y, z) and at time t having an x-component of velocity

$$u=f(x, y, z, t) \tag{1.7}$$

At time $t+\delta t$, i.e., a short interval δt later, this element will have moved to the position $(x+u\delta t, y+v\delta t, z+w\delta t)$ and will have an x-component $u+\delta u$.

Hence we have

$$u+\delta u=f\,(x+u\delta t,\ y+v\delta t,\ z+w\delta t,\ t+\delta t\,)$$

$$=f\,(x,y,z,t)+\left(u\frac{\partial f}{\partial x}+v\frac{\partial f}{\partial y}+w\frac{\partial f}{\partial z}+\frac{\partial f}{\partial t}\right)\delta t$$

$$+\text{higher order terms in }\delta t \tag{1.8}$$

The x-component of acceleration of the fluid at point (x, y, z) is the limit:

$$\lim_{\delta t\to 0}\frac{\delta u}{\delta t}=\frac{Du}{Dt}=\left(\frac{\partial}{\partial t}+u\frac{\partial}{\partial x}+v\frac{\partial}{\partial y}+w\frac{\partial}{\partial z}\right)u \tag{1.9}$$

The symbol

$$\frac{D}{Dt}=\frac{\partial}{\partial t}+u\frac{\partial}{\partial x}+v\frac{\partial}{\partial y}+w\frac{\partial}{\partial z}=\frac{\partial}{\partial t}+(\boldsymbol{q}\cdot\nabla) \tag{1.10}$$

is known as the total differential with respect to time or the material differential with respect to time. It gives the rate of change of a physical quantity following the path of the fluid element. In the Eulerian method, we should use this differential with respect to time when we describe the time rate of change of any quantity of the fluid. In Eq. (1.10), q is the velocity vector of the fluid which has the components u, v, and w, i.e.,

$$\boldsymbol{q}=\boldsymbol{i}u+\boldsymbol{j}v+\boldsymbol{k}w \tag{1.11}$$

and the operator del ∇ is

$$\nabla=\boldsymbol{i}\frac{\partial}{\partial x}+\boldsymbol{j}\frac{\partial}{\partial y}+\boldsymbol{k}\frac{\partial}{\partial z} \tag{1.12}$$

where $\boldsymbol{i}, \boldsymbol{j}$, and $\boldsymbol{k}$ are respectively the unit vectors in the x-, y-, and z-direction.

There is a definite relation between the Lagrangian and the Eulerian methods. For each fluid element, we have

$$\left.\begin{aligned}\frac{dx}{dt}&=u=f_1(x,y,z,t)\\ \frac{dy}{dt}&=v=f_2(x,y,z,t)\\ \frac{dz}{dt}&=w=f_3(x,y,z,t)\end{aligned}\right\} \tag{1.13}$$

If we solve Eqs. (1.13) with the initial conditions $t=t_0$, $x=x_0$, $y=y_0$ and $z=z_0$, we will have Eqs. (1.5). Hence in principle, the Lagrangian method of description can always be derived from the Eulerian method of description by the help of Eqs. (1.13). In most engineering problems, we are interested in the pressure, velocity, and other physical quantities of the fluid at certain given points in space, for instance, on the surface of a body in the fluid. Since the Eulerian method gives us these results directly, ordinarily we use the Eulerian method. In this book we always use the Eulerian method except for special cases where the Lagrangian method is more convenient.

In the Eulerian method, the basic variables are the velocity components of the fluid. We would like to know the deformation of the velocity field of a fluid flow. Any deformation of a continuum may be accomplished by two successive processes which are independent from each other if the second order quantities are neglected. The first is a simple extension or compression, that is, the normal strain, and the second is a shearing strain which measures the change of skewness of the element. The strain tensor of a fluid has the following nine components:

$$S = \begin{vmatrix} \varepsilon_x & \frac{1}{2}\gamma_{xy} & \frac{1}{2}\gamma_{xz} \\ \frac{1}{2}\gamma_{yx} & \varepsilon_y & \frac{1}{2}\gamma_{yz} \\ \frac{1}{2}\gamma_{zx} & \frac{1}{2}\gamma_{zy} & \varepsilon_z \end{vmatrix} \tag{1.14}$$

where

$$\varepsilon_x = \frac{\partial u}{\partial x}, \varepsilon_y = \frac{\partial v}{\partial y}, \varepsilon_z = \frac{\partial w}{\partial z}, \gamma_{xy} = \frac{\partial u}{\partial y} + \frac{\partial v}{\partial x} = \gamma_{yx}$$

$$\gamma_{yz} = \frac{\partial v}{\partial z} + \frac{\partial w}{\partial y} = \gamma_{zy}, \gamma_{zx} = \frac{\partial w}{\partial x} + \frac{\partial u}{\partial z} = \gamma_{xz}$$

and ε_x is x-wise normal strain, etc., and γ_{xy} is the x-y shearing strain, etc.

The vorticity $\boldsymbol{\omega}$ of the fluid is the curl of the velocity vector, i.e.,

$$\boldsymbol{\omega} = \nabla \times \boldsymbol{q} = \boldsymbol{i}\omega_x + \boldsymbol{j}\omega_y + \boldsymbol{k}\omega_z \tag{1.15}$$

where

$$\omega_x = \frac{\partial w}{\partial y} - \frac{\partial v}{\partial z}, \omega_y = \frac{\partial u}{\partial z} - \frac{\partial w}{\partial x}, \omega_z = \frac{\partial v}{\partial x} - \frac{\partial u}{\partial y}$$

and Gibbs vector notations are used.

The strain tensor of a fluid is closely related to the viscous stress tensor of the fluid which will be discussed in Chapter XVII. The vorticity is one of the most important concepts in fluid mechanics which will be studied in details later (Chapters V and XIV).

6. Dynamic similarity

In fluid dynamics, it is always desirable to find the relation between the flow patterns around affinely related bodies or geometrically similar bodies. For instance, if we test a model, we would like to know what the results will be for a full-scale body; i.e., how we can deduce the results for the full-scale body from the results of the model test. Our problem is to determine the conditions under which the forms of flows around geometrically similar bodies are themselves geometrically similar. Such similarity of flow is known as dynamic similarity[1].

We may determine the conditions for dynamic similarity from the fundamental equations of fluid dynamics. Dynamic similarity will exist when changes of scale of the units of length, time, and mass transform the fundamental equations and the boundary conditions in one case into those of the other case so that the equations remain invariant. Assuming that we have a unique solution of the system of equations, this will furnish a direct description of the circumstances for one case and a corresponding description for the other by suitable changes of the units. Obviously, dynamic similarity may be attained if the different kinds of forces acting at every point of similar positions in the two flow fields on volume elements at these points have the same ratio. Thus various laws of similarity will result from this requirement depending on the kinds of forces in effect. In each of these laws of dynamic similarity, a nondimensional parameter may be used to characterize the particular type of dynamic similarity involved. The complete set of nondimensional parameters which characterize the dynamic similarity of the important forces for a given problem may be obtained from the fundamental equations or by dimensional analysis[1,7]. The following are some of the most important dimensionless parameters in fluid dynamics:

(a) *Mach number M*. This is a measure of the compressibility of the fluid due to a high flow speed and is defined as

$$\text{Mach number } = M = \frac{\text{velocity}}{\text{sound speed}} = \frac{U}{a} \tag{1.16}$$

It is easy to show that the ratio of the variation of density of the fluid to the variation of velocity is, to the first approximation, proportional to the square of the Mach number of the flow. Hence for very small Mach number, the variation of density, i.e., the compressibility effect, due to the variation of velocity of the flow field is negligibly small and the fluid may be considered to be incompressible. For large Mach number, the effect of compressibility must be considered. When $M < 1$, the flow is called subsonic flow, and when $M > 1$, the flow is called supersonic flow. The behavior of subsonic flow differs greatly from that of supersonic flow. This shall be discussed in detail later.

(b) *Reynolds number Re*. This is the most important parameter for the fluid dynamics of a viscous fluid and is defined as follows:

$$\text{Reynolds number} = Re = \frac{\rho UL}{\mu} = \frac{UL}{\nu} \tag{1.17}$$

where μ is the coefficient of viscosity, and $\nu = \dfrac{\mu}{\rho}$ is the coefficient of kinematic viscosity. The Reynolds number is a measure of the ratio of inertial force to viscous force. When the Reynolds number of a system is small, the viscous force is predominant and the effect of viscosity is important in the whole flow field. When the Reynolds number is large, the inertial force is predominant and the effect of viscosity is important only in the narrow boundary-layer region near the solid boundary or in any other region of large variation in velocity.

(c) *Prandtl number* Pr. The Prandtl number is a measure of the relative importance of heat conduction and viscosity of a fluid and is defined as

$$Pr = \frac{c_p \mu}{\kappa} \tag{1.18}$$

where c_p is the specific heat of the fluid at constant pressure, and κ is the coefficient of thermal conductivity of the fluid. The Prandtl number may be also written as follows:

$$Pr = \frac{\mu/\rho}{\kappa/c_p\rho} = \frac{\nu}{\kappa/c_p\rho} = \frac{\textit{kinematic viscosity}}{\textit{thermal diffusivity}} \tag{1.19}$$

The value of ν shows the effect of viscosity of a fluid. If other things are the same, the smaller the value of ν, the narrower will be the region affected by viscosity. This region is known as the boundary-layer region when ν is very small[7]. While ν shows the momentum diffusivity due to viscosity effect, $\kappa/c_p\rho$ shows the thermal diffusivity due to heat conduction. The smaller the value of $\kappa/c_p\rho$, the narrower will be the region affected by heat conduction. This region is known as thermal boundary layer when $\kappa/c_p\rho$ is very small. Thus the Prandtl number shows the relative importance of heat conduction and viscosity of a fluid. Since for a gas the Prandtl number is of the order of unity, whenever the effect of viscosity is considered we must simultaneously take into account the influence of thermal conductivity of the gas.

(d) *The ratio of specific heats*. The ratio of the specific heat of a fluid at constant pressure c_p to that at constant volume c_V, i.e.,

$$\gamma = c_p/c_V \tag{1.20}$$

is a measure of the relative internal complexity of the molecules of the fluid. It is important in the study of gas dynamics.

There are many other nondimensional parameters which may be important in certain special problems of fluid dynamics. These will be discussed later. The above four parameters are the most important ones for fluid dynamics of a compressible fluid.

7. Problems

1. Find the expression of acceleration at a point in the flow field in cylindrical coordinates.

2. Find the expression of acceleration at a point in the flow field in spherical coordinates.

3. If we allow 1% variation of density of air in a flow field in which the air may be considered as an incompressible fluid, estimate the maximum velocity of the flow below which the air may be considered as an incompressible fluid. Assume that in this problem the flow is an isentropic flow and the stagnation conditions are the sea-level standard conditions of atmosphere. What will be the maximum Mach number of the flow below which the air may be considered as an incompressible fluid as the above condition is satisfied?

4. What are the definitions of the following nondimensional parameters? (See reference 4.) Discuss their significance. (a) Peclet number, (b) Froude number, (c) Grashoff number, and (d) Schmidt number.

References

1. Pai, S. I., *Modern Fluid Mechanics*, Science Press, China, 1981, distributed by Van Nostrand Reinhold Co., New York.
2. Burgers, J. M., *Flow Equations for Composite Gases*, Academic Press, New York, 1969.
3. Patterson, G. N., Mechanics of Rarefied Gases and Plasma, *UTIAS Rev.*, No. 18, Institute for Aerospace Studies, University of Toronto, 1964.
4. von Kármán, Th., Aerothermodynamics, Lecture notes given in the Department of Physics, Columbia University, New York, 1947.
5. Tsien, H. S., Superaerodynamics, *Jour. Aero. Sci.*, **13**, No. 12, Dec. 1946, pp. 653–664.
6. Kennard, E. H., *Kinetic Theory of Gases*, McGraw-Hill Book Co., Inc., New York, 1938.
7. Pai, S. I., *Viscous Flow Thoery I-Laminar Flow*, D. Van Nostrand Company, Inc., Princeton, N.J., 1956.
8. Lamb, H., *Hydrodynamics*, 6th edition, Cambridge University Press, 1932.

Chapter II

THERMODYNAMICS AND PHYSICAL PROPERTIES OF COMPRESSIBLE FLUIDS

1. General description of matter[1, 2]

The properties of matter have been described from both microscopic and macroscopic points of view since ancient times. The atomic theory of matter which is usually attributed to the Greek philosopher Democritus has been clearly expressed by Newton. Further development of the atomic theory by Dalton and others gives us the modern concept of the kinetic theory of matter. Theoretically, it should be possible to calculate the bulk properties of matter from the properties of individual particles of matter and the knowledge of the forces which the particles exert on each other. However, because of many physical and mathematical difficulties, the kinetic theory of matter in general and that of liquid in particular is still far from well developed. But the results of some kinetic theories such as those of gases and solids do give us many interesting and important results of the properties of matter.

Sometimes, one finds that the results of the kinetic theory are too detailed to be useful. We do not care about the motion of individual particles in the matter but are interested only in the resultant effects due to the motion of a large number of particles . In other words , we are interested in the macroscopic quantities only, such as pressure, temperature. density and flow velocities. Hence we should consider the macroscopic properties of matter. It is still not easy to describe the macroscopic properties of matter. It has been known for a long time that the same matter may have entire different properties at different temperature and pressure ranges. For instance, water, ice, and steam are same matter but their properties differ greatly from one another. This leads to the introduction of states or phases of matter. The states or phases of a matter are an ancient concept too, which is known as the theory of elements. Aristotel had this idea over two thousand years ago. He thought that all matter consisted of four elements: earth, water, air and fire. Of course, the meanings of these four words: earth, water, air and fire are not the same today as they were then. In our modern language, we probably should translate them as solid (earth), liquid (water), gas (air) and plasma (fire).

Hence the four states or phases of matter are really an old concept. In this chapter, we shall discuss the properties of matter according to the four states or phases of matter.

Roughly speaking, the properties of different states of a matter behave entirely differently from each other, even though there are some similarities among them, and sometimes, the transition from one state to the other may be gradual instead of being sharp. The most obvious differences of properties of different states are that (i) solid maintains its shape indefinitely if left undisturbed, (ii) liquid will take up the shape of its container up to the level of a free surface and (iii) gas and plasma cannot contain themselves with a boundary surface but take up the entire space available to it. Since liquid, gas and plasma cannot maintain their shape by themselves, we may use the term *fluid* for these three states. There are many similar properties among these three states of fluid. Furthermore, as far as the shapes of gas and plasma are concerned, plasma behaves exactly as an ordinary gas; it fills up all the space available. Hence we sometimes call plasma an ionized gas which has all the important properties of a gas, but plasma consists of electrically charged particles and is influenced greatly by electromagnetic forces. The flow field of a plasma may differ greatly from that of a neutral gas because of the interaction of the plasma with electromagnetic fields. Hence it is better to consider plasma as a state separate from a gas. In the following, when the effects of electromagnetic field are not essential, we shall consider both "neutral" gas and plasma as "gas". When the electromagnetic forces are important, we shall separate a plasma from a neutral gas. On the other hand, since liquid lies between solid and gas, some of the properties of a liquid are similar to those of a gas while the others are similar to those of a solid. This is one fact that causes the difficulty in developing the kinetic theory of a liquid.

The properties of solid, liquid and gas are different because the arrangement of molecules and the forces between molecules are different in these three states. In solids and liquids, the molecules are so close together that every one of them is subject to large forces due to the effects of its neighboring molecules. In this way, there are many similar properties between solid and liquids. Firstly, the densities of liquid and solid state of a matter do not differ appreciably. Usually, the difference in density between a liquid and a solid of the same matter is of the order of less than 10%. On the other hand, the variation of the density of a gas is very large, depending on the pressure and temperature of the gas. One may compress the gas to a high density near the value of its liquid state or expand the gas to a low density near a vacuum.

Secondly, the molecules of a solid and liquid do not fly apart easily, as is the case with a gas or a plasma, and they generally can maintain a boundary surface. Because of the close packing of molecules in a solid and liquid, the change of density due to the variation of pressure in a solid or liquid is extremely small. For instance, if we increase the pressure of water at room temperature from one atmosphere to 220 atmospheres, the density of water will be increased only by 1%. On the other hand, for most gases, the density is directly proportional to pressure at constant temperature.

Thirdly, in solids, the molecules are arranged in a regular geometrical pattern and may oscillate about at fixed positions; in liquids, the molecules are not arranged in regular order and the molecules change position frequently; in gases, the molecules are, on an average, so far apart that the forces they exert on one another can be neglected except when two of them come close together in a collision or an encounter. Because of the above facts, the kinetic theory of gases has been well developed and the kinetic theory of solids is also reasonably developed but the kinetic theory of liquid is still in the process of being developed.

The state of a matter depends on its temperature and pressure. In general, at low temperature and high pressure, the matter is in a solid state; at intermediate temperature and pressure, it is in a liquid state, while at high temperature and low pressure, it is in a gaseous state, and at very high temperature and low pressure, it may be in a plasma state. In the whole universe, most matter is in a plasma state even though on the earth only a small portion of matter is in plasma state.

For a simple substance, the transition from one state to another occurs at definite temperature and pressure. At some temperature and pressure, two states of a matter may exist side by side. Fig. 2.1 shows a phase diagram of a simple substance. In general, the range of pressures and temperatures in the phase diagram of Fig. 2.1 is very large. Hence, Fig. 2.1 is not drawn in scale but shows qualitatively the variation from one state to another.

At very high temperature, the matter is in a plasma state. The transition from a gas state to a plasma state is gradual. In other words, the degree of ionization of the gas increases gradually with temperature. Hence, we do not have any definite division line between gas and plasma.

At very low temperature and high pressure, the matter is in solid state. When the temperature increases, the solid may change into liquid along an almost vertical line *T-F* in the pressure-temperature diagram or into vapor (gas) along the curve *O-T*. The point *T* is known as a triple point at which the three states: solid, liquid and vapor, can coexist in equilibrium. At all the other points in the phase diagram of Fig. 2.1, at most only two states may coexist in equilibrium.

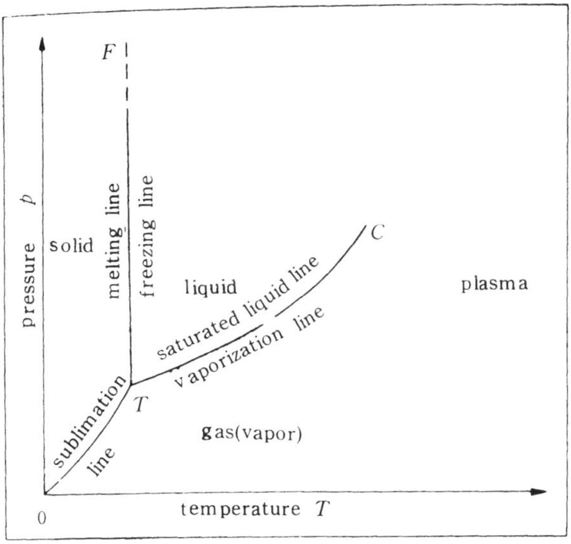

Fig. 2.1 The phase diagram of a simple or pure substance (not in scale)

The line *T-F* may be called the melting line or freezing line or fusion line at which, if we add heat to the substance, the solid will change into liquid and, if we subtract heat from the substance, the liquid will freeze into solid. This is the well known phenomenon of melting or freezing. But in general, the melting point or freezing point temperature changes with pressure even though the variation of melting temperature is small. From all present experimental results no upper limit of the point F has been found. The triple point temperature for water is nearly 0.01 ℃ greater than the melting temperature at one atmosphere pressure.

The curve *O-T* may be called sublimation line along which the solid may change directly into vapor (gas) when heat is added or the vapor will condense directly into solid when heat is subtracted.

Finally, the curve *T-C* may be called the saturated liquid line or vaporization line along which the liquid and vapor of this substance can coexist in equilibrium. The point *C* is known as the critical point beyond which there is no distinction between liquid and gas. Above the critical temperature, the surface tension of the liquid, the latent heat of vaporization, and the difference in the refractive indices of liquid and vapor all vanish. Hence, there will be no boundary between liquid and vapor. All the physical properties of liquid and gas states of the substance are identical. Hence, above the critical temperature, the substance is in a fluid state and no separation between liquid and gaseous states may be observed. The critical temperature T_c is different for different matter. For water, the critical temperature is 374.15 ℃ and the critical pressure is 218.3 atmospheres and the density is 0.32 gm/cm^3. For helium which has the lowest critical temperature among matters, the critical temperature is only 5.3 K and the critical pressure is 2.26 atmospheres.

In order to show the properties between liquid and gas in greater detail, we may draw the isothermal lines of a simple substance in the pressure volume diagram in Fig. 2.2. Let us compress a simple substance in an isothermal

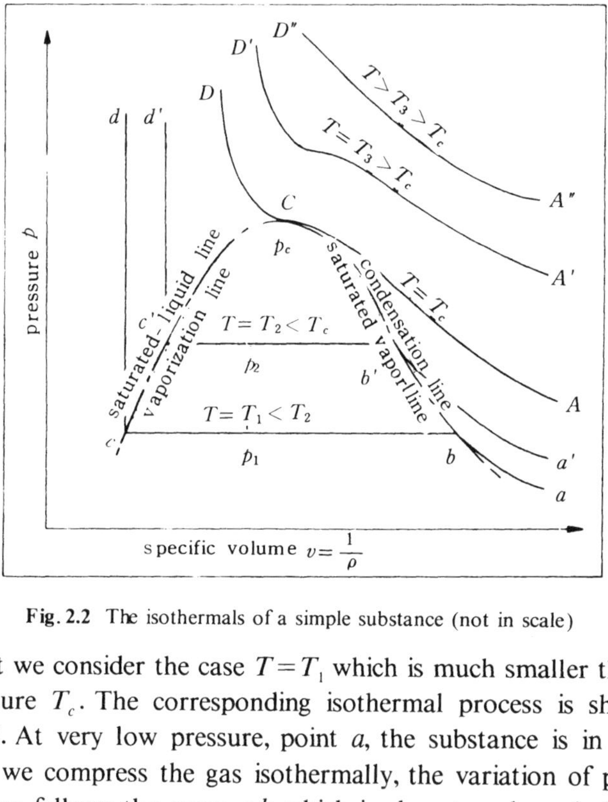

Fig. 2.2 The isothermals of a simple substance (not in scale)

process. First we consider the case $T=T_1$ which is much smaller than the critical temperature T_c. The corresponding isothermal process is shown by the curve *a-b-c-d*. At very low pressure, point *a*, the substance is in the gaseous state. When we compress the gas isothermally, the variation of pressure and specific volume follows the curve *a-b* which is close to a hyperbola. The point *b* is on saturated vapor line or simply condensation line at which the pressure and temperature have the corresponding values of a saturated vapor line given by a corresponding point on the curve *T-C* of Fig. 2.1. If we further compress the fluid isothermally, the specific volume decreases as some of the vapor changes into liquid, but the pressure remains unchanged until the point *c* at which all matter changes into liquid state. When we compress the matter from *b* to *c*, part of the vapor condenses into liquid. Hence, in the range of *b-c*, we have a mixture of liquid and vapor. Let us call the fraction of mass in vapor state the dryness fraction and denote it by λ_d. We have $\lambda_d=0$ at point *c* which corresponds to the liquid at saturated temperature T_1 under pressure

p_1, and $\lambda_d=1$ at point b which corresponds to dry saturated vapor at temperature T_1 and pressure p_1. The value λ_d varies continuously from unity to zero from point b to point c. If we repeat the isothermal process at a higher temperature $T_2 > T_1$, we have the curve a'-b'-c'-d' which has a similar shape to curve a-b-c-d but the volume change b'-c' is smaller than that of b-c. We still have $\lambda_d=0$ at point c' and $\lambda_d=1$ at point b'. If we repeat the isothermal process at still higher temperature such as the critical temperature T_c, the two points on the saturated liquid line b-c or b'-c' reduce to a single critical point C. There is no distinction between liquid and vapor. When the temperature is higher than the critical temperature T_c, we do not have sharp distinction between liquid and vapor and the matter is in a single fluid state. It is interesting to note that the vaporization line c-c'-C-b'-b in Fig. 2.2 reduces to a single line T-C in Fig. 2.1. In other words, b and c are at the same point on the T-C curve of Fig. 2.1.

In Fig. 2.2, the matter at points a, a' or A is in a gaseous state which is sometimes known as superheated vapor. It should be noted that gases are called vapors when their states are not far from the point of liquefaction. At the points b, b' and C, the matter is known as saturated vapor which marks the onset of condensation. In the region under the saturated liquid line b-b'-C-c'-c, the matter is of unsaturated vapor which is a mixture of saturated liquid and saturated vapor. The curve c-c'-C is called the saturated liquid line at which all matter is in liquid state or which marks the onset of evaporation. On the left-hand side of the curve c-c'-C, the saturated liquid line, all matter is in the liquid state. The critical temperature T_c and critical pressure p_c are different for different matter. For some of the common gases such as oxygen O_2, nitrogen N_2 hydrogen H_2 and helium He, their critical temperatures are very low[2]. In the past, one could not handle the matter at such low temperatures and one could not liquefy these gases. Hence they were called permanent gases. Now we can reduce the temperature close to absolute zero, and thus we may liquefy all the matters. Qualitatively, Fig. 2.2 represents the situation for all matters.

Even though most matter changes from the solid state to the liquid state at definite temperature so that the transition is sharp, there is matter such as glass or pitch for which the transition from solid to liquid is gradual, and there is no definite temperature of melting. For instance, glass is hard and maintains its shape at room temperature; but at high temperatures, glass will begin to soften and flow like a liquid. The transition is so gradual that no sharp definite melting temperature can be found.

Another point to mention which we discussed in Figs. 2.1 and 2.2, is the case for thermodynamic equilibrium conditions. Under the thermodynamic equilibrium condition, we may have liquid and/or gaseous states. The liquid and gaseous states of the same matter can exist side by side in equilibrium.

The liquid will have a free boundary surface and will maintain its volume. However, when the liquid of a matter is in an open container, it will evaporate, particularly when the liquid is a volatile one such as ether or gasoline. The evaporation occurs because the liquid is not under thermodynamic equilibrium condition. Similarly, solids like camphor, if left exposed, will slowly disappear into vapor by sublimation. We shall discuss the nonequilibrium conditions later.

2. Thermodynamic relations[3-7]

In the last section, we discussed qualitatively the states of a substance. For quantitative discussion, we have to use some macroscopic variables which would represent the state of a substance. For simplicity, in this section we consider a simple substance only. By simple substance, we mean a substance of a specified chemical composition. We shall discuss the case of a mixture of substances with different chemical composition later. Hence we shall not discuss in this section such problems as chemical reactions, dissociation, and ionization.

The properties of a simple substance may be obtained from any two independent state variables, i.e., temperature T, pressure p and/or specific volume v (or density $\rho = 1/v$) of the substance. It is an empirical fact that there is a functional relation between the three state variables: T, p and v, i.e.,

$$F(T, p, v) = 0 \tag{2.1}$$

From Eq. (2.1), we may solve for T, p or v and obtain:

$$T = T(p, v) \; ; \qquad p = p(T, v) \; ; \qquad v = v(T, p) \tag{2.2}$$

If we know any two of the three state variables T, p and v, the third one will be given by Eq. (2.1) or (2.2). Hence the thermodynamic state of a simple substance is determined by the values of two of the state variables. After the thermodynamic state of the simple substance is fixed, all the properties of the substance such as internal energy, viscosity, thermal conductivity, and optical refractive index are also fixed.

The essential property of a state variable is that at a definite thermodynamic state, it has a definite value no matter what changes it undergoes before it reaches this thermodynamic state. The state variables are macroscopic variables which are the average values of a large number of molecules of the substance. For the same thermodynamic state, the macroscopic variables have the same values, even though the microscopic state of the molecules may differ.

In thermodynamics, we deal with many partial derivatives of the state variables. For instance, let us consider the third equation of (2.2)

$$v = v(T, p) \tag{2.2a}$$

The total differential of Eq. (2.2a) gives

$$dv = \left(\frac{\partial v}{\partial T} \right)_p dT + \left(\frac{\partial v}{\partial p} \right)_T dp \tag{2.3}$$

where the subscript refers to the variables which are kept constant during the process. For instance, if we keep the pressure p constant ($dp=0$), the change of specific volume due to a change of temperature dT is

$$dv = \left(\frac{\partial v}{\partial T} \right)_p dT \tag{2.3a}$$

The partial derivative $(\partial v/\partial T)_p$ has been used quite often in thermodynamics in association with the coefficient of volume expansion or simply coefficient of expansion α which is defined as

$$\alpha = \frac{1}{v}\frac{dv}{dT} = \frac{1}{v}\left(\frac{\partial v}{\partial T} \right)_p \tag{2.4}$$

The coefficient of expansion α gives the ratio of volume change per degree of variation of temperature at a constant pressure process. Sometimes, one defines the coefficient of expansion α_0 in terms of some standard specific volume v_0 instead of the local specific volume v. In other words, we have

$$\alpha_0 = \frac{1}{v_0}\left(\frac{\partial v}{\partial T} \right)_p \tag{2.4a}$$

where v_0 is the specific volume at a reference temperature, e.g. 0 ℃ .

The second term on the right-hand side of Eq. (2.3) is associated with the compressibility factor of the substance which is defined as

$$k_c = -\frac{1}{v}\left(\frac{\partial v}{\partial p} \right)_T \tag{2.5}$$

Hence k_c represents the ratio of volume contraction per unit pressure at isothermal process. We may also define the compressibility factor in terms of a standard specific volume v_0 instead of local specific volume v.

Similarly, the first two equations of (2.2) give the following total differentials:

$$dp = \left(\frac{\partial p}{\partial v} \right)_T dv + \left(\frac{\partial p}{\partial T} \right)_v dT \tag{2.6}$$

$$dT=\left(\frac{\partial T}{\partial p}\right)_v dp+\left(\frac{\partial T}{\partial v}\right)_p dv \tag{2.7}$$

It is easy to find the physical significance of these partial derivatives in Eqs. (2.6) and (2.7). For instance we may define a coefficient of tension β as

$$\beta=\frac{1}{p}\left(\frac{\partial p}{\partial T}\right)_v \tag{2.8}$$

These partial derivatives are not independent of one another. For instance, Eq. (2.3) may be written as

$$\frac{dv}{dp}=\left(\frac{\partial v}{\partial p}\right)_T+\left(\frac{\partial v}{\partial T}\right)_p\frac{dT}{dp} \tag{2.9}$$

On the line of constant specific volume, $dv=0$, Eq. (2.9) gives

$$\left(\frac{\partial v}{\partial p}\right)_T+\left(\frac{\partial v}{\partial T}\right)_p\left(\frac{\partial T}{\partial p}\right)_v=0$$

or

$$\left(\frac{\partial v}{\partial p}\right)_T\left(\frac{\partial p}{\partial T}\right)_v\left(\frac{\partial T}{\partial v}\right)_p=-1 \tag{2.10}$$

This cyclic order relation (2.10) holds true for partial derivatives of any function of three variables which has continuous derivatives.

With the help of Eq. (2.10), the coefficients α, k_c and β have the relation:

$$\alpha=p\beta k_c \tag{2.11}$$

Besides the three thermodynamic variables T, p and v, we may use some other thermodynamic quantities such as internal energy U_m, enthalpy h and entropy S, to determine the thermodynamic state. It is assumed that the reader of this book is familiar with these terms. Here we just briefly mention some of the important relations for these quantities which will be used in this book.

3. Gases, vapors and liquids[8, 9]

Gases are called vapors when their state is not far removed from the vaporization line. In general, the equation of the state of vapor is more complicated than that of a gas far away from the vaporization line. The simplest case of a gas is a perfect gas or an ideal gas. We shall discuss first the properties of a perfect gas, then the properties of real gases far away from the vaporization line and finally the properties of vapor.

(i) Perfect gases

If the pressure of a gas is much smaller than its critical pressure p_c, the behavior of the gas is close to a perfect gas. A perfect gas may be defined as a gas which obeys the following laws:

(a) The internal energy per unit mass U_m depends only on the absolute temperature, i. e.,

$$U_m = U_m(T) \tag{2.12}$$

where (b) The equation of state of this gas is given by the relation

$$p = \rho R T = \frac{R_A}{m} \rho T \tag{2.13}$$

where m is the gram molecular weight of the gas, e.g. oxygen O_2, $m=32$ and R_A is the universal gas constant or the Boltzmann constant which has a value of 1.379×10^{-16} erg/ K.

(ii) Vapors and dense gases[9, 10]

The perfect gas law (2.13) may be derived from simple kinetic theory of gases in which the gas is assumed to be an aggregate of rapidly moving particles which are rigid bodies of zero diameter but of finite mass and which are constantly colliding with one another to exchange energy. The influence of the particles on each other can be conveniently neglected until they are so close together that a "collision" takes place[9]. The basic assumption for such a simple kinetic theory is that the density of the gas must be low so that the forces between the particles can be neglected until a collision takes place and that the volume occupied by the particles is negligible. When the density of the gas is not low, particularly that of a vapor near the critical condition, the forces between molecules cannot be neglected. We should include the molecular interaction all the time in the kinetic theory of a dense gas. In fact, the kinetic theory of dense gases may be considered as a first approximation for the kinetic theory of liquid[11].

(a) *Equation of state of a dense gas*[9]

From the kinetic theory of a dense gas, it has been found that the equation of state may be written as follows:

$$\frac{p}{\rho R T} = 1 + B(T)\rho + C(T)\rho^2 + \cdots \tag{2.14}$$

where $B(T)$, $C(T)$, etc. are functions of temperature only but not of density nor pressure. The coefficients $B(T)$, etc., are known as virial coefficients. Eq. (2.14) was first introduced empirically by Kammerlingh Onnes and it may be derived from the kinetic theory of gases[9]. It is evident that for a first approximation when all the virial coefficients are neglected, Eq. (2.14) reduces to the equation of state of a perfect gas (2.13).

(b) Van der Waals equation of gas

The second approximation, in which we take $B(T) = b_1 - b_2/RT$ with b_1 and b_2 as constants for a given gas and all the other virial coefficients as zeros, gives us the well known Van der Waals equation. Van der Waals equation is the first and most successful equation which gives fairly good description of fluid behavior including both gas and liquid. Hence we are going to discuss it in detail.

For Van der Waals approximations, Eq. (2.14) becomes

$$p + b_2\rho^2 = \rho RT(1 + b_1\rho) \tag{2.15}$$

If $b_1\rho$ is small, Eq. (2.15) may be written as

$$\left(p + \frac{b_2}{v^2}\right) = \frac{RT}{v - b_1} \tag{2.16}$$

Eq. (2.16) is the well known form of van der Waals equation.

The Van der Waals equation gives a fairly satisfactory description of fluid behavior including both gas and liquid phases. At high pressure or at low temperature, i.e., near the condensation region, it is sometimes necessary to use Eq. (2.16) or (2.14) instead of the simple perfect gas law (2.13) in studying the fluid flow problem.

The equation of state of vapor is very complicated. Usually special tables are used to express the relations for vapor state. The most common vapor is the steam. For instance, reference 12 shows the steam tables.

(iii) Liquids[11]

Even though in many engineering problems we assume that a liquid is an incompressible fluid, actually liquid is compressible, but the compressibility factor k_c, Eq. (2.5), is very small. For a perfect gas, the compressibility factor k_c is simply equal to $1/p$. Hence at atmospheric pressure, the compressibility factor k_c is $1/(14.7\ \text{lb/in}^2) = 1/(14.7\ \text{psi})$. However, for water at $68°\text{F} = 20°\text{C}$ and a pressure of 15 psi, the compressibility factor k_c of water is $1/(320{,}000\ \text{psi})$. Hence the compressibility factor k_c of a perfect gas is more than 20,000 times larger than that of water. But the compressibility factor k_c of water is still 100 times that of steel. Hence when we deal with very high temperature and/or very large pressure variation, we should take the compressibility effect of liquid into consideration.

The compressibility effect of a liquid, shown in Fig. 2.2, is that the liquid isothermals *c-d* and *c′-d′* are not vertical lines but are curves along which the specific volume decreases as pressure increases. The equation of state of a liquid is usually given by tables or charts. However, we may approximate these values given in tables and charts by simple equations. Some equations have been proposed in connection with underwater explosion. One is the Tate equation[4] which is as follows:

$$v(T, p) = v(T, 0)\left[1 - \frac{1}{n}\log\left(1 + \frac{p}{B}\right)\right] \tag{2.17}$$

where n is a constant and $B(T)$ is a function of temperature only. For a first approximation we may take $n=7.15$ and $B(T)=\text{constant}=3.0$ kilobars. One kilobar is equal to 14,513 lb/in^2 (psi).

In the state under the saturated liquid line and the saturated vapor line c-c'-C-b'-b of Fig. 2.2, the liquid is a mixture of liquid and vapor. For the stable conditions, the isothermals are isobars, i.e., the isothermals in Fig. 2.2 are horizontal lines. We have a two-phase flow which will be discussed in Chapter XVI. This condition may be called wetted saturated vapor and special treatment of its thermodynamic properties may be used as we shall discuss them in Chapter XVI.

4. First and second laws of thermodynamics

The first law of thermodynamics which is a law of conservation of energy may be written as:

$$dQ = dU_m + pdv \tag{2.18}$$

where dQ is the heat added per unit mass in any thermodynamic process; dU_m is the change of the internal energy per unit mass of the medium, and pdv is the amount of work done during the change of specific volume dv. In most flow problems, the heat dQ may be due to heat conduction or thermal radiation or from chemical reaction which actually involves a change of internal energy of the medium but for convenience we may regard it as an external source of heat, or from the latent heat of phase change.

The internal energy of the medium U_m is the quantity of mechanical energy or heat stored by the medium at rest. For a fluid, the internal energy of the fluid represents the kinetic and potential energies of the molecules of the fluid. We shall discuss the internal energy of a gas or a liquid in greater detail later. The internal energy per unit mass U_m is a function of the two state variables for a simple substance. Hence we may write

$$U_m = U_m(T, v) \tag{2.19}$$

The total differential of Eq. (2.19) gives

$$dU_m = \left(\frac{\partial U_m}{\partial T}\right)_v dT + \left(\frac{\partial U_m}{\partial v}\right)_T dv \tag{2.20}$$

For a constant volume process, Eqs. (2.18) and (2.20) give

$$c_v = \left(\frac{\partial Q}{\partial T}\right)_v = \left(\frac{\partial U_m}{\partial T}\right)_v \tag{2.21}$$

where c_v is known as the specific heat at constant volume. In general, c_v is a function of both temperature T and specific volume v. As we shall discuss later, for a perfect gas, c_v is a function of temperature T only.

We may introduce another thermodynamic property known as enthalpy h as follows:

$$h = U_m + pv \tag{2.22}$$

It is evident that enthalpy h is, in general, a function of both temperature T and pressure p.

The total differential of Eq. (2.22) gives

$$dh = \left(\frac{\partial h}{\partial T} \right)_p dT + \left(\frac{\partial h}{\partial p} \right)_T dp \tag{2.23}$$

The reason that we use T and p as independent variables in the case of enthalpy h is that enthalpy has a special significance in a constant pressure process as follows:

The first law of thermodynamics (2.18) may be written as

$$dQ = dU_m + d(pv) - vdp = dh - vdp \tag{2.18a}$$

From Eqs. (2.18a) and (2.23), we have

$$c_p = \left(\frac{\partial Q}{\partial T} \right)_p = \left(\frac{\partial h}{\partial T} \right)_p \tag{2.24}$$

where c_p is known as the specific heat at constant pressure. In general, c_p is a function of both temperature T and pressure p. For a perfect gas, c_p is a function of temperature only.

To show the thermodynamic state of a substance, diagram-involving enthalpy has been used often. Such a diagram-involving enthalpy is usually referred to as Mollier diagram in honor of Richard Mollier who first used such a plot in 1904.

From Eq. (2.18), we see that the heat addition dQ is not an exact differential. Physically, it is because the heat added in a process dQ depends on the path of the process. In other words, the term pdv depends on the thermodynamic process considered. It is convenient to define a thermodynamic quantity which is associated with dQ but which is a function of state variables only. This characteristic quantity is known as entropy S and is defined as

$$dS = \left(\frac{dQ}{T} \right)_{\text{rev}} \tag{2.25}$$

Since dQ is a function of two variables T and p, it is always possible mathematically to find an integrating factor to make an exact differential from dQ. In the present case, the integrating factor is $1/T$. Integration of Eq. (2.25) gives

$$S_2 - S_1 = \int_1^2 \left(\frac{dQ}{T} \right) \tag{2.26}$$

which is independent of the path from state 1 to state 2.

The second law of thermodynamics may be stated as follows: In an adiabatic process ($dQ=0$), the entropy either increases or remains unchanged, i.e.,

$$dS \geqslant 0 \tag{2.27}$$

where the upper sign (>) corresponds to an irreversible process and the lower sign (=) corresponds to a reversible process. Often the second law of thermodynamics determines whether a physical process is possible or not.

For an adiabatic reversible process, we have

$$dS = 0 \quad \text{or} \quad S = \text{constant} \tag{2.28}$$

Hence adiabatic reversible process is an isentropic process.

Now we are considering some general relations of entropy and specific heats first before we consider the specific cases of gas and liquid. From Eqs. (2.18) and (2.25), we have

$$T\,dS = dU_m + p\,dv = dh - v\,dp \tag{2.29}$$

We have immediately

$$\left(\frac{\partial S}{\partial T} \right)_v = \frac{1}{T} \left(\frac{\partial U_m}{\partial T} \right)_v \tag{2.30}$$

and

$$\left(\frac{\partial S}{\partial v} \right)_T = \frac{1}{T} \left[p + \left(\frac{\partial U_m}{\partial v} \right)_T \right] \tag{2.31}$$

Since $\dfrac{\partial^2 S}{\partial v \partial T} = \dfrac{\partial^2 S}{\partial T \partial v}$, Eqs. (2.30) and (2.31) give

$$p + \left(\frac{\partial U_m}{\partial v} \right)_T = T \left(\frac{\partial p}{\partial T} \right)_v \tag{2.32}$$

From Eqs. (2.29) and (2.32), we then have

$$dS = \frac{c_v\,dT}{T} + \left(\frac{\partial p}{\partial T} \right)_v dv \tag{2.33}$$

From Eqs. (2.22) and (2.23), we have

$$dh=c_p dT+\left(\frac{\partial h}{\partial p}\right)_T dp=c_p dT+\left[v-T\left(\frac{\partial v}{\partial T}\right)_p\right]dp \tag{2.34}$$

then Eqs. (2.29) and (2.34) give

$$dS=\frac{dh-vdp}{T}=c_p\frac{dT}{T}-\left(\frac{\partial v}{\partial T}\right)_p dp \tag{2.35}$$

From Eqs. (2.33) and (2.35), we have the relations:

$$\left(\frac{\partial c_v}{\partial v}\right)_T=T\left(\frac{\partial^2 p}{\partial T^2}\right)_v \tag{2.36}$$

and

$$\left(\frac{\partial c_p}{\partial p}\right)_T=-T\left(\frac{\partial^2 v}{\partial T^2}\right)_p \tag{2.37}$$

Now if we consider the isentropic process of Eq. (2.33), we have

$$c_v=-T\left(\frac{\partial p}{\partial T}\right)_v\Big/\left(\frac{\partial T}{\partial v}\right)_S \tag{2.38}$$

and similarly from the isentropic process of Eq. (2.35), we have

$$c_p=T\left(\frac{\partial v}{\partial T}\right)_p\Big/\left(\frac{\partial T}{\partial p}\right)_S \tag{2.39}$$

The ratio of the specific heats c_p to c_v shows the internal complexity of the molecules and it is usually denoted by the symbol γ:

$$\gamma=\frac{c_p}{c_v}=-\frac{\left(\frac{\partial v}{\partial T}\right)_p\left(\frac{\partial T}{\partial v}\right)_S}{\left(\frac{\partial T}{\partial p}\right)_S\left(\frac{\partial p}{\partial T}\right)_v}=-\frac{\left(\frac{\partial p}{\partial v}\right)_S}{\left(\frac{\partial p}{\partial T}\right)_v\left(\frac{\partial T}{\partial v}\right)_p}=\frac{\left(\frac{\partial p}{\partial v}\right)_S}{\left(\frac{\partial p}{\partial v}\right)_T} \tag{2.40}$$

where the relation (2.10) is used. In the diagram of $p-v$ curves, the ratio of specific heats γ is equal to the ratio of the slope of $(\partial p/\partial v)$ at constant entropy to that at constant temperature.

We may use the entropy S to replace the temperature as one of the three basic state variables and then we have

$$F_1(p, \rho, S)=0 \tag{2.41}$$

Eq. (2.41) is the equation of state which is identical to Eq. (2.1).

From Eq. (2.29), if we use S and v as two independent variables we then have

$$\left(\frac{\partial U_m}{\partial S}\right)_v = T \quad , \left(\frac{\partial U_m}{\partial v}\right)_S = -p \tag{2.42}$$

Similarly, if we use S and p as two independent variables, we have

$$\left(\frac{\partial h}{\partial S}\right)_T = T \quad , \left(\frac{\partial h}{\partial p}\right)_S = v = \frac{1}{\rho} \tag{2.43}$$

5. Specific heats, internal energy and enthalpy

If we add an amount of heat dQ to a fluid, its temperature will usually increase by an amount dT. We define a specific heat c_{sp} of the fluid as

$$c_{sp} = \lim_{dT \to 0} \frac{dQ}{dT} = \frac{\partial Q}{\partial T} \tag{2.44}$$

The specific head c_{sp} depends on the process in which the heat is added; in other words, when we add heat, we have to keep certain thermodynamic characteristics of the fluid unchanged and the value of c_{sp} depends on this characteristic of the fluid. For instance, two most common cases are:

$$c_v = \left(\frac{\partial Q}{\partial T}\right)_v = \text{specific heat at constant volume} \tag{2.44a}$$

$$c_p = \left(\frac{\partial Q}{\partial T}\right)_p = \text{specific heat at constant pressure} \tag{2.44b}$$

The specific heats are closely connected with the internal energy of the molecules, which depends greatly on the molecular structure. For a perfect gas, the internal energy per unit mass U_m is a function of temperature T only and we have

$$c_v = \frac{dU_m}{dT} \tag{2.45}$$

For a monatomic gas at low temperature, the internal energy of the molecules is simply the difference of their total kinetic energy due to molecular motion (ξ, η, ζ) and the mean kinetic energy of the flow (u, v, w). This type of internal energy is usually referred to as translational internal energy U_{mt} which for per unit mass is

$$U_{mt} = \frac{1}{2}\left(\overline{\xi^2} + \overline{\eta^2} + \overline{\zeta^2}\right) - \frac{1}{2}\left(u^2 + v^2 + w^2\right) = \frac{1}{2}\overline{c_a^2} = \frac{3}{2}RT \tag{2.46}$$

where c_a is the magnitude of the random velocity of a molecule. The specific heat at constant volume due to U_{mt} is simply $\frac{3}{2}R$. This part of the internal energy holds true for all kinds of gases or plasmas.

The temperature so defined is referred to as kinetic temperature.

For a polyatomic gas, the internal energy depends on the molecular structure. For simplicity, we consider the case of a diatomic gas (Fig. 2.3). The internal energy consists of six parts as follows:

$$U_m = U_{mt} + U_{mr} + U_{mv} + U_{md} + U_{me} + U_{mi} \tag{2.47}$$

where

(a) $U_{mt} = \frac{3}{2} RT$ for all kinds of gases represents the internal energy due to translational motion of the molecules. It is translational internal energy (Fig. 2.3a).

(b) U_{mr} represents internal energy due to rotational motion of the molecules. It is rotational internal energy (Fig. 2.3b). For a monatomic gas, the moment of inertia of the atom is negligibly small and we may assume $U_{mr}=0$. For a diatomic gas, the moment of inertia about the axis connecting the two atoms is negligibly small, while those about axes perpendicular to the axis connecting the atoms are not small. We may consider that there are two degrees of freedom due to rotational motion of a diatomic gas. By the principle of equipartition of energy[9], $U_{mr}=RT$ for diatomic gases or a polyatomic gas with linear molecules because there are only two degrees of freedom of rotational motion. $U_{mr} = \frac{3}{2} RT$ for polyatomic gas with nonlinear molecules because there are three degrees of freedom of rotational motion in this case.

(c) U_{mv} represents the vibrational motion between atoms in a molecule. It is vibrational internal energy (Fig. 2.3c). For each vibrational mode, there is a fundámental frequency ν_j and the corresponding internal energy is

$$U_{mv} = RT\left[\frac{h\nu_j/kT}{\exp(h\nu_j/kT) - 1}\right] \tag{2.48}$$

For complicated molecule there are a number of fundamental frequencies ν_j, the total vibrational internal energy U_{mv} is the sum of the internal energy of individual modes given by Eq. (2.48).

(d) U_{md} represents the amount of energy needed to dissociate the atoms of a molecule. It is the dissociation energy (Fig. 2.3d) (see next section).

(e) U_{me} represents the energy to excite an electron of the molecule from its ground state to a higher state. It is the electron excitation energy (Fig. 2.3e).

(f) U_{mi} represents the energy to ionize a molecule so that one or more of the electrons move away from the neighborhood of the nucleus. The molecule becomes an ion. It is the ionization energy (Fig. 2.3f).

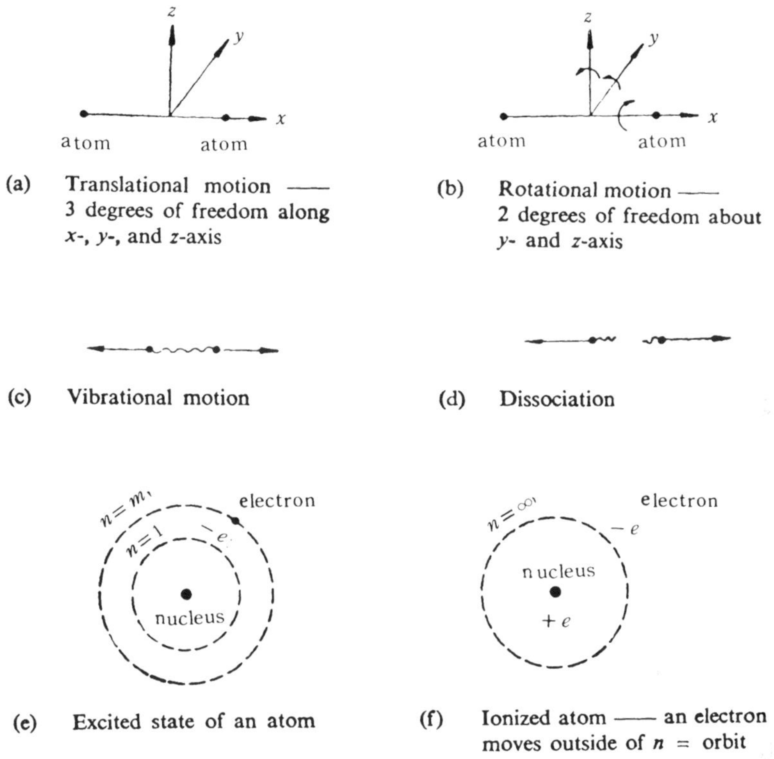

Fig. 2.3 Various modes of internal energy of a diatomic gas

Below $T = 2000$ K, the only modes of internal energy of a gas which are important are the translational energy and rotational energy. Under this condition, we have $U_m = U_{mt} + U_{mr}$ and the specific heat of a gas at constant volume is a constant, i.e.,

$$c_v = \frac{3}{2} R \text{ for monatomic gas} \tag{2.49a}$$

$$c_v = \frac{5}{2} R \text{ for diatomic gas} \tag{2.49b}$$

A gas with constant c_v is usually referred to as an *ideal gas*. When the temperature T is above 2000 K, the other modes of internal energy should be considered in the flow problem, and the specific heat at constant volume is a function of temperature. Fig. 2.4 shows a typical variation of specific heat at constant pressure of air with temperature[13-18].

Under equilibrium conditions, the temperatures T corresponding to various modes of internal energies are the same. But in non-equilibrium conditions, the temperature T corresponding to various modes of internal energy may not be the same[19]. In such cases, we may define one temperature for each mode of internal energy in the study of the flow problem of a gas. For instance, we have a temperature of vibrational mode T_v which may be different from the kinetic temperature of the gas.

We define an enthalpy H of a gas as

$$H = U_m + \frac{p}{\rho} \tag{2.50}$$

For a perfect gas, we have

$$c_p = \frac{dH}{dT} = c_v + R \tag{2.51}$$

In the fluid dynamic problems, sometimes it is more convenient to use enthalpy instead of internal energy. The ratio of c_p to c_v is usually denoted by the symbol γ, i.e.,

$$\gamma = c_p/c_v \tag{2.52}$$

The ratio γ is used to measure the relative internal complexity of the molecules.

The specific heats of gases are additive. Hence the specific heat of a plasma may be determined from our knowledge of the specific heat of gases because a plasma may be considered as a mixture of several species of gases in which electrons and ions are considered as special kinds of gas.

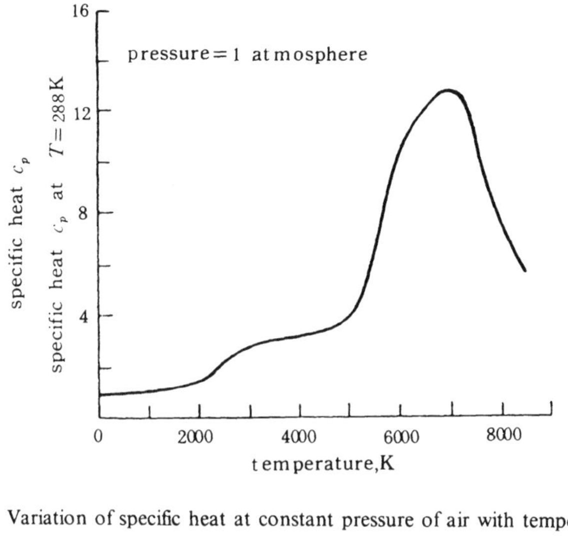

Fig. 2.4 Variation of specific heat at constant pressure of air with temperature

The specific heat of a liquid has not been well understood because the kinetic theory of liquid has not been well developed yet. However many liquids, including water and mercury, behave like a solid in that they have atomic heats about six calories/mol per atom, i.e., $c_v = 3R$. Many other liquids may have atomic heats much larger than six cal/mol.

Another interesting point of the heat capacity of liquid in fluid mechanics is that there is a definite relation between the pressure and temperature at the boiling point. In the process of evaporation or condensation, the heat may be absorbed or released. Such a heat is known as latent heat. In the study of the flow of fluids in two phases, this phenomenon should be considered.

6. Dissociation, ionization and radiation[20, 21]

Let us consider a diatomic gas such as oxygen O_2 or nitrogen N_2. At low temperature, the molecules remain unchanged even though the molecules are constantly in collision with one another. Such collisions are referred to as elastic collisions in which there is no change in internal energy of the particles, singly or collectively. At high temperature, the kinetic energy of the molecules is high. In the collisions of molecules of high kinetic energy, their internal energy may change. These are referred to as inelastic collisions[9]. During the inelastic collisions, the molecules may be excited or even dissociated. For instance, the molecule oxygen O_2 may dissociated into two oxygen atoms 2O. Of course, the atoms O may recombine into O_2. At a given temperature, an equilibrium condition may be reached such that x fraction of the molecules dissociates into atoms. Let n_0 be the number density of the molecules if there were no dissociation. For the degree of dissociation x, we have number density $(1-x)$ n_0, for molecules and number density $2xn_0$ for the atoms. Hence the total number density of the dissociating gas is $(1+x)n_0$. Since both the molecules and the atoms may be considered as perfect gases, the equation of state for the dissociating gas of degree of dissociation x is

$$p = \rho RT(1+x) \tag{2.53}$$

The degree of dissociation x in the equilibrium condition is a function of both the temperature T and the pressure p of the gas. Usually the degree of dissociation x may be expressed in terms of the equilibrium constant K for the dissociation reaction considered, i.e.,

$$\frac{4x^2}{1-x^2} = \frac{K}{p} \tag{2.54}$$

where K is a function of temperature only[20].

At very high temperatures, the collisions between particles in a gas may ionize some of the particles. The simplest case is the case of a monatomic gas. Let the degree of ionization of this gas be α. Then the equation of state for the plasma which consists of $(1-\alpha)n_0$ neutral particles, αn_0 ions and

αn_0 electrons per unit volume will be

$$p=\rho RT(1+\alpha) \tag{2.55}$$

The degree of ionization α in equilibrium condition is also a function of temperature and pressure of the gas. For singly ionized plasma of monatomic gas, the degree of ionization α is given by the Saha relation:

$$\frac{\alpha^2 p^2}{1-\alpha^2}=3.16\times 10^{-7}\times T^{5/2}\exp[-eV_i/(kT)] \tag{2.56}$$

where p is the pressure in atmospheres, T is the temperature in K, V_i is the ionization potential of the gas, and eV_i is the ionization energy in ergs.

The degree of ionization of a mixture of gases will be discussed in Chapter XVII.

In high temperature gases, photons may be emitted or absorbed by the particles of gas and they represent the thermal radiation phenomena. Except at very high temperatures the effects of thermal radiation is usually negligible in the flow problems of a fluid. However, when the temperature is very high or the density of the gas is very low, thermal radiation may have predominant influence on the flow problem of gases or plasma. In thermal equilibrium conditions, the spectral energy density of thermal radiation is given by the Planck radiation law:

$$U_\nu=\frac{8\pi h\nu^3}{c^3}\frac{1}{\exp[h\nu/(kT)]-1} \tag{2.57}$$

where ν is the frequency of the thermal radiation and c is the speed of light. The total energy density of radiation in thermal equilibrium is then:

$$E_R=\int_0^\infty U_\nu d\nu=a_R T^4 \tag{2.58}$$

where $a_R=7.67\times 10^{-15}$ erg $\cdot$ cm^{-3} $\cdot$ K^{-4} is known as the Stefan–Boltzmann constant. Since a_R is a very small number, the energy density of radiation E_R is usually negligibly small in comparison with the internal energy of a gas U_m except when the temperature T is very high and the density of the gas is very low.

7. Clausius-Clapeyron equation[3]

The latent heat of evaporation L is related with the change of specific volume (v_v-v_L) at the saturated temperature and the slope of the saturated curve

$(dp/dT)_s$ by the Clausius-Clapeyron equation which may be derived as follows:

From Eqs. (2.31) and (2.32), we have

$$\left(\frac{\partial S}{\partial v}\right)_T = \left(\frac{\partial p}{\partial T}\right)_v \tag{2.59}$$

For evaporation from saturated liquid to saturated vapor, we have

$$\left(\frac{\partial S}{\partial v}\right)_T = \frac{S_v - S_L}{v_v - v_L} = \left(\frac{dp}{dT}\right)_s \tag{2.60}$$

where subscript s refers to the value on the saturated curve $p_s = p_s(T_s)$ for the case $\lambda = 1$ gives

$$L = (v_v - v_L) T_s \left(\frac{dp}{dT}\right)_s \tag{2.61}$$

Eq. (2.61) is the Clausius-Clapeyron equation.

If the latent heat L and the specific volumes v_L and v_v are known, we may derive an equation for the saturated curve $p_s(T_s)$ from the Clausius-Clapeyron equation (2.61).

If the latent heat L is a constant and the specific volume of saturated liquid v_L is negligible in comparison with that for saturated vapor v_v, i.e., $v_L \ll v_v$ and if we may write $v_v = RT/p$, Eq. (2.61) gives

$$\ln p_s = -\frac{L}{RT_s} + \text{constant} \tag{2.62}$$

For many substance, we may write $L = a + bT$ where a and b are constants, Eq. (2.58) gives the following relation with the help of relations $v_L \ll v_v$, $v_v = RT/p$ and $L = a + bT$:

$$\ln p_s = -\frac{a}{RT_s} + \frac{b}{R} \ln T_s + \text{constant} \tag{2.62a}$$

In general, we have $v_v(T_s)$, $v_L(T_s)$ and $L(T_s)$. Eq. (2.61) gives

$$p_s - p_0 = \int_{T_0}^{T_s} \frac{L\,dT}{T(v_v - v_L)} \tag{2.63}$$

where subscript 0 refers to the values at some reference condition. The functions $L(T_s)$, $v_v(T_s)$ and $v_L(T_s)$ may be determined experimentally.

8. Transport coefficients: viscosity, heat conductivity, diffusion, electrical conductivity and absorption coefficient of thermal radiation[22, 23, 24]

Even though our main interest in this book is the study of the flow of a compressible fluid without transport phenomena, the transport phenomena do have important influence on the flow field of a compressible fluid. By transport phenomena we mean that (i) the transport of momentum by viscosity; (ii) the transport of heat energy by heat conductivity; (iii) the transport of mass by diffusion; (iv) the transport of electrical current by electrical conductivity; and (v) the transport of thermal radiation energy by the absorption of thermal radiation. Besides these laminar transport phenomena, we have also the turbulent transport phenomena about these quantities. Special treatises should be referred to in dealing these transport phenomena[1, 2, 23, 24].

However, some preliminary discussions of these transport phenomena are useful to the readers of this book for a complete understanding of the flow problems of a compressible fluid., In Chapter XVII, we shall discuss the essential features of the transport phenomena in a compressible flow which consist of (i) the discussions of various transport coefficients, i.e., the coefficient of viscosity, the coefficient of heat conductivity, the diffusion coefficient, the electrical conductivity and the absorption coefficient of thermal radiation and (ii) the influence of these transport coefficients on the flow field, particularly the concept of boundary layer flow[23]. When we study the boundary layer flows, we have to know the flow field outside the boundary layer which may be considered as the flow field without the transport phenomena and which are the main results discussed in the major portion of this book.

We may refer to the fluid without transport phenomena as the ideal compressible fluid, i.e., ideal gas or ideal plasma. In the major portion of this book, we consider mainly the flow field of an ideal compressible fluid unless otherwise specified, as those in Chapter XVII.

9. Physical properties of air[25]

One of the most important compressible fluids in practice is air. Air is a mixture of gases, ions and electrons. However, for most of the practical conditions at low altitude, the proportions of the constituents of air are so constant that air may be considered as a single gas as far as the physical properties in the study of fluid dynamics are concerned. In the present space age, we have to consider a large variation of altitudes, the earth's atmosphere can no longer be considered as a single gas for all altitudes. The composition of air varies with altitude as well as time and space. We should consider the earth's atmosphere in greater detail as follows.

It is accurate enough to divide the earth's atmosphere into various regions in which different laws of variation of pressure and temperature with altitude hold. There are two major regions in the atmosphere: One is known as homosphere which covers the altitude approximately below 85 kilometers and in which the composition of the earth's atmosphere remains the same, and the other is known as heterosphere which covers the altitude above 85 kilometers and in which the composition of the earth atmosphere changes with altitude. In each of these two main regions, we may also divide them into several subregions according to the variation of temperature with altitude as follows.

(*A*) *Homosphere*

Homosphere may be divided into three subregions: troposphere, stratosphere, and mesosphere (see Fig. 2.5).

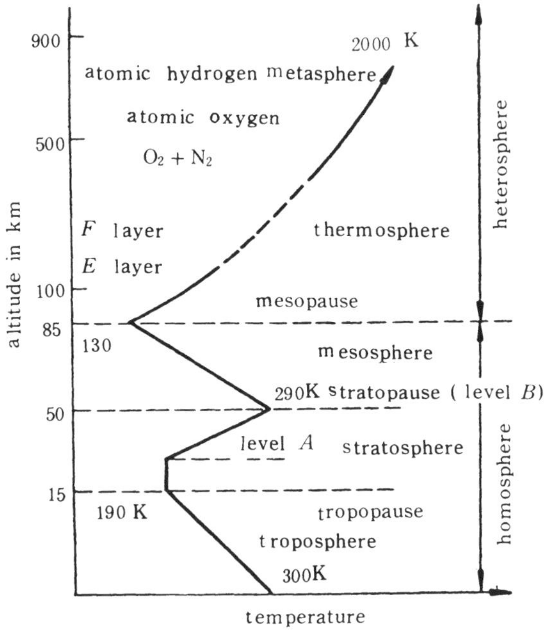

Fig. 2.5 A sketch of various regions of the earth's atmosphere (not in scale) (The temperatures are just some representative values)

(1) Troposphere. At low altitude, from sea level to an altitude of about 10 kilometers, the composition of the air is approximately constant and the temperature decreases with altitude. It is known as the troposphere. The lapse rate λ in the troposphere is almost a constant even though its value varies from place to place and from time to time. Dry atmosphere at sea level consists of 78.03% of molecular nitrogen N_2, 20.99% of molecular oxygen O_2, 0.93% of argon A, 0.030% of carbon dioxide CO_2, and 0.01% of molecular hydrogen H_2 by volume and a slight amount of neon Ne, helium He, krypton Kr, and xenon Xe. For many practical purposes, the physical properties of air are computed by assuming that the air is simply composed of 78.12% of N_2, 20.95% of O_2, and 0.93% of A at sea level by volume. For low altitude, a standard atmosphere has been adopted in order to facilitate the comparison of test date in different atmosphere conditions. The standard atmosphere for the

troposphere is based on the following assumptions[26].

(a) The air is a perfect gas with a gas constant $R=53.33$ ft/°F.

(b) The pressure of air at sea level is $p_0=29.921$ in Hg.

(c) The temperature of air at sea level is $T_0=59$°F.

(d) In the troposphere, the lapse rate is a constant of a value -0.003, 566°F/ft.

(e) The troposphere ends when the temperature reaches $T_s=-67$°F.

The lapse rate λ of the atmosphere is defined as follows:

$$\lambda=-\frac{dT}{dL}=\text{lapse rate of the atmosphere} \tag{2.64}$$

where L is the altitude and T is the temperature of the air at the altitude L. In general, the value of the lapse rate λ varies from place to place and from time to time in the atmosphere. But for a sufficiently thin layer of atmosphere, we may assume that the lapse rate λ is a constant. Hence in the standard atmosphere for the troposphere, we assume that $\lambda=-0.003566$°F/ft.

At the end of the troposphere, there is a region known as the tropopause which is a short region between troposphere and stratosphere. The height of the tropopause depends greatly on the latitude. It may be below 10 km in the polar region and above 15 km in the equatorial belt.

(2) Stratosphere. The stratosphere covers the region from the tropopause up to an altitude of about 50 km. In this region, the temperature does not decrease with increase of altitude. The temperature first remains constant and then increases with altitude. In the old definition of standard atmosphere, we assume that the temperature remains constant in the whole stratosphere. But from recent data, we know that the temperature in the stratosphere increases with altitude to a peak of the order of 270 K. Sometimes we may divide the stratosphere into two sub-regions. In the lower region, the temperature is constant up to a level A and in the upper region the temperature increases with altitude up to the stratopause which is also known as level B.

(3) Mesosphere. It is the region above the stratopause up to an altitude of about 85 km, in which the temperature decreases again with increase of altitude. The minimum temperature is of the order of 150 K. The end of mesosphere is known as the mesopause. In the mesosphere, the photochemical action is very important.

(B) Heterosphere

In the heterosphere the composition of the air varies greatly with altitude and the dissociation and ionization processes are important. The science which studies the upper atmosphere of the heterosphere is known as Aeronomy. The heterosphere may be divided into five subregions: thermosphere, ionosphere, metasphere, protosphere, and exosphere.

(1) Thermosphere and ionosphere. Above the mesopause, the temperature of the air again increases with altitude up to a maximum value of the order of 1500 K to 2000 K. This region is known as the thermosphere in which dissociation and ionization processes are important and the composition of air is no longer constant. At the lower end of the thermosphere at an altitude of about 100 km, the dissociation of oxygen takes place. Hence we cannot assume that the composition of the air remains constant. At an altitude of around 500 km, most of the molecular oxygen has dissociated and we may assume that the major constituents of the air are molecular nitrogen and atomic oxygen. At higher altitudes, the dissociation of molecular oxygen and the recombination of atomic oxygen are not the only chemical reactions. There are many other chemical reactions which take place. For instance, the atomic oxygen may attach to molecular oxygen to form ozone. The molecular nitrogen may dissociate into atomic nitrogen. The atomic oxygen and atomic nitrogen may form nitric oxide. The water vapor in the air may dissociate into atomic hydrogen and hydroxyl (OH). The carbon dioxide may dissociate into atomic oxygen and carbon monoxide. Many other chemical reactions take place.

From mesopause and up, ionization of air takes place. Even below mesopause (80 km), there are some free electrons in the atmosphere in what is known as the *D*-layer with free electrons with a number density of 10^2 to 10^4 electrons per cubic centimeter. As the altitude increases, the electron density increases to a maximum of the order of 10^5/cc at what is known as *E*-layer and which is at about 120 km. After a slight drop of electron density, the electron density increases again with altitude to a greater maximum of the order of 10^6/cc at the *F*-layer and which is at about 300 km. As the altitude further increases, the electron density decreases. The region in which there are a considerable number of free electrons is known as the ionosphere. In the ionosphere, the diffusion and photochemical and photoelectric actions are very important. The thermosphere extends a few hundred kilometers above the *F*-layer of ionosphere until the temperature of the atmosphere reaches a maximum of 1500 to 2000 K.

(2) Exosphere— metasphere and protosphere. The temperature of the atmosphere above the thermosphere remains almost constant for a considerable altitude. This isothermal region was called the exosphere because it was thought that in this region the laws of gas kinetics no longer applied. In this region, the particles will suffer few collisions and when they move upward, they may escape from the earth gravitational field. Hence the name exosphere is used. In this region, we should not consider the air as a continuum and should consider the discrete character of the particles. Hence we have to use the rarefied gasdynamics to study the motion of the particles, particularly

using the free molecule flow analysis.

In the lower portion of the exosphere, the air is still mainly un-ionized. We should use the free molecule flow of neutral particles with gravitational force as the main body force to study the dynamic process of the atmosphere. This region is known as metasphere. In the upper portion of the exosphere, the gas particles of the atmosphere are almost fully ionized and the protons are more abundant than the neutral hydrogen. This region is known as the protosphere. In the protosphere, we should consider both the gravitational force and the electromagnetic force to study the dynamic process of the atmosphere. In the exosphere, we should find out the molecular distribution function and then determine the statistical average of various properties of the atmosphere.

10. Problems

1. From the simple kinetic theory of gases, derive the equation of state of a perfect gas (2.13).

2. Show that from the simple kinetic of gases, the ratio of specific heats γ may be written as follows:

$$\gamma = \frac{c_p}{c_v} = \frac{n+2}{n} \tag{2.65}$$

where n is the number of degrees of freedom of a molecule of the gas.

3. Derive the equation of state for an anisentropic process for an ideal gas with constant γ, i.e.,

$$\frac{p}{p_0} = \left(\frac{\rho}{\rho_0}\right)^{\gamma} \exp\left(\frac{S-S_0}{c_v}\right) \tag{2.66}$$

where subscript o refers to the value at the reference state.

4. Calculate the pressure, density, and absolute temperature in the standard atmosphere at the following altitudes:

(a) 10,000 ft., (b) 20,000 ft. and (c) 50,000 ft. from the sea level.

5. For a Van der Waals gas (2.16), find the critical pressure p_c, critical specific volume v_c and the critical temperature T_c in terms of the constant b_1, b_2 and R. If we express the state variables p, v and T in terms of their critical values, for the expression of the equation of state in the reduced state variables $p_r = p/p_c$, $v_r = v/v_c$ and $T_r = T/T_c$.

6. For a real gas, the internal energy U_m depends on both the absolute temperature T and the specific volume v. Show that the Joule-Thompson effect for a real gas demonstrates that the internal energy is a function of both the temperature and the specific volume.

7. The dryness fraction λ_d is defined as the ratio of the mass of the

saturated vapor to the total mass of the mixture of saturated vapor and saturated liquid. Express the total specific volume and the enthalpy of the mixture in terms of the dryness fraction λ_d and their corresponding values of saturated vapor and saturated liquid.

8. For an ideal perfect gas in an atmosphere without wind, the equation of motion is

$$\frac{dp}{p} = -\frac{dL}{RT/g} = -\frac{dL}{H_c} \tag{2.67}$$

where g is the gravitational acceleration and $H_c = RT/g$ is the scale height. For a polytropic gas:

$$p\rho^{-n_0} = \text{constant} = K = p_1 \rho_1^{-n_0} \tag{2.68}$$

where n_0 is a constant and known as polytropic index.

Find the relation between the polytropic index n_0 and the lapse rate λ of the atmosphere.

Show that if $n_0 < \gamma$, the atmosphere is stable
if $n_0 > \gamma$, the atmosphere is unstable and
if $n_0 = \gamma$, the atmosphere is neutrally stable.

References

1. Pai, S. I, *Modern Fluid Mechnics*, Science Press, Beijing, 1981, distributed by Van Nostrand Reinhold Company, New York.
2. Pai, S. I, *Two-phase Flows*, Vieweg Verlag, Braunschweig, West Germany, 1977.
3. Schmidt, E., *Thermodynamics*, Oxford University Press, 1949.
4. Prigogine, I., and Defay, R., *Chemical Thermodynamics*, Longmans, Green and Co., London and New York, 1954.
5. Fowler, R. H., *Statistical Mechancs*, Cambridge University Press, 1936.
6. Fowler, R. H., and Guggenheim, E. A., *Statistical Thermodynamics*, Cambridge University Press, 1939.
7. Rossini, F. D. (Ed), Thermodynamics and Physics of Matter, I. High Speed Aerodynamics and Jet Propulsion, Princeton University Press, 1955.
8. Chapman, S., and Cowling, T. G., *The Mathematical Theory of Nonuniform Gases*, Cambridge University Press, 1939.
9. Hirschfelder, J. O., Curtiss, C. F., and Bird, R. B., *Molecular Theory of Gases and Liquids*, John Wiley and Sons, Inc., New York, 1954.
10. Patterson, G. N., *Molecular Flow of Gases*, John Wiley and Sons, Inc., New York, 1956.
11. Pryde, J. A., *The Liquid State*, Hutchinson University Library, London, 1966.
12. *ASME Steam Tables*, 1967.
13. Krieger, F. J., and White, W. B., The composition and thermodynamic properties of air at temperature from 500 to 8000 K and pressures from 0.00001 to 100 atmospheres, *Report* R – 149, The Rand Corp., California, 1949.
14. Hilsenrath, J., and Beckett, C. W., Thermodynamic properties of Argon free air (0.78847N_2, 0.21153O_2) to 15,000 K, *Report* 3991 Nat. Bureau of Standards, 1955.
15. Woolley, H. W., Thermodynamic functions for atomic oxygen and atomic oxygen ions, *Report*

3989 Nat. Bureau of Standards, 1955.

16. Woolley, H. W., Thermodynamic functions for atomic nitrogen and atomic nitrogen ions, *Report* 3990 Nat. Bureau of Standards, 1955.
17. Beckett, C. W., Hilsenrath, J., et al., Tables of thermal properties of gases, *Nat. Bureau of Standards Circular*, **564**, Government Printing Office, Washington, D. C., 1955.
18. Bethe, H. A., Specific heats of air to 25,000℃, *Dept. of Commerce Report*, PB-27307, 1942.
19. Gunn, J. G., Relaxation time effects in gas dynamics, *British ARC R and M*, No. 2338, 1952.
20. Gaydon, A. G., *Dissociation Energies and Spectra of Diatomic Molecules*, John Wiley and Sons, Inc., New York, 1947.
21. Lighthill, M. J., Dynamics of a dissociating gas, pt. I. Equilibrium flow, *Jour. Fluid Mech.* **2**, pt. 1, Jan. 1957, pp. 1 –32.
22. Hirschfelder, J. O., Bird, R. B., and Spotz, E. L., Viscosity and other physical properties of gases and mixtures, *ASME Trans.* **71**, 1949, pp. 921 –937.
23. Pai, S. I, *Viscous Flow Theory -I Laminar Flow*, D. Van Nostrand Company, Inc., New York, 1956.
24. Pai, S. I, *Viscous Flow Theory -II Turbulent Flow*, D. Van Nostrand Company, Inc., New York, 1957.
25. Roberts, H. E., The earth's atmosphere, Aero. *Eng. Review*, Oct. 1949, pp. 18 –31.
26. Diehl, W. S., Standard atmosphere— Tables and data, *NACA Report* No. 218, 1925.

Chapter III

ONE-DIMENSIONAL FLOW OF AN INVISCID COMPRESSIBLE FLUID[1-9]

1. Introduction

In this chapter we shall discuss the one-dimensional flow of an inviscid compressible fluid. Since the effect of viscosity is limited to regions of very large variation in velocity, such as the boundary layer or a shock front, we may neglect viscosity in the study of flow fields outside the boundary layer or the shock. This gives us a very good approximation to the real physical situation in a very large group of aerodynamic problems, as, for example, the pressure distribution over a solid body in a fluid flow. In Chapters III to XIV we shall discuss only the flow problems of an inviscid compressible fluid.

In order to bring out the essential features of the flow of a compressible fluid, we shall discuss the one-dimensional case first. We shall be concerned only with the velocity along a streamtube and the accompanying variation in pressure, density, and temperature of the fluid. To a first approximation, this is the case of the flow in a nozzle whose cross-sectional area varies slightly along its axis.

In § 2, we derive the energy equation of the one-dimensional steady flow. In flow problems of a compressible fluid, the most important parameter is the Mach number (1.16). This will be discussed in § 3 together with the definition of velocity of sound. In § 4, isentropic steady flow through a nozzle will be discussed.

In § 5, we discuss the general relations between pressure and velocity in an isentropic steady flow.

In § 6, the propagation of sound waves will be discussed, sound waves being waves of infinitesimal amplitude. In § 7, waves of finite amplitude will be discussed; particular attention will be given in the case in which the sound wave may develop into a shock wave.

In the above sections, we are concerned with the adiabatic flow of a perfect gas without thermal radiation effects. In § 8, we shall discuss the effects of heat addition and mass addition to the flow field, while in § 9, we shall consider the effects of thermal radiation on the flow field, particularly when the thermal radiation effects should be considered, and their influence on the flow field.

2. Energy equation

For one-dimensional steady flow of an inviscid compressible fluid in a channel, the momentum law gives:

$$\frac{dp}{\rho}+qdq=0 \tag{3.1}$$

where q is the velocity, p is the pressure, and ρ is the density of the fluid. If we integrate Eq. (3.1) from state 1 to state 2, we have

$$\int_1^2 \frac{dp}{\rho}+\frac{q_2^2-q_1^2}{2}=0 \tag{3.2}$$

which is the well known Bernoulli equation. It is not an energy law, because the integral in equation (3.2) depends on the path from 1 to 2. Eq. (3.2) states that the kinetic energy of unit mass of the fluid is equal to the area in the p-$(1/\rho)$ diagram (Fig. 3.1).

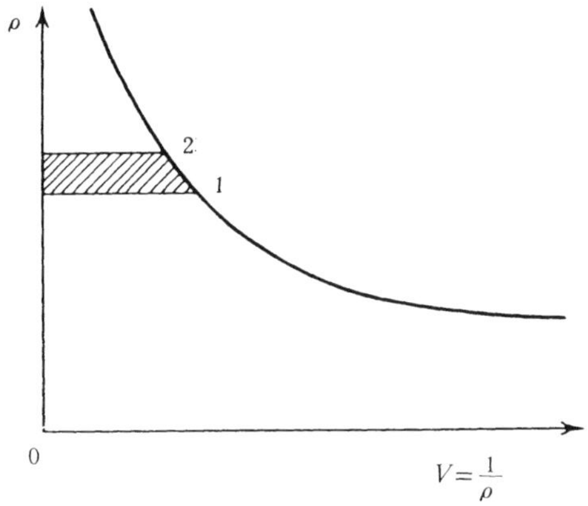

Fig. 3.1 Pressure-volume diagram

In order to derive the energy law, we use Eqs. (2.18) and (2. 50), i.e.,

$$dQ=dH-\frac{dp}{\rho}=dH+qdq=d\left(H+\frac{q^2}{2}\right) \tag{3.3}$$

For an adiabatic process, i. e., a process in which no heat is added, conducted, or radiated from the flow field, $dQ=0$. Integration of Eq. (3.3) gives

$$H+q^2/2=\text{constant}=H_0 \tag{3.4}$$

which is the energy equation for steady adiabatic flow. H_0 is known as the stagnation enthalpy of the system.

If we apply the perfect gas law (2.3) to Eq. (3.4), we have

$$\frac{q^2}{2}+\frac{\gamma}{\gamma-1}\frac{p}{\rho}=\text{constant}=\frac{\gamma}{\gamma-1}\frac{p_0}{\rho_0} \tag{3.5}$$

3. Velocity of sound and Mach number

We ask the question: Is it possible for a small pressure change in a compressible fluid to maintain itself? In answering this question, let us consider the steady flow in a nozzle of constant cross-sectional area. The conservation of mass gives

$$\rho q=\text{constant} \tag{3.6}$$

which upon logarithmic differentiation becomes

$$\frac{d\rho}{\rho}+\frac{dq}{q}=0 \tag{3.7}$$

Substituting Eq. (3.1) into Eq. (3.7) yields

$$q^2=\frac{dp}{d\rho} \tag{3.8}$$

For isentropic process, the pressure p is a function of density ρ only; hence $dp/d\rho$ is a function of p or ρ only. The velocity q defined by Eq. (3.8) may be considered as a critical velocity which satisfies both the equation of motion (3.1) and the equation of continuity (3.7), and represents the velocity of propagation in a fluid of small disturbance with its shape unchanged.

We shall call this velocity the velocity of sound and designate it by a:

$$a=\sqrt{dp/d\rho} \tag{3.9}$$

We shall show that the velocity of sound is the velocity of propagation of small disturbance in the unsteady flow problem in § 6.

If we assume an isentropic change (2.63), we have

$$a=\sqrt{\gamma p/\rho}=\sqrt{\gamma RT} \tag{3.10}$$

which shows that a is a function of T only. If p_0 and ρ_0 are the absolute pressure and density of the fluid at a stagnation point, respectively, the corresponding velocity of sound is

$$a_0=\sqrt{\gamma p_0/\rho_0} \tag{3.11}$$

The energy Eq. (3.5) may be written as follows:

$$a^2 = a_0^2 - \frac{\gamma - 1}{2} q^2 \tag{3.12}$$

Eq. (3.12) shows the variation of local velocity of sound with the velocity of the flow q. From Eq. (3.12) we have

$$q^2 = \frac{2\gamma}{\gamma - 1} \frac{p_0}{\rho_0} \left[1 - \left(\frac{p}{p_0} \right)^{\frac{\gamma - 1}{\gamma}} \right] = \frac{2a_0^2}{\gamma - 1} \left[1 - \left(\frac{p}{p_0} \right)^{\frac{\gamma - 1}{\gamma}} \right] \tag{3.13}$$

and

$$p = p_0 \left(1 - \frac{\gamma - 1}{2} \frac{q^2}{a_0^2} \right)^{\frac{\gamma}{\gamma - 1}} \tag{3.14}$$

Now let us find the velocity of the flow which is equal to the local velocity of sound. Putting $q = a$ in Eq. (3.12), we have

$$a^* = q = a_0 \sqrt{\frac{2}{\gamma + 1}} = 0.913\ a_0 \text{ (if } \gamma = 1.4) \tag{3.15}$$

where a^* is sometimes called critical velocity of sound. The corresponding pressure at $q = a$ will be the critical pressure $p^* = p_0 \left(\frac{\gamma + 1}{2} \right)^{-\frac{\gamma}{\gamma + 1}}$. We also see that the greatest attainable value of q occurs when $p = 0$. Hence from Eq. (3.14) we have

$$q_{max} = a_0 \sqrt{\frac{2}{\gamma - 1}} = 2.236\ a_0 \text{ (if } \gamma = 1.4) \tag{3. 16}$$

We conclude that the flow is in the subsonic range when

$$0 < q < a_0 \sqrt{\frac{2}{\gamma + 1}} = a^* \tag{3.17a}$$

and in the supersonic range when

$$a^* = a_0 \sqrt{\frac{2}{\gamma + 1}} < q < a_0 \sqrt{\frac{2}{\gamma - 1}} \tag{3.17b}$$

A dimensionless quantity which signifies the effect of compressibility is the Mach number which is defined as

$$M = q/a \tag{3.18}$$

For subsonic flow, $M<1$, and for supersonic flow, $M>1$. Sometimes we write $M^*=q/a^*$. It is easy to show that

$$M^{*2}=\frac{q^2}{a^{*2}}=\frac{(\gamma+1)M^2}{(\gamma-1)M^2+2} \tag{3.19}$$

$$M^2=\frac{2M^{*2}}{(\gamma+1)-(\gamma-1)M^{*2}} \tag{3.20}$$

The relation between the pressure and velocity at different Mach numbers will be discussed in § 5.

From Eq. (3.1) we have

$$\frac{dp}{dq}=-q\rho$$

and

$$\frac{d^2p}{dq^2}=-\rho-q\frac{d\rho}{dp}\frac{dp}{dq}=-\rho\left(1-\frac{q^2}{a^2}\right)=-\rho(1-M^2)\lesseqgtr 0$$

if

$$M\lesseqgtr 1 \tag{3.21}$$

Fig. 3.2 shows the relation between the pressure and velocity of the flow, where the point of inflection occurs at $M=1$, and the curvatures of the curve for subsonic and supersonic flows have opposite signs as given by Eq. (3.21). It shows that the behavior of subsonic flow is different from that of supersonic flow. We shall discuss this point in detail in later chapters.

4. One-dimensional steady flow in a nozzle

For one-dimensional steady flow of a compressible fluid in a nozzle, the conservation of mass gives

$$\rho qA=\text{constant} \tag{3.22}$$

where A is the cross-sectional area of the nozzle which is, in general, a function of x, the coordinate along the nozzle. Logarithmic differentiation of Eq. (3.22) gives

$$\frac{dA}{A}+\frac{dq}{q}+\frac{d\rho}{\rho}=0 \tag{3.23}$$

Substituting Eqs. (3.1) and (3.9) into Eq. (3.23), we have

$$\frac{dq}{dx}\frac{1}{q}\left(1-\frac{q^2}{a^2}\right)=-\frac{1}{A}\frac{dA}{dx} \tag{3.24}$$

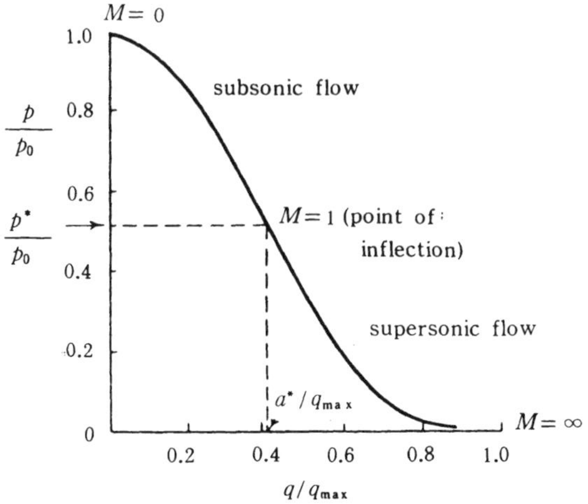

Fig. 3.2 Relation between pressure p and velocity q for isentropic flow

Substituting Eq. (3.12) into Eq. (3.24), we have

$$f(q)\frac{dq}{dx}=g(x) \tag{3.24a}$$

where $f(q)$ is a known function of q only and $g(x)$ is a known function of x only. Eq. (3.24a) can always be solved for $q(x)$ since the variables are separable, although at times graphical integration may be necessary.

Let us discuss the general behavior of the solution of Eq. (3.24). Eq. (3.24) may be rewritten as follows:

$$\frac{dq}{dx}=-\frac{\dfrac{1}{A}\dfrac{dA}{dx}}{a^2-q^2}a^2q \tag{3.25}$$

At the neck, $dA/dx=0$. Hence the maximum velocity occurs at the neck of the nozzle (since $dq/dx=0$ and $d^2q/dx^2<0$) unless $q=a$. If we plot the solution of Eq. (3.25) for a given inlet pressure and various exit pressures, we would have the results as shown in Fig. 3.3. There are two distinct regions, ONB and BNC. In the region ONB, all curves except the ONB curve have horizontal tangents, $dq/dx=0$, at the next $x=x_N$. In this region, the flow is always subsonic, except for the possible sonic velocity at the neck which corresponds to the exit pressure p_B. This region may be designated as the "venturi tube region" where the exit pressure is higher than, or equal to, the pressure p_B. If the exit is at the neck, p_B will be the critical pressure

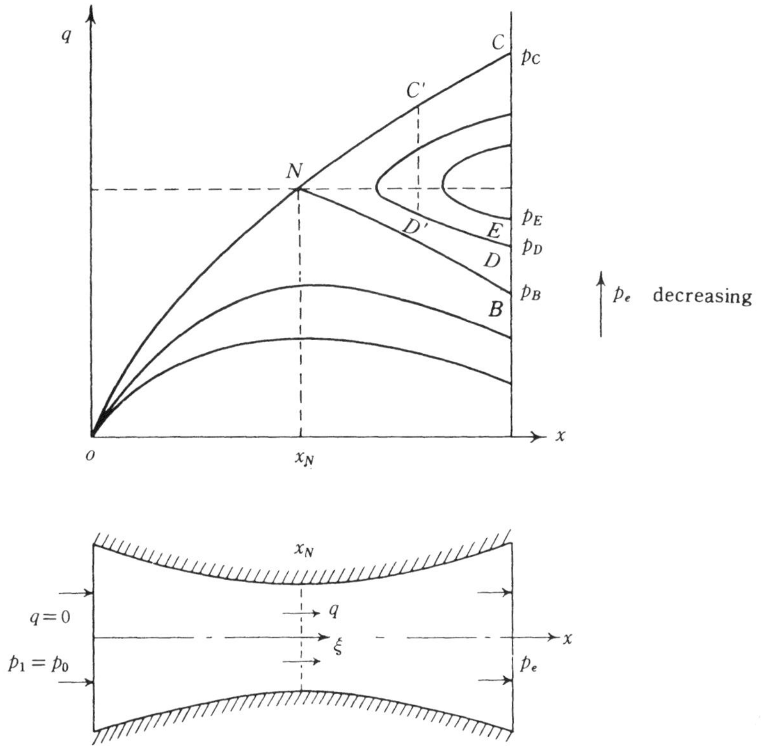

Fig. 3.3 Velocity distributions in a nozzle

$p^* = p_0\left(\frac{\gamma+1}{2}\right)^{-\frac{\gamma}{\gamma+1}}$; where p_0 is the stagnation pressure. The point N in Fig. 3.3 is a singular point of the differential Eq. (3.25). The nature of this singular point may be obtained in the following way. Let us take an origin of q-x diagram at the minimum section, and denote ξ as the distance along the nozzle axis from this origin. At a small distance from this origin, the cross-sectional area of the nozzle will be

$$A = A_{\min} + b\xi^2 + \cdots \tag{3.26}$$

where b is a constant for a given nozzle.

The velocity at the section ξ will be

$$q = a + \frac{dq}{dx}\xi + \cdots \tag{3.27}$$

Substituting Eqs. (3.26) and (3.27) into Eq. (3.25), we have at the neck, $\xi = 0$,

$$\frac{dq}{dx} = \pm a\sqrt{b/A_{\min}} \tag{3.28}$$

Consequently at the singular point N, the solution has two tangents with the slopes $\pm a\sqrt{(b/A_{\min})}$. Thus the singular point is a saddle point[8, 9, 10]. If we plot up the solution of the Eq. (3.25), we have Fig. 3.3 with two distinct regions with the saddle point at N.

In the region BNC, all the curves have vertical tangents at $q=a$, when $dA/dx \neq 0$. This region may be designated as "de Laval nozzle region" where the flow may be either subsonic or supersonic. If the flow is in the region BNC, the flow will go from the initial pressure p_0 isentropically to the exit in the nozzle of Fig. 3.3 either along ONB or ONC, depending on whether the exit pressure is p_B or p_C. If the exit pressure is higher than p_c but lower than p_B, say p_D, no continuous curve will go from the initial pressure p_0 to the exit pressure p_D. In this case, a normal shock occurs in the nozzle. The flow first follows the curve ONC up to the point C' and then it suddenly drops from C' to D' such that the pressure rise p_D'/p_C' is the pressure rise across a normal shock. From D' to D the flow again follows an isentropic process. Since there is a definite relation for the pressure rise across the normal shock (Chapter IV, § 2), the exit pressure determines the location of the normal shock such that the flow will be isentropic to the exit after the normal shock. The location of the normal shock may occur at the exit if p_E/p_C corresponds to the normal shock relation. When the exit pressure lies between the values p_E and p_C, an oblique shock occurs at the exit of the nozzle. When the exit pressure is less than p_C, expansion waves occur at the exit. When the exit pressure is lower than p_B, the flow condition from the entrance to the neck is unaffected by the change of exit condition. Hence the mass flow through the nozzle first increases as the exit pressure decreases for a given stagnation pressure until the critical pressure p_B is reached, i. e. until the velocity at the neck is sonic. Further decrease of the exit pressure will not change the mass flow; i. e., the mass flow through a given nozzle is constant if the exit pressure is below p_B.

5. Pressure and velocity relations in isentropic flow

Many practical flow problems may be approximated by isentropic flow. It is for this reason that it is of interest to find an explicit relation between the velocity and the pressure in isentropic flow. In many practical problems we consider the problem of a body moving with a constant velocity in the fluid at rest, or alternatively we consider a uniform flow passing over a body at rest.

We usually express the pressure at an arbitrary point in the field in terms of the pressure of the uniform flow. The uniform flow is called the free stream. The local velocity is also expressed in terms of the free stream velocity. Let us denote the value of any quantity, such as pressure and velocity, etc, by the subscript ∞ in the free stream, the subscript 0 at a stagnation point, and without subscript in the flow field. Then by Eq. (3.14), we have

$$\frac{p_\infty}{p_0} = \frac{1}{\left(1+\frac{\gamma-1}{2}M_\infty^2\right)^{\gamma(\gamma-1)}} \tag{3.29}$$

where $M_\infty = q_\infty / a_\infty$.

From Eqs. (3.14) and (3.29), we have

$$\frac{p}{p_\infty} = \left[1+\frac{\gamma-1}{2}M_\infty^2\left(1-\frac{q^2}{q_\infty^2}\right)\right]^{\frac{\gamma}{\gamma-1}} \tag{3.30}$$

which is an exact relation between the pressure p and velocity q in isentropic flow. In many practical problems, Eq. (3.30) may be simplified by proper approximations. The following are some of these approximations:

(a) *Small free stream Mach number*. When M_∞ is small, we may expand Eq. (3.30) in power series of M_∞^2 and have

$$\begin{aligned}\frac{p}{p_\infty} &= 1+\frac{\gamma}{2}\left(1-\frac{q^2}{q_\infty^2}\right)M_\infty^2+\frac{\gamma}{8}M_\infty^4\left(1-\frac{q^2}{q_\infty^2}\right)^2 \\ &\quad +\frac{\gamma(2-\gamma)}{48}M_\infty^6\left(1-\frac{q^2}{q_\infty^2}\right)^3+\cdots\end{aligned} \tag{3.31}$$

It is customary to define a pressure coefficient C_p as follows:

$$C_p = \frac{p-p_\infty}{\frac{\rho_\infty}{2}q_\infty^2} \tag{3.32}$$

Substituting Eq. (3.31) into Eq. (3.32), we have

$$\begin{aligned}C_p &= \left(1-\frac{q^2}{q_\infty^2}\right)\left[1+\frac{M_\infty^2}{4}\left(1-\frac{q^2}{q_\infty^2}\right)+\frac{2-\gamma}{24}M_\infty^4\left(1-\frac{q^2}{q_\infty^2}\right)^2\right. \\ &\quad \left.+\frac{(2-\gamma)(3-2\gamma)}{192}M_\infty^6\left(1-\frac{q^2}{q_\infty^2}\right)^3+\cdots\right]\end{aligned} \tag{3.33}$$

In incompressible fluid flow, $M_\infty = 0$; using this, Eq. (3.33) becomes

$$C_p = 1 - \frac{q^2}{q_\infty^2}$$

Therefore the factor in the bracket represents the compressibility effect.

(b) *Small difference between the local velocity and the free stream velocity*. If the local velocity differs slightly from the free stream velocity, we may write

$$q = q_\infty + \Delta q \tag{3.34}$$

where $\Delta q \ll q_\infty$.

Now we expand Eq. (3. 30) in power series of $(\Delta q / q_\infty)$ and have

$$\begin{aligned} \frac{p}{p_\infty} &= \left\{1 + \frac{\gamma - 1}{2} M_\infty^2 \left[1 - \left(1 + \frac{\Delta q}{q_\infty}\right)^2\right]\right\}^{\frac{\gamma}{\gamma-1}} \\ &= \left\{1 - (\gamma - 1) M_\infty^2 \left[\frac{\Delta q}{q_\infty} + \frac{1}{2}\left(\frac{\Delta q}{q_\infty}\right)^2\right]\right\}^{\frac{\gamma}{\gamma-1}} \\ &= 1 - \gamma M_\infty^2 \left[\frac{\Delta q}{q_\infty} + \frac{1}{2}\left(\frac{\Delta q}{q_\infty}\right)^2\right] + \frac{\gamma M_\infty^4}{2}\left(\frac{\Delta q}{q_\infty}\right)^2 \\ &\quad + \cdots \end{aligned} \tag{3.35}$$

The pressure coefficient (3.32) for this case becomes

$$C_p = -2\frac{\Delta q}{q_\infty}\left[1 + \frac{1}{2}(1 - M_\infty^2)\frac{\Delta q}{q_\infty} + \cdots\right] \tag{3.36}$$

If the terms $(\Delta q / q_\infty)^2$ and $M_\infty^2 (\Delta q / q_\infty)^2$ are small and negligible, Eq. (3.36) becomes

$$C_p = -2(\Delta q / q_\infty) \tag{3.37}$$

This simplified formula has been widely used in the linearized theory of compressible flow where the free stream Mach number M_∞ is not very large.

(c) *Small disturbance Δq but large free stream Mach number M_∞*. In this case, the term $M_\infty^2 (\Delta q / q_\infty)^2$ is no longer small, but $(\Delta q / q_\infty)^2$ is still negligible when compared with $\Delta q / q_\infty$, and the simple formula (3.37) does not hold. We may drop the term $(\Delta q / q_\infty)^2$ in the second line of Eq. (3.35) and have the approximate formula

$$\frac{p}{p_\infty} = \left\{1 - (\gamma - 1) M_\infty^2 \frac{\Delta q}{q_\infty}\right\}^{\frac{\gamma}{\gamma-1}} \tag{3.38}$$

The corresponding formula for the pressure coefficient is

$$C_p = \frac{2}{\gamma}\frac{1}{M_\infty^2}\left\{\left[1-(\gamma-1)M_\infty^2\frac{\Delta q}{q_\infty}\right]^{\frac{\gamma}{\gamma-1}}-1\right\} \tag{3.39}$$

This formula may be used in hypersonic flow over a thin body where the disturbance velocity is small in comparison with the free stream velocity (q_∞) and the Mach number M_∞ is very large.

6. Nonsteady one-dimensional flow[6]. Sound wave

We now shall consider nonsteady one-dimensional isentropic flow. The fundamental equations are (cf. Chapter V):

$$\frac{\partial u}{\partial t}+u\frac{\partial u}{\partial x} = -\frac{1}{\rho}\frac{\partial p}{\partial x} \tag{3.40a}$$

$$\frac{\partial \rho}{\partial t}+\frac{\partial \rho u}{\partial x} = 0 \tag{3.40b}$$

$$p = p(\rho) \tag{3.40c}$$

Eq. (3.40) is obtained by assuming that all quantities depend only on the time t and a single coordinate x. The velocity vector has the direction of the x-axis. Hence $q=u$, $v=w=0$. Here we assume that the fluid is a barotropic gas, i. e. the pressure is a function of density only. Eq. (3.40c) replaces the equation of state and that of energy. For isentropic flow, Eq. (3.40c) has the form of Eq. (2.63) with $s=s_0$.

First we shall investigate the case of a small disturbance, i. e. the case where the velocity u and its derivatives are small, and the density ρ can be represented as

$$\rho = \rho_0 + \rho' \tag{3.41}$$

where ρ_0 is the density of the fluid in the undisturbed state and ρ' is the change of density. We assume that $\rho' \ll \rho_0$.

If we neglect higher-order terms, such as $u\dfrac{\partial u}{\partial x}$ in comparison with the first-order term, such as $\dfrac{\partial u}{\partial t}$, Eqs. (3.40a) and (3.40b) become, respectively,

$$\frac{\partial u}{\partial t} \cong -\frac{a^2}{\rho}\frac{\partial \rho}{\partial x} \cong -\frac{a_0^2}{\rho_0}\frac{\partial \rho}{\partial x} \tag{3.42a}$$

$$\frac{\partial \rho}{\partial t} = -\rho_0\frac{\partial u}{\partial x} \tag{3.42b}$$

where $a^2 = \frac{dp}{d\rho}$, and $a_0 = \sqrt{\gamma p_0 / \rho_0}$ is the sound velocity in the undisturbed fluid. If u is eliminated from Eq. (3.42), we have

$$\frac{\partial^2 \rho}{\partial t^2} = a_0^2 \frac{\partial^2 \rho}{\partial x^2} \tag{3.43a}$$

or, if we eliminate ρ, we have

$$\frac{\partial^2 u}{\partial t^2} = a_0^2 \frac{\partial^2 u}{\partial x^2} \tag{3.43b}$$

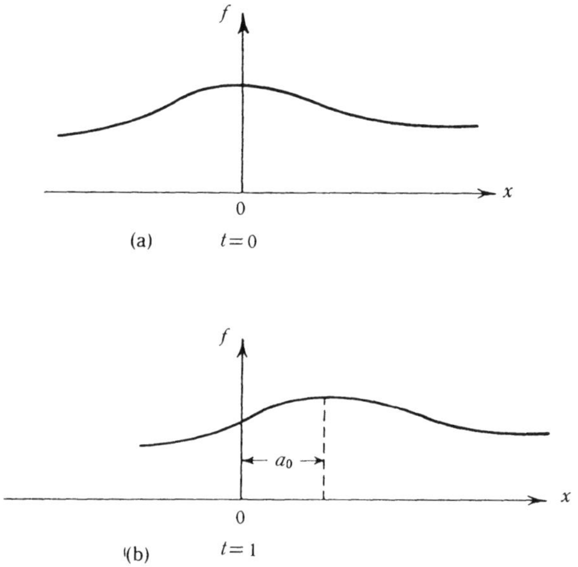

Fig. 3.4 Sound wave propagation

This is the well known one-dimensional wave equation whose general solution is

$$\rho = f(x - a_0 t) + g(x + a_0 t) \tag{3.44}$$

where f and g are arbitrary functions[11].

Let us investigate the significance of Eq. (3.44). First we consider the solution $f(x - a_0 t)$. We suppose that at $t=0$, we have the density distribution $\rho = f(x)$ shown in Fig. 3.4(a). One second later, we have $\rho = f(x - a_0)$, i.e. the density distribution has been shifted along the x-axis a distance a_0 without suffering any distortion (Fig. 3.4b). The interpretation is that the density distribution is propagated without distortion at a velocity a_0 to the right. Similarly the solution $g(x + a_0 t)$ represents a wave propagation to the left.

Since we are considering only an infinitesimal disturbance, the velocity of propagation depends only on the fluid and is independent of the nature of disturbance. This is precisely the case for sound waves. If the disturbance is finite, this is not true. We shall discuss the finite disturbance in § 7.

If we introduce a diagram of time t plotted against the spatial coordinate x (Fig. 3.5), along the lines $x-a_0t$=constnat, the solution $f\,(x-a_0t)$ = constant, while along the lines $x+a_0t$=constant, the solution $g\,(x+a_0t)$ = constant. These lines are called the characteristics of the differential equation (3.43). We shall discuss this point in some detail in Chapter XI.

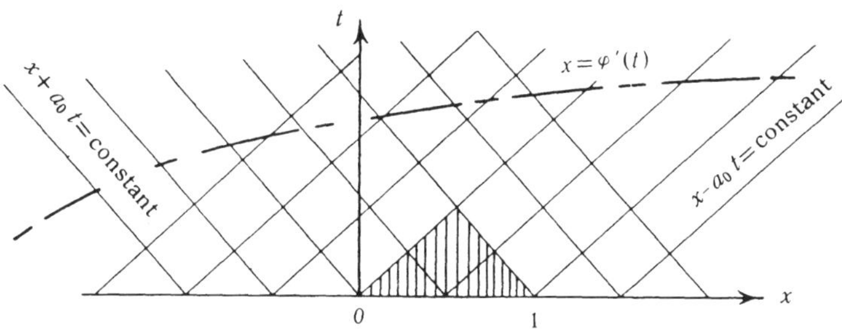

Fig. 3.5 x-t diagram of one-dimensional nonsteady flow

If we know the value of ρ and $\partial\rho/\partial t$ for all values of x from $-\infty$ to $+\infty$ for $t=0$, we know the value of the density distribution at all later times t and all points x. It is easy to show that

$$\begin{aligned} f' &= \frac{df(x)}{dx} = \frac{1}{2}\left(\frac{\partial\rho}{\partial x}\right)_{t=0} + \frac{1}{2a_0}\left(\frac{\partial\rho}{\partial t}\right)_{t=0} \\ g' &= \frac{dg(x)}{dx} = \frac{1}{2}\left(\frac{\partial\rho}{\partial x}\right)_{t=0} - \frac{1}{2a_0}\left(\frac{\partial\rho}{\partial t}\right)_{t=0} \end{aligned} \qquad (3.45)$$

Since we know the values of ρ and $\partial\rho/\partial t$ at $t=0$ for all x, we may calculate $(\partial\rho/\partial x)_{t=0}$ and $(\partial\rho/\partial t)_{t=0}$, and hence f' and g', from Eq. (3.45). Finally f, g, and ρ can be determined.

If we know ρ and $\partial\rho/\partial t$ at $t=0$ for the segment $0\leqslant x\leqslant l$ only, the density distribution is completely determined only in the shaded triangular region shown in Fig. 3.5. To interpret this, let us consider an infinite long pipe. If we know ρ and $\partial\rho/\partial t$ at all points along the pipe at $t=0$, we know the density at all times. However, if we only know ρ and $\partial\rho/\partial t$ in the stretch $0\leqslant x\leqslant l$ at $t=0$, then at later times waves from outside this segment will begin to affect this segment. These outside influences can travel only with velocity a_0, i. e.

along the lines $x-a_0t=$constant and $x+a_0t=$constant. Since we can determine the density only at points unaffected by the outside influences, it is easy to see that the region ρ is known as triangular region shown in Fig. 3.5.

We shall next show that if ρ, $\partial\rho/\partial x$ and $\partial\rho/\partial t$ are given along an arbitrary line $x=\varphi(t)$, the solution $\rho(x, t)$ can be continued from this initial value, i. e. the higher differential quotients can be determined except when $dx/dt=\pm a_0$, which is, in the present case, $x-a_0t=$constant and $x+a_0t=$ constant.

Let us consider a line element dx and dt along the line $x=\varphi(t)$. We denote

$$\rho_t=\frac{\partial\rho}{\partial t} \text{ and } \rho_x=\frac{\partial\rho}{\partial x}$$

Along the line $x=\varphi(t)$, the increments of ρ_t and ρ_x are

$$\begin{aligned} d\rho_t&=\frac{\partial^2\rho}{\partial t^2}\,dt+\frac{\partial^2\rho}{\partial x\partial t}\,dx \\ d\rho_x&=\frac{\partial^2\rho}{\partial x\partial t}\,dt+\frac{\partial^2\rho}{\partial x^2}\,dx \end{aligned} \tag{3.46}$$

Substituting Eq. (3.43) into Eq. (3.46) gives

$$\begin{aligned} d\rho_t&=a_0^2\frac{\partial^2\rho}{\partial x^2}\,dt+\frac{\partial^2\rho}{\partial x\partial t}\,dx \\ d\rho_x&=\frac{\partial^2\rho}{\partial x\partial t}\,dt+\frac{\partial^2\rho}{\partial x^2}\,dx \end{aligned} \tag{3.47}$$

Eliminating $\dfrac{\partial^2\rho}{\partial x^2}$ from Eq. (3.47), we have

$$\frac{\partial^2\rho}{\partial x\partial t}=\frac{d\rho_t\,dx-a_0^2d\rho_x dx}{(dx)^2-a_0^2(dt)^2} \tag{3.48}$$

Obviously we can calculate $\dfrac{\partial^2\rho}{\partial x\partial t}$ from the given data along $x=\varphi(t)$ except when $dx=\pm a_0dt$, i.e. the inclination of the line is equal to $\pm a_0$. Similar calculation may be performed for other higher-order derivatives. These lines $dx=\pm a_0dt$ are the characteristics of the differential Eq. (3.43), and they are exceptional among all lines which we can draw in the $x-t$ plane. We are not able to continue our solution beyond the characteristics, although ρ and its first derivatives are known along the characteristics. The theory of characteristics will be discussed further in Chapter XI.

7. Waves with finite amplitude. Formation of a shock[1]

If the velocity of the disturbances is not small, we have to solve the nonlinear equation (3.40). In general, it is very difficult to solve this equation, particularly in analytical form. However there is a powerful mathematical tool, the method of characteristics, which may be used to solve this equation. This will be discussed in Chapter XI. Now we shall find a simple analytic solution of equation (3.40). From the results of small disturbances, we see that the differential equations for density and velocity are the same. Hence we may attempt to find a solution of the nonlinear equations (3.40) such that the density is a function of velocity u only.

$$\rho = f(u) \tag{3.49}$$

where the function f will be determined from Eq. (3.40). Substituting Eq. (3.49) into Eq. (3.40), the first two equations give respectively,

$$\frac{\partial u}{\partial t} + u\frac{\partial u}{\partial x} = -\frac{a^2}{f} f' \frac{\partial u}{\partial x} \tag{3.50a}$$

$$\frac{\partial u}{\partial t} + u\frac{\partial u}{\partial x} = -\frac{f}{f'} \frac{\partial u}{\partial x} \tag{3.50b}$$

where $f' = d\rho/du$. If Eq. (3.49) is a solution of Eq. (3.40), the two equations (3.50a) and (3.50b) must be the same; hence we have

$$\frac{a^2 f'}{f} = \frac{f}{f'}$$

or

$$\frac{d\rho}{\rho} = \pm\frac{du}{a} \tag{3.51}$$

Integration of Eq. (3.51) gives

$$u = \pm\int_{\rho_0}^{\rho} \frac{a d\rho}{\rho} \tag{3.52}$$

where subscript 0 refers to the stagnation condition. For isentropic flow, we have

$$u = \pm\frac{2}{\gamma - 1}(a - a_0) \tag{3.53}$$

Now if we substitute Eq. (3.51) into Eq. (3.40), we have

$$\begin{aligned} \frac{\partial u}{\partial t} + (u \pm a)\frac{\partial u}{\partial x} &= 0 \\ \frac{\partial \rho}{\partial t} + (u \pm a)\frac{\partial \rho}{\partial x} &= 0 \end{aligned} \tag{3.54}$$

It is easy to verify by direct differentiation that the general solutions of Eq. (3.54) are

$$u=F_1[x-(u\pm a)t],\ \rho=F_2[x-(u\pm a)t] \tag{3.55}$$

where F_1 and F_2 are arbitrary functions. In order to show the significance of Eq. (3.55), let us consider the solution

$$u=F_1[x-(u+a)t] \tag{3.56}$$

This solution means that the variation of u with respect to a point moving with velocity $u+a$ is zero, i.e., the disturbance is propagated with an instantaneous velocity $(u+a)$. If the velocity u is much smaller than the local sound velocity, we have $u=F_1(x-at)$ which is the case of sound wave discussed in § 6.

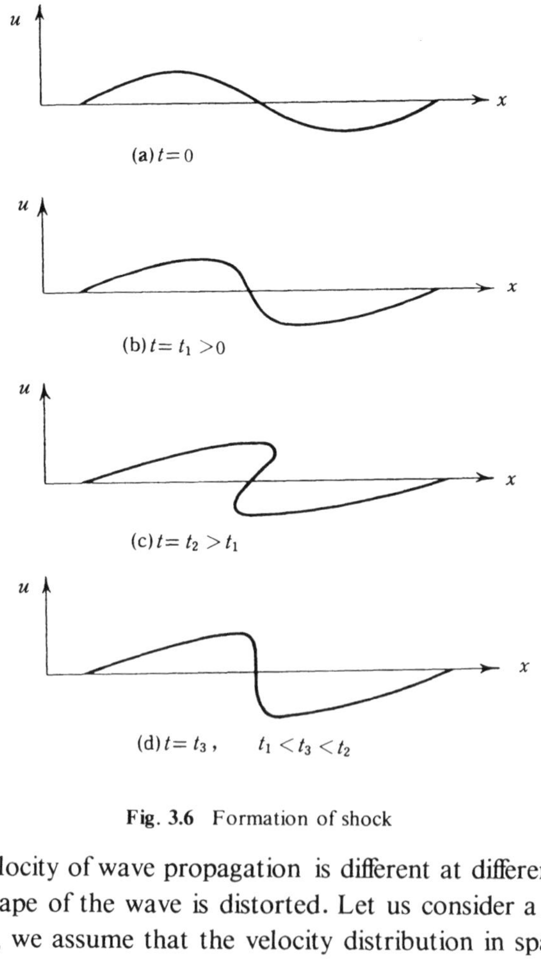

Fig. 3.6 Formation of shock

Since the velocity of wave propagation is different at different points in the flow field, the shape of the wave is distorted. Let us consider a case shown in Fig. 3.6. At $t=0$, we assume that the velocity distribution in space is given by

Fig. 3.6(a). At a time t_1 later, the crest has a tendency to overtake the trough, because the velocity of propagation at crest is larger than that at the trough. We have then Fig. 3.6(b). If we follow the mathematical solution (3.56) indefinitely at certain later time $t_2 > t_1$, we will have a picture of Fig. 3.6c in which we have three values of velocity at a given space x. This is an impossible condition for a longitudinal wave such as the sound wave because the velocity and the density must be single valued functions of position. Actually before the condition Fig. 3.5(c) is reached, a shock wave forms, and we have a large change in density and velocity in a very narrow region in which the fluid no longer behaves like a barotropic gas. We have the conditions sketched in Fig. 3.6(d). Under the assumption of an inviscid fluid, we may consider the shock wave as a surface of discontinuity which will be discussed in Chapter IV. Inside the shock wave, the effect of viscoscity must be considered; this will be discussed in Chapter XVII.

In the case of transverse waves, such as water waves, the case of Fig. 3.6c is possible; in fact, waves of this type, known as breakers, furnish a great deal of sport for beach swimmers. For water, it can be shown that the velocity of propagation of waves in shallow water of depth h is $a=\sqrt{gh}$, where g is the gravitational acceleration. In this case when the velocity of the wave u is large, the crests overtake the troughs, eventually resulting in an overturning phenomena as sketched in Fig. 3.6(c).

For the solution (3.56), the local sound velocity is

$$a = a_0 + \frac{\gamma - 1}{2}\, u \tag{3.57}$$

As the velocity of the flow u increases, the local velocity of sound increases. As a result, both the local temperature and pressure increase. Hence the wave is a compression wave. Thus we have shown that a compression wave will develop into a shock.

On the other hand, the solution $u=F_1[x-(u-a)t]$ represents an expansion wave if $F_1[x-(u+a)t]$ represents a compression wave. It is easy to show that such a wave will smooth itself out as the wave propagates.

8. One-dimensional steady flow with heat addition and mass addition[12–14]

So far we have considered only the adiabatic cases without mass addition. The analysis may be easily extended to the case with heat addition which is important in practical problems, such as those involving combustion and with mass addition where the mass may be introduced through the wall of the nozzle by injection or suction. Let us consider the problem discussed in § 4, of the one-dimensional flow in a nozzle now modified to take account of the heat addition dQ and the mass addition dm. The fundamental equations are then

$$dm = d(\rho A q) \tag{3.58a}$$

$$\rho q dq = -dp \tag{3.58b}$$

$$dQ = c_v dT + p\, d(1/\rho) \tag{3.58c}$$

$$p = \rho\, RT \tag{3.58d}$$

where dQ is the heat introduced into the system corresponding to the change in temperature dT and a change in density $d\rho$. In general, dQ may be considered as a constant or as a function of the position x in the nozzle. dm is the mass introduced in the system and dm may be considered as a constant or as a function of the axial distance of the nozzle. In the adiabatic case without mass addition, we have $dQ=0$ and $dm=0$ and then Eqs. (3.58) reduce to those corresponding equations in § 4.

From Eqs. (3.58), we may obtain the following differential relations from which the effect of heat addition and mass addition may be obtained.

$$\frac{dq}{q} = \frac{1}{(1-M^2)}\left(\frac{dQ}{H} + \frac{dm}{m} - \frac{dA}{A}\right) \tag{3.59a}$$

$$\frac{dp}{p} = \frac{-\gamma M^2}{(1-M^2)}\left(\frac{dQ}{H} + \frac{dm}{m} - \frac{dA}{A}\right) \tag{3.59b}$$

$$\frac{d\rho}{\rho} = \frac{-1}{(1-M^2)}\left(\frac{dQ}{H} + \frac{dm}{m} - \frac{dA}{A}\right) + \frac{dm}{m} - \frac{dA}{A} \tag{3.59c}$$

$$\frac{dT}{T} = \frac{1-\gamma M^2}{(1-M^2)}\left(\frac{dQ}{H} + \frac{dm}{m} - \frac{dA}{A}\right) + \frac{dA}{A} - \frac{dm}{m} \tag{3.59d}$$

$$\frac{dM^2}{M^2} = \frac{1+\gamma M^2}{(1-M^2)}\left(\frac{dQ}{H} + \frac{dm}{m} - \frac{dA}{A}\right) + \frac{dm}{m} - \frac{dA}{A} \tag{3.59e}$$

where $m=\rho q A$, $M=q/a$, $a=$local sound speed$=(\gamma RT)^{1/2}$, $H=$enthalpy per unit mass and $\gamma=$ratio of specific heats $=c_p/c_v$.

From Eqs. (3.59), we may easily obtain the changes in velocity q, pressure p, density ρ, temperature T and the Mach number of the flow M resulting from the heat addition dQ and the mass addition dm.

It is interesting to know that if we define an effective area variation dA/A' as follows:

$$\frac{dA'}{A'} = \frac{dA}{A} - \frac{dm}{m} \tag{3.60}$$

Eqs. (3.59) in terms of dA'/A' become exactly the same form of those without mass addition. Hence the effect of mass addition is the same as some modification of area variation. The new physical phenomenon in this section is mainly the effect of heat addition.

For simplicity, let us consider the case of a nozzle of uniform cross section, i. e. $dA=0$ or $dA'=0$. Here we have two distinguishable cases: (a) the subsonic case $M<1$ and (b) the supersonic case $M>1$.

(a) When the heat is added to a subsonic flow, i. e. $M<1$, from Eqs. (3.59) with $dA=dm=0$, we see that q increases, p and ρ decrease, while T increases or decreases according to $M<1/\sqrt{\gamma}$ or $M>1/\sqrt{\gamma}$. It is interesting to note that when the local Mach number is between $1/\sqrt{\gamma}$ and 1, the temperature of the flow decreases when the heat is being added, because the increase of kinetic energy is overbalancing the heat added. The local Mach number is increasing continuously toward $M=1$ as the heat is continually being added to the system.

We see that when $dQ\neq 0$, we have an infinite gradient in all variable q, p, etc. at $M=1$. Actually this means that our analysis ceases to be valid at $M=1$. In other words, when the heat is being added to the flow of a nozzle of constant cross section, the flow will always approach the sonic state but cannot become supersonic.

(b) When the heat is added to a supersonic flow, i. e. $M>1$, from Eqs. (3.59) with $dA=dm=0$, we see that q decreases, and p, ρ and T increase, while M decreases. Here again we have an infinite gradient in all variables q, p, etc. When the heat is being continually added to a supersonic flow in a nozzle of constant cross section, the flow will always approach the sonic state but cannot become subsonic.

For a nozzle of variable cross section area $dA\neq 0$ and/or with mass addition $dm\neq 0$, the effect of heat addition is much more complicated. We may have the transition from subsonic flow to supersonic flow with heat addition. Without heat addition, the transition from subsonic flow to supersonic flow occurs at $dA'=0$ as shown in § 4. But with heat addition $dQ=dQ(x)\neq 0$, the transition from subsonic flow to supersonic flow does not occur at $dA'=0$ but at the location $dQ/H-dA'/A'=0$. We shall discuss some cases like this in § 10 of this chapter. Furthermore, a similar problem occurs in magnetogasdynamics which will be discussed in Chapter XV, § 5.

9. One-dimensional flow in radiation gasdynamics[15–19]

In many new technological developments such as reentry of space vehicles, fission and fusion reactions and others, the temperature of the gas is so high and the density of the gas is so low that thermal radiation becomes a very important factor in the determination of the flow field. A complete analysis of such a high temperature flow field should be based on a study of the gasdynamic field and the thermal radiation field simultaneouly. The term "radiation gasdynamics"[15, 16] has been used for this new branch of gasdynamics. There are two groups of thermal radiation effects: one involves the radiation energy density and the radiation pressure, and the other is due to the radiative heat flux which may be considered as a new transport phenomenon in addition to the well-known transport phenomena of ordinary gasdynamics, i. e. viscosity and heat conductivity. Most of the current literature of radiation gasdynamics concerns only the radiative heat flux[16, 18, 19] and little has been done about the effects of radiation energy density and radiation pressure because the temperature in many practical problems such as reentry of space vehicles from Mars is not high enough that radiation pressure is still negligible. When the temperature reaches 10^5 K or higher, the radiation energy density and radiation pressure become important. It is also of academic interest to find out the essential features of the effects of radiation energy density and radiation pressure on the flow field of a very high temperature gas. In this section, we study only the effects of radiation energy density and radiation pressure while in Chapter XVII § 12, we shall study the effects of radiative heat flux. We consider an optically thick gas with such a high temperature and low pressure that the radiation pressure number R_p is defined as follows:

$$R_p = \frac{\text{radiation pressure}}{\text{gas pressure}} = \frac{a_R T^4}{3p} = \frac{a_R}{R}\frac{T^3}{\rho} \tag{3.61}$$

where the expression of radiation pressure $p_R = a_R T^4/3 = E_R/3$, i.e. Eq. (2.58) and the perfect gas law (2.13) have been used.

Fig. 3.7 shows some typical variation of R_p with temperature and gas pressure. We consider in this section that R_p is not negligibly small. It is easy to see from Fig. 3.7 that in ordinary gasdynamic problems the value of R_p is really negligible.

When the radiation pressure number is not negligible, we should replace the gas pressure by the total pressure which is the sum of gas pressure and radiation pressure in the equation of motion and that of energy and add the radiation energy density as an additional internal energy, i. e. E_R/ρ should be added to the gas internal energy U_m. For inviscid and thermal radiating gas, the equation of state may be written as Ref. 17.

It is easy to show that the equation of state of a radiating gas in equlibrium is as follows:

$$\frac{p}{p_o} = \left(\frac{\rho}{\rho_o}\right)^{\gamma} \exp\left(\frac{S - S_o}{c_v}\right) \exp[4(\gamma - 1)(R_{po} - R_p)] \tag{3.62}$$

where subscript o refers to values at a reference state. If both R_{po} and R_p are negligibly small, Eq. (3.62) reduces to the formula of ordinary gasdynamics, i. e. Eq. (2.63). If R_p is very large, the equation of state (3.62) may be written approximately as

$$S - S_o = \frac{4}{3a_R}\left(\frac{T^3}{\rho} - \frac{T_o^3}{\rho_o}\right) \tag{3.63}$$

For isentropic flow, $S = S_o$, Eq. (3.63) may be considered as a special case of Eq. (3.62) with effective γ or 4/3 but the difference of behavior of Eq. (3.62) and (3.63) should be noticed.

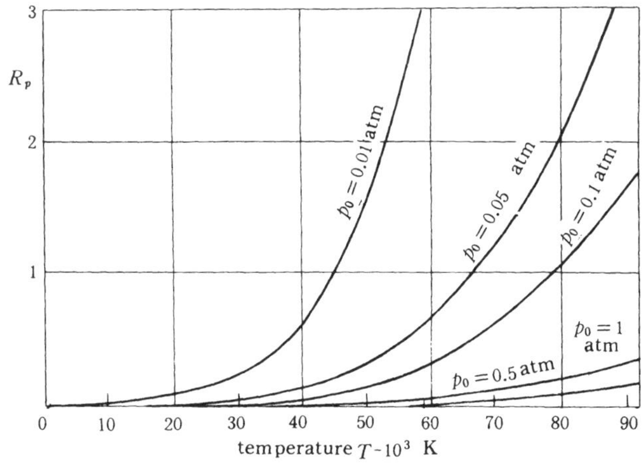

Fig. 3.7 Radiation pressure number R_p as a function of temperature T and gas pressure p_g

For isentropic flow $S = S_o =$ constant, the following relations of the variation of state variables of a radiating gas hold:

$$\frac{dT}{d\rho} = \frac{T}{\rho}\,\frac{(\gamma-1)(1+4R_p)}{1+12(\gamma-1)R_p} \tag{3.64}$$

$$\frac{dp}{d\rho} = \frac{p}{\rho}\,\frac{\gamma+16(\gamma-1)R_p}{1+12(\gamma-1)R_p} \tag{3.65}$$

$$\frac{dp_R}{dT} = 4\frac{p_R}{T} \tag{3.66}$$

and

$$\frac{d(p+p_R)}{d\rho} = \frac{p}{\rho}\,\frac{\gamma+20(\gamma-1)R_p+16(\gamma-1)R_p^2}{1+12(\gamma-1)R_p} = C_R^2 \tag{3.67}$$

where C_R is the effective sound speed of a radiating gas which is the speed of propagation of small disturbance in an ideal radiating gas and which is a function of both temperature T and density ρ, because the radiation pressure number R_p is a function of both T and ρ. Of course, for isentropic case, ρ may be considered as a function of temperature T, i.e. $\rho = \rho(T)$ but the function $\rho(T)$ still depends on some reference density and we cannot say that the sound speed C_R depends only on temperature because $\rho(T)$ depends on some reference density. Hence we conclude that the sound speed of a radiating gas depends on both temperature and density of the gas. Only in the limiting case, $R_p \to 0$, $C_R = \sqrt{\gamma RT}$ and C_R is a function of T only. For very large R_p, $R_p \to \infty$, we have

$$C_R = \frac{2}{3}\sqrt{a_R}\ \frac{T^2}{\sqrt{\rho}} = \sqrt{\frac{4}{3}RT}\sqrt{R_{p_o}} \tag{3.68}$$

where $R_{p_0} = p_{R_o}/p_o$ is a reference value of the radiation pressure number R_p. Since R_{p_o} is a constant, at a first glance of Eq. (3.68), one may attempt to draw the conclusion that C_R depends on temperature T only which behaves like a gas with effective value of ratio of specific heats $\gamma_e = 4/3$. However, the constant R_{p_o} depends on both the temperature T_o and the density ρ_o of the gas. Hence even when $R_p \to \infty$, we cannot draw the conclusion that C_R is independent of the density of the gas.

For isoenergetic radiation gas, the energy equation may be written as

$$c_p T + \frac{4}{3}\frac{E_R}{\rho} + \frac{1}{2}q^2 = \frac{\gamma}{\gamma - 1}\frac{p}{\rho} + 4\frac{p_R}{\rho} + \frac{1}{2}q^2 = \text{constant} \tag{3.69}$$

Eq. (3.69) may be written as follows:

$$\frac{q^2}{2} + \frac{C_R^2}{\gamma_e - 1} = \frac{C_{R_o}^2}{\gamma_{eo} - 1} \tag{3.70}$$

where

$$\gamma_e = \frac{\gamma + 4(\gamma - 1)R_p}{1 + 3(\gamma - 1)R_p} = \text{effective } \gamma \text{ of a radiating gas} \tag{3.71}$$

and

$$C_R = \sqrt{\gamma_e \frac{p + p_R}{\rho}} = \text{effective sound speed of a radiating gas} \tag{3.72}$$

and subscript o refers to the values at a stagnation point $q = 0$.

Eq. (3.72) is another way to state effective sound speed of a radiating gas which is practically the same as that given by Eq. (3.67). The results of Eqs. (3.67) and (3.72) are equal as shown in Fig. 3.3.

The maximum possible valocity q_m of a radiating gas from a given stagnation condition T_o and ρ_o is

$$q_m = \sqrt{\frac{2}{\gamma_{eo} - 1}}\ C_{R_o} \tag{3.73}$$

The maximum possible velocity q_m is a function of both the stagnation temperature T_o and the stagnation density ρ_o instead of a function of T_o only in the case of ordinary gasdynamics, i.e.,

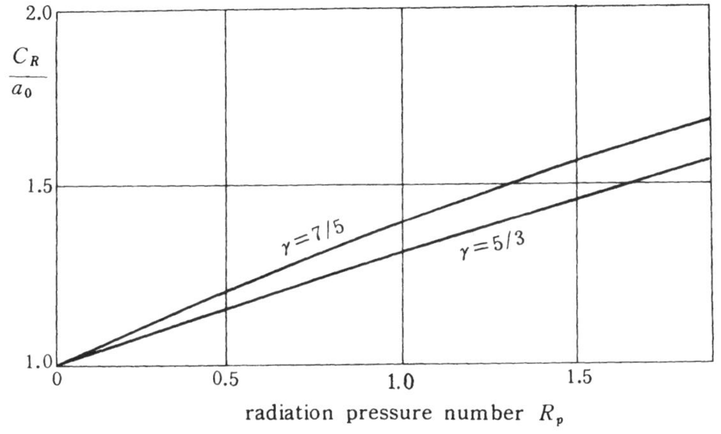

Fig. 3.8 The ratio of the effective sound speed of a radiating gas C_R to the ordinary sound speed of a gas $a_o=\gamma RT$ as a function of R_p and γ

$$R_p\to 0,\ q_m=\sqrt{\frac{2}{\gamma-1}}\ \sqrt{\gamma RT_o}$$

For the other limiting case $R_p\to\infty$, we have

$$q_m(R_p\to\infty)=\sqrt{\frac{8a_R}{3}}\ \cdot\ \frac{T_o^2}{\sqrt{\rho_o}} \tag{3.74}$$

The critical effective sound speed C_R^* , which occurs at $q=q_c=C_R=C_R^*$, is

$$q_c=C_R^*=C_{RO}\sqrt{\frac{2}{\gamma_{eo}-1}}\left(\frac{\gamma_{ec}-1}{\gamma_{ec}+1}\right)^{1/2} \tag{3.75}$$

Since R_p and γ_e vary with velocity q, the effective γ_{ec} at $q=q_c$ is in general different from that at stagnation point, i. e., $\gamma_{ec}\neq\gamma_{eo}$. Only when $R_p\to 0$, $\gamma_{ec}=\gamma_{eo}$, i. e., $R_p\to 0$, $\gamma_{ec}=\gamma_{eo}=\gamma$ and $R_p\to\infty$, $\gamma_{ec}=\gamma_{eo}=4/3$.

$$R_p\to 0:\ q_c=\sqrt{\frac{2}{\gamma+1}}\ \cdot\ \sqrt{\gamma RT_o} \tag{3.76}$$

$$R_p\to\infty:\ q_c=\sqrt{\frac{8a_R}{21}}\left(\frac{T_0^2}{\sqrt{\rho_o}}\right) \tag{3.77}$$

For the one dimensional flow of a radiating gas in a nozzle of cross-sectional area $A(X)$ where X is the distance along the axis of the nozzle the equation is

$$\frac{dA}{A} + \frac{du}{u} + \frac{d\rho}{\rho} = 0 \tag{3.78}$$

where u is the axial velocity of the gas.

For isentropic flow, the equation of motion in the present case is

$$udu + C_R^2 \frac{d\rho}{\rho} = 0 \tag{3.79}$$

Eliminating $d\rho/\rho$ from Eqs. (5.1) and (5.2), we have

$$\frac{du}{dx} = -\frac{C_R^2}{u} \frac{1}{(C_R^2 - u^2)} \frac{1}{A} \frac{dA}{dx} \tag{3.80}$$

Eq. (3.80) is identical to Eq. (3.25) with C_R replacing a. Hence the general behavior of an ideal radiating gas in a nozzle is the same as that of ordinary gasdynamics except that the effective sound speed C_R should be used instead of the ordinary sound speed a.

10. Problems

1. Explain the choking of a wind tunnel from the theory of one-dimensional steady flow in a nozzle discussed in § 4.

2. Calculate the mass flow $\theta = \rho q / \rho^* q^*$ in terms of the pressure ratio (p/p_0) as well as in terms of Mach number M. Plot the mass flow θ with p/p_0 from $p/p_0 = 1$ to $p/p_0 = 0$ from a given de Laval nozzle. Here ρ^* and q^* are the value of ρ and q, respectively, at $q = a$.

3. Find the analytic relation between the velocity q in a de Laval nozzle and its cross-sectional area A if the gas is under an isothermal process.

4. Discuss the singular point of the following equation

$$\frac{dy}{dx} = \frac{ax + by}{cx + dy}$$

for various values of a, b, c, and d, where x and y are the variables. a, b, c, and d are constants, assuming at least one of a, b, c, and d is different from zero.

5. Discuss the solution of the one-dimensional wave equation (3.43) for the initial conditions:

At $t = 0$: $u = C = \text{constant}$ for $|x| < x_0$

$u = 0$ for $|x| > x_0$

$$\frac{\partial u}{\partial t} = 0$$

6. Discuss the solution of the one-dimensional wave equation (3.43b) for the initial conditions:

At $t=0$: $u=0$ for $|x|<x_0$

$u=U\cos x$ for $|x|>x_0$

$$\frac{\partial u}{\partial t}=0$$

7. Derive the one-dimensional wave equation similar to Eq. (3.43) in cylindrical coordinates where the only velocity component different from zero is the radial component. All the variables depend only on the radial coordinate and the time.

8. Derive the one-dimensioal wave equation similar to Eq. (3.43) in spherical coordinates where the only velocity component different from zero is the radial component. All the variables depend only on the radial coordinate and the time. Compare the general solution of this equation with that of Eq. (3.43).

9. Discuss the one-dimensional steady flow of an inviscid fluid in a nozzle with two minimum cross sections for various exit pressures below a certain critical pressure such that immediately behind the first minimum section the flow is supersonic.

10. If we consider the heat added dQ in Eq. (3.58c) consisting of two parts: dQ_1 due to heat conduction across the nozzle wall and dQ_2 due to chemical reaction, i.e., $dQ_2=Q_0dn$, where Q_0 is the heat generated by the chemical reaction per mole increase of the substance, derive the corresponding differential formulas similar to Eq. (3.59) for the variation of velocity q, pressure p, density ρ, temperature T, Mach number M, and change in mole dn in terms of the change in heat dQ, change in mole dn and change in cross sectional area dA.

11. For a Van der Waals gas with the following assumptions:

(i) $c_v=a_1+a_2T+a_3T^2$ and

(ii) $p=\dfrac{\rho RT}{1-b_1\rho}-b_2\rho^2$

where the constants a_2, a_3, b_1 and b_2 are small numbers.

Find the following first order differential relations with respect to the small constants a_2, etc:

$$T(\rho),\quad p(\rho),\quad q^2(T,p,\rho) \text{ and } a^2(p,\rho)$$

12. Find the expression $\rho(T)$, $p(T)$, $\rho(p)$ and $T(p)$ for the isentropic expansion of a Van der Waals gas to the same order of magnitude as those in problem 11.

13. Find the expression $T(M^2)$ for isentropic expansion of a Van der Waals gas to the same order of magnitude as those in problem 11.

14. Discuss the unsteady one-dimensional flow of a Van der Waals gas when the radiation pressure is not negligible.

15. Using the Runge-Kutta method, compute the steady isothemal flow of van der Waals gas[20] through the nozzle without shock wave. The nozzle is defined by

$$A(x)=\begin{cases} 11.057x^3-8.707x^2+1.8, & 0\leqslant x\leqslant x^* \\ -11.197x^3+25.614x^2-17.636x+4.819, & x^*<x\leqslant 0.7 \\ 1.7640x-0.050711, & 0.7<x<13 \end{cases}$$

$$T=1{,}403{,}001{,}100\,\mathrm{K}$$

p_0 (entrence pressure) $=0.1$, 3.5, 20, 50, 200, and 400 N/m^2. Compare the results with those of the perfect gas.

References

1. Courant, R., and Friedrichs, K. O., *Supersonic Flow and Shock Waves*, Interscience Publishers, Inc., New York, 1948.
2. Oswatitsch, K., *Gasdynamik*, Wien, Springer Verlag, 1952.
3. St. Venant, B., and Wantzel, L., Mémoires et experiences sur l'écoulement de l'eir determine par des differences de pressions considérables, *Jour. de l'École Poly.* **27**, 1839, pp. 85–122.
4. Reynolds, O., On the flow of gases, *Phil. Mag.* (5) **21**, 1886, pp. 185–199.
5. Shapiro, H., and Hawthorne, W., The mechanics and thermodynamics of steady one-dimensional gas flow, *Jour. Appl. Mech.* **14**, 1947, pp. 317–336.
6. Riemann, B., Ueber die Fortpflanzung ebener Luftwellen von endlicher Schwingungsweite, *Ges. Werke*, 2 Aufl., 1892, pp. 156–181.
7. Mark, H., Zur Theorie der Zylinder-und Kugelwellen in reibungsfreien Gasen und Flüssigkeiten, *Ann. Physik* (5) **XLI**, 1942, pp. 61–88.
8. Pai, S. I., *Fluid Dynamics of Jets*, D. Van Nostrand Company, Inc., Princeton, N. J., 1954.
9. Busemann, A., *Gasdynamik*, Part 1, **IV** of Handbuch der Experimental Physik, Akademische Verlag, Leipzig, 1931.
10. Ince, E. L., *Ordinary Differential Equations*, Dover Publications, Inc., New York, 1956.
11. Sommerfeld, A., *Partial Differential Equations in Physics*, Academic Press, Inc., New York, 1949.
12. Hiks, B. L., Montgomery, D. J., and Wasserman, R. H., The one-dimensional theory of steady compressible fluid flow in ducts with friction and heat addition, *NACA*, *TN* No. 1336, 1947.
13. Foa, J. V., and Rudinger, G., On the addition of heat to a gas flowing in a pipe at subsonic speed, *Jour. Aero. Sci.* **16**, No. 2, Feb. 1949, pp. 84–94.
14. Chambre, P., and Lin, C. C., On the steady flow of a gas through a tube with heat exchange or chemical reaction, *Jour. Aero. Sci.* **13**, 1946, pp. 537–542.
15. Pai, S.I, *Modern Fluid Mechanics*, *Science Press*, Beijing, 1981 distributed by Van Nostrand Reinhold Company, New York.
16. Pai S.I, *Radiation Gas Dynamics*, Springer-Verlag New York Inc., 1966.
17. Pai, S. I., Inviscid Flow of Radiation Gasdynamics, *Jour. of Math & Phys. Sci.* **3**, No. 4, pp. 361–370, 1969.
18. Pai, S. I. & Tsao, C.K., A Uniform flow of radiative gas over a flat plate, *Proc. of 3rd Intern. Heat Transfer Conf.*, Chicago, 1966.
19. Scala, S. M. and Sampson, D. H., Heat transfer in hypersonic flow with radiation and chemical

reaction. *Supersonic flow, Chemical Processes and Radiative transfer*, Pergamon Press, pp. 319 – 354, 1966.
20. Tsien, H. S., One-dimensional flows of a gas characterized by Van der Waals equation of state, *Jour. Math. and Phys.* **25**, 1946, pp. 301 – 324.

Chapter IV

SHOCK WAVES

1. Introduction

In Chapter III, §4, we saw that the flow in the nozzle can not be isentropic if the exit pressure is between p_B and p_C, i. e. in the de Laval nozzle region. We said that in such cases, the flow in the nozzle exhibits discontinuity in pressure and density, called a shock wave. In this chapter we shall study the relations between the quantities in front of and behind a shock. Although it was Riemann[1] who first conceived the possibility of such a jump , the theory of a shock was developed in great detail by Rankine[2] and Hugoniot.[3] We shall discuss, in § 2, the Rankine-Hugoniot relations across a normal shock for an ideal gas with constant specific heats.

Actually the shock wave is not a surface of discontinuity but a very narrow region in which a large variation of pressure and velocity occurs. In order to study fully this shock region, viscous effects must be considered, as is done in Chapter XVII. It is possible to estimate the thickness of a shock wave approximately. From this approximate estimate we may show that the extent of the shock-wave region is very small and may be considered as a surface of discontinuity for many practical problems. We shall show this result in § 3.

For an actual gas, the shock-wave relations differ somewhat from the ideal case discussed in § 2, particularly at very high temperatures. We shall discuss in § 4 deviations from the ideal case due to such effects as relaxation, dissociation, and ionization.

Shock waves appear in many compressible flow problems. Often the flow direction is not perpendicular to the shock-wave front. Such is the case of an oblique shock. We shall discuss the relation of the flow variables across an oblique shock in § 5 for an ideal gas. There is a simple graphical relation for the velocity components across an oblique shock known as the shock polar which will be discussed in § 6.

Problems associated with shock waves at boundaries, either solid wall or free surface, will be briefly discussed in § 7.

The behavior of shock wave will be influenced if heat is added to the flow (§ 8). One illustration of this case is the condensation shock which occurs when moisture in the air condenses. This will be discussed in § 9. Another example, which will be discussed briefly in § 10, is the detonation wave due to a very rapid release of heat by explosion.

In §11, shock waves in radiation gasdynamics where radiation pressure and radiation energy density are not negligible will be considered. Finally, in § 12, the concept of artificial viscosity will be introduced for capturing the shock waves in inviscid flow.

2. Normal shock wave in an ideal gas [4,5]

Let us consider a normal shock to be a surface of discontinuity in velocity, pressure, density, and temperature of the fluid. We shall find the relations between these quantities in front of and behind the shock. For simplicity, we choose the coordinate system such that the shock is stationary and the fluid moves through it; under these conditions the flow is steady (Fig.4.1). We assume that the fluid is an ideal gas, so that the perfect gas law (2.13) holds both in front of and behind the shock, and the specific heats are constants. We assume that the velocities are perpendicular to the shock. Our problem is to

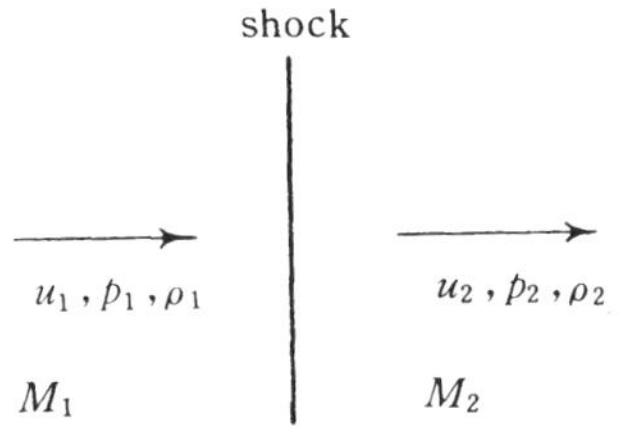

Fig. 4.1 Normal shock

find the velocity u_2, pressure p_2, and density ρ_2 behind the shock if the corresponding values in front of the shock are given. We have the following three relations:

(a) Equation of continuity

$$\rho_1 u_1 = \rho_2 u_2 \tag{4.1}$$

(b) Equation of momentum

$$\rho_1 u_1 (u_1 - u_2) = p_2 - p_1 \tag{4.2}$$

(c) Equation of energy

$$\frac{\gamma}{\gamma-1}\frac{p_1}{\rho_1} + \frac{u_1^2}{2} = \frac{\gamma}{\gamma-1}\frac{p_2}{\rho_2} + \frac{u_2^2}{2} \tag{4.3}$$

where subscript 1 refers to the values in front of the shock and subscript 2, values behind the shock.

Eliminating u_2 from Eqs. (4.1) and (4.2), we have

$$\Delta p = p_2 - p_1 = \rho_1 u_1^2 \frac{\rho_2 - \rho_1}{\rho_2} = \rho_1 u_1^2 \frac{\Delta\rho}{\rho_2} \tag{4.4}$$

If we use Eqs.(4.3) and (4.4) to eliminate u_1, we have

$$\frac{p_2-p_1}{\rho_2-\rho_1}=\frac{\Delta p}{\Delta \rho}=\gamma\frac{p_1+p_2}{\rho_1+\rho_2} \tag{4.5}$$

This is known as the Rankine-Hugoniot relation for the shock. If the strength of the shock is very small, the differences in the states in front of and behind the shock is very small and Eq. (4.5) becomes

$$\frac{dp}{d\rho}=\gamma\frac{p}{\rho} \tag{4.6}$$

Thus for small disturbances, the fluid behaves as in an isentropic process.

From Eqs. (4.4) and (4.5), we have

$$u_1=\sqrt{\gamma\left(\frac{p_1+p_2}{\rho_1+\rho_2}\right)\frac{\rho_2}{\rho_1}} \tag{4.7}$$

Form Eq. (4.7), we see that a very weak shock wave travels with the speed of sound, for in this case, $\rho_1 \cong \rho_2$ and $u_1=\sqrt{\gamma p_1/\rho_1}$. For a very strong shock, $p_2 \gg p_1$, $\rho_2 \gg \rho_1$, we have $u_1=\sqrt{\gamma p_2/\rho_1}$ which is larger than the speed of sound both front of the shock $a_1=\sqrt{\gamma p_1/\rho_1}$ and that behind the shock $a_2=\sqrt{\gamma p_2/\rho_2}$.

We may find the pressure jump, density jump, etc., across the shock in terms of the incoming Mach number $M_1=u_1/a_1$. From Eq. (4.2), we have

$$\frac{u_2}{u_1}=1-\frac{1}{\gamma M_1^2}\left(\frac{p_2}{p_1}-1\right) \tag{4.8}$$

From Eq. (4.3) we have

$$1+\frac{\gamma-1}{2}M_1^2=\frac{p_2}{p_1}\frac{u_2}{u_1}+\frac{\gamma-1}{2}M_1^2\left(\frac{u_2}{u_1}\right)^2 \tag{4.9}$$

From Eqs. (4.8) and (4.9), we have

$$\left(\frac{p_2}{p_1}\right)^2-\frac{2}{\gamma+1}(1+\gamma M_1^2)\frac{p_2}{p_1}-\frac{2}{\gamma+1}\left(\frac{\gamma-1}{2}-\gamma M_1^2\right)=0 \tag{4.10}$$

Eq. (4.10) gives two solutions: one solution is

$$\frac{p_2}{p_1}=1$$

which is a trival solution and which means that nothing happens in a uniform flow. The other solution is

$$\frac{p_2}{p_1} = \frac{2\gamma}{\gamma+1} M_1^{\,2} - \frac{\gamma-1}{\gamma+1} \tag{4.11}$$

which gives the pressure jump across a normal shock.

The other relations of the quantities behind the shock to those in front of the shock can be easily found as follows:

$$\frac{u_2}{u_1} = \frac{\rho_1}{\rho_2} = \frac{\gamma-1}{\gamma+1} + \frac{2}{(\gamma+1) M_1^{\,2}} \tag{4.12}$$

$$\frac{T_2}{T_1} = \frac{p_2}{p_1}\frac{\rho_1}{\rho_2} = \frac{[2\gamma M_1^{\,2} - (\gamma-1)]\,[(\gamma-1) M_1^{\,2} + 2]}{(\gamma+1)^2 M_1^{\,2}} \tag{4.13}$$

$$M_2^{\,2} = \frac{1 + \dfrac{\gamma-1}{2} M_1^{\,2}}{\gamma M_1^{\,2} - \dfrac{\gamma-1}{2}} \tag{4.14}$$

Another important condition across a normal shock, known as Prandtl's relation, is:

$$u_1 u_2 = \left(\frac{u_2}{u_1}\right) u_1^{\,2} = \frac{u_1^{\,2}}{M_1^{\,2}} \frac{1 + \dfrac{\gamma-1}{2} M_1^{\,2}}{\dfrac{\gamma+1}{2}} = a_1^{\,2} \frac{T_0/T_1}{\dfrac{\gamma+1}{2}} = \frac{a_0^{\,2}}{\dfrac{\gamma+1}{2}} = a^{*2} \tag{4.15}$$

where a^* is the critical velocity of sound for $M = u/a = 1$. From Eq. (4.15), we see that, if the velocity of the flow in front of the normal shock u_1 is greater than the corresponding speed of sound, i. e., supersonic flow, the velocity of the flow behind of the normal shock u_2 is less than the corresponding speed of sound, i. e. subsonic flow.

In the above analysis no restriction was imposed on the value of M_1; it could be greater than one or less than one. However, according to the second law of thermodynamics, there is a restriction.

From Eq. (2.66), we may write the entropy S as follows:

$$S - S_0 = c_v \log\left[\left(\frac{p}{p_0}\right)\left(\frac{\rho_0}{\rho}\right)^{\gamma}\right] \tag{4.16}$$

Eq. (4.16) may be applied to the flow in front of and behind the shock. Hence we have

$$\Delta S = S_2 - S_1 = c_v \log\left[\left(\frac{p_2}{p_1}\right)\left(\frac{\rho}{\rho_2}\right)^{\gamma}\right] \tag{4.17}$$

If we substitute the relations (4.11) and (4.12) into Eq. (4.17), we have a relation between ΔS and M_1 which is shown in Fig. 4.2. From Fig. 4.2, we have

$$\text{(a) } M_1 > 1, \quad \Delta S > 0, \qquad \text{(b) } M_1 < 1, \quad \Delta S < 0$$

The entropy S is increased across a shock when the oncoming flow is supersonic and decreased when it is subsonic. In our analysis we consider only the adiabatic case. According to the second law of thermodynamics, the entropy must increase and cannot decrease in an adiabatic process; hence case (b) is not possible. In an adiabatic process, shock waves can occur only in a supersonic flow.

Let us examine the change in entropy in the case of a weak shock. We take the oncoming Mach number M_1 as a value different slightly from unity, i.e.

$$M_1^2 = 1 + m \tag{4.18}$$

where $|m| \ll 1$ and m may be positive or negative.

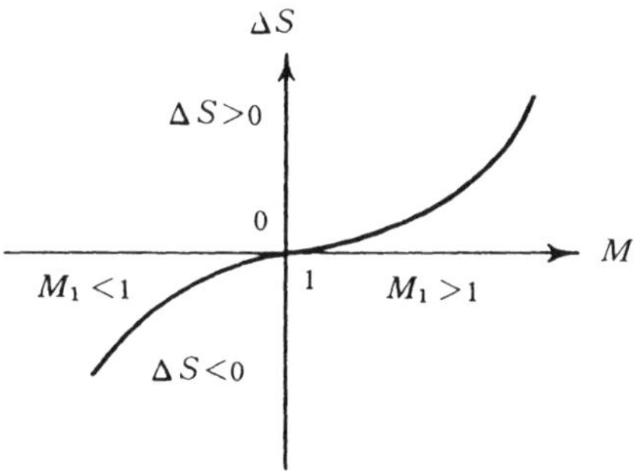

Fig. 4.2 Entropy change across a normal shock

Substituting Eq. (4.18) into Eq. (4.11) gives

$$\frac{p_2}{p_1} = 1 + \frac{2\gamma m}{\gamma + 1} \tag{4.19}$$

Combining Eq. (4.18) and Eq. (4.12) yields

$$\frac{\rho_1}{\rho_2} = \left(1 + \frac{\gamma - 1}{\gamma + 1} m\right)(1 + m)^{-1} \tag{4.20}$$

Substituting Eqs. (4.19) and (4.20) into Eq. (4.17) and developing the result in powers of m, we have

$$\Delta S = c_v\left(\log\frac{p_2}{p_1} + \gamma\log\frac{\rho_1}{\rho_2}\right) = c_v\gamma\,\frac{2(\gamma-1)m^3}{3(\gamma+1)^2} + O(m^4) \quad (4.21)$$

Eq. (4.21) shows again that only the case, where $m > 0$ and $\Delta S > 0$, is physically possible. Since $m > 0$ means $p_2 > p_1$, we can have only compression shocks. Eq. (4.21) also shows that the change of entropy across a weak shock is a third-order quantity with respect to the shock strength. The latter may be characterized by the jump of the pressure and is proportional to m in Eq. (4.19). This is the reason why, in the problem of weak shocks, the flow may be considered as isentropic in the first approximation.

3. Thickness of a shock wave

In the last section we considered the shock as a surface of discontinuity. Actually the shock represents a very narrow region in which a large change in velocity and the state of the fluid takes place. In order to see what occurs inside the shock, we must take the viscosity into consideration. We shall discuss the flow inside the shock in Chapter XVII. In this section we are going to estimate roughly the length of this shock region. It will be shown that the thickness of the shock is so thin that for many practical purposes the shock may be considered as a surface of discontinuity.

In the shock region it is to be expected that the viscous stress is of the same order of magnitude as the pressure. Taking δ as a measure of the thickness of the shock (Fig. 4.3), the viscous stress normal to the shock is approximately

$$\sigma_x \cong \mu\frac{u_1 - u_2}{\delta} \quad (4.22)$$

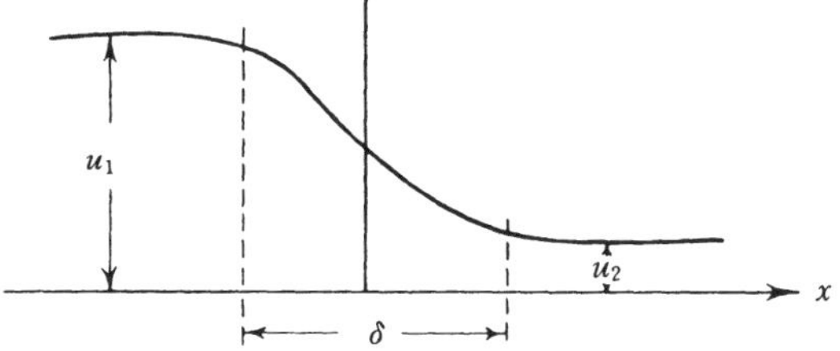

Fig. 4.3 Shock-wave thickness

where μ is the coefficient of viscosity. The viscous stress given by Eq. (4.22) is of the same order of magnitude as the pressure in the flow. As a first approximation we may write

$$\mu\frac{u_1 - u_2}{\delta} = p_2 - p_1 \quad (4.23)$$

If we substitute the normal shock relation (4.22) into Eq. (4.23), we have

$$\delta = \frac{\mu}{\rho_1} \frac{1}{u_1} \tag{4.24}$$

For air under standard conditions, $\dfrac{\mu}{\rho_1} = \dfrac{1}{6380}$ ft/sec and $\gamma = 1.4$. Furthermore, assuming $M_1 = 2.0$ which corresponds to $u_1 = 2240$ ft/sec, Eq. (4.24) gives

$$\delta = 1 \times 10^{-6} \text{ inch}$$

which shows that the shock is indeed a very thin layer. Thus our assumption that shock may be considered as a surface of discontinuity is fairly well justified for many practical problems. A more accurate formula given by Taylor and Maccoll[5] for shock wave thickness is

$$\delta = \frac{1}{u_1 - u_2} \text{ cm}$$

where u_1 and u_2 are in cm/sec. We shall discuss the thickness of shock waves in more detail in Chapter XV.

4. Shock waves in a perfect gas [6–12]

In § 2, we considered the shock wave in an ideal gas and found some simple expression for the jumps of pressure, density, etc., across a shock. The corresponding relations in real gases differ considerably from these ideal relations, particularly at high temperatures, i.e. when the shock is strong, say $M_1 > 5$ and T_1 is room temperature. At high temperatures, the specific heats are no longer constant because of the excitation of vibrational energy. Since

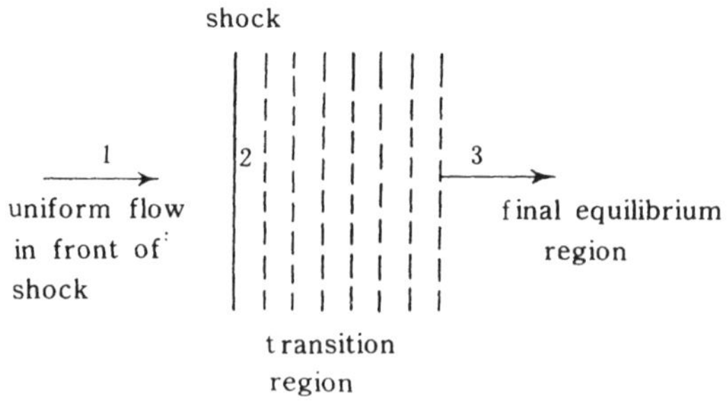

Fig. 4.4 Shock wave with effects of relaxation time

the vibrational energy reaches its equilibrium value relatively slowly, the time of relaxation has important influences on the shock structure. Furthermore at high temperature, dissociation and ionization phenomena may occur. These also will affect the shock relations. We shall discuss briefly the effects of relaxation time, dissociation, and ionization on shock-wave relation in the following sections.

(a) *Effects of relaxation time.* In the coordinate system of Fig. 4.4, the normal shock relations are

$$
\begin{aligned}
\rho_1 u_1 &= \rho_3 u_3 = \rho_2 u_2 \\
p_1 + \rho_1 u_1^2 &= p_3 + \rho_3 u_3^2 \cong p_2 + p_2 u_2^2 \\
\frac{u_1^2}{2} + H_1 &= \frac{u_3^2}{2} + H_3 \cong \frac{u_2^2}{2} + H_2
\end{aligned}
\tag{4.25}
$$

where $H = U_m + p/\rho$ = enthalpy per unit mass. Eq. (4.25) holds true for both the ideal gas and the real gas. The only difference between these cases lies on the relation of H with temperature T and the equation of state.

For a perfect gas, one should use the expression of U_m given in Chapter II, § 5, i.e., $U_m = U_{mt} + U_{mr} + U_{mv}$, in Eq. (4.25) to determine the variables of the final state. The values so obtained give the final equilibrium state values. In actual flow conditions, it takes considerable time to reach such equilibrium values. Qualitatively the structure of shock including the effects of relaxation time is as follows (Fig. 4.4).

At first there is a very sharp shock front which is of the order of a few mean free path lengths. The state 2 immediately behind this shock front corresponds to the situation in which the vibrational energy is unaltered and in which the translational and rotational energies are completely adjusted. In other words, we mav find the variables for state 2 from Eq. (4.25) by putting $U_m = U_{mt} + U_{mr}$ only. After state 2, there is a transition region in which the vibrational energy is more or less continuously excited until it has reached the equilibrium condition 3 which corresponds to the values obtained from Eq. (4.25) by setting $U_m = U_{mt} + U_{mr} + U_{mv}$. The flow in the transition region is rather complicated. It depends on the composition of the fluid and the temperature range. This area is one of the most interesting current research problems in compressible fluid flow. Some detailed discussions of shock wave structure with relaxation are given in Chapter XVII, § 15.

(b) *Effects of dissociation and ionization.* At high temperatures, dissociation and ionization phenomena may occur in the flow field. In this case, Eq. (4.25) still holds true between the conditions in front and behind the shock, but now both the effective molecular weight m and the enthalpy H are functions of both pressure and temperature. If one has the properties of the fluid in terms of pressure and temperature, as those given in Refs. 13 and 14 of Chapter II, one may easily compute the final equilibrium conditions across the shock from the known values in front of the shock. As there is no simple formula for summarizing the thermodynamic properties of a gas which includes the effects of dissociation and ionization, one must use an iteration method to compute the final equilibrium conditions. One way to calculate these variables will be given below.

From Eqs. (4.25) and (2.13), we have the following relation:

$$2(H_3-H_1)=\frac{p_1}{p_3}\left(\frac{p_3}{p_1}-1\right)\left(\frac{p_3}{p_1}+\frac{m_1}{m_3}\frac{T_3}{T_1}\right)\frac{R'T_1}{m_1} \tag{4.26}$$

Solving (p_3/p_1) from (4.26), we have

$$\frac{p_3}{p_1}=\frac{(H-mt)+\sqrt{(H-mt)^2+4mt}}{2} \tag{4.27}$$

where $m = m_1/m_3$, $H=\dfrac{2m_1(H_3-H_1)}{R'T_1}+1$, $t=T_3/T_1$, and R' is the universal gas constant.

The positive sign is used in Eq. (4.27) because the pressure must be positive. Since, in general, both H and m are functions of pressure and temperature, we must use iteration procedures to find (p_3/p_1). For a given temperature ratio t, we may first assume a pressure ratio; from this temperature ratio t and the assumed pressure ratio, we may compute H and obtain H and m from some thermodynamic tables such as those given in Ref. 13 of Chapter II. Substituting the values of H, m, and t into Eq. (4.27), we obtain a new pressure ratio. This pressure ratio is then used instead of our assumed value and the process repeats. If the second pressure ratio agrees with the first one, it is the required result. Otherwise, the procedure is repeated until the desired accuracy is obtained. In general, the convergence of the procedure is rapid.

For an ideal gas consisting of diatomic molecules, $\gamma = 1.4$, Eq. (4.12) shows that the maximum density ratio $\rho_2/\rho_1=(\gamma+1)/(\gamma-1)=6$. For a perfect gas, it was found that the density ratio across a shock may reach values much larger than six for a diatomic gas such as air.

5. Oblique shock in an ideal gas

A more general case of shock wave is the oblique shock where the incident velocity is at an angle to the shock front. In the analysis of an oblique shock, we observe that only the normal component of the velocity is discontinuous; thus the pressure, density, and temperature jumps across the shock are determined in the same manner as for a normal shock; while the tangential component of the velocity is continuous across the shock.

Let us consider the oblique shock shown in Fig. 4.5. The velocity vector u_1 in front of the shock makes a shock angle α with the shock front. The velocity vector u_2 makes a shock angle β with the shock front. Since the tangential velocity components to the shock front are the same for both the velocity vectors u_1 and u_2, we have

$$w=u_1\cos\alpha=u_2\cos\beta \tag{4.28a}$$

The corresponding normal velocity components in front of and behind the shock are, respectively,

$$v_1 = u_1 \sin \alpha \text{ and } v_2 = u_2 \sin \beta \tag{4.28b}$$

These normal velocity components should be used in place of the normal shock relation (§ 2) to calculate the corresponding jumps in pressure, etc.,

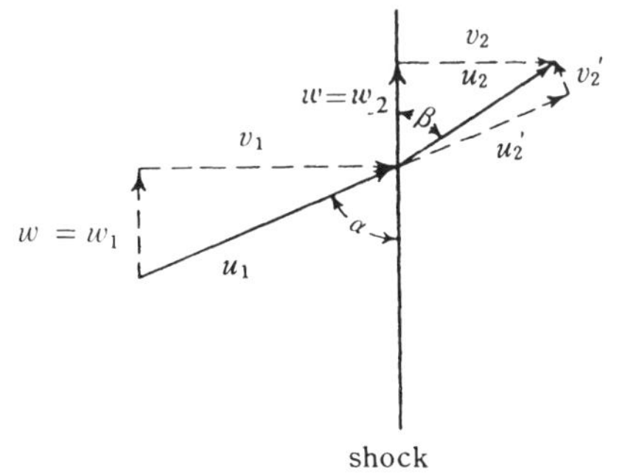

Fig. 4.5 Oblique shock

for an oblique shock. For instance, the following relations may be easily obtained:

$$\frac{p_2}{p_1} = \frac{2\gamma}{\gamma+1} M_1^2 \sin^2 \alpha - \frac{\gamma-1}{\gamma+1}$$

$$\frac{\rho_1}{\rho_2} = \frac{v_2}{v_1} = \frac{\tan \beta}{\tan \alpha} = \frac{\gamma-1}{\gamma+1} + \frac{2}{\gamma+1} \frac{1}{M_1^2 \sin^2 \alpha} \tag{4.29}$$

$$M_2^2 \sin^2 \beta = \frac{1 + \frac{\gamma-1}{2} M_1^2 \sin^2 \alpha}{\gamma M_1^2 \sin^2 \alpha - \frac{\gamma-1}{2}}$$

where $M_1 = \frac{u_1}{a_1}$, $M_2 = \frac{u_2}{a_2}$, and γ is assumed to be constant.

Since the tangential velocity component is not changed but the normal component does change across an oblique shock, the direction of the flow will change in general across an oblique shock. The angle of flow deflection across an oblique shock is $\theta = \tan^{-1}\left(\frac{v_2'}{u_2'}\right)$. Using the second relation of Eq. (4.29), we have

$$\tan\theta = \tan(\alpha-\beta) = \frac{\tan\alpha - \tan\beta}{1+\tan\alpha\tan\beta} = \frac{\tan\alpha - \dfrac{\tan\beta}{\tan\alpha}\tan\alpha}{1+\tan^2\alpha\left(\dfrac{\tan\beta}{\tan\alpha}\right)}$$

$$= \frac{M_1^2\sin 2\alpha - 2\cot\alpha}{M_1^2(\gamma+\text{con } 2\alpha)+2} \qquad (4.30)$$

It is easy to see that if the angle of flow deflection is zero, we have

$$M_1^2\sin 2\alpha - 2\cot\alpha = 0 \qquad (4.31)$$

or (1) $\cos\alpha = 0$, $\alpha = 90°$ which is the case of normal shock and (2) $\alpha = \sin^{-1}\left(\frac{1}{M_1}\right)$ which is the Mach angle corresponding to M_1 or a shock of infinitesimal strength.

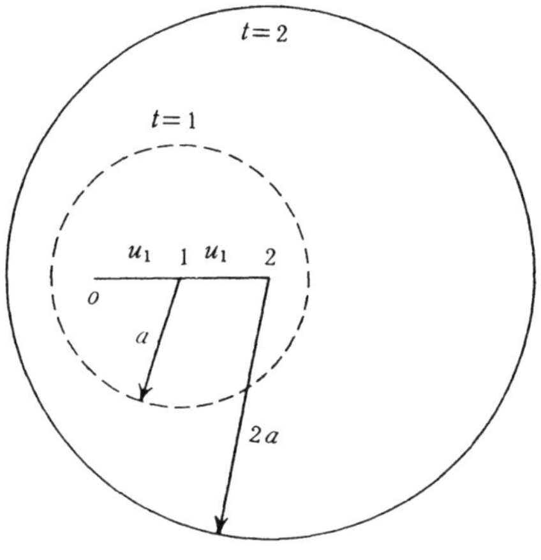

(a) Subsonic flow

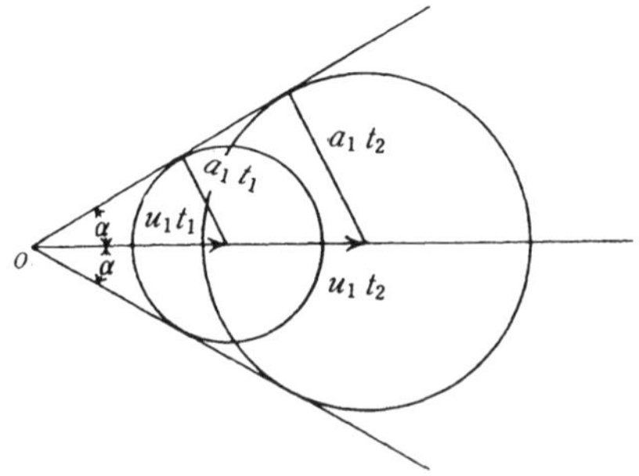

(b) Supersonic flow $\alpha = \sin^{-1}\frac{a_1}{u_1} = \sin^{-1}\frac{1}{M_1}$

Fig. 4.6 Propagation of small disturbances in compressible fluid flow

If we consider a very small disturbance center moving with the flow velocity u_1, the disturbance will propagate out with respect to the center with a speed of the velocity of sound a_1. When the velocity of the flow is less than the velocity of the sound, we shall have the picture of Fig. 4.6a. In this case, after time t from the starting position, the center of disturbance has traveled a distance u_1t but the spherical wave has a radius a_1t Since a_1t is larger than u_1t, eventually the disturbance will cover the whole space. When the velocity of the flow u_1 is larger than the velocity of sound a_1, we shall have the picture of Fig. 4.6b. Here $a_1t < u_1t$, and the disturbance will be confined to the conical region where the semi-angle of the cone is the Mach angle. Thus we have quite different flow patterns for subsonic and supersonic flow fields.

Since there are two values of the shock angle giving zero flow deflection, it appears as if a maximum flow deflection point for a given M_1 might exist. If we plot the shock angle α against θ for various M_1, we obtain Fig. 4.7, which

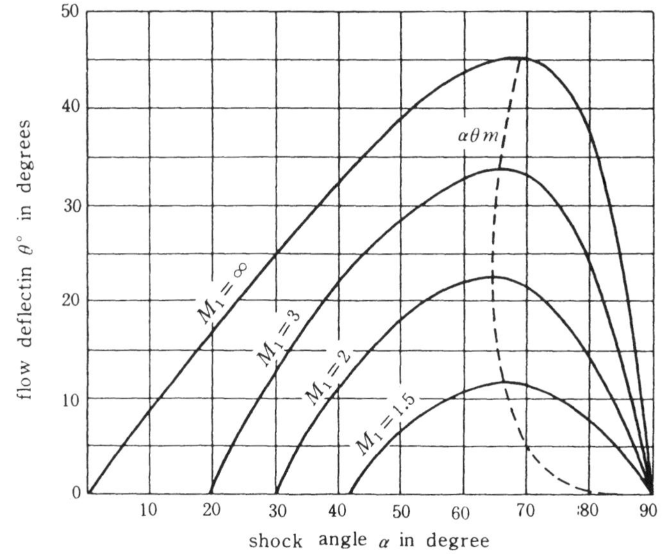

Fig. 4.7 Flow deflections across various oblique shocks

shows not only that there is a maximum flow deflection for a given M_1, but also that in general we have two possible shock configurations for a given M_1 and θ.

We may apply the above results to the case of a wedge in a supersonic flow of Mach number M_1. We assume that a wedge of semi-vertex angle θ is placed symmetrically with respect to a uniform supersonic flow u_1. When the flow reaches the wedge, it will deflect an angle θ. Now if the angle θ is less than the corresponding maximum angle of deflection across an oblique shock, the flow condition may be represented by an oblique shock attached to the leading edge of the wedge (Fig. 4.8a) so that the flow after the shock is parallel to the surface of the wedge and the pressure over the wedge is then constant.

For a given value of M_1 there are two possible shock configurations for a given wedge: the strong shock with a larger shock angle and a larger entropy increase and the weak shock with a smaller shock angle and a smaller entropy increase. Experimentally one generally observes the weak shock on wedges of finite dimensions. However there is still some possibility for the occurrence of the strong shock. This point has been studied by many authors, but no definite results have been obtained yet.

If the semi-vertex angle of the wedge is larger than the maximum flow deflection across an oblique shock, an attached shock on the wedge will not satisfy the boundary condition. In this case, we have a curved detached shock located a short distance in front of the wedge (Fig. 4.8b). The flow behind the

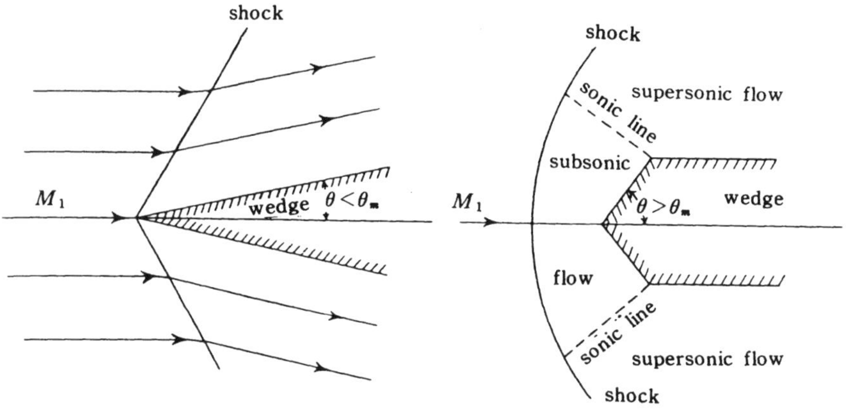

(a) Attached shock case (b) Detached shock case

Fig. 4.8 Supersonic flow over a wedge

detached shock and near the line of symmetry is subsonic. Far a way from the wedge, the flow is supersonic. Hence the shock angle is 90° on the line sysmmetry and gradually decreases outward. The detached shock problem, which is very complicated, will be further discussed in Chapter XIV.

6. Shock polar

There is a simple graphical representation of the velocity components behind oblique shocks for a given u_1 or M_1, known as the shock polar.

We first derive a generalized formula for Eq. (4.15). From Eq. (4.29), we have

$$v_1 v_2 = v_1^2\left(\frac{v_2}{v_1}\right) = \frac{v_1^2}{M_1^2 \sin^2\alpha}\left(\frac{\gamma-1}{\gamma+1} M_1^2 \sin^2\alpha + 1\right)$$

$$= a^{*2} - \frac{\gamma-1}{\gamma+1} w^2 \tag{4.32}$$

When $\alpha = 90°$, Eq. (4.32) is reduced to Eq. (4.15), because $w = 0$.

Now we plot Eq. (4.32) in a hodograph plane in which the axis of abscissa is in the direction of u_1 (Fig. 4.9). Then the velocity vector u_2 has a component u_2' in the direction of u_1 and another component v_2' in the direction normal to u_1. Hence we have the following relations (see Fig. 4.10):

$$w = u_1 \cos\alpha = u_2' \cos\alpha + v_2' \sin\alpha$$

$$v_1 = u_1 \sin\alpha; \quad v_2 = u_1 \sin\alpha - \frac{v_2'}{\cos\alpha} \tag{4.33}$$

$$\tan\theta = \frac{v_2'}{u_2'}, \qquad \tan\alpha = \frac{u_1 - u_2'}{v_2'}$$

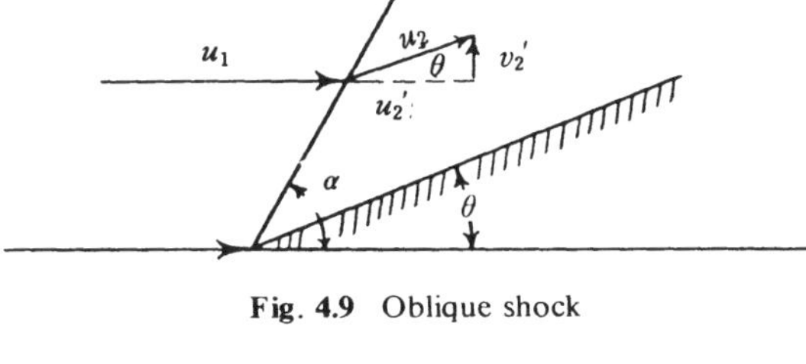

Fig. 4.9 Oblique shock

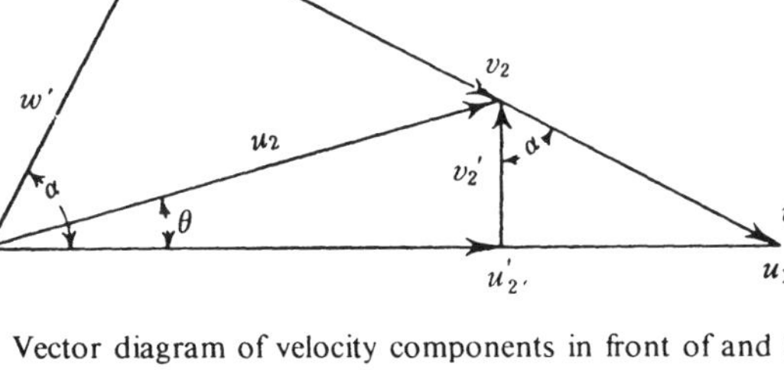

Fig. 4.10 Vector diagram of velocity components in front of and behind an oblique shock

Subs ituting Eq. (4.33) into Eq. (4.32) and simplifying, we have

$$(u_1 - u_2')^2\left(u_2 - \frac{a^{*2}}{u_1'}\right) = v_2'^2\left(\frac{a^{*2}}{u_1'} + \frac{2}{\gamma+1} u_1 - u_2'\right) \tag{4.34}$$

For a given u_1 and $a^* = a_0\sqrt{\dfrac{2}{\gamma+1}}$, if we plot Eq. (4.34) in the hodograph plane (Fig. 4.11), we have the shock polar which is a curve known as the Folium of Descartes. From Eq. (4.34), we see that

(a) When $v_2' = \pm\infty$,

$$u_2' = \frac{a^{*2}}{u_1} + \frac{2}{\gamma+1}u_1 = OA$$

(b) When $v_2' = 0$,

$$u_1 = u_2' = OP$$

which corresponds to a Mach wave, i. e., a shock wave of infinitesimal strength, or

$$u_2' = \frac{a^{*2}}{u_1} = OQ$$

which corresponds to a normal shock.

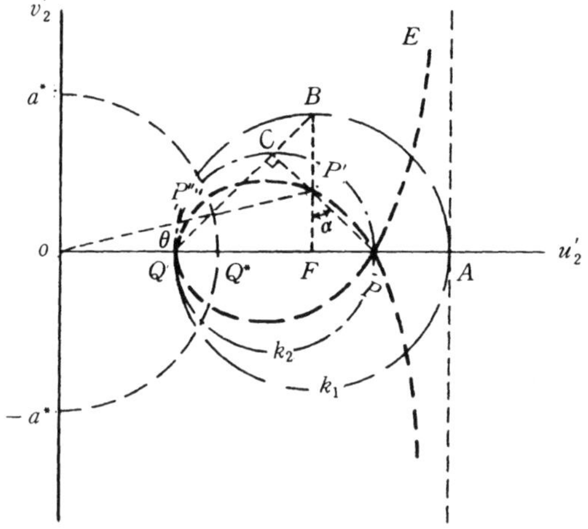

Fig. 4.11 Shock polar

The portion of the shock polar for $u_2' > u_1$, i.e., *PE* in Fig. 4.11, should be discarded because it represents an expansion shock which violates the second law of thermodynamics. Ordinarily the shock polar refers to the curve *PP′Q*, i.e., the portion of the curve given by Eq. (4.34) with $u_2' < u_1$. For a given M_1 (or u_1) and θ, two solutions of u_2' and v_2' are obtained corresponding to the points *P′* and *P″* in Fig. 4.11. *P′* represents the weak shock case and *P″* the strong shock case. These two solutions coincide only when $\theta = \theta_{max}$. The sonic circle with radius a^* divides the shock polar into two parts: one has supersonic flow behind the shock, and the other has subsonic flow behind the shock.

The following is a simple graphical method for constructing the shock polar for any given u_1 and a^*.

From the given values of u_1 and a^*, we can calculate OQ and OA in Fig. 4.11. We then draw a circle k_2 with diameter QP through the points Q and P, and another circle k_1 with diameter QA through the points Q and A. From any arbitrary point B on the circle k_1, draw the line QB. Draw a perpendicular line PC to BQ. Draw another perpendicular line BF to the axis QP. The intersection point $P'$of PC and BF is a point on the required shock polar. Repeat this process for other points on the circle k_1. We thus have as many points on the shock polar as we wish.

If the point P' is a point on the shock polar, $\overline{P'F}=v_2'$; from the geometry, we have

$$\overline{P'F}:\overline{FP}=\overline{QF}:\overline{FB}$$

$$v_2'^{\,2}=\overline{P'F}^2=\frac{\overline{FP}^2\cdot\overline{QF}^2}{\overline{FB}^2}=\frac{\overline{FP}^2\cdot\overline{QF}^2}{\overline{QF}\cdot\overline{FA}}=\overline{FP}^2\frac{\overline{QF}}{\overline{FA}} \quad (4.35)$$

Since $\overline{FP}=u_1-u_2'$, $\overline{QF}=u_2'-\dfrac{a^{*2}}{u_1}$, $\overline{FA}=OA'-u_2'$. Eq. (4.35) is identical to Eq. (4.34). We see now $\overline{P'F}=v_2'$.

7. Reflections of shock waves from solid or free boundary[13,14]

If a shock wave meets the boundary between adjacent media, in general a reflected disturbance occurs in the first medium and a new shock wave with another pressure jump originates in the second medium. Problems of this sort may be solved by using the shock polar discussed above.

If an oblique shock of not too large strength is incident on a rigid solid wall, we have regular shock reflection as shown in Fig. 4.12(a). The incoming

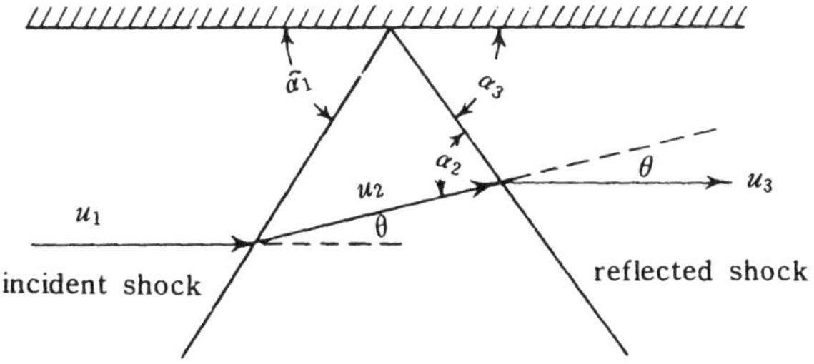

(a) Regulor shock reflection

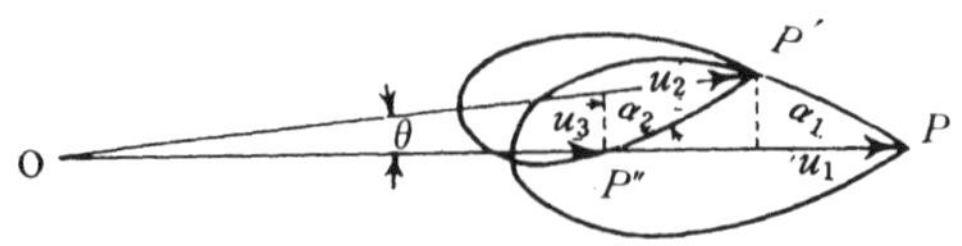

(b) Shock polars in regular shock reflection

Fig. 4.12 Regular shock reflection

stream of velocity u_1 will be deflected by angle θ in passing through the incident shock 1. The flow behind incident shock 1 with velocity u_2 will be deflected back by the same angle of θ in passing through the reflected shock 2. The strength of the reflected shock 2 is determined by the velocity u_2 and the flow deflection, which in turn are determined by the incoming stream velocity u_1 and the incident shock angle α_1. The strength of the reflected shock may be calculated from the shock polar shown in Fig. 4.12(b). From the incoming stream u_1 and a^*, we can draw the shock polar 1 on the vector u_1. From the strength of the shock, we can determine the flow behind the shock 1, i.e., OP'. From OP' and a^*, we can draw another shock polar 2. Because of the boundary condition that the wall is a straight line, the final flow OP'' must be in the same direction as OP. Hence the intersection of the second shock polar and the axis OP gives the final velocity u_3, i.e. OP'', and the shock angle α_2.

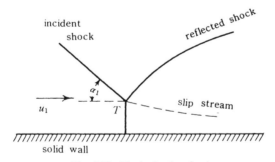

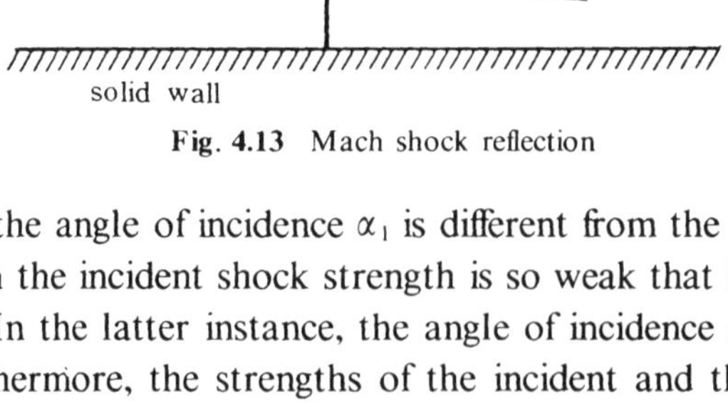

Fig. 4.13 Mach shock reflection

In general, the angle of incidence α_1 is different from the angle of reflection α_3, except when the incident shock strength is so weak that it degenerates into a Mach wave. In the latter instance, the angle of incidence is equal to that of reflection. Furthermore, the strengths of the incident and the reflected Mach waves in the case of a solid straight wall are the same.

If the incident wave is very strong, i. e., the incident shock angle α_1 is too steep, then the shock polar 2 will not intersect the axis OP. It is then not possible here to have a regular shock reflection. In this case, Mach shock reflection, as shown in Fig. 4.13, will occur. For a straight incident shock, there will be a curved reflected shock together with a normal shock near the wall. In addition, there will be a surface of discontinuity, known as a slip stream, trailing from the triple point T. The theory of regular shock reflection has been completely worked out and checks fairly well with experimental results. However, the problem of Mach shock reflection is not yet completely solved.

When an oblique shock impinges on a free boundary i.e. a surface of constant pressure such as the boundary of a jet, the shock may reflect as another shock or a fan of expansion waves[4]. The formulation of reflection and refraction of shock waves at a gaseous interface is not difficult, but an exhaustive analysis is rather cumbersome due to algebraic complexity.[14]

8. Shock waves in an ideal gas with heat addition

In the last few sections, we considered shocks in adiabatic flows only. If heat is added to the flow, the flow pattern will be different from that in adiabatic flow. Let us now consider the normal shock wave discussed in § 2 with heat addition. For simplicity we may assume that the heat added per unit mass between the states 1 and 2 is h, a constant. The equations of continuity and of momentum, (4.1) and (4.2), respectively, hold true here. The energy equation (4.3) should be changed to the following from:

$$\frac{u_1^2}{2} + \frac{\gamma}{\gamma-1}\frac{p_1}{\rho_1} + h = \frac{u_2^2}{2} + \frac{\gamma}{\gamma-1}\frac{p_2}{\rho_2} \tag{4.36}$$

In order to find the shock relations with heat addition, we must solve Eqs. (4.1), (4.2), and (4.36). For instance, the Prandtl relation (4.15) in the present case becomes

$$\frac{a_1^{*2}}{u_1 u_2} = 1 + \frac{q^2}{u_2(u_2-u_1)} \tag{4.37}$$

where $h = \dfrac{\gamma+1}{2(\gamma-1)} q^2$, and a_1^* is the critical sound speed for flow in front of the shock and in general $a_1^* \neq a_2^*$. When $q = 0$, Eq. (4.37) reduces to Eq. (4.15), and $a_1^* = a_2^*$.

For the velocity behind the normal shock, the following relation may be easily obtained:

$$m_2 = \frac{1}{2}\left(m_1 + \frac{1}{m_1}\right) \pm \sqrt{\left[\frac{1}{2}\left(m_1 - \frac{1}{m_1}\right)\right]^2 + k^2} \tag{4.38}$$

where

$$m_1 = \frac{u_1}{a_1^*}, \quad m_2 = \frac{u_2}{a_2^*}, \quad k = \frac{q}{a_1^*} = \sqrt{\frac{h}{c_p T_1}}$$

We have two solutions for the shock with heat addition. The first solution is developed from the trivial solution of shock with zero strength, i.e.,

$$(m_2)_1 = \frac{1}{2}\left(m_1 + \frac{1}{m_1}\right) + \sqrt{\left[\frac{1}{2}\left(m_1 - \frac{1}{m_1}\right)\right]^2 + k^2} \quad (4.39)$$

If $k = 0$, $m_1 = m_2$, i.e., $u_1 = u_2$. If k is very small, we have

$$(m_2)_1 \cong m_1 - \frac{k^2}{m_1 - \frac{1}{m_1}} + \cdots \quad (4.39a)$$

This represents a condensation shock (see § 9).

The second solution is developed from the conventional normal shock, i.e.

$$(m_2)_2 = \frac{1}{2}\left(m_1 + \frac{1}{m_1}\right) - \sqrt{\left[\frac{1}{2}\left(m_1 - \frac{1}{m_1}\right)\right]^2 + k^2} \quad (4.40)$$

If $k = 0$, $m_2 = 1/m_1$. If k is very small, we have

$$(m_2)_2 \cong \frac{1}{m_1} + \frac{k^2}{m_1 - \frac{1}{m_1}} + \cdots \quad (4.40a)$$

In this case, it may be difficult to distinguish this shock from a conventional shock. On the other hand, if k is very large, we have a detonation wave (see § 10).

9. Condensation shock[15, 16]

One example of a shock wave with heat addition is the condensation shock. In the actual flow of air, there is normally a certain amount of moisture present which is in the vapor state. If this vapor condenses, heat is released to the gas in an amount equal to the latent heat of vaporization of the condensation vapor. The shock-wave relations will change as shown in the last section.

Although the condensation process is rather complicated. we shall briefly discuss it here.

The partial pressure of the vapor is called the vapor pressure and is denoted by p_v. A vapor is in equilibrium with an infinite plane surface of its liquid when the vaopr pressure is equal to a value known as saturation pressure p_s. The saturation pressure p_s is given by the Clausius-Clapeyron equation (26.1). As a first approximation, we assume that the latent heat L is a constant for wate vapor. Then the integration of.Eq. (2.61) gives

$$\frac{p_s}{p_{s0}} = \exp\left[\frac{L}{RT_0}\left(1-\frac{T_0}{T}\right)\right] \tag{4.41}$$

where the subscript 0 refers to the reservoir conditions.

The relative humidity r is defined as

$$r = p_v/p_s$$

A necessary condition for condensation is $r = 1$, but this is not sufficient as we shall see presently. Let us calculate the relative humidity in the flow terms of the relative humidity in the reservoir r_0. We assume that the flow is an isentropic flow so that

$$\frac{T}{T_0} = \left(\frac{p}{p_0}\right)^{\frac{\gamma-1}{\gamma}} \tag{4.42}$$

As the velocity of the flow increases, its pressure p decreases, but in the absence of condensation the specific humidity (weight of water vapor per unit weight of air) remains unaltered and hence the ratio p_v/p remains constant. Then we have

$$\frac{r}{r_0} = \frac{p_v/p_s}{p_{v0}/p_{s0}} = \frac{p}{p_0}\frac{p_{s0}}{p_s} = \frac{p}{p_0}\exp\left[\frac{L}{RT}\left(\frac{p}{p_0}\right)^{\frac{\gamma-1}{\gamma}} - 1\right] \tag{4.43}$$

From Eq. (4.43), we see that the value of r increases very rapidly as the Mach number of the flow increases. For instance, if T_0 is 100°F, $\gamma = 1.4$ at $M = 1$, r/r_0 is more than 20. In order to keep $r < 1$, r_0 must be smaller than 0.05. This shows that if $r = 1$ were really a correct criterion for condensation, it would be impossible to have even moderately supersonic flow without condensation.

Fortunately there are two effects which tend to raise the condensation value of r very materially. The first is that the Clausius-Clapeyron equation (2.61) is based on the assumption that the vapor is in equilibrium with and condenses on an infinite plane surface of liquid. Actually the condensation will occur on small droplets. The effective relative humidity for droplets decreases with the size of the droplets due to the effect of surface tension of the droplets. There is a critical droplet size below which condensation cannot occur. Actually the condensation is a statistical phenomenon. Whether the droplet will exceed its critical value depends on probability considerations. When condensation does occur it is extremely rapid and sudden. The second effect is that the regions of large supersonic velocities are usually very limited in extent, for example, in a wind tunnel, and gas particles remain there only very short periods of time before returning to regions of higher pressure and temperature. Condensation may not be able to occur in such a short period. It may be concluded from these arguments that the condensation phenomenon is rather complicated. Much theoretical and experimental work remains to be done before it is completely understood.

If the water vapor does condense, two possible shock configurations may occur according to the simple analysis of § 8. Usually we refer to a condensation shock as that corresponding to $(m_2)_1$ [Eq. (4.39)]. The thickness of this shock is rather large.

The second shock $(m_2)_2$ which may be called a shock with condensation, differs only slightly from a conventional normal shock.

Further discussion of the condensation shock as a two-phase flow problem — a mixture of a liquid and its own vapor and a gas-will be given in Chapter XVI, § 3.

10. Detonation wave[17-20]

A second example of a shock wave with heat addition is the detonation wave which arises from the rapid transformations of explosive material. The theory of detonation has been thoroughly investigated by Becker.[18] In the theory, the explosive is assumed to be contained in an infinitely long rigid tube designed so that there is no heat conduction through the walls. The detonation is assumed to appear steady to an observer who moves with the appropriate velocity D along the tube. All state variables are supposed to depend only on the distance along the tube and not on the distance from the axis of the tube. At sufficiently large distances from the zone of chemical reaction, the state variables have constant values. Viscous friction at the wall of the tube is also neglected.

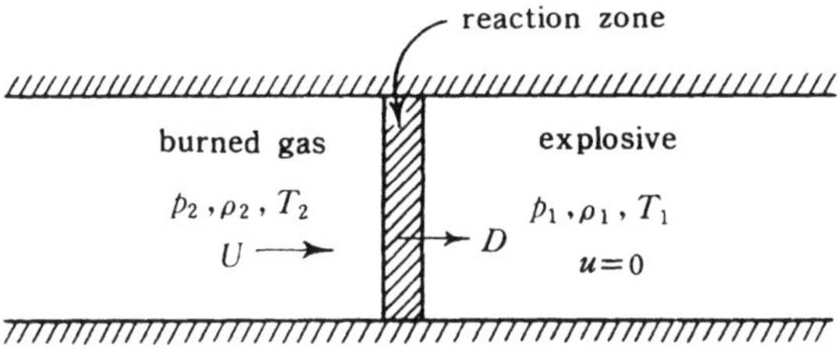

Fig. 4.14 Idealized pattern of detonation wave

Fig. 4.14 shows the idealized pattern of detonation. The conservation theorems of mass, momentum, and energy give the following relations:

$$D\rho_1 = (D-U)\rho_2 \tag{4.44}$$

$$D^2\rho_1 + p_1 = (D-U)^2\rho_2 + p_2 \tag{4.45}$$

$$U_{m1} + \frac{D^2}{2} + \frac{p_1}{\rho_1} = U_{m2} + \frac{(D-U)^2}{2} + \frac{p_2}{\rho_2} \tag{4.46}$$

where p denotes the pressure; ρ, the density; U_m, the internal energy per unit mass; D, the velocity with which the reaction zone travels into the explosive zone; and U, the change of flow velocity in the reaction zone. Subscript 1 refers to the values in front of the reaction zone, and subscript 2 to the values behind it.

From Eqs. (4.44) and (4.45), we have

$$D = \frac{1}{\rho_1}\sqrt{\frac{p_2 - p_1}{\dfrac{1}{\rho_1} - \dfrac{1}{\rho_2}}} \tag{4.47}$$

and

$$U = \sqrt{\left(\frac{1}{\rho_1} - \frac{1}{\rho_2}\right)(p_2 - p_1)} \tag{4.48}$$

Eq. (4.46) may be written as follows:

$$U_{m2} - U_{m1} = \frac{1}{2}(p_1 + p_2)\left(\frac{1}{\rho_1} - \frac{1}{\rho_2}\right) \tag{4.49}$$

If we assume that the gas is ideal, the change of internal energy may be written as

$$U_{m2} - U_{m1} = c_v(T_2 - T_1) - h \tag{4.50}$$

where h is the energy released in the chemical reaction. We also have the equation of state as follows:

$$p_2 = \rho_2 R_2 T_2 \tag{4.51}$$

where the gas constant R_2 may not be the same as R_1 because of the chemical reaction.

Our problem is to find D, U, p_2, ρ_2 and T_2 for a given amount of heat release h and the known state 1, i. e. p_1, ρ_1, and T_1. We have only four relations (4.47), (4.48), (4.49) and (4.51). The additional relation needed is the Chapman-Jouguet condition which may be obtained from the consideration of the $p-(1/\rho)$ diagram.

In the $p-(1/\rho)$ plane, Eq. (4.49) together with Eq. (4.50) defines a one-parameter family of curves known as Hugoniot curves. Each curve is a shock wave curve with a given value of heat released [cf. Eq. (4.36)]. A typical curve *BNGFK* is shown in Fig. 4.15. In contrast to the case of shock waves, the initial state of the explosion p_1, ρ_1, i. e. point A, does not lie on the Hugoniot curve. The final state p_2, ρ_2, however, must lie on the Hugoniot curve. The point G represents the final state of combustion at constant volume

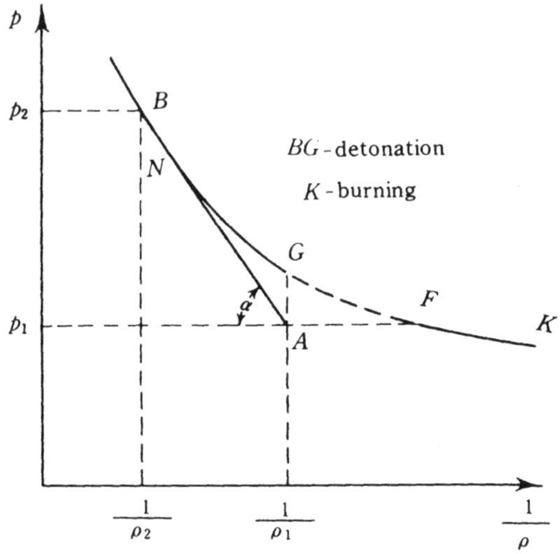

Fig. 4.15 Hugoniot curve for detonation

$\rho_1=\rho_2$, whereas the point F represents the final state of combustion at constant pressure. For an arbitrary final state B, the values of D and U are given by the following formulas:

$$D=\frac{1}{\rho_1}\sqrt{\tan\alpha}\,,\quad U=\left(\frac{1}{\rho_1}-\frac{1}{\rho_2}\right)\sqrt{\tan\alpha} \tag{4.52}$$

It should be note that for the portion of the Hugoniot curve between G and F, $\sqrt{\tan\alpha}$ is imaginary. Consequently it does not correspondto any actual process. When the final state lies on the portion BG, we have detonation; when the final state lies in the portion FK, we have normal burning.

It has been found experimentally that the detonation velocity is a constant for a given mixture of combustible gas. It is generally accepted that the detonation velocity is given by the point N which is the tangent point on the Hugoniot curve drawn from the initial state A. At this point,

$$D_N=U_N+a_2 \tag{4.53}$$

where a_2 is the sound velocity in the region 2. This condition N is known as the Chapman-Jouguet state which is the most stable condition of detonation. If we use Eq. (4.53) as the fifth condition, the final state of detonation may be obtained.

For more detailed properties of detonation waves, Refs. 17 to 20 should be consulted.

Further discussion of detonation waves as a problem of aer-thermochemistry will be given in Chapter XVII, § 14.

11. Shock waves in radiation gasdynamics[21-23]

Since shock waves are important in high speed flow where the thermal radiation is also important, it is interesting to see what the effects of thermal radiation are on the shock waves. In order to show some main effects of thermal radiation, we consider a normal shock in an optically thick medium. The main effects of the thermal radiation are the radiation pressure p_R and the radiation energy density E_R. For an inviscid radiating gas without radiative transfer, the normal shock relations (4.1) to (4.3) should be modified by the thermal radiation effects as follows:

$$\rho_1 u_1 = \rho_2 u_2 \tag{4.54a}$$

$$\rho_1 u_1 (u_1 - u_2) = p_2 + p_{R2} - p_1 - p_{R1} \tag{4.54b}$$

$$\frac{\gamma}{\gamma-1}\frac{p_1}{\rho_1} + \frac{E_{R1}}{\rho_1} + \frac{u_1^2}{2} = \frac{\gamma}{\gamma-1}\frac{p_2}{\rho_2} + \frac{E_{R2}}{\rho_2} + \frac{u_2^2}{2} \tag{4.54c}$$

where $3p_R = E_R = a_R T^4$, Eq. (2.58) and the other notations are the same as those for Eqs. (4.1) to (4.3).

In order to solve Eqs. (4.54), it is convenient to introduce the following non-dimensional variables:

$$\xi = \frac{u_2}{u_1};\ T^* = \frac{RT_2}{u_1^2};\ T_1^* = \frac{1}{\gamma M_1^2};\ M_1 = \frac{u_1}{a_1}$$
$$a_1 = (\gamma R T_1)^{1/2};\ R_p = p_R/p \tag{4.55}$$

With these non-dimensional variables, Eqs. (4.54) give an equation of ξ as follows:

$$(\xi - 1)\left[\xi - \frac{8R_{p2} + r^2 + 1}{7R_{p2} + r^2}(1 + R_{p1}) f T_1^* + (1 + R_{p1})\right] = (\xi - 1)(\xi - \xi_2) = 0 \tag{4.56}$$

where $r^2 = \dfrac{\gamma+1}{\gamma-1}$; $f = \dfrac{\xi - g(R_{p2})}{\xi - 1}$; $g(R_{p2}) = \dfrac{(R_{p2}+1)(8R_{p1}+r^2+1)}{(R_{p1}+1)(8R_{p2}+r^2+1)}$

There are two roots of Eq. (4.56). The root $\xi = 1$ represents the velocity of the original flow, i.e. no shock. The other root $\xi = \xi_2$ represents the velocity behind a normal shock. The formal expression of ξ_2 is

$$\xi_2 = \frac{\gamma_e - 1}{\gamma_e + 1} + \frac{2\gamma_e p_t^*}{\gamma_e + 1} \tag{4.57}$$

where

$$\gamma_e = \frac{4(\gamma-1)R_{p2}+\gamma}{3(\gamma-1)R_{p2}+1} = \text{effective ratio of specific heats in radiation gasdynamics} \tag{4.58}$$

$$p_t^* = (1+R_{p1})f(R_{p2})\,T_1^* = \text{effective value of } T_1^* \text{ in radiation gasdynamics} \tag{4.59}$$

when $R_{p2} = 0$, $\gamma_e = \gamma$ and $p_t^* = T_1^*$ we have the Rankine-Hugoniot relation across a normal shock in ordinary gasdynamics, Eq. (4.12). When R_{p2} is very large, $\gamma_e = 4/3$ for all values of γ.

Since both γ_e and p_t^* are functions of R_{p2} and R_{p2} depends on ξ_2, we have to find ξ_2 for a given set of initial conditions T_1^* and R_{p1} by the method of successive approximation.

It is interesting to find the values of ξ_2 for a few limiting cases:

(a) *Low temperature case*. If the temperatures both in front of and behind the normal shock are not too high, we have $R_{p1} = R_{p2} = 0$. Hence $\gamma_e = \gamma$ and $p_t^* = T_1^*$, Eq. (4.57) is identical to the normal shock relation in ordinary gasdynamics, Eq. (4.12).

(b) *Weak shock in a high temperature gas*. If the temperature of the gas is initially very high, R_{p1} is then not negligible. If in addition, the shock wave strength is weak R_{p2} will be approximately equal to R_{p1}. Hence in Eq. (4.57), we may write $\gamma_e = \gamma_{e1}$ and $p_t^* = p_{t1}^*$. The effects of thermal radiation on the uniform state behind a weak shock in this case are:

(i) The value of γ is replaced by the effective value γ_{e1}, i.e.

$$\gamma_{e1} = \frac{4(\gamma-1)R_{p1}+\gamma}{3(\gamma-1)R_{p1}+1} \tag{4.60}$$

and (ii) the value of T_1^* is replaced by p_{t1}^*, i.e. the gas pressure is replaced by the total pressure which is the sum of the gas pressure and the radiation pressure. When the shock strength is infinitesimally small, we have

$$u_1^2 = \gamma_{e1}\frac{p_1 + p_{R1}}{\rho_1} = C_R^2 \tag{4.61}$$

This formula (4.61) is another way to define a radiation sound speed C_R which is identical to that given by Eq. (3.67), i. e. Eq. (3.72).

(c) *Very strong shock in a cold gas*. In this case, $R_{p1} \ll 1$ but $R_{p2} \gg 1$, we have then

$$\xi_2 = \frac{1}{7} + \frac{27}{70M_1^2} \tag{4.62}$$

where we take $\gamma = 5/3$. Without thermal radiation effect, the limiting value of ξ_2 when $M_1 \gg 1$ depends on the value of γ, but with thermal radiation effects when $M_1 \gg 1$, the limiting value of $\xi_2 = 1/7$ for all the values of γ. Fig. 4.16 shows the variation of ξ_2 with the effective Mach number $M_{e1} = u_1/C_{R1}$. In general, the effects of thermal radiation on the velocity field is not very large.

From Eqs. (4.54) with the help of Eq. (4.55), we have an equation for the temperature T_2 behind the normal shock as follows:

$$T^{*4} + A^{-1} T^* - A^{-1} B = 0 \tag{4.63}$$

where

$$A^{-1} = \frac{T_1^{*3}}{R_{p1}\xi_2} > 0$$

and

$$B = [(1 + R_{p1}) T_1^* + 1]\ \xi_2 - \xi_2^2 > 0$$

For $R_{p1} \gg 1$, $A^{-1} T^* \ll T^{*4}$. Hence Eq. (4.63) becomes

$$\frac{T_2}{T_1} = \frac{T^*}{T_1^*} \cong \frac{(A^{-1}B)^{1/4}}{T_1^*} \cong \left[1 + \frac{8}{7}\left(M_{e1}^2 - 1\right)\right]^{1/4} \tag{4.64}$$

where M_{e1} is the effective shock wave Mach number, defined by the equation

$$M_{e1}^2 = \frac{\gamma[1 + 12(\gamma - 1) M_1^2 R_{p1}]}{\gamma + 20(\gamma - 1) R_{p1} + 16(\gamma - 1) R_{p1}^2} \tag{4.65}$$

When $R_{p1} \gg 1$, we have

$$M_{e1}^2 = \frac{3\gamma M_1^2}{4R_{p1}} \tag{4.66}$$

For very large M_{e1}, Eq. (4.64) becomes

$$\frac{T_2}{T^1} = 1.033\ M_{e1}^{1/2} \tag{4.67}$$

Without radiation effect, it is well known that at very high shock Mach number, the temperature ratio across a shock increases with the square of the shock Mach number M_1. Here we have shown that if the radiation effects are included at very high shock Mach number, the temperature across a shock wave increases only with the square root of the effective Mach number M_{e1}.

For finite R_{p1} (not very large compared to unity, but still with $A^{-1}T^* \ll T^{*4}$), we have

$$\frac{T_2}{T_1} = \frac{T^*}{T_1^*} \cong \frac{(A^{-1}B)^{1/4}}{T_1^*} \cong \left\{\frac{1 + R_{p1}}{R_{p1}} + \frac{6(M_{e1}^2 - 1)[\gamma + 20(\gamma - 1) R_{p1} + 16(\gamma - 1) R_{p1}^2}{7R_{p1}[1 + 12(\gamma - 1) R_{p1}]}\right\}^{1/4} \tag{4.68}$$

It was found that Eq. (4.68) gives very accurate results when $R_{pl} \geqslant 1$. The accurate values of temperature ration T_2/T_1 have been calculated and are shown in Fig. 4.17.

12. Artificial viscosity

In the investigation of the inviscid flow with shock waves, the shock waves are considered as surfaces of discontinuity. The partial differential equations governing the inviscid flow require boundary conditions connecting the values of velocity, pressure, density and the like on the two sides of each shock surface. The necessary boundary conditions are supplied by the Rankine-Hugoniot equations, but their application is complicated because the position of the shock surfaces is not known in advance and is governed by the differential equations and boundary conditions themselves. In consequence, the treatment of shock waves requires lengthy computations of trial and error procedures. This method is known as shock-fitting method. We shall discuss another method.

We know that shock waves are continuous phenomena when the viscosity of fluid is considered (see Chapter XVII § 16). The partial differential equations of viscous flow can be applied to the entire flow field including the shock waves. However, the thickness of the physical shock waves is too small to be resolved on an affordable computational mesh. It is impractical to compute flow with shock waves by the equations of physical viscous motion. Moreover, we in general need neither the detail structure of shock waves nor the viscous terms of governing equations outside of the shock waves.

Von Neumann and Richtmyer[24] introduced artificial viscosity terms into the inviscid flow equations so as to give shock waves a thickness comparable to the spacing of the points of the computational mesh. The artificial viscosity dies out automatically when the local flow gradients are small in comparison with those in the shock waves. Then the differential equations of the artificial viscous flow can be used for the entire field just as though there were no shock waves at all. In the numerical results obtained, the thickness of the shock layers is small in comparison with other physically relevant dimensions of the system and the shock waves are immediately evident as near-discontinuities and across which velocity, pressure, etc. have very nearly correct jumps.

In this section we shall apply the method of artificial viscosity to steady one-dimensional flows and show its suitability by analytical solution. The equations of continuity, motion, and energy for steady one-dimensional adiabatic flows are written in divergence form

$$\frac{\partial}{\partial x}(\rho u)=0 \tag{4.69}$$

$$\frac{\partial}{\partial x}(\rho u^2 + p + q) = 0 \tag{4.70}$$

$$\frac{\partial}{\partial x}\left[\left(\rho c_v T + \frac{\rho u^2}{2} + p + q\right)u\right] = 0 \tag{4.71}$$

where q is the artificial viscous normal stress or viscous pressure. We shall show that the expression

$$q = -c^2 \Delta x^2 \rho \left|\frac{\partial u}{\partial x}\right| \frac{\partial u}{\partial x} \tag{4.72}$$

meets the requirements. Here Δx is the interval length used in the numerical computations and c is dimensionless constant near unity. The physical viscous normal stress is $-\frac{4}{3}\mu\frac{\partial u}{\partial x}$ where μ is the physical viscosity coefficient. In comparison with this, the expression (4.72) denotes a nonlinear dissipative mechanism which is effective in the shock layer and negligible elsewhere. Eqs. (4.69) to (4.71) can be applied to shock waves which are now considered as continuous phenomena.

Integrating equations (4.69) to (4.71) we obtain

$$\rho u = C_1 \tag{4.73}$$

$$\rho u_2 + p + q = C_2 \tag{4.74}$$

$$\left(\rho c_v T + \frac{\rho u^2}{2} + p + q\right)u = C_3 \tag{4.75}$$

where C_1, C_2, C_3 are constants.

Let the initial and final values be denoted by

$$\text{As } x \to -\infty,\ u \to u_1,\ p \to p_1,\ q \to 0 \tag{4.76}$$

$$\text{As } x \to \infty,\ u \to u_2,\ p \to p_2,\ q \to 0 \tag{4.77}$$

Then Eq. (4.73) to (4.75) give

$$\rho_1 u_1 = \rho_2 u_2 \tag{4.78}$$

$$\rho_1 u_1^2 + p_1 = \rho_2 u_2^2 + p_2 \tag{4.79}$$

$$\left(\rho_1 c_v T_1 + \frac{\rho_1 u_1^2}{2} + p_1\right)u_1 = \left(\rho_2 c_v T_2 + \frac{\rho_2 u_2^2}{2} + p_2\right)u_2 \tag{4.80}$$

Eq. (4.78) to (4.80) are the equations of Rankine-Hugoniot and are seen to be independent of the amount and form of the dissipation, provided that $q \to 0$ as $x \to \pm\infty$. The physical reason for this is that the Rankine-Hugoniot equations are direct consequence of the conservation laws of mass, momentum, and energy for adiabatic flows. These laws (4.69) to (4.71) require that in a shock a certain amount of mechanical energy be converted irreversibly into

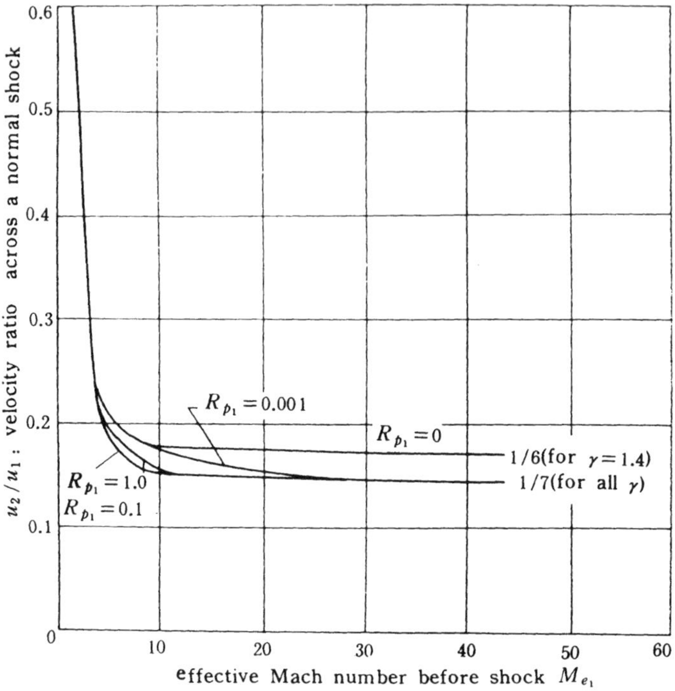

Fig. 4.16 Velocity ratio across a normal shock in an ideal gas with and without radiation effect

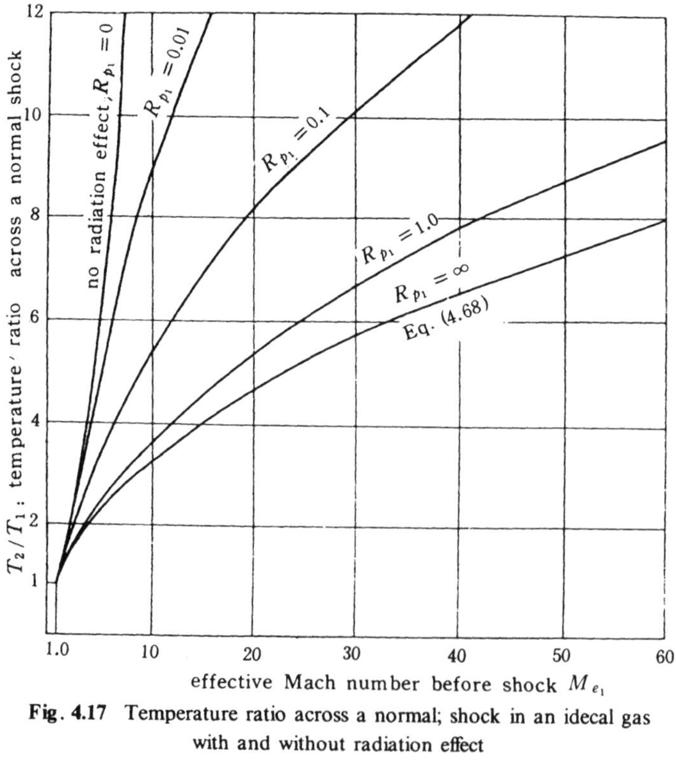

Fig. 4.17 Temperature ratio across a normal; shock in an idecal gas with and without radiation effect

heat. In the steady adiabatic state, the motion adjusts itself, in the shock layer, until precisely the amount of work done against the artificial viscous pressure q is converted into heat.

To investigate the shape of the shock, we look for solutions satisfying

$$\frac{\partial u}{\partial x} \leqslant 0 \tag{4.81}$$

This is normally the case for a shock in the flow of the x direction (Fig. 4.18). Then Eq. (4.72) can be written

$$q = c^2 \Delta x^2 \rho \left(\frac{du}{dx} \right)^2 \tag{4.82}$$

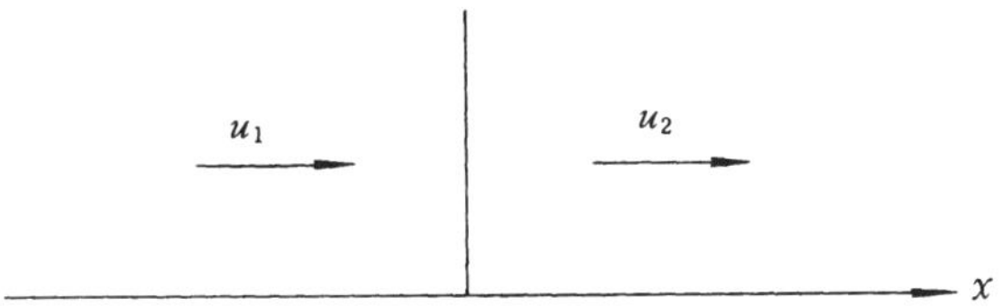

Fig. 4.18 Flow with shock

Eliminating $(p+q)$ from Eqs. (4.74) and (4.75) and using the equation of state $p = \rho RT$, we obtain

$$\frac{\rho u^3}{2} - \frac{pu}{\gamma - 1} = C_2 u - C_3 \tag{4.83}$$

From Eqs. (4.73), (4.74) and (4.83), we get

$$qu = -\frac{\gamma+1}{2} C_1 u^2 + \gamma C_2 u - (\gamma - 1) C_3 \tag{4.84}$$

The right member of Eq. (4.84) is a quadratic in u that vanishes for $u = u_1$, and for $u = u_2$, so that, clearly

$$qu = \frac{\gamma+1}{2} C_1 (u_1 - u)(u - u_2) \tag{4.85}$$

With the aid of Eq. (4.73), Eq. (4.85) gives

$$q = \frac{\gamma+1}{2} \rho (u_1 - u)(u - u_2) \tag{4.86}$$

Combining Eqs. (4.82) and (4.86), we obtain

$$c^2 \Delta x^2 \left(\frac{du}{dx} \right)^2 = \frac{\gamma+1}{2} (u_1 - u)(u - u_2) \tag{4.87}$$

Solving (4.87) for u, we have

$$u = \frac{u_1 + u_2}{2} - \frac{u_1 - u_2}{2} \sin \frac{x}{x_0} \tag{4.88}$$

where

$$x_0 = \sqrt{\frac{2}{\gamma+1}}\ c\Delta x \tag{4.89}$$

Because of our initial assumption (4.81) that $\frac{\partial u}{\partial x} \leqslant 0$, we can use only a half wave of the solution (4.88), but this half wave can be pieced together with two other particular solutions

$$u \equiv u_1, \quad \text{and} \quad u \equiv u_2 \tag{4.90}$$

to make the composite continuous solution depicted in Fig. 4.19. x_0 is a measure of the thickness of the shock, and is of order Δx, provided c is of order

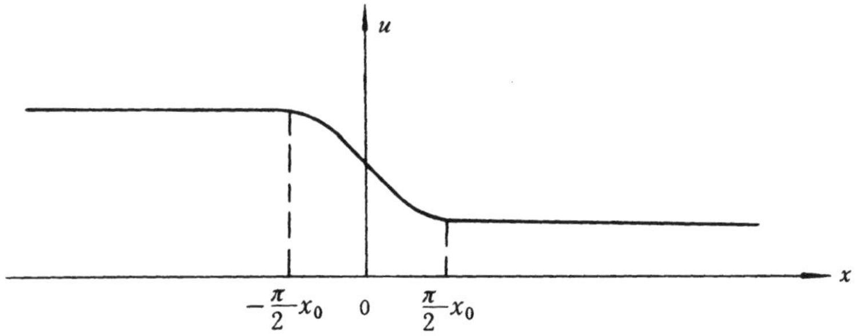

Fig. 4.19 Velocity distribution in shock

unity, independent of the strength of the shock and conditions ahead of it, by Eq. (4. 89). Throughout most of the system q is negligible in comparison with the ordinary pressure p, because the factor Δx^2 in Eq. (4.72), but in the shock layer q becomes comparable with p because of the abnormally large value of $\left|\frac{\partial u}{\partial x}\right|$ there. Expression (4.72) thus meets all requirements.

The method of artificial viscosity is here applied only to one-dimensional flows, but it is equally suited to the study of more complicated flows; where, indeed, shock calculations by shock-fitting method would ordinarily be prohibitively difficult. The method of artificial viscosity is known as the shock-capturing method. The practical numerical solution of shock waves by the artificial viscosity will be presented in Chapter X .

13. Problems

1. Show the detailed derivation of the Rankine-Hugoniot relation (4.5). Show also that Eq. (4.5) may be written as follows:

$$\frac{1}{2}(p_1 + p_2)\left(\frac{1}{\rho_1} - \frac{1}{\rho_2}\right) = \frac{1}{\gamma-1}\left(\frac{p_1}{\rho_1} - \frac{p_2}{\rho_2}\right)$$

2. Derive the expression for the maximum shock angle in terms of Mach number M_1 in front of the shock.

3. Consider that a shock wave travels in an ideal gas which is at rest in front of the shock, toward a rigid wall which is parallel to the shock wave. A reflected wave occurs after the original shock hits the wall. If $P = p_2/p_1$ is the pressure ratio of the original shock wave, and $P' = p_3/p_2$ is the pressure ratio across the reflected shock, show that

$$P' = \frac{(3\gamma-1)P-(\gamma-1)}{(\gamma-1)P+(\gamma+1)}$$

From the above general expression show that for a weak shock wave the pressure jump is doubled in the reflection.

4. Find the expression for the pressure jump across a normal shock in a van der Waals gas whose internal energy is given by equation

$$U_m = a_1 T + \frac{a_2 T^2}{2} + \frac{a_3 T^3}{3}$$

where a_1, a_2, a_3 are constants.

5. Find the ratio of the pressure p_t measured by the pitot tube in a uniform supersonic stream and the static pressure p of the uniform stream in terms of the Mach number M of the uniform stream.

6. Calculate the oblique shock wave relations with thermal radiation effects i.e. radiation pressure and radiation energy density.

References

1. Riemann, B., Ref. 6 of Chapter III.
2. Rankine, W. J. M., On the thermodynamic theory of finite longitudinal disturbance, *Phil. Trans. Roy. Soc. London* **CLX**, 1870, pp. 277 –288.
3. Hugoniot, H., Mémoire sur la propagation du mouvement dans les corps et spécialement dans les gases parfaits, *Jour. de l'École polyt. Cahier* **57**, 1887, pp. 1 –97, Cahier **58**, 1889, pp. 1 – 125.
4. Pai, S. I., *Fluid Dynamics of Jets*, D. Van Nostrand Company, Inc., Princeton, N. J., 1954.
5. Taylor, G. I., and Maccoll, J. W., *The Mechanics of Compressible Fluids*, Division H, Aerodynamic theory, **IV**, edited by W. F. Durand, Julius Springer, 1935.
6. Gunn, M. A., Relaxation time effects in gas dynamics, *British ARC R and M* No. 2338, 1952.
7. Howarth, L., *Modern Developments in Fluid Dynamics*, **I**, High speed flow, Chapter IV, Oxford University Press, 1953.
8. Courant, R. and Friedrichs, K. O., *Supersonic flow and shock waves*, Interscience Publishers, Inc., New York, 1948.
9. Oswatitsch, K., *Gasdynamik*, Wien, Springer Verlag, 1952.
10. Bethe, H. A., and Teller, E. Deviations from thermal equilibrium in shock waves, *Aberdeen Proving Ground B. R. L. Report* X-117, 1945.
11. Meyer, Th., Ueber zweidimensionale Bewenungsvorgaenge in einem Gas, das mit Ueberschallgeschwindigkeit stroemt, Dissertation Goettingen 1908, *Mitteilungen ueber Forschungsarbeiten auf dem Gebiete des Ing.Wesens* (edited by VDI), No. 62 (1908).

12. Busemann, A., *Verdichtungsstoesse in ebenen Gasstroemungen, Vortraege aur dem Gebiet der Aerodynamik und verwandter Gebiele* (Aachen 1929), edited by A. Giles, L. Hopf, and Th. V. Kármán, 1930, p. 162.
13. Bleakney, W., and Taub, A. H., Interaction of shock waves, *Revs. of Modern Phys.* **21**, No. 4, Oct. 1949, pp. 584-605.
14. Polachek, H., and Seeger, R. J., On shock –wave phenomena re-fraction of shock waves as a gaseous interface, *Phys, Rev.* **84**, No. 5, 1951, pp. 922- –929.
15. Hermann, R., Condensation shock waves in supersonic wind tunnel nozzles, *Luftfar* (LFF **19**, No. 6, 1942, pp. 201-209); also *RTP Trans.* No. 1581.
16. Oswatitsch, K., Kondensationsstoesse in Laval-Duesen, *ZAMM* **22**, 1942, pp. 1-14; also *RTP Trans.* No. 1905.
17. Döring, W., and Burkhardt, G., Beitraege zur Theorie der Detonation. FB 1939 (1944). Translation into English as Contributions to the Theory of Detonation, *Tech. Report* No. F-TS-1227-IA (GDAM-A9-T –46), May 1949, Wright –Patterson Air Force Base, Air Material Command.
18. Becker, R., Stosswelle and detonation, *Zeitschrift fuer Physik*, **8**, 1921-1922, pp. 321-362; also *NACA*, *TM* No. 505 and No. 506, 1929.
19. Jost, W., *Explosion and Combustion Processes in Gases*, McGraw-Hill Book Company, Inc., New York, 1946.
20. Lewis, B., and von Elbe, G., *Combustion, Flames and Explosions of Gases*, Academic Press, Inc., New York, 1951.
21. Pai, S. I., *Radiation gasdynamics*, Springer Verlag, New York, Inc., 1966.
22. Pai, S. I., Inviscid Flow of Radiation Gasdynamics, *Loui. Math. & Phy. Sci.* **3**, No. 4, pp. 361 –370, 1969.
23. Pai, S. I., *Modern Fluid Mechanics*, Science Press, Beijing, 1981, distributed by Van Nostrand Reinhold Co., N. Y..
24. Von Neumann, J. and Richtmyer, R. D., A method for the numerical calculation of hydrodynamic shocks, *Jour. of Appl. Phy.* **21**, No. 3, March 1950, pp. 232-237.

Chapter V

FUNDAMENTAL EQUATIONS OF THE DYNAMICS OF A COMPRESSIBLE INVISCID, NON-HEAT-CONDUCTING AND RADIATING FLUID[1-5]

1. Introduction

In the analysis of one-dimensional flow, we are actually considering only the average values of the velocity, pressure, etc., over certain sections. In order to study the details of the flow patterns, we have to find the velocity, pressure, etc., at every point in space. Hence we study the three-dimensional flow and consider all three components of velocity together with the pressure, density and temperature of the fluid as functions of the three spatial coordinates x, y, and z and time t. We shall derive the fundamental equations for three-dimensional flow and discuss some general properties of these equations.

In this chapter we shall consider only inviscid, non-heat-conducting and radiating fluid because the effects of viscosity, heat conductivity and radiative transfer are usually negligible except near a solid boundary or inside a shock. The effects of the transport phenomena will be discussed in Chapter XVII. We shall also consider only the non-electrically conducting fluid only in this chapter. The flow of an electrically conducting fluid will be discussed in Chapter XV.

We discuss the equation of state including the radiation effects in § 2; the equation of continuity as well as the equation of diffusion, the equation of continuity of a mixture of gases in § 3; the equations of motion in § 4 and the equation of energy in § 5. Kelvin's theorem, one of the most important theorems in an ideal fluid will be discussed for a compressible fluid in § 6. This theorem is the basis for the assumption of irrotational flow discussed in § 7.

All real fluid flows are rotational, and the rotationality of the flow may be characterized by vorticity. In § 8, we discuss the vortex motion and the Helmholtz's theorem for a compressible fluid. The equations of two-dimensional and axially symmetrical isoenergetic rotational flow are treated in § 9, and adiabatic flow is discussed in § 10.

A general discussion of various methods of solution of these fundamental equations is given in § 11. The characteristics of the unsteady Euler equations are discussed in § 12. More knowledge on the theory of characteristics will be given in Chapter XI. In § 13, the Euler differential equations are written into the conservation form. By use of the conservation form, the discontinuities which possibly exist in the inviscid flow are discussed in § 14 and the concept of solution for differential equations is generalized to include the discontinuous solution in § 15.

In § 16 some simple flows of a compressible fluid are treated. In § 17, we consider the general orthogonal coordinates while in § 18, the moving coordinates will be considered. Sometimes it is convenient to use vector notations for the fundamental equation of fluid mechanics. In § 19, the vector notations will be summarized and explained.

2. Equation of state

The equation of state is the relation between the pressure p, density ρ, and temperature T of the fluid considered. In general we may write

$$p=F(\rho, T) \tag{5.1}$$

where F is a function that was discussed in Chapter II, § 2.

In this book, we shall always consider a perfect gas except when otherwise specified. For this, the equation of state is

$$p=nR_A T \tag{5.2}$$

where n is the total number of molecules per unit volume under the pressure p and temperature T, and R_A is the universal gas constant. Eq. (5.2) holds true for a single perfect gas or a perfect mixture of several gases. For a single perfect gas or a perfect mixture of gases with constant composition, Eq. (5.2) may be written as

$$p=\rho \frac{R_A}{m} T=\rho RT \tag{5.2a}$$

where m is the mass of a molecule of the perfect gas or an equivalent mass for the case of a perfect mixture of gases and R is the gas constant. It should be noticed that if the concentration of the mixture of the gases changes and if Eq. (5.2a) is used as the equation of state, the values of m and R must be changed accordingly. For instance, under ordinary conditions, the value of m and R for air may be considered as a constant, e.g., $R=53.33$ ft/°F for air. However, at high temperature, when the phenomena of dissociation and ionization occur, the value of R must be recalculated.

For non-isentropic process with radiation effects, the equation of state of an ideal gas is given in Eq. (3.62). Since in most of the ordinary situations, the radiative effects are negligible, i.e., $R_p \to 0$ (see Fig. 3.7), we shall neglect the radiation effects unless otherwise specified.

In ordinary gasdynamics with $R_p \to 0$, Eq. (3.62) becomes

$$\frac{p}{p_0} = \left(\frac{\rho}{\rho_0}\right)^{\gamma} \exp\left(\frac{S - S_0}{c_v}\right) \tag{5.3}$$

We shall use this equation for rotational flow. For irrotational flow, we shall demonstrate that the entropy has the same constant value for all points in the field and Eq. (5.3) becomes

$$\frac{p}{p_0} = \left(\frac{\rho}{\rho_0}\right)^{\gamma} \tag{5.4}$$

where the subscript 0 refers to some reference state.

3. Equation of continuity and equation of diffusion

The second fundamental equation, the equation of continuity, is based upon the principle of conservation of mass. Let u, v, w, and ρ be the x-, y- and z-velocity components and the density of the fluid, respectively, at a point $P(x, y, z)$ in space at any given time t. Let us draw a parallelepiped with sides dx, dy, and dz from P as shown in Fig. 5.1. The mass flowing out of this parallelepiped in x-direction is

$$\left[\rho u + \frac{\partial \rho u}{\partial x} dx\right] dydz - \rho u dydz = \frac{\partial \rho u}{\partial x} dxdydz \tag{5.5}$$

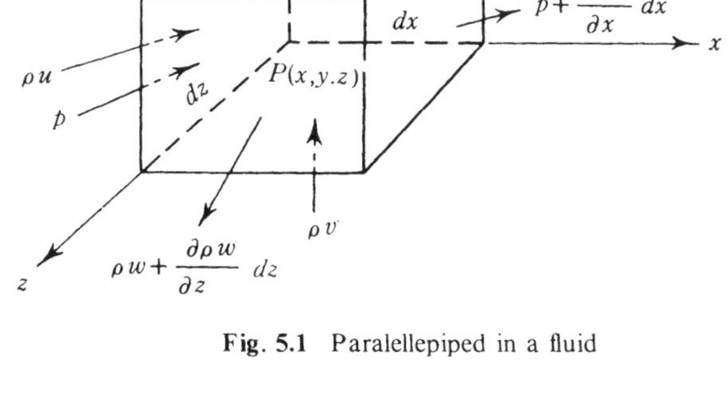

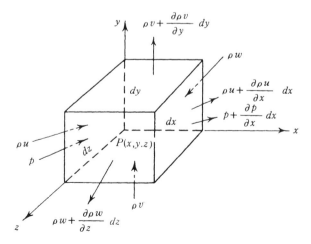

Fig. 5.1 Paralellepiped in a fluid

We have similar expressions for the mass flow in the y- and the z-directions. The total amount of mass flowing out of the parallelepiped is then

$$\left(\frac{\partial \rho u}{\partial x} + \frac{\partial \rho v}{\partial y} + \frac{\partial \rho w}{\partial z} \right) dxdydz \tag{5.5a}$$

By the conservation of mass, the expression of (5.5a) must be equal to the time rate of decrease of mass in the parallelepiped. Accordingly, we have

$$\left(\frac{\partial \rho u}{\partial x} + \frac{\partial \rho v}{\partial y} + \frac{\partial \rho w}{\partial x} \right) dxdydz = -\frac{\partial \rho}{\partial t} dxdydz$$

or

$$\frac{\partial \rho}{\partial t} + \frac{\partial \rho u}{\partial x} + \frac{\partial \rho v}{\partial y} + \frac{\partial \rho w}{\partial z} = 0 \tag{5.6}$$

In vector notation, Eq. (5.6) becomes

$$\frac{\partial \rho}{\partial t} + \nabla \cdot (\rho \boldsymbol{q}) = 0 \tag{5.6a}$$

where $\boldsymbol{q} = \boldsymbol{i}u + \boldsymbol{j}v + \boldsymbol{k}w$ is the velocity vector (see § 19). Eq. (5.6) or (5.6a) is known as the equation of continuity.

Most of the problems in which we are interested are for two-dimensional or axially symmetrical steady flows. For these two cases the equation of continuity may be written

$$\frac{\partial \rho u}{\partial x} + \frac{\partial \rho v}{\partial y} + \delta \frac{\rho v}{y} = 0 \tag{5.7}$$

where $\delta = 0$ for the two-dimensional case, and $\delta = 1$ for the axially symmetrical case. Here x and y denote the distance along and perpendicular to the flow axis, and u and v are the respective velocity components in these directions.

From Eq. (5.7), we can define a stream function ψ such that

$$\frac{\partial \psi}{\partial y} = \rho u y^{\delta}, \quad \frac{\partial \psi}{\partial x} = -\rho v y^{\delta} \tag{5.8}$$

It is clear that Eq. (5.7) is automatically satisfied by the stream function ψ. In some analyses, it is more convenient to consider the stream function as the unknown quantity.

For the surface $\psi =$ constant, we have

$$d\psi = 0 = \frac{\partial \psi}{\partial x} (dx)_{\psi} + \frac{\partial \psi}{\partial y} (dy)_{\psi}$$

or

$$\left(\frac{dy}{dx} \right)_{\psi} = -\frac{\partial \psi / \partial x}{\partial \psi / \partial y} = \frac{v}{u} \tag{5.9}$$

where the subscript ψ means along the curve ψ=constant. Eq. (5.9) shows that the slope of the curve ψ=constant is the same as the direction of the velocity vector. Consequently ψ=constant lines are streamlines and this is why ψ is called the stream function.

In many problems of fluid flow, the fluid is a mixture of several species, e. g. air is a mixture of oxygen, nitrogen and other gases. For the flow of a mixture of gases, new variables, in the form of concentration of any species in the mixture and the diffusion velocities of the species enter our problem. For a mixture of N-species, we may treat our flow problems by multi-fluid theory (Chapter XVI § 7) in which we consider the partial variables of each species, such as the velocity vector $\boldsymbol{q}_s$, temperature T_s, pressure p_s and density ρ_s for the sth species where $s=1,2,\cdots$, or N. We may derive the equation of continuity for each species in the same manner as that for the mixture as a whole, Eq. (5.6), i.e.

$$\frac{\partial \rho_s}{\partial t}+\frac{\partial \rho_s u_s}{\partial x}+\frac{\partial \rho_s v_s}{\partial y}+\frac{\partial \rho_s w_s}{\partial z}=\sigma_s \tag{5.10}$$

where σ_s is the source term for the sth species which depends on the chemical reactions and ionization. The sum of all the source terms is zero, i.e.

$$\sum_{s=1}^{N}\sigma_s=0 \tag{5.11}$$

The total density of the mixture is simply

$$\rho=\sum_{s=1}^{N}\rho_s \tag{5.12}$$

and the velocity of the mixture is defined as

$$\boldsymbol{q}=\frac{1}{\rho}\sum_{s=1}^{N}\rho_s\boldsymbol{q}_s \tag{5.13}$$

With the help of Eqs. (5.11) to (5.13), the sum of all the N-equation (5.10) gives us Eq. (5.6) which may be considered as the equation of continuity of the mixture.

If we use Eq. (5.10) in our analysis, we have to know the partial velocity u_s^i, i.e. $u_s^1=u_s$, $u_s^2=v_s$ and $u_s^3=w_s$. The partial velocity is determined by the equation of motion of the sth species. Ordinarily, the differences between the partial velocity and the velocity of the mixture as a whole are not large. Hence we may use the diffusion velocity and diffusion coefficients in our analysis. The diffusion velocity w_s^i is defined as

$$w_s^i=u_s^i-u^i \tag{5.14}$$

When the diffusion velocity is small, we may use the following approximate formula to express the diffusion velocity in terms of the diffusion coefficients:

$$\rho_r w_r^i = \sum_{q \neq r} \rho D_{qr} \frac{\partial c_q}{\partial x^i} + D_p \frac{\partial \ln p}{\partial x^i} - D_T \frac{\partial \ln T}{\partial x^i} \tag{5.15}$$

where the number concentration of species r, c_r, is the ratio of the number density of the rth species to the number density of the mixture as a whole n, i. e., $n = \sum_{r=1}^{N} n_r$ and

$$c_r = \frac{n_r}{n} \tag{5.16}$$

The mass concentration k_r of the rth species is

$$k_r = \frac{\rho_r}{\rho} = \frac{m_r n_r}{\rho} \tag{5.17}$$

where m_r is the mass of a particle or molecule of the rth species. D_{qr} is the diffusion coefficient between the rth and qth species in the mixture. For a binary mixture, the first term on the right-hand side of Eq. (5.15) becomes

$$(k_r w_r^i)_c = -D_{qr} \frac{\partial c_r}{\partial x^i} \tag{5.18}$$

The binary diffusion coefficient D_{qr} is one of the most important diffusion coefficient in fluid dynamics which will be further discussed in Chapter XVII § 4 and § 10. D_p is known as the coefficient of pressure diffusion and D_T is the coefficient of thermal diffusion. Both D_p and D_T are usually negligibly small in most of the fluid dynamical problems. The diffusion velocity w_r^i may depend on other physical phenomena. For instance, the diffusion velocity of charged particles in a plasma depends mainly on the electromagnetic field and we should add the corresponding terms on the right-hand side of Eq. (5.15). We shall discuss the diffusion of charged particles in Chapters XV and XVII.

Substituting Eqs. (5.14) and (5.15) in Eq. (5.10) and using the equation of continuity (5.6) we have the diffusion equation as follows:

$$\frac{Dk_r}{Dt} = -\sum_{q \neq r} \frac{\partial}{\partial x^j} \left(\rho D_{qr} \frac{\partial c_q}{\partial x^j} \right) - \frac{\partial}{\partial x^j} \left(D_p \frac{\partial \ln p}{\partial x^j} \right) + \frac{\partial}{\partial x^j} \left(D_T \frac{\partial \ln T}{\partial x^j} \right) + \sigma_r \tag{5.19}$$

For the flow without transfer phenomena or source, Eq. (5.19) gives

$$k_r = \text{constant} \tag{5.20}$$

4. Equation of motion

We consider here the flow of an inviscid fluid without body forces. Only the inertial forces and pressure forces will be included. Consider again the parallelelpiped (Fig. 5.1). By Newton's second law of motion in the x-direction, we have

$$\rho \frac{Du}{Dt} dxdydz = -\frac{\partial p}{\partial x} dxdydz$$

or

$$\rho \frac{Du}{Dt} = -\frac{\partial p}{\partial x} \tag{5.21a}$$

where D/Dt is the total derivative given in Eq. (1.10), i.e.

$$\frac{D}{Dt} = \frac{\partial}{\partial t} + u\frac{\partial}{\partial x} + v\frac{\partial}{\partial y} + w\frac{\partial}{\partial z} = \frac{\partial}{\partial t} + \boldsymbol{q} \cdot \nabla \tag{5.22}$$

Similarly the equations of motion in the y- and z-directions are

$$\rho \frac{Dv}{Dt} = -\frac{\partial p}{\partial y} \tag{5.21b}$$

$$\rho \frac{Dw}{Dt} = -\frac{\partial p}{\partial z} \tag{5.21c}$$

In vector notation, the equation of motion (5.10) may be written as follows:

$$\rho \frac{D\boldsymbol{q}}{Dt} = \rho\left[\frac{\partial \boldsymbol{q}}{\partial t} + (\boldsymbol{q} \cdot \nabla)\boldsymbol{q}\right] = -\nabla p \tag{5.23}$$

Using the vector identity [see § 19, Eq. (5.176)]

$$\nabla\left(\frac{1}{2} q^2\right) = \frac{1}{2} \nabla(\boldsymbol{q} \cdot \boldsymbol{q}) = (\boldsymbol{q} \cdot \nabla)\boldsymbol{q} + \boldsymbol{q} \times (\nabla \times \boldsymbol{q}) \tag{5.24}$$

Eq. (5.23) becomes

$$\frac{\partial \boldsymbol{q}}{\partial t} + \nabla\left(\frac{1}{2} q^2\right) - \boldsymbol{q} \times \boldsymbol{\omega} = -\frac{1}{\rho} \nabla p \tag{5.25}$$

where $\boldsymbol{\omega}$ is known as the vorticity and is defined by

$$\boldsymbol{\omega} = \nabla \times \boldsymbol{q} = \boldsymbol{i}\left(\frac{\partial w}{\partial y} - \frac{\partial v}{\partial z}\right) + \boldsymbol{j}\left(\frac{\partial u}{\partial z} - \frac{\partial w}{\partial x}\right) + \boldsymbol{k}\left(\frac{\partial v}{\partial x} - \frac{\partial u}{\partial y}\right) \tag{5.26}$$

A further discussion of the properties of vorticity will be given in § 7 and § 8.

In the above equations of motion, the radiation effects are neglected. If we include the thermal radiation effects, we should replace the pressure p with the total pressure $p_t = p + p_R$ and all the other terms remain unchanged.

5. Equation of energy

Since we consider only inviscid and no-heat-conducting fluids, the conservation of energy gives the energy equation as [cf. Eq. (2.18)]

$$dQ = TdS = dU_m + pd\frac{1}{\rho} = dh - \frac{1}{\rho}dp \tag{5.27}$$

which, in vector notation, becomes

$$T\nabla S = \nabla h - \frac{1}{\rho}\nabla p \tag{5.28}$$

In the discussion of energy, it is sometimes convenient to introduce a stagnation enthalpy h_0 defined as

$$h_0 = h + \frac{1}{2}q^2 \tag{5.29}$$

This represents the sum of heat energy h and kinetic energy fluid per unit mass. A knowledge of the time rate of change of the stagnation enthalpy may be gotten by the following procedure. From Eq. (5.29), we have

$$\frac{Dh_0}{Dt} = \frac{Dh}{Dt} + \frac{D}{Dt}\left(\frac{1}{2}q^2\right) = \frac{Dh}{Dt} + \boldsymbol{q}\cdot\frac{D\boldsymbol{q}}{Dt} \tag{5.30}$$

The energy equation (5.27) may be written as

$$T\frac{DS}{Dt} = \frac{Dh}{Dt} - \frac{1}{\rho}\frac{Dp}{Dt} = \frac{Dh}{Dt} - \frac{1}{\rho}\left[\frac{\partial p}{\partial t} + (\boldsymbol{q}\cdot\nabla)p\right] \tag{5.27a}$$

Combining Eq. (5.27a) and Eq. (5.30) gives

$$\begin{aligned}\frac{Dh_0}{Dt} &= T\frac{DS}{Dt} + \frac{1}{\rho}\frac{\partial p}{\partial t} + \boldsymbol{q}\cdot\left\{\frac{D\boldsymbol{q}}{Dt} + \frac{1}{\rho}\nabla p\right\} \\ &= T\frac{DS}{Dt} + \frac{1}{\rho}\frac{\partial p}{\partial t} = \frac{DQ}{Dt} + \frac{1}{\rho}\frac{\partial p}{\partial t}\end{aligned} \tag{5.31}$$

which shows the time rate change of stagnation enthalpy with heat addition. For adiabatic flow, the heat added is zero, i.e. $dQ=0$, and we have

$$\frac{Dh_0}{Dt} = \frac{\partial h_0}{\partial t} + (\boldsymbol{q}\cdot\nabla)h_0 = \frac{1}{\rho}\frac{\partial p}{\partial t} \tag{5.32}$$

and

$$\frac{DS}{Dt} = \frac{\partial S}{\partial t} + (\boldsymbol{q}\cdot\nabla)S = 0 \tag{5.33}$$

Eqs. (5.32) and (5.33) are the energy equation for the three-dimensional adiabatic flow of a compressible fluid. It is interesting to note that for steady flow we have

$$(\boldsymbol{q} \cdot \nabla) h_0 = 0 \text{ and } (\boldsymbol{q} \cdot \nabla) S = 0 \tag{5.34}$$

These relations state that both the stagnation enthalpy and entropy are constant along any stream line for steady adiabatic flow. It should be noted that for nonsteady flow, the stagnation enthalpy is not a constant along a stream line but is dependent on the time rate of change of pressure.

In the above equations of energy, we neglect the thermal radiation effects. If the thermal radiation effects are included, we should repalce the pressure p with the total pressure $p_t = p + p_R$ and the internal energy per unit mass U_m with the total internal energy per unit mass $U_{mR} = U_m + E_R/\rho$ and the enthalpy per unit mass h with the enthalpy including the thermal radiation effects $h_R = U_{mR} + p_p/\rho$. The other terms remain the same.

In the following sections, we shall neglect the thermal radiation effects unless otherwise specified. The extension of the results without thermal radiation to the cases with thermal radiation effects is straightforward. We shall put them in the section of problems so that the students may easily derive those formulas with thermal radiation effects.

6. Kelvin's theorem

We now introduce an important concept of fluid dynamics, the circulation. The circulation Γ is defined as the line integral of velocity along a closed curve:

$$\Gamma = \oint_c \boldsymbol{q} \cdot d\boldsymbol{l} \tag{5.35}$$

where $\boldsymbol{q}$ is the velocity vector, $d\boldsymbol{l}$ is the differential of arc length of the closed curve c, and $\boldsymbol{q} \cdot d\boldsymbol{l}$ is the scalar product of these two vectors. We would like to know the time rate of change of circulation for a closed fluid curve where, by "fluid curve", we mean a curve which always consists of the same particles of fluid. Let c be a closed fluid curve. We have then

$$\frac{D\Gamma}{Dt} = \oint_c \frac{D\boldsymbol{q}}{Dt} \cdot d\boldsymbol{l} + \oint_c \boldsymbol{q} \cdot \frac{Dd\boldsymbol{l}}{Dt} = \oint_c \frac{D\boldsymbol{q}}{Dt} d\boldsymbol{l} + \oint_c \boldsymbol{q} \cdot d\boldsymbol{q}$$

$$= \oint_c \frac{D\boldsymbol{q}}{Dt} d\boldsymbol{l} + \frac{1}{2} \oint_c dq^2 = \oint_c \frac{D\boldsymbol{q}}{Dt} d\boldsymbol{l} = \oint_c \left(-\frac{1}{\rho} \nabla p \right) \cdot d\boldsymbol{l} \tag{5.36}$$

By Stokes' theorem

$$\oint_c \boldsymbol{B} \cdot d\boldsymbol{l} = \oiint_A (\nabla \times \boldsymbol{B}) \cdot d\boldsymbol{A} \tag{5.37}$$

Eq. (5.36) becomes

$$\frac{D\Gamma}{Dt} = -\oiint_A \left[\nabla \times \left(\frac{1}{\rho} \nabla p\right)\right] \cdot d\boldsymbol{A} \tag{5.38}$$

where $\boldsymbol{A}$ is the vector area enclosed by the closed fluid curve c. Since

$$\nabla \times \left(\frac{1}{\rho} \nabla p\right) = \frac{1}{\rho} \nabla \times \nabla p + \nabla\left(\frac{1}{\rho}\right) \times \nabla p = \nabla\left(\frac{1}{\rho}\right) \times \nabla p$$

$$= \nabla \times (h - T\nabla S) = -\nabla T \times \nabla S$$

because

$$\nabla \times \nabla p = \nabla \times \nabla h = \nabla \times \nabla S = 0$$

Eq. (5.38) may be written as follows:

$$\frac{D\Gamma}{Dt} = -\oiint_A \left[\nabla\left(\frac{1}{\rho}\right) \times \nabla p\right] \cdot d\boldsymbol{A} = \oiint_A (\nabla T \times \nabla S) \cdot d\boldsymbol{A} \tag{5.38a}$$

Eq. (5.38a) shows that the circulation along a closed fluid curve is in general not a constant even when there are no viscous forces. When the stagnation enthalpy h_0 and the entropy S are not constant in space, $D\Gamma/Dt$ will not be zero.

However, for so-called barotropic fluids where the pressure is a function of density only, i.e. $p = p(\rho)$, we have

$$\nabla \times \left(\frac{1}{\rho} \nabla p\right) = \nabla \times \left(\frac{1}{\rho} \frac{dp}{d\rho} \nabla \rho\right) = \nabla \times \nabla f(\rho) = 0$$

and then

$$\frac{D\Gamma}{Dt} = 0 \tag{5.39}$$

which is Kelvin's theorem. This states that, for a barotropic fluid, the circulation along a closed fluid curve remains constant in time. For an incompressible fluid, ρ = constant and $\nabla \times \frac{1}{\rho} \nabla p = \frac{1}{\rho} \nabla \times \nabla p = 0$, so that Kelvin's theorem holds good here also. For a compressible fluid, the Kelvin's theorem holds only when $p = p(\rho)$. Since in general for compressible fluid $p = p(\rho, S)$, Kelvin's theorem will not hold good if S is not a constant in the whole flow field, e.g., for the flow behind a curved shock.

7. Irrotational motion

If we apply Stokes′theorem (5.37) to Eq. (5.35), we have

$$\Gamma=\oint_{c} \boldsymbol{q} \cdot d\boldsymbol{l}=\oiint_{A} \boldsymbol{\omega} \cdot d\boldsymbol{A} \tag{5.40}$$

where $\boldsymbol{\omega}=\nabla\times\boldsymbol{q}$ = vorticity.

If $\boldsymbol{\omega}=\mathbf{0}$, we say that the flow is irrotational. If the whole flow field is irrotational at any given time, the circulation in the whole flow field at this time will be zero by Eq. (5.40) because $\boldsymbol{\omega}=\mathbf{0}$. If Kelvin′s theorem applies, the circulation Γ will be zero everywhere and at all subsequent times. As a result, the vorticity will be zero at all times. Furthermore, when the flow starts from rest, initially the vorticity is everywhere zero, and the flow will be irrotational. Thus a very large class of physically important isentropic flows of an inviscid fluid are irrotational; this is the reason that irrotational flow has been extensively investigated.

For irrotational flow, there exists a velocity potential φ such that

$$\boldsymbol{q}=\nabla\varphi \tag{5.41}$$

Since

$$\boldsymbol{\omega}=\nabla\times\boldsymbol{q}=\nabla\times\nabla\varphi=\mathbf{0}$$

the condition irrotationality is satisfied. The relations between the velocity potential and the velocity components (u, v, w) are

$$u=\frac{\partial\varphi}{\partial x}, \quad v=\frac{\partial\varphi}{\partial y}, \quad w=\frac{\partial\varphi}{\partial z} \tag{5.41a}$$

Substituting Eq. (5.41) into the equation of motion (5.25) yields

$$\nabla\left(\frac{\partial\varphi}{\partial t}+\frac{q^2}{2}+\int\frac{dp}{\rho}\right)=0$$

or

$$\frac{\partial\varphi}{\partial t}+\frac{q^2}{2}+\int\frac{dp}{\rho}=f(t) \tag{5.42}$$

The arbitrary function of time $f(t)$ may be absorbed into the velocity potential φ without altering the relation (5.41). If this is done, we have

$$\frac{\partial\varphi}{\partial t}+\frac{q^2}{2}+\int\frac{dp}{\rho}=\text{constant}=B_0 \tag{5.42a}$$

which is known as Bernoulli′s theorem; B_0 is known as Bernoulli′s constant.

Differentiating Eq. (5.42a) with respect to t, we have

$$\frac{\partial^2\varphi}{\partial t^2} + \boldsymbol{q} \cdot \frac{\partial \boldsymbol{q}}{\partial t} + a^2 \frac{1}{\rho} \frac{\partial \rho}{\partial t} = 0 \qquad (5.43)$$

where $a = \sqrt{\dfrac{dp}{d\rho}}$ = local sound speed.

Now the equation of continuity (5.6) may be written as

$$\frac{1}{\rho} \frac{\partial \rho}{\partial t} + \nabla^2\varphi + \frac{1}{\rho} \boldsymbol{q} \cdot \nabla\rho = 0$$

but

$$\boldsymbol{q} \cdot \frac{1}{\rho} \nabla\rho = \frac{1}{a^2} \boldsymbol{q} \cdot \frac{1}{\rho} \nabla p = \frac{1}{a^2} \boldsymbol{q} \cdot \left\{ -\frac{\partial \boldsymbol{q}}{\partial t} - (\boldsymbol{q} \cdot \nabla)\boldsymbol{q} \right\}$$

so that Eq. (5.43) becomes

$$\frac{1}{a^2} \frac{\partial^2\varphi}{\partial t^2} + \frac{2}{a^2} \boldsymbol{q} \cdot \frac{\partial \boldsymbol{q}}{\partial t} = \nabla^2\varphi - \frac{1}{a^2} \boldsymbol{q} \cdot [(\boldsymbol{q} \cdot \nabla)\boldsymbol{q}] \qquad (5.44)$$

or

$$\left(1 - \frac{u^2}{a^2}\right)\frac{\partial^2\varphi}{\partial x^2} + \left(1 - \frac{v^2}{a^2}\right)\frac{\partial^2\varphi}{\partial y^2} + \left(1 - \frac{w^2}{z^2}\right)\frac{\partial^2\varphi}{\partial z^2}$$

$$-2\frac{uv}{a^2} \frac{\partial^2\varphi}{\partial x \partial y} - 2\frac{vw}{a^2} \frac{\partial^2\varphi}{\partial y \partial z} - 2\frac{wu}{a^2} \frac{\partial^2\varphi}{\partial z \partial x}$$

$$= \frac{1}{a^2} \frac{\partial^2\varphi}{\partial t^2} + \frac{2u}{a^2} \frac{\partial^2\varphi}{\partial x \partial t} + \frac{2v}{a^2} \frac{\partial^2\varphi}{\partial y \partial t} + \frac{2w}{a^2} \frac{\partial^2\varphi}{\partial z \partial t} \qquad (5.44a)$$

This is the differential equation for the velocity potential φ. Here the flow is isentropic and the velocity of sound a is related to the velocity q by the relation (3.12).

For incompressible fluid, $a = \infty$, Eq. (5.44) becomes

$$\nabla^2\varphi = \frac{\partial^2\varphi}{\partial x^2} + \frac{\partial^2\varphi}{\partial y^2} + \frac{\partial^2\varphi}{\partial z^2} = 0 \qquad (5.45)$$

For the two-dimensional steady flow of a compressible fluid, Eq. (5.44a) becomes

$$\left(1 - \frac{u^2}{a^2}\right)\frac{\partial^2\varphi}{\partial x^2} - 2\frac{uv}{a^2} \frac{\partial^2\varphi}{\partial x \partial y} + \left(1 - \frac{v^2}{a^2}\right)\frac{\partial^2\varphi}{\partial y^2} = 0 \qquad (5.46)$$

8. Vortex motion. Helmholtz's theorem

In more general fluid flow, the vorticity is not zero everywhere and the fluid elements have the components of angular velocity.

$$\xi = \frac{1}{2}\left(\frac{\partial w}{\partial y} - \frac{\partial v}{\partial z}\right) = \frac{1}{2}\,\omega_x,\ \eta = \frac{1}{2}\left(\frac{\partial u}{\partial z} - \frac{\partial w}{\partial x}\right) = \frac{1}{2}\,\omega_y,$$

$$\zeta = \frac{1}{2}\left(\frac{\partial v}{\partial x} - \frac{\partial u}{\partial y}\right) = \frac{1}{2}\,\omega_z$$

where u, v, w are, respectively, the x-, y- and z-velocity components at the point (x, y, z). It should be noted that it is possible that some portions of the fluid may possess rotation while others move irrotationally.

Vortex lines are lines drawn in the fluid which at every point coincide with the instantaneous axis of rotation of the corresponding fluid elements. Vortex filaments or vortices are portions of the fluid bounded by vortex lines drawn through every point of an infinitesimal closed curve. The boundary of a vortex is called a vortex tube.

We would like to investigate the behavior of vortices in fluid flow. To find such relation we take the curl of equation of motion (5.25)

$$\begin{aligned}\nabla \times \frac{D\boldsymbol{q}}{Dt} &= \nabla \times \frac{\partial \boldsymbol{q}}{\partial t} + \nabla \times \nabla\left(\frac{1}{2}q^2\right) - \nabla \times (\boldsymbol{q} \times \boldsymbol{\omega}) \\ &= \frac{\partial}{\partial t}\nabla \times \boldsymbol{q} - (\boldsymbol{\omega} \cdot \nabla)\boldsymbol{q} + (\boldsymbol{q} \cdot \nabla)\boldsymbol{\omega} - \boldsymbol{q}(\nabla \cdot \boldsymbol{\omega}) + \boldsymbol{\omega}(\nabla \cdot \boldsymbol{q}) \\ &= \frac{\partial \boldsymbol{\omega}}{\partial t} + (\boldsymbol{q} \cdot \nabla)\boldsymbol{\omega} - (\boldsymbol{\omega} \cdot \nabla)\boldsymbol{q} + \boldsymbol{\omega}(\nabla \cdot \boldsymbol{q}) = -\nabla \times \left(\frac{1}{\rho}\nabla p\right)\end{aligned}$$

which is

$$\begin{aligned}\frac{D\boldsymbol{\omega}}{Dt} &= (\boldsymbol{\omega} \cdot \nabla)\boldsymbol{q} - \boldsymbol{\omega}(\nabla \cdot \boldsymbol{q}) - \nabla\frac{1}{\rho}\nabla p \\ &= (\boldsymbol{\omega} \cdot \nabla)\boldsymbol{q} - \boldsymbol{\omega}(\nabla \cdot \boldsymbol{q}) + \nabla T \times \nabla S \end{aligned} \qquad (5.47)$$

If the fluid is barotropic, i.e., $p = p(\rho)$, we have

$$\frac{D\boldsymbol{\omega}}{Dt} = (\boldsymbol{\omega} \cdot \nabla)\boldsymbol{q} - \boldsymbol{\omega}(\nabla \cdot \boldsymbol{q}) = (\boldsymbol{\omega} \cdot \nabla)\boldsymbol{q} + \frac{\boldsymbol{\omega} D\rho}{\rho Dt} \qquad (5.48)$$

Eqs. (5.47) and (5.48) give the time rate of change of vorticity of the fluid with respect to an observer who is moving with the fluid. One important conclusion is that, for barotropic fluids, if at one instant $\boldsymbol{\omega} = \boldsymbol{0}$, the flow will be irrotational for all subsequent times. For instance, for shock-free flow, the motion of a body in a uniform stream will be irrotational.

For a barotropic fluid, the vortex lines move with the fluid. This can be proved as follows.

We consider a surface S composed of vortex lines at time t. The circulation in any closed fluid curve C on the surface S is zero because $\boldsymbol{\omega} \cdot d\boldsymbol{A}$ is always zero [cf. Eq. (5.36)]. At a later time $t+\delta t$, the particles that formed the surface S will lie on another surface S' and the circulation in it is still zero because of Eq. (5.39). This is true for all such closed fluid curves on S'. The surface S' must be also composed of vortex lines. Hence any surface composed of vortex lines, as it moves with the fluid, continues to be composed of vortex lines. The intersection of two such surfaces must always be a vortex line. Thus we have the theorem that for barotropic fluid the vortex lines move with the fluid. The theorem is true also for incompressible fluids.

Eq. (5.48) is known as Helmholtz's theorem. It states that the vorticity for barotropic fluid moves with the fluid and that the change of vorticity is due to the distortion of the vortex tube as given by Eq. (5.48). The second part of the above statement may be seen more clearly by considering a simple case in which $\omega_x \neq 0$, $\omega_y = \omega_z = 0$. We consider a small vortex tube which at $t=0$, we assume, has cross-sectional area A, length L, and vorticity ω_x. At a later time $t=\delta t$, this same tube has cross-sectional area A', length L', and vorticity ω_x'. By Kelvin's theorem we have

$$\omega_x A = \omega_x' A' \tag{5.49}$$

The length of the tube at time δt is

$$L \quad = L + \frac{\partial u}{\partial x} L\delta t = L\left(1 + \frac{\partial u}{\partial x}\delta t\right) \tag{5.50}$$

The mass in the tube at $t=0$ is equal to that at $t=\delta t$; thus

$$AL = A'L'\left(\rho + \frac{D\rho}{Dt}\delta t\right) \tag{5.51}$$

Hence

$$\frac{A}{A'} = \frac{L'}{L}\left(1 + \frac{1}{\rho}\frac{D\rho}{Dt}\delta t\right) \cong 1 + \left(\frac{\partial u}{\partial x} + \frac{1}{\rho}\frac{D\rho}{Dt}\right)\delta t \tag{5.52}$$

Substituting Eq. (5.52) into Eq. (5.49) results in

$$\omega_x' - \omega_x = D\omega_x = \omega_x\left(\frac{\partial u}{\partial x} + \frac{1}{\rho}\frac{D\rho}{Dt}\right)\delta t$$

or

$$\frac{D\omega_x}{Dt} = \omega_x\left(\frac{\partial u}{\partial x} + \frac{1}{\rho}\frac{D\rho}{Dt}\right) \tag{5.53}$$

Eq. (5.53) is identically the x-component of Eq. (5.48) if we put $\omega_x = \omega_x$, $\omega_y = \omega_z = 0$. Thus Eq. (5.53) or (5.48) represents the change of vorticity due to the stretching of the vortex tube.

9. Two-dimensional and axially symmetrical steady isoenergetic rotational flow[6-10]

For rotational flow, such as the flow behind a curved shock, the velocity potential does not exist and we cannot use Eq. (5.44). In rotational flow, it is sometimes convenient to use the stream function ψ for two-dimensional or axial symmetric flow as defined in Eqs. (5.8). We shall develop the differential equation for the stream function ψ as follows:

For two-dimensional or axially symmetric flow, only one component of the vorticity $\boldsymbol{\omega}$ is different from zero. Let us denote this component of vorticity by $\boldsymbol{\omega}$. For steady flow, the stagnation enthalpy h_0 and the entropy S are functions of stream function ψ. From Eqs. (5.25) and (5.28), we have for steady flow

$$\boldsymbol{q} \times \boldsymbol{\omega} = T\nabla S - \nabla h_0 \tag{5.54}$$

which for the two-dimensional or axially symmetrical case becomes

$$q\omega = T\frac{dS}{dn} - \frac{dh_0}{dn} \tag{5.55}$$

where q is the resultant velocity, and n is the normal to the streamline.

For isoenergetic flow, h_0 is a constant, and using Eq. (5.8) we then have

$$\omega = y^{\delta}\rho T\frac{dS}{d\psi} \tag{5.56}$$

Eq. (5.55) shows that vorticity depends on the spatial variation of stagnation enthalpy and entropy. For isoenergetic flow, vorticity is a function of the derivative of entropy. For isentropic flow, the vorticity is zero. Hence isentropic flow is irrotational flow.

We have now

$$\omega = \frac{\partial v}{\partial x} - \frac{\partial u}{\partial y} \tag{5.57}$$

then

$$\rho\omega = \rho\frac{\partial v}{\partial x} - \rho\frac{\partial u}{\partial y} = \frac{\partial \rho v}{\partial x} - \frac{\partial \rho u}{\partial y} - \left(v\frac{\partial \rho}{\partial x} - u\frac{\partial \rho}{\partial y}\right) \tag{5.58}$$

In an isoenergetic rotational flow, the entropy is nonuniform, and the pressure is a function of density and entropy given by Eq. (5.3). Thus

$$v\frac{\partial \rho}{\partial x} - u\frac{\partial \rho}{\partial y} = \frac{1}{a^2}\left(v\frac{\partial p}{\partial x} - u\frac{\partial p}{\partial y}\right) + \frac{1}{c_p}\frac{dS}{d\psi}\left[\left(\frac{\partial \psi}{\partial x}\right)^2 + \left(\frac{\partial \psi}{\partial y}\right)^2\right]\frac{1}{y^{\delta}}$$

$$= -\frac{1}{y^{\delta}}\left[\frac{u^2}{a^2}\frac{\partial^2 \psi}{\partial x^2} + \frac{2uv}{a^2}\frac{\partial^2 \psi}{\partial x \partial y} + \frac{y^2}{a^2}\frac{\partial^2 \psi}{\partial y^2}\right]$$

$$+\frac{1}{c_p}\frac{dS}{d\psi}\left[\left(\frac{\partial\psi}{\partial x}\right)^2+\left(\frac{\partial\psi}{\partial y}\right)^2\right]\frac{1}{y^\delta} \tag{5.59}$$

$$a^2=\left(\frac{\partial p}{\partial\rho}\right)_S=\frac{\gamma p}{\rho} \tag{5.60}$$

Combining Eqs. (5.56), (5.58), (5.59) and (5.8) and rearranging them, we have

$$\left(1-\frac{u^2}{a^2}\right)\frac{\partial^2\psi}{\partial x^2}-\frac{2uv}{a^2}\frac{\partial^2\psi}{\partial x\partial y}+\left(1-\frac{v^2}{a^2}\right)\frac{\partial^2\psi}{\partial y^2}-\frac{\delta}{y}\frac{\partial\psi}{\partial y}$$

$$=-\rho^2y^{2\delta}T\frac{dS}{d\psi}-\frac{1}{c_p}\frac{dS}{d\psi}\left[\left(\frac{\partial\psi}{\partial x}\right)^2+\left(\frac{\partial\psi}{\partial y}\right)^2\right]$$

$$=-\frac{y^\delta}{\gamma R}\frac{a}{M}\frac{dS}{dn}[1+(\gamma-1)M^2] \tag{5.61}$$

where $M=q/a$. This is the fundamental differential equation for the stream function ψ for the rotational isoenergetic flow of a compressible fluid. If Eq. (5.55) is used, a more general form is obtained. Eq. (5.61) and Eq. (5.3) should be considered simultaneously with the energy equation:

$$\frac{T}{T_0}=1-\frac{\gamma-1}{2}\frac{1}{\rho^2a_0^2y^{2\delta}}\left[\left(\frac{\partial\psi}{\partial x}\right)^2+\left(\frac{\partial\psi}{\partial y}\right)^2\right] \tag{5.62}$$

where the subscript 0 refers to the stagnation values. If the variation of entropy with streamline $dS/d\psi$ is specified, we may try to solve for the stream function ψ from Eq. (5.61) together with Eqs. (5.62) and (5.3). However, in actual cases, $dS/d\psi$ is not specified a priori but depends on the solution itself. For instance, in the case of detached shock, $dS/d\psi$ depends on the shape of the shock which has to be fixed by matching the solution of stream function ψ from Eq. (5.61) to the shock relation. It can thus be seen that the problem is very complicated.

Crocco has defined another stream function for rotational flow. He has obtained a slightly different form for the differential equation of the stream function[6, 7].

For two-dimensional irrotational flow, Eq. (5.61) becomes

$$\left(1-\frac{u^2}{a^2}\right)\frac{\partial^2\psi}{\partial x^2}-\frac{2uv}{a^2}\frac{\partial^2\psi}{\partial x\partial y}+\left(1-\frac{v^2}{a^2}\right)\frac{\partial^2\psi}{\partial y^2}=0 \tag{5.63}$$

which is identical in form to the corresponding differential equation for velocity potential (5.46).

10. Diabatic flow[8, 9]

In diabatic flow, heat is added to the flow, i. e., $dQ \neq 0$. The study of diabatic flow is important for combustion aerodynamics and meteorology. The fundamental equations are (5.2), (5.6), (5.25), and (5.31). Since heat is added to the flow, by Eq. (5.31), the stagnation enthalpy h_0 in general will not be a constant. By Eq. (5.54), we see that the variation of h_0 will introduce vorticity in the flow. Hence in general in diabatic flow we have rotational flow. In adiabatic flow, we may have a surface of discontinuity, such as shock wave. In diabatic flow, we may also have a surface of discontinuity, such as a detonation wave or a flame front. Across such a surface of discontinuity we must apply the relations of conservation of mass, momentum, and energy in a manner similar to that done in shock relations.

In two-dimensional or axially symmetrical flow, we may derive a differential equation for the stream function ψ for diabatic flow similar to Eq. (5.61) except that $dh_0/dn \neq 0$. Simple diabatic flow shall be discussed in Chapter XIV.

11. Methods of solution of compressible flow problems

In general for compressible flow problems of an inviscid perfect fluid, we should solve the fundamental equations (5.2), (5.6), (5.25), and (5.31) for $\mathbf{q}$, p, ρ, and T fitted to proper initial and boundary conditions. Since these equations are nonlinear, and there is no general method of solution for arbitrary initial and boundary conditions, it is impossible to solve the general problem exactly. For practical problems, we use approximate methods. We shall review briefly various approximate methods in this section. They will be discussed in greater detail in the following chapters. Even in the approximate methods we have to make further assumptions in order to simplify the fundamental equations. One assumption used is that the flow is adiabatic, so that $dQ = 0$. In this book we shall discuss mainly adiabatic flow. However many of the methods discussed may be applied to flows with heat addition. Such flows are important in problems involving combustion[8, 9].

The problems which have been most extensively investigated are those involving two-dimensional and axially symmetrical flows. The following are some of the most important approximate methods used to solve the compressible flow problems.

(a) *Method of small perturbations*[10]. Many of the practical problems deal with the uniform flow passing over a thin body. Since the thickness ratio of the body is small, we may express the solution in power series of the thickness. In this way we may linearize the fundamental equations. Fortunately in most cases, the results of the first-order linearized theory agree with the experimental results very well. For this reason, the linearized theory of the first order has been extensively used. Occasionally second-order theory is used. We shall discuss the theory of small perturbation in Chapter VI.

(b) *Rayleigh-Janzen method*[11, 12]. The theory of small perturbations is good only for a body with small thickness. For an arbitrary body some other method should be used. If the Mach number of the flow is not large, we may express the solution of the problem in power series of Mach number M^2. The zeroth-order solution will be that for incompressible fluid and the higher order terms correspond to the correction due to compressibility. We shall discuss this method in Chapter VII.

(c) *Hodograph method*[13]. For two-dimensional steady irrotational flow, we may transform the fundamental equations (5.46) and (5.63) into linear equations by using the velocity components (u, v) as the independent variables instead of the space coordinates (x, y). This is the hodograph method and will be discussed in Chapter VIII.

(d) *Exact solutions of fundamental equations*. Another way to study the properties of the solution of the fundamental equations of compressible flow is to find the exact solutions of these equations without prescribing the boundary conditions. The exact solutions so obtained are usually simple[14] but may give us some insight into the properties of the solutions of compressible fluid flow. We shall discuss some of these in Chapters IX and XIII. In Chapter IX we shall also discuss a method of finding the exact solutions of the two-dimensional irrotational steady flow for arbitrary boundary conditions.

(e) *Method of characteristics*[15–19]. The method of characteristics is one of the most important tools in the analysis of steady supersonic flow, particularly for problems with two independent variables. Prandtl and Busemann were the first to establish the well known graphical method for the treatment of steady plane, irrotational, isentropic supersonic flow by the method of characteristics[15]. The method has since been generalized for motion with vorticity and for flow with axial symmetry. We shall discuss the method of characteristics in Chapters XI, XIII and XIV.

(f) *Variational method*[20, 21]. In the variational method, we do not solve the differential equation but formulate a variational integral and find the approximate solution of the problem by the Rayleigh-Ritz method. We shall discuss this method in Chapter VII.

(g) *Computational fluid mechanics*. With the development of high speed computers, the numerical solution of fluid mechanics has been well developed during the last thirty years, particularly the finite difference method. It has been developed as a new branch of investigation of fluid dynamics, the computational fluid dynamics, which lies between the classical experimental fluid mechanics and the classical theoretical fluid mechanics. Because the computational fluid dynamics becomes an important supplement to the classical experimental fluid dynamics and classical theoretical fluid dynamics, in this edition of our book, considerable efforts have been made to add the essential features of computational fluid dynamics. The numerical methods used in the solutions of fluid mechanics may be divided into two main groups: (i) the finite difference method which may be called the computational fluid mechanics in a narrow sense and (ii) the panel method which is based on the principle of superposition. The basic concepts of the finite difference method will be given in Chapter VII and the further developments are discussed in Chapters X, XIII, XIV and XVII. We shall discuss the panel method in Chapters VI and XII.

12. Characteristics of unsteady Euler equations

Characteristics behaviors are an intrinsic property of the partial differential equations. The solution methods for the partial differential equations depend substantially on the characteristics properties. In this section, we study the characteristics of the differential equation system for unsteady, inviscid, non-heat-conducting and adiabatic flows, known as unsteady Euler equations. Euler equations generally consist of the equations of continuity, motion and energy, expressed by Eqs. (5.6), (5.23) and (5.32). In calculating the characteristics, it is convenient to replace the energy equation (5.32) with the entropy equation (5.33). Characteristics are invariant for equivalent differential equation systems. Eqs. (5.32) and (5.33) are equivalent. For simplicity, we consider one-dimensional flow. The governing differential equations are

Equation of continuity $$\rho_t + u\rho_x + \rho u_x = 0 \tag{5.64}$$

Equation of motion $$u_t + uu_x + \frac{1}{\rho} p_x = 0 \tag{5.65}$$

Equation of entropy $$p_t + up_x + \rho a^2 u_x = 0 \tag{5.66}$$

Eq. (5.66) may be obtained from $\frac{D}{Dt}\left(\frac{p}{\rho^\gamma}\right) = 0$ with the aid of Eq. (5.64).

The characteristics are defined as the curves on the plane (t, x), across which the solutions may have discontinuous first order derivatives. Consider a curve C on (t, x) plane. Its equation is

$$\varphi(t, x) = 0 \tag{5.67}$$

Given ρ, u and p along C, we want to extend the solutions into the neighbourhood of the curve C.

Along C, we have

$$d\rho = \rho_t dt + \rho_x dx \tag{5.68}$$

By Eq. (5.67), (5.68) may be written as

$$d\rho = \rho_t dt + \rho_x \lambda dt \tag{5.69}$$

where

$$\lambda = \frac{dx}{dt} = -\varphi_t / \varphi_x \tag{5.70}$$

Eq. (5.69) gives

$$\rho_t = -\lambda \rho_x + \frac{d\rho}{dt} \tag{5.71}$$

Similarly, we have

$$u_t = -\lambda u_x + \frac{du}{dt} \tag{5.72}$$

$$p_t = -\lambda p_x + \frac{dp}{dt} \tag{5.73}$$

where $\frac{d\rho}{dt}$, $\frac{du}{dt}$ and $\frac{dp}{dt}$ along C are given.

Using Eqs. (5.71) to (5.73) to eliminate ρ_t, u_t and p_t in Eqs. (5.64) to (5.66), we obtain

$$(u-\lambda)\rho_x + \rho u_x = f_1 \tag{5.74}$$

$$(u-\lambda)u_x + \frac{1}{\rho} p_x = f_2 \tag{5.75}$$

$$(u-\lambda)p_x + a^2 \rho u_x = f_3 \tag{5.76}$$

where f_1, f_2 and f_3 are known functions.

The necessary and sufficient condition for ρ_x, u_x and p_x being indeterminate or infinite is

$$\begin{vmatrix} u-\lambda & \rho & 0 \\ 0 & u-\lambda & 1/\rho \\ 0 & a^2\rho & u-\lambda \end{vmatrix} = 0$$

or

$$(u-\lambda)\left[(u-\lambda)^2 - a^2\right] = 0 \tag{5.77}$$

Solving for λ, we obtain

$$\lambda = u, \quad u \pm a \tag{5.78}$$

Thus there exist three real and different characteristics for the equation system (5.64) to (5.66)

$$\frac{dx}{dt} = u, \ u \pm a \tag{5.79}$$

and the equation system (5.64) to (5.66) is of hyperbolic type. It is important to see that the unsteady Euler equations are of single type—hyperbolic type, no matter which the flow speed is (subsonic or supersonic). Therefore, the solution method can be established once for the entire speed ranges. For this reason, steady flow problems are usually considered as unsteady flow problems. The steady flow solution is obtained by solving the unsteady Euler differential equations with the steady boundary conditions and arbitrary initial conditions until the unsteady flow solution approaches a limit as time $t \to \infty$. This method is known as time dependent method for steady flow problems.

More information about the theory of characteristics will be given in Chapters XI and XIV.

The first-order differential equations (5.64) to (5.66) are nonlinear, however, they are linear in the first-order derivatives and are thus called quasi-linear equations.

13. Conservation forms of Euler equations

In studying one-dimensional flow with shock waves in Chapter IV § 12, we have seen that the governing differential equations ought to be written in divergence form in order to have correct jump conditions across the shock waves. This is true in general. In this section, we shall write the general differential equations of unsteady inviscid, non-heat-conducting, adiabatic flow into divergence or conservation form.

The equation of continuity (5.6) is already in divergence form

$$\frac{\partial \rho}{\partial t} + \frac{\partial(\rho u)}{\partial x} + \frac{\partial(\rho v)}{\partial y} + \frac{\partial(\rho w)}{\partial z} = 0 \qquad (5.80)$$

Equations of motion (5.21a) to (5.21c) may be put into divergence form with the aid of (5.80).

$$\frac{\partial(\rho u)}{\partial t} + \frac{\partial}{\partial x}(\rho u^2 + p) + \frac{\partial}{\partial y}(\rho uv) + \frac{\partial}{\partial z}(\rho uw) = 0 \qquad (5.81)$$

$$\frac{\partial(\rho v)}{\partial t} + \frac{\partial}{\partial x}(\rho uv) + \frac{\partial}{\partial y}(\rho v^2 + p) + \frac{\partial}{\partial z}(\rho vw) = 0 \qquad (5.82)$$

$$\frac{\partial(\rho w)}{\partial t} + \frac{\partial}{\partial x}(\rho uw) + \frac{\partial}{\partial y}(\rho vw) + \frac{\partial}{\partial z}(\rho w^2 + p) = 0 \qquad (5.83)$$

Equation of energy (5.32) is similarly put into divergence form,

$$\frac{\partial(\rho e)}{\partial t} + \frac{\partial}{\partial x}[(\rho e + p)u] + \frac{\partial}{\partial y}[(\rho e + p)v] + \frac{\partial}{\partial z}[(\rho e + p)w] = 0 \quad (5.84)$$

where e is specific total energy,

$$e=E+\frac{u^2+v^2+w^2}{2}$$

Here E is specific internal energy.

Equation of entropy (5.33) in divergence form becomes

$$\frac{\partial(\rho s)}{\partial t}+\frac{\partial}{\partial x}(\rho s u)+\frac{\partial}{\partial y}(\rho s v)+\frac{\partial}{\partial z}(\rho s w)=0 \tag{5.85}$$

In the above Eqs. (5.80) to (5.85), only five equations are independent. The equation system is closed by an equation of state $p=p(\rho, E)$. For ideal gas whose ratio of specific heat is γ, we have

$$p=(\gamma-1)\rho E \tag{5.86}$$

Eqs. (5.81) to (5.84) can be written in a concise divergence form,

$$\frac{\partial U}{\partial t}+\frac{\partial F}{\partial x}+\frac{\partial G}{\partial y}+\frac{\partial H}{\partial z}=0 \tag{5.87}$$

where U, F, G and H are vector functions. For the conservation of mass, momentum and energy, they are

$$U=\begin{bmatrix}\rho\\ \rho u\\ \rho v\\ \rho w\\ \rho e\end{bmatrix},\quad F=\begin{bmatrix}\rho u\\ \rho u^2+p\\ \rho uv\\ \rho uw\\ (\rho e+p)u\end{bmatrix},\quad G=\begin{bmatrix}\rho v\\ \rho uv\\ \rho v^2+p\\ \rho vw\\ (\rho e+p)v\end{bmatrix},\quad H=\begin{bmatrix}\rho w\\ \rho uw\\ \rho vw\\ \rho w^2+p\\ (\rho e+p)w\end{bmatrix} \tag{5.88}$$

F, G and H are composite functions of U. For simplicity, we consider one-dimensional cases. Let

$$U=\begin{bmatrix}U_1\\ U_2\\ U_3\end{bmatrix},\qquad F=\begin{bmatrix}F_1\\ F_2\\ F_3\end{bmatrix} \tag{5.89}$$

where $U_1=\rho$, $U_2=\rho u$, $U_3=\rho e$.

$$F_1=U_2,\ F_2=(\gamma-1)U_3+\frac{3-\gamma}{2}\frac{U_2^2}{U_1}$$
$$F_3=\frac{\gamma U_2 U_3}{U_1}-\frac{\gamma-1}{2}\frac{U_2^3}{U_1^2} \tag{5.90}$$

By the differentiation rule of the composite function,

$$\frac{\partial F}{\partial x}=\frac{\partial F}{\partial U}\frac{\partial U}{\partial x} \tag{5.91}$$

where $\frac{\partial F}{\partial U}$ is the Jacobian matrix,

$$\frac{\partial F}{\partial U}=\begin{bmatrix} 0 & 1 & 0 \\ -\frac{3-\gamma}{2}\frac{U_2^2}{U_1^2} & (3-\gamma)\frac{U_2}{U_1} & \gamma-1 \\ -\frac{\gamma U_2 U_3}{U_1^2}+(\gamma-1)\frac{U_2^3}{U_1^3} & \frac{\gamma U_3}{U_1}-\frac{3}{2}(\gamma-1)\frac{U_2^2}{U_1^2} & \frac{\gamma U_2}{U_1} \end{bmatrix} \qquad (5.92)$$

From Eqs. (5.90) we see that the flux vector $F(U)$ is a homogeneous function of degree one in U which means that $E(dU)=dE(U)$. Thus we have from Eqs. (5.90) and (5.92)

$$F=\frac{\partial F}{\partial U}U \qquad (5.93)$$

This may be verified by simply multiplying the indicated matrices. Similar relations exist for two- and three-dimensional cases.

Now we study the integral form of the Euler equations. Take a volume V in the physical field of flow and integrate Eqs. (5.87) in V. We obtain by an application of Gauss' divergenc theorem

$$\frac{\partial}{\partial t}\int_V U dV+\int_S \boldsymbol{w}\cdot\boldsymbol{n} dS=0 \qquad (5.94)$$

where

$$W=F\boldsymbol{i}+G\boldsymbol{j}+H\boldsymbol{k} \qquad (5.95)$$

S is the surface enclosing the volume V, n is outward unit normal of the surface S. In fact we may write out the integral equations (5.94) directly as the expressions of conservation laws. They denote that the time-rate increments of mass, momentum and energy in V are balanced by the convection flux of the corresponding quantities with appropriate terms due to pressure acting on S. And we may derive Eq. (5.87) by another application of Gauss′ theorem.

Indeed, the integral equations (5.94) are more fundamental statements than the differential equation (5.87), since they hold whether or not the flow variables can be differentiated with respect to the spatial variables (x, y, z). Therefore, the Euler integral equations may have discontinuous solutions. The numerical discretization of Euler integral equations is known as finite-volume method, whereas the numerical discretization of Euler differential equations is called finite-difference method.

14. Discontinuous solutions of steady Euler equations

The Euler integral equations can be applied to flow with discontinuities.

In this section we shall study the possible discontinuities in steady inviscid non-heat-conducting adiabatic flow. The governing equations are from Eq. (5.94)

$$\int_S \boldsymbol{W} \cdot \boldsymbol{n} dS = 0 \tag{5.96}$$

Consider two-dimensional case for simplicity. Let C be a curve of discontinuity in the flow plane (x, y). Consider a unit arc length along C (Fig. 5.2). Take a closed contour S enclosing the unit length of C and infinitely close to

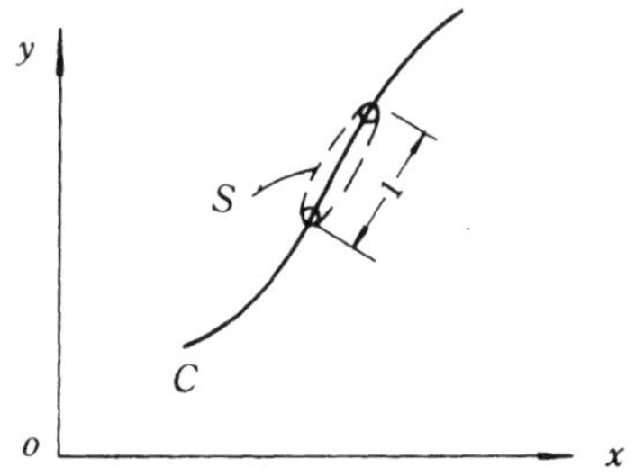

Fig. 5.2 Discontinuity curve

to the two sides of C. By applying Eqs. (5.96), we obtain

$$\begin{aligned} &\rho_1 q_{n1} - \rho_2 q_{n2} = 0 \\ &\rho_1 q_{n1}^2 + p_1 - \rho_2 q_{n2}^2 - p_2 = 0 \\ &\rho_1 q_{n1} q_{t1} - \rho_2 q_{n2} q_{t2} = 0 \\ &(\rho_1 e_1 + p_1) q_{n1} - (\rho_2 e_2 + p_2) q_{n2} = 0 \end{aligned} \tag{5.97}$$

where q_n and q_t are normal and tangential components of velocity along C, subscripts 1 and 2 denote two sides of the discontinuity curve C.

In general, there exist two kinds of discontinuities:

(i) If $q_{n1} \neq 0$, we have

$$\begin{aligned} \rho_1 q_{n1} &= \rho_2 q_{n2} \\ \rho_1 q_{n1}^2 + p_1 &= \rho_2 q_{n2}^2 + p_2 \\ q_{t1} &= q_{t2} \\ (\rho_1 e_1 + p_1) q_{n1} &= (\rho_2 e_2 + p_2) q_{n2} \end{aligned} \tag{5.98}$$

These are the Rankine-Hugoniot relations for adiabatic shock waves.

(ii) If $q_{n_1} = 0$, we have

$$\begin{aligned} q_{n1} &= 0 \\ p_1 &= p_2 \end{aligned} \tag{5.99}$$

and

$$\rho_1 \neq \rho_2,\ q_{t1} \neq q_{t2},\ e_1 \neq e_2$$

These are relations for contact surfaces, e.g. vortex sheets.

There are a number of incisive differences between shock waves and contact surfaces.

(i) There is no convection flux across the contact surface and the flow. One side of the surface cannot be determined by the flow on the other side. Across the shock wave, there is a convection flux.

(ii) The jump conditions (5.98) for shock waves have to be supplemented by the entropy condition in order to make them yield a unique solution of the discontinuity (see Chapter IV § 2). For the contact surfaces, there is no "entropy" condition.

(iii) In the physical world, the thickness of a shock wave is very thin and for most practical purposes the shock wave may be considered as a discontinuity. The vortex sheet always thickens after its generation due to viscosity and generally it can harely be considered as a discontinuity.

The development of methods which solve the Euler equations and the application of these methods to the prediction of vortical flow about delta wings and slender bodies at a high angle of attack has been an area of active research. It has also been the source of considerable controversy. The controversy centers upon the degree of realism that the inviscid model provides in describing the actual physics of vortical flow.

15. Weak solutions of quasi-linear differential equations

Discontinuous inviscid flow cannot be the solution of the Euler differential equations in the original sense, since the flow variables are not differentiable with respect to the space variables. The concept of solution for differential equations may be generalized to include the discontinuous solution. We shall give an intrinsic characterization of such concept in this section[22].

Consider the quasi-linear system of conservation laws in divergence form with two independent variables, x and t.

$$L(u) = p_t(x, t, u) + q_x(x, t, u) + r(x, t, u) = 0 \tag{5.100}$$

where p, q and r are twice continuously differentiable function vectors of their arguments x, t in domain G and of function vector u in some domain. To define weak solutions for system (5.100), we consider arbitrary smooth test functions ζ in subdomain R of G vanishing outside R. We multiply $L(u)$ by ζ, integrate over R and obtain

$$\iint_R \zeta L(u)\, dx dt = 0 \tag{5.101}$$

For smooth u, Gauss'theorem yields

$$\iint_R (p\zeta_t + q\zeta_x - r\zeta)\,dxdt = 0 \tag{5.102}$$

Conversely, if Eq. (5.102) holds for a function vector u with continuous first derivatives and for all admissible test functions ζ in all subdomains R of G, then Eq. (5.102) yields Eq. (5.101) by another application of Gauss'theorem. We hence can conclude that $L(u) = 0$.

We now give the generalization of the concept of solution: we allow the function vector u and its first-order derivatives to be piecewise continuous, i. e. to possess jump discontinuities along piecewise smooth curve C. Such a function vector u is called a weak solution of the equation $L(u) = 0$ in G if Eq. (5.102) is satisfied for all admissible test functions ζ and all subdomains R of G.

Assuming the discontinuous solution u of $L(u) = 0$ to be regular in all domains not containing C, we shall derive relations for the jumps of u. Suppose C divides a domain R into two parts, R_1 and R_2. Integrate Eq. (5.101) by parts separately in R_1 and R_2. Since in both these domains $L(u) = 0$, and since $\zeta = 0$ on the boundary of R, we obtain

$$\int_C \zeta(\varphi_t [p] + \varphi_x [q])\,ds = 0 \tag{5.103}$$

Here φ_x and φ_t denote the direction cosines of the normal to C, s is arc length along C and $[p]$, $[q]$ are the jumps of p and q across C. In Eq. (5.103), ζ is arbitrary on C and hence we obtain the jump conditions across C,

$$\varphi_t [p] + \varphi_x [q] = 0 \tag{5.104}$$

For example, consider steady two-dimensional Euler equations. The jump condition (5.104) gives

$$\begin{aligned}
&[\rho u]\cos(n,x) + [\rho v]\cos(n,y) = 0 \\
&[\rho u^2 + p]\cos(n,x) + [\rho uv]\cos(n,y) = 0 \\
&[\rho uv]\cos(n,x) + [\rho v^2 + p]\cos(n,y) = 0 \\
&[(\rho e + p)u]\cos(n,x) + [(\rho e + p)v]\cos(n,y) = 0
\end{aligned} \tag{5.105}$$

where [ ] denotes the jump across discontinuity curve C. It is seen that the jump conditions (5.105) coincide with Eq. (5.97) given by the integral Euler equations.

Now we study weak solutions of linear hyperbolic differential equations. The linear system may be considered as a special case of quasi-linear system.

For linear system, let

$$p=A(x,t)u,\quad q=B(x,t)u,\quad r=C(x,t)u \tag{5.106}$$

We have from Eq. (5.100)

$$L(u)=Au_t+Bu_x+(A_t+B_x+C)u=0 \tag{5.107}$$

We call a function vector u a weak solution of $L(u)=0$, if it is piecewise continuous with piecewise continuous first derivatives in domain G, and if

$$\iint_R u(A\zeta_t+B\zeta_x-C\zeta)\,dxdt=0 \tag{5.108}$$

is satisfied for all admissible test functions ζ in all subdomains R of G.

We shall show that the curves of discontinuity for linear system are necessarily characteristics of the system. From Eq. (5.104), the jump conditions for linear system are across discontinuity curve C,

$$(\varphi_t A+\varphi_x B)[u]=0 \tag{5.109}$$

Under the assumption that the jump $[u]$ is different from zero, this implies

$$\varphi_t A+\varphi_x B=0 \tag{5.110}$$

Hence the discontinuity curve C is a characteristic curve.

There are a number of incisive differences between weak solutions of quasilinear and linear differential equations.

(i) In the linear case, discontinuity curves C are characteristics. For nonlinear conservation laws, however, the relations for the discontinuities and for the slope of the curve C are not separated but interlocked. Thus in the quasilinear case, discontinuity curves C are not characteristics.

(ii) Solutions of linear equations are discontinuous only if the prescribed data are discontinuous. In contrast, solutions of quasilinear equations with smooth (even analytic) initial data can develop discontinuities after a finite time interval has elapsed.

(iii) The jump conditions for the linear case suffice to determine a unique discontinuous solution to a given discontinuous initial (or mixed initial and boundary) value problem, e.g. wave motion initiated by an impulse is a discontinuous solution. The jump conditions for the quasi-linear case may have to be supplemented, e.g. by the entropy condition for shock waves, in order to make them yield a unique solution of the corresponding problem.

(iv) Different systems of conservation laws may be equivalent to systems of differential equations, i.e. smooth solutions of one system are also smooth solutions of the other. But a discontinuous solution of one system need not (and in general will not) be a solution of the other. A striking example is the system consisting of the equations of conservation of mass, momentum and energy on one hand, and the system of the conservation of mass, momentum and entropy on the other hand.

16. Some simple types of compressible flow

Before we discuss the general methods of solution for the compressible flow problem, we will discuss two simple types of compressible flow which illustrate the large difference between the flow of compressible and incompressible fluids.

(a) *Sources and Sinks*. Sources or sinks represent the type of flow in which the streamlines are radial lines from a vertex 0 and the only velocity component which differs from zero is the radial component (Fig. 5.3). The radial velocity component q_r, the pressure, and the density of the fluid are functions only of the radial distance r from the origin. In this case the equation of continuity (5.6) gives

$$\frac{d}{dr}(\rho q_r) = -\varepsilon \frac{\rho q_r}{r} \tag{5.111}$$

which, upon integration, becomes

$$r^{\varepsilon} = \frac{\text{constant}}{\rho q_r} \tag{5.112}$$

where $\varepsilon = 1$ for two-dimensional flow, and $\varepsilon = 2$ for three-dimensional flow. It is easy to show that for source or sink flow, the mass flowing through concentric

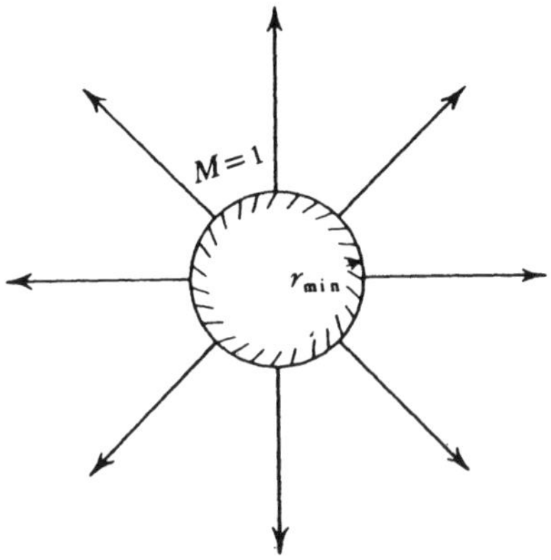

Fig. 5.3 Source (or sink) flow in a compressible fluid

spheres in three-dimensional flow or through constant circles in two-dimensional flow, is constant; this fact defines the strength of the source or sink.

If we express the mass flow ρq_r in terms of the local Mach number,

$M=\frac{q_r}{a}$, Eq. (5.112) becomes

$$\left(\frac{r}{r_{\min}}\right)^{\varepsilon}=\frac{1}{M}\sqrt{\left(\frac{\gamma-1}{\gamma+1}M^2+\frac{2}{\gamma+1}\right)^{\frac{\gamma+1}{\gamma-1}}} \tag{5.113}$$

in which $r=r_{\min}=a$ minimum radius when local Mach number $M=1$. Source or sink flow in a compressible fluid exists only outside the minimum radius. The stronger the source is, the smaller will be the value of $r_{\min}$. The flow pattern of a source or a sink behaves as a divergent channel. As we saw in Chapter III, for such a channel we can have either pure supersonic flow or pure subsonic flow with sonic velocity at the minimum section. We cannot have transonic flow for this case.

In incompressible flow, the source or sink starts from a point.

(b) *Vortex*. Another simple flow pattern is a vortex flow, whose streamlines are concentric circles about a point 0 (Fig. 5.4). The only component of velocity different from zero is the tangential velocity component q_θ

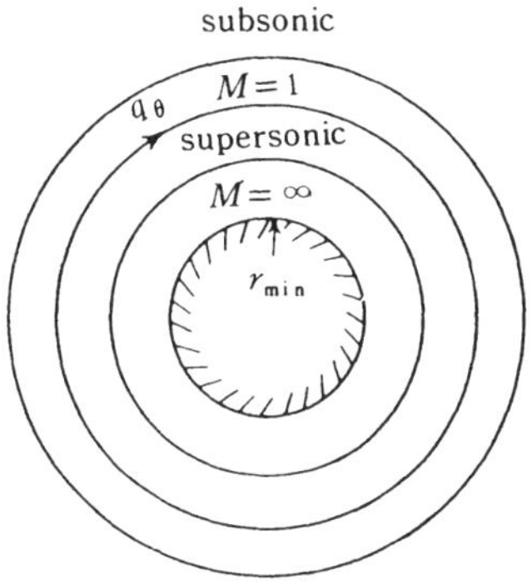

Fig. 5.4 Vortex flow in a compressible fluid

which is a function only of radial distance r. If we assume that the flow is irrotational, then

$$rq_\theta=\text{constant} \tag{5.114}$$

If we express the velocity q_θ in terms of local Mach number $M=\frac{q_\theta}{a}$, Eq. (5.114) becomes

$$r=\frac{\text{constant}}{q_\theta}=\frac{\text{constant}}{M^*}$$

or

$$\left(\frac{r}{r_{\min}}\right)^2=1+\frac{2}{(\gamma-1)M^2} \tag{5.115}$$

At the minimum radius r_{min}, the local Mach number is infinite. Hence vortex flow exists only outside of the region of minimum radius r_{min}. The local Mach number decreases from infinity at $r=r_{min}$ continuously as r increases. At $r=\infty$, $M=0$.

The corresponding flow in an incompressible fluid is a point vortex. The combination of the above two types of flow gives a spiral flow.

17. General orthogonal coordinates

In our previous discussions, we have used the simple Cartesian coordinates. For many problems, it may be convenient to use systems of curvilinear coordinates instead of Cartesian coordinates. In fluid mechanics, the most convenient curvilinear coordinates are the orthogonal ones. In this section, we are going to write down some general properties of orthogonal curvilinear coordinates which will be useful in the analysis of fluid mechanics.

Let three independent orthogonal families of surfaces be

$$f_1(x, y, z) = a_1;\ f_2(x, y, z) = a_2;\ f_3(x, y, z) = a_3 \tag{5.116}$$

where x, y, and z are the Cartesian coordinates. The surfaces $a_1=$constant, $a_2=$constant, and $a_3=$constant form an orthogonal system. The values of a_1, a_2, and a_3 may be used as coordinates of a point in space. The relations between the two system of coordinates x, y, z and a_1, a_2, a_3 may be given by Eqs. (5.116) or by the relations:

$$x=x(a_1, a_2, a_3);\quad y=y(a_1, a_2, a_3);\quad z=z(a_1, a_2, a_3) \tag{5.117}$$

The element of length $ds=(dx^2+dy^2+dz^2)^{1/2}$ may be expressed in terms of the orthogonal coordinates a_1, a_2, a_3 as follows:

$$(ds)^2=h_1^2(da_1)^2+h_2^2(da_2)^2+h_3^2(da_3)^2 \tag{5.118}$$

where

$$\begin{aligned} h_1&=\left[\left(\frac{\partial x}{\partial a_1}\right)^2+\left(\frac{\partial y}{\partial a_1}\right)^2+\left(\frac{\partial z}{\partial a_1}\right)^2\right]^{1/2}\\ h_2&=\left[\left(\frac{\partial x}{\partial a_2}\right)^2+\left(\frac{\partial y}{\partial a_2}\right)^2+\left(\frac{\partial z}{\partial a_2}\right)^2\right]^{1/2}\\ h_3&=\left[\left(\frac{\partial x}{\partial a_3}\right)^2+\left(\frac{\partial y}{\partial a_3}\right)^2+\left(\frac{\partial z}{\partial a_3}\right)^2\right]^{1/2} \end{aligned} \tag{5.119}$$

From the above expressions, we see that the elements of length at a point (a_1, a_2, a_3) in the direction of increasing a_1, a_2, and a_3 are respectively h_1da_1, h_2da_2, and h_3da_3.

Let $\boldsymbol{q}$ be a vector of components q_{a1}, q_{a2}, and q_{a3} in the direction of a_1,

a_2, and a_3 respectively, i.e.

$$\boldsymbol{q}=\boldsymbol{i}_1\boldsymbol{q}_{a1}+\boldsymbol{i}_2\boldsymbol{q}_{a2}+\boldsymbol{i}_3\boldsymbol{q}_{a3} \tag{5.120}$$

where $\boldsymbol{i}_1$, $\boldsymbol{i}_2$, and $\boldsymbol{i}_3$ are respectively the unit vector in the direction of a_1, a_2, and a_3 .

The gradient of a scalar function φ in the orthogonal curvilinear coordinates is

$$\nabla\varphi=\boldsymbol{i}_1\frac{1}{h_1}\frac{\partial\varphi}{\partial a_1}+\boldsymbol{i}_2\frac{1}{h_2}\frac{\partial\varphi}{\partial a_2}+\boldsymbol{i}_3\frac{1}{h_3}\frac{\partial\varphi}{\partial a_3} \tag{5.121}$$

The divergence of a vector $\boldsymbol{q}$ is

$$\nabla\cdot\boldsymbol{q}=\frac{1}{h_1h_2h_3}\left\{\frac{\partial}{\partial a_1}(q_{a1}h_2h_3)+\frac{\partial}{\partial a_2}(q_{a2}h_3h_1)+\frac{\partial}{\partial a_3}(q_{a3}h_1h_2)\right\} \tag{5.122}$$

The curl of a vector $\boldsymbol{q}$ is

$$\nabla\times\boldsymbol{q}=\boldsymbol{i}_1\omega_1+\boldsymbol{i}_2\omega_2+\boldsymbol{i}_3\omega_3 \tag{5.123}$$

where

$$\omega_1=\frac{1}{h_2h_3}\left\{\frac{\partial}{\partial a_2}(q_{a3}h_3)-\frac{\partial}{\partial a_3}(q_{a2}h_2)\right\}$$

$$\omega_2=\frac{1}{h_3h_1}\left\{\frac{\partial}{\partial a_3}(q_{a1}h_1)-\frac{\partial}{\partial a_1}(q_{a3}h_3)\right\}$$

$$\omega_3=\frac{1}{h_1h_2}\left\{\frac{\partial}{\partial a_1}(q_{a2}h_2)-\frac{\partial}{\partial a_2}(q_{a1}h_1)\right\}$$

the Laplacian of a scalar function φ is

$$\nabla^2\varphi=\frac{1}{h_1h_2h_3}\left\{\frac{\partial}{\partial a_1}\left(\frac{h_2h_3\partial\varphi}{h_1\partial a_1}\right)+\frac{\partial}{\partial a_2}\left(\frac{h_3h_1\partial\varphi}{h_2\partial a_2}\right)\right.$$
$$\left.+\frac{\partial}{\partial a_3}\left(\frac{h_1h_2\partial\varphi}{h_3\partial a_3}\right)\right\} \tag{5.124}$$

The most common curvilinear coordinates are as follows:

(1) Cylindrical coordinates. These coordinates are related with circular cylinders with radius r and angular coordinate θ and axial coordinate z. The relation between the cylindrical coordinates (r, θ, z) and the Cartesian coordinates (x, y, z) are:

$$x=r\cos\theta,\ y=r\sin\theta,\ z=z \tag{5.125}$$

If we take

$$a_1=r,\ a_2=\theta,\ a_3=z \tag{5.126}$$

we have

$$h_1=1,\ h_2=r,\ h_3=1 \tag{5.127}$$

Substituting Eqs. (5.126) and (5.127) into Eqs. (5.121) to (5.124), we have the gradient of φ, divergence of q, curl of q, and Laplacian of φ in cylindrical coordinates respectively.

(2) Spherical coordinates. These coordinates are related to spheres with radius r and two angular coordinates θ and φ such that

$$x=r\ \sin\theta\cos\varphi,\ y=r\ \sin\theta\sin\varphi,\ z=r\ \cos\theta \tag{5.128}$$

If we take

$$a_1=r,\ a_2=\theta,\ a_3=\varphi \tag{5.129}$$

we have

$$h_1=1,\qquad h_2=r,\quad h_3=r\ \sin\theta \tag{5.130}$$

By the help of Eqs. (5.129) and (5.130), we may easily find the fundamental equations in spherical coordinates.

(3) Elliptical coordinates. These coordinates are related with elliptical cylinders such that

$$x=c\xi,\quad y=c[(\xi^2-1)(1-\eta^2)]^{1/2},\ z=z \tag{5.131}$$

It is easy to show that ξ=constant is a cylinder of elliptical cross section and η=constant is hyperbolic cylinder of two sheets. The factor c is a constant. If we take

$$a_1=\xi,\ a_2=\eta,\ a_3=z \tag{5.132}$$

we have

$$h_1=c\left(\frac{\xi^2-\eta^2}{\xi^2-1}\right)^{1/2},\quad h_2=c\left(\frac{\xi^2-\eta^2}{1-\eta^2}\right)^{1/2},\quad h_3=1 \tag{5.133}$$

Eqs. (5.132) and (5.133) give the corresponding values for the elliptical coordinates. If we investigate the flow around an elliptical cylinder in a flow field, it is convenient to use the elliptical coordinates because on the boundary of the elliptical cylinder, we have simply ξ=constant.

18. Moving coordinates

It is sometimes convenient in special problems to employ a system of reference axes which is in motion. Let U, V, W be the x-, y-, and z-component of velocity of the origin of the references axes and ω_x, ω_y, and ω_z, the angular velocities of the frame of the reference axes. Let u, v, w be the absolute velocities of the fluid at a point (x, y, z) rigidly connected to the frame of reference axes and u', v', w', the velocities of the fluid at the same point relative to the frame. We have

$$\left.\begin{aligned} u &= U + u' - y\omega_z + z\omega_y \\ v &= V + v' - z\omega_x + x\omega_z \\ w &= W + w' - x\omega_y + y\omega_x \end{aligned}\right\} \tag{5.134}$$

If we consider the increase in mass in a small rectangular element of volume attached to the frame of reference axes with the center at (x, y, z) where ρ is the density of the fluid, we obtain the equation of continuity from the conservation of mass without any mass source [Eq. (5.6)] as follows:

$$\frac{\partial \rho}{\partial t} + \frac{\partial \rho u'}{\partial x} + \frac{\partial \rho v'}{\partial y} + \frac{\partial \rho w'}{\partial z} = \frac{\partial \rho}{\partial t} + \nabla \cdot (\rho \boldsymbol{q}') = 0 \tag{5.135}$$

Hence the equation of continuity has the same form as that for the case where the reference axes are fixed in space.

For the expression of the acceleration [Eq. (5.21)], we shall show that the expressions in the moving axes differ from those in the fixed axes.

Let A be the absolute velocity in the direction fixed in space whose direction cosines referred to the moving axes are l_A, m_A, n_A, i.e., $A = l_A u + m_A v + n_A w$. In time δt, the coordinates of the fluid particle at (x, y, z) will increase by an amount $(u', v', w')\,\delta t$ so that to a first approximation, $l_A u$ will become

$$(l_A + \delta l_A)\left\{u + \left(\frac{\partial u}{\partial t} + u'\frac{\partial u}{\partial x} + v'\frac{\partial u}{\partial y} + w'\frac{\partial u}{\partial z}\right)\delta t\right\}$$

whence we get

$$\begin{aligned} \frac{DA}{Dt} &= \frac{Dl_A}{Dt}u + \frac{Dm_A}{Dt}v + \frac{Dn_A}{Dt}w + l_A\left(\frac{\partial u}{\partial t} + u'\frac{\partial u}{\partial x} + v'\frac{\partial u}{\partial y}\right. \\ &\quad \left. + w'\frac{\partial u}{\partial z}\right) + m_A\left(\frac{\partial v}{\partial t} + u'\frac{\partial v}{\partial x} + v'\frac{\partial v}{\partial y} + w'\frac{\partial v}{\partial z}\right) \\ &\quad + n_A\left(\frac{\partial w}{\partial t} + u'\frac{\partial w}{\partial x} + v'\frac{\partial w}{\partial y} + w'\frac{\partial w}{\partial z}\right) \end{aligned} \tag{5.136}$$

But since l_A, m_A, n_A are direction cosines referring to the moving axes of a line in space, therefore

$$\left.\begin{aligned}\frac{Dl_A}{Dt}-m_A\omega_z+n_A\omega_y=0\\ \frac{Dm_A}{Dt}-n_A\omega_x+l_A\omega_z=0\\ \frac{Dn_A}{Dt}-l_A\omega_y+m_A\omega_x=0\end{aligned}\right\}\tag{5.137}$$

and substituting Eqs. (5.137) into Eq. (5.136), we have

$$\begin{aligned}\frac{DA}{Dt}=&l_A\left(\frac{\partial u}{\partial t}-v\omega_z+w\omega_y+u'\frac{\partial u}{\partial x}+v'\frac{\partial u}{\partial y}+w'\frac{\partial u}{\partial z}\right)\\ &+m_A\left(\frac{\partial v}{\partial t}-w\omega_x+u\omega_z+u'\frac{\partial v}{\partial x}+v'\frac{\partial v}{\partial y}+w'\frac{\partial v}{\partial z}\right)\\ &+n_A\left(\frac{\partial w}{\partial t}-u\omega_y+v\omega_x+u'\frac{\partial w}{\partial x}+v'\frac{\partial u}{\partial y}+w'\frac{\partial w}{\partial z}\right)\end{aligned}\tag{5.138}$$

In general curvilinear coordinates such as cylindrical or spherical coordinates, at every instance we consider that the origin of the moving axes is coincident with that of the fixed axes and that the moving axes have no linear velocities but only angular velocities. Hence

$$x=y=z=U=V=W=0,\ u=u',\quad v=v',\quad w=w'\tag{5.139}$$

From Eqs. (5.139), we see that the absolute accelerations of the fluid particle which should be used in the equations of motion in the moving coordinates are as follows:

$$\left.\begin{aligned}\left(\frac{DA}{Dt}\right)_x=\frac{Du'}{Dt}-v'\omega_z+w'\omega_y\\ \left(\frac{DA}{Dt}\right)_y=\frac{Dv'}{Dt}-w'\omega_x+u'\omega_z\\ \left(\frac{DA}{Dt}\right)_z=\frac{Dw'}{Dt}-u'\omega_y+v'\omega_x\end{aligned}\right\}\tag{5.140}$$

where

$$\frac{D}{Dt}=\frac{\partial}{\partial t}+u'\frac{\partial}{\partial x}+v'\frac{\partial}{\partial y}+w'\frac{\partial}{\partial z}$$

In cylindrical coordinates, we have

$$\begin{aligned}&u'=q_r,\ v'=q_\theta,\ w'=q_z,\ \omega_x=\omega_y=0\\ &\omega_z=\frac{d\theta}{dt}=\frac{q_\theta}{r},\ dx=dr,\ dy=rd\theta,\ dz=dz\end{aligned}\tag{5.141}$$

Substituting Eqs. (5.141) into Eqs. (5.140), we have the acceleration terms in the cylindrical coordinates as follows:

$$\left.\begin{aligned} &\text{The radial acceleration} = \frac{Dq_r}{Dt} - \frac{q_\theta^2}{r} \\ &\text{The tangential acceleration} = \frac{Dq_\theta}{Dt} + \frac{q_r q_\theta}{r} \\ &\text{The axial acceleration} = \frac{Dq_z}{Dt} \end{aligned}\right\} \tag{5.142}$$

where

$$\frac{D}{Dt} = \frac{\partial}{\partial t} + q_r \frac{\partial}{\partial r} + \frac{q_\theta}{r}\frac{\partial}{\partial \theta} + q_z \frac{\partial}{\partial z}$$

Similar expressions may be obtained for the general orthogonal coordinates. The corresponding expressions for the angular velocities are

$$\begin{aligned} \omega_x &= \frac{1}{h_2}\left(\frac{\partial h_3}{\partial a_2}\right)\frac{da_3}{dt} - \frac{1}{h_3}\left(\frac{\partial h_2}{\partial a_3}\right)\frac{da_2}{dt} \\ \omega_y &= \frac{1}{h_3}\left(\frac{\partial h_1}{\partial a_3}\right)\frac{da_1}{dt} - \frac{1}{h_1}\left(\frac{\partial h_3}{\partial a_1}\right)\frac{da_3}{dt} \\ \omega_z &= \frac{1}{h_1}\left(\frac{\partial h_2}{\partial a_1}\right)\frac{da_2}{dt} - \frac{1}{h_2}\left(\frac{\partial h_1}{\partial a_2}\right)\frac{da_1}{dt} \end{aligned} \tag{5.143}$$

$$u' = q_{a1},\ v' = q_{a2},\ w' = q_{a3},\ dx = h_1 da_1,\ dy = h_2 da_2,\ dz = h_3 da_3$$

19. Vector notation[23]

In this section, we shall explain the vector notation used in previous sections and review some vector analysis which shall prove useful in deriving the fundamental equations.

A scalar quantity is one which is specified completely by its magnitude. For instance, temperature, density, volume, etc., are scalar quantities. A vector is one which is specified completely by its magnitude and direction. For instance, velocity, temperature gradient, etc., are vector quantities. In the ordinary three-dimensional space, a vector has three components. Any vector $\boldsymbol{q}$ with components u, v, and w along the x-, y-, and z-axes may be expressed as follows:

$$\boldsymbol{q} = \boldsymbol{i}u + \boldsymbol{j}v + \boldsymbol{k}w \tag{5.144}$$

where $\boldsymbol{i}$, $\boldsymbol{j}$, and $\boldsymbol{k}$ are, respectively, unit vectors in the direction of the x-, y-, and z-axes. The addition of two vectors $\boldsymbol{q}$ and $\boldsymbol{q}_1$ gives

$$\boldsymbol{q}+\boldsymbol{q}_1=(\boldsymbol{i}u+\boldsymbol{j}v+\boldsymbol{k}w)+(\boldsymbol{i}u_1+\boldsymbol{j}v_1+\boldsymbol{k}w_1)$$
$$=\boldsymbol{i}(u+u_1)+\boldsymbol{j}(v+v_1)+\boldsymbol{k}(w+w_1) \tag{5.145}$$

If s is a scalar quantity, the multiplication of $\boldsymbol{q}$ by s gives

$$s\boldsymbol{q}=\boldsymbol{i}su+\boldsymbol{j}sv+\boldsymbol{k}sw \tag{5.146}$$

The scalar or dot product of two vectors $\boldsymbol{q}$ and $\boldsymbol{q}_1$ is defined as follows:

$$\boldsymbol{q}\cdot\boldsymbol{q}_1=qq_1\cos\theta=\boldsymbol{q}_1\cdot\boldsymbol{q} \tag{5.147}$$

where q and q_1 are the magnitude of $\boldsymbol{q}$ and $\boldsymbol{q}_1$, respectively, and θ is the angle between these two vectors.

From Eq. (5.147) it is easy to show that

$$\boldsymbol{i}\cdot\boldsymbol{i}=\boldsymbol{j}\cdot\boldsymbol{j}=\boldsymbol{k}\cdot\boldsymbol{k}=1$$

$$\boldsymbol{i}\cdot\boldsymbol{j}=\boldsymbol{j}\cdot\boldsymbol{i}=\boldsymbol{j}\cdot\boldsymbol{k}=\boldsymbol{k}\cdot\boldsymbol{j}=\boldsymbol{k}\cdot\boldsymbol{i}=\boldsymbol{i}\cdot\boldsymbol{k}=0$$

Using this result, Eq. (5.147) may be written as

$$\boldsymbol{q}\cdot\boldsymbol{q}_1=uu_1+vv_1+ww_1 \tag{5.147a}$$

We also have $\boldsymbol{a}\cdot(\boldsymbol{q}+\boldsymbol{q}_1)=\boldsymbol{a}\cdot\boldsymbol{q}+\boldsymbol{a}\cdot\boldsymbol{q}_1$ where $\boldsymbol{a}$ is a vector. The scalar product of a force vector $\boldsymbol{F}$ and a distance vector $\boldsymbol{d}$ gives the work done W, i.e.

$$\boldsymbol{F}\cdot\boldsymbol{d}=W$$

The vector or cross product of two vectors $\boldsymbol{q}$ and $\boldsymbol{q}_1$ is defined as follows:

$$\boldsymbol{q}\times\boldsymbol{q}_1=\varepsilon qq_1\sin\theta=-\boldsymbol{q}_1\times\boldsymbol{q} \tag{5.148}$$

where ε is a unit vector in the direction perpendicular to both $\boldsymbol{q}$ and $\boldsymbol{q}_1$.

From Eq. (5.148) we have

$$\boldsymbol{i}\times\boldsymbol{i}=\boldsymbol{j}\times\boldsymbol{j}=\boldsymbol{k}\times\boldsymbol{k}=0$$

$$\boldsymbol{i}\times\boldsymbol{j}=-\boldsymbol{j}\times\boldsymbol{i}=\boldsymbol{k},\quad \boldsymbol{j}\times\boldsymbol{k}=-\boldsymbol{k}\times\boldsymbol{j}=\boldsymbol{i}$$

$$\boldsymbol{k}\times\boldsymbol{i}=-\boldsymbol{i}\times\boldsymbol{k}=\boldsymbol{j}$$

Using this result, Eq. (5.148) may be written as

$$\boldsymbol{q}\times\boldsymbol{q}_1=\begin{vmatrix}\boldsymbol{i} & \boldsymbol{j} & \boldsymbol{k}\\ u & v & w\\ u_1 & v_1 & w_1\end{vmatrix}$$
$$=\boldsymbol{i}(vw_1-wv_1)+\boldsymbol{j}(wu_1-uw_1)+\boldsymbol{k}(uv_1-vu_1) \tag{5.148a}$$

We also have

$$\boldsymbol{a}\times(\boldsymbol{q}+\boldsymbol{q}_1)=\boldsymbol{a}\times\boldsymbol{q}+\boldsymbol{a}\times\boldsymbol{q}_1$$

The area A formed by two radius vectors $\boldsymbol{r}$ and $\boldsymbol{r}_1$ is

$$\boldsymbol{A}=\boldsymbol{r}\times\boldsymbol{r}_1$$

whose direction is the normal to the surface.

The scalar triple product of three vectors $\boldsymbol{r}$, $\boldsymbol{q}$, and $\boldsymbol{q}_1$ is defined as

$$\boldsymbol{r}\cdot\boldsymbol{q}\times\boldsymbol{q}_1=\begin{vmatrix} r_1 & r_2 & r_3 \\ u & v & w \\ u_1 & v_2 & w_1 \end{vmatrix}=\boldsymbol{r}\times\boldsymbol{q}\cdot\boldsymbol{q}_1=[\boldsymbol{r}\boldsymbol{q}\boldsymbol{q}_1] \tag{5.149}$$

where $\boldsymbol{r}=\boldsymbol{i}r_1+\boldsymbol{j}r_2+\boldsymbol{k}r_3$. The volume is a scalar triple product of three radius vectors.

The following are some useful formulas for various products of vectors:

$$\boldsymbol{A}\times(\boldsymbol{B}\times\boldsymbol{C})=\boldsymbol{A}\cdot\boldsymbol{C}\boldsymbol{B}-\boldsymbol{A}\cdot\boldsymbol{B}\boldsymbol{C} \tag{5.150}$$

$$(\boldsymbol{A}\times\boldsymbol{B})\cdot(\boldsymbol{C}\times\boldsymbol{D})=\boldsymbol{A}\cdot\boldsymbol{B}\times(\boldsymbol{C}\times\boldsymbol{D})=\boldsymbol{A}\cdot\boldsymbol{C}\boldsymbol{B}\cdot\boldsymbol{D}-\boldsymbol{A}\cdot\boldsymbol{D}\boldsymbol{B}\cdot\boldsymbol{C} \tag{5.151}$$

$$(\boldsymbol{A}\times\boldsymbol{B})\times(\boldsymbol{C}\times\boldsymbol{D})=(\boldsymbol{A}\times\boldsymbol{B}\cdot\boldsymbol{D})\boldsymbol{C}-(\boldsymbol{A}\times\boldsymbol{B}\cdot\boldsymbol{C})\boldsymbol{D} \tag{5.152}$$

If the vector $\boldsymbol{q}$ is a continuous function of a scalar s, the derivative of $\boldsymbol{q}$ with respect to s is

$$\frac{d\boldsymbol{q}}{ds}=\lim_{\Delta s\to 0}\frac{\Delta\boldsymbol{q}}{\Delta s}=\lim_{\Delta s\to 0}\frac{\boldsymbol{q}(s+\Delta s)-\boldsymbol{q}(s)}{\Delta s} \tag{5.153}$$

If we write $\boldsymbol{q}(s)=\boldsymbol{i}u(s)+\boldsymbol{j}v(s)+\boldsymbol{k}w(s)$, then

$$\frac{d\boldsymbol{q}}{ds}=\boldsymbol{i}\frac{du}{ds}+\boldsymbol{j}\frac{dv}{ds}+\boldsymbol{k}\frac{dw}{ds} \tag{5.154}$$

$$\frac{d\boldsymbol{q}\cdot\boldsymbol{q}_1}{ds}=\boldsymbol{q}\cdot\frac{d\boldsymbol{q}_1}{ds}+\frac{d\boldsymbol{q}}{ds}\cdot\boldsymbol{q}_1 \tag{5.155}$$

$$\frac{d}{ds}(\boldsymbol{q}\times\boldsymbol{q}_1)=\frac{d\boldsymbol{q}}{ds}\times\boldsymbol{q}_1+\boldsymbol{q}\times\frac{d\boldsymbol{q}_1}{ds} \tag{5.156}$$

If $\boldsymbol{q}$ is a function of more than one scalar, such as

$$\boldsymbol{q}(x,y,z,t)=\boldsymbol{i}u(x,y,z,t)+\boldsymbol{j}v(x,y,z,t)+\boldsymbol{k}w(x,y,z,t) \tag{5.157}$$

then

$$\frac{\partial\boldsymbol{q}}{\partial x}=\boldsymbol{i}\frac{\partial u}{\partial x}+\boldsymbol{j}\frac{\partial v}{\partial x}+\boldsymbol{k}\frac{\partial w}{\partial x}$$

$$\frac{\partial u\boldsymbol{q}}{\partial x}=u\frac{\partial\boldsymbol{q}}{\partial x}+\boldsymbol{q}\frac{\partial u}{\partial x} \tag{5.158}$$

Vector integration of a vector $\boldsymbol{q}(s)$ with respect to s gives

$$\int\boldsymbol{q}(s)\,ds=\boldsymbol{Q}(s)+\text{constant} \tag{5.159}$$

or

$$\int_a^b\boldsymbol{q}(s)\,ds=\boldsymbol{Q}(b)-\boldsymbol{Q}(a) \tag{5.159a}$$

If $\boldsymbol{A}$ is a constant vector, then

$$\begin{aligned}\int \boldsymbol{A} \cdot \boldsymbol{q}(s)\,ds &= \boldsymbol{A} \cdot \int \boldsymbol{q}(s)\,ds \\ \int \boldsymbol{A} \times \boldsymbol{q}(s)\,ds &= \boldsymbol{A} \times \int \boldsymbol{q}(s)\,ds\end{aligned} \tag{5.160}$$

If a curve c in space is specified by a radius vector $\boldsymbol{r}(s)$, the line integral along this curve of any vector $\boldsymbol{q}$ is defined as

$$\Gamma = \int_c \boldsymbol{q} \cdot d\boldsymbol{r} = \int_c \boldsymbol{q} \cdot \boldsymbol{t}ds \tag{5.161}$$

where $\boldsymbol{t}$ is the unit vector in the direction of the curve $c.\boldsymbol{r} = \boldsymbol{t}s$.

If the area of a surface S is specified by a vector $\boldsymbol{A} = \boldsymbol{n}A$, the surface integral of any vector $\boldsymbol{q}$ over the area S is defined as

$$Q = \iint_S \boldsymbol{q} \cdot dA = \iint_S \boldsymbol{q} \cdot \boldsymbol{n}dA \tag{5.162}$$

where $\boldsymbol{n}$ is the unit normal vector of the surface S.

Now we define an operator del as follows:

$$\nabla = \boldsymbol{i}\frac{\partial}{\partial x} + \boldsymbol{j}\frac{\partial}{\partial y} + \boldsymbol{k}\frac{\partial}{\partial z} \tag{5.163}$$

where x, y, and z are the Cartesian coordinates. The radius vector r is defined by

$$\boldsymbol{r} = \boldsymbol{i}x + \boldsymbol{j}y + \boldsymbol{k}z \tag{5.164}$$

We consider a scalar function $T(x, y, z)$. On the surface of $T(x, y, z)$ = constant, we have a point P represented by the radius vector $\boldsymbol{r}$, and on a

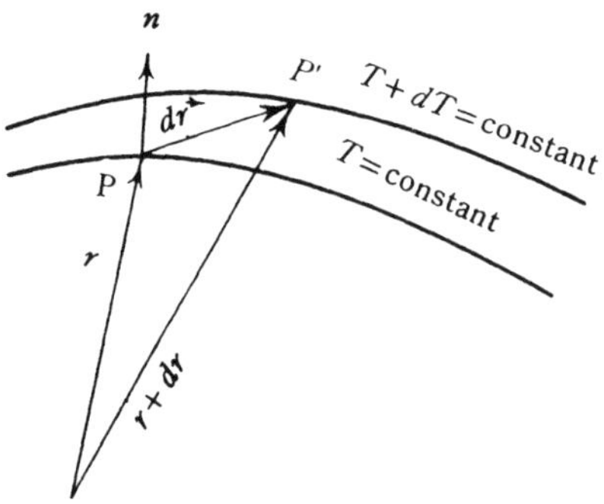

Fig. 5.5 Gradient of a scalar function T

neighboring surface of $T + dT$=constant, we have a point P' represented by the radial vector $\boldsymbol{r} + d\boldsymbol{r}$ (see Fig. 5.5). We have

$$dT = \frac{\partial T}{\partial x} dx + \frac{\partial T}{\partial y} dy + \frac{\partial T}{\partial z} dz$$

$$= (\boldsymbol{i}dx + \boldsymbol{j}dy + \boldsymbol{k}dz) \cdot \left(\boldsymbol{i}\frac{\partial T}{\partial x} + \boldsymbol{j}\frac{\partial T}{\partial y} + \boldsymbol{k}\frac{\partial T}{\partial z} \right)$$

$$= d\boldsymbol{r} \cdot \nabla T \quad (5.165)$$

Let $\boldsymbol{n}$ be the unit normal vector to the surface T=constant, then

$$dn = \boldsymbol{n} \cdot d\boldsymbol{r} \quad (5.166)$$

and

$$dT = \frac{dT}{dn} dn = d\boldsymbol{r} \cdot \boldsymbol{n} \frac{dT}{dn} \quad (5.167)$$

Comparing Eq. (5.165) with Eq. (5.167), we have

$$\nabla T = \boldsymbol{n} \frac{dT}{dn} = \text{gradient of } T = \text{grad } T \quad (5.168)$$

where dT/dn is the greatest spatial rate of change of T.

The divergence of a vector function $\boldsymbol{q}$ (5.157) is defined as

$$\nabla \cdot \boldsymbol{q} = \frac{\partial u}{\partial x} + \frac{\partial v}{\partial y} + \frac{\partial w}{\partial z} \quad (5.169)$$

The curl of a vector function q (5.157) is defined as

$$\nabla \times \boldsymbol{q} = \boldsymbol{i}\left(\frac{\partial w}{\partial y} - \frac{\partial v}{\partial z} \right) + \boldsymbol{j}\left(\frac{\partial u}{\partial z} - \frac{\partial w}{\partial x} \right) + \boldsymbol{k}\left(\frac{\partial v}{\partial x} - \frac{\partial u}{\partial y} \right) \quad (5.170)$$

If $\boldsymbol{q}$ is the gradient of a scalar function φ, i.e., $\boldsymbol{q} = \nabla\varphi$, then

$$\nabla \times \boldsymbol{q} = \nabla \times \nabla\varphi = 0 \quad (5.171)$$

If the vector $\boldsymbol{q}$ is the curl of a vector function $\boldsymbol{A}$, then

$$\nabla \cdot \boldsymbol{q} = \nabla \cdot \nabla \times \boldsymbol{A} = 0 \quad (5.172)$$

The following are some other useful formulas:

$$\nabla \cdot (u\boldsymbol{q}) = \boldsymbol{q} \cdot \nabla u + u\nabla \cdot \boldsymbol{q} \quad (5.173)$$

$$\nabla \times (u\,\boldsymbol{q}) = u\nabla \times \boldsymbol{q} + (\nabla u) \times \boldsymbol{q} \quad (5.174)$$

$$\nabla \cdot (\boldsymbol{q} \times \boldsymbol{q}_1) = -\boldsymbol{q} \cdot (\nabla \times \boldsymbol{q}_1) + \boldsymbol{q}_1 \cdot (\nabla \times \boldsymbol{q}) \quad (5.175)$$

$$\nabla(\boldsymbol{q} \cdot \boldsymbol{q}) = 2(\boldsymbol{q} \cdot \nabla)\boldsymbol{q} + 2\boldsymbol{q} \times (\nabla \times \boldsymbol{q}) \quad (5.176)$$

$$\nabla \times (\boldsymbol{q} \times \boldsymbol{q}_1) = -(\boldsymbol{q} \cdot \nabla)\boldsymbol{q}_1 + \boldsymbol{q}(\nabla \cdot \boldsymbol{q}_1) + (\boldsymbol{q}_1 \cdot \nabla)\boldsymbol{q} - \boldsymbol{q}_1(\nabla \cdot \boldsymbol{q}) \quad (5.177)$$

20. Problems

1. If the lines of motion are curves on the surface of cones having their vertices at the origin and the axis of z for a common axis, find the equation of continuity in this case where the velocity component normal to the surface of a cone is zero.

2. Find the equation of continuity in spherical coordinates.

3. Find the equations of motion, equation of continuity, and equation of energy of inviscid and nonheat-conducting fluid in cylindrical coordinates.

4. Prove the Stokes′ theorem (5.37).

5. The velocity components of a flow field are as follows:

$$u=\frac{ax+by}{x^2+y^2}, \quad v=\frac{ay+bx}{x^2+y^2}, \quad w=0$$

where a and b are constants. Investigate the nature of the motion of the fluid. Include the circulation around a closed curve enclosing the origin, the divergence, vorticity of the flow, etc.

6. Define a stream function ψ for the one-dimensional unsteady compressible fluid flow so that the equation of continuity is automatically satisfied. Find the differential equation for this stream function for adiabatic flow.

7. Show that the vector equation (5.176) is true.

8. Show that

$$[\boldsymbol{pqr}]\,[\boldsymbol{abc}] = \begin{vmatrix} \boldsymbol{p}\cdot\boldsymbol{a} & \boldsymbol{p}\cdot\boldsymbol{b} & \boldsymbol{p}\cdot\boldsymbol{c} \\ \boldsymbol{q}\cdot\boldsymbol{a} & \boldsymbol{q}\cdot\boldsymbol{b} & \boldsymbol{q}\cdot\boldsymbol{c} \\ \boldsymbol{r}\cdot\boldsymbol{a} & \boldsymbol{r}\cdot\boldsymbol{b} & \boldsymbol{r}\cdot\boldsymbol{c} \end{vmatrix}$$

9. Rederive the equations of motion (5.12) and equation of energy (5.22) if the effect of decrease of gravity in major distance from the earth center is taken into account.

10. If we assume that the pressure-density relation is approximated by the following relation (see Chapter VIII, § 5),

$$p=A+\frac{B}{\rho}$$

where A and B are constants, find the corresponding energy equation and the Bernoulii equation. What is the corresponding velocity of sound in this case?

11. Derive the differential equation for the stream function for two-dimensional steady diabatic flow.

12. Show that the characteristics of the two-dimensional unsteady Euler equations are

$$dt=1\Big/\left(\frac{U}{dx}+\frac{V}{dy}\right)$$

and

$$dt = 1 \Big/ \left[\frac{U}{dx} + \frac{V}{dy} \pm a \left(\frac{1}{dx^2} + \frac{1}{dy^2} \right)^{\frac{1}{2}} \right]$$

13. Show that the characteristics of the three-dimensional unsteady Euler equations are

$$dt = 1 \Big/ \left(\frac{U}{dx} + \frac{V}{dy} + \frac{W}{dz} \right)$$

and

$$dt = 1 \Big/ \left[\frac{U}{dx} + \frac{V}{dy} + \frac{W}{dz} \pm a \left(\frac{1}{dx^2} + \frac{1}{dy^2} + \frac{1}{dz^2} \right)^{\frac{1}{2}} \right]$$

14. Prove that for two-dimensional unsteady Euler equations in divergence form, the Jacobian matrices are

$$\frac{\partial F}{\partial U} = \begin{bmatrix} 0 & 1 & 0 & 0 \\ -\frac{3-\gamma}{2}\frac{U_2^2}{U_1^2} + \frac{\gamma-1}{2}\frac{U_3^2}{U_1^2} & (3-\gamma)\frac{U_2}{U_1} & -(\gamma-1)\frac{U_3}{U_1} & \gamma-1 \\ -\frac{U_2U_3}{U_1^2} & \frac{U_3}{U_1} & \frac{U_2}{U_1} & 0 \\ -\frac{\gamma U_4 U_2}{U_1^2} + (\gamma-1)\frac{U_2^2+U_3^2}{U_1^3}U_2 & \frac{\gamma U_4}{U_1} - \frac{\gamma-1}{2}\frac{3U_2^2+U_3^2}{U_1^2} & -(\gamma-1)\frac{U_2U_3}{U_1^2} & \frac{\gamma U_2}{U_1} \end{bmatrix}$$

$$\frac{\partial G}{\partial U}=\begin{bmatrix} 0 & 0 & 1 & 0 \\ -\dfrac{U_2U_3}{U_1^2} & \dfrac{U_3}{U_1} & \dfrac{U_2}{U_1} & 0 \\ \dfrac{\gamma-1}{2}\dfrac{U_2^2}{U_1^2}-\dfrac{3-\gamma}{2}\dfrac{U_3^2}{U_1^2} & -(\gamma-1)\dfrac{U_2}{U_1} & (3-\gamma)\dfrac{U_3}{U_1} & \gamma-1 \\ -\dfrac{\gamma U_4U_3}{U_1^2}+(\gamma-1)\dfrac{U_2^2+U_3^2}{U_1^3}U_3 & -(\gamma-1)\dfrac{U_2U_3}{U_1^2} & \dfrac{\gamma U_4}{U_1}-\dfrac{\gamma-1}{2}\dfrac{U_2^2+3U_3^2}{U_1^2} & \dfrac{\gamma U_3}{U_1} \end{bmatrix}$$

and

$$F=\frac{\partial F}{\partial U}U, \quad G=\frac{\partial G}{\partial U}U .$$

15. Derive the Jacobian matrices for three-dimensional unsteady Euler equations in divergence form, $\dfrac{\partial F}{\partial U}$, $\dfrac{\partial G}{\partial U}$, and $\dfrac{\partial H}{\partial U}$, and prove that $F=\dfrac{\partial F}{\partial U}U$, $G=\dfrac{\partial G}{\partial U}U$ and $H=\dfrac{\partial H}{\partial U}U$.

16. Prove that the characteristics of the linear differential equations (5.107) are (5.110).

17. Reexamine the Kelvin's theorem (5.38) by including the effects of radiation pressure and radiation energy density.

References

1. Busemann, A., Gasdynamik, *Handbuch der Experimental physik*, **4**, pt I, pp. 343–460, Leipzig, 1931.
2. Ackeret, J., Gasdynamik, *Handbuch der Physik*, 7, Kap, 5.
3. Pai, S. I., *Fluid Dynamics of Jets*, D. Van Nostrand Company, Inc., Princeton, N. J., 1954.
4. Lamb, H., *Hydrodynamics*, 6th edition, Cambridge University Press, 1932.
5. Pai, S. I., *Modern Fluid Mechanics*, Science Press, Beijing, 1981, distributed by Van Nostrand Reinhold Co. N. Y.
6. Vazsonyi, A., On rotational gas flows, *Quart. Appl. Math.* **3**, No. 1, April 1945, pp. 29–37.
7. Crocco, L., Eine neus Stromfunktion fuer die Erforschung der Bewegung der Gas mit Rotation, *ZAMM* **17**, 1937, pp. 1–7.
8. Gross, R. A., and Esch, R., Low speed combustion aerodynamics, *Jet Propulsion*, March-

April 1954, pp. 95 – 101.

9. Fourth Symp. (international) on Combustion (combustion and detonation waves) held at M. I. T., Sept. 1 – 5, 1952, The Williams and Wilkins Company, Baltimore, Md., 1953.
10. Sears, W. R., General theory of high speed aerodynamics, **6** of *High Speed Aerodynamics and Jet Propulsion*, Princeton University Press, 1954, chapters C, D, and E.
11. Rayleigh, Lord, On the flow of compressible fluid past an obstacle, *Phil. Mag.* (6) **32**, 1916, pp. 1 – 6.
12. Janzen, O., Beitrag zu einer Theorie der stationaeren Stroemung kompressibler Fluessigkeiten, *Phys. Z.* **XIV**, 1913, pp. 639 – 643.
13. Chang, C. C., General considerations of problems in compressible flow using the hodograph method, *NACA*, *TN* No. 2582, Jan. 1952.
14. Ringleb, F., Exacte Loesungen der Differentialgleichungen einer adiabatischen Gasstroemung, *ZAMM* **20**, 1940, pp. 185 – 198.
15. Prandtl, L., and Busemann, A., *Naeherungsverfahren zur zeichnerischen Ermittlung von ebenen Stroemungen mit Ueberschallgeschwindigkeit*, Stodola-Festschrift Zuerich, 1929, pp. 499 – 509.
16. Ramsey, A. S., *Hydrodynamics*, G. Bell and Sons, Ltd., London, 1935.
17. von Mises, R., Notes on mathematical theory of compressible fluid flow, Harvard University, Grad. School of Eng., *Special Pub.* No. 2, 1949.
18. Hicks, B. L., Diabatic flow of a compressible fluid, *Quart. Appl. Math.* **6**, No. 3, 1948, pp. 221 – 237.
19. Hicks, B. L., On the characterization of fields of diabatic flow, *Quart. Appl. Math.* **6**, No. 4, 1949, pp. 405 – 416.
20. Wang, C. T., Two dimensional subsonic compressible flows past arbitrary bodies by the variational method, *NACA*, *TN* No. 2326, March 1951.
21. Southwell, R. V., *Relaxation Methods in Theoretical Physics*, Oxford Clarendon Press. 1946.
22. Courant, R., Methods of Mathematical Physics, **II** *Partial Differential Equations*, *Chap.* V, § 9, pp. 486 – 490, Inter Science Publisher, 1962.
23. Wills, A. P., *Vector Analysis with an Introduction to Tensor Analysis*, Prentice-Hall., Inc., 1931.

Chapter VI

METHOD OF SMALL PERTURBATIONS, LINEARIZED THEORY

1. Introduction

The method of small perturbations was first used to solve the problem of the flow around a thin body in a uniform stream of a compressible fluid. If the thickness ratio δ of the thin body is much smaller than unity, we may develop the velocity potential of the irrotational flow as a power series of the thickness ratio[1]. For two-dimensional flows without a stagnation point, this power series expansion is valid. On the other hand, for two-dimensional flows which exhibit a stagnation point, terms of $\delta^n \log \delta$ appear in this expansion of the velocity potential. Since this kind of singularity does not appear explicitly in the expansion of series of stream function, it is more convenient to solve this latter problem using the stream function instead of the velocity potential.

The method of small perturbations may be further generalized. Consider the case of a two-dimensional steady rotational flow[2,3]. For this, we write the stream function ψ

$$\psi\ (x, y) = \psi_0(x, y) + \psi_1(x, y) + \psi_2(x, y) + \cdots \tag{6.1}$$

where $\psi_0\ (x, y)$ is a specified solution for the irrotational motion corresponding to the problem considered, and ψ_1 is the first-order perturbation of ψ_0. Of course, the simplest form of ψ_0 is that of uniform flow of constant velocity U_0, i.e., $\psi_0 = \rho_0 U_0 y$. For the purpose of defining the iteration procedure, the function ψ_{n+1} is to be regarded as being small as compared with the preceding function ψ_n; the derivatives are held to show similar relationships. Thus, the total index fixes the order of the terms; for example ψ_2 is of the same order as ψ_1^2 or $\psi_0\psi_2$.

In § 2, we shall use this iteration method to find the differential equations for the various ψ_n's for two-dimensional steady rotational flow. In this section, the detailed derivations shall be given.

In § 3, we shall give the first-order linearized equation for the velocity potential with the basic flow as a uniform stream. This is the fundamental equation which has been extensively used in high-speed aerodynamics.

In § 4, the boundary conditions for the linearized theory will be discussed.

In § 5, we shall use the first-order linearized equation of stream function to study subsonic two-dimensional irrotational steady flow. This analysis gives the important Prandtl-Glauert rule for the correction of velocity and pressure coefficient due to compressibility effects.

In § 6, we shall discuss two-dimensional irrotational steady supersonic flow by means of linearized equation. The well-known Ackeret formula will be given.

In § 7, we shall study the compressible flow over an infinite waveshaped wall and point out the essential difference between subsonic and supersonic flows.

In § 8, the Fourier integral method of solving the subsonic flow over a wall of arbitrary shape will be discussed briefly.

In § 9, a general discussion of the linearized theory of three-dimen-sional irrotational flow will be given. This theory will be further presented in chapter XII.

For both transonic and hypersonic flows, the linearized theory based on a uniform stream fails. The small perturbation theory of both transonic and hypersonic flows will be discussed in Chapter X, as the famous similarity laws will be in the same chapter.

In § 10, higher-order approximations will be discussed briefly.

The linearized theory of rotational flows will be discussed in Chapter XIV.

In § 11, the panel method, a powerful numerical method for solving the linearized flow problems is discussed for the two-dimensional Laplace equation. The panel method for supersonic flow and unsteady flow problems will be given in Chapter XII.

2. Two-dimensional steady rotational flow[2,3]. Basic equations for iteration

The fundamental equation for the stream function is [see Eq. (5.61)]

$$\left(1-\frac{u^2}{a^2}\right)\frac{\partial^2\psi}{\partial x^2}-\frac{2uv}{a^2}\frac{\partial^2\psi}{\partial x\partial y}+\left(1-\frac{v^2}{a^2}\right)\frac{\partial^2\psi}{\partial y^2}=$$

$$-\rho\omega\left\{1+\frac{\gamma-1}{\rho^2a^2}\left[\left(\frac{\partial\psi}{\partial x}\right)^2+\left(\frac{\partial\psi}{\partial y}\right)^2\right]\right\} \quad (6.2)$$

If we substitute Eq. (6.1) into Eq. (6.2), we may find the differential equations for ψ_n by collecting terms of the same order of n. Before we can collect the terms of the same order we must express the density ρ and sound velocity a in terms of the perturbation stream function ψ_n's and the entropy perturbation which is, in general, assumed to be of the same order of magnitude as the velocity perturbation.

We write

$$\frac{\partial\psi_0}{\partial y}=\rho_0 u_0, \quad \frac{\partial\psi_0}{\partial x}=-\rho_0 v_0 \tag{6.3}$$

and

$$q_0^2=u_0^2+v_0^2 \tag{6.4}$$

where u_0 and v_0 are the x-and y-component of velocity of the basic flow, and ρ_0 is the density of the basic flow. In general u_0, v_0, and ρ_0 are known functions of x and y; the subscript 0 refers to values of the basic flow.

For isoenergetic flow, the velocity of sound is given by the following equation:

$$a^2+\frac{\gamma-1}{2}(u^2+v^2)=a_0^2+\frac{\gamma-1}{2}q_0^2 \tag{6.5}$$

From the equation of state (5.3), we have

$$\frac{a^2}{a_0^2}=\frac{T}{T_0}=\frac{p}{p_0}\frac{\rho_0}{\rho}=\left(\frac{\rho}{\rho_0}\right)^{\gamma-1}e^{\frac{S-S_0}{C_v}} \tag{6.6}$$

Combining Eqs. (6.6) and (6.5) we have

$$\left(\frac{\rho}{\rho_0}\right)^{\gamma-1}e^{\frac{S-S_0}{C_v}}+\frac{\gamma-1}{2}\left(\frac{\rho_0}{\rho}\right)^2 M_0^2\left[\frac{\rho^2}{\rho_0^2}\frac{u^2+v^2}{q_0^2}\right]=1+\frac{\gamma-1}{2}M_0^2 \tag{6.7}$$

where $M_0=q_0/a_0$=the Mach number corresponding to ψ_0 which is, in general, a function of x and y.

Let us write

$$1+\beta=\frac{1}{\rho_0^2 q_0^2}\left[\left(\frac{\partial\psi}{\partial x}\right)^2+\left(\frac{\partial\psi}{\partial y}\right)^2\right]=\frac{\rho^2}{\rho_0^2}\frac{u^2+v^2}{q_0^2} \tag{6.8}$$

and

$$\frac{\rho_0}{\rho}=1+\zeta \tag{6.9}$$

Substituting Eqs. (6.8) and (6.9) into Eq. (6.7), we have $(1+\zeta)^{-(\gamma-1)e^A}+$

$$\frac{\gamma-1}{2} M_0^2 (1+\zeta)^2 (1+\beta) = 1 + \frac{\gamma-1}{2} M_0^2 \tag{6.10}$$

where $A = (S - S_0)/c_v$ which is of the same order of magnitude as the first-order perturbation stream function ψ_1.

Solving for β from Eq. (6.10), we have

$$\beta = \frac{1}{(1+\zeta)^2} \left\{ \frac{2}{(\gamma-1)M_0^2} \left[1 - (1+\zeta)^{-(\gamma-1)e^{A}} \right] + 1 \right\} - 1 \tag{6.11}$$

Expanding the right-hand side of Eq. (6.11), a power series of ψ_n and A, and collecting terms of the same order, we obtain

$$\beta = -\{2\zeta + B[A - (\gamma-1)\zeta]\} + \left\{ 3\zeta^2 + 2\zeta B[A - (\gamma-1)\zeta] - B\left[\frac{A^2}{2} + \frac{\gamma(\gamma-1)}{2}\zeta^2 - (\gamma-1)\zeta A \right] \right\} + \cdots \tag{6.12}$$

where $B = \dfrac{2}{(\gamma-1)M_0^2}$.

From Eqs. (6.1) and (6.8) we have

$$\beta = 2\frac{u_0}{\rho_0 q_0^2} \frac{\partial \psi_1}{\partial y} - \frac{2v_0}{\rho_0 q_0^2} \frac{\partial \psi_1}{\partial x} + \cdots \tag{6.13}$$

We may now solve for ζ from Eqs. (6.12) and (6.13) in terms of A and the ψ_n's to get

$$\zeta = \frac{M_0^2}{1-M_0^2} \frac{u_0}{\rho_0 q_0^2} \frac{\partial \psi_1}{\partial y} - \frac{M_0^2 v_0}{(1-M_0^2)\rho_0 q_0^2} \frac{\partial \psi_1}{\partial x} + \frac{A}{(\gamma-1)(1-M_0^2)} + \cdots \tag{6.14}$$

With the help of Eq. (6.14), we can calculate the velocity ratios in terms of the ψ_n's and A:

$$\frac{u^2}{a_2} = \frac{u_0^2}{a_0^2} \left\{ 1 + \frac{2}{\rho_0 u_0} \left[1 + \frac{M_0^2(\gamma+1)u_0^2}{(1-M^2)_0 q_0^2} \right] \frac{\partial \psi_1}{\partial y} - \frac{v_0 M_0^2(\gamma+1)}{\rho_0(1-M_0^2)q_0^2} \frac{\partial \psi_1}{\partial x} + \frac{2A}{(\gamma-1)(1-M_0^2)} \left(1 + \frac{\gamma-1}{2} M_0^2 \right) + \cdots \right\} \tag{6.15}$$

$$\frac{v^2}{a^2}=\frac{v_0^2}{a_0^2}\left\{1+\frac{u_0M_0^2(\gamma+1)}{\rho_0(1-M_0^2)q_0^2}\frac{\partial\psi_1}{\partial y}-\frac{2}{\rho_0v_0}\left[1+\frac{M_0^2(\gamma+1)v_0^2}{(1-M_0^2)2q_0^2}\right]\frac{\partial\psi_1}{\partial x}\right.$$
$$\left.+\frac{2A}{(\gamma-1)(1-M_0^2)}\left(1+\frac{\gamma-1}{2}M_0^2\right)+\cdots\right\} \tag{6.16}$$

$$\frac{uv}{a^2}=\frac{u_0v_0}{a_0^2}\left\{1+\frac{M_0^2(\gamma+1)u_0^2}{(1-M_0^2)q_0^2}\frac{\partial\psi_1}{\partial y}-\frac{1}{\rho_0v_0}\left[1+\frac{M_0^2(\gamma+1)v_0^2}{(1-M_0^2)q_0^2}\frac{\partial\psi_1}{\partial y}\right]\right.$$
$$\left.+\frac{2A}{(\gamma-1)(1-M_0^2)}\left(1+\frac{\gamma-1}{2}M_0^2\right)+\cdots\right\} \tag{6.17}$$

Since we assume that the basic flow is irrotational, we may expand the vorticity as follows:

$$\omega=\omega_1+\omega_2+\omega_3+\cdots \tag{6.18}$$

where $\omega_1\sim\psi_1$, $\omega_2\sim\psi_2$, etc., and

$$\frac{\rho}{\rho_0}\omega=\omega_1-(\xi\omega_1-\omega_2)+\cdots \tag{6.19}$$

Furthermore, from Eqs. (6.6) and (6.8) to (6.11),

$$\frac{1}{\rho^2a^2}\left[\left(\frac{\partial\psi}{\partial x}\right)^2+\left(\frac{\partial\psi}{\partial y}\right)^2\right]=\frac{1}{\rho_0^2a_0^2}\left[\rho_0^2q_0^2+2\frac{\partial\psi_0}{\partial y}\frac{\partial\psi_1}{\partial y}+2\frac{\partial\psi_0}{\partial x}\frac{\partial\psi_1}{\partial x}\right.$$
$$+2\rho_0^2q_0^2\zeta+\frac{\gamma-1}{2}\frac{M_0^2}{1-M_0^2}\rho_0^2q_0^2\left(\frac{2u_0}{\rho_0q_0^2}\frac{\partial\psi_1}{\partial y}\right.$$
$$\left.\left.-\frac{2v_0}{\rho_0q_0^2}\frac{\partial\psi_1}{\partial x}+\frac{2A}{\gamma-1}\right)+\cdots\right] \tag{6.20}$$

Substituting Eqs. (6.1) and (6.15) to (6.20) into Eq. (6.2) and collecting terms of the same order, we obtain the differential equations for ψ_n's. For the zeroth-order equation

$$\left(1-\frac{u_0^2}{a_0^2}\right)\frac{\partial^2\psi_0}{\partial x^2}-\frac{2u_0v_0}{a_0^2}\frac{\partial^2\psi_0}{\partial x\partial y}+\left(1-\frac{v_0^2}{a_0^2}\right)\frac{\partial^2\psi_0}{\partial y^2}=0 \tag{6.21}$$

which is indeed the irrotational flow equation for ψ_0 [cf. Eq. (5.63)].

The first-order perturbation differential equation for stream function ψ_1 is

$$\left(1-\frac{u_0^2}{a_0^2}\right)\frac{\partial^2\psi_1}{\partial x^2}-\frac{2u_0v_0}{a_0^2}\frac{\partial^2\psi_1}{\partial x\partial y}+\left(1-\frac{v_0^2}{a_0^2}\right)\frac{\partial^2\psi_1}{\partial y^2}$$

$$-\frac{u_0^2}{a_0^2}\frac{\partial^2\psi_0}{\partial x^2}\left[\left(Ku_0+\frac{2}{\rho_0u_0}\right)\frac{\partial\psi_1}{\partial y}-Kv_0\frac{\partial\psi_1}{\partial x}+A^*\right]$$

$$-\frac{2u_0v_0}{a_0^2}\frac{\partial^2\psi_0}{\partial x\partial y}\left[\left(Ku_0+\frac{1}{\rho_0u_0}\right)\frac{\partial\psi_1}{\partial y}-\left(Kv_0+\frac{1}{\rho_0v_0}\right)\frac{\partial\psi_1}{\partial x}+A^*\right]$$

$$-\frac{v_0^2}{a_0^2}\frac{\partial^2\psi_0}{\partial y^2}\left[Ku_0\frac{\partial\psi_1}{\partial y}-\left(Kv_0+\frac{2}{\rho_0v_0}\right)\frac{\partial\psi_l}{\partial x}+A^*\right]$$

$$=-\rho_0\omega_1[1+(\gamma-1)M_0^2] \tag{6.22}$$

where

$$K=\frac{(\gamma+1)M_0^2}{(1-M_0^2)\rho_0q_0^2}\ .\ A^*=\frac{2A}{(\gamma-1)(1-M_0^2)}\left(1+\frac{\gamma-1}{2}M_0^2\right)$$

If the basic flow is a uniform flow of velocity U_0=constant, $M_0=\dfrac{U_0}{a}$ =constant, Eq. (6.22) becomes

$$\left(1-\frac{U_0^2}{a_0}\right)\frac{\partial^2\psi_1}{\partial x^2}+\frac{\partial^2\psi_1}{\partial y^2}=-\rho_0\omega_1[1+(\gamma-1)M_0^2] \tag{6.23}$$

an equation which was first derived by Sears.[5]

For the case of uniform basic flow, the second-order differential equation of the perturbed stream function ψ_2 is

$$\left(1-\frac{U_0^2}{a_0^2}\right)\frac{\partial^2\psi_2}{\partial x^2}+\frac{\partial^2\psi_2}{\partial y^2}=\frac{U_0^2}{a_0^2}\frac{\partial^2\psi_1}{\partial x^2}\left[\left(KU_0+\frac{2}{\rho_0U_0}\right)\frac{\partial\psi_1}{\partial y}+A^*\right]$$

$$-\frac{2U_0}{\rho_0a_0^2}\frac{\partial^2\psi_1}{\partial x\partial y}\frac{\partial\psi_1}{\partial x}-\rho_0(-\zeta\omega_1+\omega_2)[1+(\gamma-1)M_0^2]$$

$$-\rho_0\omega_1(\gamma-1)M_0^2\left[\left(\frac{2U_0}{q_0^2}+KU_0\right)\frac{\partial\psi_1}{\partial y}+A^*\right] \tag{6.24}$$

The interesting result is that all the differential equations for ψ_n's are linear. The equation for ψ_{n+1} depends on the functions ψ_0 to ψ_n. Thus we may solve the complete problem by successive approximation. However in many practical problems, the first-order equation will give sufficiently good results.

Occasionally, the second-order equation may be required. These equations for ψ should be solved for the proper boundary conditions; these will be discussed in § 4.

3. Three-dimensional irrotational flow. Linearized first-order equation

The method of iteration also may be used to linearize the equation involving the velocity potential (5.44). Here we shall consider only the first-order perturbation equation with a basic flow of uniform velocity U_0, higher-order equations being obtained by similar processes.

Let the velocity potential be

$$\varphi = U_0(x+\varphi_1) \tag{6.25}$$

where φ_1 is the perturbed velocity potential (assumed to be small). Substituting Eq. (6.25) into Eq. (5.44) and collecting terms of the same order φ_1, we obtain the differential equation for φ_1

$$\frac{\partial^2\varphi_1}{\partial x^2}+\frac{\partial^2\varphi_1}{\partial y^2}+\frac{\partial^2\varphi_1}{\partial z^2}=\frac{1}{a_0^2}\left(\frac{\partial^2\varphi_1}{\partial t^2}+2U_0^2\frac{\partial^2\varphi_1}{\partial x\partial t}+U_0^2\frac{\partial^2\varphi_1}{\partial x^2}\right) \tag{6.26}$$

in which a_0 is the sound speed of the basic flow corresponding to the velocity U_0. In a steady flow, Eq. (6.26) becomes

$$(1-M_0^2)\frac{\partial^2\varphi_1}{\partial x^2}+\frac{\partial^2\varphi_1}{\partial y^2}+\frac{\partial^2\varphi_1}{\partial z^2}=0 \tag{6.27}$$

$M_0=U_0/a_0$ being the Mach number of the basic uniform flow.

Eq. (6.26) is derived for a coordinate system attached to the body moving in the fluid with a velocity U_0. If we refer to a coordinate system at rest with respect to the fluid, the corresponding equation for the perturbation velocity potential may be obtained from Eq. (6.26) by the Galilian transformation:

$$x'=x-U_0t,\ y'=y,\ z'=z,\text{ and } t'=t,\ \varphi_1(x',y',z',t')=\varphi_1(x,y,z,t) \tag{6.28}$$

where the primes refer to the new coordinate system at rest with respect to the fluid.

Combining Eqs. (6.28) and (6.26) we see that

$$\frac{\partial^2\varphi_1}{\partial x'^2}+\frac{\partial^2\varphi_1}{\partial y'^2}+\frac{\partial^2\varphi_1}{\partial z'^2}=\frac{1}{a_0^2}\frac{\partial^2\varphi_1}{\partial t'^2} \tag{6.29}$$

Eq. (6.29) is the well-known wave equation with the velocity of wave propagation a_0. It is interesting to note that for small disturbances, the perturbation velocity potential always satisfies the simple wave equation (6.29). Since this simple wave equation has been extensively studied, we may apply the known results of this latter problem to our problem of compressible flow.[7]

4. Boundary conditions

Before we attempt to solve the linearized equation (6.29) or (6.26) we must say something about the boundary conditions (cf. Chapter I, § 4) which should be consistent with the linearization performed on the basic equation. The boundary conditions are: (1) at infinity, the disturbance velocity produced by the thin body should be finite (or zero in subsonic case); (2) the surface of the body must be a streamline, i.e., the velocity over the surface must be tangential to the surface. For the second boundary condition, approximations are used in the linearized theory.

Let us consider a two-dimensional steady flow. If the solid body has an ordinate for the upper surface y_u such that

$$y_u = \delta Y(x), \quad 0 \leqslant x \leqslant c \tag{6.30}$$

where δ is the thickness ratio of the body which is assumed to be small or of the order of magnitude of ψ_1, $Y(x)$ is of the order of unity and c is the chord of the body. The condition that the surface is a streamline gives

$$\psi\,[x, \delta Y(x)] = 0 \tag{6.31}$$

Now, for simplicity we define a nondimensional stream function such that

$$\frac{\partial \psi}{\partial y} = \frac{\rho}{\rho_0}\,\frac{u}{U_0}, \quad \frac{\partial \psi}{\partial x} = -\frac{\rho}{\rho_0}\,\frac{v}{U_0} \tag{6.32}$$

The stream function ψ with a uniform basic flow U_0 may be written as follows:

$$\psi = \psi_0 + \psi_1 + \psi_2 + \cdots = y + \psi_1 + \cdots \tag{6.33}$$

On the surface of the body, we have

$$y_u + \psi_1 + \cdots = \delta Y(x) + \psi_1\,[x, \delta Y(x)] + \cdots = 0$$

If we develop ψ_1 as a power series of δ, we obtain

$$\delta Y(x) + \psi_1(x, 0) + O(\psi_2) = 0 \tag{6.34}$$

For the first-order perturbation stream function ψ_1, the boundary condition is

$$\psi_1(x, 0) = -\delta Y(x) \tag{6.35}$$

It is interesting to observe that in the linearized theory, the boundary condition for two-dimensional flow must be satisfied only on the x-axis but not on the body. This is not so for the axially symmetrical flow.

If the perturbation velocity potential φ_1 of Eq. (6.25) is used, the boundary condition that the velocity over the surface is equal to the slope of the surface gives

$$\frac{dy_u}{dx}=\frac{v}{u}=\frac{v_1}{U_0+u_1}\cong\frac{v_1}{U_0}=\left(\frac{\partial\varphi_1}{\partial y}\right)_{y=0} \tag{6.36}$$

where the subscript 1 refers to the perturbed value, i.e.,

$$\frac{\partial\varphi_1}{\partial x}=\frac{u_1}{U_0},\ \frac{\partial\varphi_1}{\partial y}=\frac{v_1}{U_0} \tag{6.37}$$

5. Two-dimensional subsonic irrotational steady flow. Prandtl-Glauert rule [4–6]

We consider a uniform stream of velocity U_0 and Mach number $M_0 < 1$ passing over a thin body. The flow is irrotational, two dimensional and steady. The equation for first-order perturbed stream function (6.22) reduces to the following simple form

$$(1-M_0^2)\frac{\partial^2\psi_1}{\partial x^2}+\frac{\partial^2\psi_1}{\partial y^2}=0 \tag{6.38}$$

For incompressible flow, $M_0=0$, Eq. (6.38) becomes

$$\frac{\partial^2\psi_1}{\partial x^2}+\frac{\partial^2\psi_1}{\partial y^2}=0 \tag{6.39}$$

which is the Laplace equation; the method of solution for this is well known.

For subsonic compressible flow with $M_0 < 1$, we may use the transformation:

$$x=\xi,\ y=\frac{1}{\sqrt{1-M_0^2}}\eta \tag{6.40}$$

so that Eq. (6.38) is transformed into the Laplace equation, i.e.,

$$\frac{\partial^2\psi_1}{\partial \xi^2}+\frac{\partial^2\psi_1}{\partial \eta^2}=0 \tag{6.41}$$

If $\psi_1^{(i)}(x, y)$ be the incompressible solution with the boundary condition

$$\psi_1^{(i)}(x, 0)=-\delta Y(x)$$

then

$$\psi_1^{(i)}(\xi, \eta)=\psi_1(x, y) \tag{6.42}$$

is the compressible solution for the same body.

We would like to know the compressibility effect on the flow, e.g., the effect of compressibility on the pressure coefficient which is defined as [cf. Eq. (3.37)]

$$C_p = \frac{p - p}{\frac{\rho_0}{2} U_0^2} = -2\frac{u_1}{U_0} \tag{6.43}$$

From Eq. (6.32), we have

$$\frac{\partial \psi_1}{\partial y} = \frac{u_1}{U_0} + \frac{\Delta\rho}{\rho_0} \tag{6.44}$$

where

$$\rho = \rho_0 + \Delta\rho, \; u = U_0 + u_1$$

From the energy equation,

$$\frac{\gamma p}{(\gamma - 1)\rho} + \frac{u^2}{2} = \frac{\gamma p_0}{(\gamma - 1)\rho_0} + \frac{U_0^2}{2} \tag{6.45}$$

or, for the first-order quantity,

$$\frac{\Delta\rho}{\rho_0} = -M_0^2 \frac{u_1}{U_0} \tag{6.46}$$

Finally,

$$\frac{\partial \psi_1}{\partial y} = (1 - M_0^2)\frac{u_1}{U_0} \tag{6.47}$$

For incompressible fluid, we have on the surface

$$C_{P_i} = -2\left(\frac{\partial \psi_1^{(i)}}{\partial y}\right)_{y=0} = -2\left(\frac{\partial \psi_1^{(i)}}{\partial \eta}\right)_{\eta=0} \tag{6.48}$$

and for compressible fluid, we have

$$C_p = -\frac{2}{(1 - M_0^2)}\left(\frac{\partial \psi_1}{\partial y}\right)_{y=0} = -\frac{2}{\sqrt{1 - M_0^2}}\left(\frac{\partial \psi_1^{(i)}}{\partial \eta}\right)_{\eta=0} \tag{6.49}$$

Comparing Eq. (6.48) with (6.49), we obtain the well known Prandtl-Glauert rule:

$$C_p = \frac{C_{Pi}}{\sqrt{1 - M_0^2}} \tag{6.50}$$

which states that the local pressure coefficient on a body in a subsonic flow of Mach number M_0 is equal to $(1/\sqrt{1 - M_0^2})$ times the corresponding pressure coefficient for the same body in an incompressible flow. This rule is, of course, valid only for a thin body. It holds for $M_0 < 0.5$. For higher subsonic velocities, the correction formulas discussed in Chapter VIII will give better results. For a more accurate analysis, it may be shown that the compressibility effect depends not only on M_0 but also on the shape of the body. The effect of the shape will be more important when M_0 is larger. We shall discuss this point further in Chapter VIII. Another fact which should be mentioned is that the simple formula of (6.50) holds only for a thin body in a stream of infinite extent. For a complicated boundary condition, such as an airfoil in a wind tunnel, no such simple formula exists.[7]

The region of influence of a body in a compressible flow is larger than that of the same body in an incompressible flow, since for the same disturbed velocity, the y-distance in compressible fluid y_c is equal to $(1/\sqrt{1-M_0^2})$ times that of incompressible flow y_i.

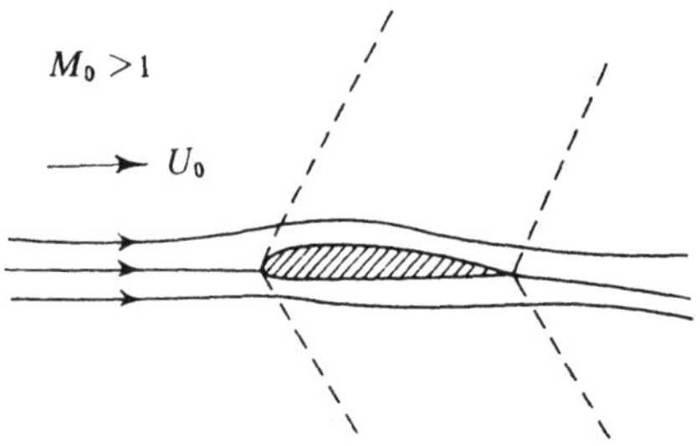

Fig. 6.1 Supersonic flow over an airfoil

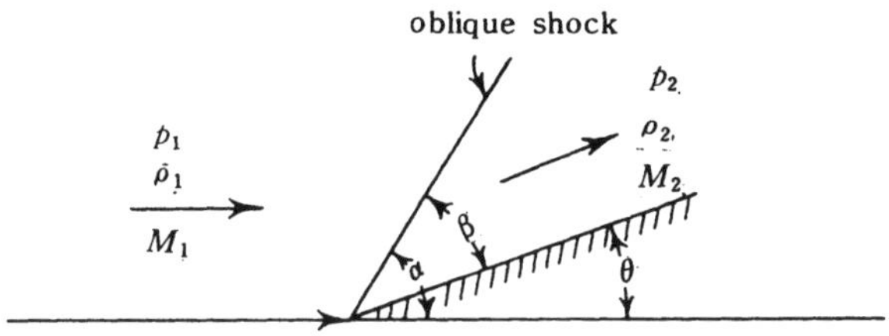

Fig. 6.2 Supersonic flow over a wedge

6. Two-dimensional supersonic irrotational steady flow. Ackeret formula[8]

Now we consider a supersonic uniform stream of velocity U_0 and Mach number $M_0 > 1$ passing over a thin body as shown in Fig. 6.1. If the flow is compressed through deflection by the body, a shock wave will occur. It is known that across the shock the entropy of the flow increases. Since the body is curved, the flow deflection is not a constant and the shock is not straight but curved. Behind the curved shock the flow will be rotational with a nonuniform distribution of entropy. Let us examine the entropy change across an oblique shock with a flow deflection angle θ (Fig. 6.2). From Eq. (4.16)

$$\frac{\Delta S}{c_V} = \log \frac{p_2}{p_1} + \gamma \log \frac{\rho_1}{\rho_2} \tag{6.51}$$

where p_2/p_1 is given by Eq. (4.11) and ρ_1/ρ_2, by Eq. (4.12), the oblique shock relations.

Since we assume that the flow deflection θ is small and is of the same order of magnitude as the thickness ratio δ of the body, the change in density of the fluid across the shock is also of the same order of magnitude as δ. We may write

$$\frac{\rho_1}{\rho_2} = 1 - \zeta \tag{6.52}$$

and then the oblique shock relation

$$\frac{p_2}{p_1} = \frac{1 + \dfrac{\gamma - 1}{2}\zeta}{1 - \dfrac{\gamma - 1}{2}\zeta} \tag{6.53}$$

Substituting Eqs. (6.52) and (6.53) into Eq. (6.51), we have

$$\frac{\Delta S}{c_V} = \frac{\gamma(\gamma^2 - 1)}{2}\zeta^3 + O(\zeta^4) \tag{6.54}$$

Hence ΔS is of the third order of ζ. Since ΔS is of the same order of magnitude of the vorticity ω, the vorticity ω is also of the third order in ζ. In the present problem, ζ is of the same order of magnitude as δ or ψ_1. Hence in the expansion of the vorticity ω of Eq. (6.18), $\omega_1 = \omega_2 = 0$, and only ω_3 and higher-order terms are different from zero. Hence for thin body with weak shock in supersonic flow, the vorticity is negligible in the first- and second-order theories. In these cases, the flow may be considered as irrotational. However, if the shock is not weak, i.e., θ is not of the same order of magnitude as ψ_1, the change of entropy may be of the same order of magnitude as ψ_1. In this case the rotationality of the flow must be considered. The subject of rotationality will be discussed in Chapter XIV.[2]

For a two-dimensional supersonic flow over a thin body at a small angle of attack, the fundamental equation of perturbed stream function ψ_1 is still Eq. (6.38) but with $M_0 > 1$. Eq. (6.38) is now a simple wave equation [cf. Eq. (3.43)] whose solution may be written as follows:

$$\psi_1(x, y) = f(x - my) + g(x + my) \tag{6.55}$$

where $m = \sqrt{M_0^2 - 1}$. Since we assume that no disturbance comes from infinity, $g(x+my)=0$. Furthermore from the boundary condition (6.35), we have

$$\psi_1(x,0)=f(x)=-\delta Y(x) \tag{6.56}$$

Hence the perturbed stream function is

$$\psi_1(x,y)=-\delta Y(x-my) \tag{6.57}$$

The pressure coefficient on the surface of the body is then

$$C_P=\frac{2}{M_0^2-1}\left(\frac{\partial\psi_1}{\partial y}\right)_{y=0}=\frac{2\delta}{\sqrt{M_0^2-1}}Y'(x) \tag{6.58}$$

where $Y'(x)=dY/dx=$slope of the surface of the body. Eq. (6.58) is the well-known Ackeret formula for supersonic flow, which indicates that the pressure coefficient in supersonic flow depends on the local slope of the surface.

7. Flow past an infinite wave-shaped wall

We have discussed the effect of compressibility on the pressure coefficient on a thin body. Now we shall examine the whole flow pattern of some typical two-dimensional steady compressible flows. We consider a uniform stream

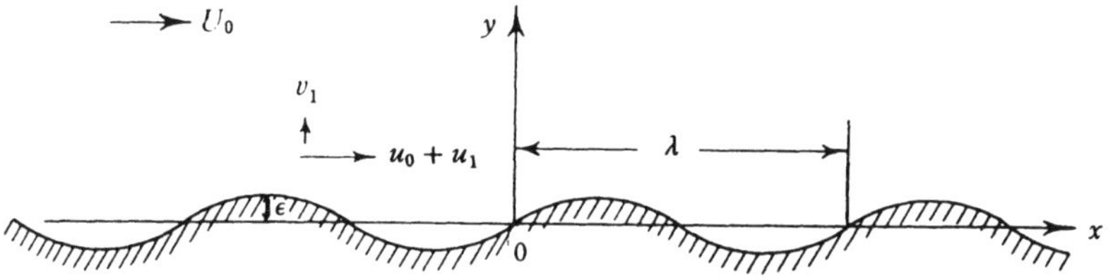

Fig. 6.3 Compressible flow over an infinite wave-shaped surface

of velocity U_0 and Mach number M_0 passing over an infinite wave-shaped wall whose ordinate is given by the following formula

$$y_w=\varepsilon\sin\alpha x \tag{6.59}$$

where $\alpha=2\pi/\lambda$ and λ is the wave length of the wall. We assume that $\varepsilon\ll\lambda$ and $\alpha\varepsilon\ll 1$, so that the first-order perturbation equation holds in this case (Fig. 6.3).

On the boundary of the wall, the inclination of the flow must be equal to the slope of the surface and then we have

$$\left(\frac{v_1}{U_0+u_1}\right)_{\text{wall}}\cong\left(\frac{v_1}{U_0}\right)_{y=0}=\left(\frac{dy_w}{dx}\right)=\varepsilon\alpha\cos\alpha x \tag{6.60}$$

Since for this problem the flow is irrotational, we may use the velocity potential equation (6.27) to solve this problem. We have

$$(1-M_0^2)\frac{\partial^2\varphi_1}{\partial x^2}+\frac{\partial^2\varphi_1}{\partial y^2}=0 \qquad (6.61)$$

with the boundary condition

$$\left(\frac{\partial\varphi_1}{\partial y}\right)_{y=0}=\varepsilon\alpha\cos\alpha x \qquad (6.62)$$

$$\varphi_1(x,\infty)\neq\infty$$

where the perturbed velocity potential φ_1 is defined by Eq. (6.25). The boundary condition at $y=\infty$ depends on whether the uniform flow is supersonic $M_0>1$ or subsonic $M_0<1$. We shall discuss these two cases separately.

(a) *Subsonic uniform flow.* $M_0<1$. If we write $1-M_0^2=\beta^2$, Eq. (6.61) becomes

$$\frac{\partial^2\varphi_1}{\partial x^2}+\frac{1}{\beta^2}\frac{\partial^2\varphi_1}{\partial y^2}=0 \qquad (6.63)$$

We may use the method of separation of variables to solve Eq. (6.63) by writing

$$\varphi_1=\Phi_1(x)\Phi_2(y) \qquad (6.64)$$

Substituting Eq. (6.64) into Eq. (6.63). we have

$$\frac{1}{\Phi_1}\frac{d^2\Phi_1}{dx^2}=-k^2=-\frac{1}{\beta^2\Phi_2}\frac{d^2\Phi_2}{dy^2} \qquad (6.65)$$

where k is an arbitrary constant, which will be determined later from the boundary condition. From Eq. (6.65),

$$\Phi_1=A_1\cos kx+B_1\sin kx$$

and

$$\Phi_2=Ae^{-\beta ky}+B_2e^{+\beta ky}$$

or

$$\Phi_1=(A_2e^{-\beta ky}+B_2e^{+\beta ky})(A_1\cos kx+B_1\sin kx) \qquad (6.66)$$

where A_1, A_2, B_1 and B_2 are arbitrary constants to be determined from the boundary conditions. At $y=\infty$, $\varphi_1\neq\infty$; hence $B_2=0$. At $y=0$,

$$\left(\frac{\partial\varphi_1}{\partial y}\right)_{y=0}=-A_2\beta k(A_1\cos kx+B_1\sin kx)=\varepsilon\alpha\cos\alpha x \qquad (6.67)$$

Hence $B_1=0$, $k=\alpha$, and $-A_1A_2\beta=\varepsilon$. The perturbed velocity potential φ_1 is then

$$\varphi_1 = -\frac{\varepsilon}{\sqrt{1-M_0^2}} e^{-\alpha\sqrt{1-M_0^2}\,y} \cos \alpha x \tag{6.68}$$

and the perturbed velocity components are

$$\frac{u_1}{U_0} = \frac{\partial \varphi_1}{\partial x} = \frac{\varepsilon\alpha}{\sqrt{1-M_0^2}} e^{-\alpha\sqrt{1-M_0^2}\,y} \sin \alpha x$$

and (6.69)

$$\frac{v_1}{U_0} = \frac{\partial \varphi_1}{\partial y} = \varepsilon\alpha\, e^{-\alpha\sqrt{1-M_0^2}\,y} \cos \alpha x$$

The maximum perturbed velocity ratio is

$$\left(\frac{u_1}{U_0}\right)_{\max} = \frac{\varepsilon\alpha}{\sqrt{1-M_0^2}} \ll 1$$

Hence (6.70)

$$\varepsilon\alpha \ll \sqrt{1-M_0^2}$$

Thus, for our solution to remain valid as the Mach number M_0 increases toward unity, the amplitude of the wave ε for a given wave length must decrease toward zero.

(b) *Supersonic uniform flow.* $M_0 > 1$. If we write $M_0^2 - 1 = m^2$, Eq. (6.61) becomes

$$\frac{\partial^2 \varphi_1}{\partial x^2} - \frac{1}{m^2}\frac{\partial^2 \varphi_1}{\partial y^2} = 0 \tag{6.71}$$

which is the simple wave equation whose solution is

$$\varphi_1 = f_1(x - my) + f_2(x + my) \tag{6.72}$$

If we restrict ourselves to the condition that no disturbance comes from infinity, $f_2(x + my) = 0$. Then from the boundary condition (6.62)

$$\left(\frac{\partial \varphi_1}{\partial y}\right)_{y=0} = -mf_1'(x) = \varepsilon\alpha \cos \alpha x \tag{6.73}$$

where $f_1' = df_1/dx$. By integration

$$f_1(x) = -\frac{\varepsilon}{m} \sin \alpha x \tag{6.74}$$

Hence the perturbed velocity potential φ_1 in a supersonic stream is

$$\varphi_1 = f_1(x - my) = -\frac{\varepsilon}{m} \sin \alpha(x - my) \tag{6.75}$$

and the velocity components are

$$\frac{u_1}{U_0} = \frac{\partial \varphi_1}{\partial x} = -\frac{\varepsilon\alpha}{\sqrt{M_0^2-1}} \cos \alpha\,(x-my)$$
$$\frac{v_1}{U_0} = \frac{\partial \varphi_1}{\partial y} = \varepsilon\alpha \cos \alpha\,(x-my) \tag{6.76}$$

Flow patterns for various Mach number M_0 are shown in Fig. 6.4. For incompressible flow, the magnitude of the disturbance decreases rapidly from the surface. As the Mach number increases, the rate of decrease of the magnitude of the disturbance reduces. At $M_0=1$, the disturbance will propagate laterally without decrease of magnitude.

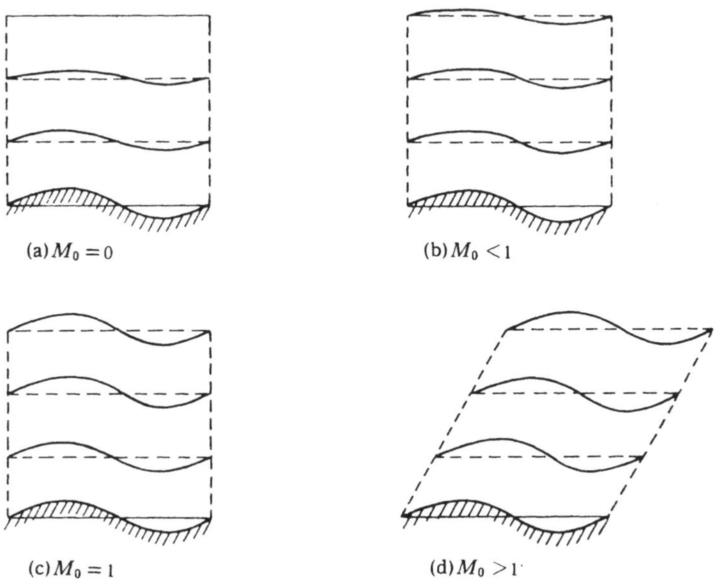

Fig. 6.4 Flow patterns over an infinite wave-shaped surface at various free stream Mach numbers M_0

For supersonic flow, the disturbance will propagate without distortion along the Mach lines.

On the surface of the wall, the pressure coefficients are

$$C_{P\,\text{sub.}} = -\frac{2\varepsilon\alpha}{\sqrt{1-M_0^2}} \sin \alpha x\,,\ \text{for subsonic flow}$$

$$C_{P\,\text{sup.}} = \frac{2\varepsilon\alpha}{\sqrt{M_0^2-1}} \cos \alpha x\,,\ \text{for supersonic flow}$$

This shows that subsonic flow is of a different nature from that of supersonic flow. For $M_0 < 1$, C_p is proportional to sin αx ; while for $M_0 > 1$, C_p is proportional to cosαx. The variation of the maximum pressure coefficient $C_{P\,max}$ with free stream Mach number M_0 is sketched in Fig. 6.5. At $M_0 = 1$, $C_{Pmax} = \infty$ which means that the linearized theory does not hold at $M_0 = 1$. Actually the pressure coefficient is finite at sonic speed. At $M_0 = \infty$, the linearized theory gives zero pressure coefficient which also demonstrates that the linearized theory fails for hypersonic flow. These points will be discussed further in Chapter X.

8. Subsonic two-dimensional steady flow over wall of arbitrary shape [9, 10]. Fourier integral method

The method of solution of Eq. (6.72) for supersonic flow is applicable to walls of arbitrary shape, although the method of solution of Eq. (6.66) for subsonic flow is not directly applicable to a wall of arbitrary shape. However, this solution may be generalized so that it furnishes the solution for a subsonic flow over a wall of arbitrary shape. This generalization is accomplished using the technique of Fourier integrals.

Before we try to discuss the solution for an arbitrary wall, we shall briefly review the Fourier integral formula. An arbitrary function $f(x)$

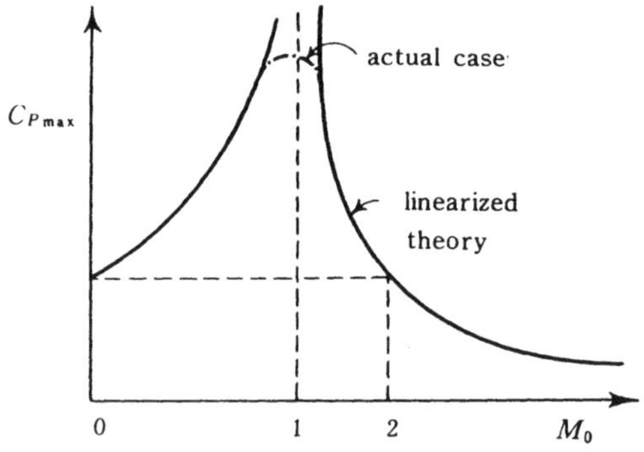

Fig. 6.5 Variation of maximum pressure coefficient $C_{P\,max}$ with free stream Mach number M_0

defined in an interval $-c \leqslant x \leqslant c$ may be represented by the Fourier series:

$$f(x) = \frac{a_0}{2} + \sum_{n=1}^{\infty}\left(a_n \cos\frac{n\pi x}{c} + b_n \sin\frac{n\pi x}{c}\right) \tag{6.77}$$

where

$$a_n = \frac{1}{c}\int_{-c}^{c} f(\xi)\cos\frac{n\pi\xi}{c}\,d\xi,\quad b_n = \frac{1}{c}\int_{-c}^{c} f(\xi)\sin\frac{n\pi\xi}{c}\,d\xi \tag{6.78}$$

provided that certain conditions regarding the continuity, differentiability, etc., are satisfied.[11]

The Eqs. (6.77) and (6.78) may be combined into the following form:

$$f(x) = \frac{1}{2c} \int_{-c}^{c} f(\xi)d\xi \left[1 + 2\sum_{n=1}^{\infty} \cos \frac{n\pi}{c} (\xi - x) \right]$$

$$= \frac{1}{2\pi} \int_{-c}^{c} f(\xi)d\xi \sum_{n=-\infty}^{\infty} \frac{\pi}{c} \cos \frac{n\pi}{c} (\xi - x) \tag{6.79}$$

Now if the region in which $f(x)$ is defined is extended to infinity, i.e., $c \to \infty$, we should take the limiting form of Eq. (6.79). Let $\pi/c = \Delta\lambda$, then

$$\lim_{\Delta\lambda \to 0} \sum_{n=-\infty}^{\infty} \Delta\lambda \varphi(n\Delta\lambda) = \int_{-\infty}^{\infty} \varphi(\lambda)d\lambda \tag{6.80}$$

Thus we have

$$\lim_{c \to \infty} \sum_{n=-\infty}^{\infty} \frac{\pi}{c} \cos \frac{n\pi}{c} (\xi - x) = \int_{-\infty}^{\infty} \cos \lambda (\xi - x)d\lambda \tag{6.80a}$$

In this limiting case, Eq. (6.79) becomes

$$f(x) = \frac{1}{2\pi} \int_{-\infty}^{\infty} f(\xi)d\xi \int_{-\infty}^{\infty} \cos \lambda (\xi - x)d\lambda \tag{6.81}$$

Since $\cos \lambda(\xi - x)$ is an even function of λ, Eq. (6.81) may be written as follows:

$$f(x) = \frac{1}{\pi} \int_{-\infty}^{\infty} f(\xi)d\xi \int_{0}^{\infty} \cos \lambda (\xi - x)d\lambda \tag{6.82}$$

If the function $f(x)$ belongs to the Lebesque class L[12], we may inter change the order of integral of Eq. (6.82) and have

$$f(x) = \frac{1}{\pi} \int_{0}^{\infty} d\lambda \int_{-\infty}^{\infty} f(\xi) \cos \lambda (\xi - x)d\xi \tag{6.83}$$

Eq. (6.83) is the Fourier integral formula, which we shall use in our analysis of subsonic flow over a wall of arbitrary shape.

For a wall of arbitrary shape whose ordinate is

$$y_w = F(x) \tag{6.84}$$

the boundary condition at $y = 0$ is

$$\left(\frac{\partial \varphi_1}{\partial \eta}\right)_{\eta=0} = \frac{U_0}{\beta} F'(x) = \frac{U_0}{\beta\pi} \int_0^{\infty} d\lambda \int_{-\infty}^{\infty} F'(\xi) \cos \lambda(\xi - x) d\xi \quad (6.85)$$

where $\eta = \beta y = y\sqrt{1-M_0^2}$ .Eq. (5.83) is used for $F'(x) \left[F'(x) = \frac{dF}{dx} \right]$.

We want an expression for φ_1 to satisfy Eq. (6.85). Since the differential equation (6.63) of φ_1 is linear, the method of superposition may be used to find a more complicated solution of Eq. (6.63). The elementary solution of Eq. (6.63) is given in Eq. (6.66) with $B_2=0$ because of the boundary condition at infinity. This elementary solution may be written as

$$\varphi_{1i} = A_i e^{-k_i \eta} \cos k_i(\xi_i - x) \quad (6.86)$$

where A_i, k_i, and ξ_i are arbitrary constants. If we put $\xi_i = \xi$, and $A_i = f_1(\lambda) f_2(\xi)\, d\lambda d\xi$ and sum over an infinite number of these elementary solutions for $0 \leqslant \lambda \leqslant \infty$, $-\infty \leqslant \xi \leqslant \infty$, we have in the limit the following expression for the perturbed velocity potential:

$$\varphi_1(x, \eta) = \int_0^{\infty} f_1(\lambda) d\lambda \int_{-\infty}^{\infty} e^{-\lambda\eta} f_2(\xi) \cos \lambda(\xi - x) d\xi \quad (6.87)$$

This is the expression which will be used for subsonic flow over a wall of arbitrary shape. The arbitrary functions f_1 and f_2 may be determined from the boundary conditions. At $y=\eta=0$, we have from Eq. (6.87)

$$\left(\frac{\partial \varphi_1}{\partial \eta}\right)_{\eta=0} = -\int_0^{\infty} \lambda f_1(\lambda) d\lambda \int_{-\infty}^{\infty} f_2(\xi) \cos \lambda(\xi - x) d\xi \quad (6.88)$$

Comparing Eq. (6.88) with Eq. (6.85), we find that if

$$f_1(\lambda) = \frac{1}{\lambda}, f_2(\xi) = -\frac{U_0}{\beta\pi} F'(\xi) \quad (6.89)$$

the velocity potential Eq. (6.87) satisfies the boundary condition. Hence for a wall of arbitrary shape $F(x)$, the perturbed velocity potential φ_1 is

$$\varphi_1(x, \eta) = -\frac{U_0}{\beta\pi} \int_0^{\infty} \frac{d\lambda}{\lambda} \int_{-\infty}^{\infty} e^{-\lambda\eta} F'(\xi) \cos \lambda(\xi - x) d\lambda \quad (6.90)$$

The pressure coefficient C_p is

$$C_p = \frac{2}{\beta\pi} \int_{-\infty}^{\infty} \frac{F'(\xi)(\xi - x)}{(\xi - x)^2 + \eta^2} d\xi \quad (6.91)$$

On the surface of the wall, we have

$$C_p = \frac{2}{\pi\sqrt{1-M_0^2}} \int_{-\infty}^{\infty} \frac{F'(\xi)}{\xi - x} d\xi \quad (6.92)$$

9. Three-dimensional irrotational flow. Linearized theory

We shall discuss three-dimensional potential flow based on linearized theory in detail in Chapter XII. Here we shall make only a few general remarks. The fundamental Eq. (6.26) or Eq. (6.29) is linear; the method of superposition is always applicable for the linearized theory. If the fundamental solution of the simple wave equation (6.29) is known, as for the solution for source or doublet, the general solution of the problem may be obtained by the integration of the source and doublet distributions for given initial and boundary conditions. This procedure is applicable to both supersonic and subsonic flow whether steady or not.

For steady irrotational flow, the simpler equation (6.27) may be used. If the flow is subsonic, Eq. (6.27) may be reduced to the Laplace equation by the Prandtl-Glauert transformation. The Prandtl-Glauert rule may be extended to the three-dimensional case, but in this extension special attention must be paid to the boundary conditions. This will be discussed in Chapter XII. For supersonic flow, Eq. (6.27) may be reduced to the form of a two-dimensional wave equation, i.e.,

$$\frac{\partial^2 \varphi_1}{\partial x^2} = \frac{1}{m^2}\left(\frac{\partial^2 \varphi_1}{\partial y^2} + \frac{\partial^2 \varphi_1}{\partial z^2}\right) \tag{6.93}$$

Since the solution of a two-dimensional wave equation is more complicated than both that of a one-dimensional wave equation and that of a three-dimensional wave equation, the solution for three-dimensional steady supersonic flow (corresponding to the two-dimensional wave equation) is much more complicated than that of the two-dimensional steady supersonic flow (corresponding to the one-dimensional wave equation).

For the one-dimensional wave equation

$$\frac{\partial^2 \varphi_1}{\partial t^2} = a_0^2 \frac{\partial^2 \varphi_1}{\partial x^2} \tag{6.94}$$

the general solution is simply

$$\varphi_1 = f\,(x - a_0 t) + g\,(x + a_0 t) \tag{6.95}$$

where f and g are arbitrary functions.

For the two-dimensional wave equation, there is no simple general solution such as that of Eq. (6.95). If we consider the case for which the solution is a function of time t and radial distance $r=\sqrt{x^2+y^2}$ only, the differential equation becomes

$$\frac{\partial^2 \varphi_1}{\partial t^2} = a_0^2\left[\frac{\partial^2 \varphi_1}{\partial r^2} + \frac{1}{r}\frac{\partial \varphi_1}{\partial r}\right] \tag{6.96}$$

and the general solution of Eq. (6.96) may be written as follows:

$$\varphi_1 = \frac{1}{a_0} \int_0^\infty f\left(t - \frac{r}{a_0} \cosh q\right) dq \tag{6.97}$$

where f is an arbitrary function.

For the three-dimensional wave equation, if we consider only solutions which are functions of time t and radial distance $R=\sqrt{x^2+y^2+z^2}$, the differential equation becomes

$$\frac{\partial^2\varphi_1}{\partial t^2} = a_0^2\left(\frac{\partial^2\varphi_1}{\partial R^2} + \frac{2}{R}\frac{\partial\varphi_1}{\partial R}\right) = \frac{a_0^2}{R^2}\frac{\partial}{\partial R}\left(R^2\frac{\partial\varphi_1}{\partial R}\right) \tag{6.98}$$

and the general solution is

$$\varphi_1 = \frac{f(R-a_0t)}{R} + \frac{g(R+a_0t)}{R} \tag{6.99}$$

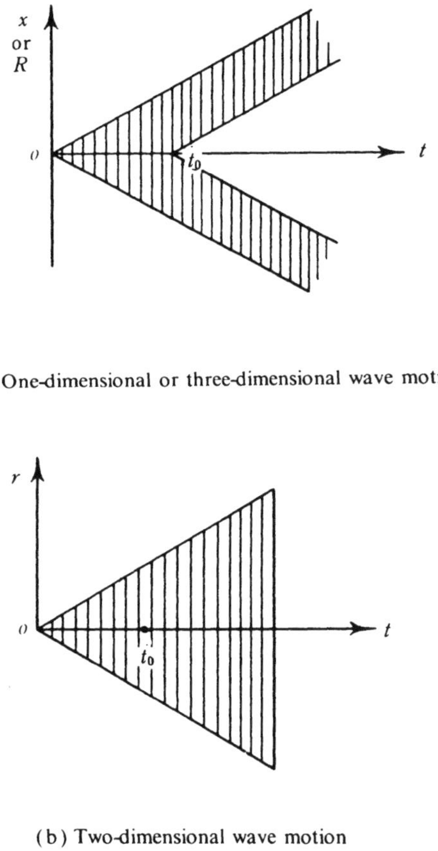
x
or
R
0
t_0
t
(a) One-dimensional or three-dimensional wave motion
r
0
t_0
t
(b) Two-dimensional wave motion

Fig. 6.6 x-t diagram of wave motion

where f and g are arbitrary functions. Thus in the three-dimensional case we again obtain simple solutions.

The meaning of these solutions may be seen from the following examples: in the one-dimensional case, if we imagine a disturbance at $x=0$, which lasts from $t=0$ to $t=t_0$ we have the x-t diagram of Fig. 6.6a, in which the shaded area represents the disturbed region.

The regions both in front of and behind the chevron-shaped regions are regions of no disturbance. In the three-dimensional case, the picture is quite similar except that the intensity of the chevron belt decreases as R increases.

In the two-dimensional case, the corresponding picture is that of Fig. 6.6b. In this case only the region ahead of the wave is undisturbed. There is no undisturbed region behind the wave.

10. Higher approximations[11-16]

The linearized first-order theory discussed in previous sections holds good for very thin bodies, i.e., for bodies with very small thickness to chord ratio. If the thickness ratio of a body is not very small, higher-order approximations are needed for good results.

For subsonic flows, Hantzsche[4] and Kaplan[11] studied the case of elliptic cylinders up to the third approximation, i.e., φ_3. We shall not discuss the details of the solution as it is quite tedious and long. If the thickness ratio of the body is very large, it is not advisable to use the linearized theory discussed in this chapter, and some other method should be applied. In Chapter VII we shall discuss some which are particularly suitable for subsonic flows.

For supersonic flows, there is a problem where the first-order linearized theory fails even for a very thin body. In order to obtain the correct wave pattern, some of the second-order terms should be retained in the first-order equation which is then effectively of the first order. Belonging to this class are the similarity laws of transonic and hypersonic flows. In transonic and hypersonic flows (Chapter X), the second order terms will affect the pressure distribution on the thin body. For ordinary supersonic flow, the second-order terms will not affect the pressure distribution on the surface of the thin body, this value being given by ordinary linearized first-order theory. However, the second-order terms in the differential equations are needed to describe the firstorder wave structure. The reason is due to the fact that at large distance from the body, the error in the first-order linearized theory accumulates so that the approximated characteristics curves will be far away from the actual characteristics curves of the exact solution.

In supersonic flow, we have a powerful method, the method of characteristics, which gives the exact flow pattern for a body regardless of its thickness ratio. The method of characteristics will be discussed in Chapter XI.

11. Panel methods for Laplace equation[17]

The subsonic and supersonic small perturbation with the basic flow as a uniform stream is governed by a linear equation for the velocity potential. Linear partial differential equation can be solved by method of superposition. Bodies around which uniform streams flow are represented by distributions of source, doublet, vortex and the like singularities whose strengths are adjusted to satisfy the flow-tangency boundary conditions required by the geometry of the particular configuration. The superposition of the fundamental solutions yields a linear integral equation of the unknown strengths of singularities. Panel methods are a powerful set of techniques for solving the linear integral equations. In contrast to the finite difference method, the panel methods do not solve the flow field throughout the domain. Instead, they solve the singularity strength distribution on or within the body boundary. Panel methods are also called finite boundary element methods.

The basic idea underlying the use of panel methods will be discussed in this section. For simplicity we consider the two-dimensional flow of an incompressible, inviscid, irrotational fluid around body C with free stream velocity U (Fig. 6.7). The governing equation for the

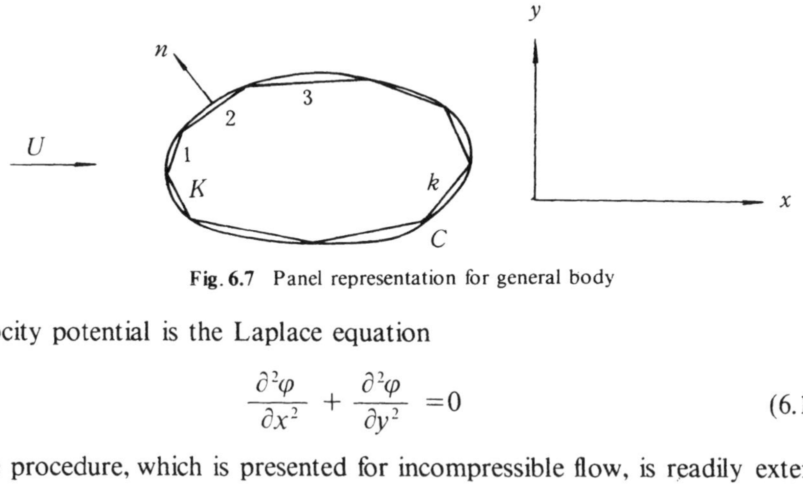

Fig. 6.7 Panel representation for general body

velocity potential is the Laplace equation

$$\frac{\partial^2 \varphi}{\partial x^2} + \frac{\partial^2 \varphi}{\partial y^2} = 0 \tag{6.100}$$

The procedure, which is presented for incompressible flow, is readily extended to the linear subsonic regime by means of the Prandtl-Glauert transformation (6.40). Let us distribute source singularities along the boundary C. The strength of the source distribution $f(\xi, \eta)$ is defined as the volume flow issued from a unit length at point (ξ, η) of C per unit time. Superposing the velocity potentials of the distributed sources, we obtain

$$\varphi(x, y) = \frac{1}{2\pi} \oint_C f(\xi, \eta) \ln r \, dl \tag{6.101}$$

where

$$r=[(x-\xi)+(y-\eta)]^{1/2}$$

dl is an element length of the boundary C.

The general solution (6.101) satisfies the boundary conditions at infinity, i.e. the perturbation velocity vanishes at infinity. The boundary condition of the body C is

$$\frac{\partial \varphi}{\partial n} = -U \cos(U, n) \tag{6.102}$$

where n is normal to C.

Substituting Eq. (6.101) into Eq. (6.102), we obtain an integral equation for f

$$\frac{1}{2\pi} \oint_C f(\xi, \eta) \frac{\partial}{\partial n} (\ln r) dl = -U \cos(U, n) \tag{6.103}$$

Now we divide the body boundary C into a large number of small panels, $k=1, 2, \cdots K$ (Fig. 6.7). Each panel defines an area over which the source strength is held constant. Some advanced panel methods assume other distributions, e.g. polynomial, and the corresponding numerical computation becomes more complex. Thus the line integral in Eq. (6.103) can be carried out along each panel. Prior to the integration, we select a control point on each panel, at which the boundary condition (6.103) is applied. In this case, the control point may be selected at the midpoint of each panel, denoted by j. In calculating the integral in Eq. (6.103), we ought to exercise caution as k approaches j, that is, when $r \to 0$. This particular term on the left-hand side of Eq. (6.103) expresses the normal velocity component induced at the jth control point due to the source distribution on the j th panel. By use of the definition of source strength, the normal velocity component is $f_j/2$. Therefore, the integral equation may be approximated by the algebraic equation system

$$\frac{f_j}{2} + \frac{1}{2\pi} \sum_{\substack{k=1 \\ (k \neq j)}}^{K} f_k \int_{C_k} \frac{\partial}{\partial n_i} (\ln r_{jk}) dl_k = -U \cos(U, n_j) \tag{6.104}$$

$$j=1, 2, \cdots, K$$

where

$$r_{jk} = [(x_j - x_k)^2 + (y_j - y_k)^2]^{1/2} \tag{6.105}$$

n_j is normal to the jth panel, dl_k is an element length of the kth panel C_k.

In applying the panel methods, the most difficult part is the evaluation of the integral in Eq. (6.104). In general, it is convenient to view the integral as an aerodynamic influence coefficient and write the system of algebraic equations in the compact form

$$[a_{jk}]\{f_k\} = -\{U \cos(U, n_j)\} \tag{6.106}$$

where $[a_{jk}]$ is the aerodynamic influence coefficient matrix whose element a_{jk} is the normal velocity component at control point j induced by the source panel k of unit strength distribution.

$$a_{jk}=\begin{cases} \dfrac{1}{2\pi}\displaystyle\int_{C_k}\dfrac{\partial}{\partial n_j}(\ln r_{jk})dl_k\ , & k\neq j \\ 1/2\ , & k=j \end{cases} \tag{6.107}$$

and { } denotes column vector.

An example demonstrating the procedure for generating the required algebraic equation is in order. As farasthe incompressible flow is concerned, the above method may be used for arbitrary bodies.

Suppose we wish to solve for the pressure distribution on a circular cylinder of unit radius in incompressible flow using the method of source panels. The cylinder is to be represented by eight panels of equal length and the configuration is shown in Fig. 6.8.

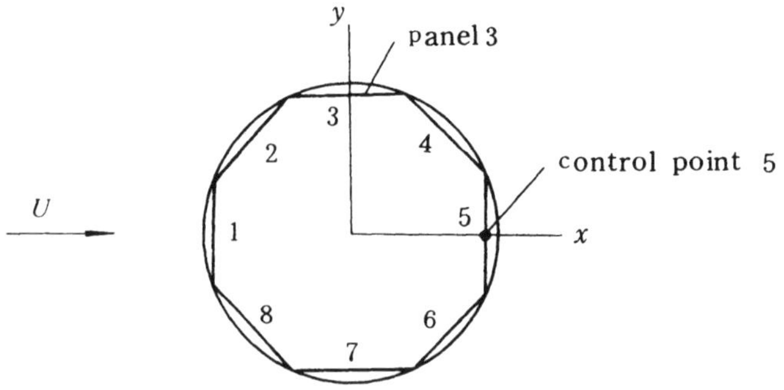

Fig. 6.8 Panel representation of cylinder

To demonstrate the application of Eq. (6.107) Eq. we elect to compute a_{53}. which represents the normal velocity component at the control point of panel 5 due to a unit source strength on panel 3. The integrand function of Eq. (6.107) is most easily developed by using the vector dot product and may be written

$$\frac{\partial}{\partial n_j}(\ln r_{jk})=\nabla_j(\ln r_{jk})\cdot \boldsymbol{n}_j \tag{6.108}$$

where $\boldsymbol{n}_j$ is unit vector normal to panel j.
By use of Eq. (6.105), we have

$$\nabla_j(\ln r_{jk})=\frac{(x_j-x_k)\boldsymbol{i}+(y_j-y_k)\boldsymbol{j}}{(x_j-x_k)^2+(y_j-y_k)^2} \tag{6.109}$$

For the case of $j=5$ and $k=3$,

$x_5=0.9239,\qquad y_5=0,\qquad \boldsymbol{n}_5=\boldsymbol{i},\qquad y_3=0.9239,$

x_3 is an integration variable between the limits ± 0.3827.
This reduces the integral required to the form

$$\int_{C_3} \frac{\partial}{\partial n_5}(\ln r_{53})\, dl_3 = \int_{-0.3827}^{0.3827} \frac{0.9239 - x_3}{x_3^2 - 1.848x_3 + 1.707}\, dx_3 = 0.4018$$

A comparison of the eight panel solution with the analytically derived pressure coefficient is shown in Fig. 6.9. Clearly the panel scheme provides

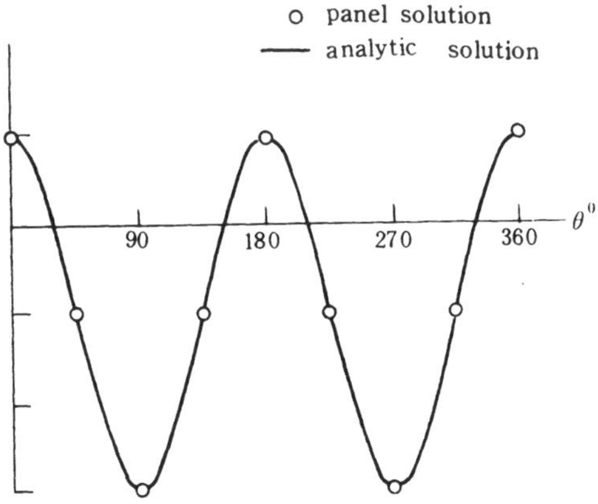

Fig. 6.9 Pressure coefficient for circular cylinder

a very accurate numerical solution for this case.

The matrix $[a_{jk}]$ is symmetric, i.e. $a_{jk} = a_{kj}$ and the solution for the f_j 's must be such that

$$\sum_{j=1}^{K} f_j = 0$$

This requirement is an obvious result of the requirement that we have a closed body.

We have used the method of source panels in our example to demonstrate the techniques of applying the scheme. We could use doublets or dipoles to construct bodies as well as vortex panels. Clearly we must include circulation if we are concerned with lifting bodies.

Panel methods represent a powerful scheme for solving flow problems governed by linear partial differential equations. They have received extensive use in industry. In Chapter XII, we shall discuss panel methods for three-dimensional subsonic and supersonic small perturbation flows.

12. Problems

1. Derive the second-order approximation of velocity potential for the the-

ory of small perturbations based on a uniform flow. Discuss the boundary conditions and the method of solution (cf. Ref. 6).

2. Find the relation between the perturbed velocity components and the perturbed stream function and perturbed entropy for the first-order linearized theory of rotational flow.

3. In flow past a very thin airfoil the velocity at any point differs slightly from a uniform velocity U_0 in the direction of the axis of x. The equation for for the perturbed velocity distribution is given by Eq. (6.27). Find the line on a two-dimensional airfoil for subsonic flow on the supposition that the circulation around the airfoil is C.

4. Compute the lift, drag, and moment coefficients for an airfoil of rhomboidal cross section in supersonic two-dimensional steady flow at an angle of attack α.

5. Compute the lift, drag, and moment coefficients in two-dimension supersonic steady flow for an airfoil of sickle form (i.e. consisting of two arcs of a circle).

6. A jet of uniform velocity U_0 is discharged in to a medium at rest. If the density of the jet at the exit of the nozzle is slightly different from that for the surrounding medium at rest, derive the linearized equation for the density in the jet. Assume that the flow is two-dimensional and steady. Discuss the density distribution in the jet if (a) the flow is subsonic $M_0 = U_0/a_0 < 1$ and (b) supersonic $M_0 = U_0/a_0 > 1$.

7. Show that Eq. (6.97) is a solution of Eq. (6.96). Discuss the nature of this solution. Is there another solution for Eq. (6.96)? If so, what is it?

References

1. Kaplan, C., Effect of compressibility at high subsonic velocities on the lifting force acting on an elliptic cylinder, *NACA, TN* No. 118, 1946.
2. Pai, S. I., On the flow behind an attached curved shock, *Jour. Aero. Sci.* **19**, No. 11, Nov. 1952, pp. 734 – 742.
3. Sears. W. R., The linear perturbation theory for rotational flow. *Jour. Math. and Phys.* **28**, No. 4 July 1950, pp. 268 – 271.
4. Hantzsche, W., Die Prandtl Glauertsche Naeherung als Grundlage fuer ein Iterationsverfahren zur Berechnung kompressibler Unterschall stroemungen, *ZAMM* **23**, No. 4, 1943, pp. 185 – 199.
5. Prandtl. L., Ueber Stroemungen, deren Geschwindigkeiten mit der Schallgeschwindigkeit vergleichbar sind, *Jour. Aero. Res. Inst. Univ. of Tokyo*, No. 6, 1930, p. 14.
6. Glauert, H., The effect of compressibility on the lift of an airfoil, *British ARCR and M* No. 4 1135, 1927.
7. Tsien, H. S., and Lees. L., The Glauert-Prandtl approximations for subsonic flows in a compressible fluid, *Jour. Aero. Sci.* **12**, No. 2, 1945, pp. 173 – 187.
8. Ackeret, J., Ueber Luftkraefte bei sehr grossen Geschwindigkeit insbesondere bei ebenen Stroemungen, *Helvetia Phys. Acta*, **1**, fasc, 5, 1928, pp. 301 – 322.
9. Byerly. W. E., *Fourier Series and Spherical, Cylindrical and Ellipsoidal Harmonics*, Ginn and Company, Boston, 1893.

10. Kampe de Feriet, J., Mathematical Methods used in the statistical theory of turbulence: ha rmonic analysis, *Lecture Series* No. 1, Inst. for Fluid Dynamics and Appl. Math., University of Maryland, 1950 – 1951.
11. Kaplan, C., The flow of a compressible fluid past a circular arc profile, *NACA Report* No. 794, 1949.
12. Baker, B. B., and Copson, E. T., *Huggen's Principle*, Oxford University Press, 1939.
13. Hayes, W. D., Pseudotransonic similitude and first-order wave structure, *Jour. Aero. Sci.* **21**, No. 11, Nov. 1954, pp. 721 – 730.
14. Lighthill, M. J., Higher approximations, Sec. E of General theory of high speed aerodynamics, edited by W. R. Sears. Vol. 6 of *High Speed Aerodynamics and Jet Propulsion*, Princeton University Press, 1954.
15. Sears. W. R., Small perturbation theory, Sec. C of General theory of high speed aerodynamics, edited by W. R. Sears, **6** of *High Speed Aerodynamics and Jet Propulsion*, Princetorn University Press, 1954.
16. Ward, G. N., *Linearized Theory of Steady High Speed Flow*, Cambridge University Press, 1955.
17. Hess, J. L. and Smith, A. M. O., Calculation of potential flow about arbitrary bodies, *Progress in Aeronautical Sciences*, **8**, Pergamon, New York, 1967, pp. 1 – 138.

Chapter VII

TWO-DIMENSIONAL SUBSONIC STEADY POTENTIAL FLOW

1. Introduction

The method of solution discussed in Chapter VI is valid only for bodies of small thickness ratio. If the thickness of the body is not small, approximations of an order higher than the first are required to obtain accurate results. The resulting computation proves to be extremely laborious. Consequently, fora body of larger thickness ratio, othen methods of solution for the compressible fluid flow problem are to be preferred. One method, particularly suitable for low subsonic flow, is the Rayleigh-Janzen method,[1, 2] in which the velocity potential is developed in a power series of the Mach number. This method will be discussed in § 2 to § 5.

The Rayleigh-Janzen method is not suitable for flows at a high Mach number because higher-order terms are required; furthermore, the computation for higher approximations is very complicated. Even for a body of such a simple shape as a circular cylinder, computation beyond the third approximation would be too laborious to be carried out. Another method which has been used in solving for high subsonic flow about a body of relatively large thickness ratio is the variational method.[3-5] Here, one attempts to satisfy the boundary conditions exactly and the differential equation approximately. In the variational method, we do not solve the differential equation but formulate a variational integral and find the approximate solution of the problem by the Rayleigh-Ritz method. We shall discuss this method in § 6 and § 7. In establishing Ritz's proof, it is necessary that the integral has either a maximum or minimum. Originally the variational problem was applied to subsonic flow in which the variational integral has an extreme. However, this method can be extended to the case of transonic flow so long as the flow is irrotational. The extension to transonic flow will be briefly discussed in § 7. The basic concepts in the finite difference methods will be given in § 8. The resulting difference equation systems are solved by the relaxation method[6] in § 9 and § 10. Another method for solving subsonic steady two-dimensional compressible flow problems involves the use of electrical analogies. This will be discussed in § 11.

2. Rayleigh-Janzen method

The method of developing the velocity potential φ in powers of M_∞^2 was first given by Janzen[2] and Rayleigh[1] and has been further studied by Poggi,[7] Imai,[8] Kaplan,[9] and others. In the problems considered, we always assume that a uniform stream of velocity U and Mach number $M_\infty = U/a_\infty$ at infinity is passing over a body whose thickness ratio is not necessarily small. In this case, we may write the velocity potential φ for steady flow in the following form:

$$\varphi = U(\varphi_0 + M_\infty^2 \varphi_1 + M_\infty^4 \varphi_2 + \cdots) \tag{7.1}$$

where φ_0, φ_1, etc., are functions of spatial coordinates (x, y, z) which do not contain M_∞ and have the dimensions of a length. In the following, we shall consider the two-dimensional case only so that all the variables are also independent of z. The differential equation for the velocity potential φ is given by Eq. (5.35)

$$\frac{\partial^2 \varphi}{\partial x^2} + \frac{\partial^2 \varphi}{\partial y^2} = \frac{1}{a^2}\left(u^2 \frac{\partial^2 \varphi}{\partial x^2} + 2uv \frac{\partial^2 \varphi}{\partial x \partial y} + v^2 \frac{\partial^2 \varphi}{\partial y^2} \right) \tag{7.2}$$

From Eqs. (7.1) and (7.2) we may find the differential equations for the φ_m's. However, we should first find the relation between u, v, and a in terms of the φ_m's.

From the definition of the velocity potential we have

$$\frac{u}{U} = \frac{1}{U}\frac{\partial \varphi}{\partial x} = \frac{\partial \varphi_0}{\partial x} + M_\infty^2 \frac{\partial \varphi_1}{\partial x} + M_\infty^4 \frac{\partial \varphi_2}{\partial x} + \cdots \tag{7.3}$$

$$\frac{v}{U} = \frac{1}{U}\frac{\partial \varphi}{\partial y} = \frac{\partial \varphi_0}{\partial y} + M_\infty^2 \frac{\partial \varphi_1}{\partial y} + M^4 \frac{\partial \varphi_2}{\partial y} + \cdots \tag{7.4}$$

For irrotational flow, the process is isentropic; hence the relation between the local sound speed a and that for the uniform stream a_∞ is [cf. Eq. (3.12)]

$$a^2 + \frac{\gamma - 1}{2} q^2 = a_\infty^2 + \frac{\gamma - 1}{2} U^2$$

or

$$\frac{a^2}{a_\infty^2} = 1 - \frac{\gamma - 1}{2} M_\infty^2 \left[\left(\frac{q}{U} \right)^2 - 1 \right] \tag{7.5}$$

where $q^2 = u^2 + v^2$. Substituting Eq. (7.1) into Eq. (7.5) we have

$$\frac{U^2}{a^2}=\frac{U^2}{a_\infty^2}\,\frac{a_\infty^2}{a^2}=\frac{M_\infty^2}{1-\frac{\gamma-1}{2}M_\infty^2\left[\left(\frac{\partial\varphi_0}{\partial x}\right)^2+\left(\frac{\partial\varphi_0}{\partial y}\right)^2+\cdots-1\right]} \tag{7.6}$$

We now substitute Eqs. (7.1), (7.3), (7.4), and (7.6) into Eq. (7.2) and collect terms of the same powers of M_∞^2. From this the differential equations for the various φ_m's are obtained. For example, the zeroth-order terms give

$$\frac{\partial^2\varphi_0}{\partial x^2}+\frac{\partial^2\varphi_0}{\partial y^2}=0 \tag{7.7}$$

which is the equation of the velocity potential for incompressible flow. Apparently, then, the zeroth-order approximation of compressible flow is the corresponding incompressible flow.

For the first-order terms, we see that

$$\frac{\partial^2\varphi_1}{\partial x^2}+\frac{\partial\varphi_1}{\partial y^2}=\left(\frac{\partial\varphi_0}{\partial x}\right)^2\frac{\partial^2\varphi_0}{\partial x^2}+2\frac{\partial\varphi_0}{\partial x}\,\frac{\partial\varphi_0}{\partial y}\,\frac{\partial^2\varphi_0}{\partial x\partial y}+\left(\frac{\partial\varphi_0}{\partial y}\right)^2\frac{\partial^2\varphi_0}{\partial y^2} \tag{7.8}$$

an equation which is linear in φ_1. Since the method of iteration is used in solving this equation, φ_0 is assumed to be known when we solve for φ_1.

Similarly, the differential equation of mth order term is given as

$$\frac{\partial^2\varphi_m}{\partial x^2}+\frac{\partial^2\varphi_m}{\partial y^2}=f\,(\varphi_0,\varphi_1\cdots,\varphi_{m-1}) \tag{7.9}$$

which is linear with respect to φ_m ; φ_0 to φ_{m-1} are assumed to be known when we solve for φ_m. It may be seen that the Rayleigh-Janzen method is another technique for linearizing the basic equation of compressible flow.

After obtaining the differential equations, we must find the boundary conditions corresponding to these φ_m's. For our problem, we assume that (i) the velocity components at infinity are $u=U$, $v=0$, and (ii) on the surface of the body, the velocity vector must be tangential to the solid boundary. Our original boundary conditions are:

At infinity:

$$\frac{\partial\varphi}{\partial x}=1,\quad \frac{\partial\varphi}{\partial y}=0 \tag{7.10}$$

At the solid surface:

$$\frac{\partial \varphi}{\partial n}=0$$

where n is the normal to the surface of the body.

Combining Eqs. (7.1) and (7.10), we find the boundary conditions for the φ_m's to be:

(a) For φ_0,

At infinity: $$\frac{\partial \varphi_0}{\partial x}=1, \quad \frac{\partial \varphi_0}{\partial y}=0 \tag{7.11}$$

At the solid surface $$\frac{\partial \varphi_0}{\partial n}=0$$

(b) For φ_m, $m \geqslant 1$

At infinity: $$\frac{\partial \varphi_m}{\partial x}=\frac{\partial \varphi_m}{\partial y}=0 \tag{7.12}$$

At the solid surface: $$\frac{\partial \varphi_m}{\partial n}=0$$

3. Solution by means of complex variables

The mathematical theory of two-dimensional fluid flow has been very extensively developed through the use of the theory of complex variables. Let us define a complex variable z as:

$$z=x+iy=re^{i\theta} \tag{7.13}$$

where x and y are the Cartesian coordinates of the two-dimensional flow $i=\sqrt{-1}$, $r^2=x^2+y^2$, and $\theta=\tan^{-1}\frac{y}{x}$. The complex variable z should not be confused with the third coordinate in a three-dimensional flow.

Let $F(z)$ be an analytic function of z. We may write

$$F(z)=\varphi_0+i\psi_0 \tag{7.14}$$

where φ_0 and ψ_0 are real functions of the real variables x and y. Since $F(z)$ is an analytic function of z, its derivative dF/dz must be unique regardless of the path taken in the complex plane. Thus $F(z)$ of Eq. (7.14) must satisfy the Cauchy-Reimann conditions[10]

$$\frac{\partial \varphi_0}{\partial x}=\frac{\partial \psi_0}{\partial y}, \quad \frac{\partial \varphi_0}{\partial y}=-\frac{\partial \psi_0}{\partial x} \tag{7.15}$$

From Eq. (7.15) it is easy to show that both φ_0 and ψ_0 satisfy Laplace's equation (7.7). We may identify the real part of any analytic function of z as the velocity potential φ_0 and thus apply the theory of complex variables to our flow problem. In this case, ψ_0 may be interpreted as the corresponding stream function.

We shall compute our independent variables x and y in terms of z and $\bar{z}$, where the bar denotes the complex conjugate, i.e., $\bar{z}=x-iy$. Specifically,

$$x=\frac{z+\bar{z}}{2}, \quad y=\frac{z-\bar{z}}{2i} \tag{7.16}$$

In terms of F and $\bar{F}$, we have

$$\varphi_0=\frac{F+\bar{F}}{2} \tag{7.17}$$

If we write

$$\varphi_1(x,y)=\varphi_1(z,\bar{z}) \tag{7.18}$$

the differential equation (7.8) becomes

$$\frac{\partial^2\varphi_1}{\partial z\partial\bar{z}}=\frac{1}{8}\left(\frac{dF}{dz}\right)^2\frac{d^2\bar{F}}{d\bar{z}^2}+\frac{1}{8}\left(\frac{d\bar{F}}{d\bar{z}}\right)^2\frac{d^2F}{dz^2} \tag{7.19}$$

or

$$\frac{\partial^2\varphi_1}{\partial z\partial\bar{z}}=\mathrm{Re}\left[\frac{1}{4}\left(\frac{dF}{dz}\right)^2\frac{d^2\bar{F}}{d\bar{z}^2}\right]=\mathrm{Re}\left[\frac{1}{4}\left(\frac{d\bar{F}}{d\bar{z}}\right)^2\frac{d^2F}{dz^2}\right] \tag{7.19a}$$

where Re denotes the real part of the complex function.

Integrating Eq. (7.19a) first with respect to $\bar{z}$ and then with respect to z, we have

$$\varphi_1=\mathrm{Re}\left\{\frac{1}{4}\frac{dF}{dz}\int\left(\frac{d\bar{F}}{d\bar{z}}\right)^2 d\bar{z}+f_1(z)+f_2(\bar{z})\right\} \tag{7.20}$$

where $f_1(z)$ and $f_2(\bar{z})$ are arbitrary functions determined by the boundary condtion (7.10). Since we are interested in the real part of Eq. (7.20), either $f_1(z)$ or $f_2(\bar{z})$ is sufficient to determine the solution. It is, however, more convenient to retain both in practical calculation. For a body of arbitrary shape, we first use a conformal transformation to transform it into a simple shape, such as a circular cylinder, for which the complex potential is known. From this we may easily carry out the ensuing calculations.[8] In the following we shall give a simple example for a uniform flow over a circular cylinder without circulation.

4. Flow around a circular cylinder

The complex potential of a uniform flow over a circular cylinder without circulation is

$$F=\varphi_0+\psi_0=z+\frac{1}{z}=\left(r+\frac{1}{r}\right)\cos\theta+i\left(r-\frac{1}{r}\right)\sin\theta \tag{7.21}$$

where the surface of the cylinder is at $r=1$. In this case, the normal to the surface is the radial direction, i.e., $n=r$. On the surface, the normal velocity component is

$$\left(\frac{\partial\varphi_0}{\partial r}\right)_{r=1}=\left(1-\frac{1}{r^2}\right)\cos\theta=0 \tag{7.22}$$

Furthermore it is easy to see that, on the surface of the cylinder, the stream function $\psi_0=0$.

From Eq.(7.21), we have

$$\bar{F}=\bar{z}+\frac{1}{\bar{z}} \tag{7.23}$$

Substituting Eqs. (7.21) and (7.23) into Eq. (7.20) gives

$$\varphi_1=\mathrm{Re}\left[\frac{1}{4}\left(1-\frac{1}{z^2}\right)\left(\bar{z}+\frac{2}{\bar{z}}-\frac{1}{3}\frac{1}{\bar{z}^3}\right)+f_1(z)+f_2(\bar{z})\right] \tag{7.24}$$

Now we will fix $f_1(z)$ and $f_2(\bar{z})$ so that the boundary conditions (7.12) for $m=1$ are satisfied. It is easy to show that if we take

$$f_2(\bar{z})=-\frac{1}{4}\left(\bar{z}+\frac{2}{\bar{z}}-\frac{1}{3}\frac{1}{\bar{z}^3}\right),$$

$$f_1(z)=\frac{13}{12z}+\frac{1}{12z^3} \tag{7.25}$$

the boundary conditions (7.12) are satisfied. Hence the velocity potential φ_1 is

$$\varphi_1=\cos\theta\left(\frac{13}{12r}-\frac{1}{2r^3}+\frac{1}{12r^5}\right)+\cos 3\theta\left(\frac{1}{12r^3}-\frac{1}{4r}\right) \tag{7.26}$$

On the surface of the cylinder, only the tangential velocity component q_θ is different from zero. The tangential velocity component on the surface of the cylinder, for $0\leqslant\theta\leqslant\pi$, is

$$q_\theta=U\left[2\sin\theta+M_\infty^2\left(\frac{2}{3}\sin\theta-\frac{1}{2}\sin 3\theta\right)+O(M_\infty^4)\right] \tag{7.27}$$

The maximum value of $q_{\theta m}$ occurs at $\theta=\pi/2$ and is

$$q_{\theta m}=U\left[2+\frac{7}{6}M_{\infty}^{2}+0(M_{\infty}^{4})\right] \tag{7.28}$$

5. Higher-order approximations

For high M_{∞}, the higher order terms of φ_m are needed in order to obtain accurate results. However, the computation for higher-order terms φ_m becomes rather complicated and, furthermore, the convergence of the series of Eq. (7.1) for large M_{∞} is uncertain. As a result, for most problems only the first approximation φ_1 has been calculated. However, higher-order approximations have been worked out successfully by Imai and his associates without consideration of the convergence of the series (7.1).[11]

From Eq. (7.8), we see that for the first-order approximation the properties of the fluid do not enter because the differential equation for φ_1 is independent of γ, the ratio of specific heats. For terms of order higher than the first, the ratio of the specific heats γ does enter the differential equation of φ_m.

The pressure coefficient defined by Eq. (3.33) may be expressed in terms of the φ_m's. We write

$$\frac{q}{U}=q_0+M_{\infty}^{2}q_1+M_{\infty}^{4}q_2+\cdots \tag{7.29}$$

where

$$q_0^2=\left(\frac{\partial\varphi_0}{\partial x}\right)^2+\left(\frac{\partial\varphi_0}{\partial y}\right)^2 \quad\text{and}\quad q_m^2=\left(\frac{\partial\varphi_m}{\partial x}\right)^2+\left(\frac{\partial\varphi_m}{\partial y}\right)^2$$

Substituting Eq. (7.29) into Eq. (3.33), we have

$$C_p=(1-q_0^2)+M_{\infty}^{2}\left[\frac{1}{4}(1-q_0^2)^2-2q_0q_1\right]$$
$$+M^4\left\{\frac{2-\gamma}{24}(1-q_0^2)^3-q_0q_1(1-q_0^2)-(2q_0q_2-q_1^2)\right\}+\cdots \tag{7.30}$$

which shows that the pressure coefficient can be calculated to the same order of accuracy as the velocity potential φ_m.

6. The variational method

In the variational method, an approximate solution of irrotational compressible flow problem is found not by solving the governing nonlinear differential equations, but by satisfying a certain variational integral and the boundary conditions of the problem. Hargreaves[3] first showed that when the hydrodynamic equations are satisfied, the integrand in the variational principle is a linear function of the pressure; Bateman[4] was the first one to apply the variational method to inviscid compressible flow; this method has been studied further by Wang and his associates.[5, 12]

Consider the steady subsonic irrotational flow of a compressible fluid in which a velocity potential φ exists such that

$$u_i = \partial\varphi/\partial x_i \tag{7.31}$$

Bateman showed that for a finite domain of the subsonic steady irrotational compressible flow, the first variation of the integral

$$I = \int_V p dV = \max \tag{7.32}$$

is equivalent to the equation of continuity. Here p is the pressure of the flow field, and V is the volume of the domain. Eq. (7.32) may be alternatively written as

$$\delta I = \delta \int_V p dV = 0 \tag{7.33}$$

Eq. (7.32) may be proved as follows: since for irrotational flow, the process is isentropic, the pressure p of the flow field may be written as

$$p = B\,(q_m^2 - q^2)^{\frac{\gamma}{\gamma-1}} \tag{7.34}$$

where B and q_m are constants for any given flow problem.

Substituting Eq. (7.34) into Eq. (7.33) and noting that the constant B is irrelevant to the variational problem, we have

$$\delta I = \delta \int_V (q_m^2 - q^2)^{\frac{\gamma}{\gamma-1}} dV = 0 \tag{7.35}$$

where $q^2 = \left(\dfrac{\partial\varphi}{\partial x_i}\right)\left(\dfrac{\partial\varphi}{\partial x_i}\right)$, and $i = 1, 2, 3$.

Carrying out the first variation and simplifying,[5] we have

$$\int_V \frac{\partial}{\partial x_i}\left(\rho\delta\varphi \frac{\partial\varphi}{\partial x_i}\right) dV - \int_V \delta\varphi \frac{\partial}{\partial x_i}\left(\rho \frac{\partial\varphi}{\partial x_i}\right) dV = 0 \tag{7.36}$$

Applying Gauss′ theorem to the first integral of Eq. (7.36), we have

$$-\int_S \rho\delta\varphi \frac{\partial\varphi}{\partial n} dS - \int_V \delta\varphi \frac{\partial}{\partial x_i}\left(\rho \frac{\partial\varphi}{\partial x_i}\right) dV = 0 \tag{7.37}$$

where S is the surface of the domain V, and n is the inner normal on the surface S.

The first integral in Eq.(7.37) is zero because the boundary conditions in fluid dynamics are either (a) $\partial\varphi/\partial n=0$ on the boundary surfaces S, or (b) $\partial\varphi=0$ on the boundary surface. Boundary condition (a) corresponds to the case that S is a stream surface, whereas boundary condition (b) corresponds to the case that S is a constant pressure surface.

The second integral of Eq. (7.37) is zero because of the equation of continuity.

Since Eq. (7.37) is equivalent to Eq. (7.33), we thus prove that Eq. (7.33) is true and that the integral (7.32) is a maximum.

The variational integral (7.32) cannot be applied directly to the case of infinite domain. For an infinite domain, Wang[5] showed that the following variational integral should be used instead of integral (7.32):

$$I=\int_V [p(\varphi)-p(\varphi_1)]\,dV+K=\text{maximum} \tag{7.38}$$

where φ_1 is the velocity potential of the corresponding incompressible flow, $\varphi=\varphi_1+\varphi_2$ is the required velocity potential of the compressible flow, and K is a constant depending on the form assumed for φ_1 and φ_2.

7. Rayleigh-Ritz method

In the variational method, we transform the nonlinear, differential equation into a variational integral and then use the Rayleigh-Ritz method to obtain an approximate expression for the velocity potential for any given boundary value problem. In the Rayleigh-Ritz method, we choose a sequence of functions all of which satisfy the boundary conditions, such that

$$\varphi_2=\sum_{i=1}^{m}(a_iF_i+b_iG_i+\cdots) \tag{7.39}$$

where F_i and G_i are functions which satisfy the boundary conditions of the problem studied, and a_i and b_i are undetermined constants. Substituting Eq. (7.39) into Eq. (7.38) and setting the first variation of I to be zero, we obtain a set of simultaneous algebraic equations for the a_i and b_i etc. These constants may be calculated by solving these algebraic equations, i.e.,

$$\frac{\partial I}{\partial a_i}=0\,,\quad \frac{\partial I}{\partial b_i}=0,\ \text{etc.} \tag{7.40}$$

where $i=1,2,\cdots,m$.

As an example, let us consider the problem of a uniform flow over a circular cylinder without circulation discussed in § 4 again. In this case, we have

$$\varphi_1 = U\left(r + \frac{1}{r}\right)\cos\theta \tag{7.41}$$

To satisfy the boundary conditions, we assume that

$$\varphi_2 = \sum_{m=1}^{M}\sum_{n=1}^{N}\left[A_{mn}R_m(r)\cos n\theta + B_{mn}R_m(r)\sin n\theta\right] \tag{7.42}$$

where

$$R_m(r) = \frac{1}{mr^m} - \frac{1}{(m+2)\,r^{m+2}}$$

A_{mn} and B_{mn} are the undetermined constants. Since we consider the case of zero circulation so that the flow is symmetrical with respect to θ, all of the B_{mn} are zero. K of Eq. (7.38) is now equal to $4(q_m^2 - U^2)\,U\pi A_{11}$.

Wang has determined the constants A_{mn} in Ref. 5. The accuracy of the method depends on the terms used in Eq. (7.42). In general, the computation is rather lengthy.

For bodies of arbitrary shape, it is rather difficult to find simple analytic functions F_i. G_i, etc., to satisfy the boundary conditions. This difficulty may be overcome if the technique of conformal transformation is used.

In the Rayleigh-Ritz method, the integral must have either a maximum or a minimum. Thus the method, strictly speaking, is applicable to subsonic flow only. Wang has shown that the variational method also may give a good approximate result for transonic flow prior to the onset of a compression shock.[12]

8. Basic concepts in finite difference methods

The fundamental equations for compressible flow are nonlinear. There is no general method for expressing the solutions in analytic form for nonlinear differential equations. For this reason finite difference methods are often applied to the fundamental equations to obtain quantitative results for particular compressible flow problems. In the finite difference approach, derivatives are approximated by differences resulting in an algebraic representation of the partial differential equation. Thus, a problem involving calculus is transformed into an algebraic problem.

The nature of the resulting algebraic system depends on the character of the problem posed by the original partial differential equations. There are three different types of the partial differential equations: the elliptic type, e.g. Laplace equation for the linearized first-order perturbed velocity potential in steady subsonic flow; the hyperbolic type, e.g. wave equation for the first-order perturbed velocity potential in supersonic flow, and the parabolic type, e. g. the heat conduction equation which will occur in the study of boundary layer flow (Chapter XVII). In a steady subsonic flow field, each point is influenced by all other points in the entire flow field. Elliptic problems usually result in a system of algebraic equations which must be solved simultaneously throughout the problem domain in conjunction with specified boundary values. In supersonic flow field there exists domain of dependence. Hyperbolic and parabolic problems result in algebraic equations which can usually be solved one or several at a time in conjunction with specified initial values. The former problems are called equilibrium problem, and the latter two problems are called marching problems.

Several considerations determine whether the solution so obtained will be a good approximation to the exact solution of the original partial differential equation. Among these considerations are truncation error, consistency, and stability, all of which will be discussed in the present section.

(a) *Truncation error.* One of the first steps to be taken in establishing a finite-difference procedure for solving a partial differential equation is to replace the continuous problem domain by a discretized mesh or grid. As an example, suppose that we wish to solve a Laplace equation for which $\varphi(x, y)$ is the dependent variable in the rectangular domain. We establish a uniform grid on the domain by replacing $\varphi(x, y)$ by $\varphi(i\Delta x, j\Delta y)$, where $i=1, 2, \cdots, I$, $j=1, 2, \cdots, J$. If we think of $\varphi_{i,j}$ as $\varphi(x, y)$ then (Fig. 7.1)

$$\varphi_{i+1,j}=\varphi\ (x+\Delta x, y), \quad \varphi_{i-1,j}=\varphi\ (x-\Delta x, y)$$

$$\varphi_{i,j+1}=\varphi\ (x, y+\Delta y), \quad \varphi_{i,j-1}=\varphi\ (x, y-\Delta y)$$

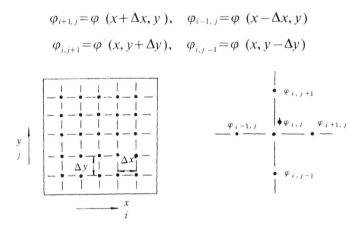

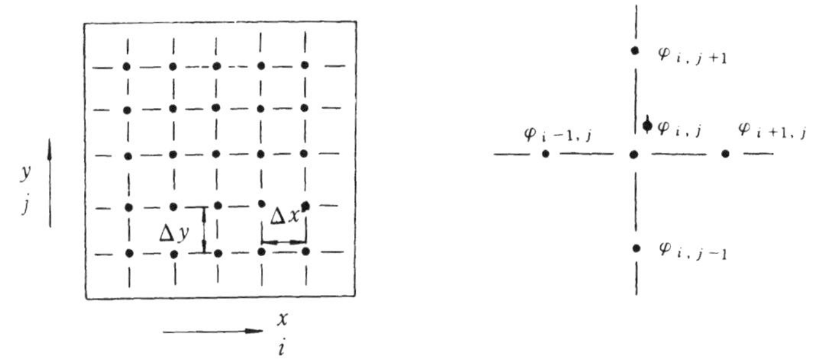

Fig. 7.1 Finite-difference grid

The idea of a finite-difference representation for a derivative can be introduced by recalling the definition of the derivative. Consider the central difference at point (x, y),

$$\frac{\partial \varphi}{\partial x} = \lim_{\Delta x \to 0} \frac{\varphi\ (x+\Delta x, y) - \varphi\ (x-\Delta x, y)}{2\Delta x} \tag{7.43}$$

Here, if φ is continuous, it is expected that the partial difference quotient on the right-hand side will be a reasonable approximation to $\frac{\partial \varphi}{\partial x}$ for a sufficiently small but finite Δx. The difference approximation can be put on a more formal basis through the use of a Taylor-series expansion. Developing Taylorseries expansions for $\varphi\ (x \pm \Delta x, y)$ about (x, y) gives

$$\varphi\ (x \pm \Delta x, y) = \varphi\ (x, y) \pm \frac{\partial \varphi}{\partial x} \Delta x + \frac{\partial^2 \varphi}{\partial x^2} \frac{\Delta x^2}{2\,!} \pm \frac{\partial^3 \varphi}{\partial x^3} \frac{\Delta x^3}{3\,!} + \cdots \tag{7.44}$$

By subtracting the two equations in (7.44) and rearranging we obtain

$$\frac{\partial \varphi}{\partial x} = \frac{\varphi\ (x+\Delta x, y) - \varphi\ (x-\Delta x, y)}{2\Delta x} - \frac{\partial^3 \varphi}{\partial x^3} \frac{\Delta x^2}{6} + \cdots \tag{7.45}$$

Switching to the i, j notation for brevity,

$$\frac{\partial \varphi}{\partial x} = \frac{\varphi_{i+1,j} - \varphi_{i-1,j}}{2\Delta x} + \left[-\frac{\partial^3 \varphi}{\partial x^3} \frac{\Delta x^2}{6} + \cdots \right] \tag{7.46}$$

Here, $(\varphi_{i+1,\,j} - \varphi_{i-1,\,j})\,/2\Delta x$ is the finite central difference representation for $\frac{\partial \varphi}{\partial x}$. The truncation error is defined as the difference between the partial derivative and its finite-difference representation, i.e. the quantity in brackets where only the leading term has been written out. The truncation error for the central difference representation is of the second order of Δx, i.e. $O(\Delta x^2)$. Thus the central difference representation is second-order accurate.

Similarly, we have the central difference representation for $\frac{\partial^2 \varphi}{\partial x^2}$

$$\frac{\partial^2 \delta}{\partial x^2} = \frac{\varphi_{i+1,j} - 2\varphi_{i,j} + \varphi_{i-1,j}}{\Delta x^2} + \left[-\frac{\partial^4 \varphi}{\partial x^4} \frac{\Delta x^2}{12} + \cdots \right] \tag{7.47}$$

It is also second-order accurate.

Using also the central difference representation for $\frac{\partial^2 \varphi}{\partial y^2}$, we may approximate the Laplace equation by a difference equation

$$\frac{\varphi_{i+1,j}-2\varphi_{i,j}+\varphi_{i-1,j}}{\Delta x^2}+\frac{\varphi_{i,j+1}-2\varphi_{i,j}+\varphi_{i,j-1}}{\Delta y^2}=0 \tag{7.48}$$

However, there are truncation errors associated with the difference representations used in Eq. (7.48). If we include all these truncation errors, we obtain

$$\frac{\partial^2\varphi}{\partial x^2}+\frac{\partial^2\varphi}{\partial y^2}=\frac{\varphi_{i+1,j}-2\varphi_{i,j}+\varphi_{i-1,j}}{\Delta x^2}+\frac{\varphi_{i,j+1}-2\varphi_{i,j}+\varphi_{i,j-1}}{\Delta y^2}$$
$$+\left[-\frac{\partial^4\varphi}{\partial x^4}\frac{\Delta x^2}{12}-\frac{\partial^4\varphi}{\partial y^4}\frac{\Delta y^2}{12}+\cdots\right] \tag{7.49}$$

The quantity in brackets in Eq. (7.49) is identified as the truncation error for the finite difference equation (7.48) of the Laplace equation and is defined as the difference between the partial differential equation and the difference approximation to it. The order of the truncation error in this case is $O(\Delta x^2, \Delta y^2)$ and the difference equation (7.48) is second-order accurate. It is noted that the truncation errors associated with all derivatives in any partial differential equation should be obtained by expanding about the same point (i,j in the above discussion).

From Eq. (7.49) it is seen that the difference equation (7.48) is equivalent to a new partial differential equation

$$\frac{\partial^2\varphi}{\partial x^2}+\frac{\partial^2\varphi}{\partial y^2}+\frac{\partial^4\varphi}{\partial x^4}\frac{\Delta x^2}{12}+\frac{\partial^4\varphi}{\partial y^4}\frac{\Delta y^2}{12}+\cdots=0 \tag{7.50}$$

rather than the original partial differential equation. Note that in Eq. (7.50) only the leading terms have been written out utilizing Taylor series expansions. Eq. (7.50) is called the modified differential equation of the difference equation (7.48). The solution of difference equation is in fact the solution of its modified differential equation rather than the original differential equation.

(b) *Consistency*. A finite-difference representation of a partial differential equation is said to be consistent if the difference between the partial differential equation and its difference representation vanishes as the mesh is refined. This should always be the case if the order of the truncation error vanishes under grid refinement. Evidently consistency is a necessary condition for a difference equation to be an approximate representation of the original differential equation

An example of a questionable scheme is the DuFort-Frankel differencing of the heat equation,

$$\frac{\partial \varphi}{\partial t}=\alpha \frac{\partial^2 \varphi}{\partial x^2} \tag{7.51}$$

Using a central-difference representation for the time derivative and a modified central difference for the spatial derivative, we have

$$\frac{\varphi_j^{n+1}-\varphi_j^{n-1}}{2\Delta t}=\frac{\alpha}{\Delta x^2}(\varphi_{j+1}^n-\varphi_j^{n+1}-\varphi_j^{n-1}+\varphi_{j-1}^n) \tag{7.52}$$

Here the time variation is indicated by a superscript. For the difference equation (7.52) the leading terms in the truncation error are

$$\frac{\alpha}{12}\frac{\partial^4 \varphi}{\partial x^4}\Delta x^2-\alpha\frac{\partial^2 \varphi}{\partial t^2}\left(\frac{\Delta t}{\Delta x}\right)^2-\frac{1}{6}\frac{\partial^3 \varphi}{\partial t^3}\Delta t^2 \tag{7.53}$$

If

$$\lim_{\Delta t,\,\Delta x\to 0}\frac{\Delta t}{\Delta x}=0 \tag{7.54}$$

the DuFort-Frankel scheme (7.52) is consistent with the heat equation (7.51). If Δt and Δx were to approach zero at the same rate such that $\Delta t/\Delta x=\beta$, then the DuFort-Frankel scheme (7.52) is consistent with another differential equation

$$\frac{\partial \varphi}{\partial t}+\alpha\beta^2\frac{\partial^2 \varphi}{\partial t^2}=\alpha\frac{\partial^2 \varphi}{\partial x^2} \tag{7.55}$$

which is of the hyperbolic type.

(c) *Stability*. Any computer used in numerical solution has finite accuracy. All numerical values are rounded to a finite number of digits in the arithmetic operations. These errors are called round-off errors. A stable numerical scheme is one for which the round-off errors are not permitted to grow in the sequence of numerical procedures as the calculation proceeds from one marching step to the next. Thus, numerical stability is a concept applicable in the strict sense only to marching problems, i.e. hyperbolic and parabolic problems. If the round-off errors grow while computations are being performed, the numerical solution would be false and even blow out. The question of stability is usually answered by using a Fourier analysis.[13] This method is also referred to as a von Neumann analysis. Fourier analysis can be applied only to linear equations. Nonlinear equations have to be treated approximately.

For example, we consider the second-order wave equation

$$\frac{\partial^2 \varphi}{\partial t^2} = a^2 \frac{\partial^2 \varphi}{\partial x^2} \tag{7.56}$$

where a is the wave speed. This equation has the characteristics

$$x + at = \text{constant} = C_1$$

and

$$x - at = \text{constant} = C_2 \tag{7.57}$$

Using a backward-difference representation for the time derivative and a central-difference representation for the spatial derivative, we can approximate the wave equation by (Fig. 7.2)

$$\frac{\varphi_j^n - 2\varphi_j^{n-1} + \varphi_j^{n-2}}{\Delta t^2} = a^2 \frac{\varphi_{j+1}^{n-1} - 2\varphi_j^{n-1} - \varphi_{j-1}^{n-1}}{\Delta x^2} \tag{7.58}$$

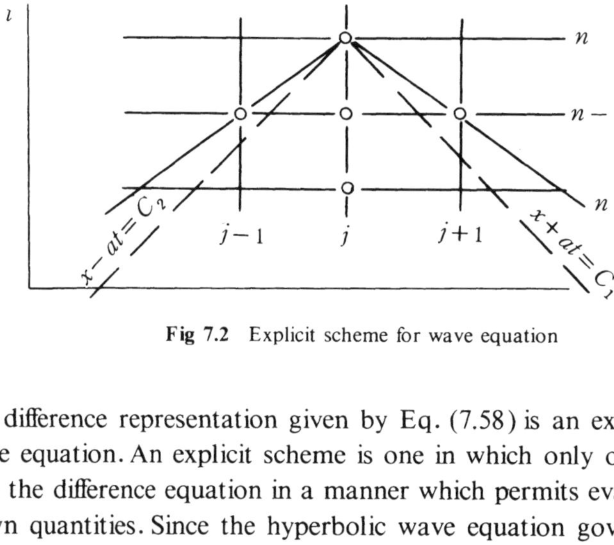

Fig 7.2 Explicit scheme for wave equation

The difference representation given by Eq. (7.58) is an explicit scheme for the wave equation. An explicit scheme is one in which only one unknown appears in the difference equation in a manner which permits evaluation in terms of known quantities. Since the hyperbolic wave equation governs a marching problem for which the initial values must be specified, φ's at the time levels $n-1$ and $n-2$ can be considered as known. If the spatial derivative term in the wave equation is approximated by φ's at the n-time level, three unknowns appear in the difference equation and the procedure is known as implicit, indicating that the algebraic formulation requires the simulataneous solution of several equations involving the unknowns.

Consider the finite-difference equation (7.58). Let ε represent the error in the numerical solution due to round-off errors. The numerical solution actually computed may be written $\varphi + \varepsilon$, where φ is the exact solution of the difference equation. This computed numerical solution must satisfy the difference equation. Substituting $\varphi + \varepsilon$ into Eq. (7.58) yields

$$\frac{\varphi_j^n+\varepsilon_j^n-2\varphi_j^{n-1}-2\varepsilon_j^{n-1}+\varphi_j^{n-2}+\varepsilon_j^{n-2}}{\Delta t^2}=a^2\frac{\varphi_{j+1}^{n-1}+\varepsilon_{j+1}^{n-1}-2\varphi_j^{n-1}-2\varepsilon_j^{n-1}+\varphi_{j-1}^{n-1}+\varepsilon_{j-1}^{n-1}}{\Delta x^2} \tag{7.59}$$

Since the exact solution φ must satisfy the difference equation, the same is true of the error, i.e.

$$\frac{\varepsilon_j^n-2\varepsilon_j^{n-1}+\varepsilon_j^{n-2}}{\Delta t^2}=a^2\frac{\varepsilon_{j+1}^{n-1}-2\varepsilon_j^{n-1}+\varepsilon_{j-1}^{n-1}}{\Delta x^2} \tag{7.60}$$

In this case, the exact solution φ and the error ε must both satisfy the same difference equation. This means that the numerical error and the exact numerical solution both possess the same growth property in time. Any perturbation of the numerical values at arbitrary time level will either be prevented from growing for a stable scheme or grow without bound for an unstable scheme.

Since the difference equation is linear, superposition may be used and we may examine the behavior of a single term of the Fourier series. Consider the term

$$\varepsilon\ (t,x)=e^{\alpha t}e^{i\beta x} \tag{7.61}$$

where β is any real value, α is a function of β and may be complex, $i=\sqrt{-1}$.

Substituting Eq. (7.61) into (7.60), we obtain

$$e^{\alpha\Delta t}=A\pm\sqrt{A^2-1} \tag{7.62}$$

$$\text{where } A=1-2\nu^2\sin\frac{\beta\Delta x}{2} \tag{7.63}$$

$\nu=a\Delta t/\Delta x$ is called the Courant number.

Since $\varepsilon_j^n=e^{\alpha\Delta t}\varepsilon_j^{n-1}$ for any β in the solution for error, it is clear that if $|e^{\alpha\Delta t}|\leqslant 1$, a general Fourier component of error will not grow from one time step to the next. $e^{\alpha\Delta t}$ is the growth (amplification) factor for the error. Therefore,

$$\left|A\pm\sqrt{A^2-1}\right|\leqslant 1 \tag{7.64}$$

Solving the inequality (7.64), we obtain

$$\frac{a\Delta t}{\Delta x}\leqslant 1 \tag{7.65}$$

which is the stability requirement for the explicit scheme (7.58). This numerically places a constraint on the size of the time step relative to the size of the spatial mesh spacing. This is called the Courant-Friedrichs-Lewy (CFL) condition.[14]

A physical interpretation of the results provided by the CFL condition (7.65) for hyperbolic equation is important. An analytic solution at a point (t, x) depends upon data contained between the characteristics (7.57) which intersect that point as sketched in Fig. 7.2. The region between the characteristics is the domain of dependence for the partial differential equation (7.56). The domain of dependence for the explicit difference equation (7.58) is the shadowed region in Fig. 7.2. The CFL condition (7.56) requires that the domain of dependence for the difference equation may include more than, but not less than, the domain of dependence for the differential equation. Thus, the CFL requirement makes sense from a physical point of view. One would also expect the numerical solution to be degraded if too much unnecessary information is included by allowing $a\,\Delta t/\Delta x$ to become much greater than unity. This is in fact what occurs numerically. The best results for hyperbolic equations using the simple explicit scheme are obtained with Courant numbers near unity.

Clearly, the influence of boundary conditions is not included in this analysis. In fact, the von Neumann stability analysis assumes that we have imposed periodic boundary conditions. If the influence of boundary conditions on stability is desired, we must use the matrix notation, referred to as the matrix method.[15] The basic idea of the matrix method is also analyzing a typical term of a Fourier series. The matrix method can also be used to evaluate stability for system of equations.

9. Successive line over relaxation (SLOR)

Let us consider the solution to Laplace equation on a rectangular domain with Dirichlet boundary conditions (Fig. 7.1). For each grid point in the interior of the domain we can write the difference equation (7.48) so that our problem is one of solving the system of $(I-2)\times(J-2)$ simultaneous linear algebraic equations for the $(I-2)\times(J-2)$ unknown $\varphi_{i,j}$'s. Methods for solving system of linear algebraic equations can be readily classified as either direct or iterative. Direct methods give the solution in a finite and predeterminable number of operations using an algorithm which is often quite complicated and immensely time-consuming. Iterative methods consist of a repeated application of an algorithm which is usually quite simple. They yield the exact answer only as a limit of a sequence, but, if the iterative procedure converges, we can come to an approximate answer in a finite but usually not predeterminable number of operations. One of the most efficient and useful iterative methods for large system of equations is the successive line over relaxation (SLOR) incorporated with the Gauss-Seidel iteration. We will discuss the SLOR in this section.

In the SLOR, we can choose either rows or columns for grouping unknowns. If we agree to start at the left end of the rectangular domain and sweep rightward by columns (Fig. 7.1), we can write as follows for the genetal point

$$\frac{\varphi^{n-1}_{i+1,j}-2\varphi^n_{i,j}+\varphi^n_{i-1,j}}{\Delta x^2}+\frac{\varphi^n_{i,j+1}-2\varphi^n_{i,j}+\varphi^n_{i,j-1}}{\Delta y^2}=0 \qquad (7.66)$$

where the superscript n denotes iteration level. If we study this equation carefully, we observe that only three unknowns are present since $\varphi^n_{i-1,j}$ is known from either the left boundary conditions if we apply the equation to the first column of unknowns, or the solution already obtained at the n level from the left column. We have chosen to evaluate $\varphi_{i+1,j}$ at $n-1$ iteration level rather than the n level in order to obtain just three unknowns in the equation so that the efficient tridiagonal algorithm can be used.

The SLOR procedure for a general system of algebraic equations is to: (1) make initial guesses for all unknowns except the first column in this case. (2) solve the system of $J-2$ simultaneous algebraic equations for the $J-2$ unknowns on each column, using the guessed values initially and the most recently computed values, i.e. the Gauss-Seidel iteration, thereafter for the other unknowns in the equation. (3) apply over relaxation before moving to the next column, Let $\overline{\varphi}^n_{i,j}$ denote the newly calculated values by the tridiagonal algorithm, i.e.

$$\frac{\varphi^{n-1}_{i+1,j}-2\overline{\varphi}^n_{i,j}+\varphi^n_{i-1,j}}{\Delta x^2}+\frac{\overline{\varphi}^n_{i,j+1}-2\overline{\varphi}^n_{i,j}+\overline{\varphi}^n_{i,j-1}}{\Delta y^2}=0 \qquad (7.67)$$

The over-relaxated values are

$$\varphi^n_{i,j}=\varphi^{n-1}_{i,j}+\omega\left(\overline{\varphi}^n_{i,j}-\varphi^{n-1}_{i,j}\right) \qquad (7.68)$$

where ω is relaxation parameter. When $\omega<1$, $\varphi^n_{i,j}$ is under-relaxed and lies between $\overline{\varphi}^n_{i,j}$ and $\varphi^{n-1}_{i,j}$. When $\omega>1$, $\varphi^n_{i,j}$ is over-relaxed and lies beyond the range of $\overline{\varphi}^n_{i,j}$ and $\varphi^{n-1}_{i,j}$. (4) repeat iteratively the solution of the equations in this manner until the changes in the unknowns between successive iterations become sufficiently small.

(a) *Convergence of SLOR.* To study the convergence behavior of the SLOR incorporated with Gauss-Seidel iteration, we use an approach due to Garabedian.[16] Viewing the iteration level n as an artificial time t, the backward difference representation may be used for

$$\frac{\partial\varphi}{\partial t}=\frac{\varphi^n_{i,j}-\varphi^{n-1}_{i,j}}{\Delta t} \qquad (7.69)$$

The modified differential equation of the difference equation (7.67) combined with Eq. (7.68) can be written as

$$\varphi_{xx}+\varphi_{yy}-\frac{\Delta t}{\Delta x}\varphi_{tx}-\left(\frac{2}{\omega}-1\right)\varphi_t=0 \tag{7.70}$$

This is an artificial time-dependent differential equation for $\varphi\ (x,y,t)$. The solution of the SLOR behaves like the solution of this differential equation if Δx, Δy and Δt are small.

Since Eq. (7.70) is linear, we may decompose $\varphi\ (x,y,t)$

$$\varphi\ (x,y,t)=\Phi\ (x,y)+\varphi_1\ (x,y,t) \tag{7.71}$$

where $\Phi\ (x,y)$ is the solution of Laplace equation with Dirichlet boundary conditions, $\varphi_1\ (x,y,t)$ is the relaxation error.

The relaxation error $\varphi_1\ (x,y,t)$ satisfies the same differential equation (7.70). The initial values for $\varphi_1\ (x,y,t)$ are

$$\varphi_1\ (x,y,0)=\varphi\ (x,y,0)-\Phi\ (x,y) \tag{7.72}$$

where $\varphi\ (x,y,0)=\varphi^{o}_{i,j}$, the guessed initial values for $\varphi^{n}_{i,j}$.
The boundary conditions for $\varphi_1\ (x,y,t)$ are $\varphi_1=0$.

Eq. (7.70) can be solved by method of variable separation after a variable transformation,

$$\begin{aligned}\xi&=x\\ \eta&=y\\ \tau&=t+\frac{1}{2}\frac{\Delta t}{\Delta x}x\end{aligned} \tag{7.73}$$

Eq. (7.70) becomes

$$\varphi_{\xi\xi}+\varphi_{\eta\eta}-\frac{\Delta t^2}{4\Delta x^2}\varphi_{\tau\tau}-\left(\frac{2}{\omega}-1\right)\frac{\Delta t}{\Delta x^2}\varphi_\tau=0 \tag{7.74}$$

which is of the hyperbolic type.
Thus, $\varphi_1\ (\xi,\eta,\tau)$ can be written as

$$\varphi_1\ (\xi,\eta,\tau)=F\ (\tau)\cdot\ G\ (\xi,\eta) \tag{7.75}$$

Substituting Eq. (7.75) into Eq. (7.74), we obtain the differential equations for $F\ (\tau)$ and $G\ (\xi,\eta)$.

The differential equation for $G\ (\xi,\eta)$ is

$$\frac{\partial^2 G}{\partial\xi^2}+\frac{\partial^2 G}{\partial\eta^2}=kG \tag{7.76}$$

where k is a constant.
By use of Green's formula and the zero boundary conditions, we can prove[17] that k must be negative, i. e. $K=-k_m^2$ where $m=1, 2,\cdots$. Let

$$k_1^2<k_2^2<\cdots<k_{m-1}^2<k_m^2<\cdots \tag{7.77}$$

k_m^2's are called eigenvalues and the corresponding solutions of (7.76), G_m's, are called eigenfunctions.

The differential equation for $F(\tau)$ is

$$\frac{1}{4}\left(\frac{\Delta t}{\Delta x}\right)^2 F''+\left(\frac{2}{\omega}-1\right)\frac{\Delta t}{\Delta x^2}F'+k_m^2F=0 \tag{7.78}$$

The solution of Eq. (7.78) is

$$F(\tau)=A_m e^{-p_m\tau}+B_m e^{-q_m\tau} \tag{7.79}$$

where A_m and B_m are constants,

$$p_m=\frac{2}{\Delta t}\left(\frac{2}{\omega}-1\right)(1-r_m)$$

$$q_m=\frac{2}{\Delta t}\left(\frac{2}{\omega}-1\right)(1+r_m) \tag{7.80}$$

$$r_m=\sqrt{1-\frac{k_m^2\,\Delta x^2}{\left(\frac{2}{\omega}-1\right)^2}}$$

Thus, the relaxation error function $\varphi_1(\xi,\eta,\tau)$ is

$$\varphi_1(\xi,\eta,\tau)=\sum_{m=1}^{\infty}(A_m e^{-p_m\tau}+B_m e^{-q_m\tau})\,G_m(\xi,\eta) \tag{7.81}$$

It is clear that if the real parts of p_m and q_m are positive, $\lim\limits_{\tau\to 0}\varphi_1(\xi,\eta,\tau)=0$. This requires that

$$0<\omega<2 \tag{7.82}$$

This is the convergence condition of the SLOR for Laplace equation. The problem now is to choose a relaxation parameter ω such that the time dependent analytical solution converges as rapidly as possible to the desired steady state solution. We observe that the convergence speed depends essentially on the leading term in the solution expression (7.81) as τ is large, i.e.

$$A_1 e^{-p_1\tau}\,G_1(\xi,\eta) \tag{7.83}$$

An optimum ω can be obtained by setting the radical r_1 in the p_1 expression (7.80) equal to zero. Thus,

$$\omega = \frac{2}{1+k_1\ \Delta x} \tag{7.84}$$

The optimum relaxation parameter approaches a value of 2 as the finite difference mesh is refined. Substituting Eq. (7.84) into the p_1 expression (7.80), we obtain

$$p_1 t = 2k_1\ \Delta x\ n \tag{7.85}$$

For a fixed degree of convergence, i. e. $p_1 t$=constant, we have the required iteration number $n = O(1/\Delta x)$.

To illustrate the acceleration effects of the optimum relaxation parameter, let us compare the case $\omega = 1$. In this case,

$$p_1 t \simeq k_1^2 \Delta x^2 n \tag{7.86}$$

For fixed degree of convergence, the required iteration number $n = O(1/\Delta x^2)$, which is larger by one order of $1/\Delta x$.

It can be seen that in the artificial time-dependent differential equation (7.70) the term φ_t has damping effects if the coefficients of φ_t and φ_{xx} have opposite signs.

(b) *Stability of SLOR*. Since the SLOR algorithm for Laplace equation is a marching procedure, the question of numerical stability appears. By use of the von Neumann method, consider a general term of Fourier series

$$e^{\alpha t} e^{i(\beta_1 x + \beta_2 y)} \tag{7.87}$$

where wave numbers β_1 and β_2 are any real values, α is a function of β_1 and β_2 and may be complex, $i = \sqrt{-1}$.

Substituting expression (7.87) into the difference equation (7.67) combined with Eq. (7.68), we obtain the growth factor for the error in the numerical solution due to round-off errore

$$e^{\alpha\Delta t} = \left[\frac{e^{i\beta_1 \Delta x} - 2\left(1 - \dfrac{1}{\omega}\right)}{\Delta x^2} - \frac{4\left(1 - \dfrac{1}{\omega}\right)\sin^2 \dfrac{\beta_2 \Delta y}{\Delta y^2}}{\Delta y^2} \right] \Bigg/ \left[\frac{\dfrac{2}{\omega} - e^{i\beta_1 \Delta x}}{\Delta x^2} + \frac{\dfrac{4}{\omega}\sin^2 \dfrac{\beta_2 \Delta y}{2}}{\Delta y^2} \right] \tag{7.88}$$

The stability requirement $\mid e^{\alpha\Delta t} \mid \leqslant 1$ yields

$$0<\omega\leqslant 2 \tag{7.89}$$

This is the stability condition for the SLOR solution of Laplace equation which is slightly different from the convergence condition (7.82).

(c) *Fourier analysis of SLOR error.* Some insight into the convergence of relaxation may be obtained by a Fourier analysis of the relaxation error. We consider the SLOR solution of Laplace equation in a square domain $0\leqslant x\leqslant L$, $0\leqslant y\leqslant L$. Let $\Delta x=\Delta y=h$, and $L/h=2N$. The mth Fourier component of error is defined as having m waves on L, i.e. wave length$=L/m$. The wave frequency is m/L which denotes the number of waves per unit length. The wave number is $2\pi m/L$ or $m\pi/Nh$ which denotes the circular angle of waves per unit length. The mth Fourier component of the SLOR error may be expressed in the form,

$$e^{\lambda t+i\frac{m\pi}{Nh}(x+y)} \tag{7.90}$$

where λ is a function of wave number $(m\pi/Nh)$, $i=\sqrt{-1}$.
In a linear problem, the error satisfies the same governing equation as the dependent variable. Substituting Eq. (7.90) into the difference Eq. (7.67) combined with Eq. (7.68) and using $\Delta x=\Delta y=h$, we obtain the growth factor for the SLOR-error component[18]

$$e^{\lambda\Delta t}=\left[2\left(1-\frac{1}{\omega}\right)\left(\cos\frac{m\pi}{n}-2\right)+e^{i\frac{m\pi}{N}}\right]\Big/\left[\frac{2}{\omega}\left(2-\cos\frac{m\pi}{N}\right)-e^{-i\frac{m\pi}{N}}\right] \tag{7.91}$$

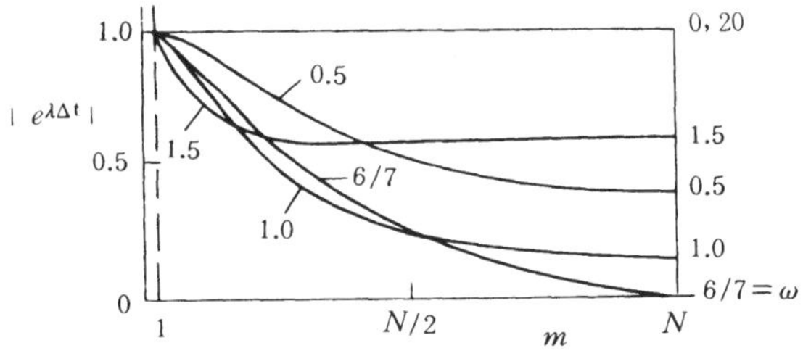

Fig. 7.3 Growth factor for SLOR error

The growth factor of the SLOR error over the entire frequency spectrum is shown in Fig. 7.3, where $m=1$ corresponds to the lowest frequency and $m=N$, the highest frequency. It is seen that the SLOR can provide efficient elimination for the high frequency components of error but not for the low frequency components of error. They are the low frequency components of error which slow down finally the convergence of the SLOR.

We notice that the SLOR works very well on high frequency components of error. Take a sequence of grids ranging from very coarse to very fine. Of course, the final solution is obtained from the finest grid. SLOR is used on each grid to remove the high frequency components of error supported on the grid. A desirable aspect of this approach is that the high frequency components of error on the coarse grids are actually the low frequency components of error existing in the finest grid. Because these troublesome low frequency components of error on the finest grid are efficiently dealt with on coarse grids, very little computational work is expended in removing them from the solution. Thus a tremendous convergence rate enhancement is obtained. This convergence acceleration technique is known as the multigrid iteration method.[19] It is enjoying a wave of popularity that has included applications in a host of different areas for numerical solutions.

Besides the SLOR, there are other efficient iterative methods for large system of simultaneous linear algebraic equations, e.g. the alternating direction implicit method. We will present the latter method in numerical solutions of transonic flow (Chapter X).

10. Numerical solution of subsonic steady potential flow

The differential equation for the two-dimensional subsonic steady velocity potential is given by Eq. (5.46), i.e.

$$(a^2-u^2)\frac{\partial^2\varphi}{\partial x^2}-2uv\frac{\partial^2\varphi}{\partial x\partial y}+(a^2-v^2)\frac{\partial^2\varphi}{\partial y^2}=0 \qquad (7.92)$$

This equation is a quasi-linear partial differential equation of second order, i.e. equation linear in second-order partial derivatives but nonlinear with respect to the first derivative and the function itself. If the local Mach number $M=\sqrt{u^2+v^2}/a$ is less than unity, this equation is of the elliptic type (cf. Chapter XI). For the elliptic equation, we solve the equation for given boundary values over a closed boundary and use the central difference representations for all derivatives in numerical solution. The resulting algebraic equations for (7.92) are nonlinear and thus have to be solved by iteration methods. We use the φ's at the $n-1$ iteration level for the coefficients a^2-u^2, $-2uv$, and a^2-v^2 in the difference equations of φ at the n iteration level, i.e. the Jacobi iteration. Then we solve the linearized algebraic system by the same iterative method for Laplace equation, i.e. the SLOR incorporated with the Gauss-Seidel iteration. Using Jacobi iteration rather than Gauss-Seidel iteration for the coefficients usually yields better chance for convergence.

Convergence and stability analyses of the above iterative method have not been completed in strict nonlinear sense. In order to perform linear analyses, we may treat the coefficients in Eq. (7.92) as local constants. Thus the

Garabedian convergence and von Neumann stability analyses can be applied locally. The results of the locally linearized analyses usually provide guidelines to nonlinear equations, although strictly speaking they are not valid for nonlinear problems.

Of course, there are peculiarities in nonlinear cases. For example, over-relaxation is usually appropriate for numerical solutions to Laplace equation. Under-relaxation is sometimes called for in elliptic problems when the equations are nonlinear. Occasionally, for nonlinear problems, under-relaxation is even observed to be necessary for convergence. In general, under-relaxation appears to be most appropriate when the convergence at a point is taking on an oscillatory pattern and tending to overshoot the apparent final solution.

Two numerical examples are given as follows. Fig. 7.4 portrays the results in pressure distribution solving the full-potential equation (7.92)[20] (dotted line) and the Euler equations (solid line) for free stream Mach number $M_\infty = 0.2$ around a circular cylinder without circulation. The numerical solution of Euler equations will be discussed in Chapter XIV. Fig 7.5 gives numerical solutions of the full-potential equation[21] and the transonic small perturbation equation (cf. Chapter X) for subcritical flow past an NACA 0012 airfoil at a 2° angle of attack.

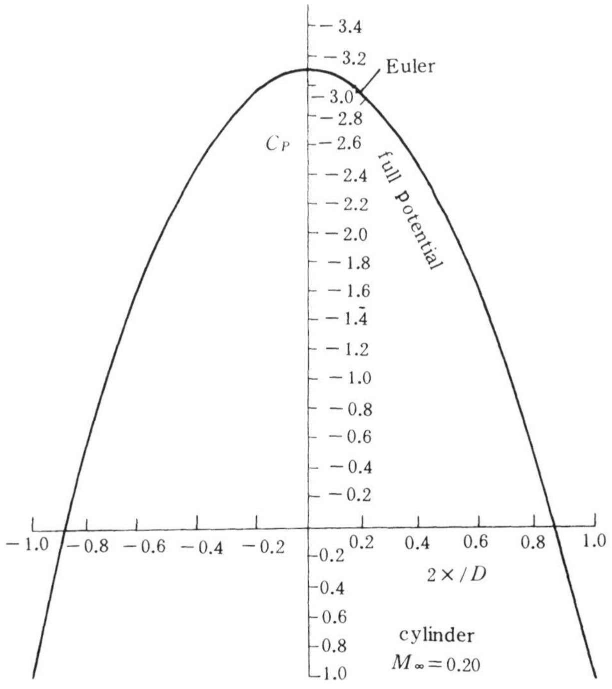

Fig. 7.4 Pressure coefficient C_p on a cylinder

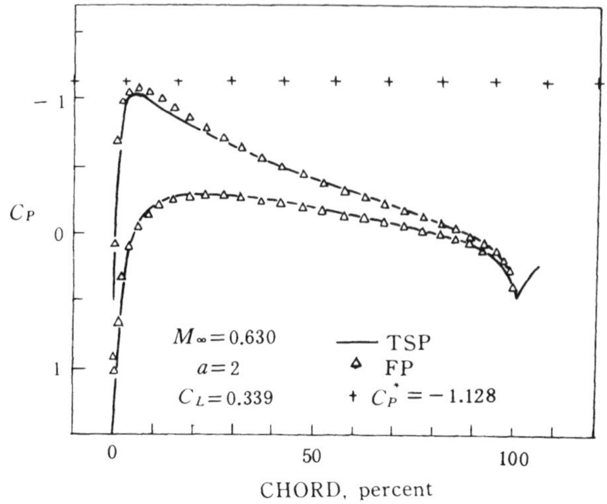

Fig. 7.5 C_p distribution on airfoil NACA 0012

In both examples, Neumann boundary conditions $\dfrac{\partial \varphi}{\partial n}=0$, where n is normal to the body surface, are fitted into the difference equations of (7.92) written at the grid points on the body. In Reference 21, the flow field exterior to the airfoil is conformally mapped into a unit circle and Kutta condition is applied to the point corresponding to the trailing edge.

11. Electrical analogies

The fundamental equation for the velocity potential φ (7.2) for a compressible flow may be solved by an iteration method. In the iteration method, we first solve the equation

$$\frac{\partial^2 \varphi}{\partial x^2}+\frac{\partial^2 \varphi}{\partial y^2}=0 \tag{7.93}$$

From the solution of Eq. (7.93), we may calculate the right-hand terms of Eq. (7.2) as a function of x and y, say, $f(x, y)$. We then solve the Poisson equation:

$$\frac{\partial^2 \varphi}{\partial x^2}+\frac{\partial^2 \varphi}{\partial y^2}=f(x, y) \tag{7.94}$$

This process is repeated until the new values obtained are sufficiently close to the previous approximation.

There is another method of iteration. At first we assume that the density of the compressible fluid is a constant, i.e., $\rho=\rho_0=$constant, and solve the Laplace equation (7.93) for the problem investigated. The results give us a velocity distribution u_0 (x, y) and v_0 (x, y), for the whole flow field investigated. From these values of velocity distribution, u_0 and v_0 and Bernoulli's equation (5.42) with the help of equation of state (5.4) and $\partial\varphi/\partial t=0$, we may calculated the density distribution of the flow field $\rho=\rho_1(x, y)$. For the next approximation, we solve the equation

$$\frac{\partial \rho_1 u_1}{\partial x}+\frac{\partial \rho_1 v_1}{\partial y}=0 \tag{7.95}$$

where ρ_1 is known. We repeat this process as before.

G.I.Taylor[22] introduced an electrical basin by means of which these calculations were performed electrically. Consider a tank filled with a conducting liquid. The bottom of the tank is of an insulating material. Portions of the sides of the tank are conductors and the rest are insulators. The depth of the liquid is variable. The axes of x and y are parallel to the surface of the liquid.

If V be the electric potential; W, the electric flux function; f and g the x-and y-components of the currents, respectively; t, the depth of the liquid; and σ, the specific resistance of the liquid, Ohm's law and the electric continuity law state that

$$\begin{aligned}\frac{1}{\sigma}\frac{\partial V}{\partial x}&=-f\,,\quad tf=\frac{\partial W}{\partial y}\\ \frac{1}{\sigma}\frac{\partial V}{\partial y}&=-g\,,\quad tg=-\frac{\partial W}{\partial x}\end{aligned} \tag{7.96}$$

However, the fundamental equations of a compressible flow may be written in the following manner:

$$\begin{aligned}\frac{\partial \varphi}{\partial x}&=u\,,\quad \rho u=\frac{\partial \psi}{\partial y}\\ \frac{\partial \varphi}{\partial y}&=v\,,\quad \rho v=-\frac{\partial \psi}{\partial x}\end{aligned} \tag{7.93}$$

In comparing Eqs. (7.96) and (7.97), two analogies are possible:

(a) We may associate the electric potential with the negative velocity potential and the electric flux function with the steam function, i. e., $V \sim -\varphi$ and $W \sim \psi$. From Eqs. (7.96) and (7.97) we have then,

$$u \sim \sigma f \, , v \sim \sigma g \, , \rho u \sim tf \, , \rho v \sim tg$$

These will be compatible if $\rho \sim t/\sigma$.

At the portions of wall that are conductors, the electric potential V is constant; and at the portions that are insulators, the electric flux function W is constant. Consequently, in this analogy, we see that insulators correspond to ψ =constant, or streamlines, and conductors to φ =constant, or equipotential lines. The iteration process may be carried out in the following manner:

The operator of the tank starts out with a uniform depth which corresponds to a constant density. He measures the components of electric currents f and g by electric meters. Our analogy gives the components of velocity u and v from the measured values of f and g. From Bernoulli's equation and these values of u and v, the new density variation ρ_1 (x, y) may be obtained. The bottom of the tank is molded according to this mew density distribution. The test is repeated and a second value for the density will be obtained. The process is repeated until sufficient convergence of the result is assured.

In general, the density will decrease rather than increase, and this means that insulating material must be added to the bottom of the tank. This is usually difficult to do. For this reason a second analogy may be developed.

(b) In this analogy, we associate the electric potential with the stream function and the electric flux function with the velocity potential, i. e. $V \sim \psi$ and $W \sim \varphi$. From Eqs. (7.96) and (7.97), we have

$$-u \sim tg \, , v \sim tf \, , \rho u \sim -\sigma g \, , \rho v \sim \sigma f$$

These relations will be compatible if $\rho \sim \sigma / t$.

For this analogy, we see that conductors (V=constant) correspond to streamlines (ψ =constant) and the depth is inversely proportional to the density. The operating method is similar to that of the first analogy except that insulators and conductors are interchanged, i.e., for the first analogy, the body tested is made of insulator, whereas in the second analogy, the body should be made of conductor. Now in the second analogy the bottom of the tank is scooped out instead of built up. This is usually a more convenient process.

The case of a uniform flow over a circular cylinder (§ 4) had been tested in an electric tank. The tests converged only if the free Mach number M_∞ was less than 0.45. For large velocities, each repetition required more scooping of the bottom of the tank than did the previous approximation.

12. Problems

1. Find the explicit form of the differential equations for φ_m of Eq. (7.9) for the cases:

(a) $m=2$, i.e., φ_2, and (b) $m=3$, i.e., φ_3.

2. Find the velocity potential of a uniform flow over a circular cylinder with circulation from the first-order equation φ of the Rayleigh-Jansen method.

3. Calculate the values of A_{mn} of Eq. (7.42) by taking $M=N=2$.

4. Show that

$$\frac{\partial^2\varphi}{\partial x^2}=\frac{-\varphi_{i+2,j}+16\varphi_{i+1,j}-30\varphi_{i,j}+16\varphi_{i-1,j}-\varphi_{i-2,j}}{12\Delta x^2}+\frac{\partial^6\varphi}{\partial x^6}\frac{\Delta x^4}{90}+O(\Delta x^6)$$

$$\frac{\partial\varphi}{\partial x}=\frac{\varphi_{i,j}-\varphi_{i-2,j}}{2\Delta x}+\frac{\partial^2\varphi}{\partial x^2}\Delta x+O(\Delta x^2)$$

$$\frac{\partial^2\varphi}{\partial x^2}=\frac{\varphi_{i,j}-2\varphi_{i-1,j}+\varphi_{i-2,j}}{\Delta x^2}+\frac{\partial^3\varphi}{\partial x^3}\Delta x+O(\Delta x^2)$$

$$\frac{\partial\varphi}{\partial x}=\frac{3\varphi_{i,j}-4\varphi_{i-1,j}+\varphi_{i-2,j}}{2\Delta x}+\frac{\partial^3\varphi}{\partial x^3}\frac{\Delta x^2}{3}+O(\Delta x^3)$$

$$\frac{\partial^2\varphi}{\partial x^2}=\frac{2\varphi_{i,j}-5\varphi_{i-1,j}+4\varphi_{i-2,j}-\varphi_{i-3,j}}{\Delta x^2}+\frac{\partial^4\varphi}{\partial x^4}\frac{11\Delta x^2}{12}+O(\Delta x^3)$$

$$\frac{\partial^2\varphi}{\partial x\partial y}=\frac{\varphi_{i,j}-\varphi_{i-1,j}-\varphi_{i,j-1}+\varphi_{i-1,j-1}}{\Delta x\Delta y}+\frac{\partial^3\varphi}{\partial x\,\partial y}\frac{\Delta x}{2}+\frac{\partial^3\varphi}{\partial x\partial y^2}\frac{\Delta y}{2}+O(\Delta x^2,\Delta y^2)$$

5. Write out the implicit scheme for the wave equation (7.56) and show that it is unconditionally stable.

6. The Richardson scheme for the heat equation (7.51) is

$$\frac{\varphi_j^{n+1}-\varphi_j^{n-1}}{2\Delta t}=\frac{\alpha}{\Delta x^2}(\varphi_{j+1}^n-2\varphi_j^n+\varphi_{j-1}^n)$$

Find the modified differential equation and study the numerical stability.

7. Show that if the SLOR for Laplace equation is incorporated with Jacobi iteration, the convergence condition is $\omega>0$ and the stability condition is $0<\omega\leqslant 1$.

References

1. Rayleigh, Lord, On the flow of compressible fluid past an obstacle, *Phil. Mag.* (6) **32**, 1916, pp. 1–6.
2. Janzen, O., Beitrag zu einer Theorie der stationaeren Stroemung kompressibler Fluessigkeiten, *Phys. Z* **14**, 1913, pp. 639–643.
3. Hargreaves, R., A pressure-integral as kinetic potential, *Phil. Mag.* (6) **16**, No. 3, Sept. 1908, pp. 436–444.
4. Bateman, H., Notes on a differential equation which occurs in the two-dimensional motion of a compressible fluid and the associated variational problems, *Proc. Roy. Soc. London* **A-125**, No. 799, Nov. 1, 1929, pp. 598–618.
5. Wang, C. T., Variational method in the theory of compressible fluid, *Jour. Aero. Sci.* **15**, No. 11, Nov. 1948, pp. 675–685.
6. Emmons, H. W., The numerical solution of compressible fluid flow *NACA*, *TN* No. 932, May 1944.
7. Poggi. L., Campo di velocita in una corrente piana di fluido compressible, *L'Aerotechnica*, **12**, p. 1579 (1932), and **14**, p. 532 (1934).
8. Imai, I., On the subsonic flow of a compressible fluid past a general Joukowki profile, *Rept. Aero. Res. Inst. Tokyo Imp. Univ.* No. **216**, 1941.
9. Kaplan, C., Two-dimensional subsonic flow past elliptic cylinders, *NACA Report* No. 624, 1938.
10. Whittaker, E. T., and Watson, G. N., *Modern Analysis*, Cambridge University Press, 4th edition, 1940.
11. Imai, I., On the flow of a compressible fluid past a circular cylinder, II *Proc. of the Phys. Math. Soc., Japan*, 3rd series **23**, No. 3, March 1941, pp. 180–193.
12. Wang, C. T., and Chou, P. C., Application of variational methods to transonic flows with shock waves, *NACA*, *TN* No. 2539, Nov. 1951.
13. O'Brien, G.G., Hyman, M.A., and Kaplan, S., A study of the numerical solutions of partial differential equations, *Jour. Math. Phys.* **29**, 1950, pp. 223–251.
14. Courant, R., Friedrichs, K.O., and Lewy, H., Über die Partiellen Differenzengleichungen der Mathematischen Physik, Math. Annalen, 100, 1928, pp. 32–74. (Translated to: On the partial difference equations of mathematical physics, *IBM Jour. Res.* Dec. 11, 1967, pp. 215–234.)
15. Ames, W. F., *Numerical Methods for Partial Differential Equations*, Academic Press, New York, 1977.
16. Garabedian, P. R., Estimation of the relaxation factor for small mesh size, *Math. Tables & Other Aids to Comp.* **10**, 1956, pp. 183–185.
17. Courant, R., and Hilbert, D., *Methods of Mathematical Physics*, I, Chapter V, § 5, pp. 297–298, Interscience Publishers, 1962.
18. Luo, S. J., Potential equation for transonic steady flow with large longitudinal disturbance and line relaxation, Proc. 2nd Asian Cong. *Fluid Mech.* 1983, Beijing, pp. 225–232.
19. Brandt, A., Multi-level adaptive solution to boundary value problems, *Math. Comp.* **31**, April 1977, pp. 333–390.
20. Schmidt, W., Jameson, A., and Whitfield, D., Finite-volume solutions to the Euler equations in transonic flow, *Jour. Aircraft*, **20**, No. 2, Feb. 1983, pp. 127–133.
21. Sells, C.C.L., Plane subcritical flow past a lifting airfoil, *RAE TR* 67146, June 1967, Royal Aircraft Establishment, England.
22. Taylor, G. I., and Sharman, C. F., A mechanical method for solving problems of flow in compressible fluids, *British ARC R and M* No. 1195, 1928; also *Proc. Roy. Soc. London* A–**121**, 1928, pp. 194–217.

Chapter VIII

HODOGRAPH METHOD[1-4] AND RHEOGRAPH METHOD[5-7]

1. Introduction

The fundamental equations for the velocity components of a compressible flow in terms of the space coordinates are nonlinear because the coefficients of the derivatives of the velocity components are themselves functions of the velocity. Since there is no general method for solving the nonlinear differential equations, it is necessary, in the theoretical investigation of compressible flow problems to linearize the fundamental equations. For two-dimensional steady irrotational flow, the fundamental equations can be linearized by inverting the roles of the dependent and the independent variables, i.e., by expressing the space coordinates or their equivalent in terms of the velocity components. Since the streamline in the plane with velocity components u and v as coordinates, i.e., hodograph plane, is known as hodograph, the method with u and v or q and θ as independent variables is called the hodograph method. The quantity q is the magnitude of the velocity vector, and θ is the angle between the velocity vector and the x-axis, i.e.,

$$u=q\cos\theta,\ v=q\sin\theta \tag{8.1}$$

It is usually possible to sketch the hodograph for a given problem investigated. For instance, for a compressible flow of a barotropic gas discharged from a tank of straight walls, Fig. 8.1a shows the flow in the physical plane, whereas Fig. 8.1b shows the corresponding hodographs. At infinity (point A in the tank) the velocity is zero, hence point A is at the origin of the hodograph plane. For the solid wall, AB, the x-component of velocity is zero, and the hodograph AB is along the v-axis in the hodograph plane. BC is a free streamline along which the magnitude of velocity vector is constant; hence BC is a circular arc with radius q_c in the hodograph plane. Similarly the hodograph $AB'C'$ may be drawn with negative v-components. For all the streamlines inside of ABC and $AB'C'$, the hodographs lie inside the semicircle $ABCB'$. Once the hodographs are known, it is not difficult to find the corresponding solution in the hodograph plane because the fundamental equation for two-dimensional irrotational flow is linear and the method of superposition may be used. The hodograph method cannot be used to linearize the fundamental equations of axially symmetrical flow or of rotational flow. As a result, the hodograph method has been extensively used only for two-dimensional steady irrotational flow.

There are several methods for transforming the fundamental equations of irrotational flow in the physical plane to the corresponding equations in the hodograph plane. In § 2, we discuss the Legendre transformation; in §3, the Molenbroek-Chaplygin transformation will be discussed.

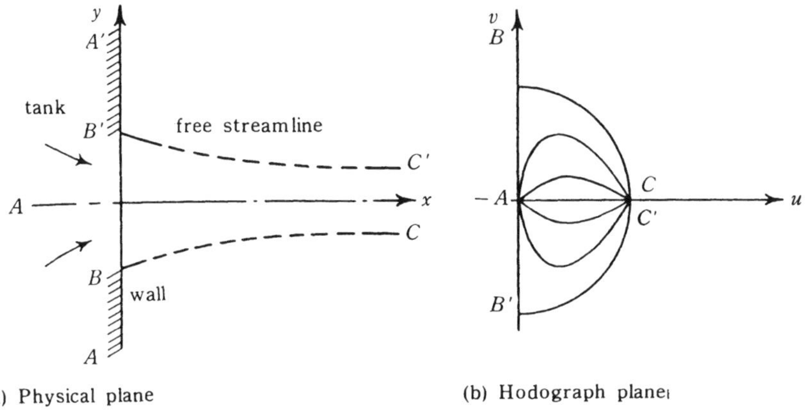

Fig. 8.1 Flow from a tank with straight walls

In §4, we shall discuss the formulation of the problem in the hodograph plane. The general solution of the hodograph method, which is still a difficult mathematical problem, will be briefly discussed in Chapter IX. In the present chapter, we shall discuss only some of the approximate solutions obtainable by the hodograph method. One of the most important approximate solutions is the well-known von Kármán-Tsien method; this will be discussed in detail in §5 to §7. Other approximate solutions obtained by the hodograph method will be discussed in §8.

The hodograph method is based upon the use of the variables (q, θ) or (u, v) as the independent variables of the problem. This is possible only if q and θ are not related to each other. If q is a function of θ, the hodograph method cannot be applied. The latter circumstance is the degenerate case or, as it is sometimes called, the "lost solution". It turns out to be one of the most important types of supersonic flow. The "lost solution" will be discussed in §9.

There is another transformation which was first used by Christianowitsch and which may also transform the fundamental equations of two-dimensional inviscid compressible flow into a system of linear equations. In this transformation[5–7], the Bertrami coordinates are used instead of the velocity components u and v or q and θ as independent variables. With the velocity components as independent variables, the study of compressible flow problems is known as hodograph method while with Bertrami coordinates s and t as independent variables, it is known as Rheograph method. Some very interesting results of compressible flow problems have been studied by Sobieczky[8, 9] by Rheograph method. We shall discuss the Rheograph method in § 10 and § 11.

In §12, some numerical solutions of hodograph method will be disucssed.[10]

2. Legendre transformation [1, 11]

The Legendre transformation, which is a special case of the contact transformation used in the theory of differential equations, is based on the idea that any given curve may be characterized as the envelope of a family of straight lines. Consider a function $\varphi\ (x)$ which is a curve in the φ-x plane (Fig. 8.2). Let us represent a family of straight lines by the slope p and intercept X and assume that X is a function of p, (see Fig. 8.2). It is easily seen that

$$\varphi = X + px \quad \text{or} \quad X = \varphi - px \tag{8.2}$$

In this way we represent the curve by a family of straight lines. For two neighboring points in the φ-x plane, we have

$$dX = d\varphi - pdx - xdp \tag{8.3}$$

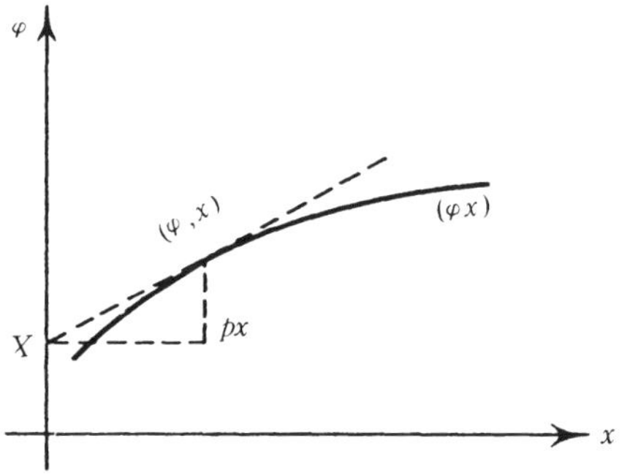

Fig. 8.2 Legendre transformation

Assuming that these two neighboring points are on the curve $\varphi\ (x)$ or $X\ (p)$

$$d\varphi = \frac{d\varphi}{dx}\, dx = pdx \tag{8.4}$$

Combining Eqs. (8.4) and (8.3) gives

$$\frac{dX}{dp} = -x \tag{8.5}$$

Usually the sign of X is taken to be the negative of that of our choice and then the first derivative gives the value x. The importance of this transformation lies in the fact that the resulting equation $X = X\ (p)$ is much simpler than the original equation of $\varphi = \varphi\ (x)$. If we can solve the equation of $X\ (p)$, then Eq. (8.5) gives $p = p\ (x)$ and $X = X\ (x)$. From Eq. (8.2) we have $\varphi = \varphi\ (x)$.

The above idea may be applied to the case of two independent variables. In this application it is called the Legendre transformation. In two-dimensional steady irrotational flow, we have a velocity potential φ which is a function of the spatial coordinates x and y, such that

$$u = \frac{\partial \varphi}{\partial x}, \quad v = \frac{\partial \varphi}{\partial y} \tag{8.6}$$

We shall now characterize the surface $\varphi\ (x, y)$ by means of the family of tangent planes represented by the two slopes (u, v) and the intercept X on the φ-axis. In a manner similar to the one-dimensional case discussed, we obtain

$$\varphi = X + ux + vy, \quad X = \varphi - ux - vy \tag{8.7}$$

and

$$dX = d\varphi - udx - xdu - vdy - ydv \tag{8.8}$$

where X is known as the Legendre potential.

If the two neighboring points lie on the surface $\varphi\ (x, y)$,

$$d\varphi = udx + vdy \tag{8.9}$$

Thus

$$dX = -xdu - ydv \tag{8.10}$$

and

$$x = -\frac{\partial X}{\partial u}, \quad y = -\frac{\partial X}{\partial v} \tag{8.11}$$

The differential equation for φ is

$$\left(1 - \frac{u^2}{a^2}\right)\frac{\partial^2 \varphi}{\partial x^2} - \frac{2uv}{a^2}\frac{\partial^2 \varphi}{\partial x \partial y} + \left(1 - \frac{v^2}{a^2}\right)\frac{\partial^2 \varphi}{\partial y^2} = 0 \tag{8.12}$$

[cf. Eq. (5.35)].

Now we shall transform the differential equation (8.12) for $\varphi\ (x, y)$ into the differential equation for $X\ (u, v)$ with the help of Eqs. (8.8) and (8.11). If the Jacobian

$$J=\frac{\partial x}{\partial u}\frac{\partial y}{\partial v}-\frac{\partial x}{\partial v}\frac{\partial y}{\partial u}=\frac{\partial^2 X}{\partial u^2}\frac{\partial^2 X}{\partial v^2}-\left(\frac{\partial^2 X}{\partial u\partial v}\right)^2 \qquad (8.13)$$

is not zero, it is simple to show that Eq. (8.12) may be transformed into the following form

$$\left(1-\frac{u^2}{a^2}\right)\frac{\partial^2 X}{\partial v^2}+\frac{2uv}{a^2}\frac{\partial^2 X}{\partial u\partial v}+\left(1-\frac{v^2}{a^2}\right)\frac{\partial^2 X}{\partial u^2}=0 \qquad (8.14)$$

When $J=0$, this transformation fails (see § 9).

Eq. (8.14) is linear in the dependent variable and its derivatives, whereas Eq. (8.12) is nonlinear in the first derivatives. Eq. (8.14) is thus much easier to solve than Eq. (8.12). However, the solution of Eq. (8.14) will be in the hodograph plane and it is very difficult to find a solution in this plane which satisfies given boundary conditions in the physical plane. We shall discuss this point further in § 4.

In the Legendre transformation, the physical interpretation of X is by no means clear. Furthermore, the velocity potential φ is unknown and the transformation of the boundary from the physical to the hodograph plane is undetermined. Usually it is more convenient to use the stream function ψ in place of the Legendre potential X. This method of linearization, which is known as the Molenbroek-Chaplygin transformation, will be discussed below.

3. Molenbroek-Chaplygin transformation [2-3]

The stream function is defined by [cf. Eq. (5. 8)]

$$\frac{\partial\psi}{\partial y}=\frac{\rho}{\rho_0}u,\quad \frac{\partial\psi}{\partial x}=-\frac{\rho}{\rho_0}v \qquad (8.15)$$

where ρ_0 is a constant and is the density at a reference point.

Bernoulli's equation is [cf. Eq. (5.31a)]

$$\frac{1}{2}q^2+\int\frac{dp}{\rho}=\text{constant} \qquad (8.16)$$

From the definitions of the velocity potential φ and the stream function ψ, we have

$$d\varphi=\frac{\partial\varphi}{\partial x}dx+\frac{\partial\varphi}{\partial y}dy=q\ (\cos\theta\ dx+\sin\theta\ dy)$$

$$d\psi = \frac{\partial \psi}{\partial x} dx + \frac{\partial \psi}{\partial y} dy = \frac{\rho}{\rho_0} q \, (-\sin\theta \, dx + \cos\theta \, dy) \tag{8.17}$$

Solving for dx and dy from Eq. (8.17) gives

$$\begin{aligned} dx &= \frac{\cos\theta}{q} d\varphi - \frac{\rho_0}{\rho} \frac{\sin\theta}{q} d\psi \\ dy &= \frac{\sin\theta}{q} d\varphi + \frac{\rho_0}{\rho} \frac{\cos\theta}{q} d\psi \end{aligned} \tag{8.18}$$

Now in the hodograph plane, we consider the velocity potential φ and the stream function ψ as functions of (u, v) or (q, θ). We then have

$$d\varphi = \frac{\partial \varphi}{\partial q} dq + \frac{\partial \varphi}{\partial \theta} d\theta, \; d\psi = \frac{\partial \psi}{\partial q} dq + \frac{\partial \psi}{\partial \theta} d\theta \tag{8.19}$$

Substituting Eq. (8.19) into Eq. (8.18), we find the expressions for dx and dy as functions of q and θ, as

$$\begin{aligned} dx &= \left(\frac{\cos\theta}{q} \frac{\partial \varphi}{\partial q} - \frac{\rho_0}{\rho} \frac{\sin\theta}{q} \frac{\partial \psi}{\partial q} \right) dq + \left(\frac{\cos\theta}{q} \frac{\partial \varphi}{\partial \theta} - \frac{\rho_0}{\rho} \frac{\sin\theta}{q} \frac{\partial \psi}{\partial \theta} \right) d\theta \\ dy &= \left(\frac{\sin\theta}{q} \frac{\partial \varphi}{\partial q} + \frac{\rho_0}{\rho} \frac{\cos\theta}{q} \frac{\partial \psi}{\partial q} \right) dq + \left(\frac{\sin\theta}{q} \frac{\partial \varphi}{\partial \theta} + \frac{\rho_0}{\rho} \frac{\cos\theta}{q} \frac{\partial \psi}{\partial \theta} \right) d\theta \end{aligned} \tag{8.20}$$

or in terms of a complex variable

$$\begin{aligned} dz &= dx + i dy = \frac{\partial z}{\partial q} dq + \frac{\partial z}{\partial \theta} d\theta \\ &= e^{i\theta} \left[\left(\frac{1}{q} \frac{\partial \varphi}{\partial q} + i \frac{\rho_0}{\rho} \frac{1}{q} \frac{\partial \psi}{\partial q} \right) dq + \left(\frac{1}{q} \frac{\partial \varphi}{\partial \theta} + i \frac{\rho_0}{\rho} \frac{1}{q} \frac{\partial \psi}{\partial \theta} \right) d\theta \right] \end{aligned} \tag{8.21}$$

From Eq. (8.21)

$$\frac{\partial z}{\partial q} = e^{i\theta} \left(\frac{1}{q} \frac{\partial \varphi}{\partial q} + i \frac{\rho_0}{\rho} \frac{1}{q} \frac{\partial \psi}{\partial q} \right)$$

$$\frac{\partial z}{\partial \theta}=e^{i\theta}\left(\frac{1}{q}\frac{\partial \varphi}{\partial \theta}+i\frac{\rho_0}{\rho}\frac{1}{q}\frac{\partial \psi}{\partial \theta}\right) \tag{8.22}$$

We differentiate the first expression $\partial z/\partial q$ with respect to θ and the second expression $\partial z/\partial \theta$ with respect to q and obtain

$$\frac{\partial^2 z}{\partial \theta \partial q}=e^{i\theta}\left[\left(\frac{1}{q}\frac{\partial^2 \varphi}{\partial \theta \partial q}-\frac{\rho_0}{\rho}\frac{1}{q}\frac{\partial \psi}{\partial q}\right)+i\left(\frac{\rho_0}{\rho}\frac{1}{q}\frac{\partial^2 \psi}{\partial \theta \partial q}+\frac{1}{q}\frac{\partial \varphi}{\partial q}\right)\right]$$

$$\frac{\partial^2 z}{\partial q \partial \theta}=e^{i\theta}\left\{\left(\frac{1}{q}\frac{\partial^2 \varphi}{\partial \theta \partial q}-\frac{1}{q^2}\frac{\partial \varphi}{\partial \theta}\right)+i\left[\frac{\rho_0}{\rho}\frac{1}{q}\frac{\partial^2 \psi}{\partial \theta \partial q}\right.\right. \tag{8.23}$$

$$\left.\left.+\frac{d}{dq}\left(\frac{\rho_0}{\rho}\frac{1}{q}\right)\cdot\frac{\partial \psi}{\partial \theta}\right]\right\}$$

Since $\dfrac{\partial^2 z}{\partial q \partial \theta}=\dfrac{\partial^2 z}{\partial \theta \partial q}$, Eq. (8.23) gives

$$\begin{aligned}\frac{\rho_0}{\rho}\frac{\partial \psi}{\partial q}&=\frac{1}{q}\frac{\partial \varphi}{\partial \theta}\\ \frac{\partial \varphi}{\partial q}&=q\frac{d}{dq}\left(\frac{\rho_0}{\rho}\frac{1}{q}\right)\cdot\frac{\partial \psi}{\partial \theta}\end{aligned} \tag{8.24}$$

Noting that

$$q\frac{d}{dq}\left(\frac{\rho_0}{\rho}\frac{1}{q}\right)=-(1-M^2)\frac{\rho_0}{\rho}\frac{1}{q} \tag{8.25}$$

where $M=\dfrac{q}{a}=\dfrac{q}{\sqrt{dp/d\rho}}$ =local Mach number of the flow, Eq. (8.24) becomes

$$\begin{aligned}\frac{\rho_0}{\rho}\frac{\partial \psi}{\partial q}&=\frac{1}{q}\frac{\partial \varphi}{\partial \theta}\\ -\frac{\rho_0}{\rho}(1-M^2)\frac{1}{q}\frac{\partial \psi}{\partial \theta}&=\frac{\partial \varphi}{\partial q}\end{aligned} \tag{8.26}$$

Eq. (8.26) is the fundamental equation of the hodograph method based on Molenbroek-Chaplygin transformation.

From Eq. (8.26) we may derive a fundamental equation for the velocity potential or the stream function. If we eliminate the velocity potential φ from

Eq. (8.26), we find the equation for stream function of compressible flow in the hodograph plane to be

$$\frac{\rho_0}{\rho} q \frac{\partial}{\partial q}\left(\frac{\rho_0}{\rho} q \frac{\partial \psi}{\partial q}\right)+\left(\frac{\rho_0}{\rho}\right)^2(1-M^2)\frac{\partial^2 \psi}{\partial \theta^2}=0 \tag{8.27}$$

It should be noted that the equation for the stream function is not the same as that for the velocity potential in the hodograph plane (q, θ), whereas these equations are the same in the physical plane (x, y). Since it is usually easy to sketch the streamlines in the hodograph plane for any given practical problem, it is often convenient to use the equation for stream function ψ to investigate compressible flow problems.

4. Formulation of the problem in hodograph plane

There are two different approaches to investigating compressible flow problems in the hodograph plane. From the purely theoretical point of view, we may find the exact solution of the fundamental equation (8.27) in the hodograph plane without prescribing the boundary conditions. After the solution is obtained, we transform it to the physical plane and investigate the type of flow field obtained. The results are very interesting. We shall discuss an example of this approach in Chapter IX.

For an engineering problem, we wish to find the flow field corresponding to a given boundary condition in the physical plane. We must transform this condition in the physical plane to a corresponding boundary condition in the hodograph plane. If the boundary condition in the physical plane is simple, it is possible to find the more or less exact boundary condition in the hodograph plane, and we may be able to find the required solution in the hodograph plane. For instance, for the jet problem of Fig. 8.1, we know the hodograph $ABCB'$ if the velocity on the free streamline BC is given. We may easily solve this problem according to the boundary condition $ABCB'$. In general, however, the solution is not always so simple. For instance, the most interesting problem in practice is the flow over a body placed in a uniform stream of velocity U (Fig. 8.3a). We may roughly sketch the corresponding hodographs of this problem in Fig. 8.3b. At infinity, the velocity is a constant $u=U$, $v=0$. Hence the point a represents all points at infinity in the physical plane. If we follow the x-axis from minus infinity to the stagnation point b of the body, we follow the u-axis in the hodograph plane from point a to the origin b. Then along the surface of the body, we know the direction of the velocity vector but not its magnitude. We can therefore only sketch approximately the hodograph bc whose velocity component v vanishes at the point c. Since we assume that the body is symmetrical with respect to y-axis, the hodograph cd is symmetrical to *bc* with respect to *u*-axis. All the streamlines in the upper half plane in the *x-y* plane are transformed into the hodograph inside the closed curve *bcd*. Hence the whole upper half plane in the *x-y* plane is transformed inside the closed curve *bcd*. However we do not know the curve *bcd* exactly before we have our required solution of $\psi(x, y)$. For this problem, the only thing which we know definitely is that there is a singular point at $q=U$, $\theta=0$. We can find only a solution similar to the hodograph sketched. We cannot predict whether a solution of equation (8.27) is satisfactory until we have calculated the physical coordinates (x, y). If we have the solution $\psi\ (q, \theta)$, the physical coordinates may be obtained by integration of the expression:

$$dz=dx+idy$$

$$=\frac{\rho_0}{\rho}\frac{e^{i\theta}}{q}\left\{-\left[(1-M^2)\frac{1}{q}\frac{\partial\psi}{\partial\theta}-i\frac{\partial\psi}{\partial q}\right]dq+\left(q\frac{\partial\psi}{\partial q}+i\frac{\partial\psi}{\partial\theta}\right)d\theta\right\} \tag{8.28}$$

If the boundary conditions are satisfied, the solution $\psi(q, \theta)$ is satisfactory. Thus the problem is not capable of being directly attacked.

This is one of the main difficulties of the hodograph method. Another difficulty is the existence of the singular point at $q=U$, $\theta=0$. We shall discuss all these in Chapter IX. It is desirable to find some simple approximate method for easily constructing a solution for the stream function in the hodograph plane which will satisfy the desired physical boundary conditions in the *x-y* plane. The best known method for doing this is the von Kármán-Tsien approximation.

5. Von Kármán-Tsien approximation [1, 12]

Eq. (8.27), although linear, is still difficult to solve. However from the results of incompressible flow, we may find some approximate solutions of Eq. (8.27). For incompressible flow, $M=0$ and $\rho=\rho_0$, Eq. (8.27) becomes

$$q_i\frac{\partial}{\partial q_i}\left(q_i\frac{\partial\psi_i}{\partial q_i}\right)+\frac{\partial^2\psi_i}{\partial\theta^2}=0 \tag{8.29}$$

where the subscript i refers to the value in the corresponding incompressible flow. If we use the transformation

$$\Lambda_i=\int\frac{dq_i}{q_i} \tag{8.30}$$

Eq. (8.29) reduces to the well-known Laplace equation, i.e.,

$$\frac{\partial^2 \psi_i}{\partial \Lambda_i^2} + \frac{\partial^2 \psi_i}{\partial \theta^2} = 0 \tag{8.31}$$

In a similar manner, we may transform Eq. (8.27) into the Laplace equation if we put

$$\Lambda = \int \frac{\rho_0}{\rho} \frac{dq}{q} \tag{8.32}$$

and

$$\left(\frac{\rho_0}{\rho}\right)^2 (1 - M^2) = 1 \tag{8.33}$$

Then

$$\frac{\partial^2 \psi}{\partial \Lambda^2} + \frac{\partial^2 \psi}{\partial \theta^2} = 0 \tag{8.34}$$

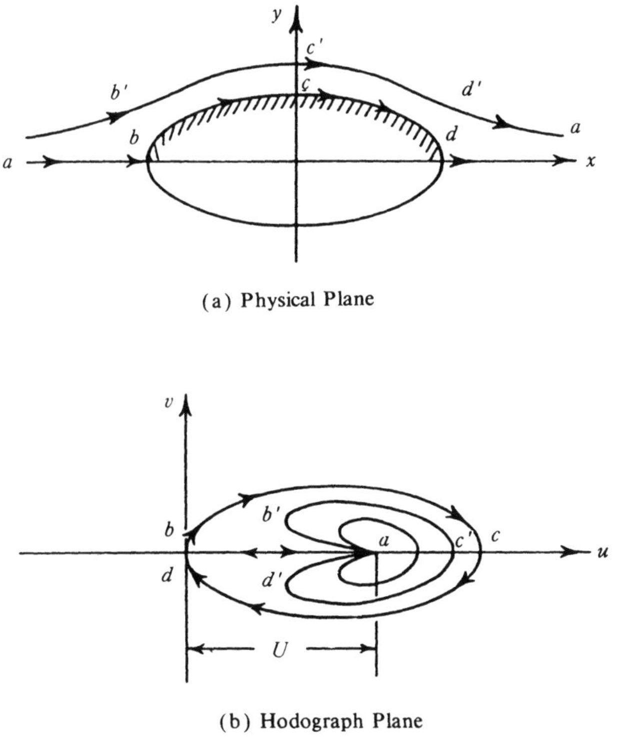

Fig. 8.3 Flow over a body in a uniform stream U

First we must find out what we mean by Eq. (8.33). For irrotational flow, we know that pressure is a function of density only, i.e. $p=p(\rho)$, but the function $p(\rho)$ is not specified. It is possible to find a function $p(\rho)$ such that Eq. (8.33) is satisfied. Eq. (8.33) may be written as follows:

$$\left[1-\left(\frac{\rho}{\rho_0}\right)^2\right]\frac{dp}{d\rho}=q^2 \tag{8.33a}$$

Differentiating Eq. (8.33a) with respect to ρ gives

$$\left[1-\left(\frac{\rho}{\rho_0}\right)^2\right]\frac{d^2p}{d\rho^2}-2\left(\frac{\rho}{\rho_0}\right)\frac{1}{\rho_0}\frac{dp}{d\rho}=2q\frac{dq}{d\rho} \tag{8.35}$$

From Bernoulli's equation (8.16) we have

$$q\frac{dq}{d\rho}+\frac{1}{\rho}\frac{dp}{d\rho}=0 \tag{8.36}$$

Eliminating $dq/d\rho$ from Eqs. (8.35) and (8.36) gives

$$\left[1-\left(\frac{\rho}{\rho_0}\right)^2\right]\left(\frac{d^2p}{d\rho^2}+\frac{2}{\rho}\frac{dp}{d\rho}\right)=0 \tag{8.37}$$

Eq. (8.37) gives the conditions under which Eq. (8.33) is true. One condition is

$$\rho=\rho_0=\text{constant} \tag{8.38}$$

This corresponds to incompressible flow.

The other condition is

$$\frac{d^2p}{d\rho^2}+\frac{2}{\rho}\frac{dp}{d\rho}=0 \tag{8.39}$$

By integrating Eq. (8.39) twice with respect to ρ, we have

$$p=A-\frac{B}{\rho} \tag{8.40}$$

where A and B are constants of integration. If the p-ρ relation of a gas satisfies Eq. (8.40), Eq. (8.34) is exact. But no real gas is known to have this particular p-ρ relation. However, as first pointed out by Demtchenko,[11] the straight line in p-$1/\rho$ plane (Fig. 8.4), i.e., Eq. (8.40), can be made a tangent to the true isentropic curve of the real gas, and thus serves as a good approximation to the real gas for conditions near the point of tangency. This

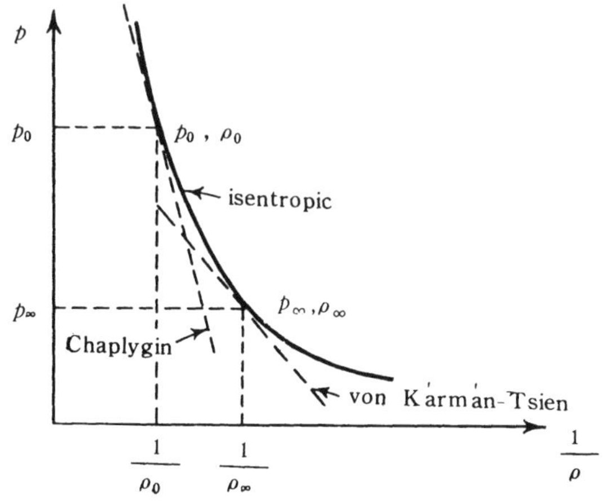

Fig. 8.4 Pressure-density relation in the von Kármán-Tsien approximation

approximation was known to Chaplygin[3] who chose the stagnation point as the point of tangency. Such an approximation is good only for very small Mach numbers. In the calculation of the flow about a body in a uniform stream, von Kármán and Tsien suggested that the point of tangency should be made at the point corresponding to the uniform stream conditions. Thus a good approximation is expected for a thin body in a uniform stream. They worked out the correction formulas for the pressure and velocity over a thin body. These are the well-known von Kármán-Tsien approximations.[12]

In the von Kármán-Tsien approximation, the point of tangency is the point corresponding to the free stream condition, i.e.,

$$B=\rho^2a^2=\rho_0^2a_0^2=\rho_\infty^2 a_\infty^2 \tag{8.41}$$

and

$$p=p_\infty=\rho_\infty^2a_\infty^2\left(\frac{1}{\rho_\infty}-\frac{1}{\rho}\right) \tag{8.42}$$

where the subscript 0 refers to the value at stagnation point; subscript ∞, free stream conditions.

The corresponding Bernoulli's equation (8.16) with the relation (8.42) becomes

$$\left(\frac{\rho_0}{\rho}\right)^2=1+\frac{q^2}{a_0^2} \tag{8.43}$$

From Eqs. (8.41) and (8.43), we have

$$\left(\frac{\rho}{\rho_0}\right)^2 = 1 - \frac{q^2}{a^2} = 1 - M^2 \tag{8.44}$$

Eq. (8.43) shows that the density of the fluid decreases as the velocity q increases. This is true for a real gas. Eq. (8.41), however, requires the velocity of sound a to increase with increasing q. This is in contradiction with the properties of a real fluid. Eq. (8.44) shows that the local Mach number M actually increases with q because ρ decreases as q increases. Since $(\rho/\rho_0)^2$ is always positive, the local Mach number must remain subsonic, and there can never be any transonic flow in this approximation. It is also interesting to note that this approximation may be applied to the case where the flow is entirely supersonic.[5]

One of the most important contributions of the von Kármán-Tsien method is that it provides a method for constructing the solution ψ of compressible fluid flow which will satisfy approximately the desired physical boundary conditions in the physical plane (x, y). If we want to investigate the flow field around a given body, we first find the corresponding incompressible solution in the physical plane, i.e., $\psi_1(x, y)$. We then transform this incompressible solution to the hodograph variables Λ_i and θ, i.e., $\psi_i(\Lambda_i, \theta)$. The corresponding compressible solution ψ is then

$$\psi = \psi_i(\Lambda, \theta) \tag{8.45}$$

That is, we simply replace Λ_i in the incompressible solution by Λ. Since ψ_1 satisfies Eq. (8.31), ψ satisfies Eq. (8.34). Furthermore, if we put the restriction that

$$q \to q_i \text{ as } M_\infty \to 0 \tag{8.46}$$

then

$$\psi(q, \theta) \to \psi_i(q_i\, \theta) \text{ as } M_\infty \to 0 \tag{8.47}$$

We thus obtain the compressible solution ψ with a parameter M_∞. Of course, as the parameter M_∞ changes, the shape of the body considered changes. One may fail to obtain exactly the desired body but the deviation will not be great.

Now let us determine the meaning of Eqs. (8.46) and (8.47).

By definition [Eqs. (8.30) and (8.32)],

$$\Lambda = \Lambda_i = \log q_i = \int \frac{\rho}{\rho_0}\frac{dq}{q} = \int \frac{dq}{q\sqrt{1 + \dfrac{q^2}{a_0^2}}} \tag{8.48}$$

Integration of Eq. (8.48) gives

$$q_i = \frac{Cq}{1+\sqrt{1+\frac{q^2}{a_0^2}}} \tag{8.49}$$

where C is a constant of integration. In the limit, $a_0 \rightarrow \infty$, $q = q_i$ and $C = 2$.

From Eq. (8.49),

$$q = \frac{4a_0^2 q_i}{4a_0^2 - q_i^2} \tag{8.50}$$

Let U be the free stream velocity of the compressible flow and U_i, the free stream velocity of the incompressible flow. Then

$$U = \frac{4a_0^2 U_i}{4a_0^2 - U_i^2} \tag{8.51}$$

From Eqs. (8.50) and (8.51)

$$\frac{q}{U} = \frac{q_i}{U_i} \frac{1-\lambda}{1-\lambda\left(\frac{q}{U_i}\right)^2} \tag{8.52}$$

where

$$\lambda = \frac{U_1^2}{4a_0^2} = \left[\frac{U/a_0}{1+\sqrt{1+\left(\frac{U}{a_0}\right)^2}}\right]^2 = \frac{M_\infty^2}{[1+\sqrt{1-M_\infty^2}]^2} \tag{8.53}$$

Eq. (8.52) gives the relation of the velocity ratio in compressible flow to that of the corresponding incompressible flow in terms of the parameter M_∞, the free stream Mach number.

Similarly, the pressure coefficient in compressible flow in terms of the pressure coefficient at a corresponding point in the incompressible flow is

$$C_P = \frac{C_{Pi}}{\sqrt{1-M_\infty^2} + \frac{M_\infty^2}{1+\sqrt{1-M_\infty^2}} \frac{C_{Pi}}{2}} \tag{8.54}$$

which is the well-known von Kármán-Tsien formula for pressure correction. As C_{Pi} tends to zero, it reduces to the Prandtl-Glauert formula (6.50).

Eqs. (8.52) and (8.53) give respectively the velocity and pressure correction for corresponding points. Let (x, y) be the point in compressible flow and (x_i, y_i) be the corresponding point in incompressible flow. We would like to find the relation between $z=x+iy$ and $z_i=x_i+iy_i$.

For incompressible fluids, the complex potential is a function of the conjugate complex velocity, i.e.,

$$\varphi_i+i\psi_i=F\ (\overline{w}_i) \tag{8.55}$$

where

$$\overline{w}_i=u_i-iv_i=q_ie^{i\theta}=\frac{dF}{dz_i} \tag{8.56}$$

in which the bar refers to the complex conjugate quantity.

In the von Kármán-Tsien approximation, we use

$$\varphi+i\psi=F\ (\overline{w}_i) \tag{8.57}$$

$$\varphi-i\psi=\overline{F}\ (w_i)$$

where $w_i=q_ie^{i\theta}$ and q_i is connected to the compressible velocity q by Eq. (8.50). Now we may calculate the spatial coordinates from Eq. (8.18) in terms of the incompressible velocity q_i. We obtain

$$\begin{aligned} dx &= \frac{\cos\theta}{4a_0^2q_i}(4a_0^2-q_i^2)d\varphi-\frac{\sin\theta}{4a_0^2q_i}(4a_0^2+q_i^2)d\psi \\ dy &= \frac{\sin\theta}{4a_0^2q_i}(4a_0^2-q_i^2)d\varphi+\frac{\cos\theta}{4a_0^2q_i}(4a_0^2+q_i^2)d\psi \end{aligned} \tag{8.58}$$

or

$$dz+dx+idy=\frac{dF}{q_ie^{-i\theta}}-\frac{U_i^2}{4a_0^2}\left(\frac{qe^{i\theta}}{U_i}\right)\frac{d\overline{F}}{q_ie^{i\theta}}=dz_i-\lambda\left(\frac{1}{U}\frac{d\overline{F}_i}{d\overline{z}_i}\right)^2d\overline{z}_i \tag{8.59}$$

Integration of Eq. (8.59) gives

$$z=z_i-\lambda\int\left(\frac{d\overline{G}}{d\overline{z}_i}\right)^2d\overline{z}_i \tag{8.60}$$

with $F\ (z_i)=U_iG\ (z_i)$.

Eq. (8.60) is the formula for coordinate correction which has again a parameter λ or M_∞. If we compare boundaries in the compressible and incompressible flows, we see that there is a relative distortion which is proportional to the compressibility factor λ and depends on the incompressible flow potential function. To a first approximation, we may put $z=z_i$, i.e., the solid boundaries in the compressible and incompressible flows have the same shape. In practice, the pressure correction formula (8.54) is assumed to give the relation between C_p at a point on a body in compressible flow and C_{Pi} at the same point on the same body in an incompressible flow. With this interpretation, the agreement of the von Kármán-Tsien theoretical results with experimental data is very good up to the condition where the local velocity of sound is reached. We shall discuss further the reason for such an application in the next section.

6. Flow around an elliptical cylinder

Now we apply the von Kármán-Tsien method to the study of the subsonic flow over a family of elliptical cylinders whose incompressible flow solution is easily obtained by the application of Joukowsky transformation.

Using the Joukowsky transformation

$$z_i = \zeta + \frac{1}{\zeta} \tag{8.61}$$

an elliptical cylinder in the incompressible physical plane z_i may be transformed into a circular cylinder of radius r_0 in ζ-plane (Fig. 8.5). For a circle in the ζ-plane of redius r_0 and center at origin, we have

$$\zeta_c = r_0 e^{i\theta} \tag{8.62}$$

Substituting Eq. (8.62) into Eq. (8.61), we find the coordinates of the cylinder in z_i-plane to be

$$x_{ic} = \left(r_0 + \frac{1}{r_0}\right)\cos\theta\,,\; y_{ic} = \left(r_0 - \frac{1}{r_0}\right)\sin\theta \tag{8.63}$$

Eliminating θ from Eq. (8.63) gives

$$\frac{x_{ic}^2}{\left(r_0 + \frac{1}{r_0}\right)^2} + \frac{y_{ic}^2}{\left(r_0 - \frac{1}{r_0}\right)^2} = 1 \tag{8.64}$$

This is the equation of an ellipse whose major axis is $2\left(r_0 + \frac{1}{r_0}\right)$ and minor axis, $2\left(r_0 - \frac{1}{r_0}\right)$. The thickness ratio of the ellipse is

$$\delta_i = \frac{y_{ic\max}}{x_{ic\max}} = \frac{r_0^2 - 1}{r_0^2 + 1} \tag{8.65}$$

If we consider the case of a uniform flow of velocity U in the direction of major axis passing over the elliptical cylinder, the complex potential in ζ-plane

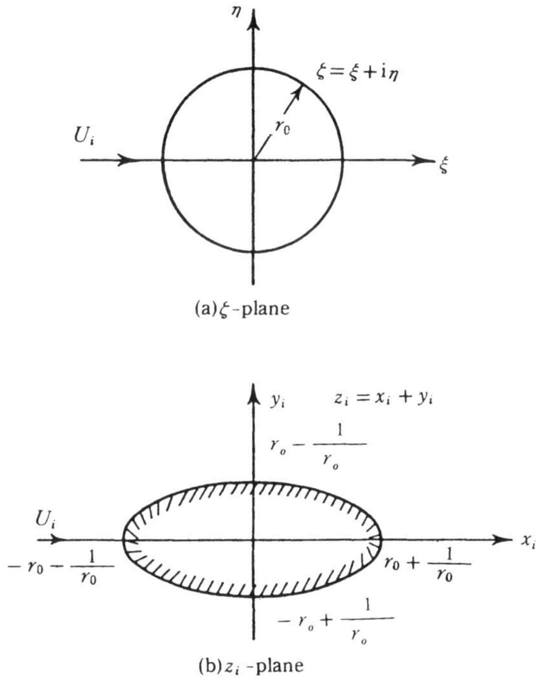

Fig. 8.5 Flow over an elliptical cylinder

is given by

$$G(\zeta) = \zeta + \frac{r_0^2}{\zeta} \tag{8.66}$$

Since

$$\frac{d\bar{G}}{d\bar{z}_i} = \frac{d\bar{G}}{d\bar{\zeta}} \bigg/ \frac{d\bar{z}_i}{d\bar{\zeta}} \tag{8.67}$$

Eq. (8.60) gives the coordinates for the corresponding compressible flow as

$$z = \left(\zeta + \frac{1}{\zeta}\right) - \lambda\left[\bar{\zeta} + \frac{r_0^4}{\bar{\zeta}} + \frac{(r_0^2 - 1)^2}{2}\log\frac{\bar{\zeta} - 1}{\bar{\zeta} + 1}\right] \tag{8.68}$$

On the surface of the elliptical cylinder, we substitute Eq. (8.62) in Eq. (8.68) and have

$$z_c = r_0(\cos\theta + i\sin\theta) + \frac{1}{r_0}(\cos\theta - i\sin\theta)$$

$$-\lambda\Bigg[r_0(\cos\theta - i\sin\theta) + r_0^3(\cos\theta + i\sin\theta)$$

$$+\frac{(r_0^2-1)^2}{2}\log\frac{(r_0\cos\theta - 1) - ir_0\sin\theta}{r_0\cos\theta + 1 - ir_0\sin\theta}\Bigg] \tag{8.69}$$

The thickness ratio of the elliptical cylinder in compressible flow is then

$$\delta = \frac{y_{c\max}}{x_{c\max}} = \delta_i \frac{1-\lambda\left[r_0^2 - \dfrac{r_0(r_0^2-1)}{2}\tan^{-1}\dfrac{2r_0}{r_0^2-1}\right]}{1-\lambda\left[r_0^2 - \dfrac{r_0(r_0^2-1)^2}{2(r_0^2+1)}\log\dfrac{r_0+1}{r_0-1}\right]} \tag{8.70}$$

It is easy to show that $\delta > \delta_i$ because of $r_0 > 1$. The thickness ratio of the body in compressible flow is always larger than the corrres-ponding body in incompressible flow. Therefore if we ignore this increase in thickness ratio from the incompressible flow to the compressible flow, we would have overestimated the effect of compressibility. However, in the von Kármán-Tsien approximation, a tangent is used to replace the true isentropic pressure-density relation. This results in an underestimation of the effect of compressibility. These two effects compensate each other. In practice, the pressure and the velocity correction formulas (8.54) and (8.52) are applied for the same surface points on the same body without the coordinate correction as suggested in the last section.

7. Improvements of the von Kármán-Tsien approximation[13–15]

If there is circulation around the body in the incompressible flow, the coordinate correction formula (8.60) will not yield a closed contour for the solid body in the compressible flow. Lin[13] improved the von Kármán-Tsien method so that this defect was removed. In Lin's method, let the solid body in incompressible flow be P_0. The incompressible solution is given by the complex potential $F(z_i)$, and the complex velocity is $w(z_i)$. Lin chose a function $k(z_i)$, regular in the region R_0 exterior to P_0 and including the point at infinity, having no root in R_0 such that

$$\left|\frac{1}{2}w(z_i)\right| < |k(z_i)| < \infty \quad \text{on } P_0 \tag{8.71}$$

and that

$$\oint k(z_i)dz_i - \frac{1}{4}\oint\frac{\overline{w^2(z_i)}}{k(z_i)}dz_i = 0 \tag{8.72}$$

where the integration is performed along any contour enclosing P_0. Then

$$z = x + iy = \int k(z_i)dz_i - \frac{1}{4}\overline{\int \frac{w^2(z_i)}{k(z_i)}\,dz_i} \tag{8.73}$$

$$\frac{2qe^{i\theta}}{1+\sqrt{1+q^2}} = \frac{w(z_i)}{k(z_i)} \tag{8.74}$$

$$\varphi + i\psi = F\ (z_i) \tag{8.75}$$

gives the parametric representation of a compressible flow past a body P in the x-y plane with z_i as parameter, where P has the same general analytic nature as the original profile P_0.

This improvement makes the von Kármán-Tsien theory complete.

Poritsky[14] proposed a generalization of von Kármán-Tsien's method by using a series of straight-line segments to achieve a closer approximation to the true isentropic pressure-density relation. His method is much more complicated than von Kármán-Tsien's original method because one has to join the different regions of the solution corresponding to the different straight-line segments in the pressure density diagram. Kármán-Tsien's method may also be applied to supersonic flow.[15]

8. Other approximations to the hodograph equation. Transonic flow

The von Kármán-Tsien approximation cannot be applied to the case of transonic flow, because the hodograph equation in their approximation is reduced to either the Laplace equation or the wave equation. For transonic flow, the hodograph equation should be of the mixed type. In order to get a good approximation, the simplified hodograph equation should also be of the mixed type. To discuss other approximations, we return to the original fundamental equation (8.27). If we introduce a new variable σ such that

$$d\sigma = -\frac{\rho}{\rho_0}\frac{dq}{q} \quad \text{or} \quad \sigma = -\int_{a^*}^{q}\frac{\rho}{\rho_0}\frac{dq}{q} \tag{8.76}$$

Eq. (8.27) becomes

$$K\ (\sigma)\frac{\partial^2\psi}{\partial\theta^2} + \frac{\partial^2\psi}{\partial\sigma^2} = 0 \tag{8.77}$$

with

$$K\ (\sigma) = (1-M^2)\left(\frac{\rho_0}{\rho}\right)^2 \tag{8.78}$$

In general, $K\ (\sigma)$ is a complicated function of σ. In various approximations, one uses simple form of $K(\sigma)$ to replace the exact expression of $K(\sigma)$ for true isentropic flow. The following are a few of the approximations which have been investigated.[16, 17]

(a) *Von Kármán-Tsien approximation.* In this approximation, we simply put $K(\sigma)=1$. It was Molenbroek who verified that $K(\sigma)$ is approximately equal to unity for the local Mach number M_∞ lying between 0 and 0.5. This approximation is valid only for purely subsonic flow ($K=+1$) or purely supersonic flow ($K=-1$).

(b) *Tricomi equation and transonic flow.*[18–20] At sonic speed $\sigma=0$. Near the sonic line, $K(\sigma)$ is very close to σ with a suitable choice of the scale. For transonic flow, then, a good approximation may be obtained if we put $K(\sigma)=\sigma$. The fundamental equation (8.77) becomes

$$\sigma\frac{\partial^2\psi}{\partial\theta^2}+\frac{\partial^2\psi}{\partial\sigma^2}=0 \tag{8.79}$$

This is known as the Tricomi equation. It was first investigated thoroughly by Tricomi from a mathematical point of view and has been used extensively in the investigation of transonic flow. This equation is of the mixed type. For $\sigma>0$, it is of elliptical type, whereas for $\sigma<0$, it is of hyperbolic type. We shall discuss it a little more in Chapter X.

(c) *Second-order approximation.* The Tricomi equation is good near the sonic lines. In order to increase the range of validity, the following approximation may be used,

$$K(\sigma)=\frac{\sigma}{A+B\sigma} \tag{8.80}$$

where A and B are constants chosen so that the approximate flow will match the exact solution at both the sonic line and at the free stream velocity. Alternatively, they may be chosen so that both the value and the first derivative of the approximate solution at the sonic line match those of the exact solution. Since we have two constants to be chosen, this approximation is sometimes called the second-order approximation.

Of course we may extend this approximation to the third approximation by putting

$$K(\sigma)=\frac{\sigma(1+A\sigma)}{(B+C\sigma)^2} \tag{8.81}$$

where A, B, and C are constants to be chosen so that the pressure-density relation in the approximation is in good agreement to the true isentropic relation.

The general solution of Eq. (8.79) and of those based on the approximations (8.80) and (8.81) will be discussed in Chapter IX where the general exact solution of Eq. (8.27) is discussed.

Other simplified methods of solution have been also worked out by some authors. For instance, Imai used successfully the *WKB* method to simplify Eq. (8.77).[21–23]

9. Lost solution[24]

In the hodograph method we use the variables (q, θ) or (u, v) instead of the spatial coordinates (x, y) as independent variables. This is possible only when q and θ are not related to each other. If they are related, we cannot use the hodograph method because we have only one variable. In fact this is the degenerate case which is not included in our previous treatment. It is sometimes called the lost solution. This solution is one of the most important types of flow from a practical point of view. It is the well-known Prandtl-Meyer expansion for supersonic flow around a corner.

Mathematically, the condition for the failure of hodograph transformation from $q\ (x, y)$, $\theta\ (x, y)$ to $x\ (q, \theta)\ y\ (q, \theta)$ is given by the vanishing of the Jacobian j, i.e.,

$$j=\frac{\partial(q,\theta)}{\partial(x,y)}=\frac{\partial q}{\partial x}\frac{\partial\theta}{\partial y}-\frac{\partial q}{\partial y}\frac{\partial\theta}{\partial x}=0 \tag{8.82}$$

If the velocity magnitude q is a function of angle θ, i.e., $q=q\ (\theta)$, Eq. (8.82) becomes

$$j=\frac{\partial(q,\theta)}{\partial(x,y)}=\frac{dq}{d\theta}\left(\frac{\partial\theta}{\partial x}\frac{\partial\theta}{\partial y}-\frac{\partial\theta}{\partial y}\frac{\partial\theta}{\partial x}\right)\equiv 0 \tag{8.82a}$$

so that the hodograph transformation fails.

Let us consider the functions $q\ (x, y)$ and $\theta\ (x, y)$ and their inversion $x\ (q, \theta)$ and $y\ (q, \theta)$, defining the Jacobian j as

$$j=\frac{\partial q}{\partial x}\frac{\partial\theta}{\partial y}-\frac{\partial q}{\partial y}\frac{\partial\theta}{\partial x}=\begin{vmatrix}\dfrac{\partial q}{\partial x} & \dfrac{\partial q}{\partial y}\\[2mm] \dfrac{\partial\theta}{\partial x} & \dfrac{\partial\theta}{\partial y}\end{vmatrix} \tag{8.83}$$

By definition

$$dq=\frac{\partial q}{\partial x}dx+\frac{\partial q}{\partial y}dy,\quad d\theta=\frac{\partial\theta}{\partial x}dx+\frac{\partial\theta}{\partial y}dy \tag{8.84}$$

From Eq. (8.84), we have

$$jdx=\frac{\partial\theta}{\partial y}dq-\frac{\partial q}{\partial y}d\theta=j\left(\frac{\partial x}{\partial q}dq+\frac{\partial x}{\partial\theta}d\theta\right) \tag{8.85}$$

$$-j dy = \frac{\partial \theta}{\partial x} dq - \frac{\partial q}{\partial x} d\theta = -j\left(\frac{\partial y}{\partial q} dq + \frac{\partial y}{\partial \theta} d\theta \right)$$

If $j \neq 0$, Eq. (8.85) gives

$$\frac{\partial q}{\partial x} = j\frac{\partial y}{\partial \theta}, \quad \frac{\partial q}{\partial y} = -j\frac{\partial x}{\partial \theta}$$
$$\frac{\partial \theta}{\partial x} = -j\frac{\partial y}{\partial q}, \quad \frac{\partial \theta}{\partial y} = j\frac{\partial x}{\partial q} \tag{8.86}$$

Only when the Jacobian j is not equal to zero may we substitute the relation (8.86) into the equation of $q\ (x, y)$ and $\theta\ (x, y)$ to obtain a corresponding equation of $x\ (q, \theta)$ and $y\ (q, \theta)$. The hodograph transformation fails whenever $j = 0$.

Similarly, if we want to transform back from $x\ (q, \theta)$ and $y\ (q, \theta)$ to the variables $q\ (x, y)$ and $\theta\ (x, y)$, the Jacobian J

$$J = \frac{\partial\ (x, y)}{\partial\ (q, \theta)} \tag{8.87}$$

must not be zero.

The lost solution $q\ (\theta)$ is the Prandtl-Meyer expansion flow of supersonic speed around a corner. The basic equations for the irrotational two-dimensional flow may be written as

$$\left(1 - \frac{u^2}{a^2}\right)\frac{\partial u}{\partial x} - \frac{uv}{a^2}\left(\frac{\partial v}{\partial x} + \frac{\partial u}{\partial y}\right) + \left(1 - \frac{v^2}{a^2}\right)\frac{\partial v}{\partial y} = 0$$
$$\frac{\partial v}{\partial x} - \frac{\partial u}{\partial y} = 0 \tag{8.88}$$

If we put $q = q\ (\theta)$ into Eq. (8.88) we have

$$\left[(1 - M^2)\cos\theta \frac{dq}{d\theta} - q\sin\theta\right]\frac{\partial \theta}{\partial x} + \left[(1 - M^2)\sin\theta\frac{dq}{d\theta} + q\cos\theta\right]\frac{\partial \theta}{\partial y} = 0$$
$$\left[\sin\theta\frac{dq}{d\theta} + q\cos\theta\right]\frac{\partial \theta}{\partial x} + \left[-\cos\theta\frac{dq}{d\theta} + q\sin\theta\right]\frac{\partial \theta}{\partial y} = 0 \tag{8.89}$$

where $M = \frac{q}{a}$.

If $\frac{\partial \theta}{\partial x} \neq \frac{\partial \theta}{\partial y} \neq 0$, the determinant formed from the coefficients of the two equations of (8.89) must be zero. Hence we have

$$\begin{vmatrix} (1-M^2)\cos\theta\dfrac{dq}{d\theta}-q\sin\theta & (1-M^2)\sin\theta\dfrac{dq}{d\theta}+q\cos\theta \\ \sin\theta\dfrac{dq}{d\theta}+q\cos\theta & -\cos\theta\dfrac{dq}{d\theta}+q\sin\theta \end{vmatrix} \tag{8.90}$$

$$=(1-M^2)\left(\frac{dq}{d\theta}\right)^2+q^2=0$$

or

$$\frac{dq}{d\theta}=\pm\frac{q}{\sqrt{M^2-1}} \tag{8.91}$$

This relation between q and θ happens to be the same relation as the characteristics in the hodograph plane. It shows that this solution is real only for supersonic flow $M>1$. We shall discuss it more in Chapters IX and XI.

10. Christianowitsch transformation and Betrami relations[5, 6, 7]

Christianowitsch introduced a new variable Λ such that

$$\begin{aligned} M\leqslant 1: d\Lambda &= \sqrt{(1-M^2)}\,\frac{dq}{q}\ ;\ \Lambda=\int_{a^*}^{q}\sqrt{(1-M^2)}\,\frac{dq'}{q'} \\ M\geqslant 1: d\Lambda &= \sqrt{(M^2-1)}\,\frac{dq}{q}\ ;\ \Lambda=\int_{a^*}^{q}\sqrt{(M^2-1)}\,\frac{dq'}{q'} \end{aligned} \tag{8.92}$$

Substituting Eqs. (8.92) into Eqs. (8.26), we have

$$\begin{aligned} M\leqslant 1: &\ \frac{\partial\Lambda}{\partial\varphi}+\frac{\rho}{\sqrt{(1-M^2)}}\frac{\partial\theta}{\partial\psi}=0;\ \frac{\rho}{\sqrt{(1-M^2)}}\frac{\partial\Lambda}{\partial\psi}-\frac{\partial\theta}{\partial\varphi}=0 \\ M\geqslant 1: &\ \frac{\partial\Lambda}{\partial\varphi}-\frac{\rho}{\sqrt{(M^2-1)}}\frac{\partial\theta}{\partial\psi}=0;\ \frac{\rho}{\sqrt{(M^2-1)}}\frac{\partial\Lambda}{\partial\psi}-\frac{\partial\theta}{\partial\varphi}=0 \end{aligned} \tag{8.93}$$

For further simplification of Eqs. (8.93), we use the Betrami relations

$$\begin{aligned} M\leqslant 1: &\ \frac{\partial\varphi}{\partial s}=K\frac{\partial\psi}{\partial t}\ ;\ \frac{\partial\varphi}{\partial t}=-K\frac{\partial\psi}{\partial s}\ ;\ K(\Lambda)=\frac{\sqrt{(1-M^2)}}{\rho} \\ M\geqslant 1: &\ \frac{\partial\varphi}{\partial s}=K\frac{\partial\psi}{\partial t}\ ;\ \frac{\partial\varphi}{\partial t}=K\frac{\partial\psi}{\partial s}\ ;\ K(\Lambda)=\frac{\sqrt{(M^2-1)}}{\rho} \end{aligned} \tag{8.94}$$

Eqs. (8.94) may be considered as the definition of the new variables s and t. Substituting Eqs. (8.94) into Eqs. (8.93), we have

$$\begin{aligned} M\leqslant 1: &\ \frac{\partial\Lambda}{\partial s}+\frac{\partial\theta}{\partial t}=0\ ;\ \frac{\partial\Lambda}{\partial t}-\frac{\partial\theta}{\partial s}=0 \\ M\geqslant 1: &\ -\frac{\partial\Lambda}{\partial s}+\frac{\partial\theta}{\partial t}=0\ ;\ \frac{\partial\Lambda}{\partial t}-\frac{\partial\theta}{\partial s}=0 \end{aligned} \tag{8.95}$$

Now we have very simple Eqs. (8.95) for the two-dimensional steady potential flow. But we have different equations for the subsonic region from those of the supersonic region.

From Eqs. (8.95), we have the characteristics for the system of Eqs. (8.95) as follows:

$$\begin{aligned} M \leqslant 1 &: \xi = s + it,\ \bar{\eta} = s - it \\ M \geqslant 1 &: \xi = s - t,\ \eta = s + t \end{aligned} \tag{8.96}$$

where $i = \sqrt{-1}$. In the supersonic region we have the real hyperbolic characteristics ξ and η while in the subsohic region, we have the complex elliptical characteristics. It is possible to use the numerical solution of the complex characteristics of Eqs. (8.96) to calculate the shockless transonic potential flow over an airfoil[10] as we shall discuss in § 12.

11. Sobieszky's exact solutions of two-dimensional potential flow by rheograph method[8, 9]

Sobieczky used a Beltrami's transformation of the coordinates to the new variables s and t and obtained a Cauchy-Riemann system of equations from the gasdynamic equation and the equation of irrotationality. The study of the compressible flow problems with the velocity components u and v or q and θ as independent variables is known as the Hodograph method, while that with Beltrami coordinates s and t as independent variables is known as Rheograph method. In both the hodograph method and the Rheograph method, the fundamental equations are linear and the method of superposition is applicable. We may easily find the elementary solutions of the fundamental equations in the hodograph method or the Rheograph method and then by the method of superposition, we may construct useful solution from various elementary solutions. We are going to follow Sobieczky to use the Rheograph method and to find some known solution in this section. In Chapter X, we shall discuss the application of Sobieczky's solutions for the transonic flow of small disturbances.

Instead of Eqs. (8.94), we consider the following Betramic equations:

$$M \lessgtr 1 : \frac{\partial x}{\partial s} = \sqrt{(\mp u_1)}\ \frac{\partial \bar{y}}{\partial t}\ ;\ \frac{\partial x}{\partial t} = \sqrt{(\mp u_1)}\ \frac{\partial \bar{y}}{\partial s} \tag{8.97}$$

For small disturbance, Eqs. (8.95) may be reduced to the Cauchy-Riemann system as follows:

$$M \lessgtr 1 : \frac{\partial \Lambda_1}{\partial s} \pm \frac{\partial v_1}{\partial t} = 0\ ;\ \frac{\partial \Lambda_1}{\partial t} - \frac{\partial v_1}{\partial s} = 0 \tag{8.98}$$

where $$M \lessgtr 1 : \Lambda_1 = \mp \frac{2}{3}\,(+u_1)^{3/2} \tag{8.99}$$

For small disturbance, we may express the various quantities in power series of the thickness ratio τ as follows:

$$\Lambda=\Lambda_1\tau+\cdots\cdots;\ \theta=\nu_1\tau+\cdots$$
$$\varphi=x+\cdots;\ \psi=\bar{y}+\cdots;\ K=\sqrt{(\mp u_1)}\ +\cdots \tag{8.100}$$

There are many ways to solve this system of Eqs. (8.97) and (8.98). For example, we may take a solution which is closely related to an incompressible solution such that we put

$$\Lambda_1\sim\ u_1\ ;\ s\sim\ x;\ t\sim\ \bar{y}$$

i.e., a source or a vortex distribution on $t=0$. with this approximation, we get the function $u_1\ (s,t)$ and then with this u_1, we solve the linear equations (8.97). In general, the solution of the supersonic case is easy to obtain while that of the subsonic case is rather difficult to obtain.

On the other hand, Sobieczky's method is to take a very simple solution of Eqs. (8.98) and then to solve Eqs. (8.97) for various elementary solutions. In his method, Sobieczky put

$$M\lessgtr 1:\ \Lambda_1=\mp s\ ;\ \nu_1=t \tag{8.101}$$

we thus have the linear Rheograph equation and obtain the following equations instead of Eqs. (8.97)

$$M\lessgtr 1:\ \frac{\partial x}{\partial s}\ -s^{1/3}A_1\ \frac{\partial\bar{y}}{\partial t}\ =0\ ;\ \frac{\partial x}{\partial t}\ \pm s^{1/3}A_1\ \frac{\partial\bar{y}}{\partial s}\ =0 \tag{8.102}$$

where $A_1=(3/2)^{1/3}$.

Now we are looking for the solution of Eqs. (8.102). In principle, the present analysis is similar to that for the small disturbance approximation. From Eqs. (8.94)of small disturbance equations, we try to obtain φ and ψ as functions s and t or at least as function of Λ and θ. With the known functions of φ and ψ, we obtain the physical coordinates x and y as function of Λ and θ or q and θ by simple integration. The main difficulty lies in the solution of the small disturbance equations.

Sobieczky solved Eqs. (8.102) by the following well-known method. He first introduced the polar coordinates r and z such that

$$M\lessgtr\ 1:\ r=\sqrt{(t^2\pm s^2)},\ z=t/r \tag{8.103}$$

and then put

$$x=r^{\nu+2/3}f_\nu(z);\quad A_1\bar{y}=r^{\nu+1/3}\,g_\nu(z) \tag{8.104}$$

Substituting Eqs. (8.103) and (8.104) into Eqs. (8.102), we obtain the following ordinary differential equations for $f\ (z)$ and $g\ (z)$:

$$\left(\nu+\frac{1}{3}\right)zg_\nu+(1-z^2)g'_\nu=(1-z^2)^{1/3}\left[\left(\nu+\frac{2}{3}\right)f_\nu-zf_\nu'\right]$$
$$\left(\nu+\frac{2}{3}\right)zf_\nu+(1-z^2)f'_\nu=(1-z^2)^{1/3}\left[\left(\nu+\frac{1}{3}\right)g_\nu-zg_\nu'\right] \quad (8.105)$$

where the prime means the derivative with respect to z. Eqs. (8.105) are hypergeometric differential equations and they have closed solutions for the following values of ν :

$$3\nu=0, \quad \pm1, \ \pm2, \ \pm3, \ \cdots\cdots . \quad (8.106)$$

The differential Eqs. (8.105) are the same for both the supersonic and the subsonic cases $M \lessgtr 1$. Therefore, we do not expect any difficulty at the sonic line.

Sobieczky[8] gave 13 different closed solutions explicitly in this paper. Here we give 5 of them in Tab. 8.1 from which we shall use these solutions in the studying of transonic flow in Chapter X.

Table 8.1 Some closed solutions of Eqs . (8.105)

ν	x	$A_1\bar{y}$
-2	$-r^{-4/3}(1+z)^{2/3}(2-3z)$ $+(1-z)^{2/3}(2+3z)$	$-r^{5/3}(1+z)^{1/3}(1-3z)$ $-(1-z)^{1/3}(1+3z)$
0	$-r^{2/3}(1+z)^{2/3}+(1-z)^{2/3}$	$-2r^{1/3}(1+z)^{1/3}-(1-z)^{1/3}$
$1/3$	$-t$	$\pm\frac{3}{2}s^{2/3}$
$4/3$	$t^2\mp\frac{3}{2}s^2$	$\mp 3s^{2/3}t$
$5/3$	$-\frac{3}{2}s^{4/3}t$	$-t^2\pm\frac{3}{4}s^2$

12. Numerical solution of hodograph and rheograph methods

Garabedian and Korn[10] used the complex characteristics of Eq. (8.96) to calculate numerically the shockless transonic potential flow over an airfoil. Fig. 8.6 shows the upper half of a symmetric airfiol with a peaky pressure distribution of the Pearcey type[25] near the nose. Mach lines are drawn in the supersonic zome and their physically meaningless analytic continuations inside the body are included to indicate where limiting lines occur. Only isolated points on the plot of the pressure coefficient C_P are given in the subsonic regime because of the restricted number of paths of integration used in the computation there.

In Fig. 8.7 we present a lifting airfoil that is shock-free at $M=0.75$ and $C_L=0.63$. It has very much the appearance of the Whitcomb upside-down wing[26] because it is flat on top and has a concave lower surface near the trailing edge. An unusually gradual decay of the pressure coefficient along the rear of the upper surface should make the airfoil especially effective in suppressing turbulent boundary layer separation and consequently eliminating global perturbation by the unwanted wake.

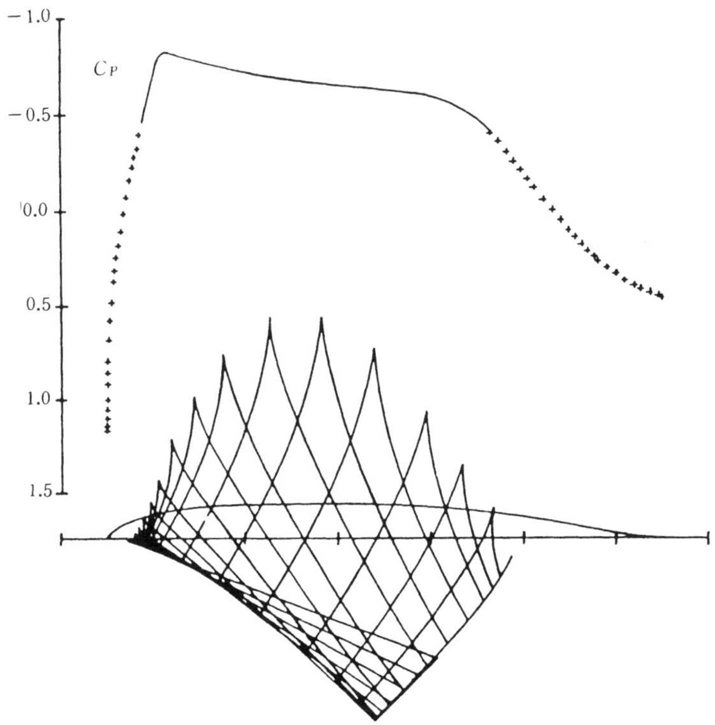

Fig. 8.6 Symmetric Transonic Airfoil at $M=0.8$, $t/c=0.13$

In the numerical solution, the following observations have been obtained. It is known that for shockless supersonic potential flow, there is no neighborhood solution. Any small disturbance in the shockless solution would produce shock front. The question is whether there is any neighborhood solution in the transonic flow case. The question has not been rigorously answered. For

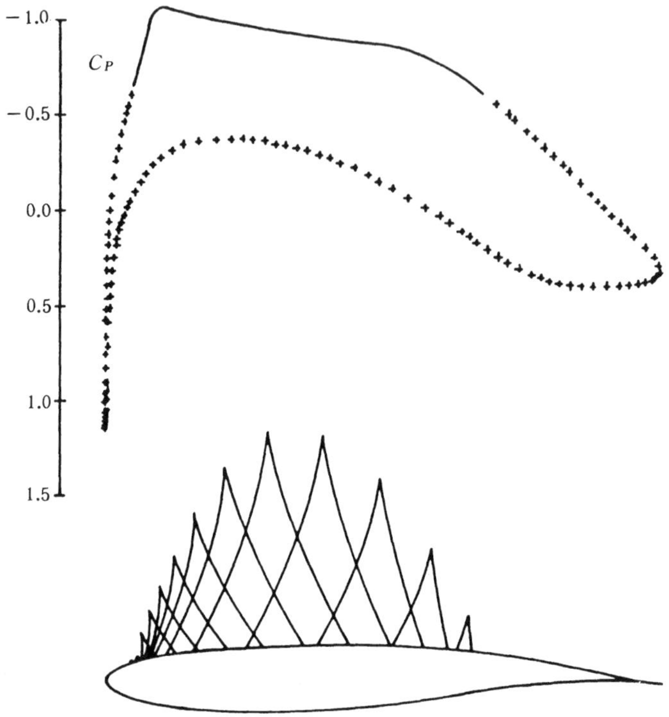

Fig. 8.7 Lifting Shockless Airfoil at $M = 0.75$, $C_L = 0.63$, $t/c = 0.12$

practical purpose, it is reasonable to assume that when the freestream Mach number is greater than a critical value say

$$M_{\infty s} = \frac{M_{\infty c} + 1}{2} \tag{8.107}$$

there will be no shockless solution at all for conventional airfoil. When M_∞ is less than $M_{\infty s}$, shockless transonic flow solution may exist. There will be no neighborhood solution in the sense that if for a given airfoil, we change the freestream Mach number slightly from the shockless solution, shock wave would occur. However, if we change the freestream Mach number from the shockless solution slightly and at same time, we modify the shape of the airfoil properly, it is possible to find another shockless transonic potential flow solution.

By indirect method, we give the velocity or pressure as a function of the flow inclination θ first and then find the corresponding shape of the airfoil from this given velocity or pressure distribution. However, it is not always possible to find the proper airfoil for a given distribution of pressure. Sometimes, we may find negative thickness for this corresponding airfoil or the shape of the airfoil will not be closed.

13. Problems

1. Derive the equation of the velocity potential in the hodograph plane, i.e., in terms of q and θ.

2. Show that the von Kármán-Tsien approximation (8.40) may be used in purely supersonic flow.

3. Show the detailed derivation of the pressure correction formula (8.54).

4. Use Lin's improvement formula to study the flow over a Joukowsky airfoil with a circulation.

5. Show that the thickness ratio δ of an elliptical cylinder in compressible flow is larger than the corresponding thickness ratio δ_i of the incompressible case for the two special cases:

$$(a)\ \delta_i \ll 1 \text{ and } (b)\ \delta_i \sim 1$$

6. Show that the approximation of von Kármán-Tsien pressure-density relation (8.40) underestimates the effect of compressibility.

7. Briefly show how to join the different regions of the solution in Poritsky's method.

8. Find the fundamental solutions of Tricomi equation (8.79).

9. Use the von Kármán-Tsien method to find the solution in compressible flow for a jet issuing from a straight wall with an opening where the Mach number on the surface of the jet is one.

10. Show that if we use the variables

$$\Psi = K^{\frac{1}{2}}\psi, \quad \tau = -\int_{a^*}^{q} (1-M^2)^{\frac{1}{2}} \frac{dq}{q}$$

equation (8.77) may be transformed into:

$$\frac{\partial^2 \Psi}{\partial \tau^2} + \frac{\partial^2 \Psi}{\partial \theta^2} = k\Psi$$

where $k = K^{-\frac{1}{2}} d^2 K^{\frac{1}{2}} / d\tau^2$. This is the fundamental equation in the W.K.B. method of Imai (Reference 21).

11. Find some exact solutions of Sobieczky's Eqs. (8.105) other than those given in Tab. 8.1.

References

1. von Kármán, Th., Compressibility effects in aerodynamics, *Jour. Aero. Soc.* **8**, July 1941, pp. 337–356.

2. Molenbroek, P., Ueber einige Bewegungen eines Gases bei Annahme eines Geschwin-

digkeitspotentials, *Archiv. der Math. und Phys.* II **9,** 1890, pp. 157 – 195.
3. Chaplygin, S., Gas jets, *Scientific Memoirs*, Moscow University, 1902, translated as *NACA, TM* No. 1063, 1944.
4. Busemann, A., Hodographenmethode der Gasdynamik, *ZAMM* **17,** 1937, pp. 73 – 79.
5. Oswatitsch, K., Special problems of inviscid steady transonic flows. *Int Jour. Eng. Sci.* **20,** No. 4, April 1982, pp. 497 – 540.
6. Christianowitsch, S. A., Flow over bodies at high subsonic velocities. *Tr. Joukowski central Aero-Hydrody. Inst* No. 481, 1940.
7. Ferrari, C. and Tricomi, F. G., *Transonic Aerodynamics*, Academic Press, New York, London, 1968.
8. Sobieczky, H, Exakte Loesungen der ebenen gasdynamischen Gleichungen in *Z. f. Flugwiss* **19**, pp. 197 – 214, 1971.
9. Sobieczky, H., Tragende Schnabelprofile in Stossfreier Schallanstroemung. *ZAMP* **26**, pp. 819 – 830, 1975.
10. Garabedian, P.R. and Korn, K. G., *Numerical solutions of partial differential equations*, Part II, pp. 253 – 271. Academic Press, New York, 1971.
11. Demtchenko, B., Quelques problems d'hydrodynamique bi-dimensionelle des fluides compressible, *C. R. Ac. Sci.* **194**, 1932; also *P. S. T. Ministere del'Air* No. **144**, 1939.
12. Tsien, H. S, Two-dimensional subsonic flow of compressible fluids, *Jour. Aero. Sci* **6**, No. 10, 1939, pp. 399 – 407.
13. Lin, C. C., On the extension of the von Kármán-Tsien method to two-dimensional subsonic flows with circulation around closed profiles, *Quart. Appl. Math.* **4,** 1946, pp. 291 – 297.
14. Poritsky, H., Polygonal approximations method in the hodograph plane, *Proc. First Sym. of Appl. Math*, 1949, American Math. Soc., New York.
15. Coburn, N., The Kármán-Tsien pressure-volume relation in the two-dimensional supersonic flow of compressible fluids, *Quart. Appl. Math.* 3, No 2, July 1945, pp. 106 – 116.
16. Chang, C. C., General consideration of problems in compressible flow using the hodograph method, *NACA, TN* 2582, Jan, 1952.
17. Temple, G., and Yarwood, T., The approximate solution of the hodograph equation for compressidle flow, *British Report* No. S.M.E. 3201, R.A.E. 1942.
18. Tricomi, F., On linear partial differential equations of the second order of mixed types, *Trans.* A9-T-26, Grad. Div. Appl. Math., Brown University, 1923.
19. Weinstein, A., Transonic flow and generalized axially symmetrical potential theory, *Naval Ord. Lab. Report* No. 1132, July 1950.
20. Guderley, G., and Yoshihara, H., The flow over a wedge profile at Mach number one, *Tech. Report* No. 5783, Air Material Command, Army Air Force, July 1949.
21. Imai, I., Application of the W.K.B. metho to the flow of a compressible fluid I, *Jour. Math. Phys.* **28,** 1948, p. 173.
22. Imai, I., and Hasimoto, H., Application of the W.K.B. method to the flow of a compressible fluid II, *Jour. Math. Phys.* **28**, 1950, pp. 205 – 214.
23. Imai, I., On a refinement of the transonic approximation theory, *Jour. Phys. Soc, Japan* **9**, 1954, pp. 1009 – 1020.
24. Meyer, Th., Ueber zweidimensionale Bewgungsvorgange in einem Gas, das mit Ueberschallgeschwindigkeit stroemt, *Verein der Deut. Ingen.*, Forsch. No. 62, 1908.
25. Pearcey, H.H., The aerodynamic design of section shapes for swept wings, *Advan. Aero. Sci.*, **3**, pp. 277 – 322, 1962.
26. Whitcomb, R.T., The upside-down wing, *Science News Item in Time Magazine*, Feb. **21**, 1969, p. 66.

Chapter IX

EXACT SOLUTIONS OF TWO-DIMENSIONAL ISENTROPIC STEADY FLOW EQUATIONS

1. Introduction

In this chapter, we will discuss the exact solutions of the two-dimensional isentropic steady flow equations. First the Prandtl-Meyer flow,[1] corresponding to the lost solution of the hodograph method, will be discussed (§ 2). This solution represents a supersonic flow over a corner.

In § 3, we shall discuss a simple exact solution of transonic flow known as the Ringleb solution.[2] This solution exhibits an important transonic flow phenomenon known as the limiting line.[3] It also yields a possible reason for the breakdown of isentropic flow. The limiting line will be discussed further in § 4.

In § 5, the general exact solution of the hodograph equation (8.27) will be discussed.[4–10] The application of this general method of solution to the problem of a subsonic jet[4,5] will be discussed in §6. Because of its mathematical complexity, the application of this general method of solution to the flow around a cylinder will be discussed only briefly in §7. References for the exact solution of the flow around a cylinder are given so that those who are interested may pursue the matter further.

In § 8. we shall discuss some exact solution of rheograph method according to Sobieczky for the transonic flow.

2. Prandtl-Meyer flow [1]

The lost solution discussed in Chapter VIII, § 9, is sometimes known as the Prandtl-Meyer solution. Here,

$$\frac{dq}{d\theta} = \pm \frac{q}{\sqrt{M^2-1}} \tag{9.1}$$

Now for lines of constant θ and constant q in the physical plane (x, y), we have

$$d\theta = \frac{\partial \theta}{\partial x}\, dx + \frac{\partial \theta}{\partial y}\, dy = 0 \tag{9.2}$$

For such lines, the slope is

$$\left(\frac{dy}{dx}\right)_{\theta,q}=-\frac{\dfrac{\partial\theta}{\partial x}}{\dfrac{\partial\theta}{\partial y}}=\frac{q\sin\theta-\cos\theta\dfrac{dq}{d\theta}}{\dfrac{dq}{d\theta}\sin\theta+q\cos\theta}=\text{function of }\theta \tag{9.3}$$

If we combine Eqs. (9.1) and (9.3), we have

$$\left(\frac{dy}{dx}\right)_{\theta,q}=\frac{\tan\theta\mp\dfrac{1}{\sqrt{M^2-1}}}{1\pm\dfrac{\tan\theta}{\sqrt{M^2-1}}}=\frac{\tan\theta\mp\tan\beta}{1\pm\tan\beta\tan\theta}=\tan(\theta\mp\beta) \tag{9.4}$$

From Eq. (9.3) we see that the lines of constant q and θ are straight lines in physical plane, and from Eq. (9.4) we see that these straight lines of constant q and θ make an angle with the velocity vector equal to the local Mach angle $\beta=\sin^{-1}\dfrac{1}{M}$. In other words, these lines of constant q and θ are Mach lines, their significance will be discussed further in Chapter XI.

Eq. (9.1) may be written

$$d\theta=\pm\frac{\sqrt{M^2-1}}{q}\,dq=\pm\frac{\sqrt{M^2-1}}{2}\,d\log\left(\frac{q^2}{a_0^2}\right)$$

$$=\pm\frac{\sqrt{M^2-1}}{2}\,d\log\left(\frac{M^2}{1+\dfrac{\gamma-1}{2}M^2}\right)$$

$$=\pm\frac{1}{2}\frac{1}{\sqrt{M^2-1}}\left[\frac{\gamma+1}{2}\frac{1}{\left(1+\dfrac{\gamma-1}{2}M^2\right)}-\frac{1}{M^2}\right]dM^2 \tag{9.5}$$

Integration of Eq. (9.5) gives

$$\theta=\text{constant}\pm\left[\sqrt{\frac{\gamma+1}{\gamma-1}}\tan^{-1}\sqrt{\frac{\gamma-1}{\gamma+1}(M^2-1)}-\tan^{-1}\sqrt{M^2-1}\right] \tag{9.6}$$

If we let $\theta=0$ when $M=1$, the constant in Eq. (9.6) is then equal to zero and Eq. (9.6) becomes

$$\theta=\pm\left[\sqrt{\frac{\gamma+1}{\gamma-1}}\tan^{-1}\sqrt{\frac{\gamma-1}{\gamma+1}(M^2-1)}-\tan^{-1}\sqrt{M^2-1}\right] \tag{9.7}$$

Eq. (9.7) is tabulated in Refs. 11 and 12. The sign of Eq. (9.7) is chosen so that (starting from $M=1$) the flow turns larger and larger angles as the local Mach number increases. The maximum value of angle of turn is obtained when the flow expands into a vacuum, i.e., $M \to \infty$. Therefore,

$$\theta_{max}=\left(\sqrt{\frac{\gamma+1}{\gamma-1}}-1\right)\frac{\pi}{2} \tag{9.8}$$

so that for $\gamma=1.4$, $\theta_{max}=130.3°$.

Eq. (9.7) gives the flow condition for a uniform supersonic flow expanded around a corner (Fig. 9.1). If the original flow is of Mach number $M_1=1$, Eq. (9.7) gives the angle of turn $\theta=\theta_c$ and the corresponding local Mach number $M_2=M$ after expansion. If the original flow is of Mach number $M_1>1$, we may use Eq. (9.6) to determine the proper constant or we may still use the tabulated values of Eq. (9.7) by taking the angle of turn as the difference of θ at these two Mach numbers M_2 and M_1, i.e., $\theta=\theta_{M_2}-\theta_{M_1}$. This

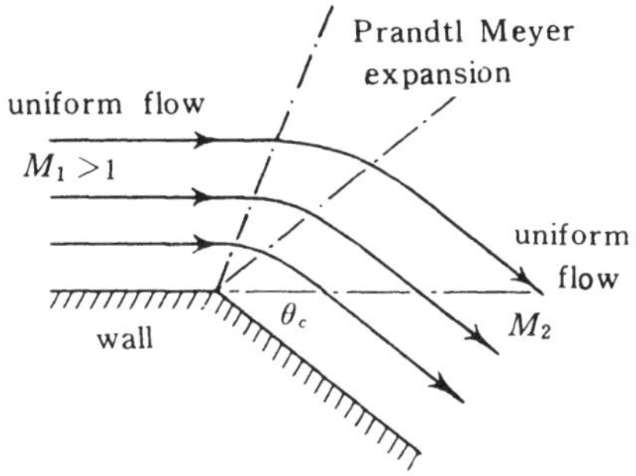

Fig. 9.1 Flow around a corner

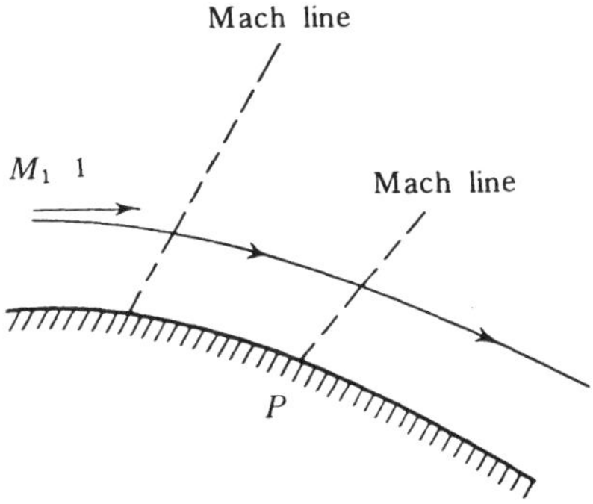

Fig. 9.2 Supersonic flow over a curved surface

type of flow is generally known as Prandtl-Meyer flow.

Prandtl-Meyer flow may be also used to find the pattern of a supersonic flow expanded continuously over any curved surface (Fig. 9.2). In this case, at any point P the local Mach number is determined from the original Mach number M_1 and the angle of turn θ_c which is the angle between the original flow direction and the tangent to the surface at the point P. This method of finding the Mach number at point P is closely associated with the method of characteristics, a powerful method for investigating the supersonic flow field. The method of characteristics will be discussed in detail in Chapter XI. This is the reason why the Prandtl-Meyer flow is one of the most important types of flow from a practical point of view.

The Prandtl-Meyer flow can be applied only in the case of an expansion of a supersonic flow. In a compression, a shock wave usually occurs. This case will be discussed in Chapter XI and also in § 8.

3. Flow with 180° turn. Ringleb solution

Ringleb[2] found that

$$\psi = la_0^2 \frac{\cos\theta}{q} \tag{9.9}$$

is an exact solution of equation (8.27). The quantity l is a constant, and a_0 is the sound velocity at the stagnation point, as well as a constant for a given

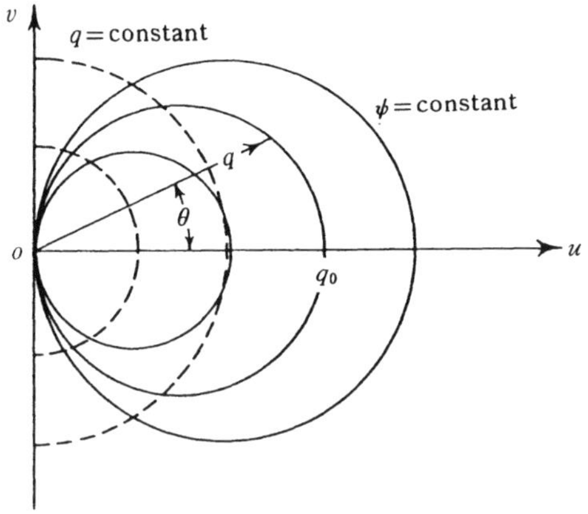

Fig. 9.3 Ringleb solution in hodograph plane

problem. By direct substitution into Eq (8.27) it is easy to show that Eq. (9.9) is an exact solution.

For any streamline, the velocity from the Ringleb solution is a maximum when $\theta = 0$. We may write

$$q = \frac{la_0^2}{\psi} \cos\theta = q_0 \cos\theta \tag{9.10}$$

where $q_0 = la_0^2/\psi$ is the maximum velocity for any given streamline, and q_0 is a function of the stream function ψ. The streamlines are circles in the hodograph plane with diameter q_0; their centers are on the u-axis, and they pass through the origin (Fig. 9.3). The streamlines of the Ringleb solution are smooth curves in the hodograph plane. To find the streamlines in the physical plane (x, y), we calculate the spatial coordinates (x, y) corresponding to the hodograph variables q and θ. Substituting Eq. (9.9) into Eq. (8.21) gives

$$dz = -la_0^2 \frac{\rho_0}{\rho} \frac{e^{i\theta}}{q} \left\{ -\left[(1-M^2) \frac{\sin\theta}{q^2} - i \frac{\cos\theta}{q^2} \right] dq + \left[\frac{\cos\theta}{q} + i \frac{\sin\theta}{q^2} \right] d\theta \right\} \tag{9.11}$$

It is convenient to use the variable τ defined by

$$\tau = \frac{\gamma - 1}{2} \frac{q^2}{a_0^2} = \frac{q^2}{q_{\max}^2} \tag{9.12}$$

instead of the velocity q. τ is the square of the ratio of the velocity q to the maximum possible velocity $q_{\max}$ for a given a_0. Hence the value of τ varies from 0 to 1.

Since Eq. (9.9) is an exact solution of Eq. (8.27) for any pressure and density relation, we may assume any $p(\rho)$ in the analysis. One plausible assumption is that the gas is perfect and is undergoing an isentropic process. Then the density ratio (ρ_0/ρ) and the local Mach number M may be expressed in terms of τ according to the isentropic relation of a perfect gas with constant γ, i.e.,

$$\frac{\rho_0}{\rho} = \frac{1}{\left(1 - \frac{\gamma-1}{2} \frac{q^2}{a_0^2}\right)^{\frac{1}{(\gamma-1)}}} = (1-\tau)^{-\frac{1}{\gamma-1}} \tag{9.13}$$

$$1 - M^2 = \frac{1 - \frac{\gamma+1}{\gamma-1}\tau}{1-\tau} \tag{9.14}$$

Substituting Eqs. (9.13) and (9.14) into Eq. (9.11) and simplifying yield

$$dz = -\frac{l}{4} i \frac{d\tau}{\tau(1-\tau)^{\gamma/(\gamma-1)}} + \frac{\gamma-1}{4} lid \left[\frac{e^{i2\theta}}{\tau(1-\tau)^{1/(\gamma-1)}} \right] \tag{9.15}$$

Integration of Eq. (9.15) gives

$$z=x+iy=\frac{\gamma-1}{4}\frac{ile^{i2\theta}}{\tau(1-\tau)^{1/(\gamma-1)}}-i\frac{l}{4}\int\frac{d\tau}{\tau(1-\tau)^{\gamma/(\gamma-1)}} \tag{9.16}$$

From Eq. (9.16) we have finally the spatial coordinates

$$x=-\frac{\gamma-1}{4}\frac{l\sin 2\theta}{\tau(1-\tau)^{1/(\gamma-1)}}$$
$$y=\frac{\gamma-1}{4}\frac{l\cos 2\theta}{\tau(1-\tau)^{1/(\gamma-1)}}+y_0(\tau) \tag{9.17}$$

where

$$y_0(\tau)=-\frac{l}{4}\int\frac{d\tau}{\tau(1-\tau)^{\gamma/(\gamma-1)}} \tag{9.18}$$

If $\gamma=1.4$, Eq. (9.18) can be expressed in terms of elementary functions as follows:

$$\frac{y_0(\tau)}{l}=\frac{1}{2}\tanh^{-1}\sqrt{1-\tau}-\frac{1}{10(1-\tau)^{5/2}}-\frac{1}{6(1-\tau)^{3/2}}-\frac{1}{2(1-\tau)^{1/2}} \tag{9.18a}$$

By eliminating θ from Eq. (9.17) we have

$$\left[\frac{x}{l}\right]^2+\left[\frac{y-y_0(\tau)}{l}\right]^2=\left[\frac{\gamma-1}{4}\frac{1}{\tau(1-\tau)^{1/(\gamma-1)}}\right]^2 \tag{9.19}$$

Eq. (9.19) shows that lines of constant velocity of τ are circles in the physical plane. These circles have a radius equal to $\frac{\gamma-1}{4}\frac{1}{\tau(1-\tau)^{1/(\gamma-1)}}$ and are centered at $x=0$, $y=y_0$.

Before discussing the streamlines of the compressible flow in the physical plane, we shall first examine the corresponding case in incompressible flow. For incompressible flow, a_0 tends to infinity. However we may adjust the scale l in Eq. (9.9) so that la_0^2 tends to unity. We may find the limiting case as $\tau \to 0$, $x \to x_i$ and $y \to y_i$ from Eq. (9.17) as follows:

$$x_i=-\frac{1}{2}\frac{\sin 2\theta}{q^2}$$
$$y_i=\frac{1}{2}\frac{\cos 2\theta}{q^2} \tag{9.20}$$

where we have shifted the origin of the physical plane by a constant in x_i, so that the curves $q=$constant are concentric circles about the origin in the physical plane. The streamlines of the incompressible flow in both hodograph and physical planes are shown in Fig. 9.4. This is the flow with a 180° turn.

The corresponding streamlines of a compressible fluid in the physical plane (Fig. 9.5) are of a similar nature. Here we observe the new phenomenon of breaks in the streamlines. If we plot circles of constant q/a_0, in the x-y plane, the center of the circle moves downward and the radius of these circles has a minimum at $\tau=\gamma-1/\gamma$. For streamlines along which the flow is always subsonic, such as circle A in the hodograph plane, we always have smooth streamlines in the physical plane such as streamline A and those above streamline A. If we decrease the value of ψ so that the hodographs are larger circles in u-v plane, we may reach the case of a streamline such as B, along which the flow is partly subsonic and partly supersonic. If the maximum velocity q_0 is not very large, we still will have smooth streamlines in the physical plane. If we further increase q_0, a critical value will be reached, correspon-

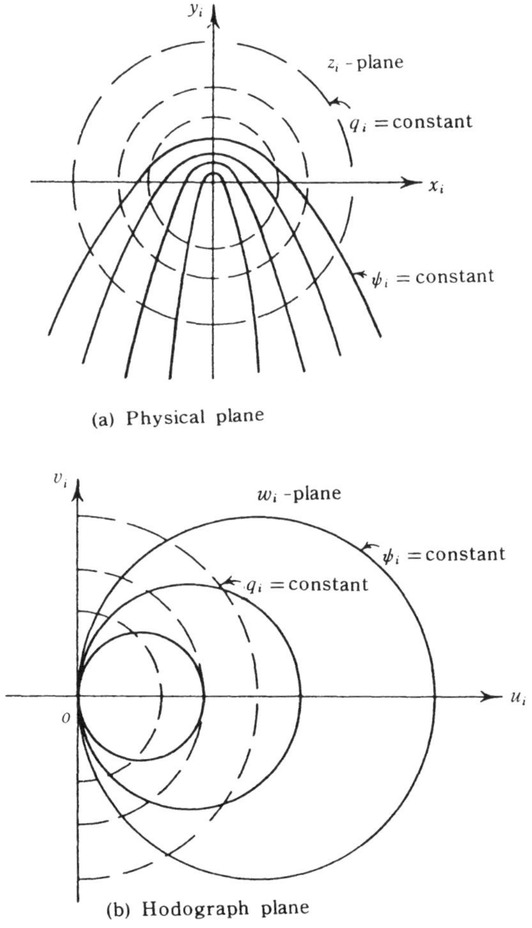

Fig. 9.4 Incompressible fluid flow of 180° turn

ding to the critical streamline C on which there are two critical points P and Q. At these points the pressure gradient is infinite. The streamline C is the last smooth streamline in the physical plane. Further increase of maximum velocity q_0 of a streamline will yield a cusped streamline such as that shown in streamline D. The streamline turns backward and the flow cannot be continued further into the region beyond the locus of the cusp. The locus of these cusp points is called the limiting line which cannot be crossed by a continuous isentropic flow. The critical points P and Q correspond to a velocity of $q/a_0=1.2$; the corresponding q_0 for the streamline C is about $1.67a_0$.

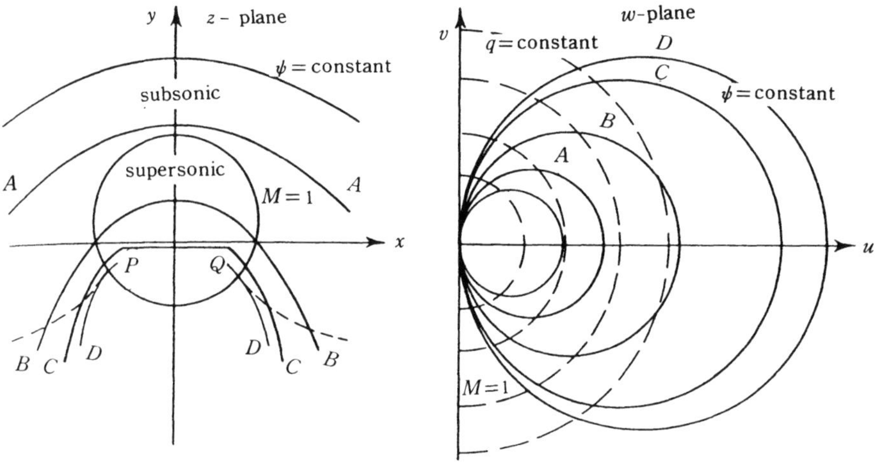

(a) Physical plane (b) Hodograph plane

Fig. 9.5 Ringleb Solution in both physical and hodograph planes

The discovery of a limiting line from the Ringleb solution is one of the most important results of such simple exact solution. We shall discuss this further in the next section.

4. Limiting line [3]

A limiting line appears when two curves of constant q intersect or are tangent in the physical plane so that there is a change in the fluid velocity but no change in the physical coordinates. Now we shall consider this condition in a general manner. Along a streamline ψ =constant, we have

$$d\psi = \frac{\partial\psi}{\partial q}\,dq + \frac{\partial\psi}{\partial\theta}\,d\theta = 0 \tag{9.21}$$

The change of complex coordinate z along a streamline is, from Eq. (8.28),

$$(dz)_\psi = -\frac{\rho_0}{\rho}\frac{e^{i\theta}}{q}\left\{\left[(1-M^2)\frac{1}{q}\frac{\partial\psi}{\partial\theta} - i\frac{\partial\psi}{\partial q}\right]dq + \left[q\frac{\partial\psi}{\partial q} + i\frac{\partial\psi}{\partial\theta}\right]\right.$$
$$\left.\times\frac{\partial\psi/\partial q}{\partial\psi/\partial\theta}\,dq\right\} \tag{9.22}$$
$$= -\frac{\rho_0}{\rho}\frac{e^{i\theta}}{\left(\dfrac{\partial\psi}{\partial\theta}\right)}\left[\left(\frac{\partial\psi}{\partial q}\right)^2 + (1-M^2)\left(\frac{1}{q}\frac{\partial\psi}{\partial\theta}\right)^2\right]dq$$

If along the streamline, there is a change in the fluid velocity but no change in the physical coordinate or alternatively an infinite pressure gradient, i.e., $(dq/dz)_\psi = \infty$, we have from Eq. (9.22)

$$\left(\frac{\partial\psi}{\partial q}\right)^2 + (1-M^2)\left(\frac{1}{q}\frac{\partial\psi}{\partial\theta}\right)^2 = 0 \tag{9.23}$$

which is the condition for the appearance of a limiting line. Eq. (9.23) shows that the limiting line can appear only when $M > 1$, i.e., only in supersonic regions.

From Eqs. (9.21) and (9.23), we have

$$\left(\frac{dq}{d\theta}\right) = \pm\frac{q}{\sqrt{M^2-1}} \tag{9.24}$$

This means that whenever the slope of the streamline in the hodograph plane satisfies Eq. (9.24), a limiting line occurs. This is exactly the differential form of the characteristics of the hodograph plane (cf. Chapter XI). The characteristics are real for supersonic flow; hence a limiting line can again occur in supersonic flow. A point on the limiting line exists whenever the streamline is tangent to the characteristic curve.

For the Ringleb solution, the condition for the limiting line is

$$\frac{\cos^2\theta}{q^4} + (1-M^2)\frac{\sin^2\theta}{q^4} = 0$$

or

$$1 - M^2\sin\theta = 0 \tag{9.25}$$

Since $\cos\theta = q/q_0$, Eq. (9.25) gives

$$\frac{q^2}{a_0^2} = \frac{\gamma+1}{4}\frac{q_0^2}{a_0^2} \pm \sqrt{\left(\frac{\gamma+1}{4}\frac{q_0^2}{a_0^2}\right)^2 - \frac{q_0^2}{a_0^2}} \tag{9.26}$$

Since for each streamline q_0 is a constant, there are in general two values of q_0/a_0 for which the limiting line condition is satisfied. For the last smooth streamline, we have

$$\left(\frac{\gamma+1}{4}\frac{q_0^2}{a_0^2}\right)^2-\frac{q_0^2}{a_0^2}=0 \quad \text{or} \quad \frac{q_0}{a_0}=\frac{4}{\gamma+1}=1.667\ (\gamma=1.4)$$

The local Mach number at this point is $q/a=1.580$. This shows that smooth transonic flow is possible only with a maximum Mach number of 1.580. Beyond this Mach number, a shock wave will occur.

Another interesting relation for the occurrence of the limiting line may be obtained by considering the Jacobian J in the transformation of the stream function ψ and velocity potential φ from the physical to the hodograph plane.

$$J=\frac{\partial\varphi}{\partial q}\frac{\partial\psi}{\partial\theta}-\frac{\partial\varphi}{\partial\theta}\frac{\partial\psi}{\partial q} \tag{9.27}$$

If we substitute the hodograph relation (8.26) into Eq. (9.27), we obtain

$$J=-\frac{\rho_0}{\rho}q\left[(1-M^2)\left(\frac{1}{q}\frac{\partial\psi}{\partial\theta}\right)^2+\left(\frac{\partial\psi}{\partial q}\right)^2\right] \tag{9.28}$$

Comparing Eq. (9.28) with Eq. (9.23), we see that the condition for the occurrence of the limiting line is

$$J=0 \tag{9.29}$$

Whenever the functional Jacobian J vanishes, we have a limiting line.

The physical reason for the occurrence of the limiting line is that our assumption of isentropic flow breaks down in the forbidden region. In this forbidden region, presumably a shock wave occurs and the flow is no longer isentropic. For adiabatic flow, a shock wave occurs only in the supersonic flow. This again shows that the limiting line occurs only in supersonic flow.

5. General exact solution of hodograph equation [6-10]

The Ringleb solution is a special case of the solution of the stream function Eq, (8.27). It is of interest to find the general solution of Eq. (8.27) satisfying arbitrary boundary conditions. The general solution has been obtained by Chaplygin in his now famous memoir,[4] "Gas Jets". It has been further developed by many authors in applications of a uniform flow over a solid body.[6-10] We shall briefly discuss these in the next three sections.

Eq. (8.27) may be written as follows:

$$q^2\frac{\partial^2\psi}{\partial q^2}+q(1+M^2)\frac{\partial\psi}{\partial q}+(1-M^2)\frac{\partial^2\psi}{\partial\theta^2}=0 \tag{9.30}$$

Chaplygin[4] tried to find the fundamental solutions of Eq. (9.30) similar to that for classical potential theory for incompressible fluids. For incompressible flow, $M=0$, the elementary solution of Eq. (9.30) is

$$\psi_i = q^n \cos n\theta \quad [\text{or } q^n \sin n\theta] \tag{9.31}$$

where n is an integer from $-\infty$ to ∞. Since Eq. (9.30) is linear, the method of superposition is applicable. The general solution of Eq. (9.30) for incompressible flow may be written as follows:

$$\psi_i = \sum_{n=-\infty}^{\infty} q^n (A_n \cos n\theta + B_n \sin n\theta) \tag{9.32}$$

where the coefficients A_n and B_n are to be determined by the boundary conditions of the problem investigated.

For compressible flow, since the Mach number M is a function of q only, we may generalize the solution (9.31) into the following form:

$$\psi = q^n f_n(q) \cos n\theta \quad [\text{or } \; q^n f_n(q) \sin n\theta] \tag{9.33}$$

where $f_n(q)$'s may be considered as correction factors for compressibility effect and approach unity as M approaches zero. Substituting Eq. (9.33) into Eq. (9.30) gives a differential equation for f_n as

$$q^2 \frac{d^2 f_n}{dq^2} + (2n+1+M^2) q \frac{df_n}{dq} + n(n+1) M^2 f_n = 0 \tag{9.34}$$

It is convenient to use the variable τ defined in Eq. (9.12) instead of q in Eq. (9.34). With this, Eq. (9.34) becomes

$$4\tau^2 \frac{d^2 f_n}{d\tau^2} + 2\tau\,[2(n+1) + M^2] \frac{df_n}{d\tau} + n(n+1) M^2 f_n = 0 \tag{9.35}$$

If we substitute the relation for M^2 in terms of τ [Eq. (9.14)], we finally have the differential equation of f_n

$$\tau(1-\tau) \frac{d^2 f_n}{d\tau^2} + \left[(n+1) - \left(n+1 - \frac{1}{\gamma-1} \right) \tau \right] \frac{df_n}{d\tau} + \frac{n(n+1)}{2(\gamma-1)} f_n = 0 \tag{9.36}$$

This equation is now of the same form as that of the hypergeometric equation,[13] i.e.,

$$\tau(1-\tau) \frac{d^2 F}{d\tau^2} + [c_n - (a_n + b_n + 1)\tau] \frac{dF}{d\tau} - a_n b_n F = 0 \tag{9.37}$$

where a_n, b_n, and c_n are parameters. The solution to the hypergeometric equation is the hypergeometric function which may be written in the following form:

$$F = F(a_n, b_n; c_n, \tau) = 1 + \frac{a_n b_n}{c_n\ 1!}\tau + \frac{a_n(a_n+1)b_n(b_n+1)}{c_n(c_n+1)\cdot 2!}\tau^2 + \cdots \tag{9.38}$$

In comparing Eq. (9.36) with Eq. (9.37), the required function f_n is given by the hypergeometric function F if we take

$$c_n = n+1,\ a_n + b_n = n - \frac{1}{\gamma - 1},\quad a_n b_n = -\frac{n(n+1)}{2(\gamma - 1)} \tag{9.39}$$

The general solution for the compressible flow may then be written as

$$\psi = A + B\theta + \sum_{n=0}^{\infty} q^n F_n(a_n, b_n; c_n, \tau)\,[A_n \cos n\theta + B_n \sin n\theta] \tag{9.40}$$

where the coefficients A, B, n, A_n, and B_n are to be determined by the Boundary conditions of the physical problems considered.

It should be noted that the same technique may be used to find the elementary solutions for the simplified equations in hodograph plane discussed in Chapter VIII, § 8. For the Tricomi equation, the hyper-geometric functions in the elementary solution reduce to the Bessel function.

Now we consider a few simple flows represented by elementary solutions of Eq. (9.34) as follows: [14]

(i) When $n=0$, we have the following two solutions of f_0 from Eq. (9.36):

$$df_0/d\tau = 0 \quad \text{or } f_0 = \text{constant} \tag{9.41}$$

or

$$\tau(1-\tau)\frac{d^2 f_0}{d\tau^2} + \left[1 - \frac{(\gamma-2)}{\gamma-1}\right]\frac{df_0}{d\tau} = 0$$

or

$$f_0 = f_0(\tau) = A\int [(1-\tau)^{1/(\gamma-1)}/\tau]\, d\tau \tag{9.42}$$

The stream function corresponding to these solutions is

$$\psi = A_0 f_0(\tau) + B_0\theta \tag{9.43}$$

If we take $A_0 = 0$, equation (9.43) represents source or sink flow [cf. Eq. (5.54)]. If we take $B_0 = 0$, Eq. (9.43) represents vortex flow [cf. Eq. (5.56)]. If both A_0 and B_0 are not zero, we have a spiral flow.

When $n=1$, we have the following two solutions for f:

$$f_1 = 1 \tag{9.44}$$

and

$$f_1 = (1-\tau)^{\gamma/(\gamma-1)} \tag{9.45}$$

Eq. (9.44) gives the Ringleb solution [cf. Eq. (9.9)], whereas Eq. (9.45) gives a solution very similar to Ringleb's Eq. (9.45) was first found by Temple and Yarwood.[15]

6. Exact solution of a subsonic gas jet [16, 17]

The general solution discussed in the last section has been used by Chaplygin to study the case of two-dimensional subsonic jet flow from a vessel with straight walls. In this case, the boundary conditions given in the physical plane are easily transformed into the conditions in the hodograph plane. As a result, reasonably simple analytic solutions may be found.

Let us consider the problem of a gas jet issuing from an infinitely wide vessel with straight walls and forming an angle θ_w with the axis of the jet as shown in Fig. 9.6 (a). On the walls AB and $A'B'$, $\theta=\theta_w$ and $\psi=\text{constant}=Q/2$, say. On the free streamlines of the jet (as, for example, BC) the velocity q of the flow is constant, say $\tau=\tau_1$; furthermore, $\psi=Q/2$. If the flow of the jet is subsonic, i.e., $\tau_1 \leqslant \dfrac{\gamma-1}{\gamma+1}$ the hodograph of this subsonic jet is shown in Fig. 9.6 (b), where q_j is the maximum velocity of the jet

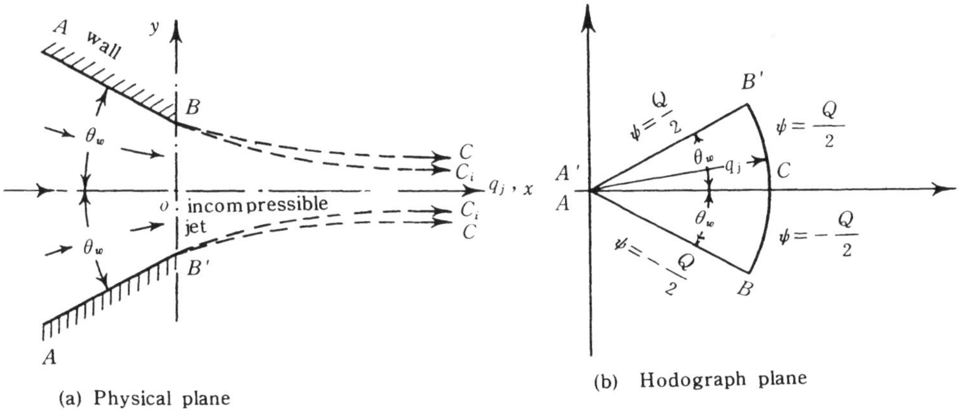

Fig. 9.6 Jet flow from vessel with inclined straight walls

which gives $\tau_1 \leqslant q_j^2/q_{\max}^2$. The solution of the incompressible jet which satisfies the streamlines $\psi=\pm Q/2$, ABC and $AB'C$ is

$$\psi_i = -\frac{Q}{\pi}\left[\frac{\theta}{\theta_c} + \sum_{n=1}^{\infty}\frac{1}{n}\left(\frac{\tau}{\tau_1}\right)^{\lambda_n/2}\sin 2n\frac{\theta}{\theta_c}\right] \tag{9.46}$$

where $\lambda_n=2n/\theta_c$, $\theta_c=2\theta_w/\pi$. It is evident that on $\theta=\pm\theta_w$, Eq. (9.46) gives $\psi=\mp Q/2$ and that $\tau=\tau_1$; Eq. (9.41) also gives $\psi=\pm Q/2$. Hence the boundary conditions are satisfied and Eq. (9.46) is the required solution of the incompressible flow.

For subsonic flow, the compressible solution of this jet may be obtained by comparing the incompressible solution with the required solution of compressible flow. We obtain the following solution for compressible flow:

$$\psi = -\frac{Q}{\pi}\left[\frac{\theta}{\theta_c} + \sum_{n=1}^{\infty}\frac{1}{n}\left(\frac{\tau}{\tau_1}\right)^{\lambda_n/2}\frac{F_n(\tau)}{F_n(\tau_1)}\sin 2n\frac{\theta}{\theta_c}\right] \tag{9.47}$$

It is evident that for $\theta=\theta_w$, Eq. (9.42) gives $\psi=\pm Q/2$, and that for $\tau=\tau_1$. Eq. (9.47) also gives $\psi=\pm Q/2$. Hence, Eq. (9.47) gives the required solution for the subsonic jet.

After the function $\psi(q,\theta)$ of Eq. (9.47) is obtained, the coordinates of the jet in the physical plane can be obtained from Eq. (8.28) by integration. This solution has been obtained by Jacob who found that the coefficient of contraction, i. e., the ratio of the width of the jet at positive infinity to the

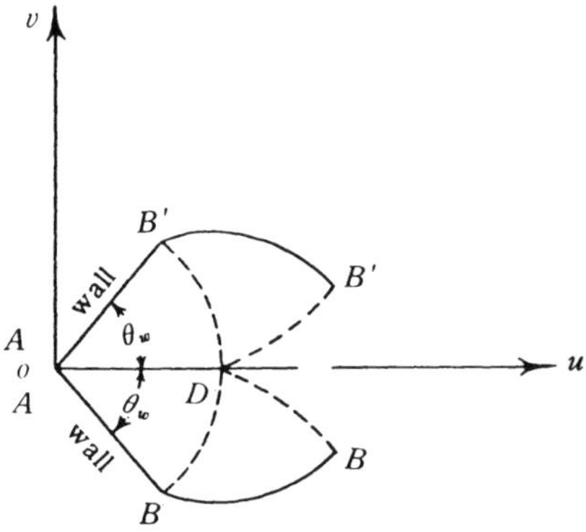

Fig. 9.7 Hodograph plane of a supersonic jet flow from vessel with inclined straight walls

opening of the vessel, for a jet of compressible fluid is larger than the corresponding value for an incompressible fluid. For an incompressible jet, the coefficient of contraction is $\dfrac{\pi}{\pi+2}=0.61$, for $\theta_w=\pi/2$; on the other hand, for a compressible jet with $\tau_1=\dfrac{\gamma-1}{\gamma+1}$ and $\theta_w=\pi/2$, the coefficient of contraction is 0.74 when $\gamma=1.4$.

For the supersonic flow of a jet from the vessel of Fig. 9.6 (a), the hodograph is different from that of Fig. 9.6(b) because, at the exit of the vessel, the velocity first reaches the local sound velocity and then increases to the supersonic speed in accordance with the Prandtl-Meyer expansion theorem. The corresponding hodograph is shown in Fig. 9.7. The curves BB and $B'B'$ are the Prandtl-Meyer expansion flow. In solving this problem, we first use Eq.(9.47) to calculate the flow up to the sonic line $B'DB$. The value of ψ on the curvilinear triangle BDB may be calculated by the method of characteristics. Frankl[16] has calculated an example of a supersonic jet with $\theta_w=\pi/2$ and found that the coefficient of contraction is larger than that of the sonic jet.

One of the difficulties in carrying out the calculations given by Eq. (9.47) is that the hypergeometric functions of all n's are not available and the original series (9.38) converges rather slowly. One improvement in this respect may be realized by transforming the series (9.38) by using the asymptotic properties of F_n for large n. The series thus obtained converges much more rapidly than the original one[6].

7. Flow around a body in a uniform stream

In practice, we often want to find the flow field around an arbitrary body in a uniform stream. In this instance, the boundary conditions in the hodograph plane generally are not known exactly. For instance, for the flow over an ellipse (shown in Fig. 8.3), we can sketch the hodograph, but, since we do not know the exact maximum velocity, we cannot find the exact solution. In general, we use the general solution (9.40) and determine the constants A, B, n, A_n, and B_n by comparison with the incompressible solution around a similar body. After the solution in the hodograph plane is obtained, we transform back into the physical plane. If the required boundary conditions are satisfied, we have obtained the required solution. Here another mathematical difficulty presents itself; there is a singular point at $q=U$, $\theta=0$, the free stream velocity. The mathematical problem of analytically continuing Eq. (9.38) beyond the circle $q=U$ is a difficult one since the point $q=U, \theta=0$ is a singular point. This problem has been solved by Lighthill[7] and Cherry,[8] whose solutions have been nicely reviewed by Pack.[6] Those who are interested in this problem should refer to the papers of Lighthill and Cherry or to the review by Pack.

8. Exact solution of rheograph method [18, 19]

We discuss one of the exact solutions of rheograph method given by Sobieczky[19] in § 11 of Chapter VIII as follows.

We consider the superposition of the solution of $\nu=1/3$ and $\nu=5/3$. The resultant solution is as follows:

$$x=-t\left(1+\frac{3}{2}s^{4/3}\right);\quad A_1\bar{y}=-t^2\pm\frac{3}{4}s^{2/3}(2+s^{4/3}) \tag{9.48}$$

(9.48)

where t represents essentially the angle of the flow inclination and s represents the magnitude of the flow velocity, Eq. (8.101). Fig. 9.8 shows the sonic line and the limiting line and the disturbance velocity u_1 at $A_1\bar{y}=9/4$. For

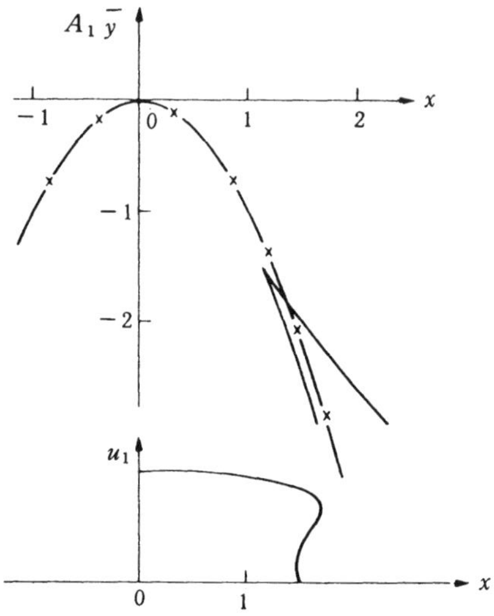

Fig. 9.8: Symmetrical local supersonic region

the sonic line, $s=0$, we have the parametric representation as follows:

$$x=-t,\ A_1\bar{y}=-t^2=-x^2 \tag{9.49}$$

The sonic line is a parabolic with y-axis as the parabola axis and with the origin as its vertex.

The conditions for the limiting line according to the general exact theory (§ 4) are

$$x_\xi=x_\eta=0 \quad \text{or} \quad x_\eta=y_\eta=0 \tag{9.50}$$

For our present case, we have

$$\begin{aligned} 2x_\xi&=x_s-x_t=0 \\ 2x_\eta&=x_s+x_y=0 \end{aligned} \tag{9.51}$$

and the limiting line is

$$2ts^{1/3}=\mp(1+\frac{3}{2}s^{4/3}) \tag{9.52}$$

with the upper sign for the upper conditions of Eqs. (9.51).

We may fit a shock front at the limiting line by Kluwick method (Ref.

20) of shock fitting. Since the present case is symmetrical with respect to the y-axis, we have one side with the compression shock while we have the opposite side with the expansion shock due to symmetry! The expansion shock is not physically possible. Hence our solution is not physically possible for $A\bar{y} < -1.5$. It is interesting to notice that for the Garabedian-Korn airfoil (Fig. 8.6), the expansion shock occurs inside the airfoil in which the flow field may be ignored for practical application. Thus the Garabedian-Korn airfoil is of practical interest.

9. Problems

1. Find the expression for the velocity potential from Eq. (8.12) for Prandtl-Meyer flow.

2. Show that in the hodograph plane the integration curve of Prandtl-Meyer flow is an epicycloid.

3. Show the detailed procedure for deriving Eq. (9.20) from Eq. (9.17) by the limiting process and evaluate the amount by which the origin has been shifted in the physical plane.

4. Plot the two following streamlines of the Ringleb solution in the physical plane:

$$\text{(a) } q_0 = 0.5a_0 \quad \text{and} \quad \text{(b) } q_0 = 2.0a_0.$$

5. Find the expression for the Legendre potential X for the jet flow shown in Fig. 9.6(a).

6. Find the expression for the coefficient of contraction of the subsonic jet shown in Fig. 9.6(a) in terms of τ_1.

7. Discuss the general solution (9.40) for the transonic flow field over a circular cylinder without circulation in a unifom stream.

8. Discuss the flow field in both physical plane and hodograph plane for the Temple-Yarwood solution (9.45).

References

1. Meyer, Th., Ueber zweidimensionale Bewegungsvorgange in einem Gas, das mit Ueberschallgeswindigkeit stroemt, *Verein der Deut. Ingen.*, Forsch. No. 62, 1908.
2. Ringleb, F., Exakte Loesungen der Differentialgleichungen einer adiabatischen Gasstroemung, *ZAMM* **20,** 1940, pp. 185 – 198.
3. Tollmien, W., Grenzlinien adiabatischer Potentialstroemungen *ZAMM* **21,** 1941, pp. 140 – 152.
4. Chaplygin, S., Gas jets, *NACA, TM* No. 1063, 1944.
5. Pai, S. I., *Fluid Dynamics of Jets,* Chapter II, D. Van Nostrand Company, Inc., Princeton, N, J., 1954.
6. Pack, D. C., Hodograph methods in compressible fluid, *Lecture Series* No. 23, Inst. for Fluid Dynamics and Appl. Math, University of Maryland, 1952.
7. Lighthill, M. J., The hodograph transformation in transonic flow, pts I-III, *Proc. Roy. Soc. London* A-**191,** No. 1026, 1947, pp. 323 – 369. Ferguson, D. F. and Lighthill, M. J., The hodograph transformation in transonic flow, pt IV, *Proc. Roy Soc. London* A-**192,** 1947, pp. 135 – 142.

8. Cherry, T. M., Flow of a compressible fluid about a cylinder, pt I, *Proc. Roy, Soc. London* A-**192**, No. 1028, 1947, pp. 45 – 79; Flow with circulation, pt II, *Proc. Roy. Soc. London* A-**196**, No. 1044, 1949, pp. 1 – 31.
9. Tsien, H. S., and Kuo, Y. H., Two-dimensional irrotational mixed subsonic and supersonic flow of a compressible fluid and the upper critical Mach number, *NACA, TN* No. 995, 1946.
10. Frankl, F., On the problems of Chaplygin for mixed sub- and supersonic flows, *NACA, TM* No. 1155, June 1947.
11. Emmons, H. W., *Gas Dynamics Tables of Air*, Dover Publications, Inc., 1947.
12. Ames Research staff: Equations, tables and charts for compressible flow, *NACA Report* No. 1135 . formerly *NACA* , *TN* No. 1428 1953.
13. Whittaker, E. T., and Watson, G. N., *Modern Analysis*, Cambridge University Press, 4th edition, 1927.
14. Chang, C. C., General consideration of problems in compressible flow using the hodograph method, *NACA, TN* No. 2582, Jan. 1952.
15. Kuo, Y. H., and Sears, W. R., Plane subsonic and transonic potential flows, Sec. F of General theory of high speed aerodynamics, Vol. 6 of *High Speed Aerodynamics and Jet Propulsion*, edited by W. R. Sears, Princeton University Press, 1954.
16. Frankl, F. I., The flow of a supersonic jet from a vessel with plane walls, *Air Material Command Tech. Report* No. F-TS-1213-IA, May 1949.
17. Jacob, C., Étude d'un jet gazeux, *C.R.Ac Sci.* **203**, 1936, p. 423.
18. Oswatitsch, K., Special problems of inviscid steady transonic flows. Int. *Jour. of Eng. Sci.* **20**, No. 4, April 1982, pp. 497 – 540.
19. Sobieczky, H., Exakte Loesungen der ebenen gasdynamischen Gleichungen in Schallnaehe, *Z. f. Flugwiss.* **19**, pp. 197 – 214, 1971.
20. Kluwick, A., Zur Ausbreitung schwacher Stoesse in dreidimensionalen instationaeren Stroemungen, *ZAMM* **51**, pp. 225 – 232, 1971.

Chapter X

TWO-DIMENSIONAL STEADY TRANSONIC AND HYPERSONIC FLOWS

1. Introduction

From our discussions in previous chapters, we see that it is extremely difficult to find exact solutions for compressible flow problems. In practice, we must use the approximate methods previously discussed. Linearization by the method of small perturbations has been extensively used because of the frequent interest in the flow around thin bodies. However, as mentioned in Chapter VI, this method of linearization fails for both the transonic and the hypersonic regions. In this chapter we shall discuss the fundamental equations of transonic (§ 2) and hypersonic (§ 17) flows for thin bodies with the help of the theory of small perturbations. These equations are nonlinear.

Since the fundamental equations for transonic and hypersonic flows are nonlinear even for thin bodies, the exact solutions of these problems are difficult to obtain. For the study of the nonlinearity effects we must rely mainly on experiment. In order to have greater usefulness of either theoretical or experimental data, it would be helpful to discover similarity laws which enable a family of possible solutions to be deduced from a single result.

von Kármán[1] was the first to obtain the similarity laws for transonic flows where the fluid velocity is very near to the velocity of sound. We shall discuss the two-dimensional transonic similarity laws in § 3. In § 4 and § 8 some problems of transonic flow will be discussed.

Considerable numerical solutions have been worked out. We discuss some of the computational gasdynamics of transonic flow in § 9 to § 16.

Tsien[2] was the first to obtain the similarity laws for hypersonic flow where the fluid velocity is much larger than the local sound velocity. We shall discuss the two-dimensional hypersonic similarity laws in § 17. In § 18 to § 21, some problems of hypersonic flows will be discussed

The numerical solution of supersonic flow will be discussed in Chapter XI § 12.

General similarity laws may be found for bodies moving in compressible fluids provided at least one of the two body dimensions perpendicular to the main flow direction is small compared to the dimension in the direction of main flow. The three-dimensional similarity laws for both transonic and hypersonic flows[3] will be discussed in Chapter XIII.

2. Transonic equation

Consider a uniform flow U which differs slightly from the critical sound speed a^*, passing over a thin body of thickness ratio δ which is much smaller than unity, i.e., $\delta \ll 1$. We assume that the velocity of the whole flow field will also be approximately equal to the sonic velocity a^*. If the higher-order terms are neglected, the flow field in the present problem is irrotational and the velocity components may be written as

$$u = a^* + \frac{\partial \varphi}{\partial x}$$
$$v = \frac{\partial \varphi}{\partial y} \tag{10.1}$$

where φ is the velocity potential of the perturbed flow. We shall find the effective first-order equation and the first-order boundary conditions for φ.

The first-order boundary conditions for φ are:

(1) At infinity, the velocity components are

$$u = U, \quad v = 0$$

or

$$\frac{\partial \varphi}{\partial x} = U - a^*, \quad v = 0 \tag{10.2}$$

(2) On the surface of the body, which is represented by

$$y = \delta h(x) \tag{10.3}$$

where $h(x)$ is of the order of unity and δ is the thickness ratio of the body, the flow must follow the surface, i.e.,

$$\frac{1}{a^*}\frac{\partial \varphi}{\partial y} = \delta \frac{dh}{dx} = \delta h' \tag{10.4}$$

at $y = \pm 0$.

The boundary condition (10.2) may be written in a more convenient form. The relation between the local sound velocity a and the velocity components of the flow is given by equation (3.12), i.e.,

$$a^2 + \frac{\gamma - 1}{2}(u^2 + v^2) = \frac{\gamma + 1}{2} a^{*2} \tag{10.5}$$

Substituting Eq. (10.1) into Eq. (10.5) and neglecting higher-order terms, we have

$$\frac{a_\infty}{a^*} \cong 1 - \frac{\gamma-1}{2}\frac{1}{a^*}\frac{\partial \varphi}{\partial x} \tag{10.6}$$

Combining Eqs. (10.6) and (10.2), we have at infinity

$$\begin{aligned} \frac{\partial \varphi}{\partial x} &= a^* \frac{2}{\gamma+1}(M_\infty - 1) \\ \frac{\partial \varphi}{\partial y} &= 0 \end{aligned} \tag{10.7}$$

where $M_\infty = U/a_\infty$ = Mach number of the uniform free stream.

From the linearized small perturbation theory (6.61), for $M_\infty = 1$, the differential equation for the perturbed velocity potential φ is

$$\frac{\partial^2 \varphi}{\partial y^2} = 0, \quad \frac{\partial \varphi}{\partial y} = f(x) \tag{10.8}$$

This shows that the disturbance in transonic flow is extended very far laterally. If it is desired to reduce transonic flow fields into one single pattern, it is necessary to contract the lateral coordinate y. Von Kármán[1] used the affine transformation:

$$x = c\xi, \ y = c(\delta\Gamma)^{-n}\eta \tag{10.9}$$

where c is the chord of the body, $\Gamma = (\gamma+1)/2$, and n is a positive exponent to be determined later. To satisfy the boundary condition (10.7), we write the perturbed velocity potential in the form

$$\varphi = ca^* \frac{1-M_\infty}{\Gamma} F(\xi, \eta) \tag{10.10}$$

The boundary condition (10.4) becomes, in nondimensional form,

$$\frac{1-M_\infty}{(\Gamma\delta)^{1-n}} \frac{\partial F}{\partial \eta} = h'(\xi) \tag{10.11}$$

at $\eta = \pm 0$.

The exact differential equation for φ is given by Eq. (5.35), i.e.,

$$(a^2 - u^2)\frac{\partial^2 \varphi}{\partial x^2} - 2uv\frac{\partial^2 \varphi}{\partial x \partial y} + (a^2 - v^2)\frac{\partial^2 \varphi}{\partial y^2} = 0 \tag{10.12}$$

Substituting Eqs. (10.1), (10.9), and (10.10) into Eq. (10.12), and neglecting higher-order terms since both δ and $1 - M_\infty$ are small quantities, we obtain the nondimensional transonic flow equation

$$\frac{\partial^2 F}{\partial \eta^2} = 2\frac{1-M_\infty}{(\Gamma\delta)^{2n}} \frac{\partial F}{\partial \xi}\frac{\partial^2 F}{\partial \xi^2} \tag{10.13}$$

The corresponding boundary condition is given by Eq. (10.11).

In order to reduce the system of the differential equation and the boundary condition so that they involve only one parameter, we should take

$$1-n=2n$$

or

$$n=1/3 \tag{10.14}$$

The differential equation for the nondimensional velocity potential F becomes

$$\frac{\partial^2 F}{\partial \eta^2}=2K\frac{\partial F}{\partial \xi}\frac{\partial^2 F}{\partial \xi^2} \tag{10.15}$$

where

$$K=\frac{1-M_\infty}{(\Gamma\delta)^{2/3}} \tag{10.16}$$

The corresponding boundary conditions are:

(1) At infinity

$$\frac{\partial F}{\partial \xi}=-1, \quad \frac{\partial F}{\partial \eta}=0 \tag{10.17}$$

(2) On the chord $\eta=0$

$$\frac{\partial F}{\partial \eta}=\frac{1}{K}h'(\xi) \tag{10.18}$$

The parameter K is known as the transonic (similarity) parameter; since $M_\infty \cong 1$, it may also be written as

$$K=\frac{1}{2}\frac{1-M_\infty^2}{(\Gamma\delta)^{2/3}} \tag{10.19}$$

For a uniform flow near $M_\infty=1$ over a series of affinely related bodies [cf. Eq. (10.3)] with different thickness ratios δ, if the values of the transonic parameters are the same, the flows are governed by one function $F(\xi,\eta)$. The system of Eqs. (10.15) to (10.18) is the transonic flow equations from which the transonic similarity laws may be derived.

3. Transonic similarity laws

In this section we shall derive the relations for the pressure coefficient, the drag coefficient, the lift coefficient, and the moment coefficient among affinely related bodies with different thickness ratios but tested under conditions where the transonic parameter is the same.

The pressure coefficient for a thin body in a uniform stream is given by Eq. (3.37), i.e.,

$$C_p = -2\frac{u-U}{U} = -2\frac{\dfrac{\partial\varphi}{\partial x} + a^* - U}{U} = -2\frac{\dfrac{\partial\varphi}{\partial x} + \dfrac{1-M_\infty}{\Gamma}a^*}{a^*\left(1 - \dfrac{1-M_\infty}{\Gamma}\right)}$$

$$\cong -2\left(\frac{1-M_\infty}{\Gamma} + \frac{1}{a^*}\frac{\partial\varphi}{\partial x}\right) \tag{10.20}$$

Substituting Eq. (10.10) into Eq. (10.20) gives

$$C_p = -\frac{2(1-M_\infty)}{\Gamma}\left(1+\frac{\partial F}{\partial\xi}\right) = -\frac{1-M_\infty^2}{\Gamma}\left(1+\frac{\partial F}{\partial\xi}\right). \tag{10.21}$$

On the surface of the body, $\dfrac{\partial F}{\partial \xi}$ is a function of K and ξ. Hence the pressure coefficient C_p on the surface of the body may be written as

$$C_p = \frac{\delta^{2/3}}{\Gamma^{1/3}}P(K,\xi) \tag{10.22}$$

where

$$P(K,\xi) = -2K\left(1 - \frac{\partial F}{\partial\xi}\right) \tag{10.23}$$

with P being a function of K and ξ only. Eq. (10.22) shows that for affinely related bodies at the same value of transonic parameter K, the pressure coefficients at the corresponding points on the surface of the bodies are proportional to the 2/3 power of their thickness ratios.

The lift coefficient of the body is obtained by integrating the pressure coefficient around the contour of the body, i.e.,

$$C_L = \oint C_p d\xi = \frac{\delta^{2/3}}{\Gamma^{1/3}}L(K) \tag{10.24}$$

where

$$L(K) = \oint P(K,\xi)d\xi \tag{10.25}$$

with $L(K)$ being a function of K only. Eq. (10.24) shows that for affinely related bodies at the same value of transonic parameter K, the lift coefficients are proportional to $\delta^{2/3}$.

Similarly, the drag coefficient of the body is

$$C_D = \oint C_p\frac{dy}{dx}dx = \oint C_p\delta h'(\xi)d\xi = \frac{\delta^{5/3}}{\Gamma^{1/3}}D(K) \tag{10.26}$$

where

$$D(K) = \oint P(K, \xi) h'(\xi) d\xi \tag{10.27}$$

with $D(K)$ being a function of K only. Eq. (10.26) shows that for affinely related bodies at the same value of transonic parameter K, the drag coefficients are proportional to $\delta^{5/3}$.

Finally the moment coefficient about a fixed point on the chord is

$$C_M(K) = \frac{\delta^{2/3}}{\Gamma^{1/3}} M(K) \tag{10.28}$$

where $M(K)$ is a function of K only. Eq. (10.28) shows that, for affinely related bodies at the same transonic parameter K, the moment coefficient about a fixed point on the chord is proportional to $\delta^{2/3}$.

Eqs. (10.22), (10.24), (10.26), and (10.28) are the transonic similarity laws for thin bodies; these have been verified experimentally.[4] These particular laws hold for Mach numbers near 1 but the results may be generalized to cover a wider range of Mach numbers.[3]

4. Slightly supersonic flow

For slightly supersonic flow, i. e., a Mach number slightly larger than one, the change in entropy is of the order δ^2 and thus, for thin bodies, the flow may be considered as irrotational; furthermore, the vorticity introduced is negligible. The fundamental solutions associated with this problem, are the oblique shock and the Prandtl-Meyer flow. The exact solutions in general may be obtained by the method of characteristics with the help of the relations of the oblique shock and Prandtl-Meyer flow; this will be discussed in Chapter XI. However it is interesting to see how the fundamental solutions of the oblique shock and the Prandtl-Meyer flow can be fitted to the transonic similarity laws involving the parameter K. These calculations were worked out by Tsien and Baron.[5]

(a) *Oblique shock*. The basic oblique shock relations are given by Eq. (4.29), i.e.,

$$\begin{aligned} \frac{p_2}{p_\infty} &= \frac{2\gamma}{\gamma+1} M_\infty^2 \sin^2\alpha - \frac{\gamma-1}{\gamma+1} \\ \frac{\tan(\alpha-\theta)}{\tan\alpha} &= \frac{\gamma-1}{\gamma+1} + \frac{2}{\gamma+1}\frac{1}{M_\infty^2 \sin^2\alpha} \end{aligned} \tag{4.29}$$

From Eq. (4.29), we have the pressure coefficient C_p:

$$C_P = \frac{4}{\gamma+1}\left(\sin^2\alpha - \frac{1}{M_\infty^2}\right) \tag{10.29}$$

From Eqs. (4.29) and (10.29), we find the relation between the angle of flow deflection θ and the pressure coefficient to be:

$$\tan\theta\left(1-\frac{C_P}{2}\right)\left(\frac{\Gamma C_P}{2}+\frac{1}{M_\infty^2}\right)^{1/2} = \frac{C_P}{2}\left[1-\left(\frac{\Gamma C_P}{2}+\frac{1}{M_\infty^2}\right)\right]^{1/2} \tag{10.30}$$

We consider the flow over thin bodies only so that the angle of flow deflection θ is small. We may take $\delta=\theta$ in Eqs. (10.19) and (10.22). Substituting these relations into Eq. (10.30) and taking the lowest order terms of θ, we have:

$$1 = \frac{P_s^2}{4}\left(2K - \frac{P_s}{2}\right) \tag{10.31}$$

where P_s is the pressure function defined by Eq. (10.23) for the case of the oblique shock. Eq. (10.31) is a cubic equation for P_s. To solve this equation, it is convenient to put

$$f = 2/P_s \tag{10.32}$$

With this, Eq. (10.31) becomes

$$f^3 - 2Kf + 1 = 0 \tag{10.33}$$

Since the flow is compressed through the oblique shock, the solution for f and P_s should be positive in order to have any physical significance. A value of $K=0.945$ is the lowest value required in Eq. (10.33) for a positive solution for f. The corresponding pressure function is $P_s=2.502$. This is the case for maximum flow deflection across an oblique shock for a given M_∞.

For larger values of K, there are two positive solutions of P_s: one for the strong shock and the other for the weak shock. The numerical values of these two pressure functions are given by Tsien and Baron.

(b) *Prandtl-Meyer flow*. In this case, the pressure coefficient may be written in differential form as

$$\frac{dC_P}{d\theta} = -\frac{2}{(M_\infty^2-1)^{1/2}\left(1-\dfrac{\Gamma}{M_\infty^2-1}C_P\right)^{1/2}} \tag{10.34}$$

Again if we take $\delta=\theta$ in Eq. (10.22), Eq. (10.34) becomes

$$\frac{1}{3} K^{5/2} \frac{d}{dK}\left(\frac{P_e}{K}\right) = \frac{1}{\sqrt{2-(P_e/K)}} \tag{10.35}$$

Integration of Eq. (10.35) gives

$$P_e = 2K - [(2K)^{2/3} + 3]^{2/3} \tag{10.36}$$

where P_e is the pressure function for Prandtl-Meyer expansion flow. It is interesting to note that for $M_\infty = 1$, $K = 0$, the pressure coefficient is

$$C_P = -\frac{(3\theta)^{2/3}}{\Gamma^{1/3}} \tag{10.37}$$

This shows that the pressure coefficient is nonlinear with respect to the angle of flow deflection in the transonic region; whereas, in Ackeret's linearized supersonic theory, the pressure coefficient is proportional to θ.

Tsien and Baron also applied the above results to calculate the aerodynamic characteristics of simple thin airfoil in slightly supersonic flow with shocks attached to the leading edge.[5]

5. Transonic flow over a wedge

The transonic flow equation (10. 13) is simpler than the original flow equation (10.12) of compressible flow but it is still nonlinear and, in general, difficult to solve. For some simple bodies, the transonic flow problem may be solved by the hodograph method. The difficulty of specifying the boundary conditions in the hodograph plane remains here even though the differential equation for transonic flow is much simpler than the original equation of compressible flow. However for wedge-shaped or flat-sided sections in a uniform flow of Mach number M equal to or greater than unity, the boundary conditions given in the physical plane may be easily transformed into those of hodograph plane. Consider the case of a double wedge section at zero angle of attack in a uniform free stream of $M_\infty = 1$. The flow patterns in both the physical and the hodograph planes are shown in Fig. 10.1. All the streamlines start at negative infinity in the physical plane, i. e., point A. In the hodograph plane, the point $A\,(q = a, \theta = 0)$ is a singular point. The upper half of the physical plane is limited by the zero streamline $\psi = 0$, i.e., ABC. It starts at minus infinity A and follows the $\theta = 0$ line to the stagnation point B. From B to C, the flow direction is constant, i.e., $\theta = \theta_0$. At point C, the velocity of the flow first reaches the sonic velocity and then increases to supersonic speed in accordance with the Prandtl-Meyer expansion theorem. At the corner, the hodograph will be represented by the Prandtl-Meyer epicycloid CD. The subsonic flow field will be affected by the supersonic flow region within CDA only. Outside this region, the supersonic flow will not affect the subsonic flow. For the transonic flow problem, we need to find only the solution satisfying the boundary condition *ABCD* in the hodograph plane.

The transonic flow equation in hodograph plane is the Tricomi equation (8.79)

$$\psi_{\sigma\sigma}+\sigma\psi_{\theta\theta}=0 \tag{10.38}$$

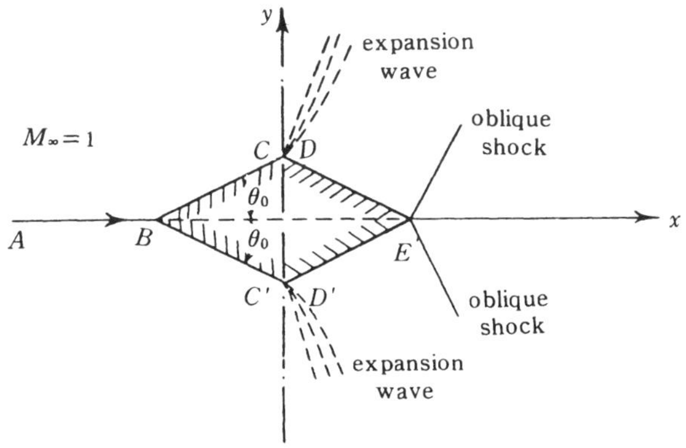

(a) Physical plane

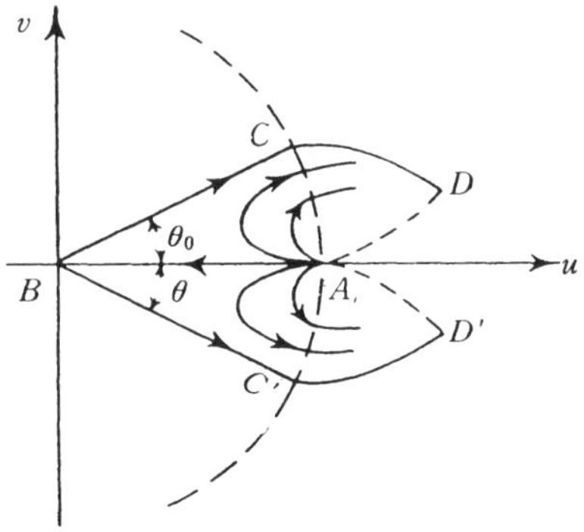

(b) Hodograph plane

Fig. 10.1 Transonic flow over a double wedge section

At the sonic line, $\sigma=0$; for subsonic flow, $\sigma>0$. In the σ-θ plane, the boundary conditions are shown in Fig. 10.2 (a). Guderley and Yoshihara[6] showed that the boundary condition of Fig. 10.2 (a) is equivalent to that of Fig. 10.2 (b) which is much easier to be satisfied. They worked out the solution of this problem with the boundary conditions of Fig. 10.2 (b) in terms of Nessel functions of 1/3 and −1/3 orders.

One of the most striking features of transonic flow is that the local Mach number distribution on an airfoil becomes independent of the free stream Mach

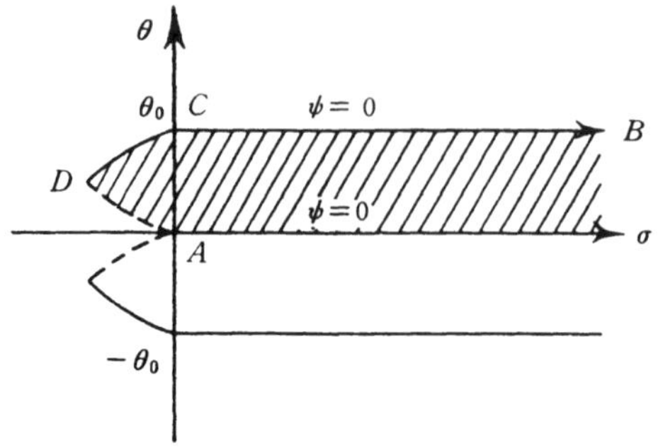

(a) Actual boundary conditions

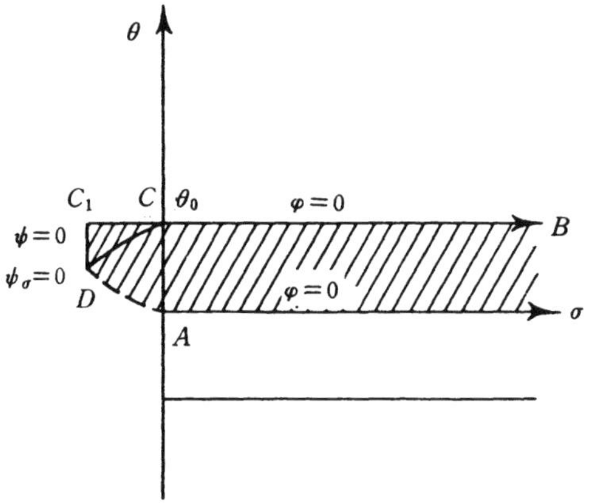

(b) Modified boundary conditions

Fig. 10.2 Boundary values on hodograph plane of a transonic flow over a double wedge section

number as the latter approaches unity. The local Mach numbers remain virtually unchanged or frozen as the free stream Mach number M_∞ varies from 0.85 to 1.24. As a result, the pressure drag coefficient is constant at the transonic region $M_\infty = 1$.[4, 7]

6. Transonic flow over airfoils [8]

The most interesting case of practice is the transonic flow over an airfoil or a wing. Let us consider a uniform stream of Mach number M_∞ passing over an airfoil. At first, the flow is low subsonic, i.e., $M_\infty^2 \ll 1$ and the whole flow field around the airfoil is subsonic. We increase M_∞ to a value $M_{\infty C}$ such that

$$M_\infty = M_{\infty C} \quad \text{if} \quad M_{\max} = 1 \tag{10.39}$$

At $M_\infty = M_{\infty C}$ the maximum Mach number $M_{\max}$ on the airfoil reaches the value of unity. We call this $M_{\infty C}$ as the subcritical Mach number. The value of $M_{\infty C}$ for a modern airplane is of the value between 0.80 and 0.85. When M_∞ is less than $M_{\infty C}$, the whole flow field is subsonic. Hence $M_{\infty C}$ may be considered as the limit of pure subsonic flow. When M_∞ is greater than $M_{\infty C}$ local supersonic region will occur.

In Fig. 10.3, we consider the case of a uniform stream of M_∞ slightly larger than $M_{\infty C}$ over a parabolic arc of zero angle of attack. There is a small local supersonic region with shock at the rear end near the maximum thickness location. We shall discuss the question whether there is a shock or not later. The line with cross $-\times-\times-$ is the sonic line which separates the subsonic and the supersonic regions. In the supersonic region, the solid curved lines are the

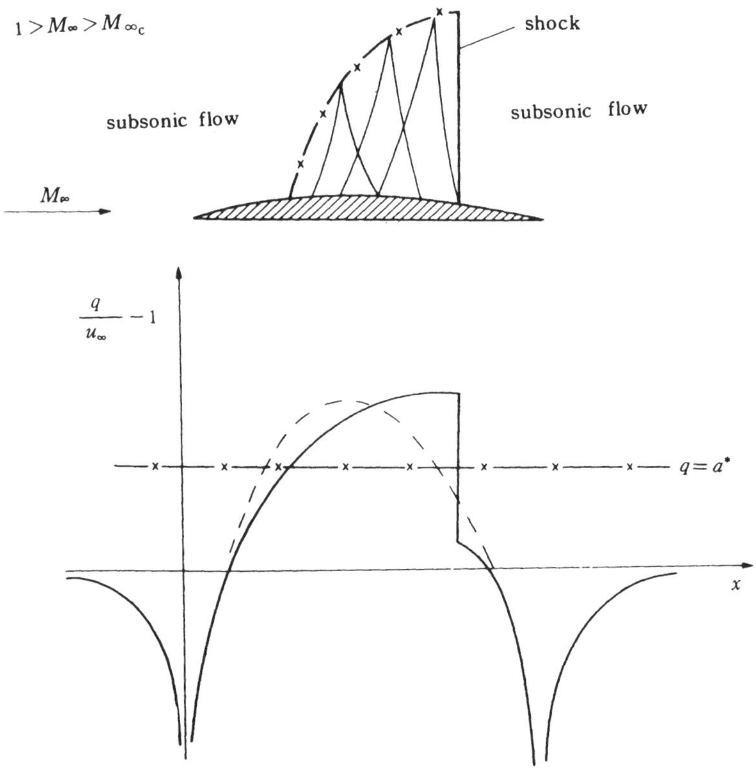

Fig. 10.3 The flow field and the surface velocity distribution of a uniform subsonic stream $M_\infty > M_{\infty C}$ over a parabolic arc airfoil

Mach lines, i. e., the hyperbolic characteristic lines. The velocity distribution along the surface of the airfoil, i.e., the parabolic arc, is shown below the airfoil, where q is the velocity on the surface of the airfoil and u_∞ is the velocity of the free stream at infinity. The sonic line $q=a^*$ is shown in Fig. 10.3

where q is the magnitude of the local velocity and a^* is the critical sound speed. The dotted curve for the velocity distribution along the airfoil surface is the result of linearized theory while the solid curve is the true transonic flow value. The location of the maximum velocity of transonic flow moves downstream from that of the linearized theory. As we increase the value of M_∞ further, the supersonic flow region increases. Fig. 10.4 shows a case when a large supersonic region exists reaching downstream of the trailing edge of the airfoil with the corresponding velocity distribution along the surface of the airfoil. In Fig. 10.4, M_∞ is still less than unity.

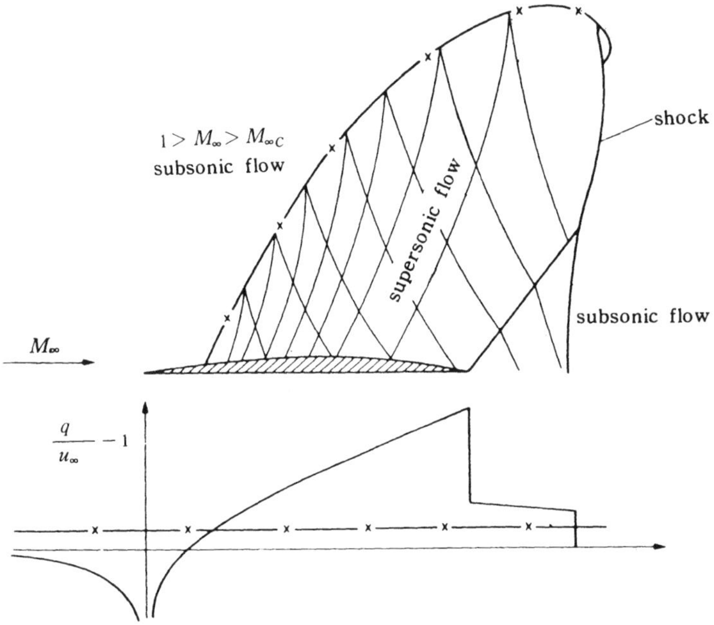

Fig. 10.4 The large supersonic flow field and the surface velocity distribution of a uniform subsonic stream of $M_\infty > M_{\infty c}$ over a parabolic arc airfoil

When we increase the Mach number of the free stream M_∞ to unity, $M_\infty = 1$, the corresponding flow field and the surface velocity distribution are shown in Fig. 10.5. We call this case of $M_\infty = 1$ as the sonic flow. There is an oblique shock wave at the trailing edge in the supersonic region. The shock wave which is at the end of the supersonic region of Fig. 10.4 moves downstream to infinity and its strength becomes zero. It is interesting to notice that when $M_\infty \cong 1$, the Mach lines and the sonic lines near the airfoil are independent of the value of M_∞. We call such a flow as Frozen flow. We shall give the exact definition of frozen flow later. We should notice that the supersonic flow field of Figs. 10.3 to 10.6 near the airfoil are quite the same as a result of the frozen flow concept.

There is a point F in Fig. 10.5 from which the left running Mach line will intersect the sonic line at infinity. From any point downstream of the point F, the left running Mach line will not intersect the sonic line. In this example, the point F is a little upstream of the location of the maximum thickness of the airfoil. Any smooth change in shape of the airfoil downstream of the point F will not influence the upstream subsonic region. The Mach line from the point F is known as the limiting Mach line.

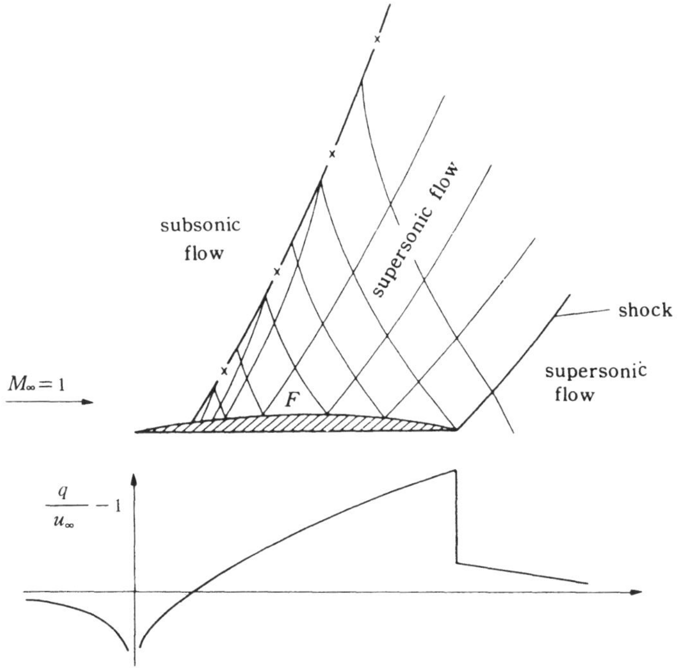

Fig. 10.5 The flow field and the surface velocity distribution of a sonic flow over a parabolic arc airfoil

When the free stream Mach number M_∞ is slightly larger than unity, the flow field and the distribution of the surface velocity on the airfoil are shown in Fig. 10.6. We have a detached or a bow shock AC in front of the airfoil. Immediately behind the detached shock, there is a subsonic region. In the supersonic flow region over the airfoil of Fig. 10.6, the Mach lines are shown. The left running limiting Mach line from the point F intersects the detached shock at point C , the end of the subsonic region behind the detached shock. The right running Mach line from the point C gives the boundary of influence of the subsonic region.

It should be noticed once more that the configuration of the sonic line near the wing for all the three cases, $M_\infty < 1$, $M_\infty = 1$, and $M_\infty > 1$, i.e., Figs. 10.4 to 10.6, are nearly similar. It was Liepmann and Bryson who first studied the frozen flow of transonic flow experimentally and introduced the concept of frozen flow.[7] Hence the special case of sonic flow $M_\infty = 1$ is very important in the study of transonic flow because it would give the whole flow field of the transonic flow around the airfoil with M_∞ slightly deviated from unity.

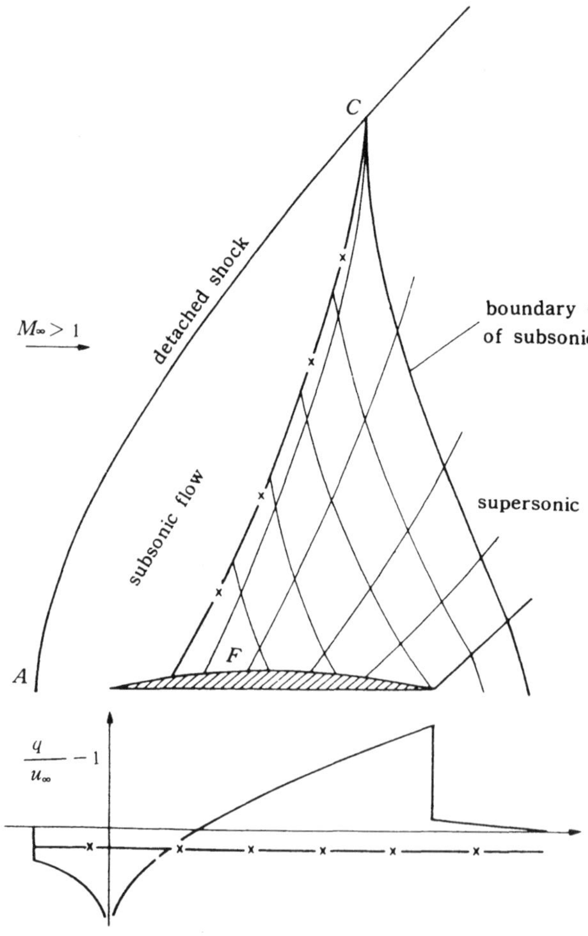

Fig. 10.6 The flow field and the surface velocity distribution of a uniform supersonic stream M_∞ slightly larger than unity over a parabolic arc airfoil

Further increase of $M_\infty > 1$, the subsonic region behind the detached shock will decrease and the boundary of influence of the subsonic region will reduce as shown in Fig. 10.7, i.e., the curve CG where G is on the surface of the airfoil. If the value of M_∞ is increased to an upper critical Mach number

M_∞^C, we have

$$M_\infty = M_\infty^C \text{ if } M_{\min} = 1 \tag{10.40}$$

the minimum Mach number $M_{\min}$ in the whole flow field is unity. The upper critical Mach number M_∞^C of a pointed airfoil is of the order of 1.3 to 1.4. For a blunt body, there is no upper critical Mach number. Hence for supersonic-hypersonic flow over a blunt body, we cannot have pure supersonic flow. For such a flow field, we always have a detached shock and a subsonic flow region behind the detached shock. We shall discuss this case in Chapter XIV.

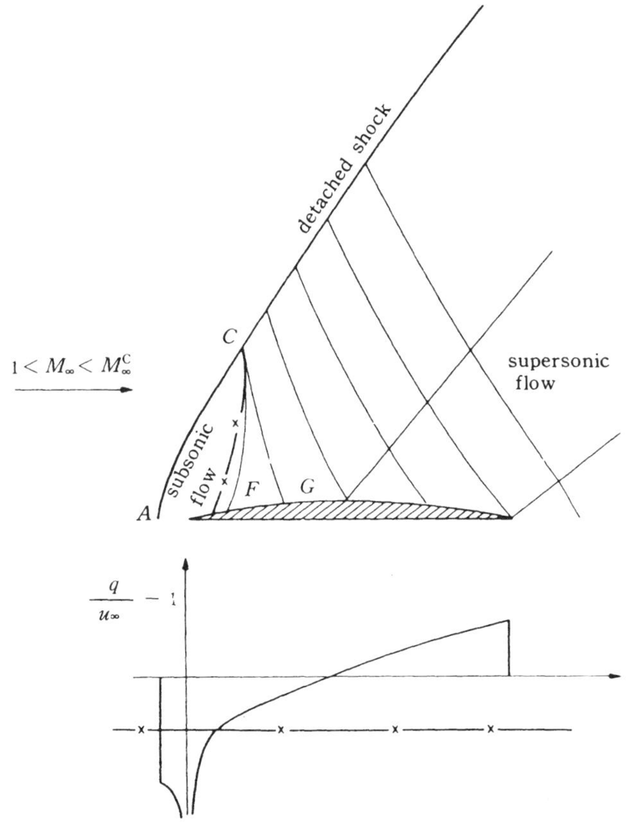

Fig. 10.7 The flow field and the surface velocity distribution of a uniform supersonic stream with Mach number slightly less than the upper critical Mach number over a parabolic arc airfoil

7. Transonic flow over an airfoil with $M_\infty = 1$

Guderley[5,6] was the first one to use the indirect method on the hodograph plane to find transonic flow over a profile without shock at $M_\infty = 1$.

It is known as Guderley profile which is a symmetrical airfoil with a cusp leading edge. Such a profile is also of interest because of the frozen flow concept of the transonic flow. Sobieczky[9, 10] has studied and extended Guderley's solution at $M_\infty=1$ by the Rheograph method. One of the interesting results of Socieczky is that he generalized Guderley's problem to the case of an airfoil with camber ω. We are going to discuss some of the Sobieczky's results in this section.

Guderley[11] used the full hypergeometric function in hodograph plane to obtain his symmetrical profile without shock at $M_\infty=1$. In general, such a calculation is tedious. However, Sobieczky has shown that the superposition of two elementary solutions $\nu=-2$ and $\nu=0$ would give the Guderley's profile. Thus, Sobieczky's method is much easier to use than Guderley's and it is possible to generalize Guderley's symmetrical profile to the case of the asymmetrical profile with a camber without too much additional work.

In Sobieczky's method (Chapter VIII § 11), the superposition of two elementary solutions $\nu=-2$ and $\nu=0$ gives

$$\begin{aligned} x &= -r^{-4/3}(1+z)^{2/3}[c_{-2}(2-3z)+c_0r^2] - r^{-4/3}(1-z)^{2/3}[c_{-2}(2+3z)+c_0r^2] \\ A_1\bar{y} &= -r^{-5/3}(1+z)^{1/3}[c_{-2}(1-3z)+2c_0r^2] + r^{-5/3}(1-z)^{1/3}[c_{-2}(1+3z)+2c_0r^2] \end{aligned} \tag{10.41}$$

where c_{-2} and c_0 are arbitrary constant for the solutions $\nu=-2$ and $\nu=0$ respectively. The absolute values of c_{-2} and c_0 are not very important. In his example, Sobieczky took $c_{-2}=1/2$ and $c_0=1/3$. The stagnation streamline for this symmetrical profile is $\bar{y}=0$, $v_1=t=0$ and therefore from the second equation of (10.41), we have

$$x=-2s^{-4/3}\ (2c_{-2}+c_0s^2) \tag{10.42}$$

Since this airfoil (Fig. 10.8) is with a cusp leading edge, there is no real stagnation point but the minimum velocity occurs at $s=-1$, $x=-1$ and $u_1=-(3/2)^{2/3}$.

The shape of this symmetrical airfoil is given by the r-z relation with $\bar{y}=0$ but $t\neq 0$, i.e.,

$$r^2\leqslant 1;\ 3r^{2/3}(1-z^2)^{1/3}=4r^2-1 \tag{10.43}$$

or

$$2t=\sqrt{(1\mp s^{2/3})(1\pm 2s^{2/3})} \tag{10.43a}$$

Eq. (10.43) or (10.43a) gives this airfoil in the Rheograph plane. From Eqs. (10.43) and the first equation of (10.41), we have for

$$x=-1:x=\mp s^{2/3}=\mp\Lambda_1^{2/3}=(2/3)^{2/3}u_1 \qquad (10.44)$$
$$v_1=1/2\sqrt{(1+x)(1-2x)}$$

The form for this Guderley's profile is then

$$h_x(x)=v_1\tau=\frac{\tau}{2}(1-2x)\sqrt{(1+x)}$$

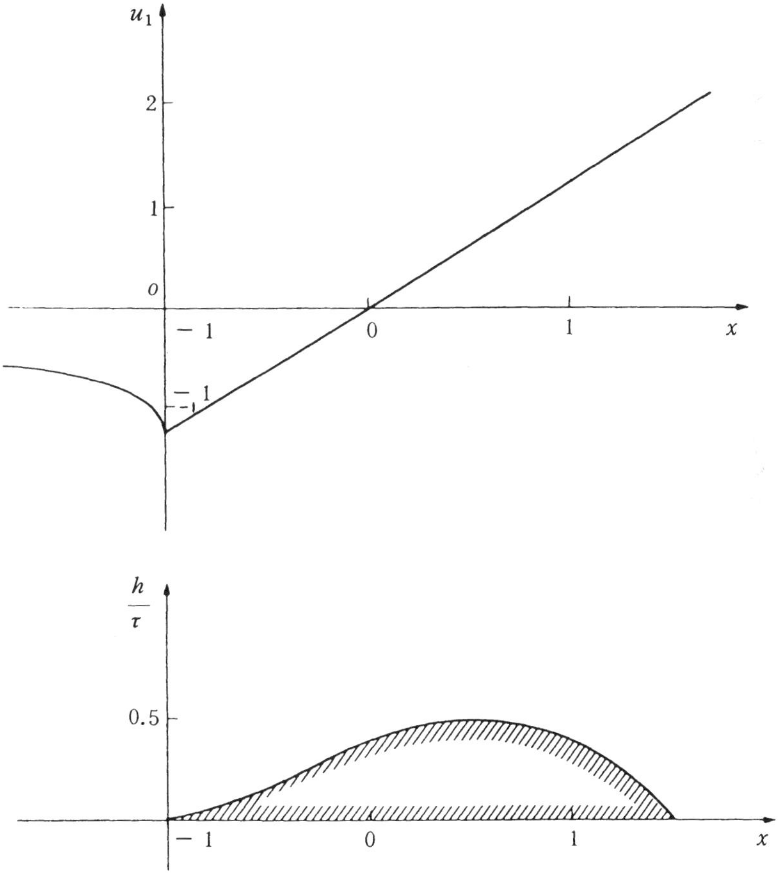

Fig. 10.8 Guderley's airfoil of Eq. (10.45) and its velocity distribution

Then

$$h(x)=\frac{\tau}{5}(1+x)^{3/2}(3-2x) \quad \text{for} \quad -1\leqslant x\leqslant 3/2 \qquad (10.45)$$

Fig. 10.8 gives the shape of Guderley's profile of Eq. (10.45) and the velocity distribution on the profile and the stagnation stream line.

The sonic line of this Guderley's profile is at $s=0$, $r=t$, $z=1$ and its shape is given from Eqs. (10.41) as follows:

$$x=\frac{2^{2/3}}{12}t^{-4/3}(1-4t^2);\quad A_1\bar{y}=\frac{2^{4/3}}{12}t^{-5/3}(1-4t^2) \tag{10.46}$$

The limiting Mach line (left running Mach line) of this Guderley's profile is

$$\xi=s-t\rightarrow 0,\ r\rightarrow 0,\ s\rightarrow t,\ z\rightarrow\infty$$

and from Eqs. (10.41)

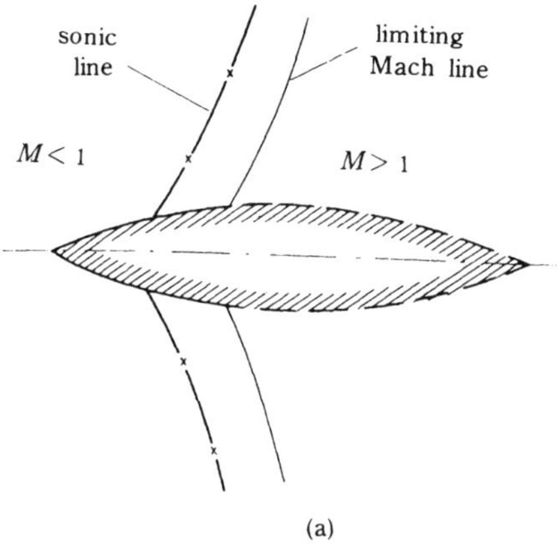

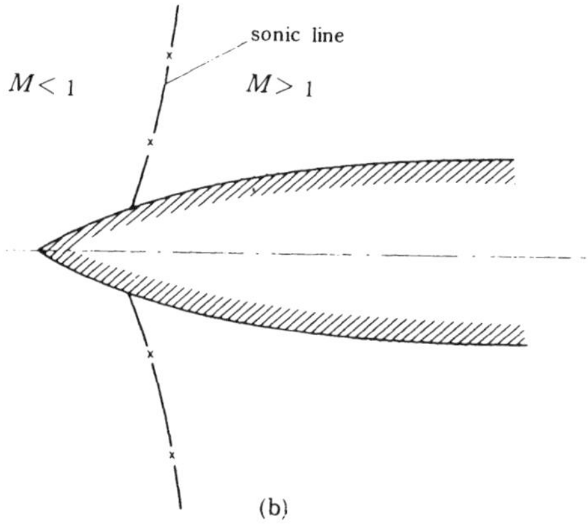

Fig. 10.9 Airfoils with and without limiting Mach line (a) body with limiting Mach line (b) Semi-infinite airfoil without limiting Mach line

$$x=\frac{5}{81}t^{-4/3}\left(1-\frac{54}{5}t^2\right);\quad A\bar{y}=\frac{4}{81}t^{-5/3}(1-27t^2) \tag{10.47}$$

and the end point of this limiting Mach line on the profile is at

$$\bar{y}=0;\; s=t=\frac{1}{9}\sqrt{3}\;, x=\frac{1}{3} \tag{10.47a}$$

It is interesting to note that the limiting Mach line occurs ahead of the maximum thickness of the airfoil. Hence the shape of this airfoil after the limiting Mach line location, i.e., $x > 1/3$, will not affect the flow in the subsonic upstream region. See Fig. 10.9(a).

Strictly speaking, in general only those changes of the profile downstream of the point $x = 1/3$ do not affect the subsonic region, which do not cause shock waves. Because shock waves can cross the limiting Mach line. Therefore, changes of the shock waves can cross the limiting Mach line. Therefore, changes of the profile of the kind of concave wedges or curvatures are excluded.

It is possible to superimpose three elementary solutions $\nu = -2$, $\nu = 0$ and $\nu = -1$ so that a semi-infinite airfoil without limiting Mach line at $M_\infty = 1$ may be obtained as shown in Fig. 10.9(b).

For an aeronautical engineer, we are very interested in the lift of an airfoil. The simplest case for a lifting airfoil is an inclined flat plate, i.e., $\tau = 0$ and $\alpha \neq 0$. Guderley[11] has calculated the flow field over an inclined flat plate as shown in Fig. 10.10. The sonic line is upstream of the leading edge of the plate and the whole flow pattern is rather complicated. There is also a sonic line at the trailing edge of the plate. However, it is possible to apply the similarity law to this problem of airfoil with zero thickness by using tan α instead of the thickness ratio. Then we have

$$C_L \sim (\tan\alpha)^{2/3}\;;\; \frac{dC_L}{d\alpha} \sim \frac{2}{3}(\tan\alpha)^{-1/3} \tag{10.48}$$

Fig. 10.11 gives the lift coefficient c_L vs. tan α where α is the angle of attack. When $\alpha = 0$, $dC_L/d\alpha = \infty$.

Sobieczky has generalized Guderley's profile at $M_\infty = 1$ to the case of airfoil with camber ω. The results are shown in Fig. 10.12. For the shockfree flow field, the angle of attack α has to change with the camber ω as shown in Fig. 10.12(a). For a given thickness ratio τ, the lift coefficient C_L is a function of the camber ω as shown in Fig. 10.12(c). Up to $\omega/\tau = 0.50$, C_L is nearly a linear function of the camber ω. In Fig. 10.12(b), some typical sonic line, limiting Mach line for the shockfree cambered profile at $M_\infty = 1$ are shown. At the trailing edge, there are shocks.

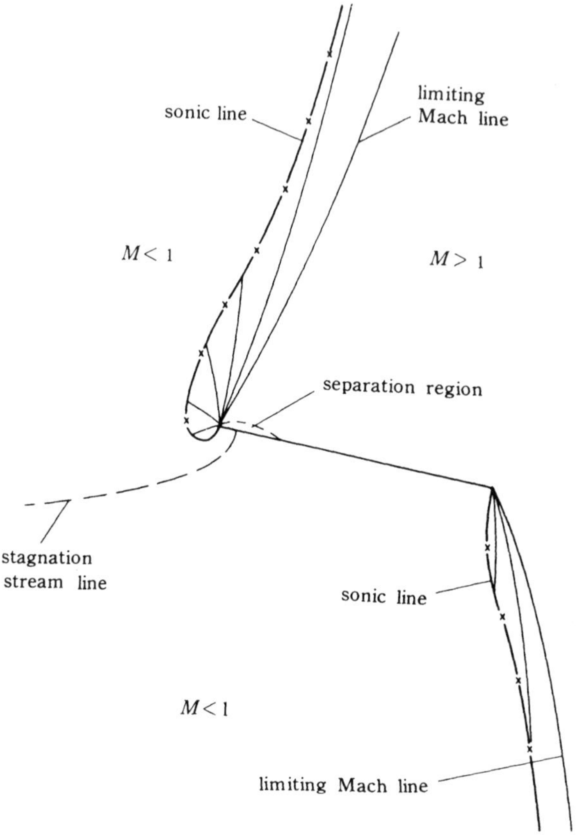

Fig. 10.10 Local supersonic region at the leading edge of an inclined flat plate

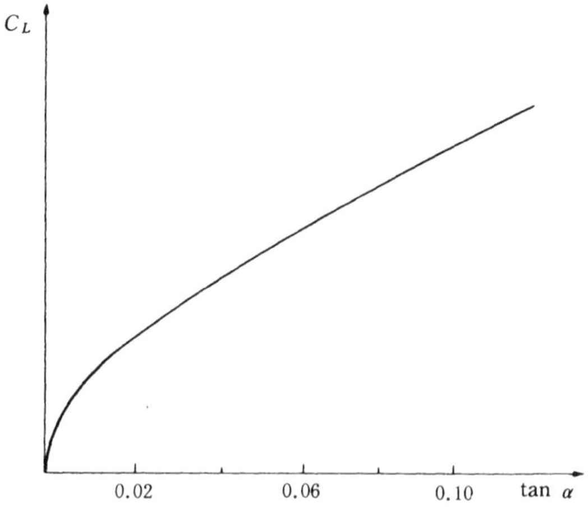

Fig. 10.11 Lift coefficient C_L vs tan α for $M_\infty = 1$ and $t=0$

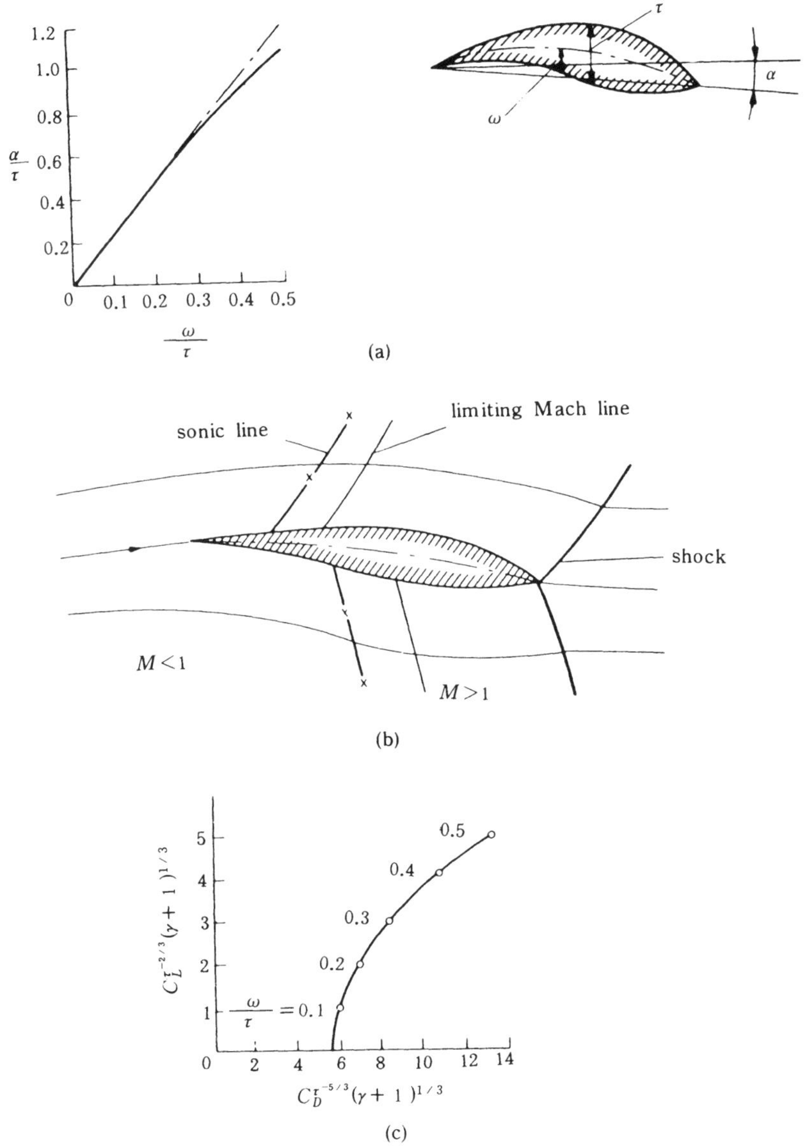

Fig. 10.12 Sobieczky's results for shockfree transonic cambered airfoil. (a) Relation between α and ω for shockfree airfoil. (b) Streamline, sonic line and limiting Mach line on a shockfree cambered airfoil. (c) Lift-drag polar curve for a shockfree cambered airfoil in the transonic flow

8. Non-lifting airfoil in the frozen supersonic region

Now we discuss in a little detail about the non-lifting airfoil in the frozen supersonic region. By frozen supersonic region, we mean that when the free stream Mach number is closed to unity, $M_\infty = 1$, the Mach lines and the sonic lines near the airfoil are the same as those given for the case $M_\infty = 1$. To show such a result, we have to consider the shock polar as well as the shape of the airfoil. Fig. 10.13 gives the reduced shock polar in the rheograph plane and some points of special interest which are listed in Tab. 10.1.

The complete boundary value problem in the rheograph plane is shown in Fig. 10.14 in which $\bar{y}=0$ gives the shape of the airfoil in the rheograph plane. Inside the profile, we have the shock polar. In order to satisfy the boundary condition at the shock, Sobieczky generalized our previous solution by replacing t by $t-m$ where m is an arbitrary constant so that

$$r=\sqrt{((t-m)^2+s^2)} \;;\; z=\frac{t-m}{r} \tag{10.49}$$

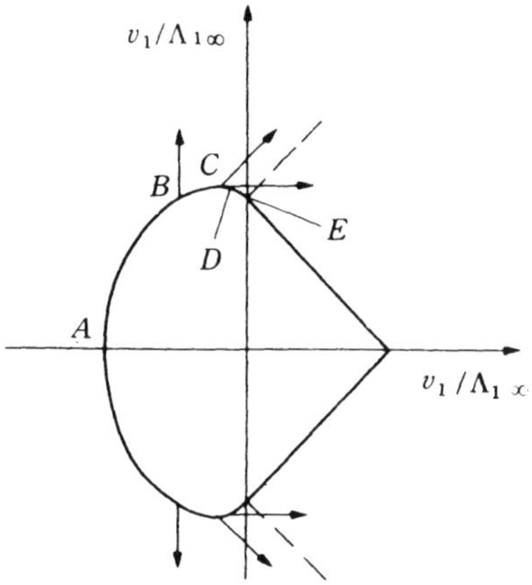

Fig. 10.13 Reduced shock polar and points of special interest

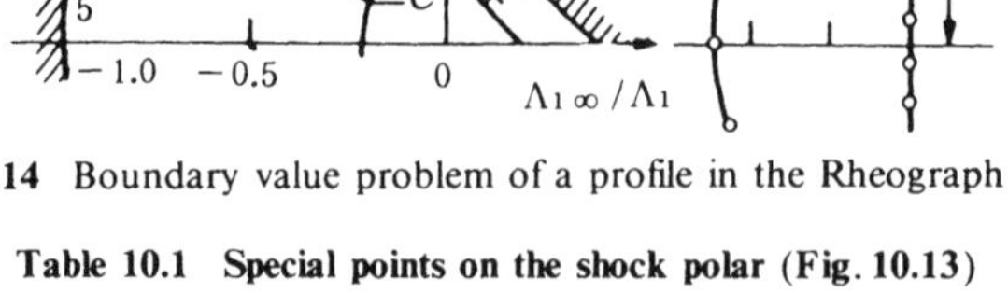

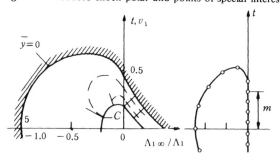

Fig. 10.14 Boundary value problem of a profile in the Rheograph plane

Table 10.1 Special points on the shock polar (Fig. 10.13)

Symbol	$\Lambda_1/\Lambda_{1\infty}$	$\hat{u}_1/u_{1\infty}$	$\hat{v}_1/\Lambda_{1\infty}$	Remark
A	-1	-1	0	Normal shock
B	$-(3/5)^{1/2}$	$-3/5$	$12\sqrt{5}/25$	No acceleration
C	$-(1/3)^{1/2}$	$-1/3$	$2\sqrt{3}/3$	Maximum v_1
D	$-(1/7)^{1/2}$	$-1/7$	$12\sqrt{21}/49$	Crocco point
E	0	0	$3\sqrt{2}/4$	Sonic point

Sobieczky tried to use these new solutions to satisfy 21 points on the shock polar. The results are shown in Fig. 10.15. In Fig. 10.15(a), we have the shock wave, sonic line, other lines of constant velocity (isotach lines), and the limiting Mach line for a slightly supersonic flow over a Guderley's profile with $u_{1\infty}=0.465$ and the corresponding sonic line, isotach lines and limiting Mach

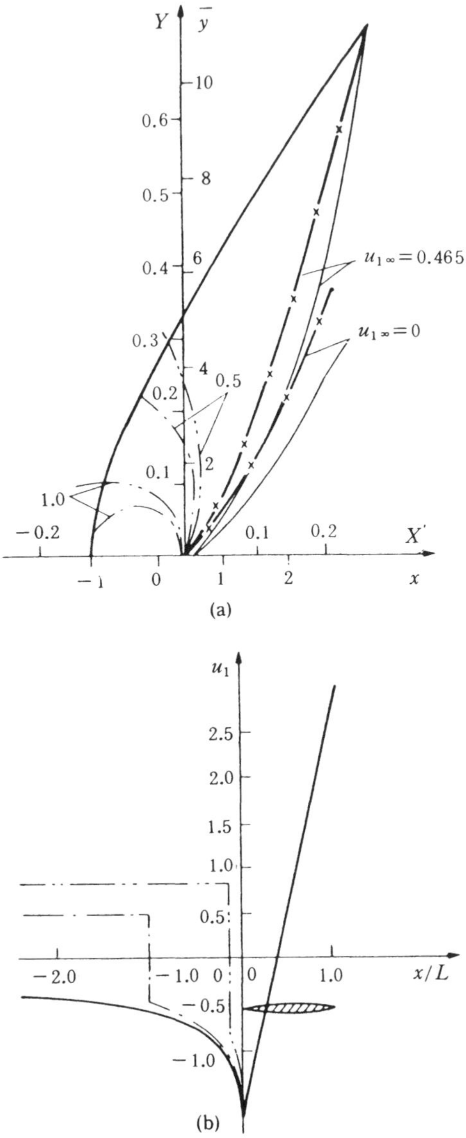

Fig. 10.15 Guderley's profile at sonic flow and slightly supersonic flow. Frozen flow concept. (a) Shockwave, lines of constant velocity and limiting Mach line for a slightly supersonic flow $u_{1\infty}=0.465$ and for sonic flow $u_{1\infty}=0$. (b) u_1 distribution for Guderley's profile at $u_1=0$ (solid line); $u_1=0.465$ ($-\cdot-$); $u_1=0.836$ ($-\cdots-$)

line for the sonic flow over a Guderley's profile with $u_{1\infty}=0$ are the same as those of slightly supersonic cases, $M_{\infty}=1$ or $u_{1\infty}=0$, but far away from the airfoil, the lines in these two cases are very different. For $M_{\infty}=1$, the bow shock is at infinity upstream. As the free stream Mach number M increases from $M_{\infty}=1$, the bow shock will come closer to the airfoil. In Fig. 10.15b, the sudden change in u_1 represents the location of the bow shock. For $u_{1\infty}=0.836$, the shock is closer to the airfoil than the case for $u_{1\infty}=0.465$. It is interesting to notice that at infinity upstream, the distribution for the subsonic case such as the Mueller-Matschet[12] with $\nu=-2$ only does not coincide with the solution of Sobieczky with shock (Fig. 10.14) even though the shock may move to infinity by reducing the strength of the shock. Similarly, the singularity of the pure subsonic solution such as the Prandtl-Glauert solution is also different from that of Guderley's solution at infinity. However the different behavior of the singularity at infinity for the pure subsonic-solution, shockfree transonic solution and transonic solution with shock seem to be not very important for practical application as in the numerical solution of the transonic flow.[13]

9. Steady transonic small perturbation equation

The full potential equation can be simplified if we consider the flow over a thin body where the uniform free stream is only slightly disturbed. The analysis of this type of transonic flow field is referred to as transonic small perturbation (TSP) theory. Besides the simplification of governing equation, a significant advantage accrues in the application of boundary conditions. Boundary conditions are applied on the slit for two-dimensional problems or on the plane for three-dimensional problems. Thus complex body-conforming grid generations are unnecessary for the application of boundary conditions. This can result in significant reductions in computer time and storage requirements, particularly in three-dimensional problems.

In order to demonstrate how the full velocity potential equation can be simplified for flow of this type, we assume that a thin body is placed in a two-dimensional flow. The body causes a disturbance of the uniform flow and the velocity components are written as

$$\left.\begin{aligned} u &= U + u' \\ v &= v' \end{aligned}\right\} \tag{10.50}$$

where the prime denotes perturbation velocity. If we let φ' be the perturbation velocity potential, then

$$\left.\begin{aligned} u &= \frac{\partial \varphi}{\partial x} = U + \frac{\partial \varphi'}{\partial x} \\ v &= \frac{\partial \varphi}{\partial y} = \frac{\partial \varphi'}{\partial y} \end{aligned}\right\} \tag{10.51}$$

Since the flow is only slightly disturbed from the free stream, we assume

$$\frac{|u'|}{U}, \frac{|v'|}{U} \ll 1 \tag{10.52}$$

As a result, the coefficients in Eq. (10.12) become

$$\left.\begin{aligned} a^2-u^2 &= a_\infty^2 - U^2 - (\gamma+1)Uu' \\ a^2-v^2 &= a_\infty^2 - (\gamma-1)Uu' \\ -2uv &= -2Uv' \end{aligned}\right\} \tag{10.53}$$

The last two coefficients can be further simplified,

$$\left.\begin{aligned} a^2-v^2 &= a_\infty^2 \\ -2uv &= 0 \end{aligned}\right\} \tag{10.54}$$

if $1/M_\infty^2 \gg |u'|/U$. This is true for transonic case. In the first coefficient a^2-u^2, the term $-(\gamma+1)Uu'$ can not be neglected, since the other term $a_\infty^2-U^2$ is also small. We have to keep both terms in order to avoid the two-dimensional equation degenerating into one-dimensional equation. Therefore, the full potential equation simplifies to

$$\left(1-M_\infty^2-\frac{\gamma+1}{U}M_\infty^2\frac{\partial\varphi'}{\partial x}\right)\frac{\partial^2\varphi'}{\partial x^2}+\frac{\partial^2\varphi'}{\partial y^2}=0 \tag{10.55}$$

The equation is called the transonic small perturbation equation. In the present approximation, the coefficient of $\frac{\partial^2\varphi'}{\partial x^2}$ can be written

$$1-M_\infty^2-\frac{\gamma+1}{U}M_\infty^2\frac{\partial\varphi'}{\partial x}=\frac{a^2-u^2}{a^2-v^2}=\frac{a^2-u^2-v^2}{a^2}=1-M^2 \tag{10.56}$$

where M is the local Mach number. This quasi-linear equation (10.55) is either elliptic or hyperbolic, depending on whether the local Mach number is less or greater than unity and thus is of a mixed type for transonic flows. At locally supersonic point, it has characteristics

$$\frac{dy}{dx}=\pm\frac{1}{\sqrt{M^2-1}} \tag{10.57}$$

which are local Mach waves. They are symmetric to coordinate axes.

For subsonic or supersonic flows where M_∞ is quite different from unity, Eq. (10.55) reduces to the linear Prandtl-Glauert equation (6.61). TSP is a good approximation for the airfoil at small angle of attack as shown in Fig. 7.5.

10. Type-dependent differences

Murman and Cole,[14] in a landmark paper treating of transonic flow, pointed out that derivatives at each mesh point in the domain of interest must be correctly treated by using type-dependent differences. Since then numerous applications of the technique have been made and many refinements of the basic method have been developed.

To understand the role of type-dependent differencing, it is instructive to consider the one-dimensional forms of the transonic small perturbation equation

$$\left(1-M_\infty^2-\frac{\gamma+1}{U}M_\infty^2\frac{\partial\varphi}{\partial x}\right)\frac{\partial^2\varphi}{\partial x^2}=0 \tag{10.58}$$

where φ is the perturbation velocity potential, i. e. the prime is omitted for brevity.

If the flow is subsonic, central differences are used for derivatives (cf. Chapter VII, § 8). If the flow is supersonic at the point of interest, the x-derivatives are treated in the upstream direction and the finite-difference representations at point (i,j) can be written as

$$\left.\begin{aligned}\frac{\partial\varphi}{\partial x}&=\frac{\varphi_i-\varphi_{i-2}}{2\Delta x}\\ \frac{\partial^2\varphi}{\partial x^2}&=\frac{\varphi_i-2\varphi_{i-1}+\varphi_{i-2}}{\Delta x^2}\end{aligned}\right\} \tag{10.59}$$

They are upwind difference representations. Substituting Eqs. (10.59) into (10.58), we obtain the finite difference equation at supersonic point (i,j)

$$\left(1-M_\infty^2-\frac{\gamma+1}{U}M_\infty^2\frac{\varphi_i-\varphi_{i-2}}{2\Delta x}\right)\frac{\varphi_i-2\varphi_{i-1}+\varphi_{i-2}}{\Delta x^2}=0 \tag{10.60}$$

Some characteristics of a difference equation are readily shown through its modified differential equation.[15] The modified differential equation of (10.60) is

$$(1-M^2)\frac{\partial^2\varphi}{\partial x^2}-(1-M^2)\Delta x\frac{\partial^3\varphi}{\partial x^3}+\frac{\gamma+1}{U}M_\infty^2\Delta x\left(\frac{\partial^2\varphi}{\partial x^2}\right)^2 \tag{10.61}$$
$$+O(\Delta x^2)=0$$

The truncation error of the upwind difference equation is $O\ (\Delta x)$. Thus, Eq. (10.60) is first-order accurate. The truncation error has important interpretations.

Since the potential equation is derived from the fundamental equations of inviscid fluid, we can identify the potential equation with equation of motion, i. e.

$$(1-M^2)\frac{\partial^2\varphi}{\partial x^2}\approx-\frac{u}{a^2}\left(\frac{Du}{Dt}+\frac{1}{\rho}\frac{\partial p}{\partial x}\right) \tag{10.62}$$

Putting (10.62) into (10.61), the modified equation becomes

$$-\frac{u}{a^2}\left[\frac{Du}{Dt}+\frac{1}{\rho}\frac{\partial p}{\partial x}-\frac{a^2}{u}(M^2-1)\Delta x\frac{\partial^2 u}{\partial x^2}-\cdots\right]=0 \tag{10.63}$$

It is seen that a leading term of the truncation error provides a positive artificial viscosity at all points where $M > 1$. Thus we can expect shock waves to form only as compressions in supersonic regions and to be captured by the numerical solution. The effect of artificial viscosity has been discussed in Chapter IV, §12. There the artificial viscosity was added to the inviscid equation of motion explicitly. Here the artificial viscosity is included implicitly through the upwind biased differencing.

Another interpretation of the truncation error for the upwind difference equation can be obtained through the identification of the potential equation with a equation of continuity. The original transonic small perturbation equation can be written

$$\frac{\partial}{\partial x}\left[1+(1-M_\infty^2)\frac{u'}{U}-\frac{\gamma+1}{2}M_\infty^2\frac{u'^2}{U^2}\right]=0 \qquad (10.64)$$

This is an approximation to the equation of continuity

$$\frac{\partial}{\partial x}\left(\frac{\rho u}{\rho_\infty U}\right)=0 \qquad (10.65)$$

Therefore, we have

$$\frac{\rho u}{\rho_\infty U_\infty}=1+(1-M_\infty^2)\frac{u'}{U}-\frac{\gamma+1}{2}M_\infty^2\frac{u'^2}{U^2} \qquad (10.66)$$

From Eq. (10.66), we obtain

$$\frac{\rho}{\rho_\infty}=1-M_\infty^2\frac{u'}{U}-\left(1+\frac{\gamma-1}{2}M_\infty^2\right)\frac{u'^2}{U^2} \qquad (10.67)$$

This is the density formula for the transonic small perturbation flow.

Now we consider the modified differential equation (10.61). It can be rewritten as

$$\begin{aligned}\frac{\partial}{\partial x}\Bigg[&1+(1-M_\infty^2)\frac{u'}{U}-\frac{\gamma+1}{2}M_\infty^2\frac{u'^2}{U^2}\\ &-\Delta x(1-M_\infty^2-\frac{\gamma+1}{U}M_\infty^2 u')\frac{\partial u'}{\partial x}\Big/U\Bigg]=0\end{aligned} \qquad (10.68)$$

Similarly we obtain

$$\begin{aligned}\frac{\rho}{\rho_\infty}=&1-M_\infty^2\frac{u'}{U}-\left(1+\frac{\gamma-1}{2}M_\infty^2\right)\frac{u'^2}{U^2}\\ &-\Delta x\left[1-M_\infty^2-(\gamma M_\infty^2+1)\frac{u'}{U}\right]\frac{\partial u'}{\partial x}\Big/U\end{aligned} \qquad (10.69)$$

This is the artificial density or compressibility provided by the upwind differ ence equation. The effect of artificial density is equivalent to the effect of artifi cial viscosity. The idea of artificial density is applied in solving the full potential equation to provide artificial viscosity in supersonic regions.[16–18]

11. Conservative and nonconservative shock schemes

In the above section we have given two of the type-dependent differences for locally subsonic and supersonic points. There are two more difference representations for the two transition points in transonic flows: sonic transition point from a subsonic region to a supersonic region and shock transition point from a supersonic region to a subsonic region. For simplicity we consider the one dimensional transonic small perturbation equation (10.58).

At sonic points, we have $M=1$. According to Eq. (10.56), the left-hand side of (10.58) vanishes. Therefore the difference representation for sonic points is zero.

At shock points, the difference representation of Eq. (10.58) is not evident. We will establish the shock difference schemes using a conservation consideration.

For conservation consideration, we rewrite the governing differential equation into conservation or divergence form

$$\frac{\partial F}{\partial x}=0 \tag{10.70}$$

where

$$F=(1-M_\infty^2)\frac{\partial \varphi}{\partial x}-\frac{\gamma+1}{2U}M_\infty^2\left(\frac{\partial \varphi}{\partial x}\right)^2 \tag{10.71}$$

The t ype-dependent difference representations for subsonic and supersonic points at grid point i are constructed.

$$\frac{\partial F}{\partial x}=\begin{cases} \dfrac{F_{i+\frac{1}{2}}-F_{i-\frac{1}{2}}}{\Delta x}, \text{ if } M<1 & (10.72) \\[2ex] \dfrac{F_{i-\frac{1}{2}}-F_{i-\frac{3}{2}}}{\Delta x}, \text{ if } M>1 & (10.73) \end{cases}$$

where the value at the cell midpoint is evaluated as

$$\left(\frac{\partial \varphi}{\partial x}\right)_{i+\frac{1}{2}}=\frac{\varphi_{i+1}-\varphi_i}{\Delta x} \tag{10.74}$$

Substituting Eq.(10.74)into (10.72)and (10.73), we obtain the central and upwind difference representations described in the above section. Putting Eqs. (10.72) and (10.73) into Eq. (10.70), we have the difference equations in conservation form.

$$F_{i+\frac{1}{2}}-F_{i-\frac{1}{2}}=0, \text{ if } M<1 \tag{10.75}$$

$$F_{i-\frac{1}{2}}-F_{i-\frac{3}{2}}=0, \text{ if } M>1 \tag{10.76}$$

We now construct the shock difference scheme using the conservation forms. Consider a transition at grid point i from supersonic flow to subsonic flow. Write out the difference equations of conservation for all grid points except point i.

$$\left.\begin{array}{lll} \vdots & & \vdots \\ i-2 & F_{i-5/2} & -F_{i-7/2}=0 \\ i-1 & F_{i-3/2} & -F_{i-5/2}=0 \\ i+1 & F_{i+3/2} & -F_{i+1/2}=0 \\ i+2 & F_{i+5/2} & -F_{i+3/2}=0 \\ \vdots & & \vdots \end{array}\right\} \tag{10.77}$$

Summing up the left-hand side of Eqs. (10.77), we obtain

$$-F_{-\infty}+F_{i-\frac{3}{2}}-F_{i+\frac{1}{2}}+F_{\infty}=0 \tag{10.78}$$

If the difference equation at shock point i is

$$F_{i+\frac{1}{2}}-F_{i-\frac{3}{2}}=0 \tag{10.79}$$

Adding Eq. (10.79) to (10.78), we get the summation of the difference equation system for the entire flow field

$$F_{\infty}-F_{-\infty}=0 \tag{10.80}$$

This means that the difference representation given by (10.79) is conservative and known as conservative shock scheme (C). Rewrite the conservative shock scheme as

$$F_{i+\frac{1}{2}}-F_{i-\frac{3}{2}}=(F_{i+\frac{1}{2}}-F_{i-\frac{1}{2}})+(F_{i-\frac{1}{2}}-F_{i-\frac{3}{2}}) \tag{10.81}$$

The conservative difference representation at the shock point i may be considered as the sum of the central and upwind differences at the point i.

If we use the central difference representation for the shock point, the resulting difference equation at the shock point i is

$$F_{i+\frac{1}{2}}-F_{i-\frac{1}{2}}=0 \tag{10.82}$$

Adding Eq. (10.82) to (10.78), we obtain

$$F_{\infty}-F_{-\infty}=F_{i-\frac{1}{2}}-F_{i-\frac{3}{2}} \tag{10.83}$$

In general, $F_{i-\frac{1}{2}}\neq F_{i-\frac{3}{2}}$. There exists a source at the shock point with the strength proportional to the right-hand side of Eq. (10.83). Thus the central difference scheme is a nonconservative shock scheme (NC) when applied at the shock wave.

In a similar manner we can verify that the zero representation for the sonic point is conservative in the transition from subsonic to supersonic flow.

12. Two-dimensional transonic small perturbation flows

In this section we will extend the type-dependent differencing of one-dimensional problems to two-dimensional problems. The steady transonic samll perturbation velocity potential equation (10.55) is considered. At subsonic points, Eq. (10.55) is elliptic and the central difference representation is used for all derivatives. At supersonic points, Eq. (10.55) is hyperbolic. The domain of dependence bounded by the characteristics (10.57) for the points of interest is symmetric about the x direction. Thus the upwind difference representation is used for the direvatives with respect to x and the central difference representation is used for $\frac{\partial^2\varphi}{\partial y^2}$.

A solution can be obtained by using either an explicit or implicit difference scheme. However, explicit schemes are not advisable in many cases because the CFL stability criterion (cf. Chapter VII, § 8) prohibits reasonable step sizes. Note that here we view the nonlinear stability problem as a locally linearized problem. Especially for transonic flow computations, an explicit solution becomes impractical. Hence it is desirable to use an implicit scheme to compute solutions. The structure of the computational molecules shown in Fig. 10.16 illustrates the correct type-dependent differencing for subsonic and supersonic points.

In the above section, we have seen that there are four possible types of flow for each grid point associated with the transonic small perturbation problems. To determine a grid point type, we may use both upwind and central difference representations for the differential expression of $1-M^2$ and compare the

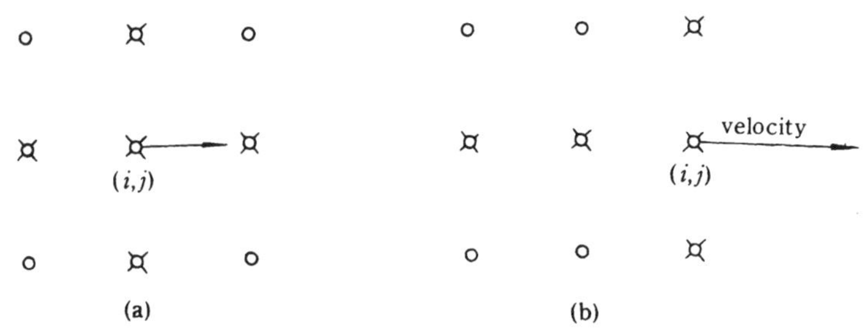

Fig. 10.16 Type-dependent differencing (a) $M<1$ (b) $M>1$

numerical values with zero. The flow type and the type-dependent term of Eq. (10.55) at grid point (i, j) are listed in Tab. 10.2, where

Table 10.2 Type and type-dependent difference

$A_{i-1,j}$	$A_{i,j}$	Type	Type-dependent difference
$\geqslant 0$	$\geqslant 0$	Subsonic	$P_{i,j}$
<0	<0	Supersonic	$P_{i-1,j}$
$\geqslant 0$	<0	Sonic	0
<0	$\geqslant 0$	Shock	$\begin{cases} P_{i,j}+P_{i-1,j}(\mathrm{C}) \\ P_{i,j}(\mathrm{NC}) \end{cases}$

$$A_{i,j}=1-M_\infty^2-\frac{\gamma+1}{U}M_\infty^2\frac{\varphi_{i+1,j}-\varphi_{i-1,j}}{2\Delta x} \tag{10.84}$$

$$P_{i,j}=A_{i,j}\frac{\varphi_{i+1,j}-2\varphi_{i,j}+\varphi_{i-1,j}}{\Delta x^2} \tag{10.85}$$

If we introduce a switch function

$$\mu_{i,j}=\begin{cases} 0, & A_{i,j}\geqslant 0 \\ 1, & A_{i,j}<0 \end{cases} \tag{10.86}$$

the four type-dependent difference equations can be expressed by a single equation Let

$$Q_{i,j}\doteq\frac{\varphi_{i,j+1}-2\varphi_{i,j}+\varphi_{i,j-1}}{\Delta y^2} \tag{10.87}$$

and use the conservative shock scheme. The single expression for four type-dependent difference equations is

$$P_{i,j}+Q_{i,j}-\mu_{i,j}P_{i,j}+\mu_{i-1,j}P_{i-1,j}=0 \tag{10.88}$$

The type-dependent difference equation (10.88) is conservative, as its modified differential equation can be written into divergence form. The last two terms in (10.88) form the upwind difference at point (i, j) of the differential expression

$$-\Delta x\frac{\partial}{\partial x}(\mu P)+O(\Delta x^2) \tag{10.89}$$

where $P=\dfrac{\partial F}{\partial x}$ by use of Eq. (10.71).

Therefore the modified differential equation of (10.88) is

$$\frac{\partial}{\partial x}\left(F-\Delta x\mu\frac{\partial F}{\partial x}\right)+\frac{\partial}{\partial y}\left(\frac{\partial\varphi}{\partial y}\right)=O(\Delta x^2) \tag{10.90}$$

In evaluating the A's of Tab. 10.2, iterative methods have to be used, because the φ's are unknowns. For better convergence, the A term is lagged by one level in the iteration process. The resulting algebraic equations are a nonlinear system. The nonlinear algebraic system may be solved by the same iterative methods as for the solution of steady subsonic potential flow in Chapter VII § 10. By this method, the convergence at supersonic points usually behaves in an oscillatory pattern and under-relaxation at supersonic points is used. Under-relaxation is immensely time consuming. More efficient iterative methods have been obtained in recent years. We will present two of the most efficient methods for steady transonic flow computation in the following sections.

Before concluding this section, we will give some numerical examples. In Fig. 10.17, we compute the pressure coefficient distributions over the airfoil

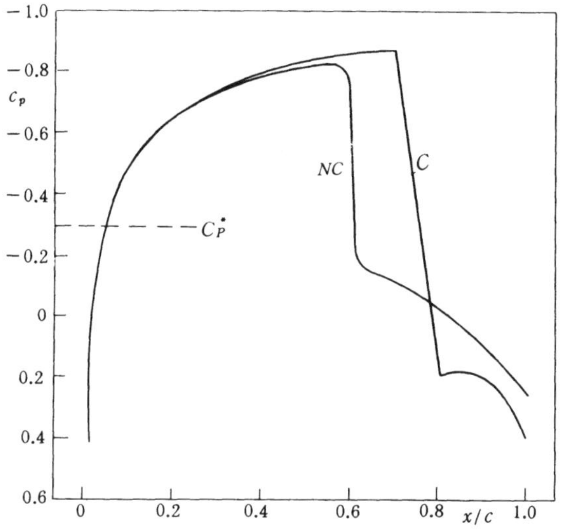

Fig. 10.17 Comparison of conservative and nonconservative shock scheme

NACA 0012 at $M_\infty = 0.85$ and angle of attack $\alpha = 0°$ using both conservative and nonconservative shock schemes. In comparison with the conservative scheme, the nonconservative shock scheme underestimates the shock strength and generally places the shock position forward on the airfoil. It is important to note that we are computing shock waves under the isentropic potential flow theory. Potential theory is useful for describing transonic flow only when shock waves are weak. As shock waves are treated approximately in potential flow theory, the nonconservative shock scheme for potential flow computation may be more useful even though the conservative scheme is mathematically most appealing.

In comparing with experimental results, we couple[19] the transonic small transverse perturbation flow computation with boundary layer integral equation algorithms to simulate the shock-boundary layer interaction. The boundary layer equations are solved by the direct method to some point upstream of the interaction or separation a nd the inverse method downstream of that point (cf. Chapter XVII). Fig. 10.18 compares computational results with experimental data for supercritical flow past a RAE2822 airfoil at $M_\infty=0.75$, $\alpha=3.19°$ and $Re=0.62\times10^7$. In the computations of Fig. 10.18, the nonconservative shock scheme is used. If the conservative shock scheme were used, the computed shock strength would be too strong and the shock location would be further downstream. The agreement between the numerical solution by the nonconservative scheme and the experimental results is excellent. The deviation of C_p on the upper surface near the blunt leading edge is possibly due to the moving of the stagnation point, resulting in a greater expansion near the nose than the numerical solution would compute.

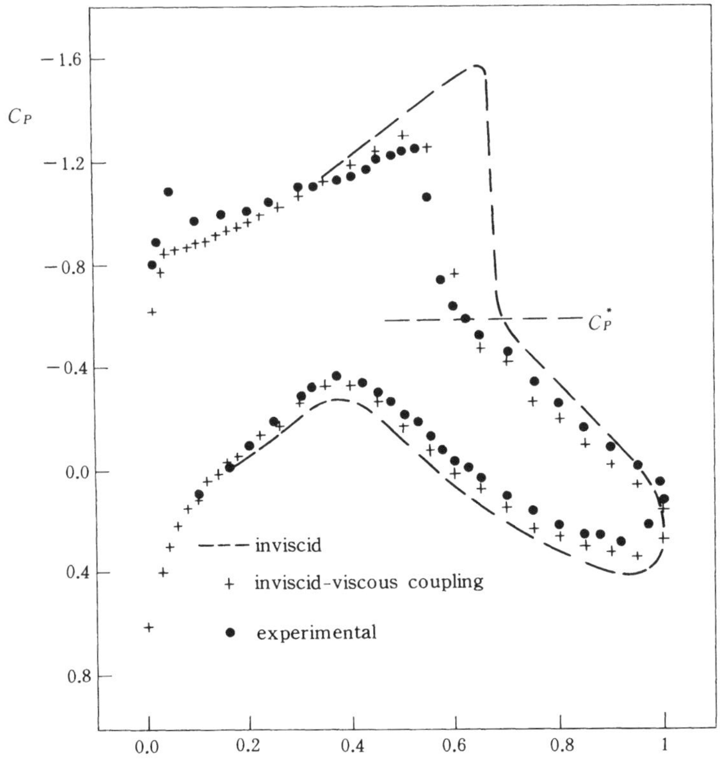

Fig. 10.18 Pressure distributions on airfoil RAE2822 $M_\infty=0.75$, $\alpha=3.19°$, $Re=0.62\times10^7$, Free Transition

13. Artificial time damping term

In order to stabilize and accelerate the SLOR algorithm for supercritical flow, Jameson[20] added an artificial time damping term $\frac{\partial^2\varphi}{\partial x\partial t}$ at supersonic points, where t denotes the iteration level n of the algorithm. We will discuss the effects of the artificial time damping term by the local linearization method.

The upwind difference scheme of Eq (10.55) at grid point (i, j) in the SLOR algorithm is

$$A^{n-1}_{i-1,j}\frac{\bar{\varphi}^n_{i,j}-2\varphi^n_{i-1,j}+\varphi^n_{i-2,j}}{\Delta x^2}+\frac{\bar{\varphi}^n_{i,j+1}-2\bar{\varphi}^n_{i,j}+\bar{\varphi}^n_{i,j-1}}{\Delta y^2}=0 \qquad (10.91)$$

incorporated with $\varphi^n_{i,j}=\omega\bar{\varphi}^n_{i,j}+(1-\omega)\varphi^{n-1}_{i,j}$. By use of artificial time following Garabedian (cf. Chapter VII, § 9), the modified differential equation of the difference equation (10.91) becomes

$$A\left[\frac{\partial^2\varphi}{\partial x^2}-\left(1-\frac{1}{\omega}\right)\frac{\Delta t}{\Delta x^2}\frac{\partial\varphi}{\partial t}\right]+\frac{\partial^2\varphi}{\partial y^2}=0 \qquad (10.92)$$

where $A=1-M^2$.

Eq. (10.92) is of the parabolic type with t as marching variable. The steady state differential equation (10.55) has x as marching variable. This is an inconsistency between the artificial time dependent equation (10.92) and the steady state equation (10.55).

We may add any artificial time dependent term to Eq. (10.92) without changing the steady state solution if the resulting artificial time dependent equation has convergent solution as $t\to\infty$. Following Jameson, we add an artificial time dependent term $\frac{\partial^2\varphi}{\partial x\partial t}$ to Eq. (10.92) and write the resulting equation in the form,

$$A\frac{\partial^2\varphi}{\partial x^2}+a\frac{\partial^2\varphi}{\partial x\partial t}+b\frac{\partial\varphi}{\partial t}+\frac{\partial^2\varphi}{\partial y^2}=0 \qquad (10.93)$$

By variable transformation,

$$\begin{aligned}\xi&=x\\ \eta&=y\\ \tau&=t-\frac{a}{2A}x\end{aligned} \qquad (10.94)$$

Eq. (10.93) becomes

$$A\frac{\partial^2\varphi}{\partial\xi^2}-\frac{a^2}{4A}\frac{\partial^2\varphi}{\partial\tau^2}+b\frac{\partial\varphi}{\partial\tau}+\frac{\partial^2\varphi}{\partial\eta^2}=0 \tag{10.95}$$

If $a\neq 0$, Eq. (10.95) is of the hyperbolic type with ξ or x as the marching variable, since $A<0$ at supersonic points. Therefore the above inconsistency is eliminated by adding the term $\frac{\partial^2\varphi}{\partial x\partial t}$.

To find the appropriate sign for the coefficient a of the term $\frac{\partial^2\varphi}{\partial x\partial t}$ in Eq. (10.93), the characteristics equations of the differential Eq. (10.93) is evaluated. The equation of the characteristic cone surface passing the origin $x=y=t=0$ is obtained

$$\left(\frac{y}{t}\right)^2=-\frac{4}{a}\left(\frac{x}{t}-\frac{A}{a}\right) \tag{10.96}$$

Eq. (10.96) is plotted out in Fig. 10.19. The characteristic cone surface

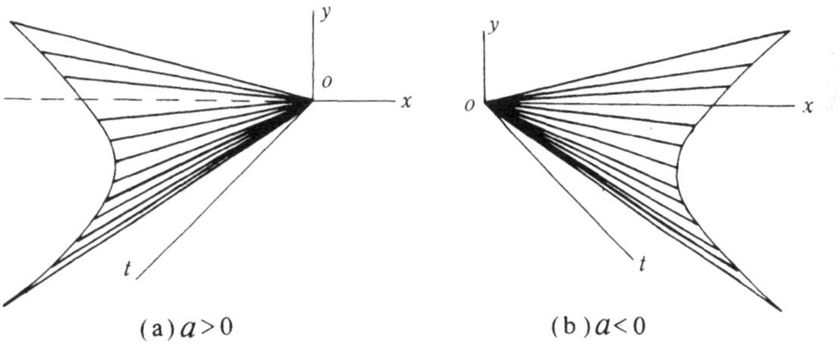

Fig. 10.19 Characteristics of equation (10.93)

including the plus x axis should be chosen. Thus, the coefficient a should be negative, i. e. the coefficients of $\frac{\partial^2\varphi}{\partial x\partial t}$ and $\frac{\partial^2\varphi}{\partial x^2}$ should have the same sign in the artificial time dependent equation (10.93).

To study the damping effects of the artificial time dependent term $\frac{\partial^2\varphi}{\partial x\partial t}$, a Fourier analysis of the relaxation error of Eq. (10.93) is made under the local linearization (cf. Chapter VII, § 9). For simplicity we may consider a normalized differential equation rather than Eq. (10.93).

$$-\frac{\partial^2\varphi}{\partial x^2}+a\frac{\partial^2\varphi}{\partial x\partial t}+\frac{\partial^2\varphi}{\partial y^2}=0 \quad (10.97)$$

Write out the upwind difference equation for (10.97) and apply the SLOR algorithm. If $\Delta x=\Delta y=h$, we obtain

$$(a-3)\bar{\varphi}^n_{i,j}+(2-a)\varphi^n_{i-1,j}-\varphi^n_{i-2,j}+\bar{\varphi}^n_{i,j+1}$$
$$+\bar{\varphi}^n_{i,j-1}-\varphi^{n-1}_{i,j}+\varphi^{n-1}_{i-1,j}=0 \quad (10.98)$$

incorporated with $\varphi^n_{i,j}=\omega\bar{\varphi}^n_{i,j}+(1-\omega)\varphi^{n-1}_{i,j}$. The growth factor for the mth Fourier component of the relaxation error can be derived

$$e^{\lambda\Delta t}=\frac{\left(2-a-2\cos\dfrac{m\pi}{N}\right)\left(1-\dfrac{1}{\omega}\right)+a(1-e^{-im\pi/N})}{\dfrac{1}{\omega}\left(a-3+2\cos\dfrac{m\pi}{N}\right)+(2-a)e^{-im\pi/N}-e^{-2im\pi/N}} \quad (10.99)$$

where $m=1, 2,\cdots, N$; $2N$=number of cells in x or y direction for the computational square region.

Eq. (10.99) is too complicated to be studied in general cases. It is seen from Fig. 7.3 that the lowest frequency component of the relaxation error yields the largest magnitude for $|e^{\lambda\Delta t}|$ over the entire frequency spectrum. If this is true in the present analysis, we may study only the case $m=1$ to determine the stability criterion for the total spectrum. Substituting $m=1$ into Eq. (10.99), we can expand the right-hand side into a power series of N^{-1} if $\omega\neq 1$.

$$|e^{\lambda\Delta t}|=1+\left(\frac{1}{2}+\frac{a\omega}{(\omega-1)(1-a)}\right)\frac{\omega}{(\omega-1)(1-a)}\left(\frac{\pi}{N}\right)^4$$
$$+O(N^{-5}) \quad (10.100)$$

If $a=0$, Eq. (10.100) reduces to

$$|e^{\lambda\Delta t}|=1+\frac{\omega}{2(\omega-1)}\left(\frac{\pi}{N}\right)^4+O(N^{-5}) \quad (10.101)$$

From Eq. (10.101), $|e^{\lambda\Delta t}|\leqslant 1$ requires

$$0<\omega<1 \quad (10.102)$$

If $a\neq 0$ and

$$a<-\frac{\omega-1}{\omega+1} \quad (10.103)$$

we find that $|e^{\lambda\Delta t}|\leqslant 1$ requires

$$\omega > 0 \tag{10.104}$$

Therefore by adding the term $a\dfrac{\partial^2\varphi}{\partial x\partial t}$ to the SLOR equation, the time consuming under-relaxation requirement (10.102) for numerical stability is relieved at least for the low frequency Fourier components of errors. The term $a\dfrac{\partial^2\varphi}{\partial x\partial t}$ has the same damping effect as the inder-relaxation and is known as the artificial time damping term or time-dependent dissipation term. The characteristics analysis and the Fourier analysis provide two conditions on the coefficient a. The conditions can be satisfied simultaneouly. Besides these conditions, the coefficient a can be chosen arbitrarily. With the artificial time damping term, over-relaxation can be used to accelerate the convergence ofthe relaxation. Numerical experiences do confirm the results of linearized analyses.

Fig. 10.20[19] compares the convergence histories of the two computations

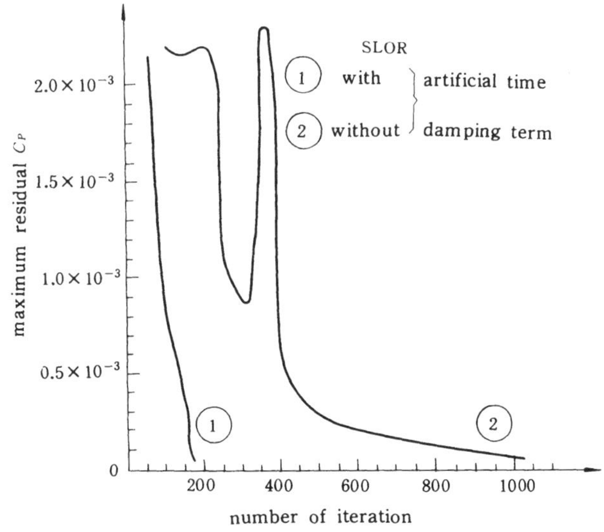

Fig. 10.20 Convergence histories of inviscid iteration NACA 0012, $M_\infty=0.8$, $\alpha=1.25°$

by SLOR with and without the artificial time damping term for a supercritical flow over a NACA0012 airfoil at $M_\infty=0.8$ and $\alpha=1.25°$. In the SLOR with the artificial time damping term, we use $a=-\dfrac{\Delta t}{\Delta x}$ and $\omega=1.7$ for $M<1$, 1.2 for $M>1$. In the SLOR without artificial time damping term, we have to use under-relaxation in order to get a convergent solution, $\omega=0.9$ for $M<1$ and 0.7 for $M>1$. The computer time for the former is about 1/5 of that for the latter.

14. Approximate factorization AF2

One technique for achieving even faster convergence than that provided by the SLOR plus artificial time damping term is to use a fully implicit scheme, that is, a scheme in which each point communicates with every other point during each iteration. This type of scheme can be constructed by using the approximate factorization (AF) or alternating direction implicit (ADI) philosophy[21].

Let us write difference equation in the form

$$L\varphi_{i,j}=0 \tag{10.105}$$

where L is the difference operator.

Any relaxation scheme can be put into a standard, two-level correction form given by

$$NC^{n}_{i,j}+\omega L\varphi^{n-1}_{i,j}=0 \tag{10.106}$$

where $C^{n}_{i,j}$ is the nth iteration correction defined by

$$C^{n}_{i,j}=\varphi^{n}_{i,j}-\varphi^{n-1}_{i,j} \tag{10.107}$$

N is the relaxation operator which determines the type of the iteration scheme, ω is the usual relaxation parameter and L $\varphi^{n-1}_{i,j}$ is the $(n-1)$ th iteration residual, which is a measure of how well the finite difference equation is satisfied by the $(n-1)$th level solution. If $C^{n}_{i,j}\to 0$, Eq. (10.106) reduces to L $\varphi^{n-1}_{i,j}=0$, then $\varphi^{n}_{i,j}$ is the solution.

In transonic flow computation, L is a nonlinear operator. The N-operator should be a linearized approximation of the L-operator. The construction of a suitable AF scheme is to factor the N-operator by using an appropriate factorization as indicated by

$$N=N_1N_2 \tag{10.108}$$

The important idea behind the factorization is that each of the factors N_1 and N_2 must involve only simple banded matrix inversions, thus reducing the computational work per iteration. And the construction of N should allow each grid point in the entire mesh to be influenced by every other grid point during each iteration, thus resulting in much faster convergence of relaxation. Both the errors associated with the linearization and factorization are removed from the solution simultaneously and automatically by iteration.

It is convenient to utilize difference operators to represent finite difference when particular forms are used repetitively. Here we define

$$\left.\begin{aligned}
\overrightarrow{\delta}_x\varphi_i &= \frac{\varphi_{i+1}-\varphi_i}{\Delta x}\\
\overleftarrow{\delta}_x\varphi_i &= \frac{\varphi_i-\varphi_{i-1}}{\Delta x}\\
\delta_{xx}\varphi_i &= \frac{\varphi_{i+1}-2\varphi_i+\varphi_{i-1}}{\Delta x^2}\\
\overleftarrow{\delta}_{xx}\varphi_i &= \frac{\varphi_i-2\varphi_{i-1}+\varphi_{i-2}}{\Delta x^2}
\end{aligned}\right\} \tag{10.109}$$

Notice that second-order difference operators can be factored

$$\left.\begin{aligned}
\delta_{xx} &= \overleftarrow{\delta}_x\overrightarrow{\delta}_x\\
\overleftarrow{\delta}_{xx} &= \overleftarrow{\delta}_x\overleftarrow{\delta}_x
\end{aligned}\right\} \tag{10.110}$$

This fact will be useful in the construction of fully implicit, approximate factorization schemes.

For the type-dependent difference equations of transonic steady small perturbation flow, the AF2 scheme can be expressed by choosing the relaxation operator N as follows.

$$\text{at } M<1\text{: N} = -\frac{1}{\sigma}\left(\sigma\overleftarrow{\delta}_x-\delta_{yy}\right)\left(\sigma-A_{i,j}^{n-1}\overrightarrow{\delta}_x\right) \tag{10.111}$$

$$\text{at } M>1\text{: N} = -\frac{1}{\sigma}\left(\sigma\overleftarrow{\delta}_x-\delta_{yy}\right)\left(\sigma-A_{i-1,j}^{n-1}\overleftarrow{\delta}_x\right) \tag{10.112}$$

where σ is a convergence acceleration parameter, cycled over a sequence of values which will be discussed below.

The N-operator has been written as the product of two factors which when multiplied out yield

$$\text{at } M<1,\quad \text{N} = A_{i,j}^{n-1}\delta_{xx}+\delta_{yy}-\sigma\overleftarrow{\delta}_x-\frac{1}{\sigma}A_{i,j}^{n-1}\overrightarrow{\delta}_x\delta_{yy} \tag{10.113}$$

$$\text{at } M>1,\quad \text{N} = A_{i-1,j}^{n-1}\overleftarrow{\delta}_{xx}+\delta_{yy}-\sigma\overleftarrow{\delta}_x-\frac{1}{\sigma}A_{i-1,j}^{n-1}\overleftarrow{\delta}_x\delta_{yy} \tag{10.114}$$

Here $A_{i,j}$ is treated as a spatial constant. These expressions contain the linearized L-operator plus error terms.

It is interesting to note that the AF2 construction generates a $\frac{\partial^2\varphi}{\partial x\partial t}$ term which provides artificial time damping to the convergence process. Regarding the superscript n as an artificial time variable t, we obtain the modified differential equation for the difference equation (10.106) combined with Eq. (10.111) or (10.112),

$$\omega\left(A\frac{\partial^2\varphi}{\partial x^2}+\frac{\partial^2\varphi}{\partial y^2}\right)+\Delta t\left[(1-\omega)\left(A\frac{\partial^3\varphi}{\partial x^2\partial t}+\frac{\partial^3\varphi}{\partial y^2\partial t}\right)\right.$$

$$\left.-\sigma\frac{\partial^2\varphi}{\partial x\partial t}-\frac{A}{\sigma}\frac{\partial^4\varphi}{2y^2\partial x\partial t}\right]=0 \qquad (10.115)$$

The coefficients of $\frac{\partial^2\varphi}{\partial x^2}$ and $\frac{\partial^2\varphi}{\partial x\partial t}$ have the same sign for ω and $\sigma>0$. Thus the AF2 has an artificial time damping effect at $M>1$.

The AF2 scheme is implemented in a two-sweep format given by the following.

$$\left.\begin{array}{l}\text{sweep 1 } (\sigma\overleftarrow{\delta}_x-\delta_{yy})f^{\,n}_{i,j}=\sigma\omega L_1\varphi^{n-1}_{i,j}\\ \text{sweep 2 } (\sigma-A^{n-1}_{i,j}\overrightarrow{\delta}_x)C^n_{i,j}=f^{\,n}_{i,j}\end{array}\right\} M<1 \qquad (10.116)$$

$$\left.\begin{array}{l}\text{sweep 1 } (\sigma\overleftarrow{\delta}_x-\delta_{yy})f^{\,n}_{i,j}=\sigma\omega L_2\varphi^{n-1}_{i,j}\\ \text{sweep 2 } (\sigma-A^{n-1}_{i-1,j}\overleftarrow{\delta}_x)C^n_{i,j}=f^{\,n}_{i,j}\end{array}\right\} M>1 \qquad (10.117)$$

where L_1 and L_2 are the linearized L-operator at $M<1$ and >1 and $f^{\,n}_{i,j}$ is an intermediate result stored at each mesh point in the finite difference mesh. In the first sweep, the f-array is obtained by solving a tridiagonal matrix equation for each y=constant line and sweeping in the positive x direction. The correction array $C^n_{i,j}$ is then obtained in the second sweep from the f-array by solving a tridiagonal matrix equation for each x=constant line in the transonic flow field.

The linearized stability of the AF2 iteration scheme may be investigated by the von Neumann method. A growth factor is obtained for subsonic points

$$e^{\lambda\Delta t}=\frac{a_1+(1-\omega)a_2}{a_1+a_2} \qquad (10.118)$$

where

$$a_1=\frac{\sigma}{\Delta x}(1-e^{-ia\Delta x})+\frac{2A}{a\Delta x\Delta y^2}(1-\cos b\Delta y)(1-e^{ia\Delta x}) \qquad (10.119)$$

$$a_2=\frac{2A}{\Delta x^2}(1-\cos a\Delta x)+\frac{2}{\Delta y^2}(1-\cos b\Delta y) \qquad (10.120)$$

a and b are real values, λ may be complex and $i=\sqrt{-1}$. Note that $\mathrm{Re}(a_1)$ and $a_2\geqslant 0$.

The numerator and denominator of the expression (10.118) have the same imaginary parts. Thus, the stability condition $|e^{\lambda\Delta t}|\leqslant 1$ is equivalent to

$$-1\leqslant \frac{\mathrm{Re}(a_1)+(1-\omega)a_2}{\mathrm{Re}(a_1)+a_2}\leqslant 1 \tag{10.121}$$

From (10.121), we obtain the stability conditions for $M<1$,

$$\sigma\geqslant 0 \tag{10.122}$$

and

$$0\leqslant\omega\leqslant 2\left[1+\frac{\mathrm{Re}(a_1)}{a_2}\right] \tag{10.123}$$

The right-hand side of (10.123) is greater than 2.

By use of the growth factor expression, we may choose a sequence of values for σ to cause the AF2 growth factor to be a minimum for various Fourier components of error. Such an optimal choice of the σ parameter may be done as follows. For instance, let $\omega=1$, then the growth factor (10.118) becomes

$$e^{\lambda\Delta t}=\frac{a_1}{a_1+a_2} \tag{10.124}$$

Take $\mathrm{Im}(a_1)=0$, we obtain

$$\sigma=\frac{1}{\Delta y}\,[2A(1-\cos b\Delta y)]^{1/2} \tag{10.125}$$

For the lowest frequency component, $b\Delta y=2\pi\Delta y/L$, where L is the length of the computational field in x or y direction, Eq. (10.125) reduces to the approximate expression,

$$\sigma\approx 2\pi\sqrt{A}\,/L \tag{10.126}$$

For the highest frequency component, $b\Delta y=\pi$, Eq. (10.125) reduces to

$$\sigma=2\sqrt{A}\,/\Delta y \tag{10.127}$$

A suitable sequence of σ's is then given by

$$\sigma_k=\sigma_H\left(\frac{\sigma_L}{\sigma_H}\right)^{\frac{k-1}{K-1}} \tag{10.128}$$

where from Eq. (10.126)

$$\sigma_L=O(1) \tag{10.129}$$

from Eq. (10.127),

$$\sigma_H=O\left(\frac{1}{\Delta y}\right) \tag{10.130}$$

$k=1, 2,\cdots, K$ and K is the number of elements in the sequence. In practise, it is well advised to optimize both end points σ_L and σ_H by trial-and-error numerical experimentation. The above analysis is only true for a linear problem. Precise estimation of the σ's for a non-linear problem is generally not possible. Instead, a repetitive application of the sequence of σ's (10.128) to the computation is used.

Fig. 10.21 compares the convergence histories of the two computations by

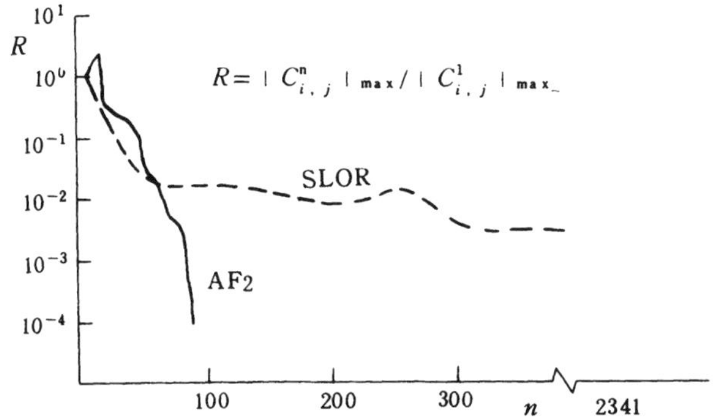

Fig. 10.21 Convergence histories of AF2 and SLOR

AF2 and SLOR for a supercritical flow over a NACA0012 airfoil at $M_\infty=0.8$ and $\alpha=0$.[22] In the AF2 computation we use $\omega=1$, $\sigma_L=0.4$, $\sigma_H=80$, $K=5$. In the SLOR computation we use $\omega=0.8$ for $M<1$ and 0.7 for $M>1$. The computer storage for the AF2 is double and the computer time for AF2 is 1/13 relative to the SLOR.

15. Numerical solution of full potential equation

The full potential equation can be written in either a nonconservative or conservative form. The nonconservative form of the potential equation for two dimensions is

$$(a^2-u^2)\frac{\partial^2\varphi}{\partial x^2}-2uv\frac{\partial^2\varphi}{\partial x\partial y}+(a^2-v^2)\frac{\partial^2\varphi}{\partial y^2}=0 \qquad (10.12)$$

where

$$u=\frac{\partial\varphi}{\partial x}, \quad v=\frac{\partial\varphi}{\partial y}$$

and a is the speed of sound which is obtained from the energy equation

$$a^2=a_\infty^2+\frac{\gamma-1}{2}(U^2-u^2-v^2)$$

The conservative form of the potential equation for two dimensions is

$$\frac{\partial(\rho u)}{\partial x}+\frac{\partial(\rho v)}{\partial y}=0 \qquad (5.7)$$

where

$$u=\frac{\partial\varphi}{\partial x}, \quad v=\frac{\partial\varphi}{\partial y}$$

and the density is calculated from the energy equation in the form

$$\rho=\rho_\infty\left(1+\frac{\gamma-1}{2}\,\frac{U^2-u^2-v^2}{a_\infty^2}\right)^{\frac{1}{\gamma-1}}$$

As differential equations, the two forms of the potential equation are equivalent, i.e. smooth solutions of one equation are also smooth solutions of the other. But a discontinuous numerical solution of one equation in general will not be a solution of the other (cf. Chapter V, § 14).

The full potential differential equation is of mixed type. It is elliptic at subsonic points and hyperbolic at supersonic points. At supersonic points, there are two families of characteristics which are symmetric Mach waves about the local velocity vector (cf. Chapter XI, §2).

Jameson[20] extended the Murman-Cole type-dependent differencing to the nonconservative full potential equation. The idea is to write the potential equation in natural coordinates as

$$(a^2-q^2)\frac{\partial^2\varphi}{\partial s^2}+a^2\frac{\partial^2\varphi}{\partial n^2}+\cdots=0 \qquad (10.131)$$

where s and n are distances along and normal to the streamlines. In the expression (10.131), the second derivative $\frac{\partial^2\varphi}{\partial s^2}$ is represented by type-dependent differences and all other derivatives are represented by central differences. The second derivative $\frac{\partial^2\varphi}{\partial s^2}$ can be written in terms of x and y as follows. By the definition of velocity potential

$$\frac{\partial^2\varphi}{\partial s^2}=\frac{\partial q}{\partial s} \qquad (10.132)$$

using the directional derivative formula

$$\frac{\partial q}{\partial s}=\frac{\partial q}{\partial x}\cos(x,s)+\frac{\partial q}{\partial y}\cos(y,s) \qquad (10.133)$$

where

$$\cos(x, s) = \frac{u}{q}, \quad \cos(y, s) = \frac{v}{q} \tag{10.134}$$

and

$$q = \sqrt{u^2 + v^2}, \tag{10.135}$$

we obtain the expression in Cartesian coordinates for $\frac{\partial^2 \varphi}{\partial s^2}$

$$\frac{\partial^2 \varphi}{\partial s^2} = \frac{1}{q^2}\left(u^2 \frac{\partial^2 \varphi}{\partial x^2} + 2uv \frac{\partial^2 \varphi}{\partial x \partial y} + v^2 \frac{\partial^2 \varphi}{\partial y^2} \right). \tag{10.136}$$

The Cartesian coordinate expression for other terms in Eq. (10.131) may be obtained by use of the potential equation (10.12), since Eqs. (10.131) and (10.12) are equivalent. Thus, Eq. (10.131) may be written in terms of x and y as

$$\begin{aligned} &(a^2 - q^2)\left(u^2 \frac{\partial^2 \varphi}{\partial x^2} + 2uv \frac{\partial^2 \varphi}{\partial x \partial y} + v^2 \frac{\partial^2 \varphi}{\partial y^2} \right) \\ &+ a^2\left(v^2 \frac{\partial^2 \varphi}{\partial x^2} - 2uv \frac{\partial^2 \varphi}{\partial x \partial y} + u^2 \frac{\partial^2 \varphi}{\partial y^2} \right) = 0 \end{aligned} \tag{10.137}$$

where the first term corresponds to the first term of Eq. (10.131). Both x and y derivative contributions to $\frac{\partial^2 \varphi}{\partial s^2}$ are retarded at supersonic points, i.e.

$$\left.\begin{aligned} \frac{\partial^2 \varphi}{\partial x^2} &= \frac{1}{\Delta x^2}(\varphi_{i,j} - 2\varphi_{i-1,j} + \varphi_{i-2,j}) \\ \frac{\partial^2 \varphi}{\partial x \partial y} &= \frac{1}{\Delta x \Delta y}(\varphi_{i,j} - \varphi_{i-1,j} - \varphi_{i,j-1} + \varphi_{i-1,j-1}) \\ \frac{\partial^2 \varphi}{\partial y^2} &= \frac{1}{\Delta y^2}(\varphi_{i,j} - 2\varphi_{i,j-1} + \varphi_{i,j-2}) \end{aligned}\right\} \tag{10.138}$$

if $u > 0$ and $v > 0$. If these signs are different, Eqs. (10.138) are replaced by similar formulas which retard the difference scheme in the proper upwind direction. Central differences are used for all other derivatives in Eq. (10.137). The above scheme is known as the rotated difference scheme.

The rotated scheme produces an artificial viscosity with leading term of the form

$$-(a^2-q^2)u^2\Delta x\frac{\partial^3\varphi}{\partial x^3} \tag{10.139}$$

This provides us with a positive artificial viscosity for all points where the flow is supersonic and we expect shock waves to form as compressions (cf. § 10).

Consider now the conservative form of the potential equation (5.7). We may use central differences for all derivatives by adding explicitly the artificial viscosity to Eq. (5.7) at supersonic points. Eq. (5.7) is equivalent to Eq. (1 0.12) for the irrotational flow. Multiplying the expression (10.139) by ρ/a^2u^2 and adding to the first term of (5.7), we obtain

$$\frac{\partial(\rho u)}{\partial x} - \frac{\rho}{a^2}(a^2-q^2)\Delta x\frac{\partial^2 u}{\partial x^2} \tag{10.140}$$

for $q>a$. Define a switch function

$$\mu=\max\left\{0,\ 1-\frac{a^2}{q^2}\right\} \tag{10.141}$$

The artificial viscosity term in (10.140) can be written for the entire flow field.

$$\frac{\rho}{a^2}\mu\, u^2\,\Delta x\frac{\partial^2 u}{\partial x^2} \tag{10.142}$$

We may replace the expression (10.142) by

$$\frac{\partial}{\partial x}\left(\frac{\rho}{a^2}\mu\, u^2\,\Delta x\frac{\partial u}{\partial x}\right) \tag{10.142a}$$

without changing the accuracy of the difference scheme. By use of equation of motion of one-dimensional flow, (10.142a) becomes

$$\frac{\partial}{\partial x}\left(-\mu\, u\,\Delta x\frac{\partial \rho}{\partial x}\right) \tag{10.143}$$

Combining (10.143) with $\dfrac{\partial(\rho u)}{\partial x}$, we have

$$\frac{\partial}{\partial x}\left[\left(\rho-\mu\,\Delta x\frac{\partial\rho}{\partial x}\right)u\right]=0 \tag{10.144}$$

Here $\left(\rho-\mu\Delta x\dfrac{\partial\rho}{\partial x}\right)$ may be interpreted as an artificial density. Therefore the effect of adding artificial viscosity is equivalent to that of using an artificial density. The technique of artificial density has been very useful for solving the conservative form of the full potential equation.[16–18]

Fig. 10.22 compares the numerical solutions[23] of the conservative full potential equation and the conservative Euler equations for a supercritical flow past a circular cylinder at $M_\infty = 0.5$. There are only minor differences at the forward part of the cylinder; however shock position as well as pressure ahead and behind the shock differ significantly between the two solutions. The

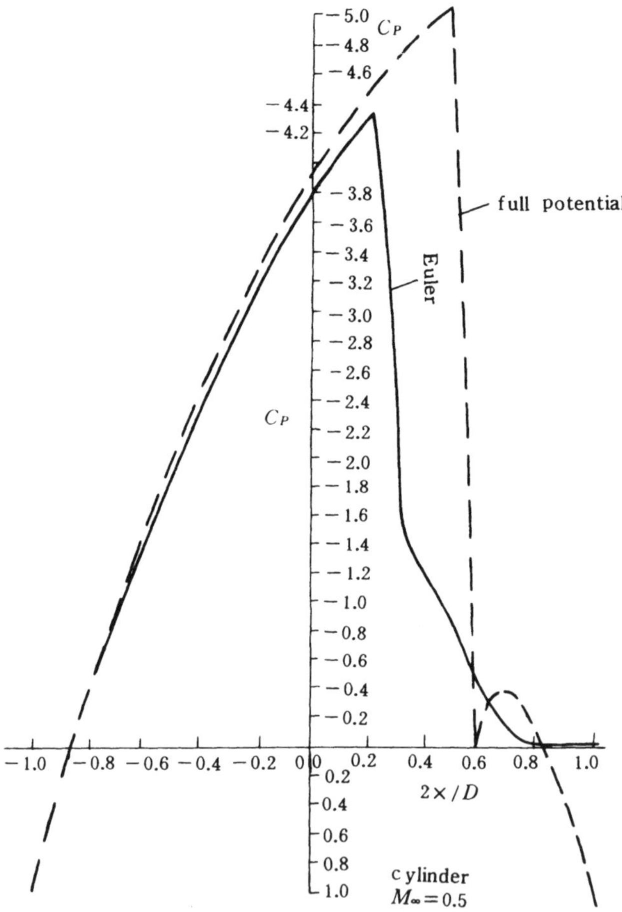

Fig. 10.22 Pressure coefficient distribution

conservative full potential equation gives a very strong shock with an incorrect position located too far downstream. The general shock relations of the conservative full potential equation will be studied in the next section. The numerical solution of Euler equation will be given in Chapter XIV.

16. Shock relations for conservative full potential equation

We will apply the weak solution theory of quasi-linear differential equations (Chapter V, § 15) to the potential flow case. The conservation laws or divergence equations for steady two-dimensional irrotational flow are

$$\frac{\partial(\rho u)}{\partial x}+\frac{\partial(\rho v)}{\partial y}=0 \tag{5.7}$$

$$\frac{\partial v}{\partial x}+\frac{\partial(-u)}{\partial y}=0 \tag{5.41}$$

$$\frac{\gamma}{\gamma-1}\frac{p}{\rho}+\frac{u^2+v^2}{2}=\text{constant} \tag{5.42a}$$

$$p/\rho^\gamma=\text{constant} \tag{5.4}$$

There are four equations for the four unknowns, u, v, p and ρ. Using the jump condition formula (5.104) of Chapter V, § 15, we obtain for the two differential equations (5.7) and (5.41)

$$[\rho u]\cos(n,x)+[\rho v]\cos(n,y)=0 \tag{10.145}$$

$$[v]\cos(n,x)+[-u]\cos(n,y)=0 \tag{10.146}$$

where [ ] denotes the jump across shock, n is normal to shock.

Eqs. (10.145) and (10.146) may be written as

$$\rho_1 q_{n1}-\rho_2 q_{n2}=0 \tag{10.147}$$

$$q_{t1}-q_{t2}=0 \tag{10.148}$$

where q_n and q_t are normal and tangential components of velocity relative to shock, subscripts 1 and 2 denote two sides of shock. The jump conditions from Eqs. (5.42a) and (5.4) are

$$\frac{\gamma}{\gamma-1}\frac{p_1}{\rho_1}+\frac{u_1^2+v_1^2}{2}-\frac{\gamma}{\gamma-1}\frac{p_2}{\rho_2}-\frac{u_2^2+v_2^2}{2}=0 \tag{10.149}$$

and

$$\frac{p_1}{\rho_1^\gamma}-\frac{p_2}{\rho_2^\gamma}=0 \tag{10.150}$$

Eqs. (10.147) to (10.150) are the jump conditions across a discontinuous curve in the full potential solution. They express the conservation of mass, tangential momentum, energy, and entropy of the flow across the discontinuous curve. Such a discontinuity may be called isentropic shock. An isentropic shock differs from the physical (adiabatic) shock in that an isentropic shock conserves entropy instead of normal momentum component. From Eq. (10.148), it can be derived that the velocity potential function $\varphi(x, y)$ is continuous across the discontinuous curve.

To illustrate the characters of isentropic shock, consider the one-dimensional flow. For one-dimensional potential flow Eqs. (10.147) to (10.150) reduce to

$$\left.\begin{aligned} \rho_1 u_1 &= \rho_2 u_2 \\ \frac{\gamma}{\gamma-1}\frac{p_1}{\rho_1}+\frac{u_1^2}{2} &= \frac{\gamma}{\gamma-1}\frac{p_2}{\rho_2}+\frac{u_2^2}{2} \\ \frac{p_1}{\rho_1^\gamma} &= \frac{p_2}{\rho_2^\gamma} \end{aligned}\right\} \tag{10.151}$$

Eliminating u_2 and ρ_2, we obtain the isentropic normal shock relation

$$M_1^2=\frac{2}{\gamma-1}\left[\left(\frac{p_2}{p_1}\right)^{\frac{\gamma+1}{\gamma}}-\left(\frac{p_2}{p_1}\right)^{\frac{2}{\gamma}}\right]\Big/\left[\left(\frac{p_2}{p_1}\right)^{\frac{2}{\gamma}}-1\right] \tag{10.152}$$

Fig. 10.23 compares the pressure ratios across normal shocks in the full potential solution [i. e. Eq. (10.152)] and in the Euler solution (i. e. the Rankine-Hugoniot relation). The full potential solution always overestimates the shock strength.

For transonic flow, $M_1 \approx 1$, we may find the explicit expression for $\frac{p_2}{p_1}$ from Eq. (10.152).

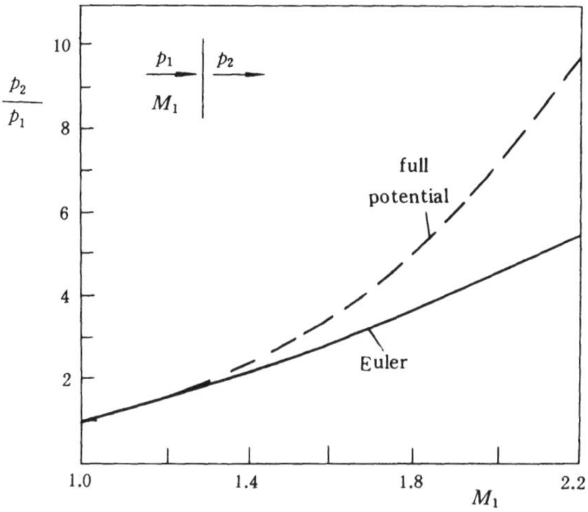

Fig. 10.23 Pressure ratio across normal shock

$$\frac{p_2}{p_1}=1+\frac{2\gamma}{\gamma+1}(M_1^2-1)+\frac{2\gamma^2}{3(\gamma+1)^2}(M_1^2-1)^2+O[(M_1^2-1)^3] \quad (10.153)$$

Compared with the exact normal shock relation (4.11), the error of the full potential solution (10.152) is $O[(M_1^2-1)^2]$ in transonic flow.

The jump of momentum flux across a isentropic shock is not zero, which can be obtained from Eqs. (10.151).

$$\frac{\Delta(p+\rho u^2)}{p_1+\rho_1 u_1^2}=\left\{\frac{p_2}{p_1}-1-\gamma M_1^2\left[1-\left(\frac{p_2}{p_1}\right)^{-\frac{1}{\gamma}}\right]\right\}\Big/(1+\gamma M_1^2) \quad (10.154)$$

where $\Delta(p+\rho u^2)=p_2+\rho_2 u_2^2-p_1-\rho_1 u_1^2$

For transonic flow, $M_1\approx 1$, Eq. (10.154) reduces to

$$\frac{\Delta(p+\rho u^2)}{p_1+\rho_1 u_1^2}=\frac{2\gamma}{(\gamma+1)^2}(M_1^2-1)^3+O[(M_1^2-1)^4] \quad (10.155)$$

The jump of the momentum flux across isentropic shock is $O[(M_1^2-1)^3]$ for the transonic case.

In the theory of weak solutions for quasi-linear equations, the jump conditions have to be supplemented with a "entropy" condition to make the weak solution unique. For the potential equation, the "entropy" condition may be derived from a consideration of wave drag. The wave drag acting on a body immersed in the isentropic potential flow may be evaluated by

$$D=\int_{\text{shock}}\Delta(p+\rho u^2)\,dy \quad (10.156)$$

where the integration contour is along the shock wave (Fig. 10.24).

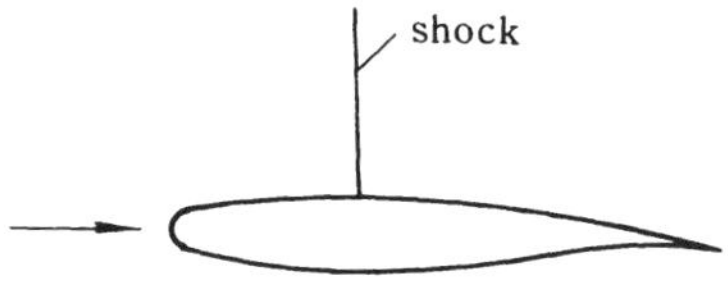

Fig. 10.24 Wave drag integral

The wave drag should not be negative. Therefore, we obtain

$$\Delta(p+\rho u^2)\geqslant 0 \quad (10.157)$$

This is the "entropy" condition for isentropic shock waves. For the transonic case, $M_1\approx 1$, using Eqs. (10.155) and (10.153), the "entropy" condition yields

$$M_1 \geqslant 1 \text{ and } p_2/p_1 \geqslant 1 \qquad (10.158)$$

Thus the "entropy" condition (10.157) causes the isentropic shocks to form as compressions.

In numerical solutions, instead of using the "entropy" condition, we may add artificial viscosity (explicitly or implicitly) to the isentropic potential equation to make the discontinuous solution unique and compressive.

It should be noted that the potential equation for transonic flow with shock of moderate strength can have nonunique solutions. The nonunique potential solutions are generally nonsense because of the entropy generation within the shocks violating the isentropic assumption upon which the theory is based.

17. Hypersonic flow equation

As in the transonic case, the linearized theory of small perturbations also fails at hypersonic speed. This failure was first mentioned in Chapter VI, § 10. Ackeret's formula (6.58) shows that the pressure coefficient and the drag coefficient tend to zero as the free stream Mach number M_∞ goes to infinity. Actually these coefficients will not be zero for hypersonic flow. Thus the linearized theory of supersonic flow fails at very large Mach numbers, even for thin bodies. However the hypersonic perturbation theory for thin bodies will give an effective first-order equation which is nonlinear.

Let us consider the case of a thin body in a uniform stream of velocity U and Mach number $M_\infty \gg 1$. The velocity components in the flow field may be written as

$$u = U + \frac{\partial \varphi}{\partial x}, \quad v = \frac{\partial \varphi}{\partial y} \qquad (10.159)$$

where φ is the velocity potential of the disturbances: $\partial\varphi/\partial\alpha$ and $\partial\varphi/\partial y$ are much smaller than U.

The exact differential equation for φ is also Eq. (10.12) with the sound veloicty given by

$$a^2 = a_\infty^2 - \frac{\gamma-1}{2}\left[2U\frac{\partial \varphi}{\partial x} + \left(\frac{\partial \varphi}{\partial x}\right)^2 + \left(\frac{\partial \varphi}{\partial y}\right)^2\right] \qquad (10.160)$$

Substituting Eq. (10.160) into (10.12) gives

$$\left\{1 - M_\infty^2 - \frac{\gamma+1}{2}\left[2M_\infty\left(\frac{1}{a_\infty}\frac{\partial \varphi}{\partial x}\right) + \left(\frac{1}{a_\infty}\frac{\partial \varphi}{\partial x}\right)^2\right]\right.$$

$$-\frac{\gamma-1}{2}\left(\frac{1}{a_\infty}\frac{\partial\varphi}{\partial y}\right)^2\Bigg\}\frac{\partial^2\varphi}{\partial x^2}-2\left(M_\infty+\frac{1}{a_\infty}\frac{\partial\varphi}{\partial x}\right)\left(\frac{1}{a_\infty}\frac{\partial\varphi}{\partial y}\right)\frac{\partial^2\varphi}{\partial x\partial y}$$

$$+\Bigg\{1-\frac{\gamma-1}{2}\left[2M_\infty\left(\frac{1}{a_\infty}\frac{\partial\varphi}{\partial x}\right)+\left(\frac{1}{a_\infty}\frac{\partial\varphi}{\partial x}\right)^2\right]$$

$$-\frac{\gamma+1}{2}\left(\frac{1}{a_\infty}\frac{\partial\varphi}{\partial y}\right)^2\Bigg\}\frac{\partial^2\varphi}{\partial y^2}=0 \qquad (10.161)$$

In the hypersonic region, the Mach lines and shock waves are very close to the body; hence the region of disturbance is very narrow laterally. In order to investigate this hypersonic region, we must expand the coordinate laterally. Tsien[2] thus put

$$x=c\xi,\ y=c\delta^n\eta \qquad (10.162)$$

where c is the chord of the body, δ is the thickness ratio which is assumed to be much smaller than unity, and n is a positive exponent to be determined later.

The appropriate form for φ was given by Tsien in accordance with the boundary condition (10.167) below:

$$\varphi=a_\infty c\frac{1}{M_\infty}f(\xi,\eta) \qquad (10.163)$$

Substituting Eqs. (10.162) and (10.163) into (10.161) and neglecting the higherorder terms for $\delta\ll 1$ and $M_\infty\delta\sim 1$, we obtain the following first-order equation for hypersonic flow

$$\left[1-(\gamma-1)\frac{\partial f}{\partial\xi}-\frac{\gamma+1}{2}\frac{1}{M_\infty^2\delta^{2n}}\left(\frac{\partial f}{\partial\eta}\right)^2\right]\frac{\partial^2 f}{\partial\eta^2}$$

$$=M_\infty^2\delta^{2n}\frac{\partial^2 f}{\partial\xi^2}+2\frac{\partial f}{\partial\eta}\frac{\partial^2 f}{\partial\xi\partial\eta} \qquad (10.164)$$

The boundary conditions for f are:

(a) Ahead of the leading edge shock, the disturbances are zero, i.e.,

$$\frac{\partial f}{\partial\xi}=\frac{\partial f}{\partial\eta}=0 \text{ at } \infty \qquad (10.165)$$

(b) At the surface of the body, the flow must follow the surface

$$\frac{\partial \varphi}{\partial y} = a_\infty M_\infty \delta h'(\xi) \tag{10.166}$$

where $h'(\xi)$ is the slope function of the surface. Substituting equation (10.163) into equation (10.166), we obtain

$$\left(\frac{\partial f}{\partial \eta}\right)_s = M_\infty^2 \delta^{1+n} h'(\xi) \tag{10.167}$$

in which the subscripts refer to the value on the surface of the body. The hypersonic flow over similar bodies can be made to depend only on one parameter if we take

$$n=1 \tag{10.168}$$

The parameter

$$k = M_\infty \delta \tag{10.169}$$

is then known as the hypersonic (similarity) parameter. The fundamental differential equation for irrotational hypersonic flow is

$$\left[1-(\gamma-1)\frac{\partial f}{\partial \xi} - \frac{\gamma+1}{2}\frac{1}{k^2}\left(\frac{\partial f}{\partial \eta}\right)^2\right]\frac{\partial^2 f}{\partial \eta^2} = k^2\frac{\partial^2 f}{\partial \xi^2} + 2\frac{\partial f}{\partial \eta}\frac{\partial^2 f}{\partial \xi \partial \eta} \tag{10.170}$$

with the boundary conditions

(a) At the surface of the body

$$\left(\frac{\partial f}{\partial \eta}\right)_s = k^2 h'(\xi) \tag{10.171}$$

(b) At infinity, ahead of leading edge shock

$$\frac{\partial f}{\partial \xi} = \frac{\partial f}{\partial \eta} = 0 \tag{10.172}$$

Eq. (10.170) was first obtained by Tsien[2]. The most interesting results are (1) that the fundamental equation for a thin airfoil is nonlinear and (2) that the flow field over affinely related bodies may be characterized by the hypersonic parameter k.

Actually the hypersonic flow is much more complicated than that described by Eq. (10.160). first, because the shock wave in hypersonic flow is very strong and the entropy variation behind the leading edge shock is by no means negligible. Hence the flow behind the shock is rotational for hypersonic flow and the velocity potential does not exist. Second, because the shock wave is very close to the surface of the body, there is interaction between the shock and the boundary layer oyer the surface. However, even when these two effects are taken into consideration, the hypersonic parameter k still characterizes the hypersonic flow. As a result Eq. (10.160) gives the correct hypersonic similarity laws.

We shall discuss the interaction of hypersonic flow with the boundary layer in Chapter XVII and the rotational hypersonic flow in Chapter XIV.

It should be noted that for hypersonic inviscid flow, the exact solution may be obtained by the method of characteristics; this shall be discussed in Chapters XI and XIV.

18. Hypersonic similarity laws

In a manner analogous to the transonic case, we may derive hypersonic similarity laws for the pressure coefficient, the drag coefficient, the lift coefficient, and the moment coefficient for affinely related bodies with different thickness ratios but tested at the same value of the hypersonic (similarity) parameter.

The pressure coeffcient for the hypersonic flow of a thin body in a uniform stream is given by Eq. (3.39) or, in terms of function f, by the expression

$$C_P = \frac{2}{\gamma M_\infty^2}\left\{\left[1-(\gamma-1)\frac{\partial f}{\partial \xi} - \frac{\gamma-1}{2}\frac{1}{k^2}\left(\frac{\partial f}{\partial \eta}\right)^2\right]^{\gamma/(\gamma-1)} - 1\right\} \quad (10.173)$$

Eq. (10.173) may be written as

$$C_p = \delta^2 P\ (k, \xi) \quad (10.174)$$

where

$$P\ (k, \xi) = \frac{2}{\gamma k^2}\left\{\left[1-(\gamma-1)\frac{\partial f}{\partial \xi} - \frac{\gamma-1}{2}\frac{1}{k^2}\left(\frac{\partial f}{\partial \eta}\right)^2\right]^{\gamma/(\gamma-1)} - 1\right\} \quad (10.175)$$

with P being a function of k and ξ. Eq. (10.174) shows that for affinely related bodies at the same value of hypersonic parameter k, the pressure coefficients at the corresponding points on the surfaces of the bodies are proportional to the square of the thickness ratio of the bodies.

Similarly, we may find the similarity laws for the lift coefficient C_L, drag coefficient C_D, and moment coefficient C_M, to be respectively,

$$C_L = \delta^2 L(k) \quad (10.176)$$

$$C_D = \delta^3 D(k) \quad (10.177)$$

$$C_M = \delta^2 M(k) \quad (10.178)$$

It is interesting to note that with increasing free stream Mach number, the pressure drag coefficient is increasingly sensitive to the thickness ratio of the thin body. For subsonic speeds, C_D is almost independent of δ; at transonic speeds, C_D is proportional to $\delta^{5/3}$, at ordinary supersonic speeds, C_D is proportional to δ^2, while at hypersonic speeds, C_D is proportional to δ^3.

19. Airfoils in hypersonic flows

Even for a thin airfoil in a uniform stream, the flow over the airfoil and behind the shock is rotational. However Linnell[24] has derived the pressure functions (10.175) for an oblique shock and for Prandtl-Meyer flow and has applied these results to calculate the pressure over the airfoil surface. Since Linnell took the boundary of the isentropic Prandtl-Meyer flow to be at the shock, the influence of the known maximum entropy change was taken into account. His result, which is a reasonable approximation, will be discussed below.

(a) *Oblique shock.* From the oblique shock relation (4.29), we have the pressure coefficient as

$$C_P = \frac{2}{\gamma M_\infty^2}\left(\frac{p}{p_\infty} - 1\right) = \frac{2}{\Gamma}\left(\sin^2\alpha - \frac{1}{M_\infty^2}\right) \tag{10.179}$$

where $\Gamma = \dfrac{\gamma+1}{2}$.

From the shock relation (4.29) we also have

$$(\tan\alpha + \cot\alpha)\tan\theta(\Gamma\sin^2\alpha) = \left(\sin^2\alpha - \frac{1}{M_\infty^2}\right)(1 + \tan\alpha\tan\theta) \tag{10.180}$$

From the similarity law (10.174) we may take here

$$C_P = \theta_s^2 P_s(k_s) \tag{10.181}$$

where θ_s is the flow deflection behind the oblique shock, and $k_s = M_\infty\theta_s$.

Substituting Eqs. (10.181) and (10.179) into Eq. (10.180), we obtain the pressure function P_s in terms of k_s as

$$P_s = \frac{\gamma+1}{2} + \sqrt{\left(\frac{\gamma+1}{2}\right)^2 + \frac{4}{k_s^2}} \tag{10.182}$$

As $M_\infty \to \infty$, we obtain the pressure coeffcient as

$$C_{Ps} \to (\gamma+1)\,\theta_s^2 \tag{10.183}$$

Thus we see that C_{Ps} tends to be a constant for a given wedge as $M_\infty \to \infty$, instead of zero as predicted by linearized theory of Ackeret.

(b) *Prandtl-Meyer expansion.* Let M_s be the Mach number of the flow immediately behind the attached shock at the leading edge of the pointed airfoil. If there is no shock, $M_s = M$. The pressure ratio for the expansion flow is then

$$\frac{p}{p_s} = \left(1 - \frac{\gamma-1}{2}\frac{M_s}{M_\infty}k_e\right)^{\frac{2\gamma}{\gamma-1}} \tag{10.184}$$

where θ_e is the angle of turn through expansion $k_e = M_\infty \theta_e$ and p_s is the pressure just behind the shock, i.e., corresponding to M. The resultant pressure coefficient on the surface of the airfoil is

$$C_P = \left(\theta_s^2 P_s + \frac{2}{\gamma M_\infty^2}\right)\left(1 - \frac{\gamma-1}{2}\frac{M_s}{M_\infty}k_e\right)^{\frac{2\gamma}{\gamma-1}} - \frac{2}{\gamma M_\infty^2} \tag{10.185}$$

where the hypersonic parameter k_e is a function of the shape of the airfoil. The lift, the drag, and the moment coefficients for an arbitrary airfoil have been worked out by Linnell. For the simple case of a flat plate at an angle of attack β with the uniform stream, we have

$$k_c = k_e = M_\infty \beta \tag{10.186}$$

20. Hayes' similitude

If one wants to find the flow field over the airfoil in hypersonic flow, one should consider the rotational flow, i. e., the effect of change of entropy. Hayes[25] showed that the problem of determining the two-dimensional hypersonic flow past a thin airfoil is analytically identical with that of unsteady one-dimensional flow. The similarity parameter is still that defined by Eq. (10.169).

The fundamental equations for two-dimensional steady inviscid rotational flow of perfect gas over a thin airfoil in a uniform stream U are

$$(U+u_1)\frac{\partial \rho}{\partial x} + v_1\frac{\partial \rho}{\partial y} + \rho\left(\frac{\partial u_1}{\partial x} + \frac{\partial v_1}{\partial y}\right) = 0 \tag{10.187a}$$

$$(U+u_1)\frac{\partial u_1}{\partial x} + v_1\frac{\partial u_1}{\partial y} + \frac{1}{\rho}\frac{\partial p}{\partial x} = 0 \tag{10.187b}$$

$$(U+u_1)\frac{\partial v_1}{\partial x} + v_1\frac{\partial v_1}{\partial y} + \frac{1}{\rho}\frac{\partial p}{\partial y} = 0 \tag{10.187c}$$

$$(U+u_1)\frac{\partial S}{\partial x} + v_1\frac{\partial S}{\partial y} = 0 \tag{10.187d}$$

$$\frac{p}{p_0} = \left(\frac{\rho}{\rho_0}\right)^{\gamma} e^{S/c_v} \tag{10.187e}$$

where u_1 and v_1 are the perturbed x-and y-velocity components, and S is the entropy.

We now use the transformations

$$x = c\xi,\ y = \frac{c}{M_\infty}\eta,\ v_1 = U\delta v'$$

$$u_1 = \frac{U\delta}{M_\infty} u',\quad p = \frac{\rho_0 U^2}{M_\infty^2} p' \tag{10.188}$$

where δ is the thickness ratio of the airfoil such that $M_\infty \delta$ is of the order of unity and $M_\infty \gg 1$.

Substituting Eq. (10.188) into (10.187) and neglecting terms of order $1/M_\infty^2$ or smaller, we have from Eqs. (10.187a) to (10.187d)

$$\frac{1}{M_\infty \delta}\frac{\partial \rho}{\partial \xi} + v'\frac{\partial \rho}{\partial \eta} + \rho\frac{\partial v'}{\partial \eta} = 0 \tag{10.189a}$$

$$\frac{1}{M_\infty \delta}\frac{\partial u'}{\partial \xi} + v'\frac{\partial u'}{\partial \eta} + \frac{\rho_0}{\rho}\frac{1}{M_\infty^2 \delta^2}\frac{\partial p'}{\partial \xi} = 0 \tag{10.189b}$$

$$\frac{1}{M_\infty \delta}\frac{\partial v'}{\partial \xi} + v'\frac{\partial v'}{\partial \eta} + \frac{\rho_0}{\rho}\frac{1}{M_\infty^2 \delta^2}\frac{\partial p'}{\partial \eta} = 0 \tag{10.189c}$$

$$\frac{1}{M_\infty \delta}\frac{\partial S}{\partial \xi} + v'\frac{\partial S}{\partial \eta} = 0 \tag{10.189d}$$

We see that the above equations are still characterized by the hypersonic parameter $k = M_\infty \delta$.

In terms of original flow quantities, Eqs. (10.189a), (10.189c) and (10.189d) become

$$U\frac{\partial \rho}{\partial x} + v_1\frac{\partial \rho}{\partial y} + \rho\frac{\partial v_1}{\partial y} = 0 \tag{10.190a}$$

$$U\frac{\partial v_1}{\partial x} + v_1\frac{\partial v_1}{\partial y} + \frac{1}{\rho}\frac{\partial p}{\partial y} = 0 \tag{10.190b}$$

$$U\frac{\partial S}{\partial x} + v_1\frac{\partial S}{\partial y} = 0 \tag{10.190c}$$

If we write $x = Ut$, Eqs. (10.190) are identical to the equations of motion governing one-dimensional unsteady inviscid anisentropic flow in the direction of y. Eq. (10.189b) determines the velocity component u' after v' and p are known.

The boundary conditions are:

(a) On the surface of the airfoil, the boundary condition, after neglecting terms of $O(1/M_\infty^2)$, becomes

$$U\frac{dy_w}{dx} - v_1 = 0 \tag{10.191}$$

where $y_w(x)$ is the shape of the airfoil.

(b) On the shock $y = y_s(x)$, after neglecting the terms of $O(1/M_\infty^2)$, we have

$$\frac{v_1}{U} = \frac{2}{\gamma+1}\left[\left(\frac{dy_s}{dx}\right)^2 - \frac{1}{M_\infty^2}\right] \Big/ \frac{dy_s}{dx} \tag{10.192}$$

$$\frac{p}{p} = \frac{2}{\gamma+1} M_\infty^2 \left(\frac{dy_s}{dx}\right)^2 - \frac{\gamma-1}{\gamma+1} \tag{10.193}$$

By the transformation $x = Ut$, the boundary condition (10.191) corresponds to that existing at a piston moving with velocity proportional to the slope of the airfoil. The boundary conditions (10.192) and (10.193) correspond to those at a normal shock in one-dimensional flow moving with a velocity which is equal to U times the slope of the shock in the two-dimensional theory. Thus we transform the two-dimensional hypersonic steady flow to a corresponding one-dimensional unsteady flow. This is known as Hayes' similitude. Of course the general analytic solution is still difficult to obtain because y_s is not known a priori. We shall discuss this point further in Chapter XIV.

21. Newtonian theory of hypersonic flow

Von Kármán[26] has pointed out an analogy between the dynamics of hypersonic flow and Newton's corpuscular theory of aerodynamics[27] when the ratio of the specific heats γ is unity. From the oblique shock relations of Chapter IV, § 5, we see that if $\gamma = 1$, the increase in density across the shock can be infinitely larger if $M_\infty \to \infty$ and $\alpha \cong \theta$. Hence, at extreme velocities, only the fluid in the immediate neighborhood of the body is affected by the presence of the body and the fluid outside this neighborhood is practically unaffected. According to Eq. (4.29), the pressure behind the oblique shock is, for $\gamma = 1$,

$$p^2 = \rho_1 U_1^2 \sin^2\alpha \cong \rho_1 U_1^2 \sin^2\theta \tag{10.194}$$

If we assume that the pressure on the surface of the body is the same as that behind the shock, the drag obtained from Eq. (10.194) will be the same as that obtained from Newton's theory. Newton's theory states that the increase in pressure is due to the inelastic reaction of the individual fluid particles which strike the surface of the body and leave it in a tangential direction. However, Busemann[28] pointed out that there is a hypersonic "boundary layer" between the shock and the surface of the body and that the pressure drop across this layer is of the same order of magnitude as the change of pressure across the shock. As a result, the resistance of the body in hypersonic flow is smaller than that given by Newton's theory. For instance, the resistance of a slender body of revolution in hypersonic flow is about one half that value given by Newton's theory.

In the case of $\gamma > 1$, the conditions are more complex because the thickness of the layer affected by the shock is finite for $M_1 = \infty$. The resistance in this case is also less than the Newtonian value.

Saenger[29] has used this concept to design the best wing and body shape for flight at infinite Mach number. This theory is not satisfactory for large but finite Mach number flight.

22. Problems

1. Show that in the linearized theory of compressible flow the similarity law may be arbitrarily chosen.

2. Find the expressions for the pressure function (10.31) for an oblique shock in slightly supersonic flow for the transonic parameter K larger than the critical value 0.945. Show that there are two values for each K, one corresponding to the weak shock case, and the other to the strong shock case.

3. Calculate the lift coefficient for a flat plate at an angle of attack 1.0° in a uniform supersonic stream for the Mach numbers from 1.0 to 1.3. Compare the results with those obtained from the Ackeret linearized theory.

4. Show that in the hodograph plane, von Kármán's transonic flow equation (10.13) is the Tricomi equation. Find the fundamental solutions of the Tricomi equation.

5. Why is there only one solution of the pressure function in hypersonic flow for the oblique shock given in Eq. (10.182).

6. Calculate the lift coefficient for a flat plate against the angle of attack at Mach numbers (a) $M_\infty = 10$ and (b) $M_\infty = 20$. Compare these results with those of the Ackeret linearized theory.

7. Derive the boundary conditions (10.192) and (10.193) from the exact relations across an oblique shock and show that the error is of the order of $1/M^2$.

8. Show the procedure by which the best wing and body shape may be designed for flight at case of infinite Mach number.

9. Prove that the critical pressure coefficient for TSP is

$$C_p^* = -\frac{2}{\gamma + 1}\,\frac{1 - M_\infty^2}{M_\infty^2}$$

and its error is $O\,[(1 - M^2)^2]$.

10. Verify that the TSP equation may be written as

$$\left(1-M_\infty^2-\frac{\gamma+1}{U}M_\infty^m\frac{\partial\varphi}{\partial x}\right)\frac{\partial^2\varphi}{\partial x^2}+\frac{\partial^2\varphi}{\partial y^2}=0$$

where m is any real value, if

$$m=\frac{2\gamma+1}{\gamma+1}$$

the error for C_p^* becomes $O\ [(1-M^2)^3]$.

11. Derive the TSP shock relation from the Rankine-Hugoniot relation

$$\left(1-M_\infty^2-\frac{\gamma+1}{U}M_\infty^2\frac{u_1^1+u_2^1}{2}\right)(u_1^1-u_2^1)^2+(v_1^1-v_2^1)^2=0$$

where u^1, v^1 are the perturbation velocity components.

12. Plot the TSP shock polar in the free stream flow $M_\infty=1.2$ and compare with the exact shock polar.

13. Derive the transonic weak normal shock relation

$$\frac{u_2^1}{U}=\frac{2(1-M_\infty^2)}{(\gamma+1)M_\infty}$$

and show that this relation is exact.

14. Derive from the TSP differential equation

$$\frac{p}{\rho_\infty U^2}=\frac{1}{\gamma M_\infty^2}-\frac{u'}{U}-\frac{1-M_\infty^2}{2}\left(\frac{u'}{U}\right)^2-\frac{v'^2}{2U^2}+\frac{\gamma+1}{6}M_\infty^2\left(\frac{u'}{U}\right)^3$$

and find the order of its error.

15. Verify that zero is a conservative sonic difference scheme for the TSP.

16. Prove that the equation of characteristic cone surface passing through the origin $x=y=0$ for the differential equation of the SLOR at $M<1$.

$$A\frac{\partial^2\varphi}{\partial x^2}+a\frac{\partial^2\varphi}{\partial x\partial t}+b\frac{\partial\varphi}{\partial t}+\frac{\partial^2\varphi}{\partial y^2}=0$$

where $A>0$ is

$$\left(\frac{y}{t}\right)^2=-\frac{4}{a}\left(\frac{x}{t}-\frac{A}{a}\right)$$

and plot out the characteristic cone surface.

17. Plot out Eq. (10.99) when $a=0$ and $\omega=0.5$, 1, 1.5 and discuss the results.

18. Show that an optimal choice of ω and σ for AF2 at subsonic points is $\omega=2$ and $\sigma_L = O(\Delta x)$, $\sigma_H = O\left(\frac{1}{\Delta x}\right)$.

19. Show that the AF2 growth factor at supersonic points is

$$e^{\lambda\Delta t} = \frac{b_1 + (1-\omega)b_2}{b_1 + b_2}$$

where

$$b_1 = \frac{1-e^{-ia\Delta x}}{\Delta x}\left[-\sigma + \frac{2A}{\sigma\Delta y^2}(1-\cos b\Delta y)\right]$$

$$b_2 = \frac{A}{\Delta x^2}(1-2e^{-ia\Delta x}+e^{-i2a\Delta x}) - \frac{2}{\Delta y^2}(1-\cos b\Delta y)$$

and use the Fourier analysis of Chapter VII, § 9c to find the stability conditions for the lowest and highest frequency components.

20. Show that an optimal choice of ω and σ for AF2 at supersonic points is $\omega=2$ and $\sigma_L = O(\Delta x)$, $\sigma_H = O\left(\frac{1}{\Delta x}\right)$.

21. Prove that in Eq. (10.131)

$$\frac{\partial^2\varphi}{\partial n^2} = 0$$

22. Prove that if the governing equations for isentropic normal shock are equation of continuity and equation of motion, there is a jump in stagnation enthalpy across the shock,

$$\frac{H_{02}-H_{01}}{H_{01}} = -\left\{1+\frac{\gamma-1}{2}M_1^2\left[1-\left(\frac{p_2}{p_1}\right)^{-\frac{2}{\gamma}}\right]-\left(\frac{p_2}{p_1}\right)^{\frac{\gamma-1}{\gamma}}\right\}\Big/\left(1+\frac{\gamma-1}{2}M_1^2\right)$$

where p_2/p_1 is related to M_1 by

$$M_1^2 = \frac{1}{\gamma}\left(\frac{p_2}{p_1}-1\right)\Big/\left[1-\left(\frac{p_2}{p_1}\right)^{-\frac{1}{\gamma}}\right]$$

For transonic flow, $M_1 \approx 1$,

$$\frac{H_{02}-H_{01}}{H_{01}} = \frac{2(3\gamma^3+4\gamma^2+7\gamma-2)}{2(\gamma+1)^3}(M_1^2-1)^3 + O[(M_1^2-1)^4]$$

$$\frac{p_2}{p_1} = 1 + \frac{2\gamma}{\gamma+1}(M_1^2-1) + \frac{2\gamma(\gamma-1)}{3(\gamma+1)^2}(M_1^2-1)^2 + O[(M_1^2-1)^3]$$

23. Prove that if the governing equations for isentropic normal shock are equation of motion and equation of energy, there is a jump in mass flux across the shock,

$$\frac{\rho_2 u_2 - \rho_1 u_1}{\rho_1 u_1} = \left[\left(\frac{p_2}{p_1}\right)^{\frac{1}{\gamma}} \frac{1+\gamma M_1^2 - p_2/p_1}{\gamma M_1^2}\right]^{\frac{1}{2}} - 1$$

where p_2/p_1 is related to M_1 by

$$M_1^2 = \left[\frac{2\gamma}{\gamma-1}\left(\frac{p_2}{p_1}\right)^{\frac{1}{\gamma}} - \frac{\gamma+1}{\gamma-1}\frac{p_2}{p_1} - 1\right] \Big/ \left\{\gamma\left[1-\left(\frac{p_2}{p_1}\right)^{\frac{1}{\gamma}}\right]\right\}$$

For transonic flow, $M_1 \approx 1$,

$$\frac{\rho_2 u_2 - \rho_1 u_1}{\rho_1 u_1} = \frac{9\gamma^3 + 153\gamma^2 - 103\gamma - 1}{48(\gamma+1)^3}(M_1^2-1)^3 + O[(M_1^2-1)^4]$$

$$\frac{p_2}{p_1} = 1 + \frac{2\gamma}{\gamma+1}(M_1^2-1) + \frac{2\gamma}{3(\gamma+1)}(M_1^2-1)^2 +$$

$$+ \frac{2}{9(\gamma+1)^3}(6 - 19\gamma + 28\gamma^2 - 7\gamma^3)(M_1^2-1)^3 + O[(M_1^2-1)^4]$$

References

1. von Kármán, Th., Similarity law of transonic flow, *Jour. Math. Phys.* **26,** 1947, pp. 182 – 190.
2. Tsien, H. S., Similarity laws of hypersonic flows, *Jour. Math. Phys.* **25,** 1946, pp. 247 – 251.
3. Pack, D. D., and Pai, S. I., Similarity laws for supersonic flows, *Quart. Appl. Math.* **11,** 1954, pp. 377 – 384.
4. Liepmann, H. W., and Bryson, A. E., Jr., Transonic flow past wedge sections, *Jour. Aero. Sci.* **17,** No. 12, Dec. 1950, pp. 745 – 755.
5. Tsien, H.S., and Baron, R. J., Airfoils in slightly supersonic flow, *Jour. Aero. Sci.* **16,** No. 1, 1949, pp. 55 – 61.
6. Guderley, G., and Yoshihara, H., The flow over a wedge profile at Mach number 1, *Jour. Aero. Sci.* **17**, No. 11, Nov. 1950, pp. 723 – 735.
7. Guderley, G., and Yoshihara, H., Two dimensional unsymmetric flow patterns at Mach number 1, *Jour. Aero. Sci.* **20,** No. 11, Nov. 1953, pp. 757 – 768.
8. Oswatitsch, K., Special problems of inviscid steady transonic flows, *Int. Jour. of Eng. Sci.* **20,** No. 4, April 1982, pp. 497 – 540.
9. Sobieczky, H., Exacte Loesungen der ebenen gasdynamischen Gleichungen in Schallnaehe, *z. f. Flugwiss.* **19,** 1971, pp. 197 – 214.
10. Sobieczky, H., Tragende Schnabelprofile in Stossfreier Schallanstroemung, *ZAMP* **26,** 1975, pp. 816 – 830.
11. Guderley, K. G., *Theory of Transonic Flow*, Pergamon Press, London, 1962.
12. Mueller, E. A., and Matschat, K. O.,*Proc. Int. Cong. Appl. Mechanics*, Springer Verlag, Verlag, Berlin, 1964, pp 1061 – 1068.
13. Zierep. J., *Theorie der Schallnahen und der Hyperschallstroemungen*, Verlag G. Braun,

Karlsruhe, 1966.

14. Murman, E.M., and Cole, J.D., Caculation of plane steady transonic flows, *AIAA Journal* **9**, No. 1, Jan. 1971, pp. 114 – 121.
15. Warming, R.F., and Hyett, B.J., The modified equation approach to the stability and accuracy analysis of finite difference methods, *J. Comp. Phys.* **14**, 1974, pp. 159 – 179.
16. Hafez, M.M., South, J.C., Jr., and Murman, E.M., Artificial compressibility methods for numerical solutions of transonic full potential equation, *AIAA Journal* **17**, No. 8, Aug. 1979, pp. 838 – 844.
17. Holst, T.L., and Ballhaus, W.F., Jr., Fast conservative schemes for the fulll potential equation applied to transonic flows, *AIAA Journal* **17**, No. 2, Feb. 1979, pp. 145 – 152.
18. Eberle, A., Transonic flow computations by finite elements, *Airfoil optimization and analysis in recent developments in theoretical and experimental fluid mechanics*, Springer Verlag, 1979.
19. Luo, S.J., Liu, P., and E, Q., Computation of transonic flow around airfoils, Presented on First World Congress on Comp. Mech. Austin, 1986.
20. Jameson, A, Iterative solution of transonic flow over airfoils and wings, including flow at Mach 1, *Comm. on Pure and Applied Math.* **27**, 1974, pp. 283 – 309.
21. Ballhaus, W.F., Jr, Jameson, A., and Albert, J., Implicit approximate factorization schemes for the efficient solution of steady transonic flow problems, *AIAA Journal* **16**, No. 6, June 1978, pp. 573 – 579.
22. Li, X. Y. Liao, Q. W., and Luo, S. J., A parallel algorithm of AF-2 scheme for plane steady transonic potential flow with small transverse disturbance, *ICAS*-88-4.7.1, 1988, pp. 1024 – 1028.
23. Schmidt, W., Jameson, A., and Whitfield, D., Finite-volume solutions to the Euler equations in transonic flow, *J. Aircraft* **20**, No. 2, Feb. 1983, pp. 127 – 133.
24. Linnell, R. D., Two dimensional hypersonic airfoils, *Jour. Aero. Sci.* **16**, No. 1, Jan. 1949, pp. 22 – 30.
25. Goldsworthy, F. A., Two dimensional rotational flow at high Mach number past thin airfoils, *QJMAM* **5**, pt. 1, March 1952, pp. 54 – 63.
26. von Kármán, Th., The problem of resistance in compressible fluids, *Proc. Fifth Volta Cong. Rome*, 1936, pp. 275 – 277.
27. Newton, I., *Principia* – Motte's translation revised, University of California Press, 1946, pp. 333, 657 – 661.
28. Busemann, A., Flussigkeits-und Gasbewung, *Handworterbuch der Naturwissenschaften, Zweite Auflage*, Gustov Fischer, Jena, 1933, pp. 266 – 279.
29. Saenger, E., *R-Antriebe, Die Deutsche Aka. der Luftfahrt.*, Berlin 1943, pp. 189 – 194.

Chapter XI

METHOD OF CHARACTERISTICS

1. Introduction

From the linearized theory of Chapter VI, we see that the nature of the differential equation for steady compressible flow is entirely different for subsonic and supersonic flows. For subsonic flow, the equation is of the elliptic type, i.e., the Laplace equation in linearized theory; whereas for supersonic flow, the equation is of the hyperbolic type, i.e., the simple wave equation in linearized theory. The exact differential equation for the steady flow of a compressible fluid has the same nature as that of the linearized theory. For steady subsonic flow, the differential equation is of the elliptic type and for steady supersonic flow, the differential equation is of the hyperbolic type.

The boundary conditions occurring in the elliptic equation are different from those in the hyperbolic equation. For the elliptic equation, the problem is to find a function throughout a field given the value of the function (or its normal derivative) over a complete boundary enclosing the field. This is known as Dirichlet's problem. We have discussed some methods of its solution in Chapter VII.

For the hyperbolic equation, the problem is to find a function throughout a field where its value and that of its normal derivative are given over a portion but not all of a boundary enclosing the field. This is called Cauchy's problem. In compressible flow problems, the differential equations for steady hypersonic flow and for nonsteady flow are hyperbolic in nature. One powerful method for finding solutions of the hyperbolic equation is the method of characteristics. We shall discuss the method of characteristics in this chapter.

The general theory of the method of characteristics for the case of two independent variables is particularly easy to visualize, and computation methods for this case have been extensively worked out. Hence in this chapter we shall restrict ourselves to cases of two independent variables. Included here are two-dimensional or axially symmetrical steady flows and unsteady flow with one spatial variable, i.e., one-dimensional unsteady flow with spherical or cylindrical symmetry.

The fundamental differential equation of compressible flow with two independent variables may be written as a single second-order differential equation, such as Eq. (5.61), or a system of simultaneous first-order partial differential equations of two dependent variables and two independent variables. We may discuss the theory of characteristics based on either of these equations. In § 2, we shall discuss the theory of characteristics of a general quasi-linear differential equation of the second order in two independent variables. In § 3, we shall discuss the theory of characteristics of two simultaneous first-order equations in two variables.

Some of the fundamental properties of characteristics will be discussed in § 4 and 5. In § 6, Cauchy initial value problems will be discussed. In § 7, 8 and 9, we apply the method of characteristics to two-dimensional irrotational supersonic flow and give the well-established numerical procedures used. In § 10, we apply this method to the design of a supersonic nozzle. In § 11, we discuss the supersonic flow in a jet in which a shock wave occurs.

In § 12, the method of characteristics will be used to fit the detached shock wave in front of a blunt body in the supersonic and hypersonic flows.

The application of the method of characteristics to axially symmetrical steady supersonic flow will be discussed in Chapter XIII, while the application of the method of characteristics to rotational flow will be discussed in Chapter XIV. The general theory of these two cases is, however, included in the general discussion of § 2 of this chapter.

Many of the ideas and techniques of the theory of characteristics of two independent variables can be carried over to the case of more than two independent variables, but the complications increase greatly. We shall discuss briefly the theory of characteristics of three variables in Chapter XIII.

2. Theory of characteristics for a second-order partial differential equation of two independent variables[1-5]

Let the function ψ of the independent variables x and y satisfy the quasi-linear differential equation

$$A\psi_{xx}+B\psi_{xy}+C\psi_{yy}+D=0 \tag{11.1}$$

where the subscript denotes partial differentiation; A, B, C, and D are functions of x, y, ψ, ψ_x, and ψ_y. Eq. (11.1) is called quasi-linear because it is linear in the derivative of the highest order, i.e., the second-order derivatives. In the gasdynamics of inviscid flow, the differential equations are all of this type, e.g , Eqs. (5.61) and (5.46).

Let

$$\eta(x, y)=\text{constant} \tag{11.2}$$

be the equation of the characteristic curves of Eq. (11.1). Let the family of the characteristic curves (11.2) be intersected by a second family of curves

$$\xi(x, y) = \text{constant} \tag{11.3}$$

No further specifications are necessary concerning this second family of curves, but in the case of hyperbolic equation, it may be chosen as the second family of characteristics.

We define the interior derivative as the derivative of a function with respect to ξ taken along η = constant, and the exterior derivative as the derivative of a function with respect to η. Thus for exterior derivatives, information concerning the variation of the function beyond η = constant is required.

We may introduce the characteristics in the following three different ways:

(a) The characteristic curves can be considered as the loci of possible small discontinuities.

(b) The characteristics are the only curves from which an integral surface can be constructed.

(c) The continuation of an integral surface beyond a characteristic may become indeterminate.

We shall discuss the characteristics from the above three different ways in the following. All give the same result.

(a) *Characteristics as loci of discontinuities of second order.* This is a most natural way for the aerodynamicist to introduce the characteristics of aerodynamic equations. In this case, the characteristics for the supersonic steady flow are the Mach lines along which small disturbances or discontinuities are propagated from the boundary into the interior of the flow field, as we shall see in § 7.

Now we assume that there is a discontinuity only in the exterior derivative of the first derivative of the function ψ of Eq. (11.1). Then only $\psi_{\eta\eta}$ is discontinuous across a characteristic curve η = constant. Since there is a jump $[\psi_{\eta\eta}]$ in $\psi_{\eta\eta}$ across η = constant, the jumps in ψ_{xx} ψ_{yx} and ψ_{yy} across η = constant are then respectively

$$\begin{aligned} [\psi_{xx}] &= [\psi_{\eta\eta}]\,\eta_x^{\,2} \\ [\psi_{xy}] &= [\psi_{\eta\eta}]\,\eta_x\eta_y \\ [\psi_{yy}] &= [\psi_{\eta\eta}]\,\eta_y^{\,2} \end{aligned} \tag{11.4}$$

because $\psi_{xx} = \psi_{\eta\eta}\eta_x^2 + 2\psi_{\xi\eta}\xi_x\eta_x + \psi_{\xi\xi}\xi_x^{\,2} + \psi_\eta\eta_{xx} + \psi_\xi\xi_{xx}$, etc.

If now the differential equation (11.1) is written for a fixed value of ξ and samll positive $(+\Delta\eta)$ and negative values $(-\Delta\eta)$ of η different from $\eta = 0$ characteristics (Fig. 11.1) (for simplicity, we may consider the characteristic $\eta = 0$ from here on). we obtain by subtraction

$$A\,[\psi_{xx}] + B[\psi_{xy}] + C[\psi_{yy}] = 0 \tag{11.5}$$

or

$$A\eta_x^2 + B\eta_x\eta_y + C\eta_y^2 = 0$$

Eq. (11.5) is the characteristic condition.

If we represent η = constant in parametric form

$$x = x(\xi), \qquad y = y(\xi) \tag{11.6}$$

then

$$\frac{\eta_x}{\eta_y} = -\frac{\dot{y}}{\dot{x}}$$

where the dot denotes the derivative with respect to ξ.

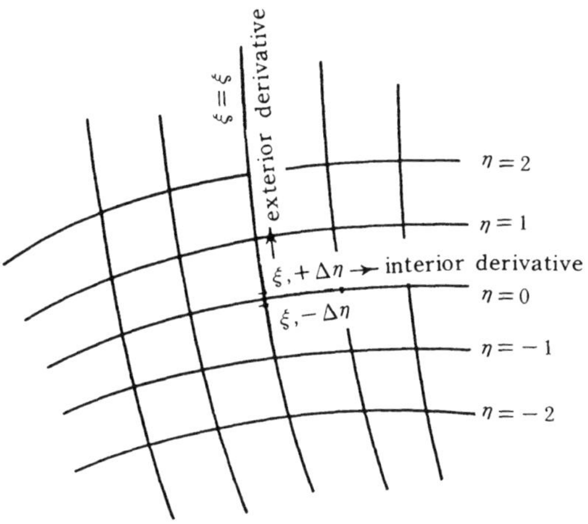

Fig. 11.1 Characteristic curves of a differential equation

Eq. (11.5) becomes

$$A\dot{y}^2 - B\dot{x}\dot{y} + C\dot{x}^2 = 0 \tag{11.7}$$

or

$$A\left(\frac{dy}{dx}\right)^2 - B\frac{dy}{dx} + C = 0 \tag{11.7a}$$

Eq. (11.7) or (11.7a) is known as the characteristic equation which gives the following two families of characteristics:

$$\left(\frac{dy}{dx}\right)_1 = \frac{B + \sqrt{B^2 - 4AC}}{2A}$$
$$\left(\frac{dy}{dx}\right)_2 = \frac{B - \sqrt{B^2 - 4AC}}{2A} \tag{11.8}$$

From the values of the functions A, B, and C of Eq. (11.1), we have the classification criteria:

B^2-4AC	*Type of equation*	*Characteristic curves*
>0	Hyperbolic	Two real familes
<0	Elliptic	Two imaginary families
$=0$	Parabolic	One real family

Since the functions A, B, C are, in general, variables,which take different values in different parts of the field, Eq. (11.1) may be of the hyperbolic type in a certain region, elliptic in the another, and parabolic in still others. In general, Eq. (11.1) is of the mixed type.

For example, let us consider the steady two-dimensional flow of compressible inviscid fluid. The fundamental equation is Eq. (5.61) or (5.63). Here we have

$$A = a^2 - u^2, \quad B = -2uv, \quad C = a^2 - v^2 \tag{11.9}$$

Hence

$$B^2 - 4AC = 4a^2 (M^2 - 1) \tag{11.10}$$

where M is the local Mach number, i. e., $M=\sqrt{u^2+v^2}/a$. We see that for supersonic flow $M > 1$, Eq. (5.61) is of the hyperbolic type; for subsonic flow $M < 1$, Eq. (5.61) is of elliptic type; and for sonic flow $M = 1$, Eq. (5.61) is of the parabolic type. In general, Eq. (5.61) is of the mixed type. It is also of interest to note that the characteristics for the rotational flow are the same as those for the corresponding irrotational flow equation (5.63). The only difference between Eqs. (5.61) and (5.63) is in D, which does not affect the characteristic equation (11.7).

From the values of Eq. (11.9), we have for supersonic flow $M > 1$ the following two families of characteristics:

$$\left(\frac{dy}{dx}\right)_{1,2} = \frac{-uv \pm a\sqrt{u^2+v^2-a^2}}{a^2-u^2} = \tan(\theta \mp \alpha) \tag{11.11}$$

where θ is the angle of the velocity vector to the x-axis and is known as the Mach angle, i. e., $\alpha=\sin^{-1}(1/M)$. Eq. (11.11) shows that the angle between the characteristic curves and the streamlines is the Mach angle. Hence the characteristics are the Mach lines in the steady two-dimensional supersonic flow.

(b) *Characteristic strips as elements of integral surfaces*. Let us consider a strip C on the surface of ψ which is associated with a curen $\eta = 0$, and which is defined parameterically by $x(\xi)$, $y(\xi)$, $\psi(\xi)$, $\psi_\xi(\xi,0)=p(\xi)$ and $\psi_\eta(\xi,0)=q(\xi)$. We want to find the conditions under which the expression on the left-hand side of Eq. (11.1), i.e.,

$$A\psi_{xx}+B\psi_{xy}+C\psi_{yy}+D \qquad (11.12)$$

can be expressed in terms of the five strip quantities $x(\xi)$, $y(\xi)$, $\psi(\xi)$, $p(\xi)$, and $q(\xi)$ and their internal derivatives. Since the coefficients A, B, C, and D depend only on the quantities x, y, ψ, ψ_ξ, and ψ_η which are given along the considered strip, we need to investigate only whether ψ_{xx}, ψ_{xy}, and ψ_{yy} can be expressed in terms of these five strip quantities and their internal derivatives.

From Eq. (11.4), we see that if

$$A\eta_x^2+B\eta_x\eta_y+C\eta_y^2 = 0 \qquad (11.13)$$

Eq. (11.12) can be expressed in terms of these five quantities and their internal derivatives. Eq. (11.13) is the characteristic condition (11.5). Only characteristic strips can be considered for the construction of an integral surface from the strips. Of course, that such a construction is indeed possible must be proved.

(c) *Indeterminate continuation of an integral surface beyond a characteristic strip.* The third definition of characteristics is obtained from the consideration of a strip of the first order on the integral surface ψ. We want to know whether the unknown second derivatives ψ_{xx}, ψ_{xy}, ψ_{yy}, and the higher derivatives along the strip of the first order can always be uniquely determined from the differential equation (11.1). If they could be determined, the function ψ might be continued beyond the strip by means of a Taylor series development.

The second derivatives r, s, and t can be determined from the following three simultaneous equations:

$$\left.\begin{aligned} A\psi_{xx}+B\psi_{xy}+C\psi_{yy} &= -D \\ \dot{x}\psi_{xx}+\dot{y}\psi_{xy} &= \psi_x \\ \dot{x}\psi_{xy}+\dot{y}\psi_{yy} &= \psi_y \end{aligned}\right\} \qquad (11.14)$$

(where the dot denotes differentiation with respect to ξ), provided that the determinant

$$\begin{vmatrix} A & B & C \\ \dot{x} & \dot{y} & 0 \\ 0 & \dot{x} & \dot{y} \end{vmatrix} = A\dot{y}^2-B\dot{x}\dot{y}+c\dot{x}^2 \qquad (11.15)$$

does not vanish. It is interesting to note that the higher derivatives, too can be uniquely determined under the same condition (11.15).

If the determinant vanishes, i.e., if the characteristic condition (11.7) is satisfied, the derivatives r, s, and t are determined, if at all, only within the solution of the homogeneous system of equations which corresponds to Eqs.

(11.14). If the characteristic condition is satisfied, Eqs. (11.14) admit a solution if the following condition

$$\begin{vmatrix} A & C & -D \\ \dot{x} & 0 & \psi_x \\ 0 & \dot{y} & \psi_y \end{vmatrix} = A\psi_x\dot{y} + C\psi_y\dot{x} + D\dot{x}\dot{y} = 0 \qquad (11.16)$$

is also satisfied. This gives the second characteristic differential equation.

The third characteristic equation is

$$\psi = \psi_x\dot{x} + \psi_y\dot{y} \qquad (11.17)$$

Eqs. (11.7), (11.16), and (11.17) are the fundamental differential equations for the characteristic method as applied to the hyperbolic type differential equation. We shall discuss the approximate solution of these equations for flow problems later.

3. Theory of characteristics for two simultaneous first-order differential equations in two variables

Let us consider the following system of quasi-linear first-order differential equations in two independent and two dependent variables :

$$\left.\begin{aligned} A_1\frac{\partial u}{\partial x} + B_1\frac{\partial u}{\partial y} + C_1\frac{\partial v}{\partial x} + D_1\frac{\partial v}{\partial y} + E_1 &= 0 \\ A_2\frac{\partial u}{\partial x} + B_2\frac{\partial u}{\partial y} + C_2\frac{\partial v}{\partial x} + D_2\frac{\partial v}{\partial y} + E_2 &= 0 \end{aligned}\right\} \qquad (11.18)$$

where $A_1, A_2, \cdots, E_2$ are functions of x, y, u, and v. If the functions $A_1, A_2 \cdots, E_2$ are functions of the independent variables x and y but not of the dependent variables u and v, the equations are linear. In general, Eqs. (11.18) are nonlinear. We say they are quasi-linear because they are linear with respect to the first derivatives of u and v. If the equations are homogeneous, i.e., $E_1 = E_2 = 0$, and if the other coefficients are functions of u, v but not of x, y explicitly, Eqs. (11.18) are said to be reducible and can be transformed into a linear system by interchanging the roles of the independent and dependent variables.

The fundamental equations of the compressible fluid for many cases may be reduced to the form of Eqs. (11. 18). For instance, the fundamental equations for steady irrotational two-dimensional or axially symmetrical flow may be written in the form of Eqs. (11.18) as follows:

$$\begin{gathered} (u^2 - a^2)\frac{\partial u}{\partial x} + uv\left(\frac{\partial u}{\partial y} + \frac{\partial v}{\partial x}\right) + (v^2 - a^2)\frac{\partial v}{\partial y} - \delta\frac{a^2 v}{y} = 0 \\ \frac{\partial v}{\partial x} - \frac{\partial u}{\partial y} = 0 \end{gathered} \qquad (11.19)$$

where $\delta = 0$ for two-dimensional flow, $\delta = 1$ for axially symmetrical flow, and y is the radial distance in axially symmetrical case. The sound speed a is a given function of the velocity $u^2 + v^2$, i.e.,

$$a^2 = a_0^2 - \frac{\gamma - 1}{2}(u^2 + v^2) \tag{11.20}$$

From Eqs. (11.19), we see that for the two-dimensional case, the equations are reducible and that for the axially symmetrical case, they are not reducible. This is well known from our consideration of the hodograph method discussed in Chapter VIII.

Now we consider the characteristics of Eqs. (11.18) in the same manner as that described in § 2(3), i.e., we consider the continuation of the functions u and y beyond a characteristic. Let us assume that along a certain curve $\sum$: $x = x(\xi)$ and $y = y(\xi)$, the values of the functions u and v are given. If we can calculate the first derivatives and higher derivatives from Eqs. (11.18) and the values of u and v on the curve $\sum$, the functions u and v might be continued beyond the curve by means of a Taylor series development.

The equations from which we can determine the four first derivatives u_x, u_y, v_x, and v_y are as follows:

$$\left.\begin{aligned} A_1 u_x + B_1 u_y + C_1 v_x + D_1 v_y &= -E_1 \\ A_2 u_x + B_2 u_y + C_2 \dot{v}_x + D_2 \dot{v}_y &= -E_2 \\ \dot{x} u_x + \dot{y} u_y &= \dot{u} \\ \dot{x} v_x + \dot{y} v_y &= \dot{v} \end{aligned}\right\} \tag{11.21}$$

We can determine u_x, etc., from Eqs. (11.21) uniquely except when

$$\begin{vmatrix} A_1 & B_1 & C_1 & D_1 \\ A_2 & B_2 & C_2 & D_2 \\ \dot{x} & \dot{y} & 0 & 0 \\ 0 & 0 & \dot{x} & \dot{y} \end{vmatrix} = (B_1 D_2 - D_1 B_2)(dx)^2 - [(A_1 D_2 - D_1 A_2) + (B_1 C_2 - C_1 B_2)]\,dxdy + (A_1 C_2 - C_1 A_2)(dy)^2 = 0 \tag{11.22}$$

Eq. (11.22) is the equation of the characteristics. The solution of Eq. (11.22) gives two families of characteristics curves

$$\left.\begin{aligned} \left(\frac{dy}{dx}\right)_1 &= C'(x, y, u, v) \\ \left(\frac{dy}{dx}\right)_2 &= C''(x, y, u, v) \end{aligned}\right\} \tag{11.23}$$

These two families of characteristics depend explicitly on the coefficients $A_1, \cdots, D_2$ of Eqs. (11.18). For instance, for two-dimensional irrotational steady flow of compressible fluid, they are the same characteristics as those given by Eq. (11.11).

Along these characteristics, the normal derivatives of u and v are indeterminate and may be discontinuous.

In a manner similar to the case discussed in § 2(3), the second condition for characteristics may be written as

$$\begin{vmatrix} B_1 & C_1 & D_1 & E_1 \\ B_2 & C_2 & D_2 & E_2 \\ \dot{y} & 0 & 0 & -\dot{u} \\ 0 & \dot{x} & \dot{y} & -\dot{v} \end{vmatrix} = 0 \tag{11.24}$$

By the help of Eq. (11.23), we have for the case $E_1 = E_2 = 0$

$$\left.\begin{aligned} \left(\frac{dv}{du}\right)_1 &= \frac{(B_1C_2-C_1B_2)C'-(B_1D_2-D_1B_2)}{(C_1D_2-D_1C_2)C'} = \Gamma' \\ \left(\frac{dv}{du}\right)_2 &= \frac{(B_1C_2-C_1B_2)C''-(B_1D_2-D_1B_2)}{(C_1D_2-D_1C_2)C''} = \Gamma'' \end{aligned}\right\} \tag{11.25}$$

Eqs. (11.25) give two distinct families of curves in the hodograph plane corresponding to the two familes of characteristics C' and C'' in the physical plane. For a reducible equation, the hodograph characteristics Γ' and Γ'' are determined in advance by Eqs. (11.18) and independent of the particular initial conditions considered. In general, the characteristics depend on both the differential equations and the initial conditions.

Consider again the two-dimensional irrotational steady flow [Eq. (11.22)]. We have for the expressions $A_1, \cdots, E_2$

$$\begin{aligned} &A_1=u^2-a^2, \quad B_1=uv, \quad C_1=uv, \quad D_1=v^2-a^2, \quad E_1=0 \\ &A_2=0, \quad B_2=-1, \quad C_2=1, \quad D_2=0, \quad E_2=0 \end{aligned} \tag{11.26}$$

The physical characteristics are then

$$\begin{aligned} C' &= \tan(\theta-\alpha) \\ C'' &= \tan(\theta+\alpha) \end{aligned} \tag{11.27}$$

which are identical to those obtained in § 2, i.e., Eq. (11.11). The corresponding hodograph characteristics are

$$\begin{aligned} \left(\frac{du}{dv}\right)_1 &= -C'' \\ \left(\frac{du}{dv}\right)_2 &= -C' \end{aligned} \tag{11.28}$$

We may integrate Eqs. (11.28) to get a universal function $f(u, v) = 0$ for the hodograph characteristics. This function is useful in the graphical method of finding the flow field of a supersonic irrotational steady two-dimensional flow. It was first used by Busemann and will be discussed in § 7.

4. Characteristic equations

From this point we shall consider only the cases with two real families of characteristics, i.e., hyperbolic type equations. From the last two sections, we see that the characteristic equations obtained from these two points of view are exactly the same. Hence for a given flow problem, we need to consider either a second-order partial differential equation or a system of two first-order differential equations. We shall have the following two families of characteristics :

$$\left.\begin{aligned} \left(\frac{dy}{dx}\right)_1 &= C'(x, y, \psi, \psi_x, \psi_y) \\ \left(\frac{dy}{dx}\right)_2 &= C''(x, y, \psi, \psi_x, \psi_y) \end{aligned}\right\} \tag{11.29}$$

The characteristic equations which hold along these characteristic curves are Eqs. (11.8), (11.16), and (11.17). For the first family of characteristics, with ξ as parameter, we have the following three relations along the characteristic C':

$$y_\xi - C' x_\xi = 0 \tag{11.30a}$$

$$AC'\psi_{x\xi} + C\psi_{y\xi} + Dy_\xi = 0 \tag{11.30b}$$

$$\psi_\xi - \psi_x x_\xi - \psi_y v_\xi = 0 \tag{11.30c}$$

For the second family of characteristics, with η as a parameter, we have another three relations along the characteristics: C:

$$y_\eta - C'' x_\eta = 0 \tag{11.31a}$$

$$AC''\psi_{x\eta} + C\psi_{y\eta} + Dy_\eta = 0 \tag{11.31b}$$

$$\psi_\eta - \psi_x x_\eta - \psi_y y_\eta = 0 \tag{11.31c}$$

If we choose ξ and η as the independent variables, we have the five unknowns $x(\xi, \eta)$, $y(\xi, \eta)$, $\psi(\xi, \eta)$, $\psi_x(\xi, \eta)$, and $\psi_y(\xi, \eta)$ and the six equations (11.30) and (11.31). However one of these six equations is automatically satisfied if the other five are satisfied. This statement may be proved as follows:

Multiplying Eq. (11.30b) by y_η, Eq. (11.31b) by y_ξ, and subtracting the resultant equations, we have

$$A(C'\psi_{x\xi}y_\eta - C''\psi_{x\eta}y_\xi) + C(\psi_{y\xi}y_\eta - \psi_{y\eta}y_\xi) = 0 \tag{11.32}$$

Dividing Eq. (11.32) by $C'C'' = C/A$ and using the relations (11.30a) and (11.31a), we obtain

$$\psi_{x\xi}x_\eta + \psi_{y\xi}y_\eta = \psi_{x\eta}x_\xi + \psi_{y\eta}y_\xi \qquad (11.33)$$

Differentiating Eq. (11.30c) with respect to η gives

$$\psi_{\xi\eta} - \psi_{x\eta}x_\xi - \psi_{y\eta}y_\xi - \psi_x x_{\xi\eta} - \psi_y y_{\xi\eta} = 0 \qquad (11.34)$$

while using Eq. (11.33), Eq. (11.34) becomes

$$\frac{\partial}{\partial \xi}(\psi_\eta - \psi_x x_\eta - \psi_y y_\eta) = 0 \qquad (11.35)$$

It follows that

$$\psi_\eta - \psi_x x_\eta - \psi_y y_\eta = h(\eta) \qquad (11.36)$$

where $h(\eta)$ is an arbitrary function of η only.

Consider a boundary curve which is intersected by the family of curves η = constant. If Eq. (11.31c) is satisfied along this boundary, the function $h(\eta)$ vanishes. Eq. (11.31c) is then also satisfied in the interior. The initial or boundary conditions must be prescribed so that Eq. (11.31c) is satisfied along the boundary. In practice the boundary is usually a streamline which may be assigned the value $\psi = 0$, so that Eq. (11.31c) is automatically satisfied.

If the value of the function ψ and its first derivatives are given along a certain curve in the region considered, we may approximately calculate the values of the function of ψ at points in the neighborhood of the given curve by finite difference methods. This is the method of characteristics. Before discussing the practical procedure, we would like to point out some fundamental concepts and properties associated with the characteristics.

5. Some fundamental properties of characteristics[2]

The first important concept of characteristics is the domain of dependence. Consider an ordinary curve K given in parametric form by $x = x_0(\xi)$, $y = y_0(\xi)$. An ordinary curve is one which intersects any single characteristic not more than once (Fig. 11.2). We assume that ψ, ψ_x, and ψ_y are given along the curve K, and that they are continuous functions. These are the initial values of the problem. Consider a segment ab on this curve. Through point a there are two characteristics C' and C''. Through point b there are also two characteristics. As shown in Fig. 11.2, the C' characteristics through a will intersect the C'' characteristic through b. The intersection point is P. The values of the functions in the curved triangle P_{ab} are uniquely determined by the initial values along the segment ab and are unaffected by the values outside ab. The segment ab of K is known as the domain of dependence of the point P. The proof of this fundamental theorem is given in Ref. 5 and will not be reproduced here.

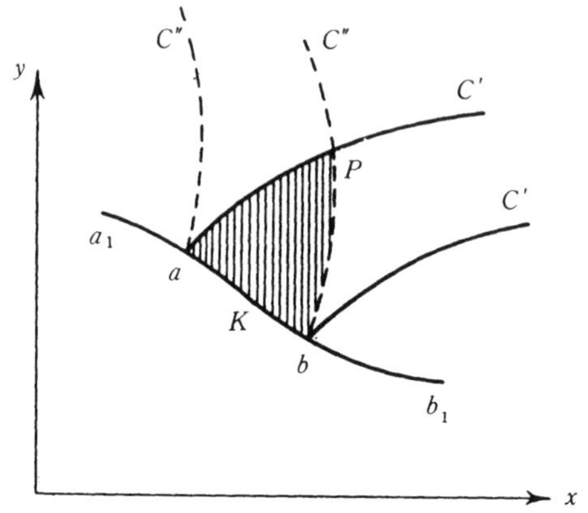

Fig. 11.2 Domain of dependence

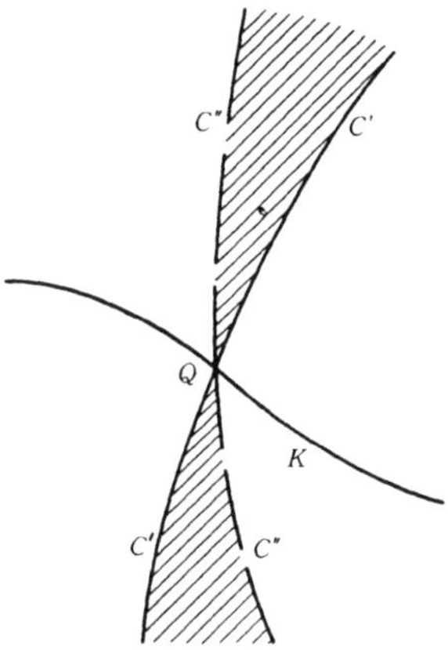

Fig. 11.3 Range of influence

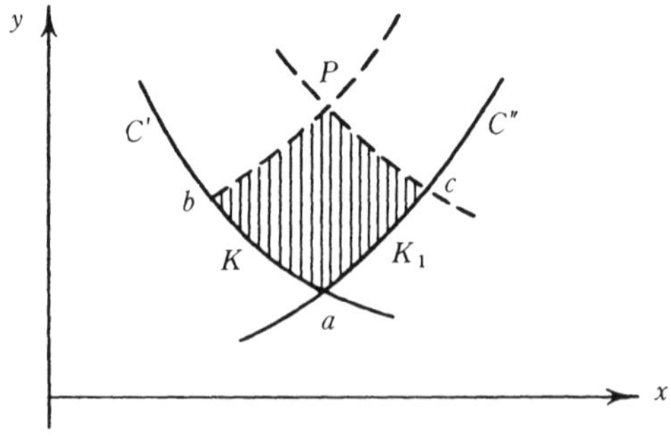

Fig. 11.4 Characteristic initial value problem

The second important concept is the range of influence of a point Q on the initial curve K. This is defined as the totality of points in the x-y plane which are influenced by the initial values at Q. It is evident that the range of influence is made up of all points P whose domains of dependence include Q. Hence the two curved triangular regions included between the two characteristics through Q are the range of influence of Q (Fig. 11.3).

If there are discontinuities in the second derivatives of ψ along an initial or boundary lines, K, they are propagated along the characteristics through the points on K where the discontinuities originate. Thus in two-dimensional steady supersonic flow, the characteristics are the Mach lines along which the discontinuities in the derivatives of velocity components propagate.

If the initial values are given along one characteristic on ly, the problem is indeterminate. However, if we are given the initial values along two characteristics K of C'' and K_1 of C', as in Fig. 11.4, the problem is uniquely determined. For instance, if ab of K and ac of K_1 are given, the value of ψ in the quadrilateral region $Pabc$ is uniquely determined.

6. Cauchy initial value problems

For the Cauchy initial value problem for the hyperbolic equation, we specify the initial values along a segment of a curve K in the x-y plane and determine the function from these initial values and the hyperbolic differential equation. If the curve K is an ordinary curve, not a characteristic (Fig. 11.2), the function ψ may be determined uniquely through a quadrilateral region enclosed by the intersecting characteristics through the endpoints of the segments. We shall discuss the detailed numerical and graphical methods for obtaining approximate solutions to such problems in §8 to §10. If the curve K is a characteristic, we must provide other data in addition to the initial values on K. Usually we may specify data on a characteristic in the second family of characteristics, such as K_1(Fig. 11.4). Since there are definite relations between the function ψ and its derivatives along the characteristics according to the characteristic equations (11.30), the initial values along the intersecting characteristics must be compatible with the characteristic relations corresponding to the differential equation. For instance, in the case of Fig. 11.4, if we give the arbitrary initial values along K, between a and b we need give only the initial value at a point c on K_1. The unknown function may then be determined uniquely in the quadrilateral region $abPc$. This is known as the characteristic initial value problem.

For a reducible equation the hodograph characteristic relation is integrable and independent of the initial value. For instance, in the case of the two-dimensional irrotational supersonic flow, it is Prandtl-Meyer expansion. Along a characteristic, we will have a relation $f(u, v) =$ constant in which an arbitrary constant may be assigned to each characteristic. The initial value at any point on the characteristic K_1 discussed in the last paragraph serves to determine the arbitrary constant. The Prandtl-Meyer flow is sometimes known as a simple wave. One of the important relations is that the flow in a region adjacent to a region of constant state (constant velocity, density, etc.) is a simple wave.

In actual computation, we usually use finite differences to replace the differentials in the characteristic equations. There are two different methods of representation: the lattice method and the field method. In both methods we replace the infinitely dense mesh of the two families of characteristics by a network of the two familes of characteristics with a finite difference in the characteristics constants (Fig. 11.5).

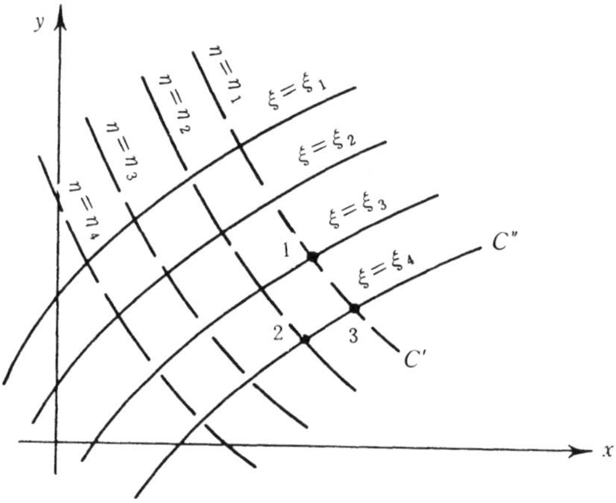

Fig. 11.5 Method of characteristics. Lattice method

In the lattice method, we consider only the values on the points of intersection of the characteristics, i.e., the lattice points. We shall find the approximate location of these lattice points and the approximate values of the values of ψ, ψ_x, and ψ_y at these points. For instance, if we know the initial values at points 1 and 2, we may calculate the values at a new point 3 of Fig. 11.5 according to the characteristic equations (11.30) and (11.31). From Eqs. (11.30a) and (11.31a) we have

$$\left.\begin{aligned} y_3-y_1&=C_1'(x_3-x_1) \\ y_3-y_2&=C_2''(x_3-x_2) \end{aligned}\right\} \tag{11.37}$$

where subscript 1 refers to the values at point 1, subscript 2 to the values at point 2, and subscript 3 to values at point 3. Eqs. (11.37) give the approximate location of point 3, i.e., x_3 and y_3. We may improve the result by using the average value of C'' at the points 1 and 3 instead of C_1'. Similar improvement may be used for C_2''.

The approximate values of ψ_{x3} and ψ_{y3} at point 3 may be obtained approximately from Eqs. (11.30b) and (11.31b) as follows:

$$\left.\begin{aligned} A_1C_1'(\psi_{x3}-\psi_{x1})+C_1(\psi_{y3}-\psi_{y1})+D_1(y_3-y_1)=0 \\ A_2C_2''(\psi_{x3}-\psi_{x2})+C_2(\psi_{y3}-\psi_{y_2})+D_2(y_3-y_2)=0 \end{aligned}\right\} \quad (11.38)$$

Finally the value of ψ may be obtained either from (11.30c) or (11.31c), i.e.,

$$\left.\begin{aligned} \psi_3-\psi_1-\psi_{x1}(x_3-x_1)-\psi_{y1}(y_3-y_1)=0 \\ \psi_3-\psi_2-\psi_{x2}(x_3-x_2)-\psi_{y2}(y_3-y_2)=0 \end{aligned}\right\} \quad (11.39)$$

In the field method, the continuous distribution of flow properties is replaced by a discontinuous patchwork in which the properties are assumed constant over each cell with discontinuous change across the boundary. For instance, in Fig. 11.6, we know the initial values at points 1,2, and 3 and extend the results to points 4 and 5. According to the field method, we assume that the flow properties in the cell I are constant, as well as those values in cell II. From the values in cells I and II, we may find the values of the flow properties in cell III.

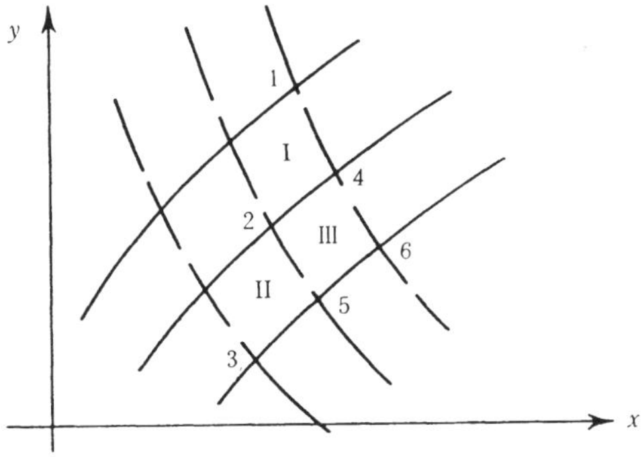

Fig. 11.6 Method of characteristics. Field method

Formulas similar to those of Eqs. (11.37) to (11.39) may be obtained from the values of the flow properties for cell III in terms of those in cells I and II from the characteristic equations (11.30) and (11.31). We shall discuss the special case of irrotational steady two-dimensional flow in § 9.

Both of the above methods have been extensively used in practice. The field method seems somewhat less clear-cut than the lattice method from the theoretical point of view but it has the conceptual advantage of emphasizing discontinuities along the disturbance wave, i.e., the Mach wave. However, practically speaking, both methods give essentially the same results.

To gain some definite ideas about the practical application of the method of characteristics, we shall discuss in the remainder of this chapter two-dimensional steady irrotational supersonic flow in detail. In § 7, we derive the characteristic equation for this special case. In § 8, the lattice method is discussed while in § 9, the field method. The application of the method of characteristics to the design of supersonic nozzle will be discussed in § 10. In § 11, we discuss the flow in a supersonic gas jet in which a shock wave will be formed. In this case, we shall show how to put the shock front in the place where the ordinary method of characteristics fails.

7. Two-dimensional steady irrotational supersonic flow

For two-dimensional steady irrotational supersonic flow, the functions A, B, C, D of Eq. (11.1) are

$$A = a^2 - u^2, \quad B = -2uv, \quad C = a^2 - v^2, \quad D = 0 \tag{11.40}$$

We may consider the function ψ as the velocity potential φ of the irrotational flow and then we have

$$\varphi_x = u, \quad \varphi_y = v \tag{11.41}$$

Furthermore, it is convenient to introduce the magnitude of velocity q and the flow-direction angle, $\theta = \tan^{-1} v/u$, instead of u and v, i.e.,

$$u = q \cos\theta, \quad v = q \sin\theta \tag{11.42}$$

Now the first characteristic equation (11.8) gives

$$C' = \left(\frac{dy}{dx}\right)_1 = \tan(\theta - \alpha), \quad C'' = \left(\frac{dy}{dx}\right)_2 = \tan(\theta + \alpha)$$
$$\alpha = \sin^{-1}(a/q) = \sin^{-1}(1/M) \tag{11.43}$$

Hence Eqs. (11.30a) and (11.31a) yield

$$dy - C'dx = 0$$
$$dy - C''dx = 0 \tag{11.44}$$

Eqs. (11.44) give the two families of Mach lines in the flow field.

The second characteristic equation (11.16) or Eqs. (11.30b) and (11.31 b), after the substitution of Eqs. (11.40)to (11.43), gives, for the first family C':

$$dq + \frac{q}{\sqrt{M^2 - 1}}\, d\theta = 0 \tag{11.45}$$

for the second family C'':

$$dq - \frac{q}{\sqrt{M^2 - 1}}\, d\theta = 0 \tag{11.46}$$

It is evident that the relations of (11.45) and (11.46) are the Prandtl-Meyer flow (9.1). For the reduced equation, this hodograph characteristic relation is integrable and independent of the initial values of the given problem. This is the reason why the two-dimensional steady flow problem is much simpler than the corresponding axially symmetrical flow problem in which the second characteristic relation depends on the initial value.

There are several integrated forms for this second characteristic equation. The following one is particularly convenient to use in the method of characteristics. It was originally given by Meyer[6].

Let us define a tangential component of velocity along the characteristic curve in the physical plane x-y as follows:

$$t = q_t = q \cos \alpha \tag{11.47}$$

Eqs. (11.45) and (11.46) become, respectively,

$$d\theta \pm d\alpha = \mp \frac{dt}{a} \tag{11.48}$$

The local velocity of sound may be written as

$$a^2 = \lambda^2 (q_m^2 - t^2) \tag{11.49}$$

where q_m is the maximum possible velocity in every given problem, and $\lambda^2 = \dfrac{\gamma - 1}{\gamma + 1}$.

Substituting Eq. (11.49) into (11.48) and integrating results in

$$\theta \pm \alpha = \mp \frac{\sin^{-1} \dfrac{t}{q_m}}{\lambda} + K_{\pm} \tag{11.50}$$

in which $K_{\pm}$ are the constants of integration which may be different for the two families of characteristics C' and C''. The constants $K_{\pm}$ may be expressed in terms of the flow inclination $\theta_{s\pm}$ at sonic speed where $q = a$, i.e.,

$$q = a, \quad t = 0, \quad \alpha = \frac{\pi}{2}, \quad \theta = \theta_{s\pm} = C_{\pm} \text{ (say)}$$

The second characteristic relation may be then written as

$$\theta \pm f(\alpha) = C_{\pm} \tag{11.51}$$

where

$$f(\alpha) = \alpha + \frac{1}{\lambda} \sin^{-1} \frac{t}{q_m} - \frac{\pi}{2}$$

$$= \alpha - \frac{\pi}{2} + \sqrt{\frac{\gamma - 1}{\gamma + 1}} \tan^{-1}\left(\sqrt{\frac{\gamma - 1}{\gamma + 1}} \cot \alpha \right) \tag{11.52}$$

and $C_{\pm}$ = constant along each C' and C'' characteristic, respectively.

The function $f(\alpha)$ is a function of Mach number M only, Hence it may be calculated once and for all for any given value of M. This greatly simplifies the actual computation.

8. Lattice method. Temple's procedure

Temple[7] used the method of lattices to carry out the computation of the method of characteristics. He simplified the computation by introducing the following notations which were first used by Busemann: [8]

Direction number $= D =$ number of degrees in θ radians

Pressure number $= P = 1000 -$ number of degrees in $f(\alpha)$ radians

$$C_+ = 1000 - 2B, \quad C_- = -1000 + 2A$$

Temple choose successive steps in $C_{\pm}$ that are made equal and integral by using a table of the pressure number P in integers as shown in Tab. 11.1. The constants for the two families of characteristics become

$$\left.\begin{aligned} P+D&=2A \text{ (for the family } C''\text{)} \\ P-D&=2B \text{ (for the family } C'\text{)} \end{aligned}\right\} \tag{11.53}$$

Table 11.1 Pressure Number and Speed Index in Supersonic Flow

P	N	M	α deg.	p/p_0
1000	0	1.000	90.00	0.528
999	2	1.081	67.62	0.479
998	4	1.133	61.98	0.450
997	6	1.176	58.22	0.425
996	8	1.219	55.15	0.402
995	10	1.258	52.65	0.382
994	12	1.294	50.62	0.364
993	14	1.330	48.75	0.346
992	16	1.365	47.10	0.330
991	18	1.400	45.58	0.314
990	20	1.435	44.17	0.299
989	22	1.469	42.90	0.285
988	24	1.503	41.70	0.271
987	26	1.537	40.60	0.258
986	28	1.571	39.55	0.246
985	30	1.605	38.55	0.234
984	32	1.638	37.62	0.221
983	34	1.672	36.73	0.211
982	36	1.706	35.88	0.201
981	38	1.741	35.07	0.190
980	40	1.775	34.27	0.181
979	42	1.810	33.53	0.171
978	44	1.845	32.82	0.162
977	46	1.880	32.13	0.154
976	48	1.915	31.48	0.146

(Continued)

P	N	M	α deg.	p/p_0
975	50	1.950	30.83	0.138
974	52	1.986	30.23	0.131
973	54	2.024	29.62	0.123
972	56	2.060	29.03	0.116
971	58	2.097	28.48	0.110
970	60	2.132	27.97	0.103
969	62	2.173	27.40	0.097
968	64	2.211	26.90	0.092
967	66	2.249	26.40	0.086
966	68	2.288	25.92	0.081
965	70	2.329	25.42	0.076
964	72	2.369	24.97	0.072
963	74	2.411	24.50	0.067
962	76	2.453	24.07	0.063
961	78	2.495	23.62	0.059
960	80	2.537	23.20	0.055
959	82	2.581	22.80	0.051
958	84	2.626	22.38	0.048
957	86	2.671	21.98	0.045
956	88	2.718	21.60	0.042
955	90	2.765	21.22	0.039
954	92	2.812	20.83	0.036
953	94	2.860	20.47	0.033
952	96	2.911	20.08	0.031
951	98	2.961	19.73	0.029
950	100	3.012	19.38	0.027
949	102	3.065	19.03	0.025
948	104	3.119	18.70	0.023
947	106	3.174	18.37	0.021
946	108	3.226	18.05	0.019
945	110	3.287	17.72	0.018
944	112	3.346	17.40	0.016
943	114	3.407	17.07	0.015
942	116	3.468	16.75	0.014
941	118	3.530	16.45	0.012
940	120	3.595	16.15	0.011
939	122	3.661	15.85	0.010
938	124	3.728	15.55	0.0099
937	126	3.798	15.27	0.0086
936	128	3.870	14.97	0.0079
935	130	3.943	14.70	0.0071
934	132	4.018	14.42	0.0064
933	134	4.095	14.13	0.0058
932	136	4.151	13.93	0.0054
931	138	4.256	13.58	0.0047
930	140	4.340	13.32	0.0042

If we choose both P and D as integers, A and B will also be integers. If the A's and B's differ by equal amounts, the computation will be greatly simplified. At any lattice point, we may use the values of A and B to specify the flow conditions, i.e.,

$$D=A-B, \quad P=A+B \tag{11.54}$$

If the initial values are given along a regular curve (Fig. 11.7), from the flow direction we have the direction number D at the lattice points, say 1 to 4; and from the Mach number M we have the pressure number P at these lattice points. The constants along the two families of characteristics through any lattice point are then obtained from Eq. (11.53). The values of P and D at lattice points outside the initial curve, such as points $1'$, $2'$, etc., may be obtained from the constant equation (11.53) along the given characteristics. For

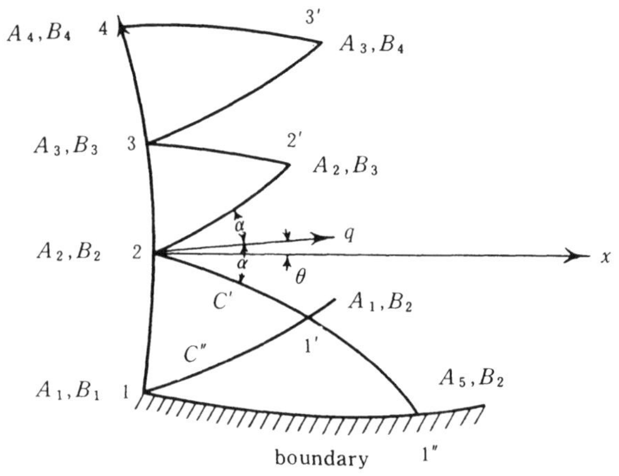

Fig. 11.7 Lattice method of Temple

instance, at the point $1'$, we have

$$P_{1'}+D_{1'}=P_1+D_1=2A_1, \quad P_{1'}-D_{1'}=P_2-D_2=2B_2 \tag{11.55}$$

where the subscripts refer to the values at that point. If we choose the mesh size such that the difference between A and B at neighboring points is integral, the actual computation labor will be greatly reduced. After the values of A and B at the new point are obtained, the location of this new point, such as point $1'$, may be determined from Eq. (11.44) in which the slopes of the characteristics C' and C'' are taken as the average values of those at points $1'$ and $1'$ for C'' and that of 2 and $1'$ for C'. In this manner, the accuracy of the computation is improved over that suggested in § 6.

Near the boundary of the flow field, the condition is always similar to that of the characteristic initial value problem. We must take the boundary condition into account. If the boundary is a solid wall, the direction number is given. If the boundary is a free surface, the pressure is constant on the boundary and the pressure number is known. We may then determine the unknown number P or D from a single characteristic. For instance, if the curve $1-1''$ is a boundary, we may determine the values of A and B from the constants along the $1'-1''$, i.e.,

$$P_{1''}-D_{1''}=P_{1'}-D_{1'} \tag{11.56}$$

The whole flow field may be determined by constructing the lattice points to cover the entire field, as shown in Fig. 11.7.

9. Field method. Puckett's procedure

In the field method, we denote each cell by a pair of numbers (a, b). We also choose an origin cell $(0,0)$. Now if we restrict ourselves to the case of flows which continually expand, both the Mach number and the function $f(\alpha)$ increase continuously downstream.

Puckett[9] writes

$$a=2\delta k', \qquad b=2\delta k'' \tag{11.57}$$

where $k'=$ number of characteristics C' crossed from $(0,0)$ cell to (a, b) cell

$k''=$ number of characteristics C'' crossed from $(0,0)$ to (a, b)

$\delta=$ constant jump in degrees in crossing a characteristic curve

If we take the origin cell as the point $\theta = 0$ and $f(0) = 0$, the flow inclination at the cell (a, b) is

$$\theta=\frac{a-b}{2} \tag{11.58}$$

The Mach number in cell (a, b) may be represented by the speed index N such that

$$N=2f(\alpha)=a+b \tag{11.59}$$

The variation of N with Mach number of the flow is also given in Tab. 11.1. If we know the initial values at any cell (a, b), the values of the index numbers at any other cell downstream may be obtained by counting the numbers of C' and C'' characteristics crossed (Fig. 11.8). In this manner, the index numbers in all of the cells in the flow field may be obtained. By Eqs. (11.58) and (11.59), the flow properties over the whole field may also be computed.

10. Design of a supersonic nozzle

In this section we shall use the two methods discussed in the last two sections to find the flow field in a supersonic nozzle. From our discussion of the flow in the nozzle of Chapter III, §4, we know that if the pressure at the exit is below the critical value, sonic velocity occurs at the throat of the nozzle, and the flow in the subsonic part of the nozzle will not be affected by the change of exit pressure. In the design of a supersonic nozzle, we shall consider only the shape of the nozzle downstream from the throat of the nozzle (Fig. 11.9). What we shall do is to find the proper shape of the nozzle so that a uniform stream of the required Mach number occurs at the exit. First the

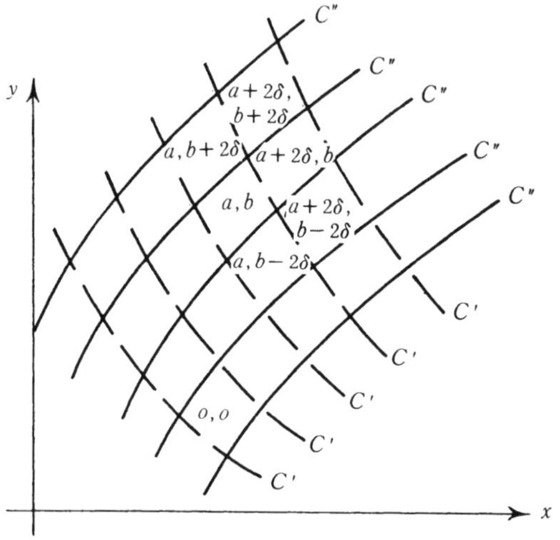

Fig. 11.8 Field method of Puckett

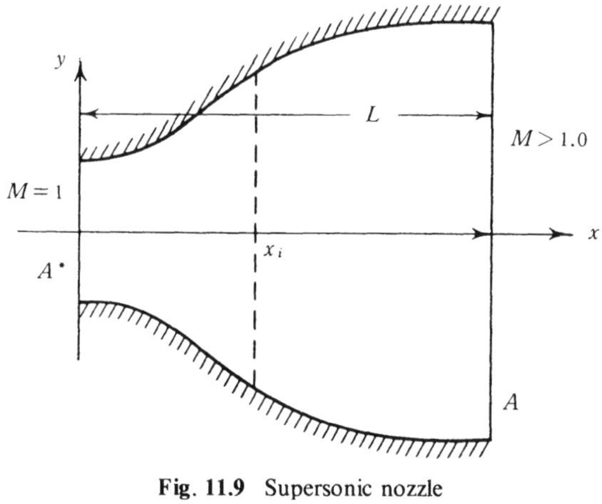

Fig. 11.9 Supersonic nozzle

ratio of the area of the exit A to that of the throat A^* may be determined from the one-dimensional theory of Chapter III, § 4, i. e, if M is the Mach number at the exit, the area ratio is

$$\frac{A}{A^*} = \frac{1}{M}\left[\frac{2}{\gamma+1}\left(1+\frac{\gamma-1}{2}M^2\right)\right]^{\frac{\gamma+1}{2(\gamma-1)}} \qquad (11.60)$$

The second step is to determine the optimum length L of the nozzle. The shorter the nozzle, the less will be the frictional effect, however the longer the nozzle, the easier is its construction. An optimum length may be chosen according to the actual working conditions. For the same exit area, the longer the nozzle, the smaller will be the inclination of the wall. The general shape of the nozzle as shown in Fig. 11.8 always has a point of inflection at $x=x_j$. Any reasonably smooth curve may be used as the shape of the wall from the throat to the point of inflection. For instance, a parabolic curve may be used. We may then use any one of the above methods (§ 8 or § 9) to find the flow in the section from the throat to the point of inflection. The shape of the nozzle beyond $x=x_i$ is determined so that at the exit the flow is parallel to the axis of the nozzle, i.e., $\theta=0$.

The inclination of the wall of the nozzle determines the length of the nozzle. There is a maximum inclination of the wall at $x=x_i$ for a given Mach number M at the exit. If the inclination of the flow of Prandtl-Meyer expan-

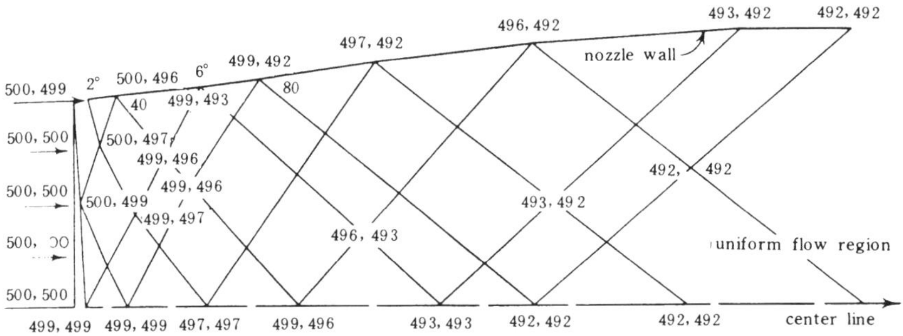

Fig. 11.10 Lattice method for supersonic nozzle design

sion at Mach number M is θ_{PM} with $\theta=0$ at sonic point, the inclination of the wall at $x=x_i$ should be

$$\theta_{x_i} \leqslant \frac{1}{2}\theta_{PM} = \frac{N_{PM}}{4} \tag{11.61}$$

In actual computations, broken straight lines segments are used to replace the actual curve of the wall in carrying out the method of characteristics. The line segments are chosen so that the flow deflection between adjoining segments differs by an integral number, e.g., $\delta=2°$. Now let us design a nozzle with the Mach number at the exit to be $M=1.638$. From Table 11.1, we see that $N=32$, which means that $\theta_{PM}=16°$. Hence the maximum deflection at $x=x_i$ is 8°.

We may design the nozzle according to the two methods discussed above:

(a) *Lattice method*. The diagram of the flow pattern is shown in Fig. 11.10. At the throat $M=1$, $\theta=0$ except at the wall where $\theta=2°$. We therefore start with $A=500$, $B=500$ at all points except the point on the wall

$A = 501$, where $B = 499$. On the axis of the nozzle θ is always equal to zero. By symmetry we may consider only the flow on the upper half of the nozzle, as shown in Fig. 11.10.

(b) *Field method.* The diagram of the flow pattern is shown in Fig. 11.11. Now we start with $a = 0$, $b = 0$, because both N and θ are zero at the throat.

The results obtained by these two methods agree quite well.

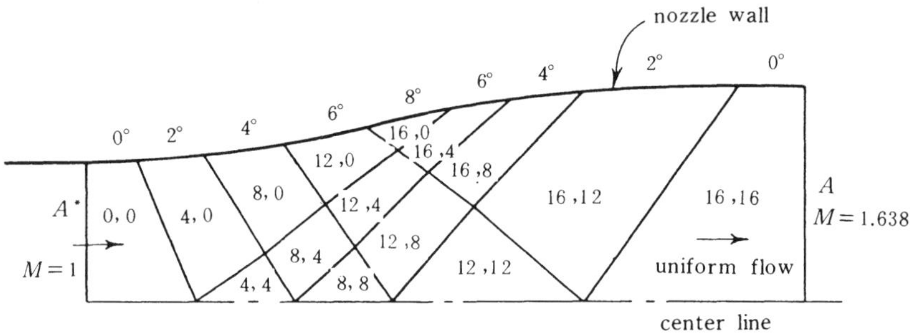

Fig. 11.11 Field method for supersonic nozzle design

11. Supersonic gas jet[10, 11]

In the flow field of a supersonic nozzle, the flow continually expands and the Mach number increases continually. In general, the flow may expand in one part of the flow field and compress in another. If the compression is large enough, a shock wave may occur. In this section we shall give an example in which both expansion and compression occur and a shock wave may develop. The problem is that of a supersonic gas jet issuing from a supersonic nozzle designed by a process similar to that discussed in § 10.

We consider a supersonic uniform steady flow issuing into a medium at rest from a properly designed nozzle. At the exit of the nozzle (Fig. 11.12), there is a uniform supersonic flow of Mach number $M > 1$. We shall assume that the pressure in the medium at rest is smaller than the static pressure at the exit of the nozzle. We may calculate the flow field in the jet by the method of characteristics discussed in § 7 and § 8 A typical case of the supersonic gas jet is shown in Fig. 11.12. At the exit aa' the flow is uniform of Mach number M_1. The lines ab and $a'b$ are the Mach lines corresponding to the Mach number M_1. The region abc is the expansion fan of Prandtl-Meyer flow where the pressure decreases from p to p_b. In the region acd the pressure is constant of the value p_b. When the Mach line cd intersects the boundary of the jet, it reflects into another Mach line of compression in a manner similar to that discussed in Chapter IV, § 8, where the pressure is constant, i. e., the pressure number on the boundary is constant. If the pressure ratio p_1/p_b is small, no curved shock will occur in the jet and the jet has periodic structure. If the pressure ratio p_1/p_b is large, a curved shock may occur in the jet. In Fig. 11.12, $p_1/p_b = 2$, we see that the consecutive characteristics from the region *de* intersect each other near $f'g'$. The intersection of characteristics of the same family may be considered as the sign of breakdown of the continuous motion and the beginning of a shock wave. We may replace the envelope of these characteristics by a shock wave. The Rankine-Hugoniot relations hold across this shock. Behind the curved shocks *fg* and $f'g'$, the flow will be rotational. It is necessary to use the method of characteristics for rotational flow for computing the flow downstream of the curved shocks. The method of characteristics for rotational flow will be discussed in Chapter XIV. When the pressure ratio p_1/p_b

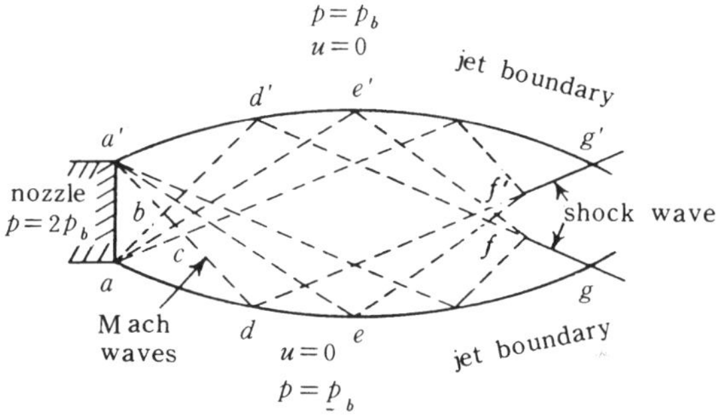

Fig. 11.12 Supersonic jet in a medium at rest. (Fig. 11.7 of Ref. 10 by D.C. Pack, courtesy of *Quarterly Journal of Mech. and Applied Math.*)

increases, the curved shocks *fg* and $f'g'$ increase in length toward the axis of the jet. At very large p_1/p_b, this shock may develop into Mach shock.

12. Shock fitting by method of characteristics

In this section we discuss numerical solution of the supersonic and hypersonic flows with detached shock waves. Consider steady supersonic flow past a blunt body as shown in Fig. 11.13. A detached shock wave is formed. In front of the detached shock, the flow is the given uniform supersonic free stream. Behind the detached shock, the central region is subsonic and beyond this region the flow becomes supersonic. As the free stream Mach number is not close to one, the flow behind the detached shock is no longer irrotational. However we may still assume the flow is inviscid and non-heat-conducting. Thus the flow can be modeled by the Euler equations. As seen from Chapter V, the unsteady Euler equations are hyperbolic in both subsonic and supersonic regions and thus are used to solve the steady flow problems. Here we sill use a shock fitting method[12] to treat the detached shock rather than the shock capturing method. If there is no other shock appearing in the flow, we may choose the entropy equation instead of the energy equation and write the unsteady Euler equation in nonconservation form.

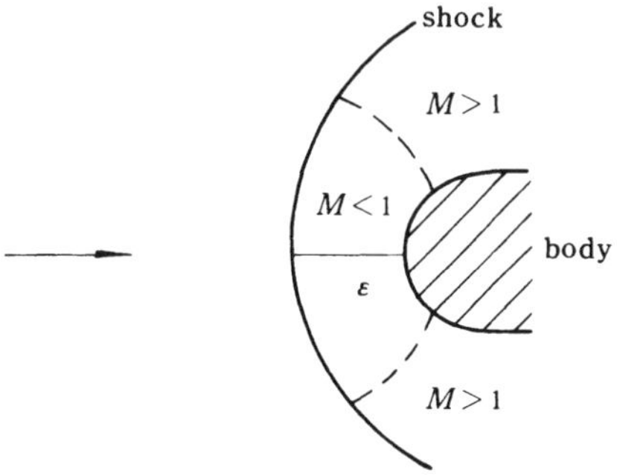

Fig. 11.13 Blunt body with detached shock

For simplicity, we consider the two-dimensional flow. The governing differential equations for the shock layer, i.e. the region between the detached shock and the body, are the following four equations

$$R_t + uR_x + vR_y + u_x + v_y = 0 \tag{11.62}$$

$$u_t + uu_x + vu_y + \frac{p}{\rho} P_x = 0 \tag{11.63}$$

$$v_t + uv_x + vv_y + \frac{p}{\rho} P_y = 0 \tag{11.64}$$

$$S_t + uS_x + vS_y = 0 \tag{11.65}$$

where

$$P = \ln(p/p_\infty), \quad R = \ln(\rho/\rho_\infty) \tag{11.66}$$

and

$$P - \gamma R = S \tag{11.67}$$

The finite-difference representations for the unsteady Euler equations will be presented in Chapter XIV. In this section, we confine our discussions to the treatment of the detached shock in the numerical solution of the blunt body problems. One can easily see the advantages in treating the detached shock by the shock-fitting methods. First of all, no point needs to be computed in front of the detached shock. Second, the mesh in the shock layer may be kept rather coarse since the flow parameters are smooth functions of the space variables. In addition, the shock is assumed as a discontinuity in the shock-fitting methods, which seems to be more realistic than one several mesh sizes thick obtained by the shock capturing methods.

We now study the shock points (Fig. 11.13). To understand the philosophy of the approach, let us consider the problem of a one-dimensional unsteady flow with a shock wave penetrating into a uniform medium, region 1 (Fig. 11.14). Suppose that the flow is known at all the values of x at $t=t_0$, and that the location and the velocity of the shock wave are to be evaluated at $t=t_1=t_0+\Delta t$, together with the conditions behind the shock. Starting at

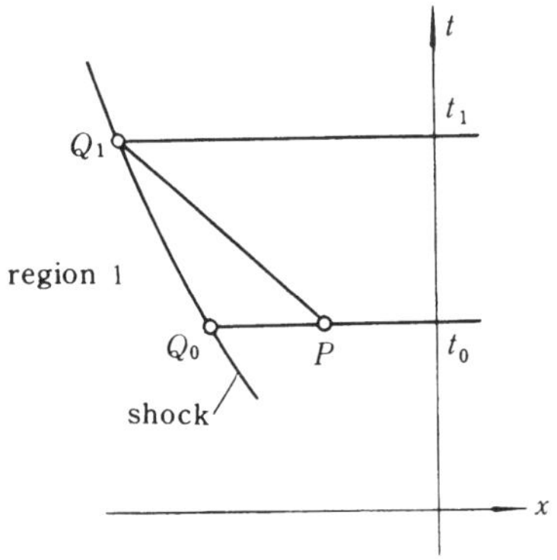

Fig. 11.14 Shock point in one-dimensional flow

Q_0 with a given slope of the shock, the approximate location of Q_1 may be found. A characteristic, defined by $dx/dt=u-a$, then is drawn backward from Q_1 and the compatibility equation is integrated along it, starting at P.

The matching of the values obtained from the Rankine-Hugoniot equations for a moving shock and of the values obtained from the compatibility equation provides the velocity of the shock at Q_1, together with the rest of the unknowns.

The same technique may be used in the present problem. Consider Fig. 11.15, where point Q_1 is the point on the shock wave at which the

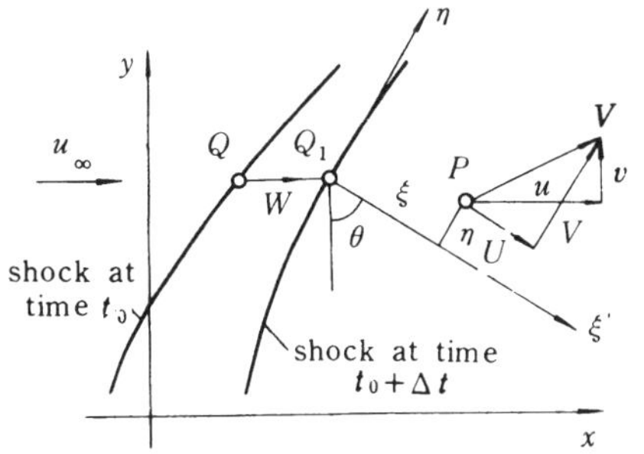

Fig. 11.15 Shock point in two-dimensional unsteady flow

values are to be calculated at time $t_0 + \Delta t$ and Q is the point on the shock wave at time t_0, having the same y. A fixed Cartesian or thogonal frame $Q_1\xi\eta$ is defined whose η axis is tangent to the shock wave at Q_1. Let U and V be the ξ and η components of the velocity vector at any point P. The governing equations in the (ξ, η, t) system may be written in the following form.

$$\left.\begin{aligned} R_t + UR_\xi + U_\xi &= -VR_\eta - V_\eta \\ U_t + UU_\xi + \frac{p}{\rho} S_\xi + \frac{\gamma p}{\rho} R_\xi &= -VU_\eta \\ S_t + US_\xi &= -VS_\eta \\ V_t + UV_\xi &= -VV_\eta - \frac{p}{\rho} P_\eta \end{aligned}\right\} \qquad (11.68)$$

It is well known that the η component of the velocity is the same on both sides of the shock. If a rigid motion defined by such a velocity, but in the negative direction of the η axis, is superimposed to the flow field, the values of V then become very small. The righthand sides of Eqs. (11.68) would actually vanish if the shock wave were rectilinear, though oblique. It thus is evident that, in the neighborhood of a shock point, the relevant parameters are ξ and U. Consequently, we consider the first three equations as quasi-one-dimensional equations, modified by forcing terms in the right-hand sides. From these three equations three characteristics are found in the (ξ, t) plane. They are defined by

$$\frac{d\xi}{dt} = U - a, \quad \frac{d\xi}{dt} = U + a, \quad \frac{d\xi}{dt} = U \qquad (11.69)$$

and have immediate interpretation in terms of a quasi-one-dimensional flow. The compatibility equation along the first of these characteristics reads

$$\frac{a}{\gamma}\frac{dp}{dt} - \frac{dU}{dt} = -\left(\frac{aV}{\gamma} P_\eta + aV_\eta - VU_\eta\right) \qquad (11.70)$$

The disturbance signals from the region between the body and the shock are propagated along the first characteristic of Eq. (11.69) and governed by Eq. (11.70) to influence the shock. Therefore, Eq. (11.70) is one of the governing equations for the shock point Q_1 at time t_1.

The other governing equations for the shock point are the Rankine-Hugoniot equations. Here the shock is moving with the velocity w (Fig. 11.15) in the x direction. Relative to the moving shock, the normal components of the flow velocities in front of and behind the shock are

$$(u_\infty - W)\sin\theta \qquad \text{and} \qquad U - W\sin\theta \qquad (11.71)$$

respectively, where θ is the angle between the shock and the x axis. The tangential components of the relative velocities in front of and behind the shock are the same, i.e.

$$(u_\infty - W)\cos\theta = V - W\cos\theta \qquad (11.72)$$

Substituting Eqs. (11.71) and (11.72) into the Rankine-Hugoniot equation (4.32) for stationary shock, we obtain

$$(u_\infty - W)(U - W\sin\theta)\sin\theta = a^{*2} - \frac{\gamma-1}{\gamma+1}(u_\infty - W)^2\cos^2\theta \qquad (11.73)$$

where a^* is given by

$$\frac{a_\infty^2}{\gamma-1} + \frac{(u_\infty - W)^2}{2} = \frac{\gamma+1}{2(\gamma-1)}a^{*2} \qquad (11.74)$$

Eliminate a^* from (11.73) and (11.74). The resulting equation is

$$U = \frac{(\gamma-1)(u_\infty - W)^2\sin^2\theta + 2a_\infty^2}{(\gamma+1)(u_\infty - W)\sin\theta} + W\sin\theta \qquad (11.75)$$

By use of Eqs. (4.29) we obtain

$$\frac{p}{p_\infty} = \frac{2\gamma}{\gamma+1}\left(\frac{u_\infty - W}{a_\infty}\right)^2\sin^2\theta - \frac{\gamma-1}{\gamma+1} \qquad (11.76)$$

and

$$\frac{\rho}{\rho_\infty} = [(\gamma+1)\frac{p}{p_\infty} + \gamma - 1]/[(\gamma-1)\frac{p}{p_\infty} + \gamma + 1] \qquad (11.77)$$

Eq. (11.72) can be written as

$$V = u_\infty\cos\theta \qquad (11.78)$$

Eqs. (11.75) to (11.78) are the Rankine-Hugoniot equations for the shock point Q_1 at time t_1.

The five equations (11.70) and (11.75) to (11.78) determine the solution at a shock point, namely U, V, p, ρ, W, whereas θ is determined by W. The system is solved numerically by iteration. In this process, the location of point Q_1 and the corresponding value of θ are computed initially from the shape of the shock and the values of W at the previous step, and then are kept constant. From an assumed W, the values of U, V, p, and ρ are computed at the shock from Eqs. (11.75) to (11.78). A characteristic with the slope $d\xi/dt = U - a$ is drawn from point Q_1 in the (ξ, t) plane (Fig. 11.16), and its intersection A with the ξ axis is found. Since point A lies in the (x, y) plane at time t_0, and the solution is known completely at this time, local values of U and a are found at A by interpolating from the values at the mesh points. A new charac-

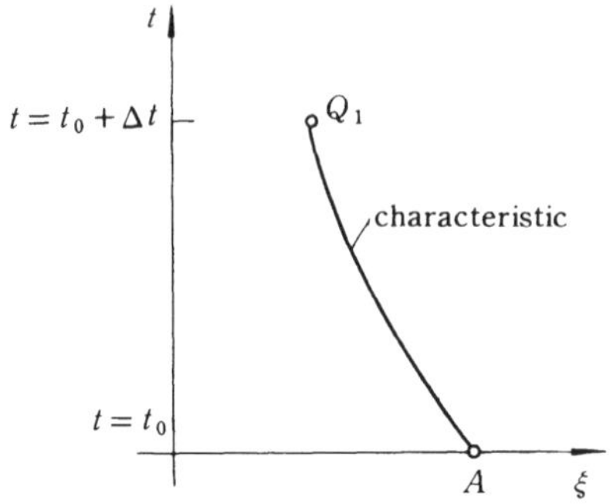

Fig. 11.16 One-dimensional characteristic

teristic then is issued from Q_1, with the slope

$$\frac{d\xi}{dt} = [(u-a)]_{\text{shock}} + (U-a)_A]/2 \tag{11.79}$$

and the process is iterated until the location of point A is stabilized. At this stage, the right-hand side of Eq. (11.70) is computed at A and at Q_1, averaged, and considered constant between A and Q_1. Eq. (11.70) can be integrated, assuming that P at the shock is the one obtained from Eq. (11.76), and a new value of U at the shock is computed. If it does not agree with the value obtained from (11.75), W is changed and the process repeated from the start. A trial-and-error technique then is used to achieve convergence on W.

The method of characteristics can also be used in determining the solution on the body surface with the boundary condition that the velocity over the surface must be tangential to the surface.

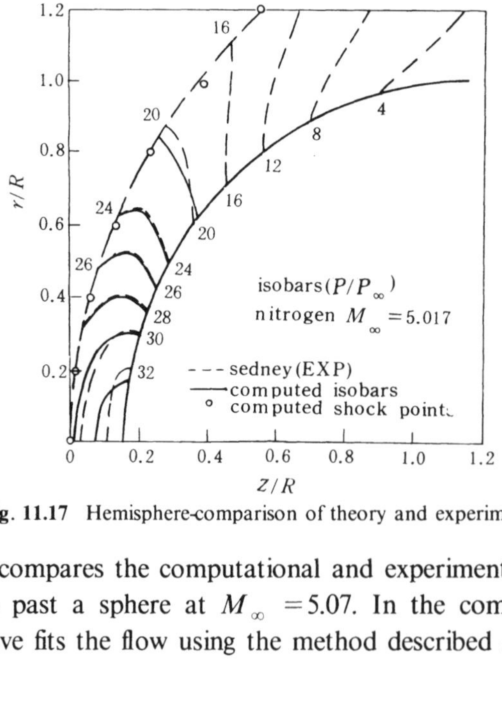

Fig. 11.17 Hemisphere-comparison of theory and experiment

Fig. 11.17[12] compares the computational and experimental results for the flow of nitrogen past a sphere at M_∞ =5.07. In the computation, the detached shock wave fits the flow using the method described above. The shock wave and the isobars obtained from experiments are drawn in dotted lines. The computed points on the shock are the dots, and the computed isobars are the solid lines. Some of the computed isobars fit the dotted lines so well that the latter are not visible. This comparison is very satisfactory, particularly in considering that the computed results have been obtained with a rather coarse mesh, which has as many lines in the y direction as dots on the shock of Fig. 11.17.

If the shock waves are embedded in the nonuniform flow region, the shock fitting methods would be very complicated to apply and one would rather use the shock capturing methods to compute the shock waves (cf. Chapters X and XIV).

13. Problems

1. Derive the system of differential equations for one-dimensional nonsteady isentropic flow. Find the characteristic equations of this system of equations by the method of § 3.

2. Define a stream function ψ for one-dimensional nonsteady isentropic flow so that the equation of continuity is automatically satisfied. Find the differential equation for this stream function. Find the characteristic equations of this differential equation by the method of § 2.

3. Derive the system of differential equations for nonsteady isentropic flow with spherical symmetry. Find the characteristic equations of this system of equations using the method of § 3.

4. Define a stream function ψ for nonsteady isentropic flow with axially symmetry so that the equation of continuity is automatically satisfied. Find the differential equation for this stream function ψ and the corresponding characteristic equations.

5. Find the characteristic equations for steady axially symmetrical supersonic flow. What are the differences between this case and the two-dimensional case?

6. Find the flow field in a wedge-shaped diffusor of vertex angle 20° with Mach number $M = 2.0$ at the entrance section. Calculate the flow field in this diffusor by the

(a) Lattice method of Temple.

(b) Field method of Puckett.

The flow is assumed to be isentropic and steady.

7. For reducible equations, the characteristics are functions of the variables u and v only. By a simple wave, we mean the integral of the characteristics in the u-v plane. For two-dimensional supersonic steady irrotational flow, the simple wave is the Prandtl-Meyer expansion wave. Find the expression of the simple wave in one-dimensional nonsteady isentropic flow.

8. Show that for the case of a reducible equation, the flow in a region adjacent to a region of constant state is a simple wave.

9. Calculate the flow field of a supersonic gas jet in a medium at rest with $M_1=2.0$ and $p_1/p_b=10$ by the method of characteristics and indicate the region at which shock wave forms.

10. Calculate the flow field of a supersonic gas jet in a supersonic uniform stream of Mach number 1.5 by the method of characteristics. The Mach number of the jet flow at the exit of the nozzle is 2.0, and the pressure ratio of p_1/p_2 is 5.

11. Find the characteristic equation of a two-dimensional steady supersonic flow when the radiation pressure and radiation energy are not negligible.

12. Verify Eq. (11.70).

13. How do you determine the solution at a body point for the blunt body problem (§ 12) by method of characteristics?

References

1. Massau, J., Mem. sur l'integration graphique des equations aux derivees partielles, Gand 1900 – 03; cf. also *Enzykl. d. Math Wiss.* **2**, 31, p. 159.
2. Meyer, R. E., The method of characteristics for problems of compressible flow involving two independent variables, Part I, General Theory, *Quart. J. Mech. Appl. Math.* **1**, pt 2, June 1948.
3. Tollmien, W., Steady two-dimensional and axially symmetrical supersonic flows, *Brown Univ. Translation* No. A9-T-1, 1948.
4. Courant, R., and Friedrichs, K. O., *Supersonic Flow and Shock Waves*, Interscience Publishers, Inc., New York, 1948.
5. Courant, R., and Hilbert, D., *Methoden der Mathematischen Physik II*, chapter 5, Verlag von Julius Springer, Berlin 1937.
6. Meyer, Th., Ref. 1 of chapter IX.
7. Temple, G., The method of characteristics in supersonic flow, *British ARC R and M* No. 2091, 1944.
8. Prandtl, L., and Busemann, A., *Naeherungsvernahren zur zeichnerischen Ermittlung von ebenen Stroemungen mit Ueberschallgeschwindigkeit* (Festschrift Professor Dr. A. Stodola zum 70 Geburtstag, Zuerich, 1929, pp. 499 – 509).
9. Puckett, A. E., Supersonic nozzle design, *Jour, App. Mech.*, Dec, 1946, p. 1 — 265.
10. Pack, D. C., On the formation of shock waves in supersonic gas jets, *Quart. J. Mech. Appl. Math.* **1**, pt 1, 1948, pp. 1 – 17.
11. Pai, S. I., *Fluid Dynamics of Jets*, D. Van Nostrand Company, Inc., Princeton, New Jersey, 1954.
12. Moretti, G. and Abbet, M., A time-dependent computational method for blunt body flows, *AIAA Journal*, **4**, No. 12, December 1966. pp. 2136 – 2141.

Chapter XII

L INEARIZED THEORY OF THREE-DIMENSIONAL POTENTIAL FLOW

1. Introduction

In many engineering problems, we are interested in the flow around a thin body in a uniform stream. In these cases, we may linearize our fundamental equations. We have discussed the linearized theory for two-dimensional cases in more detail in Chapter VI. In this chapter we shall discuss three-dimensional cases of linearized theory. We shall restrict ourselves to irrotational flows whose fundamental relation is Eq. (6.26). In § 2, we shall discuss some basic properties of this equation and also the boundary conditions for this equation which often occur in aeronautical engineering problems.

In § 3, we shall consider three-dimensional steady subsonic flow in which the wellknown Prandtl-Glauert rule is applicable. In the application of the Prandtl-Glauert rule to the three-dimensional case there is an important difference from that of the two-dimensional case discussed in Chapter VI, § 5.

One of the powerful methods of solution of the fundamental equation of the linearized theory of compressible flow is the method of source and doublet distributions. In § 4, we shall discuss the source and doublet distributions in compressible flow. Their properties are quite different in supersonic and subsonic flows. In § 5, the general expressions for forces on solid bodies are derived. In § 6, supersonic flow over a body of revolution is discussed. In § 7 to § 10, supersonic wing theory based on the source integration method will be discussed. In § 9, we also present supersonic flow over a slender body lacking symmetry.

Another important concept of supersonic flow is that of conical flow, first introduced by Busemann[1] in 1935 and greatly developed in the past twenty years. We shall briefly discuss this important type of flow in § 11.

In § 12 and § 13 the improvements of the solution are discussed.

In § 14, nonsteady three-dimensional flow will be briefly discussed.

In § 15, we discuss an approximate analysis which is known as the law of equivalence and is related to the area rule for a wing of small aspect ratio.

Finally in § 16 and § 17, the numerical solutions of linearized equations of a compressible flow will be discussed by using the panel method.

2. General discussions of the linearized theory of compressible flow

In considering the flow around a thin body in a uniform stream of velocity U, we may assume that there exists a velocity potential φ such that

$$u=U\left(1+\frac{\partial\varphi}{\partial x}\right),\quad v=U\frac{\partial\varphi}{\partial y},\quad w=U\frac{\partial\varphi}{\partial z} \tag{12.1}$$

where u, v, and w are the x-, y- and z-components of velocity, respectively, φ is the perturbation velocity potential, U is the velocity of the uniform free stream, and $\left|\frac{\partial\varphi}{\partial x}\right|, \left|\frac{\partial\varphi}{\partial y}\right|, \left|\frac{\partial\varphi}{\partial z}\right| \ll 1$.

The differential equation for φ is Eq. (6.26)

$$\frac{\partial^2\varphi}{\partial x^2}+\frac{\partial^2\varphi}{\partial y^2}+\frac{\partial^2\varphi}{\partial z^2}=\frac{1}{a_\infty^2}\left(\frac{\partial^2\varphi}{\partial t^2}+2U\frac{\partial^2\varphi}{\partial x\partial t}+U^2\frac{\partial^2\varphi}{\partial x^2}\right) \tag{12.2}$$

where a_∞ is the sound velocity of the basic flow corresponding to the velocity U.

The deviation of the local pressure p from the uniform stream pressure p_∞ is

$$\Delta p=p-p_\infty=-\rho_\infty U\left(\frac{\partial\varphi}{\partial t}+U\frac{\partial\varphi}{\partial x}\right) \tag{12.3}$$

where ρ_∞ is the density of the uniform stream U.

Eq. (12.2) is derived for a coordinate system attached to a body moving with velocity U in the fluid. If we refer to a coordinate system at rest with the fluid, the corresponding equation for the perturbation velocity potential may be obtained from Eq. (12.2) by the transformation

$$x'=x-Ut,\ y'=y,\ z'=z,\ t'=t,\ \varphi(x',\ y',\ z',\ t')=\varphi(x,y,z,t) \tag{12.4}$$

where the prime refers to the new coordinate system at rest with the fluid.

Substituting Eq. (12.4) into (12.2) gives

$$\frac{\partial^2\varphi}{\partial x'^2}+\frac{\partial^2\varphi}{\partial y'^2}+\frac{\partial^2\varphi}{\partial z'^2}=\frac{1}{a_\infty^2}\frac{\partial^2\varphi}{\partial t'^2} \tag{12.5}$$

which is the well known wave equation where the velocity of propagation is a_∞. It is interesting to note that, for small disturbances, the perturbation velocity potential always satisfies the simple wave Eq. (12.5).

Since Eq. (12.5) is linear, the method of superposition is applicable. It is therefore important to know the fundamental solutions of Eq. (12.5) for particular boundary and initial values (see § 4).

In the linearized theory, we usually consider the perturbation produced by a thin or slender body as an obstacle in the flow. It is convenient to distinguish between a thin and slender body. A thin body is one whose thickness is small in one direction only, and a slender body is one whose dimensions are small in two directions; the limit of a thin body is a surface, whereas the limit of a slender body is a line. A typical thin body is a supersonic wing, a typical slender body is a pointed body of revolution, such as a projectile.

The fundamental Eq. (12.2) or (12.5) is solved with the following initial and boundary conditions:

The first boundary condition is that of zero normal velocity on the body. For a thin body this may also be completely linearized and applied at the mean surface of the body. For a slender body however, this boundary condition may not be completely linearized in the same way because of the nature of the singularity at the axis when the solution is continued into the body.

Other boundary conditions follow from the physical picture which for the subsonic flow is different from that for supersonic flow. For subsonic flow, the disturbances tend to be zero at large distances from the body; whereas on the other hand, in supersonic flow, there is no disturbance upstream of the obstacle. We shall discuss these boundary conditions in more detail in the following sections.

The initial conditions are usually specified so that they are consistent with the boundary conditions at time $t=0$.

3. Prandtl-Glauert rule for subsonic flow[2, 3]

Now we shall consider steady three-dimensional subsonic flow. We introduce the new variables.

$$\eta=\beta y, \quad \xi=\beta z, \quad \varphi_i=k\varphi \tag{12.6}$$

where $\beta=\sqrt{1-M^2}$, $M=U/a_\infty$.

Substituting Eqs. (12.6) into Eq. (12.2), we have, for the steady case $\frac{\partial(\ )}{\partial t}=0$,

$$\frac{\partial^2\varphi_i}{\partial x^2}+\frac{\partial^2\varphi_i}{\partial \eta^2}+\frac{\partial^2\varphi_i}{\partial \xi^2}=0 \tag{12.7}$$

Eq. (12.7) shows that our fundamental equation for φ may be transformed into the equation of φ_i for the corresponding incompressible flow. Thus the Prandtl-Glauert rule of Chapter VI, § 5, may also be used in the three-dimensional case. However, in the two-dimensional case, we may choose any value of k and still obtain the same formula for the Prandtl-Glauert rule (6.50). This is not so in three dimensions.

Consider the case of a thin body. The surface of this body may be written as

$$\bar{y}=f\ (x,\ z)\ \text{or}\ \ \bar{z}=g\ (x,\ y) \tag{12.8}$$

The boundary conditions for Eq. (12.7) are

$$\frac{\bar{v}_i}{U}=\frac{\partial\bar{\varphi}_i}{\partial\eta}=\frac{k}{\beta}\frac{\partial\bar{\varphi}}{\partial y}=\frac{k}{\beta}\frac{\bar{v}}{U}=\frac{k}{\beta}\frac{\partial f}{\partial x},\ \frac{\bar{w}_i}{U}=\frac{\partial\varphi_i}{\partial\xi}=\frac{k}{\beta}\frac{\partial g}{\partial x} \tag{12.9}$$

Now if we consider $\bar{v}_i$ and w_i to be the fictitious y- and z-velocity components on the body surface in the fictitious incompressible flow with the coordinates of the surface $\bar{y}_i$ and $\bar{z}_i$ we have

$$\begin{aligned}\bar{y}_i&=\bar{\eta}=\beta\bar{y}=\beta f\ (x,\ z)\\ \bar{z}_i&=\bar{\xi}=\beta\bar{z}=\beta g\ (x,\ y)\end{aligned} \tag{12.10}$$

and

$$\frac{\bar{v}_i}{U}=\frac{\partial\bar{y}_i}{\partial x},\ \frac{\bar{w}_i}{U}=\frac{\partial\bar{z}_i}{\partial x} \tag{12.11}$$

We may also find the values of $\bar{y}_i$ and $\bar{z}_i$ by integration of Eq. (12.11) with the help of the boundary condition (12.9). We have

$$\begin{aligned}y_i&=\int\frac{v_i}{U}\,dx=\frac{k}{\beta}\int\frac{\partial f}{\partial x}\,dx=\frac{k}{\beta}f\ (x,\ z)+A(z)\\ z_i&=\int\frac{w_i}{U}\,dx=\frac{k}{\beta}\int\frac{\partial g}{\partial x}\,dx=\frac{k}{\beta}g\ (x,\ y)+B\ (y)\end{aligned} \tag{12.12}$$

Comparing Eqs. (12.10) and (12.12) shows that

$$A(z)=B(y)=0,\quad k=\beta^2 \tag{12.13}$$

Hence in three-dimensional flow, we should take $k=\beta^2$ in the Prandtl-Glauert rule. The relations between the quantities in the compressible flow and those of the corresponding incompressible flow are as follows:

$$\begin{aligned}u&=U\frac{\partial\varphi}{\partial x}=\frac{U}{\beta^2}\frac{\partial\varphi_i}{\partial x}=\frac{u_i}{\beta^2}\\ C_p&=-2\frac{u}{U}=\frac{1}{\beta^2}\left(-2\frac{u_i}{U}\right)=\frac{C_{Pi}}{\beta^2}\end{aligned} \tag{12.14}$$

$$v=U\frac{\partial\varphi}{\partial y}=\frac{U}{\beta}\frac{\partial\varphi_i}{\partial y_i}=\frac{v_i}{\beta},\quad w=U\frac{\partial\varphi}{\partial z}=\frac{U}{\beta}\frac{\partial\varphi_i}{\partial z_i}=\frac{w_i}{\beta}$$

The application of the Prandtl-Glauert rule for steady three-dimensional subsonic flow should follow the procedure:

(a) The incompressible disturbance velocity potential φ_i is found for a slenderized body obtained by contracting the lateral dimensions of the given body by the factor β.

(b) The perturbation velocity components and the pressure coefficient for the compressible flow are given by Eq. (12.14) in terms of the corresponding incompressible solution obtained in (a).

It should be noted that, if the pressure coefficient is proportional to the thickness ratio of the body, such as in the two-dimensional case, we have the simple correction formula (6.50), i.e.,

$$C_P=\frac{C_{Pi}}{\sqrt{1-M^2}}$$

In Eq. (6.50), the thickness ratio of the body in compressible fluid flow is the same as that in incompressible fluid flow. In incompressible three-dimensional flow, there is no simple relation between the pressure coefficient and the thickness ratio for a family of bodies of affinely related shape. In general, we have no simple correction formula for compressibility effect in three-dimensional flows because the correction depends both on the free stream Mach number and the shape of the body.

4. Source and doublet in compressible flow

Since the fundamental equation of the linearized theory is linear, we may apply the method of superposition. It is interesting to study some of the fundamental solutions of the differential Eq. (12.2) or (12.5). One fundamental solution is the point source. The flow field associated with the solution of a point source is different for the incompressible, subsonic, and supersonic undisturbed flow. We shall now discuss these three cases. For simplicity we shall consider a point source situated at the origin of the coordinate system.

(a) *Incompressible flow*. $M=0$, $a_\infty=\infty$. In this case, the fundamental equation (12.2) becomes

$$\frac{\partial^2\varphi}{\partial x^2}+\frac{\partial^2\varphi}{\partial y^2}+\frac{\partial^2\varphi}{\partial z^2}=0 \qquad (12.15)$$

If we consider a point source situated at the origin, the perturbed velocity potential φ must be a function of radial distance only, i. e., $\varphi=\varphi(r, t)$ and independent of the angular coordinates. Eq. (12.15) then becomes

$$\frac{d^2(r\varphi)}{dr^2}=0 \tag{12.16}$$

where $r=\sqrt{x^2+y^2+z^2}$ whose solution is

$$\varphi=-\frac{f(t)}{4\pi r} \tag{12.17}$$

where $f(t)$ is the strength of the source. For steady flow, $f(t)$ is a constant. The flow is obviously purely radial as shown in Fig. 12.1. There is a singularity at the origin. If the source is at a point (x_0, y_0, z_0), the expression r in Eq. (12.17) should be expressed as

$$r=\sqrt{(x-x_0)^2+(y-y_0)^2+(z-z_0)^2} \tag{12.18}$$

(b) *Subsonic flow*. $0<M<1$, $0<\beta=\sqrt{1-M^2}<1$. The fundamental equation is Eq. (12.2). If we use the transformation

$$x'=x,\ y'=\beta y,\ z'=\beta z,\ t'=\beta^2 t+\frac{M}{a_\infty}x \tag{12.19}$$

Eq. (12.2) becomes

$$\frac{\partial^2\varphi}{\partial x'^2}+\frac{\partial^2\varphi}{\partial y'^2}+\frac{\partial^2\varphi}{\partial z'^2}=\frac{1}{a_\infty^2}\frac{\partial^2\varphi}{\partial t'^2} \tag{12.20}$$

For the case where $\varphi(r', t')$ is a function of r' and t' only, where $r'=\sqrt{x'^2+y'^2+z'^2}$ Eq. (12.20) becomes

$$\frac{\partial^2(r'\varphi)}{\partial r'^2}=\frac{1}{a_\infty^2}\frac{\partial^2(r'\varphi)}{\partial t'^2} \tag{12.21}$$

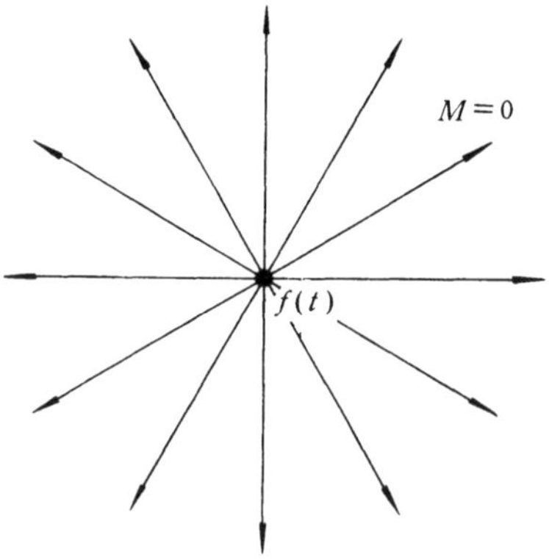

Fig. 12.1 Streamlines of a source flow in an incompressible fluid

One solution of Eq. (12.21) is

$$\varphi = -\frac{f(t'-\tau)}{4\pi r'} - \frac{g(t'+\tau)}{4\pi r'} \tag{12.22}$$

where $\tau = \frac{r'}{a_\infty}$, and f and g are arbitrary functions. Here f represents a divergent wave, and g represents a convergent wave. For the steady case, f and g are constant, and the solution becomes

$$\varphi = -\frac{k}{4\pi r'} = -\frac{k}{4\pi\sqrt{x^2+\beta^2(y^2+z^2)}} \tag{12.23}$$

In this case the equipotential surfaces are ellipsoids. The streamlines are shown in Fig. 12.2. The flow is essentially a distorted radial flow with a singularity still at the origin. Near the origin, the velocity becomes very large, and the linearized theory no longer holds.

(c) *Supersonic flow.* $M > 1$, $\beta = im$, $m^2 > 0$, $i = \sqrt{-1}$. In this case the differential equation and the fundamental solution are the same as those of subsonic flow except that β is now a pure imaginary number. Because of this fact, the flow pattern differs greatly from that of the subsonic source discussed in the last two sections. For instance for the steady case, we have

$$\varphi = -\frac{k}{4\pi r'} = -\frac{k}{4\pi\sqrt{x^2-m^2(y^2+z^2)}} \tag{12.24}$$

The velocity potential φ and its derivative are real and physically defined only for $x^2 \geqslant m^2(y^2+z^2)$, i.e., inside the Mach cone. Fig. 12.3 shows the streamlines for the supersonic source. Singularities lie on the Mach cone.

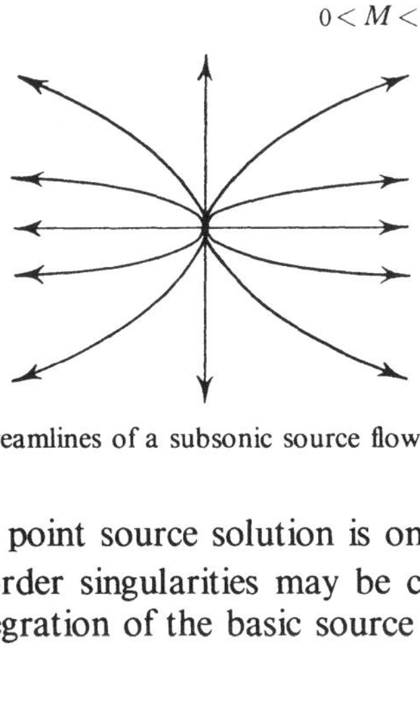

Fig. 12.2 Streamlines of a subsonic source flow in a compressible fluid

The fundamental point source solution is one of the simplest solutions of Eq. (12.2). Higher-order singularities may be constructed by superposition, differentiation, or integration of the basic source distribution discussed above.

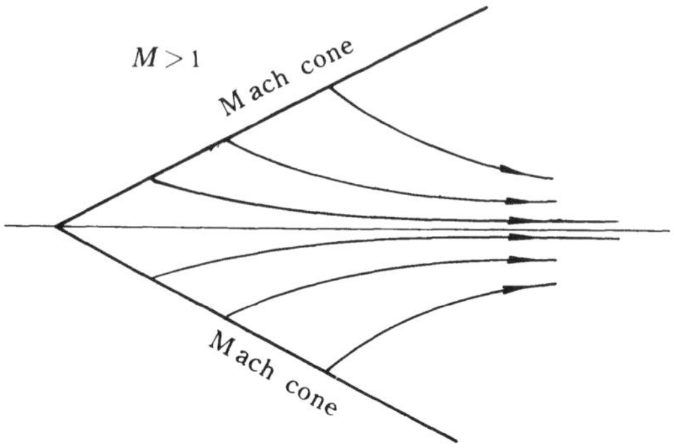

Fig. 12.3 Streamlines of a supersonic source flow in a compressible fluid

If φ_1 is a solution of the basic wave Eq. (12.20), the derivative or the integral of φ_1 with respect to any of the Cartesian coordinates is also a solution.

If φ_1 is a simple source, the derivative of φ_1 with respect to any of the coordinates, say x, gives a doublet.

If we distribute the source strength along a line, we have a line source. In a steady flow of a compressible fluid over a body of revolution, a line source along the axis of the body is usually used. This will be discussed in § 6.

We may distribute the source strength over a surface and thus obtain a surface source which may be used to study the flow over a wing. This procedure will be discussed in § 7.

5. General expressions for forces on solid bodies

In the linearized theory, it is possible to represent the flow around solid bodies by superposition of a rectilinear flow and the fundamental solutions of source and doublet flow. Before explaining such flows, we shall first discuss the forces which arise in this method. We take for a control surface, a circular cylinder of radius R whose axis is coincident with the x-axis (Fig. 12.4). The

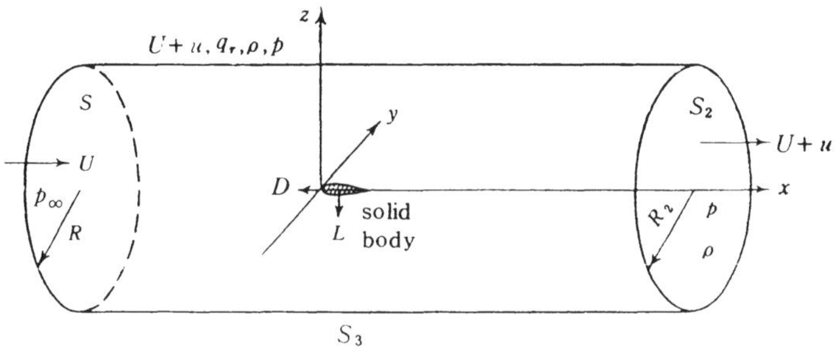

Fig. 12.4 Control surface over a solid body in a compressible fluid flow

upstream surface S_1 is taken at a position such that the flow is essentially undisturbed. For the case of supersonic flow, S_1 need only be taken to the left of the origin which is the leading edge of the body. S_2 is the downstream surface, and S_3 is the lateral surface.

Considering the conservation of the x-component of momentum we have

$$D=\int_{S_2}(p_\infty-p)\,dS_2+\int_{S_1}\rho_\infty U^2 dS_1-\int_{S_2}\rho(U+u)^2 dS_2$$
$$-\int_{S_3}\rho q_r(U+u)\,dS_3 \qquad (12.25)$$

where D is the drag force acting on the body. The drag acting on the fluid surrounding the body is shown in Fig. 12.4.

The equation of continuity may be written as

$$\int_{S_1}\rho_\infty U dS_1-\int_{S_2}\rho(U+u)\,dS_2-\int_{S_3}\rho q_r dS_3=0 \qquad (12.26)$$

Substituting Eq. (12.26) into (12.25) gives

$$D=\int_{S_2}(p_\infty-p)\,dS_2-\int_{S_2}\rho u(U+u)\,dS_2-\int_{S_3}\rho q_r u dS_3 \qquad (12.27)$$

where u is the perturbed x-velocity component, and q_r is the perturbed radial velocity component at the cylindrical surface S_3.

We now define the drag coefficient C_D and pressure coefficient C_P by the relations

$$C_D=\frac{D}{\frac{\rho_\infty}{2}U^2}, \quad C_P=\frac{p-p_\infty}{\frac{\rho_\infty}{2}U^2} \qquad (12.28)$$

Eq. (12.27) then becomes

$$C_D=-\int_{S_2}\left[C_P+2\frac{\rho}{\rho_\infty}\frac{u}{U}\left(1+\frac{u}{U}\right)\right]dS_2-2\int_{S_3}\frac{\rho}{\rho_\infty}\frac{q_r}{U}\frac{u}{U}\,dS_3 \quad (12.29)$$

which is exact. We insert the approximate expression for C_p and ρ/ρ_∞ for the perturbation theory, i.e.,

$$C_p = -2\frac{u}{U} - \frac{q_r^2 + q_\theta^2 - m^2u^2}{U^2} + O\left(\frac{q^3}{U^3}\right)$$

$$\frac{\rho}{\rho_\infty} = 1 - M^2\frac{u}{U} + O\left(\frac{q^2}{U^2}\right)$$

(q_θ=tangential perturbed velocity component) into Eq. (12.29), to yield after simplification

$$C_D = \int_{S_2} \frac{q_r^2 + q_\theta^2 + m^2u^2}{U^2}\, dS_2 - 2\int_{S_3} \frac{uq_r}{U^2}\, dS_3 + O\left(\frac{q^3}{U^3}\right) \qquad (12.30)$$

It should be noted that the drag coefficient C_D is essentially a second-order quantity in disturbed velocity components. To calculate C_D, we should include the second-order term in the expression of C_P. If we are interested in the value of C_P only, for a first approximation we need to retain the first term only in the expression of C_P, i.e., Eq. (12.57). From the source and doublet distributions we may calculate the perturbed velocity components q_r, q_θ, and u; from Eq. (12.30) the drag coefficient of the solid is obtained.

Similarly, from the consideration of the conservation of z-component of momentum we have for the expression of the lift coefficient

$$C_L = \frac{L}{\frac{\rho_\infty}{2}U^2} = 2\int_{S_3} U \sin\theta\, dS_3 - 2\int_{S_2} \frac{q_r}{U} \sin\theta\, dS_2 + O\left(\frac{q^2}{U^2}\right) \qquad (12.31)$$

It is interesting to note that the lift coefficient C_L is of the order of q/U, whereas the drag coefficient is of the order of $(q/U)^2$ where q is the perturbed velocity. We shall apply these formulas to some simple bodies.

6. Axially symmetric steady supersonic flow over a body of revolution[4-8]

We may find the field of a supersonic flow over a body of revolution at zero angle of attack by putting the source distribution along the axis of the body. We consider here only the steady case and take the source with strength $f(\xi)d\xi$ at the point $x=\xi$, $y=z=0$. By integrating over all sources along the x-axis, we obtain for the resultant perturbed velocity potential the expression

$$\varphi = \int_{-\infty}^{\xi} \frac{f(\xi)\,d\xi}{\sqrt{(x-\xi)^2 - m^2r^2}} = \int_{-\infty}^{x-mr} \frac{f(\xi)\,d\xi}{\sqrt{(x-\xi)^2 - m^2r^2}} \qquad (12.32)$$

If we calculate the velocity potential at a point in space (x, r) where r

is the radial distance from the x-axis, i. e., $r=\sqrt{y^2+z^2}$, the upper limit of the integration of equation (12.32) is $\xi_1=x-mr$. If ξ is larger than ξ_1, the contribution to the velocity potential φ from ξ_1 to ξ is imaginary; this means that the influence of those source distributions from ξ_1 to ξ does not reach the point (x, r). As we usually consider the disturbance due to a pointed slender body, $f(\xi)$ is zero for $\xi \leqslant 0$, and we may then replace the lower limit of Eq. (12.32) by 0.

It is convenient to calculate the velocity components from Eq. (12.32) with the help of the transformation

$$\xi=x-mr\cosh\sigma \tag{12.33}$$

$$d\xi=-mr\sinh\sigma d\sigma=-mr\sqrt{\cosh^2\sigma-1}\,d\sigma=\sqrt{(x-\xi)^2-m^2r^2}\,d\sigma$$

Substituting Eq. (12.33) into (12.32), we obtain

$$\varphi(x, r)=\int_0^\infty f(x-mr\cosh\sigma)\,d\sigma \tag{12.34}$$

From Eq. (12.34), we have

$$u=\frac{\partial\varphi}{\partial x}=\int_0^\infty f'(x-mr\cosh\sigma)\,d\sigma=\int_{-\infty}^{x-mr}\frac{f'(\xi)\,d\xi}{\sqrt{(x-\xi)^2-m^2r^2}}$$

$$q_r=\frac{\partial\varphi}{\partial r}=-\int_0^\infty m\cosh\sigma\cdot f'\,d\sigma=-\frac{1}{r}\int_{-\infty}^{x-mr}\frac{f'(\xi)(x-\xi)\,d\xi}{\sqrt{(x-\xi)^2-m^2r^2}} \tag{12.35}$$

where $f'(\xi)=\dfrac{df}{d\xi}$.

First we shall consider the simplest case:

$$f'(\xi)=A=\text{constant},\ f(\xi)=A\xi \tag{12.36}$$

for $\xi\geqslant 0$, and $f'=f=0$ for $\xi\leqslant 0$.
Substituting Eq. (12.36) into (12.35) yields

$$u=A\cosh^{-1}\frac{x}{mr},\quad q_r=-\frac{A}{r}\sqrt{x^2-m^2r^2} \tag{12.37}$$

Eq. (12.37) represents the flow around a cone because both u and q_r are constant along the family of cones $x/mr=$constant. At the surface of the cone of semi-vertex angle θ, we have

$$\theta=\frac{r}{x}=\frac{q_r}{U+u}\cong\frac{q_r}{U} \tag{12.38}$$

So that $A = -\dfrac{U\theta^2}{\sqrt{1-m^2\theta^2}}$

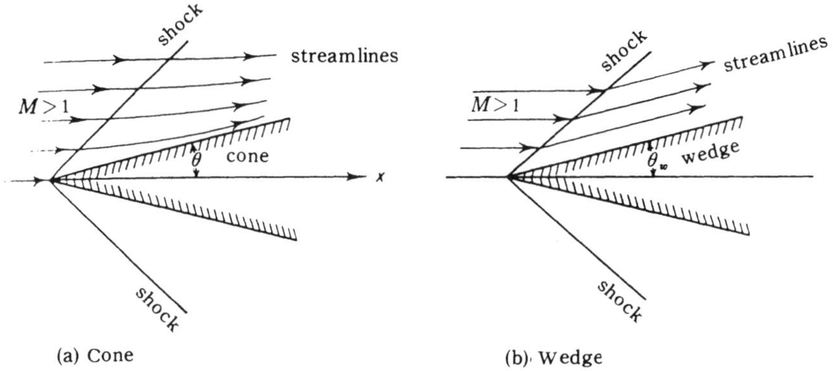

Fig. 12.5 Supersonic flow over a cone and a wedge

The perturbed velocity components are then

$$q_r = \frac{U\theta^2}{r}\,\frac{\sqrt{x^2-m^2r^2}}{\sqrt{1-m^2\theta^2}}, \quad u = -\frac{U\theta^2\cosh^{-1}\dfrac{x}{mr}}{\sqrt{1-m^2\theta^2}} \tag{12.39}$$

The flow over a cone is different from that over a two-dimensional wedge as seen by the contrasting patterns in Fig. 12.5. For the two-dmensional case of a wedge, supersonic flow behind the shock is parallel to the surface of the wedge. For the case of the cone, the flow behind the shock is not parallel.

The pressure on the surface of the cone is

$$\Delta p = p - p_\infty = -\rho U u = \frac{\rho U^2\theta^2}{\sqrt{1-m^2\theta^2}}\cosh^{-1}\left(\frac{1}{m\theta}\right) \tag{12.40}$$

When $m\theta$ is small, $\cosh^{-1}\dfrac{1}{m\theta} = \log\left(\dfrac{1}{m\theta} + \sqrt{\dfrac{1}{m\theta} - 1}\right) \cong \log\dfrac{2}{m\theta}$, then

$$\Delta p = \frac{\rho U^2}{2}\left(2\theta^2 \log\frac{2}{m\theta}\right) \tag{12.41}$$

This holds quite well if $m\theta$ is neither too large nor too small. "Too large" implies that the disturbances are too large for our approximation to be valid. "Too small" implies that the speed is too close to the speed of sound and that the transonic phenomenon enters to render our result invalid.

For a slender body of revolution of arbitrary shape, we may apply the method of superposition of the elementary solutions (12.36). For a body of revolution, shown in Fig. 12.6, we may divide the body into n parts, and put n different simple sources in these parts such that

$$f'(\xi) = A_i\,,\quad \xi_{i-1} < \xi < \xi_i \tag{12.42}$$

The perturbed velocity components at a point x, r on the surface of the body of revolution are

$$\begin{aligned} u(x_i, r_i) &= \sum_{j=1}^{i} A_j \left(\cosh^{-1} \frac{x_i - \xi_j}{mr_i} - \cosh^{-1} \frac{x_i - \xi_{j-1}}{mr_i} \right) \\ q_r(x_i, r_i) &= -\sum_{j=1}^{i} \frac{A_j}{r_i} \left[\sqrt{(x_i - \xi_j)^2 - m^2 r_i^2} - \sqrt{(x_i - \xi_{j-1})^2 - m^2 r_i^2} \right] \end{aligned} \tag{12.43}$$

The constants A_i may be determined from the boundary conditions on the body, i.e.,

$$\frac{d\bar{r}}{dx} = \frac{q_r}{U + u} = \frac{q_r}{U} \tag{12.44}$$

Since the velocity components at x_1, r_1 depend only on A_1, those at x_2, r_2 depend only on A_1 and A_2, etc., we may determine the A_i's consecutively. This method was first used by von Kármán and Moore to determine the drag of projectiles moving with supersonic velocities.[4]

We assume that the shape of the arbitrary body of revolution is given by

$$\bar{r}^2 = 2F(x) \tag{12.45}$$

The elementary solution of the cone is that $F''(x) = \dfrac{d^2F}{dx^2} = A =$ constant.

First let us consider the relation between $\bar{r}$ and F. We may assume that the body is composed of cones such that

$$F''(x) = A_i,\quad \xi_{i-1} \leqslant x \leqslant \xi_i \tag{12.46}$$

$\xi_0 = 0$; $0 \leqslant i \leqslant n$, $\xi_i < \xi_{i+1}$, $\xi_n = l$, l is the length of the body.

We have

$$F'(x) = F'(\xi_{i-1}) + A_i(x - \xi_{i-1})$$

$$F(x) = F(\xi_{i-1}) + F'(\xi_{i-1})(x - \xi_{i-1}) + \frac{A_i}{2}(x - \xi_{i-1})^2$$

The radius of the body $\bar{r}$ may be written in terms of F as

$$\bar{r}(x) = \sqrt{2F}$$

$$\frac{d\bar{r}}{dx} = \bar{r}'(x) = F'/\sqrt{2F} \tag{12.47}$$

$$\frac{d^2\bar{r}}{dx^2} = \bar{r}''(x) = \frac{1}{\sqrt{2F}}\left(F'' - \frac{F'^2}{2F}\right)$$

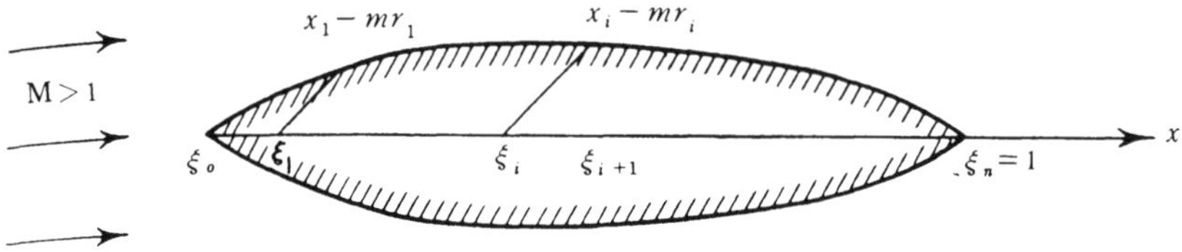

(a) A body of revolution in a supersonic flow

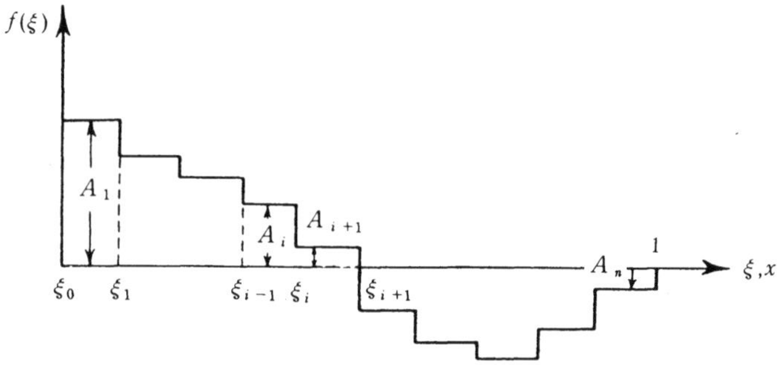

(b) Source distribution of a body of revolution

Fig. 12.6 Source distribution of a body of revolution in a supersonic flow

For a closed body, we have $r(0) = r(l) = 0$, and hence

$$F(0) = F(l) = 0 \tag{12.48}$$

For linearized theory to be valid, we require that $\bar{r}'(0) = \bar{r}'(l) = 0$, so that

$$F'(0) = F'(l) = 0 \tag{12.49}$$

Furthermore, $\bar{r}(x) > 0$, gives

$$F(x) > 0 \quad \text{for} \quad 0 < x < l \tag{12.50}$$

and

$$F''(0_+) > 0, \quad F''(l_-) > 0 \tag{12.51}$$

Conditions (12.48) to (12.51) are the necessary conditions for $F(x)$ to be used in our stepwise distributions for our analysis.

Now we consider an elementary solution corresponding to a single source element such that

$$F''(x) = 0 \text{ for } x < \xi_{i-1},\ x > \xi_i$$
$$F''(x) = A_i \text{ for } \xi_{i-1} < x < \xi_i \tag{12.52}$$

The values of F', F, and $\bar{r}$ corresponding to Eq. (12.52) are shown in Fig. 12.7. The perturbed velocity components at the point (x, r) corresponding to this source element in various regions are:

$$u_i = q_{ri} = 0 \quad \text{for } x < mr + \xi_i$$

$$\left.\begin{aligned} u_i &= -UA_i \cosh^{-1}\frac{x-\xi_i}{mr} \\ rq_r &= UA_i\sqrt{(x-\xi_{i-1})^2 - m^2r^2} \end{aligned}\right\} \text{ for } mr + \xi_{i-1} \leqslant x \leqslant mr + \xi_i$$

$$\left.\begin{aligned} u_i &= -UA_i\left(\cosh^{-1}\frac{x-\xi_{i-1}}{mr} - \cosh^{-1}\frac{x-\xi_i}{mr}\right) \\ rq_r &= UA_i\left[\sqrt{(x-\xi_{i-1})^2 - m^2r^2} - \sqrt{(x-\xi_i)^2 - m^2r^2}\right] \end{aligned}\right\} \text{ for } x > mr + \xi_i. \tag{12.53}$$

Applying Eq. (12.53) to Eq. (12.30), we have, when the higher-order terms of $O(q^3/U^3)$ are neglected,

$$C_D = -2\int_{S_3} \frac{uq_r}{U^2}\, dS_3 = -4\pi\int_{mr}^{\infty} \frac{urq_r}{U^2}\, dx \tag{12.54}$$

where the integral over the surface S_2 is zero for a closed smooth body. Since u and q_r are of opposite signs, C_D is always positive.

For the body defined by Eq. (12.45), we have on the surface of the body

$$(rq_r)_0 = UF'(x) \tag{12.55}$$

$$u_0 = \lim_{r\to 0} u(x, r)$$

$$= -U\left[F''(0)\log x + F''(x)\log\frac{2}{mr} + \int_0^x \log(x-\xi)\frac{dF''(\xi)}{d\xi}\, d\xi\right] \tag{12.56}$$

The pressure coefficient over the body of the revolution is then

$$C_P(x, 0) = -\frac{2u_0}{U} \tag{12.57}$$

while the drag coefficient of the body is

$$C_D = -4\pi\int_0^{\infty} \frac{u_0(rq_r)_0}{U^2}\, dx = 2\pi\int_0^{\infty}\int_0^{\infty} \log|x-\xi|\, F''(x)F''(\xi)\, dx\, d\xi \tag{12.58}$$

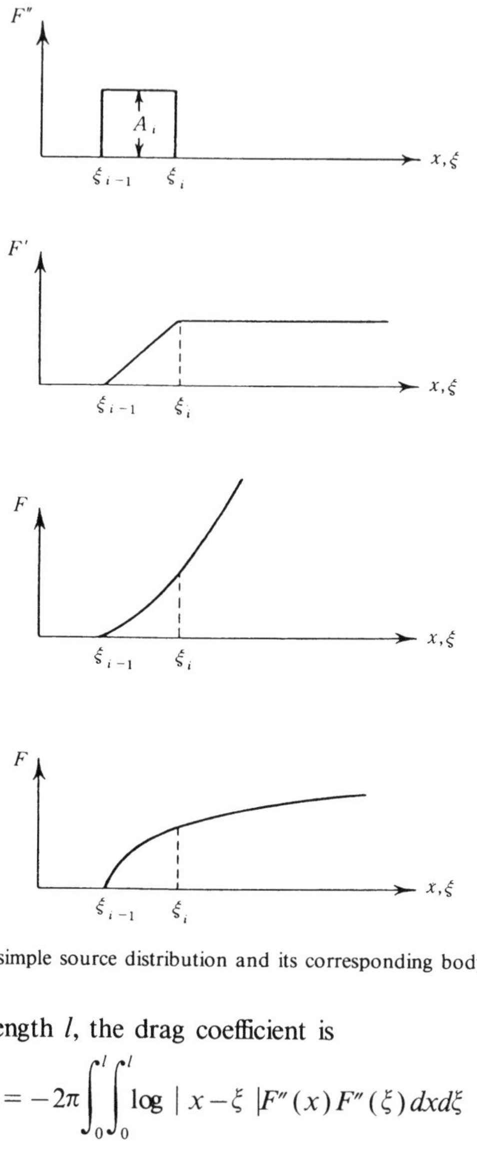

Fig. 12.7 A simple source distribution and its corresponding body shape

For a body of length l, the drag coefficient is

$$C_D = -2\pi \int_0^l \int_0^l \log |x-\xi| F''(x) F''(\xi)\, dx d\xi \tag{12.58a}$$

It is interesting to notice as was first pointed out by von Kármán[4] the remarkable analogy between the above formulas for wave drag and the Prandtl formula for induced drag of a lifting line. This analogy enables us to solve the problem of the thin shape body which gives minimum wave drag by carrying over the Prandtl result for circulation distribution giving minimum induced drag.

7. Linearized supersonic wing theory of steady flow[9-11]

In this section we shall consider the steady supersonic flow over a thin body, such as a wing of finite aspect ratio. In this case, we may use the source and doublet distributions over the wing surface. Let the equation of the wing surface be given by

$$z=\bar{z}(x, y) \tag{12.59}$$

where $\bar{z}$, $\partial\bar{z}/\partial x$, and $\partial\bar{z}/\partial y$ are all very small. The disturbed velocity components u, v, w are all very small in comparison with the uniform stream U, i. e., $\frac{|u|}{U}, \frac{|v|}{U}, \frac{|w|}{U} \ll 1$.

The boundary condition on the surface of the wing gives

$$(U+u)\frac{\partial\bar{z}}{\partial x}+v\frac{\partial\bar{z}}{\partial y}-w=0 \tag{12.60}$$

For linearized theory, Eq. (12.60) becomes

$$\frac{w(z=0)}{U}=\frac{w_0}{U}=\frac{\partial\bar{z}}{\partial x} \tag{12.61}$$

A complete discussion of all the problems of supersonic wing theory[11] is beyond the scope of this book. To illustrate the principle we shall consider the cases where the flows are symmetrical about the z-axis. Obviously, these are the cases of symmetrical wings at zero angle of attack. The results will give us the wave drag of these wings. To analyze these problems we may distribute the fundamental sources over the x-y plane in the gerion where the wing is located. Our elementary solution is

$$d\varphi(x, y, z)=-\frac{f(\xi, \eta)\,d\xi d\eta}{\sqrt{(x-\xi)^2-m^2[(y-\eta)^2+z^2]}} \tag{12.62}$$

where $f(\xi, \eta)$ is the source strength per unit area at the point ξ, η, 0. Eq. (12.62) shows that a source $f(\xi, \eta)$ at the point ξ, η, 0 gives a nonvanishing contribution to the perturbed velocity potential at another point x, y, z only if

$$(x-\xi)^2 \geqslant m^2[(y-\eta)^2+z^2]$$

i.e., the point x, y, z is within the Mach cone through the point ξ, η, 0

$$\frac{(x-\xi)^2}{(y-\eta)^2+z^2}=\cot^2\alpha=m^2=M^2-1$$

The total perturbed velocity potential at a point x, y, z is due to all the sources lying within the "fore Mach cone" through the point x, y, z (Fig. 12.8). Thus the total perturbed velocity potential may be obtained by integration of the elementary solution (12.62), i.e.,

$$\varphi(x, y, z) = -\int_0^{\xi_1} d\xi \int_{\eta_1}^{\eta_2} \frac{f(\xi, \eta)\, d\eta}{\sqrt{(x-\xi)^2 - m^2[(y-\eta)^2 + z^2]}} \tag{12.63}$$

where we choose the coordinate origin so that all the sources lie at or downstream of $\xi = 0$, i.e., $\xi < 0, f(\xi, \eta) = 0$. The points $\xi, \eta_1; \xi, \eta_2$, and $\xi_1, 0$,

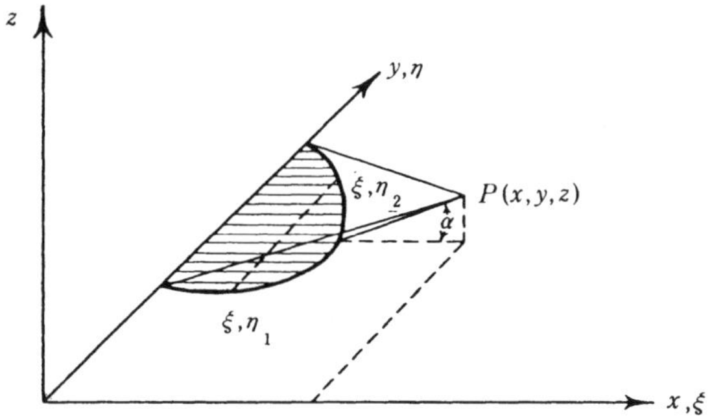

Fig. 12.8 Supersonic source distribution over a wing

are on the fore Mach cone through the point x, y, z, i.e.

$$\begin{aligned}
\eta_1 &= y - \sqrt{\frac{(x-\xi)^2}{m^2} - z^2} \\
\eta_2 &= y + \sqrt{\frac{(x-\xi)^2}{m^2} - z^2} \\
\xi_1 &= x - m\,|\,z\,|
\end{aligned} \tag{12.64}$$

In order to evaluate integral (12.63), for any particular case we must evaluate the strength of source in terms of the boundary conditions (12.61). In other words, we determine $f(\xi, \eta)$ from the value of $\partial\varphi/\partial z$ at $z = 0$. This relation may be approximately determined by the following procedure:

If we consider a point $P(x, y, z)$ very near the plane $z = 0$, the downwash depends only on the source distribution over the element $d\xi d\eta$ whose ξ, η coordinates are the same as the x, y coordinates of P. If we take the element small enough, we may replace $f(\xi, \eta)$ over it by the constant average value $F = f(x = \xi, y = \eta)$. The downwash $w(0)$ is then

$$\begin{aligned}
w(z=0) &= \lim_{z \to 0} \frac{\partial}{\partial z}\left[-F \int_{\xi_0}^{\xi_1} d\xi \int_{\eta_1}^{\eta_2} \frac{d\eta}{\sqrt{(x-\xi)^2 - m^2[(y-\eta)^2 + z^2]}} \right] \\
&= \lim_{z \to 0} \frac{\partial}{\partial z}\left[\pi F z - \frac{\pi}{m} F(x - \xi_0) \right] = \pi F \\
&= \pi f(x, y) = U \frac{\partial \bar{z}(x, y)}{\partial x}
\end{aligned} \tag{12.65}$$

Hence

$$w(x, y, 0) = \pi f(x, y)$$
$$-w(x, y, 0-) = \pi f(x, y) \tag{12.66}$$

The general expression of the perturbed velocity potential at a point P due to a symmetrical wing is

$$\varphi(x, y, z) = -\frac{U}{\pi}\int_0^{\xi_1} d\xi \int_{\eta_1}^{\eta_2} \frac{\frac{\partial \bar{z}(\xi, \eta)}{\partial \xi}\, d\eta}{\sqrt{(x-\xi)^2 - m^2[(y-\eta)^2 + z^2]}} \tag{12.67}$$

where $\bar{z}(x, y)$ is the upper surface of a symmetrical wing defined by Eq. (12.59) and ξ_1, η_1, and η_2 are given by Eq. (12.64).

8. Rectangular wing of uniform airfoil section and finite span

If the airfoil section $\bar{z}(x, y)$ and the plan form of a symmetrical wing are given, Eq. (12.67) gives the perturbed velocity potential from which the perturbed velocity components may be calculated. As a simple example, let us consider the rectangular wing of uniform airfoil section, untwisted and of span b. In this case, the function $\bar{z}(\xi, \eta)$ is independent of η, i.e.,

$$\frac{\partial \bar{z}(\xi, \eta)}{\partial \xi} = \bar{z}'(\xi) \tag{12.68}$$

Because of the symmetry of the problem we shall consider only the calculation for $z \geqslant 0$ and deduce the results for $z \leqslant 0$ from symmetry considerations. We may distinguish the three cases:

(a) $AR > 2/m$, (b) $1/m < AR < 2/m$, and (c) $AR < 1/m$, where AR is the aspect ratio of the rectangular wing $AR = b/c$, c = chord of the wing.

(a) $AR > 2/m$. In this case the Mach cones from the wing tip leading edges do not intersect within the wing. We have two distinct regions on the wing: (i) points outside the tip Mach cone, and (ii) points inside the tip Mach cone. For points outside the tip Mach cone, the results must be the same as those of two-dimensional flow because those sources which influence these points are exactly the same as those of two-dimensional flow. By direct integration of Eq. (12.67) with the condition (12.68), Ackeret's formula will be obtained (Chapter VI, § 6). We need, therefore, to study only the tip effects, i.e., the region of the wing surface lying inside the Mach cone from the tip leading edge. In the present case, the two wing tips will not influence each other and we need to consider only one tip region (Fig. 12.9a). Consider a point P (x, y) in the right-hand tip Mach cone (Fig. 12.9a). The fore cone from P includes the shaded area off the wing. If the source distribution

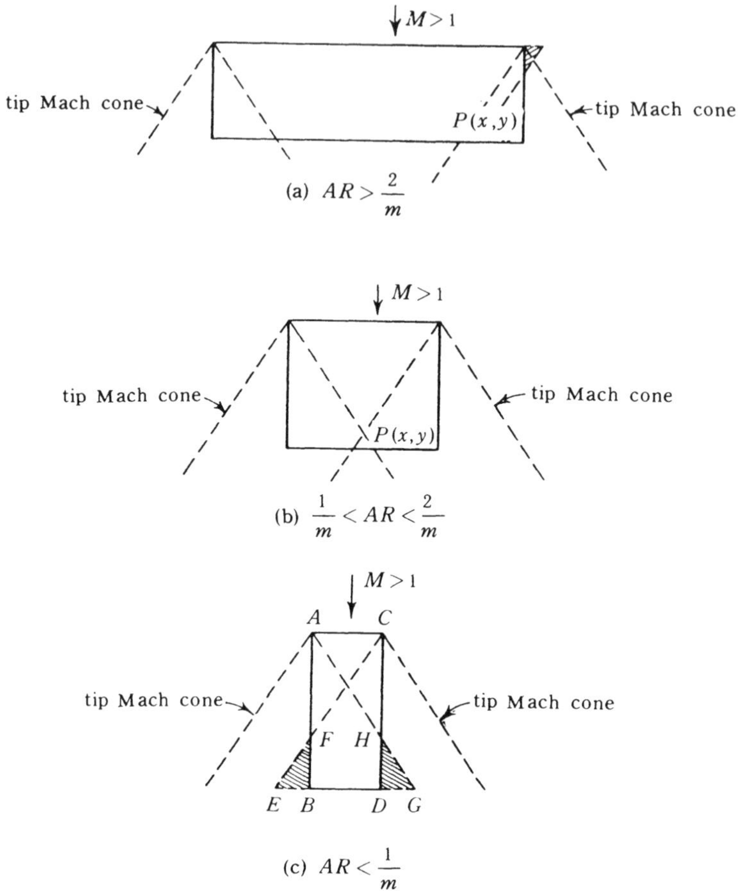

Fig. 12.9 Rectangular wings in a supersonic flow

$z'(x)$ which exists on the wing were extended to cover the shaded area, then the potential at P would have the same two-dimensional value as exists for the same y in the region between the tip Mach cone. Since the equation is linear, the method of superposition may be used; the actual velocity potential at P may be written as:

$$\varphi = \varphi_0 + \Delta\varphi \tag{12.69}$$

where φ_0 is the two-dimensional value given by the Ackeret formula, and $\Delta\varphi$ is the negative value due to the hypothetical source in the shaded area, i.e.,

$$\varphi = \frac{U}{\pi} \int_0^{\xi_1} \bar{z}'(\xi)\, d\xi \int_{b/2}^{\eta_1} \frac{d\eta}{\sqrt{(x-\xi)^2 - m^2(y-\eta)^2}}$$

$$=\frac{U}{m\pi}\int_0^{\xi_1} \bar{z}'(\xi)\cos^{-1}\frac{m\left(\frac{b}{2}-y\right)}{x-\xi}\,d\xi \qquad (12.70)$$

where

$$\xi_1=x,\ -m\left(\frac{b}{2}-y\right)$$

$$\eta_1=y+\frac{1}{m}(x-\xi)$$

We also put $z=0$, because we consider only the condition on the wing. It is of interest to determine the increment of wave drag due to the increment of velocity potential $\Delta\varphi$. The increment of drag of the wing is

$$D=\frac{2\rho_\infty U^2}{\pi}\int_0^c \bar{z}'(x)\,dx\int_{\frac{b}{2}-\frac{x}{m}}^{b/2}\left(\frac{b}{2}-y\right)dy$$

$$\times\int_0^{x-m\left(\frac{b}{2}-y\right)}\frac{\bar{z}'(\xi)\,d\xi}{(x-\xi)\sqrt{(x-\xi)^2-m^2\left(\frac{b}{2}-y\right)^2}}=0 \qquad (12.71)$$

It is interesting to note that the total drag of a symmetrical rectangular wing at zero angle of attack is the same as if the flow over all of it were two-dimensional, but the pressure distribution over the tip Mach cone region is distorted.

(b) $1/m\leqslant AR\leqslant 2/m$. In this case (see Fig. 12.9b), we must calculate the tip effects due to both tips simultaneously in the region where the two tip Mach cones overlap. Since the tip effect is linear, we may still use the method of superposition. As a result, the total wave drag is still the same as that of the two-dimensional flow but the pressure distribution is different. We have the very important result:

The total wave drag of a uniform symmetrical rectangular wing at zero angle of attack is the two-dimensional value whenever $AR\geqslant 1/m$.

(c) $AR<1/m$. In this case we may introduce a negative source distribution shown in the shaded area of Fig. 12.9(c). If the wing tip, say AB, is on the left of E, the tip effect of CD would be the same as that in case (b). As far as the wave drag of the wing is concerned, the increment of drag will be due to those negative sources in EBF and GDH, This increment of drag is

$$D=\frac{4\rho_\infty U^2}{\pi}\int_{mb}^{c}\bar{z}'(x)\,dx\int_{\frac{b}{2}-\frac{x}{m}}^{-b/2}\left(\frac{b}{2}-y\right)dy$$

$$\times \int_0^{x-m\left(\frac{b}{2}-y\right)} \frac{\bar{z}'(\xi)\,d\xi}{(x-\xi)\sqrt{(x-\xi)^2-m^2\left(\frac{b}{2}-y\right)^2}} \tag{12.72}$$

$$= \frac{4\rho_\infty U^2}{m}\int_{mb}^{c} \bar{z}'(x)\,dx \int_0^{x-mb} \bar{z}'(\xi)\,\frac{\sqrt{(x-\xi)^2-m^2b^2}}{x-\xi}\,d\xi$$

The actual variation of the drag with the aspect ratio depends on the airfoil section. But in general the drag coefficient decreases with the aspect ratio in

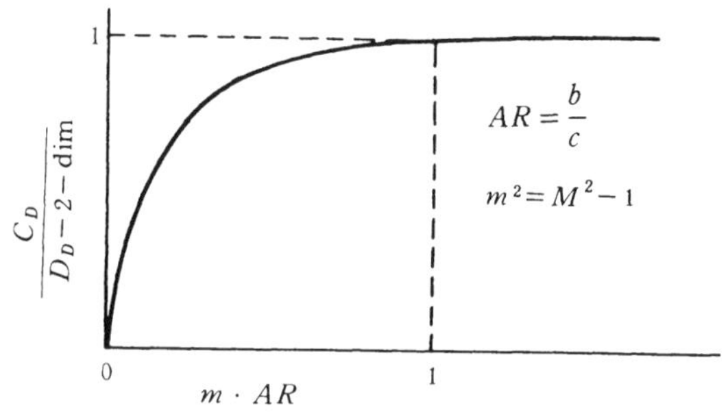

Fig. 12.10 A sketch of the variation of wave drag coefficient of a rectangular wing with its aspect ratio and the Mach number of the supersonic flow

the region $0\leqslant AR\leqslant 1/m$, as shown in Fig. 12.10 (see also Problem 5).

9. Three-dimensional steady supersonic flow over bodies without symmetry

In the last three sections we considered bodies with either axial symmetry (§ 6), or plane symmetry (§ 7 and § 8). For bodies without symmetry, the solution will be much more complicated. For such bodies, it is still convenient to divide them into two classes: slender bodies and thin bodies.

(a) *Slender bodies without axial symmetry*. This class includes bodies of revolution at an angle of attack and bodies whose cross sections are not circles. In this case it is convenient to use cylindrical coordinates (r, θ, x). The equation of the velocity potential for the steady case [cf. Eq. (12.2)] becomes

$$\frac{\partial^2\varphi}{\partial r^2}+\frac{1}{r}\frac{\partial\varphi}{\partial r}+\frac{1}{r^2}\frac{\partial^2\varphi}{\partial\theta^2}-(M^2-1)\frac{\partial^2\varphi}{\partial x^2}=0 \tag{12.73}$$

where $y=r\cos\theta$, $z=r\sin\theta$, $M=U/a$. In the present case, the velocity potential depends on the polar coordinate θ as well as r and x. The methods of solution of this problem are as follows:

(i) *The Heaviside operational method*[12]. This has been used to solve this problem. For a flow field where the x-axis is not included, the solution is quite satisfactory. This includes slender bodies of arbitrary cross section. However for the flow field including the x-axis, such as the flow inside a duct of arbitrary cross section, singularities occur when there is discontinuity in the boundary conditions. These singularities are the same as those for the flow inside a circular pipe, initially of constant cross section; then the cross section changes in such a way that the boundary of a meridian section has a discontinuous tangent. For the latter case, Chen[23] has shown that these singularities are the results of linearization. For the flow field including the x-axis, this method of solution is therefore not satisfactory. One should probably use the nonlinear solution obtained by the method of characteristics.

(ii) *The method of distribution of multipoles*. For bodies of arbitrary cross section in a uniform stream in the x-axis direction, we may distribute the multipoles along the x-axis in order to find the flow around the body in the same way as we distributed the sources for the case of bodies of revolution. The velocity potential of a multipole may be obtained by considering the source distribution on a circle of radius r' with center at the origin where the strength is proportional to $\cos n\theta$ and taking the limit of $r' \to 0$. For incompressible flow, the velocity potential of multipole of order n is of the form

$$\varphi_n = -\frac{1}{4\pi}\,\frac{r^n}{(x^2+r^2)^{n+1/2}}\,\cos n\theta \qquad (12.74)$$

When $n=0$, Eq. (12.74) gives the velocity potential of a point source while for $n=1$, it gives the velocity potential of a dipole or doublet.

(iii) *Reduced potential problem*[12]. For a slender body (not necessarily of revolution) pointed at both ends or with a pointed nose and a flat base, under certain conditions, the solution may be obtained approximately near the body by solving a two-dimensional potential problem

$$\frac{\partial^2\varphi_0}{\partial y^2} + \frac{\partial^2\varphi_0}{\partial z^2} = 0 \qquad (12.75)$$

with the boundary condition

$$\frac{\partial\varphi_0}{\partial n} = \frac{dv}{dx} \qquad (12.76)$$

where dv is the normal distance between the projections on the same plane normal to the axis of the boundaries of two sections at a distance dx apart.

(b) *Nearly plane thin bodies*. One of the classes of bodies that has been extensively investigated by aerodynamicists is that of the nearly plane thin bodies which are the wings of airplanes. We have discussed the supersonic flow over wings at zero angle of attack (§ 7 and § 8). In general, we are interested in the problem of wings at a small angle of attack. Even in the case of small angle of attack, we may use the method of source distribution on the surface of $z=0$, because, within the accuracy of linearized theory, the boundary condition of zero normal velocity on the body may be applied on the projection of the wing surface on $z=0$. Our problem is still to find the source distribution on the $z=0$ which will satisfy the proper boundary condition on the wing. In other words, the problem is solved by the use of Hadamard's solution of the general second order hyperbolic equation[11]. For $z>0$, over the part of $z=0$ for which $\xi'\leqslant x-m\,|y-\eta\,|$.

$$\varphi(x, y, z)=-\frac{1}{\pi}\iint\frac{\partial\varphi(\xi'\eta',t)}{\partial z}\frac{d\xi'\,d\eta'}{\sqrt{(x-\xi')^2-m^2[(y-\eta')^2+z^2]}} \tag{12.77}$$

The problem is divided into two parts. One φ is even and $\partial\varphi/\partial z$ is odd in z, and the other φ is odd and $\partial\varphi/\partial z$ is even. The symmetrical problem is easy to solve and it has been discussed in § 7 and 8. The antisymmetrical problem is much more complicated because boundary conditions are complicated in the antisymmetrical problems. The solution is obtained by solving an integral equation of Abel's type. This type of solution is originally due to Evvard[9] and has been further developed by others[10, 12].

For a wing or any thin body, we must distinguish between leading and trailing edges and between supersonic and subsonic edges (see Fig. 12.11). Supersonic edges are inclined to the stream direction at an angle greater than the Mach angle α, while subsonic edges are inclined at a smaller angle. At a trailing edge of a lifting surface, it is now usual to take the Kutta-Joukowski condition to be satisfied so that the velocities shall be finite; grad φ is then continuous at a subsonic trailing edge, but not necessarily so at a supersonic trailing edge. There is a trailing vortex sheet whose mean position is denoted by T. It is usual to make Prandtl's assumption that the vortex lines are exactly parallel to the undisturbed velocity U. Therefore, even in a linearized theory, neglecting the perturbation velocity normal to the sheet, the boundary conditions at the sheet are that the pressure is continuous and the velocity tangential.

For the antisymmetrical problem, the boundary conditions on the $z=0$ plane are then (i) $\partial\varphi/\partial z$ is given on the wing surface S, (ii) $\partial\varphi/\partial x=0$ on the trailing vortex sheet T, (iii) $\varphi=0$ on the rest of the plane $z=0$, and (iv) φ is continuous on $z\geqslant+0$ and grad φ is finite on the common boundary of S and T. For the symmetrical problem, the conditions (ii) to (iv) are replaced by the simple condition $\partial\varphi/\partial z=0$ on the plane $z=0$ except the surface of the wing S. Hence the symmetrical problem is given immediately by the formula (12.77).

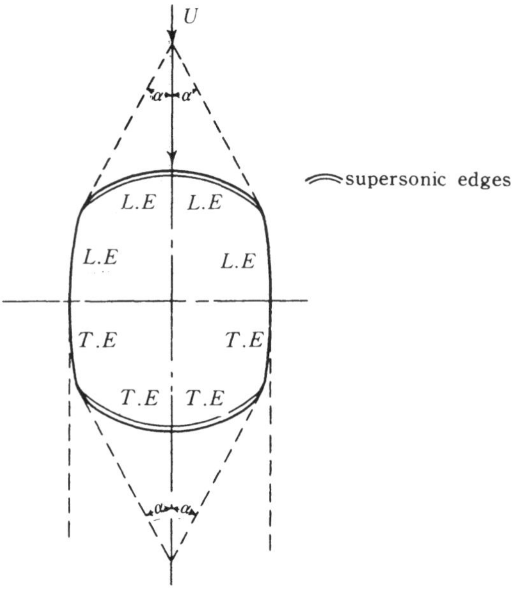

Fig. 12.11 A supersonic wing

10. Some general theorems in the linearized theory[13]

Let $\boldsymbol{U}$ be the undisturbed vector velocity, $\boldsymbol{v}$ the perturbed vector velocity, and $\boldsymbol{w}$, the vector velocity which characterizes the perturbation of the mass flux, such that

$$\rho_\infty(\boldsymbol{w}+\boldsymbol{U}) \quad =\rho\ (\boldsymbol{U}+\boldsymbol{v}) \tag{12.78}$$

where ρ is the density, and ρ_∞ the undisturbed density. In this case, the field equations are simply

$$\text{curl } \boldsymbol{v}=0, \quad \text{div } \boldsymbol{w}=0 \tag{12.79}$$

and, to a linear approximation,

$$\boldsymbol{w}=\boldsymbol{v}-\frac{\boldsymbol{U}\cdot\boldsymbol{v}}{a_\infty^2}\,\boldsymbol{U} \tag{12.80}$$

where a_∞ is the speed of sound in the undisturbed stream. The pressure perturbation is given by

$$\frac{\Delta p}{\rho_\infty}=-\boldsymbol{U}\cdot\boldsymbol{v} \tag{12.81}$$

If $\boldsymbol{v}$ and $\boldsymbol{w}$ satisfy the linearized equation and are bounded and continuous on and inside a closed surface, except that $\boldsymbol{v}$ may be discontinuous on characteristic surfaces inside S, it is easy to show that the momentum integral vanishes

$$\int_S \left(\boldsymbol{v}\boldsymbol{w} \cdot \boldsymbol{n} - \frac{1}{2} \boldsymbol{v} \cdot \boldsymbol{w}\boldsymbol{n}\right) dS = 0 \tag{12.82}$$

where $\boldsymbol{n}$ is the outward unit normal vector of the surface S.

If $\boldsymbol{v}_1$, $\boldsymbol{w}_1$ and $\boldsymbol{v}_2$, $\boldsymbol{w}_2$ are two different linearized flows, from Eq. (12.82) we have the identity

$$\int_S (\boldsymbol{v}_1\boldsymbol{w}_2 \cdot \boldsymbol{n} + \boldsymbol{v}_2\boldsymbol{w}_1 \cdot \boldsymbol{n} - \boldsymbol{v}_1 \cdot \boldsymbol{w}_2\boldsymbol{n})\, dS = 0 \tag{12.83}$$

We may use Eq. (12.83) to derive the general reversed flow theorem. Consider two boundary surfaces for thin bodies having the same mean surface $\sum$, or for a nearly plane body, the same projection on $z=0$. Let $\boldsymbol{v}_1$ and $\boldsymbol{v}_2$ be the perturbation velocities in opposite flows $\boldsymbol{U}$, $-\boldsymbol{U}$. The boundary conditions on the mean surface are

$$\boldsymbol{n} \cdot \boldsymbol{v}_1 = -\boldsymbol{U} \cdot \boldsymbol{n}_1 \text{ and } \boldsymbol{n} \cdot \boldsymbol{v}_2 = \boldsymbol{U} \cdot \boldsymbol{n}_2 \tag{12.84}$$

Multiplying Eq. (12.83) scalarly by $\boldsymbol{U}$ and using the conditions that $\boldsymbol{v}_1$ and $\boldsymbol{v}_2$ are bounded, we have

$$\int_\Sigma (\boldsymbol{U} \cdot \boldsymbol{v}_1\boldsymbol{w}_2 \cdot \boldsymbol{n} + \boldsymbol{U} \cdot \boldsymbol{v}_2\boldsymbol{w}_1 \cdot \boldsymbol{n} - \boldsymbol{v}_1 \cdot \boldsymbol{w}_2\boldsymbol{n} \cdot \boldsymbol{U})\, dS$$

$$= \boldsymbol{U} \cdot \int_\Sigma (\Delta p_1 \boldsymbol{n}_2 + \Delta p_2 \boldsymbol{n}_1)\, dS = 0 \tag{12.85}$$

This is the general formula for reversed flows. For instance, in the simplest case where the two bodies are the same so that $\boldsymbol{n}_1 = \boldsymbol{n}_2 = \boldsymbol{n}$, we have

$$\boldsymbol{U} \cdot \int_\Sigma \Delta p_1 \boldsymbol{n}\, dS = -\boldsymbol{U} \cdot \int_\Sigma \Delta p_2 \boldsymbol{n}\, dS \tag{12.86}$$

Eq. (12.86) shows that the drag forces are of the same magnitude in the two flows. Since the integrals are based on the first-order linearized perturbation pressure (12.81), the forces and moments so calculated do not include contributions from the suction forces on sharp subsonic edges. Similarly, we have the following results[8, 11, 12]:

The initial slopes of the lift curve in the two flows of the same body are the same. The rolling moment due to a given rate of yaw is equal to the yawing moment in the reversed flow due to the same rate of roll.

It should be noticed that this theorem holds for bodies which have the same mean surface or projection and which may not be of exactly the same shape. Thus the theorem is very general.

11. Supersonic conical flows[14–16]

A mathematically simple and elegant method of solution of three-dimensional supersonic steady flow in linearized theory may be found for the so-called conical flows. Conical flow solutions were first discussed by Busemann.[1] In conical flow, the velocity components u, v, w are homogeneous functions of x, y, z of degree zero and so are constant along all radial vectors through the origin. Introducing the conical coordinates (h, θ)

$$h=m\sqrt{\left(\frac{y}{x}\right)^2+\left(\frac{z}{x}\right)^2},\quad \theta=\tan^{-1}\frac{z}{x} \tag{12.87}$$

the equation for the perturbation velocity potential for steady case, i.e.,

$$(M^2-1)\frac{\partial^2\varphi}{\partial x^2}-\frac{\partial^2\varphi}{\partial y^2}-\frac{\partial^2\varphi}{\partial z^2}=0 \tag{12.88}$$

becomes

$$\left[(h^2-1)\frac{\partial^2}{\partial h^2}+\frac{1}{h}(2h^2-1)\frac{\partial}{\partial h}-\frac{1}{h^2}\frac{\partial^2}{\partial\theta^2}\right]g=0 \tag{12.89}$$

where g may be φ or any one of the perturbation velocity components u, v, w. Eq. (12.89) is of mixed type. If $h>1$, it is hyperbolic. and if $h<1$, it is elliptic, and $h=1$ corresponds to the Mach cone.

If we introduce the variable σ such that

$$\sigma=\cos^{-1}\frac{1}{h} \tag{12.90}$$

for $h>1$. Eq. (12.89) becomes

$$\frac{\partial^2 g}{\partial\sigma^2}=\frac{\partial^2 g}{\partial\theta^2} \tag{12.91}$$

If we introduce the variable s such that

$$s=-\text{sech}^{-1}\frac{1}{h} \tag{12.92}$$

for $h< 1$, Eq. (12.89) becomes

$$\frac{\partial^2 g}{\partial s^2}+\frac{\partial^2 g}{\partial \theta^2}=0 \tag{12.93}$$

We see that outside the Mach cone, the perturbed velocity components satisfy the simple wave Eq. (12.91), while inside the Mach cone, they satisfy the Laplace equation (12.93).

Inside the Mach cone, u, v, w are the real parts of three functions g_1, g_2, g_3 of the variable $\xi=e^{s+i\theta}$ and

$$g_1'(\xi):g_2'(\xi):g_s'(\xi)=\xi \tan \alpha: -(\xi^2+1): i(\xi^2-1) \tag{12.94}$$

where the prime denotes differentiation.

Similarly outside the Mach cone, u, v, w are given by the functions $f_{1,2,3}(\sigma+\theta)+F_{1,2,3}(\sigma-\theta)$, and

$$f_1':f_2':f_3' = -\tan \alpha: \cos(\sigma+\theta): \sin(\sigma+\theta) \tag{12.95}$$

$$F_1':F_2':F_3' = -\tan \alpha: \cos(\sigma-\theta): -\sin(\sigma-\theta)$$

This method has been extensively investigated[13–16] and is used to solve problems of flow over thin cones, past plane triangular wings and of superposition flow past plane polygonal airfoils, and many other problems.

This method has been generalized by Lagerstrom to include the integration of conical field solutions with different vertices and by Germain to the case when φ is a homogeneous function of degree n.

12. Poincaré-Lighthill-Kuo method[17, 18]

To improve linearized conical flow solutions, and other eases, the mathematical problem of improving the solution of a hyperbolic partial differential equation near a singular characteristics arises. A method, which is an extension of that used for an ordinary differential equation near a singular point, has been used. This method, used by Poincaré in discussing nonlinear oscillations, has been extended by Lighthill and Kuo.[18] A simple example would be

$$(x+\alpha_1 u)\frac{du}{dx}+g(x)u=r(x) \tag{12.96}$$

with α_1 small and $x=0$ as a singular point of the linearized equation. The method is to introduce a new variable z for which

$$x=z+\alpha_1 x_1(z)+\alpha_1^2 x_2(z)+\cdots \tag{12.97}$$

$$u=u_0+\alpha_1 u_1(z)+\alpha_1^2 u_2(z)+\cdots \tag{12.98}$$

Eq. (12.97) defines z and we may choose x, etc., in defining z so that the series for u converges as well near the singular point as elsewhere. This can be extended to singular characteristics of hyperbolic characteristics of hyperbolic partial differential equations. An example would be the equations:

$$\frac{\partial u}{\partial y}=\alpha_1\left(u+\frac{\partial v}{\partial y}\right)\frac{\partial u}{\partial x},\quad \frac{\partial v}{\partial x}=u \tag{12.99}$$

with α_1 small. The linearized equation $\frac{\partial u}{\partial y}=0$, for which the solution is $u=u_0\ (x)$, may have a singularity at $x=0$, say, $u_0\sim A/x^q$, where $q>0$. We write

$$\begin{aligned} x&=z+\alpha_1 x_1(z,y)+\alpha_1^2 x_2(z,y)+\cdots \\ u&=u_0(z,y)+\alpha_1 u_1(z,y)+\alpha_1^2 u_2(z,y)+\cdots \\ v&=v_0(z,y)+\alpha_1 v_1(z,y)+\alpha_1^2 v_2(z,y)+\cdots \end{aligned} \tag{12.100}$$

and $x_1(z, y)$ etc., may be found so that the expansions for u and v are valid in a region including $x=0$.

As an example, let us consider the conical flows. If $h=1$ is the Mach cone of the origin (h being equal to sec σ or sech s in Eqs. (12.90) and (12.92)), the velocity potential φ is put equal to

$$\varphi=Ux\ [1+f\ (h,\theta)] \tag{12.101}$$

The variables then change from h to H and f expandes in a series

$$h=H+h_1(\theta)+h_2(H,\theta)+h_3(H,\theta)+\cdots \tag{12.102}$$

$$f=f_1(H,\theta)+f_2(H,\theta)+f_3(H,\theta)+\cdots$$

where $f=f_1$ is the linearized solution, f_2 is quadratic in f_1 and its derivatives and so on; h is of the same order as f. Then h_1, h_2 etc., are to be found so that the expansion for f is valid in a region near $H=1$. The solutions are found separately for $H>1$ and for $H<1$. The regions of validity of the two solutions overlap; in the region of overlap, boundary conditions, appropriate to the appearance of a shock wave may be applied to determine the approximate position of the shock.

13. Improvement of solution at infinity[19]

For hyperbolic partial differential equations, approximate solutions may also be nonuniform at infinity owing to the deviation of the characteristics. As we go to infinity along one set of characteristics, the slope may be everywhere nearly the same as in linear theory, but the characteristic curve itself may be changed in position by an infinite amount. If we go to infinity along a linearized characteristic y=constant, it is sufficient, in order to improve the approximation, to replace y by a more nearly exact characteristic coordinate z, which is itself found by an expansion in y. The simplest example is as follows: consider the equation

$$\frac{\partial}{\partial r}(ur^{1/2})+b\frac{\partial}{\partial x}(ur^{1/2})+\frac{1}{2}kur^{1/2}\frac{\partial u}{\partial x}=0 \qquad (12.103)$$

For the linearized equation the solution is

$$u=\frac{1}{r^{1/2}}f(x-br) \qquad (12.104)$$

Here we suppose not that k is small, but that u is small because r is large. The full equation may be solved exactly and the solution is

$$u=-z/kr^{1/2} \qquad (12.105)$$

with

$$x=br-zr^{1/2}-h(z) \qquad (12.106)$$

where $h(x)$ is an arbitrary function of z. Hence, along z=constant,

$$\frac{dx}{dr}=b-z/2r^{1/2}=b+ku/2 \qquad (12.107)$$

and z=constant is an exact characteristic.

The above type of analysis has been applied to study the flow at a large distance from a body of revolution, where linearized theory fails. The Mach lines intersect the bow shock wave, which is thereby weakened and curved inward. The equation of the bow shock wave is found to be of the form[18]

$$x=r\cot\alpha_0-br^{1/4}-c-dr^{1/4}+\cdots \qquad (12.108)$$

and the rear shock wave of the form

$$x=r\cot\alpha_0+b_1r^{1/4}+\cdots \qquad (12.109)$$

At a constant value of r, the pressure falls approximately linearly between the shocks.

14. Nonsteady flow

For unsteady flow, sources and sinks may be used in a manner similar to that used for steady flow except that the source strength is a function of time as well as of the spatial coordinates [e.g., Eq. (12.22)]. Since the fundamental equation (12.2) is lincar, the method of superposition is applicable. Two types of problems have been extensively investigated. One is the transient behavior of a body in unsteady motion, and the other is the oscillatory motion of a body in a uniform stream[11, 20–24]

For unsteady flow, the region of validity of the linearized theory depends not only on the free stream Mach number and the thickness ratio of the body, but also on a characteristic frequency of the flow. For instance, this characteristic frequency would be the nondimensional ratio of the product of the angular frequency and the chord length to the free stream velocity in oscillatory motion; it would be the inverse of the characteristic period of time in a transient motion. We shall discuss the region of validity of linearized unsteady flow theory in Chapter XIII.

15. Law of equivalence and area rule[25]

In this section, we discuss an approximate analysis which is very useful for practical application and which is known as the law of equivalence and is related to the area rule for a wing of small aspect ratio in the transonic flow region.

For three-dimensional wing and body of small aspect ratio for transonic flow in the neighborhood of the wing and/or body, the disturbance potential φ_1 satisfies the equation:

$$\varphi_{1yy}+\varphi_{1zz}=0 \tag{12.110}$$

In Eq. (12.110) we use the assumption that the first term of the following disturbance potential φ_1 equation

$$(1-M^2)\,\varphi_{1xx}+\varphi_{1yy}+\varphi_{1zz}=0 \tag{12.111}$$

is negligible in comparison with the second and the third terms. However, in actual cases, local violation of this assumption may exist. For instance, in Fig. 12.12, at the end of the wing or at the location of discontinuity of the wing planform, this assumption is not valid. But such local violation has little influence on the main results of our following discussion of the law of equivalence or the area rule, but wrong results in the neighborhood of the region of violation are probable.

For simplicity, we consider the case of a symmetrical wing at zero angle of attack, i.e., non-lifting wing with finite thickness. The boundary condition at the wing is

$$-s(x)<z<s(x):\quad \varphi_{1y}(x,+0,\ z)=v_1(x,+0,\ z) \tag{12.112}$$

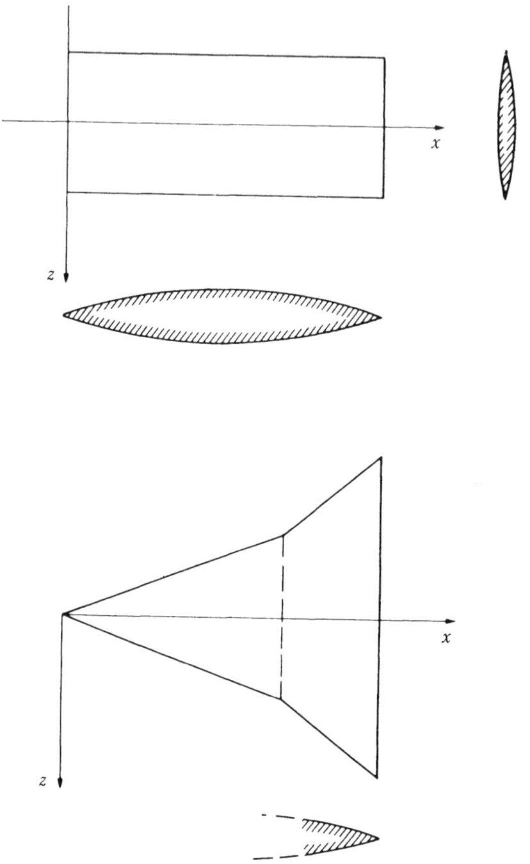

Fig. 12.12 Regions of violation that the first term of Eq. (12.111) is not negligible

where $s(x)$ is the local half-span at location x and z is the coordinates along the span.

The disturbance potential φ_1 for the boundary condition (12.112) is

$$\varphi_1(x, y, z) = \frac{1}{\pi}\int_{-S(x)}^{S(x)} v_l(x,+0,\ \xi)\ln\sqrt{(y^2+(z-\xi)^2)}\ d\xi + R(x) \quad (12.113)$$

The additional function $R(x)$ is not determined by the boundary condition of the planform of the wing, Eq. (12.112). The boundary conditions at large distance from the wing is troublesome and needs further discussion.

Since Eq. (12.110) is linear, the method of superposition is applicable. Hence, in case of a lifting wing, we may add the solution of the asymmetrical part to the results of the symmetrical wing. For the lifting wing, the boundary condition is

$$-s(x) < z < s(x): \quad \varphi_{1z}(x, +0, z) = w_l(x, +0, z) \tag{12.114}$$

and the solution of Eq. (12.110) with the boundary condition (12.114) is

$$\varphi_1(x, y, z) = -\frac{1}{\pi}\int_{-S(x)}^{S(x)} w_1(x, +0, \xi)\tan^{-1}\left(\frac{y}{z-\xi}\right)d\xi \tag{12.115}$$

Because of the asymmetry in the present case, there is no additional function of x in Eq. (12.115) because $\varphi_1(x, y, z) = \varphi_l(x, -y, z)$.

The stream lines on a symmetrical wing of finite thickness are given by the following equation:

$$w_1(x, +0, z) = \varphi_{1z} = \frac{1}{\pi}\int_{-S(x)}^{S(x)} v_1(x, +0, \xi)\frac{d\xi}{z-\xi} \tag{12.116}$$

In the near-field of a symmetrical wing of finite thickness, Eq. (12.113) may be simplified and the disturbance potential φ_1 is approximately given by the following formula:

$$\varphi_1(x, y, z) = \frac{1}{\pi}\ln\sqrt{(y^2+z^2)}\int_{-S(x)}^{S(x)} v_1(x, +0, \xi)d\xi + R(x) \tag{12.117}$$

Eq. (12.117) is also the disturbance potential for a body of revolution. Hence, we may conclude that the disturbance potential of a body of revolution with the source at a point of the x-axis is equal to the sum of sources in the plane of the wing at the same x. In Fig. 12.13, we compare the equi-potential line of a circular cone (dotted lines) and a conical wing with parabolic cross-section (solidlines). These two sets of equi-potential lines of the same cross-sectional area for these two bodies are quite close. The fact that a wing has the same function $R(x)$ as the equivalent body of revolution is known as the law of equivalence. It should be noticed that this relation holds only in the near field such that

$$|1-M_\infty^2|\sigma^2 \ll |1-M_\infty^2|(y^2+z^2) \ll 1 \tag{12.118}$$

where σ is the aspect ratio, i.e., the lateral distance is larger than the half-span but smaller than the length of the wing or the body .

From the results of Eq. (12.117), we may find an equivalent body of revolution with cross-sectional area $F(x)$ such that

$$F(x) = 2\int_{-S(x)}^{S(x)} h(x, \xi)d\xi \tag{12.119}$$

where $h(x, z)$ is the half thickness distribution of the equivalent wing. Eq. (12.119) gives a body of revolution with the same local cross-sectional area $F(x)$ as that of a corresponding wing of small aspect ratio.

For the body of revolution, we write

$$F(x) = F_{max} f(x) = \frac{\pi}{4} \tau_R^2 f(x) \tag{12.120}$$

where τ_R is the thickness ratio of the body of revolution and $f(x)$ is the distribution of the cross-sectional area of the body of revolution. For instance, in subsonic flow where a linearized gasdynamic equation is valid, the disturbance velocity potential of the body of revolution is then

$$\varphi_1(x, r) = -\frac{1}{4\pi} \int_0^1 f_x(\xi) \frac{d\xi}{\sqrt{(x-\xi)^2 + \beta^2 r^2}} \tag{12.121}$$

where $\beta^2 = 1 - M_\infty^2$. We have the disturbance parameter

$$\delta = F_{max} = C\tau_{wing}\,\sigma \tag{12.122}$$

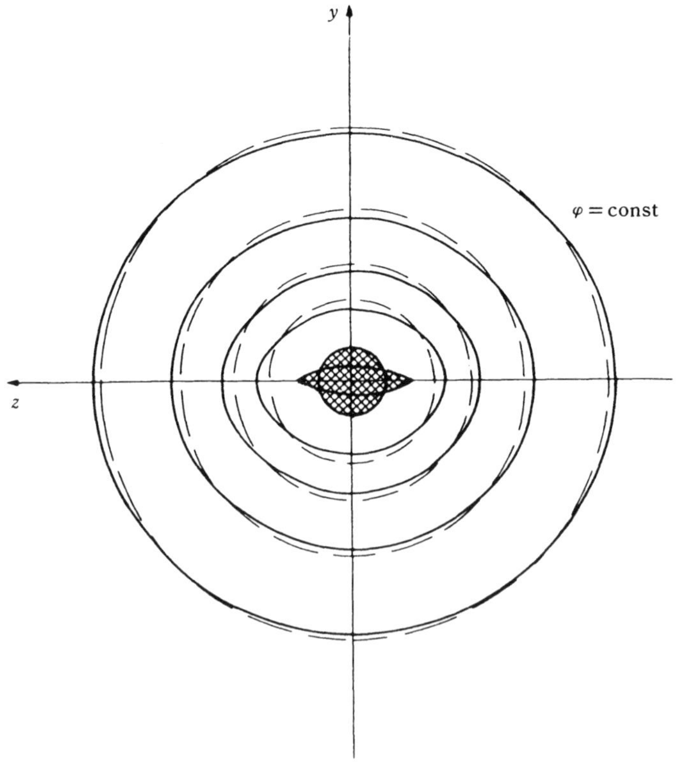

Fig. 12.13 Some equi-potential lines in a cross-section x = constant for cone with parabolic cross-section (solid lines) and cone with circular cross-section (dotted lines)

where C is a constant of the order of unity.

We may determine the additional function $R(x)$ by the following formula:

$$R(x)=\lim_{r\to 0}\left[\varphi_1(x,r)-\frac{1}{2\pi}f_x(x)\ln r\right]=-\frac{1}{4\pi}f_x(x)\ln\frac{4(1-x)x}{\beta^2}-\frac{1}{4\pi}\int_0^1\frac{f_x(\xi)-f_x(x)}{|x-\xi|}d\xi \qquad (12.123)$$

$$R_x(x)=\lim_{r\to 0}\left[u_1(x,r)-\frac{1}{2\pi}f_{xx}(x)\ln r\right]=-\frac{1}{4\pi}f_{xx}(x)\ln\frac{4(1-x)x}{\beta^2}-\frac{1}{4\pi}\int_0^1\frac{f_{xx}(\xi)-f_{xx}(x)}{|x-\xi|}d\xi \qquad (12.124)$$

In Fig. 12.14, we compare various aspect of delta wing with its equivalent body of revolution. In Fig. 12.14(a), the thickness distribution of the delta wing is compared with that of the equivalent body of revolution. In Fig. 1214 (b), the stream lines on the delta wing of $\tau=0.06$ for $0\leqslant M_\infty\leqslant\sqrt{2}$ are given. Finally in Fig. 12.14(c), the curves of constant velocity on the delta wing at four different M_∞ are given. No experimental data for u_1 is available but the numerical results based on linearized theory are available. Even though the above results are not really transonic flow, we can learn something from them with the application to the transonic problem.

The law of equivalence is not as well known as the area rule which was first found by Hayes[26] and further developed by Ward[7] and which has important practical applications. The area rule is related to the pressure drag coefficient with the cross-sectional area of the body and/or the wing. The difference of the drag coefficient of the wing C_D and that of the equivalent body of revolution C_{DR} is

$$C_D-C_{DR}=-\frac{1}{2\pi}\frac{\tau_R^2}{F_{\max}}\int_{-S(1)}^{S(1)}\int_{-S(1)}^{S(1)}f_x(1,z)f_x(1,\xi)\ln\frac{x-\xi}{h_R(1)}\,d\xi dz \qquad (12.125)$$

This formula cannot be derived here. It was first found for the linearized supersonic flow. It contains the geometrical properties of the wing only[6] and is valid also for transonic flow. If we evaluate Eq. (12.125) for a wing of cross-section of rhomboid and the equivalent circular cone, we have

$$C_{Dr}-C_{DC}=-\frac{\sigma\tau}{\pi}\left(\ln\frac{4\pi\sigma}{\tau}-3\right) \qquad (12.126)$$

In Fig. 12.15, we have a cone with rhomboid cross-section and a circular cone with the same pressure drag coefficient. In Fig. 12.16, the measured results of

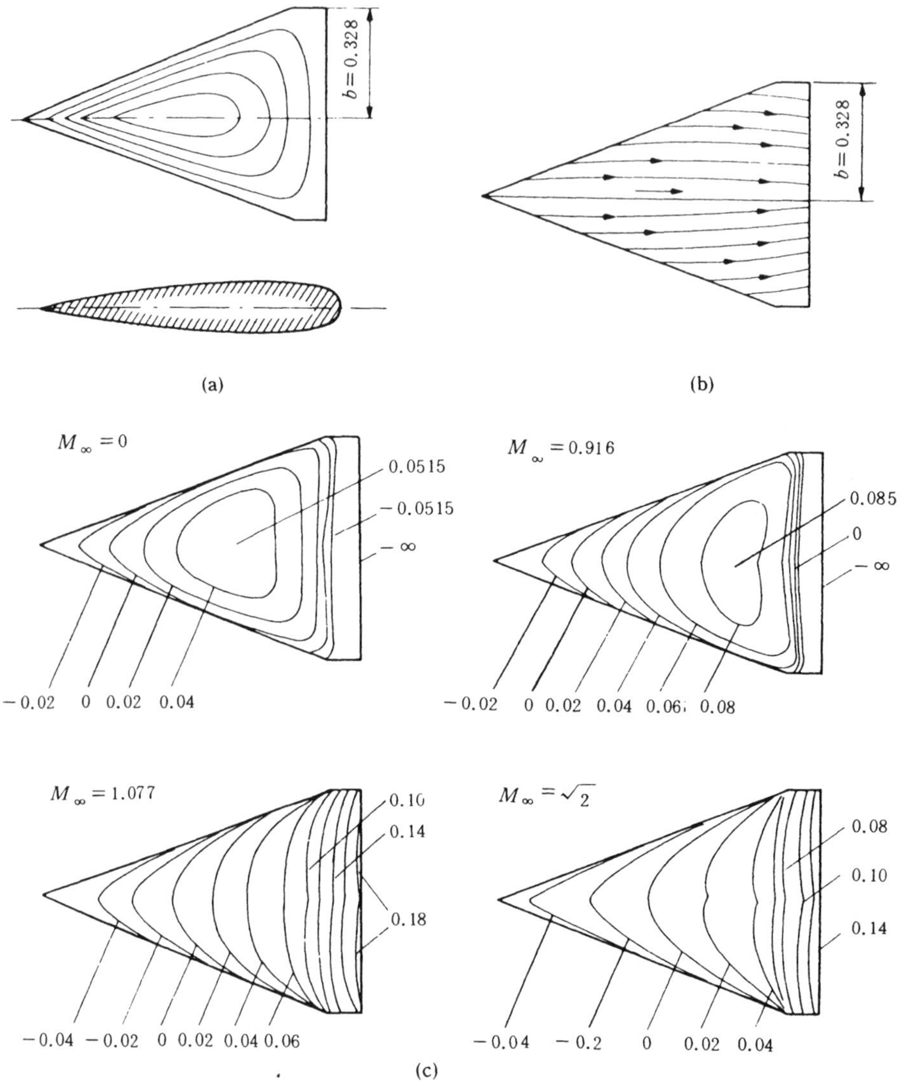

Fig. 12.14 A comparison of a delta wing and its equivalent body of revolution. (a) Thickness distribution of a delta wing and that of an equivalent body of revolution. (b) Stream lines on a delta wing with $\tau=0.06$ and $0\leqslant M_\infty \leqslant\sqrt{2}$. (c) curves of constant velocity on a delta wing at four different M_∞ for $\tau=0.06$ and $\sigma=0.328$

Whitcomb[27] for two equivalent bodies are given. In the upper diagram of Fig. 12.16, both the frictional drag and the pressure drag are included while in the lower diagram, only the pressure drag is considered. It is evident that the area rule applies to the pressure drag coefficient only.

With the help of the law of equivalence, we may compare wings of similar cross-section by affine transformation. For instance, In Fig. 12.17, we have three different cross-sections where cross-sections (a) and (b) have the same span, (b) and (c) have the same thickness and (a) and (c) have the same cross-sectional area.

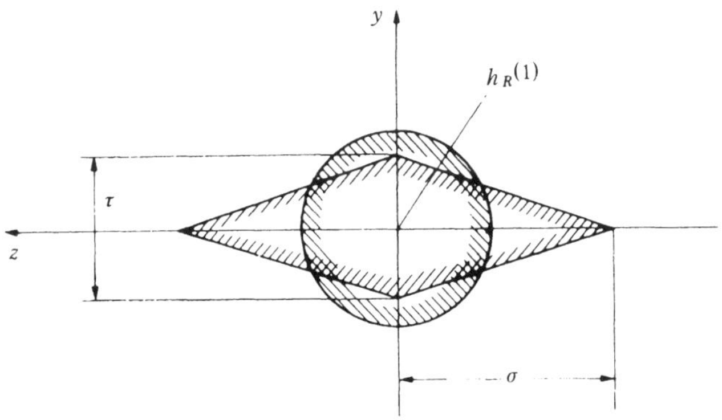

Fig. 12.15 The end cross-section of a cone with rhomboid section and that of a circular cone with the same pressure drag coefficient

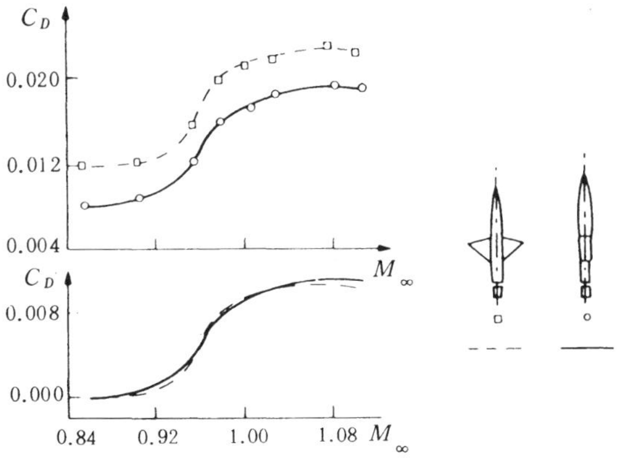

Fig. 12.16 A comparison of drag coefficients of two equivalent bodies given by Whitcomb in Ref. 29

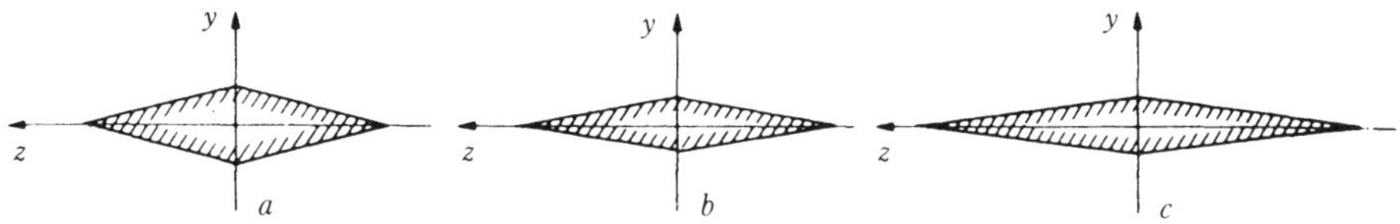

Fig. 12.17 Affine related cross-sections of three wings

16. Linearized steady supersonic flow over lifting wing

In section 7, we have studied the source method for linearized steady supersonic flow over three-dimensional wing. The effect of wing . thickness can be calculated by this method, i. e. Eq. (12.67). In general, wing camber, twist and incidence effects ought to be represented by planar vortex distributed on the mean surface of the wing. In the linearized theory of compressible flow , the discontinuity in pressure across the wing is equivalent to the discontinuity in the velocity component along free stream direction . This discontinuity in velocity may be viewed as planar vortex to simulate the effect of lifting wing . In this section, we first derive the perturbation velocity potential of the planar vortex in uniform supersonic free stream and then apply the panel method (see Chapter VI, §11) to solve the lifting problem .

We will derive the vortex potential expression from the source potential expression. From Eqs. (12.67) and (12.62) we obtain

$$\varphi(x, y, z) = -\frac{1}{\pi}\iint_S \frac{\frac{\partial\varphi}{\partial z}(\xi, \eta, 0)\, d\xi d\eta}{\sqrt{(x-\xi)^2 - m^2[(y-\eta)^2 + z^2]}} \tag{12.127}$$

where S is the area of the x-y plane within the fore Mach cone through the point (x, y, z).

Differentiating both sides of Eq. (12.127) with respect to z, we get

$$\frac{\partial\varphi}{\partial z}(x, y, z) = -\frac{1}{\pi}\frac{\partial}{\partial z}\iint_S \frac{\frac{\partial\varphi}{\partial z}(\xi, \eta, 0)\, d\xi d\eta}{\sqrt{(x-\xi)^2 - m^2[(y-\eta)^2 + z^2]}} \tag{12.128}$$

The above equation gives a relation between the downwashes at (x, y, z) and $(\xi, \eta, 0)$. We notice that the downwash function $\frac{\partial\varphi}{\partial z}(x, y, z)$ also satisfies the potential equation, i. e.

$$(1-M^2)\frac{\partial^2}{\partial x^2}\left(\frac{\partial\varphi}{\partial z}\right) + \frac{\partial^2}{\partial y^2}\left(\frac{\partial\varphi}{\partial z}\right) + \frac{\partial^2}{\partial z^2}\left(\frac{\partial\varphi}{\partial z}\right) = 0 \tag{12.129}$$

Therefore $\frac{\partial\varphi}{\partial z}$ may be considered as another solution of the potential equation. We replace $\frac{\partial\varphi}{\partial z}$ by φ in Eq. (12.128).

$$\varphi(x, y, z) = -\frac{1}{\pi}\frac{\partial}{\partial z}\iint_S \frac{\varphi(\xi, \eta, 0)\, d\xi d\eta}{\sqrt{(x-\xi)^2 - m^2[(y-\eta)^2 + z^2]}} \tag{12.130}$$

Now the integrand of (12.130) can be written in terms of $\frac{\partial \varphi}{\partial \xi}(\xi, \eta, 0)$ which is equivalent to the pressure coefficient on the wing surface.

The limits of integration are defined by the wing plane within the fore Mach cone through the point (x, y, z) (Fig. 12.18). The integral in Eq. (12.130)

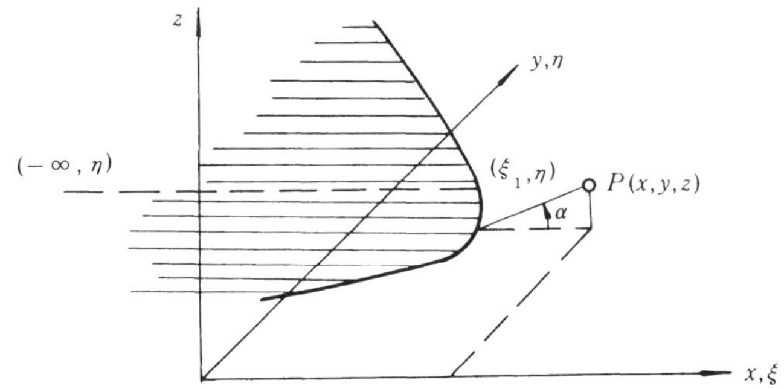

Fig. 12.18 Integration limits for supersonic wing

becomes

$$\int_{-\infty}^{\infty} d\eta \int_{-\infty}^{\xi_1} \frac{\varphi(\xi, \eta, 0)\, d\xi}{\sqrt{(x-\xi)^2 - m^2[(y-\eta)^2 + z^2]}} = I \tag{12.131}$$

where

$$\xi_1 = x - m[(y-\eta)^2 + z^2]^{1/2} \tag{12.132}$$

Integrate (12.131) by parts with respect to ξ. We obtain

$$I = \int_{-\infty}^{\infty} d\eta \left\{ -\varphi(\xi, \eta, 0) \cosh^{-1} \frac{x-\xi}{m\sqrt{(y-\eta)^2 + z^2}} \Bigg|_{-\infty}^{\xi_1} + \right.$$

$$\left. \int_{-\infty}^{\xi_1} \varphi_\xi(\xi, \eta, 0) \cosh^{-1} \frac{x-\xi}{m\sqrt{(y-\eta)^2 + z^2}}\, d\xi \right\} \tag{12.133}$$

We may assume that $\varphi(-\infty, \eta, 0) = 0$. The first term of (12.133) vanishes. Now we differentiate Eq. (12.131) with respect to z according to Eq. (12.130)

$$\frac{\partial I}{\partial z} = \int_{-\infty}^{\infty} d\eta \frac{\partial}{\partial z} \int_{-\infty}^{\xi_1} \varphi_\xi(\xi, \eta, 0) \cosh^{-1} \frac{x-\xi}{m\sqrt{(y-\eta)^2 + z^2}}\, d\xi$$

$$= \int_{-\infty}^{\infty} d\eta \left\{ \frac{\partial \xi_1}{\partial z} \left[\varphi_\xi(\xi, \eta, 0) \cosh^{-1} \frac{x-\xi}{m\sqrt{(y-\eta)^2 + z^2}} \right]_{\xi = \xi_1} + \right.$$

$$+\int_{-\infty}^{\xi_1} \varphi_\xi(\xi,\eta,0)\frac{\partial}{\partial z}\cosh^{-1}\frac{x-\xi}{m\sqrt{(y-\eta)^2+z^2}}\,d\xi\Bigg\}$$

$$=-z\int_{-\infty}^{\infty} d\eta\int_{-\infty}^{\xi_1}\frac{(x-\xi)\varphi_\xi(\xi,\eta,0)\,d\xi}{[(y-\eta)^2+z^2]\sqrt{(x-\xi)^2-m^2[(y-\eta)^2+z^2]}} \qquad (12.134)$$

Substituting Eq. (12.134) into (12.130), we obtain

$$\varphi(x,y,z)=\frac{z}{\pi}\iint_S\frac{(x-\xi)\varphi_\xi(\xi,\eta,0)\,d\xi d\eta}{[(y-\eta)^2+z^2]\sqrt{(x-\xi)^2-m^2[(y-\eta)^2+z^2]}} \qquad (12.135)$$

This is the perturbation velocity potential of planar vortex in linearized supersonic flow. Here $\varphi_\xi(\xi,\eta,0)$ is the vortex strength per unit area at the point $(\xi,\eta,0)$ and is zero except on the lifting-wing area.

Let the equation of the mean surface of the wing be given by

$$z=z_m(x,y) \qquad (12.136)$$

where z_m, $\partial z_m/\partial x$ and $\partial z_m/\partial y$ are all very small. For linearized theory, the boundary condition on the mean surface of the wing gives

$$\frac{\partial\varphi}{\partial z}(x,y,0)=U\frac{\partial z_m}{\partial x} \qquad (12.137)$$

The lifting problem of wing is to solve Eq. (12.137) for the vortex strength $\varphi_\xi(\xi,\eta,0)$ of (12.135). To do this, the panel methods of Chapter VI, § 11 will be used. We divide the wing surface into a large number of small quadrilateral panels arranged in strips parallel to the free stream so that wing edges lie on panel boundaries (Fig. 12.19). Each panel defines an area over which the vortex singularity strength is held constant. The problem

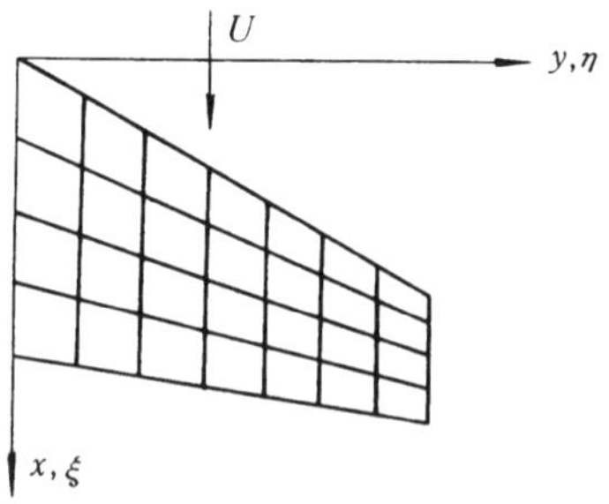

Fig. 12.19 Panel layout for wing

reduces to evaluate the integral of (12.135) with $\varphi_\xi(\xi,\eta,0)=1$ for a typical panel (Fig. 12.20).

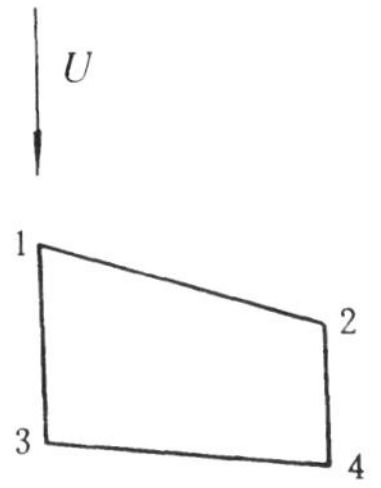

Fig. 12.20 A typical panel

The integral of (12.135) for the panel can be considered as the sum of four integrals evaluated over semi-infinite triangular regions having origins at each of the four corners (Fig. 12.21). That is the velocity potential for a

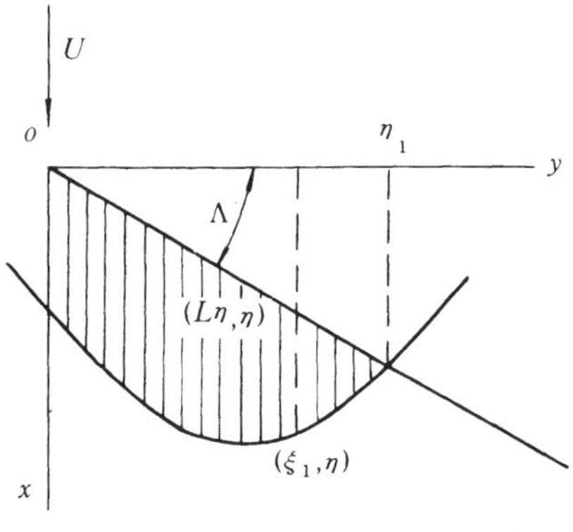

Fig. 12.21 Semi-infinite triangular region

panel is given by

$$\varphi = \varphi_1 - \varphi_2 - \varphi_3 + \varphi_4 \tag{12.138}$$

where the subscript denotes the corner of the panel (Fig. 12.20). For a typical case, the limits of the elementary integral $\varphi_k (k=1, 2, 3, 4)$ may be defined (Fig. 12.21)

$$\varphi_k(x, y, z) = \frac{z}{\pi} \int_0^{\eta_1} d\eta \int_{L\eta}^{\xi_1} \frac{(x-\xi)\,d\xi}{[(y-\eta)^2+z^2]\sqrt{(x-\xi)^2 - m^2[(y-\eta)^2+z^2]}} \tag{12.139}$$

where x, y and ξ, η are measured from corner k,

$$L = \tan\Lambda \tag{12.140}$$

Λ is leading edge sweep of the semi-infinite triangle,

$$\eta_1 = \frac{Lx - m^2y + m\sqrt{(x-Ly)^2 + (L^2-m^2)z^2}}{L^2 - m^2} \tag{12.141}$$

ξ_1 is given by Eq. (12.132).

The integral may be evaluated by elementary methods and the lengthy computation are given in reference 28. Once the velocity potential function (12.139) is obtained, the velocity potential for a single panel can be computed by the superposition of (12.138).

Associated with each panel vortex singularity used to represent the lifting surface is a control point for matching the boundary conditions given by Eq. (12.137). The aerodynamic problem is formulated as a system of linear algebraic equations relating the downwash at each control point to the unknown singularity strengths. The coefficients of this system of equations form a matrix of aerodynamic influence coefficients. The element of this matrix $a_{i,j}$ is the downwash at control point i induced by the vortex panel j of unit strength distribution. It is seen from the definition of planar vortex that

$$a_{ii} = 0 \tag{12.142}$$

In the present panel method, the downwash control point is chosen at 95% of the local panel chord through the centroid. The control point location has been determined empirically by an extensive correlation of chordwise pressure distributions for a variety of wing planforms. No theoretical argument has yet been developed to support this choice of control point.

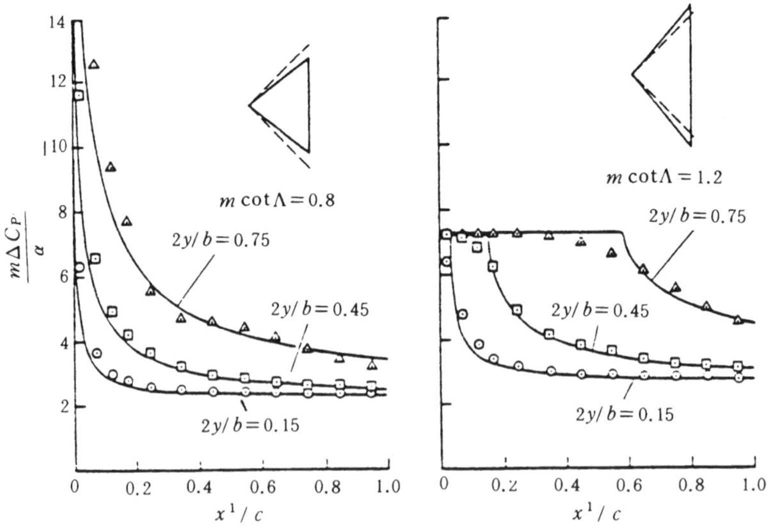

Fig. 12.22 Comparison between exact linear theory and panel method results for delta wings

The accuracy of the technique for computing pressures on isolated wings is indicated in Fig. 12.22. The panel method results are compared with exact conical flow theory for delta wings. Examples are shown for subsonic and supersonic leading-edge wings. The symbols represent the pressure coefficient jump across the wing surface given by the panel method for wing chords located at 15, 45, and 75 percent of the semispan. Here x' is the distance from the leading edge of the local chord. The solid curves were computed from conical flow theory and may be found as solutions 3 and 6 in reference 29. The results agree quite well except at the sharp ridges of the supersonic solution. This rounding of the pressures computed by this numerical method occurs because the pressures shown are actually average pressures over a wing panel.

17. Linearized unsteady subsonic flow over lifting wing

For unsteady flow, sources, doublets and vortices may be used in a manner similar to that used for steady flow except that the singularity strength is a function of time as well as of the spatial coordinates, e. g. Eq. (12.22). Since the fundamental equation (12.2) is linear, the method of superposition is applicable. In this section we will discuss the problem of unsteady subsonic flow over lifting wing. To simulate the effect of lifting surface, planar doublet distribution may be used.

The perturbation velocity potential of a point source in uniform subsonic free stream is given by Eq. (12.22). Substituting Eq. (12.19) into (12.22), we may write

$$\varphi(x, y, z, t) = -\frac{f\left[t + \dfrac{Mx}{a_\infty \beta^2} \mp \dfrac{\sqrt{x^2 + \beta^2(y^2 + z^2)}}{a_\infty \beta^2}\right]}{4\pi\sqrt{x^2 + \beta^2(y^2 + z^2)}} \tag{12.143}$$

where f is the strength of the point source located at origin (0,0,0). In this case f is a function of t only. The argument of f in (12.143) may be written as $t + \Delta t$, where

$$\Delta t = \frac{Mx}{a_\infty \beta^2} + \frac{\sqrt{x^2 + \beta^2(y^2 + z^2)}}{a_\infty \beta^2} \tag{12.144}$$

Δt is the time increment for the source affecting the point (x, y, z). For compressible flow, Δt should be negative. Therefore the negative sign in (12.144) and (12.143) has to be taken.

For a point doublet with axis parallel to z, the perturbation velocity potential may be obtained by partial differentiation of (12.143) with respect to z.

$$\varphi(x,y,z,t) = -\frac{\partial}{\partial z}\frac{f\left[t+\dfrac{Mx}{a_\infty\beta^2}-\dfrac{\sqrt{x^2+\beta^2(y^2+z^2)}}{a_\infty\beta^2}\right]}{4\pi\sqrt{x^2+\beta^2(y^2+z^2)}} \tag{12.145}$$

Here the function $f(t)$ denotes the strength of the doublet.

In the following analysis, we employ both velocity and acceleration potential. If the fluid is barotropic, there exists a function ψ such that the acceleration vector

$$\boldsymbol{a} = \nabla\psi \tag{12.146}$$

where ψ is known as acceleration potential and may be expressed in terms of pressure

$$\psi = -\int\frac{dp}{\rho} \tag{12.147}$$

With the aid of Bernoulli's theorem (5.42a), ψ is related to the velocity potential φ

$$\psi = \frac{\partial\varphi}{\partial t} + \frac{q^2}{2} \tag{12.148}$$

For small perturbation flow, Eq. (12.147) may be written as

$$\psi = -\frac{p-p_\infty}{\rho_\infty} \tag{12.149}$$

and (12.148) becomes by the use of (12.3)

$$\psi = \frac{\partial\varphi}{\partial t} + U\frac{\partial\varphi}{\partial x} \tag{12.150}$$

Therefore small perturbation acceleration potential also satisfies Eq. (12.2). For lifting wing of zero thickness, the boundary conditions on the wing plane ($z=0$) for ψ are inside the wing projection

$$\psi(x,y,0,t) = -\psi(x,y,0-,t) = \frac{\Delta p}{2\rho_\infty} \tag{12.151}$$

and outside the wing projection and on the trailing edge

$$\psi(x,y,0,t) = 0 \tag{12.152}$$

where $\Delta p = p(x,y,0-,t) - p(x,y,0+,t)$, i.e. the pressure difference across the wing surface.

Let the equation of the wing mean surface be given by

$$z = z_m(x,y,t) \tag{12.153}$$

The boundary condition on the wing mean surface for φ is

$$\frac{\partial \varphi}{\partial z}(x, y, 0, t) = \frac{\partial z_m}{\partial t} + U\frac{\partial z_m}{\partial x} \tag{12.154}$$

To relate Δp with z_m, we need to solve for both ψ and φ.

Let us consider an oscillating wing, i.e.

$$z_m(x, y, t) = \bar{z}_m(x, y)e^{i\omega t} \tag{12.155}$$

where $i = \sqrt{-1}$, ω is the angular frequency.

In this case we have

$$\varphi(x, y, z, t) = \bar{\varphi}(x, y, z)e^{i\omega t} \tag{12.156}$$

$$\psi(x, y, z, t) = \bar{\psi}(x, y, z)e^{i\omega t} \tag{12.157}$$

From Eq. (12.150)

$$\bar{\psi} = i\omega\bar{\varphi} + U\frac{\partial \bar{\varphi}}{\partial x} \tag{12.158}$$

Solving the above differential equation for $\bar{\varphi}$, we get

$$\bar{\varphi} = \frac{e^{-\frac{i\omega x}{U}}}{U}\int_{-\infty}^{x} \bar{\psi}(\lambda, y, z)e^{\frac{i\omega\lambda}{U}}\,d\lambda \tag{12.159}$$

Here we assume that $x \to -\infty$, $\bar{\varphi} = 0$.

We now construct solution $\psi(x, y, z, t)$ by superposing planar doublet distributed over the lifting surface. Let $f(\xi, \eta, t)$ be the doublet strength per unit area at the point $(\xi, \eta, 0)$. Here,

$$f(\xi, \eta, t) = \bar{f}(\xi, \eta)e^{i\omega t} \tag{12.160}$$

By the use of the fundamental solution (12.143), wh have

$$\psi(x, y, z, t) = -\frac{1}{4\pi}\iint_S \bar{f}(\xi, \eta) \times \frac{\partial}{\partial z}\frac{e^{i\omega\left\{t + \frac{M(x-\xi)}{a_\infty \beta^2} - \frac{\sqrt{(x-\xi)^2 + \beta^2[(y-\eta)^2 + z^2]}}{a_\infty \beta^2}\right\}}}{\sqrt{(x-\xi)^2 + \beta^2[(y-\eta)^2 + z^2]}}\,d\xi\, d\eta \tag{12.161}$$

where S is the wing projection on the x-y plane.

To find $\bar{f}(\xi, \eta)$ in terms of the pressure difference across the wing, we apply the boundary condition for ψ, i. e. Eq. (12.151).

The integrand of Eq. (12.161) expresses the downwash of source. If we consider a point $P(x, y, z)$ very near the plane $z=0$, the downwash depends only on the strength distribution over the neighborhood of the point $(x, y, 0)$. As $z \to 0$, we may replace the integration area S of Eq. (12.161) by a small circle, centered at $(x, y, 0)$ with radius ε (Fig. 12.23), over which $\bar{f}(\xi, \eta)$ may be

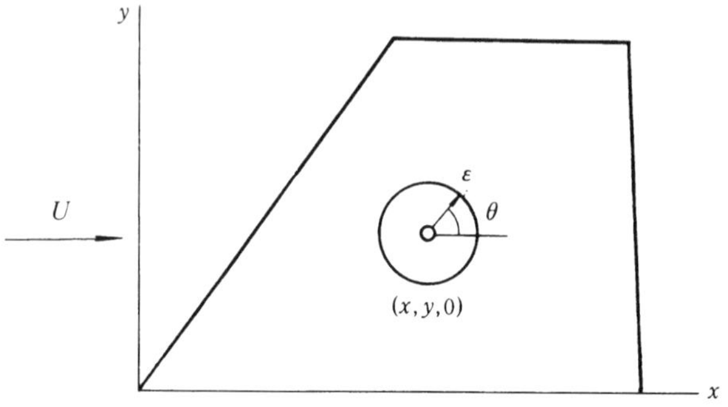

Fig. 12.23 Integral area

considered as a constant value $\bar{f}(x, y)$. Thus

$$\psi(x, y, 0, t) = \lim_{z \to 0} \frac{\bar{f}(x, y)}{4\pi} \int_0^{\varepsilon} \int_0^{2\pi} e^{i\omega\left\{t - \frac{M}{a_\infty \beta^2} r\cos\theta - \frac{\sqrt{r^2+\beta^2 z^2}}{a_\infty \beta^2}\right\}}$$

$$\times \left[\frac{1}{(r^2+\beta^2 z^2)^{3/2}} + \frac{i\omega}{a_\infty \beta^2} \frac{1}{r^2+\beta^2 z^2} \right] \beta z r dr d\theta \qquad (12.162)$$

Integrating by parts with respect to r, we obatin

$$\psi(x, y, 0, t) = \lim_{z \to 0} \frac{\bar{f}(x, y)}{4\pi} \int_0^{2\pi} e^{i\omega\left(t - \frac{|z|}{a_\infty \beta}\right)} \frac{z}{|z|} d\theta$$

$$= \frac{\bar{f}(x, y)}{2} e^{i\omega t} \qquad (12.163)$$

By the use of boundary condition (12.151), writing

$$\Delta p = \overline{\Delta p}(x, y) e^{i\omega t} \qquad (12.164)$$

we have

$$\bar{f}(x, y) = \frac{\overline{\Delta p}(x, y)}{\rho_\infty} \qquad (12.165)$$

The general expression of the acceleration potential at a point (x, y, z)

due to an oscillating lifting surface is

$$\psi(x,y,z,t)=\frac{-1}{4\pi\rho_\infty}\iint_S\overline{\Delta p}(\xi,\eta)\times$$

$$\frac{\partial}{\partial z}\frac{e^{i\omega\left\{t+\frac{M(x-\xi)}{a_\infty\beta^2}-\frac{\sqrt{(x-\xi)^2+\beta^2[(y-\eta)^2+z^2]}}{a_\infty\beta^2}\right\}}}{\sqrt{(x-\xi)^2+\beta^2[(y-\eta)^2+z^2]}}d\xi d\eta \tag{12.166}$$

where $\overline{\Delta p}$ denotes the pressure difference across the wing surface, and S is the wing projection on the $x-y$ plane.

To determine $\overline{\Delta p}(x,y)$, we now apply the boundary condition for the velocity potential φ, i.e. Eq. (12.154). Combining Eqs. (12.159) and (12.166), we obtain the general expression of φ

$$\varphi(x,y,z,t)=-\frac{1}{4\pi\rho_\infty U}\iint_S\overline{\Delta p}(\xi,\eta)e^{-\frac{i\omega(x-\xi)}{U}}$$

$$\times\frac{\partial}{\partial z}\int_{-\infty}^{x-\xi}\frac{e^{i\omega\left\{t+\frac{1}{a_\infty\beta^2}[\lambda-M\sqrt{\lambda^2+\beta^2(y-\eta)^2+\beta^2z^2}]\right\}}}{\sqrt{\lambda^2+\beta^2[(y-\eta)^2+z^2]}}d\lambda d\xi d\eta \tag{12.167}$$

Substituting Eq. (12.167) into (12.154), we get the integral equation for $\overline{\Delta p}(x,y)$. Panel methods are used to solve the integral equation for $\overline{\Delta p}(x,y)$. The difficulty is the evaluation of the integral in Eq. (12.167) over a panel with $\overline{\Delta p}(x,y)=1$. Approximations are made in the evaluation, e.g. Ref. 30.

18. Problems

1. By the Prandtl-Glauert rule, show that for subsonic flow, the maximum pressure coefficient for an ellipsoid of small thickness ratio δ is given by the following formula

$$C_{P\max}=2\delta^2\log(1.36\sqrt{1-M^2}\,\delta)$$

where M is the free stream Mach number.

2. Discuss the flow patterns of a simple doublet in compressible flows (a) $M=0$, (b) $M<1$, and (c) $M>1$, where M is the Mach number of the uniform stream upon which the linearized theory is based.

3. Work out the detailed derivation for the lift coefficient of equation (12.31).

4. Using the von Kármán-Moore method (12.43), find the flow of a uniform stream over a slender ellipsoid.

5. Calculate the wave drag of a uniform rectangular wing of the double wedge section:

$$z'(x)=\delta \qquad \text{for} \quad 0\leqslant x\leqslant c/2$$
$$z'(x)=-\delta \qquad \text{for} \quad c\geqslant x\geqslant c/2$$

c=chord of the wing in a uniform stream of Mach number M at zero angle of attack for the following cases.

(a) Aspect ratio $AR>2/m$, $m=\sqrt{M^2-1}$

(b) $1/m<AR<2/m$

(c) $AR<1/m$

6. Using the source method, calculate the drag of a rectangular wing in oscillatory motion in a uniform stream of Mach number M.

7. Show the detailed derivation of the expression for the velocity potential of a multipole given by (12.74).

8. For small harmonic oscillation of a wing, with frequency k, we may write the velocity potential $\varphi(x,y,z,t)$ in the following form:

$$\varphi(x,y,z,t)=\varphi_0(x,y,z)+\varphi_1(x,y,z)e^{ik\left(t-\frac{M^2}{1-M^2}\frac{x}{U}\right)}$$

Find the differential equations for φ_0 and φ_1 and discuss the boundary conditions for φ_0 and φ_1 in terms of the shape of the wing.

9. Derive the perturbation velocity potential of a symmetric wing at zero angle of attack and subsonic speed,

$$\varphi(x,y,z)=-\frac{U}{2\pi}\iint_S \frac{\dfrac{\partial\bar{z}(\xi,\eta)}{\partial z}\,d\xi d\eta}{\sqrt{(x-\xi)^2+\beta^2[(y-\eta)^2+z^2]}}$$

where $\bar{z}(x,y)$ is the upper surface of the wing, and S is the wing area in the x-y plane.

10. Derive the perturbation velocity potential of a lifting wing of zero thickness at subsonic speed,

$$\varphi(x,y,z)=\frac{z}{2\pi}\iint_S \frac{\varphi_\xi(\xi,\eta,0)}{(y-\eta)^2+z^2}\left\{1+\frac{x-\xi}{\sqrt{(x-\xi)^2+\beta^2[(y-\eta)^2+z^2]}}\right\}d\xi d\eta$$

where $\varphi_x(x,y,0)$ is evaluated on the upper surface of the wing, and S is the wing area in the x-y plane.

References

1. Busemann, A., Infinitesimalo Kogelige Überschallstroemung, *Deu, Akam, der Luft*, 1942–43. *NACA*, *TM* No. 1100, March 1947.

2. Gothert, B., Plane and three dimensional flow at high subsonic speeds, *NACA, TM* No. 1105, Oct. 1946.
3. Sears, W. R., A second note on compressible flow about bodies of revolution, *Quart, App. Math.* **5**, April 1947, p. 89.
4. von Kármán, Th., Supersonic aerodynamics— principles and applications, *Jour, Aero, Sci.* **14**, No. 7, July 1947, pp. 373 –402.
5. von Kármán, Th., and Moore, N. B., Resistance of slender bodies moving with supersonic velocities with special reference to projectiles, *Trans. ASME* **54**, No. 23, pp. 303 –310, 1932.
6. von Kármán, Th., The problem of resistance in compressible fluids, *Proc. Fifth Volta Congress*, R. Acad. D'Italia, Rome, pp. 210 –269, 1936.
7. Ward, G. N., Supersonic flow past slender pointed bodies, *QJMAM* **II**, 1949, pp. 75 –97.
8. Goldstein, S., Linearized theory of supersonic flow, *Lecture Series* No. 2, Inst. for Fluid Dynamics and Appl. Math., University of Maryland, 1950.
9. Heaslet, M. A., and Lomax, H., Source-sink and doublet disturbance in supersonic flow, *NACA, TR* No. 900.
10. Evvard, J. C., Use of source distribution for evaluating theoretical aerodynamics of thin finite wings at supersonic speeds, *NACA, TR* No. 951, 1950.
11. Heaslet, M. A., and Lomax, H., Supersonic and transonic small perturbation theory, Part D of General theory of high speed aerodynamics, edited by W. R. Sears, **6** of *High Speed Aerodynamics and Jet Propulsion*, Princeton University Press, 1954.
12. Ward, G. N., *Linearized Theory of Steady High Speed Flow*, Cambridge University Press, 1955.
13. Ursell, F., and Ward, G. N., On some general theorems in the linearized theory of compressible flow, *QJMAM* III, 1950, pp. 326 –348.
14. Lagerstrom, P. A., Linearized supersonic theory of conical wings, *NACA, TN* No. 1685, Jan. 1950.
15. Goldstein, S. and Ward, G. N., The linearized theory of conical field in supersonic flow, with application to plane airfoils, *Aero. Quart.* **II** pt. 1, May 1950.
16. Germain, P., General theory of conical flows and its application to supersonic aerodynamics, *NACA, TM* No. 1354, Jan. 1955.
17. Lighthill, M. J., Higher approximations, Part E of General theory of high speed aerodynamics, edited by W. R. Sears, **6** of *High Speed Aerodynamics and Jet Propulsion*, Princeton University Press. 1954.
18. Tsien, H. S., Poincaré-Lighthill-Kuo method, *Adv. in Mechanics* **IV**, 1956, Academic Press. Inc.
19. Whitham, G. B., The behavior of supersonic flow past a body of revolution far from the axis, *Proc. Roy. Soc. London* A –**201**, 1950, p. 89.
20. Kuessner, H. G., A general method for solving problems of the unsteady lifting surface theory in the subsonic range, *Jour. Aero. Sci.* **21**, No. 1, Jan. 1954, pp. 17 –26.
21. Chang, C. C., Transient aerodynamic behavior of an airfoil due to arbitrary modes of non-stationary motions in a supersonic flow, *NACA, TN* 2333, 1951.
22. Li, T. Y., and Stewart, H. J., On an integral equation in the supersonic oscillating wing theory, *Jour. Aero. Sci.* **20**, No. 10, 1953, p. 724 –726.
23. Fung, Y. C., *An Introduction to the Theory of Aeroelasticity*, John Wiley and Sons, Inc., New York, 1955.
24. Miles, J. W., Unsteady supersonic flow, *Report of ARDC and U.C.L.A.*, 1955.
25. Oswatitsch, K., Special problems of inviscid steady transonic flows, *Int. Jour. of Eng. Sci.*, **20**, No. 4, pp. 497 –540, April, 1982.
26. Hayes, W. D., Linearized supersonic Flows, *Ph. D. thesis, Cal. Inst. of Tech.* and *report N. American Aviation Inc. Co.* No. AL –222, 1947.
27. Whitcomb, R. T., A study of zero-lift dragrise characteristics of wing-body combinations near the

speed of sound, *NACA RM* L52H08, 1952.

28. Woodward, F. A., Tinoco, E. N., and Laren, J. W., Analysis and design of supersonic wing-body combinations, including flow properties in the near field, Part I: Theory and application, *NASA-CR*-73106, Aug. 1967.
29. Jones, R. T., and Cohen, D., *High Speed Wing Theory*, Princeton University Press, 1960, pp. 156 – 157.
30. Albano, E., and Rodden, W. P., A doublet-lattice method for calculating lift distributions on oscillating surfaces in subsonic flows, *AIAA Journal*, 7, No. 2, Feb. 1969, pp. 279 – 285.

Chapter XIII

NONLINEAR THEORY OF THREE-DIMENSIONAL COMPRESSIBLE FLOW

1. Introduction

In this chapter we shall discuss some analyses of three-dimensional compressible flow which include the nonlinear effects. In § 2, we shall discuss the lost solution of the exact equation for steady axially symmetric compressible flow, this being analogous to two-dimensional Prandtl-Meyer flow. This lost solution is much more restrictive than the corresponding Prandtl-Meyer solution. However, as discussed in § 2, it may be used to describe the flow over a cone at zero angle of attack.

In § 3, the method of characteristics for steady axially symmetrical irrotational flow will be discussed. The difference between this case and that of two-dimensional case will be indicated.

Like the two-dimensional cases (Chapter X), it is very difficult to find the exact solution for the nonlinear equation. Consequently, it is useful to develop similarity laws. In § 4 to §6, we shall discuss various similarity laws for transonic flow (§ 4), hypersonic flow (§ 5), and supersonic flow for a finite wing (§ 6).

In § 7, the method of characteristics with three variables will be briefly discussed.

In § 8, nonlinear effects in nonsteady flow will be discussed.

In § 9, the zonal method of the slender body theory and the finite difference method will be discussed for a three-dimensional potential flow. The numerical solution of the unsteady transonic flows will be given in § 10.

2. Supersonic flow over a cone at zero angle of attack[1-4]

The flow field around a semi-infinite cone in a uniform supersonic flow parallel to its axis (Fig. 13.1) may be described by the lost solution of compressible irrotational flow in the axially symmetrical case. The fundamental equations are

$$\left(1-\frac{u^2}{a^2}\right)\frac{\partial u}{\partial x}-\frac{uv}{a^2}\left(\frac{\partial v}{\partial x}+\frac{\partial u}{\partial r}\right)+\left(1-\frac{v^2}{a^2}\right)\frac{\partial v}{\partial r}+\frac{v}{r}=0 \qquad (13.1)$$

$$\frac{\partial v}{\partial x}-\frac{\partial u}{\partial r}=0 \qquad (13.2)$$

where x is the axial coordinate, r is the radial coordinate, u and v are respectively the x- and r-components of the velocity, and a is local speed of sound.

For the so-called lost solution, we assume that, v is a function of u only, i.e.,

$$v=v(u) \qquad (13.3)$$

Substituting Eq. (13.3) into (13.1) and (13.2), and simplifying, one obtains

$$\frac{d^2v}{du^2}\frac{\partial u}{\partial x}+\frac{1}{r}=0 \qquad (13.4)$$

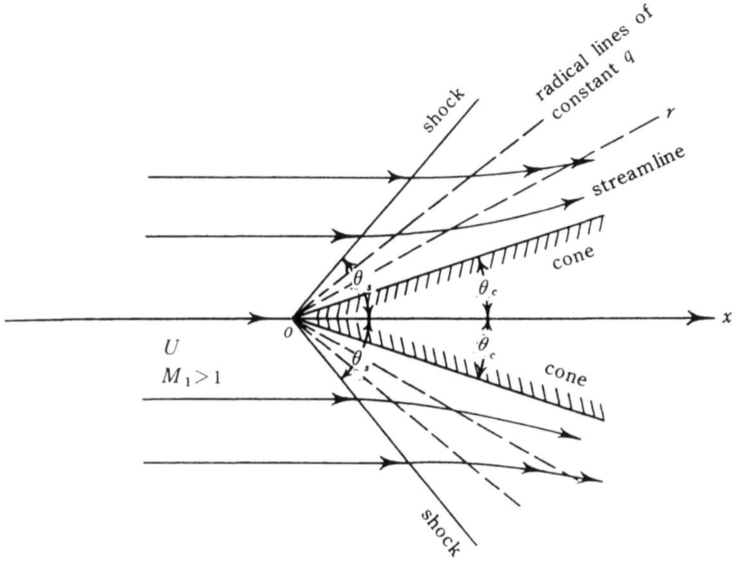

Fig. 13.1 Supersonic flow over a cone

Integrating Eq. (13.4) with respect to x gives

$$\frac{dv}{du} = \frac{f(r) - x}{r} \tag{13.5}$$

where $f(r)$ is an arbitrary function of r, still to be determined. For a surface of constant value of u

$$du = \frac{\partial u}{\partial x}(dx)_u + \frac{\partial u}{\partial r}(dr)_u = 0$$

or

$$\left(\frac{dr}{dx}\right)_u = -\frac{\partial u/\partial x}{\partial u/\partial r} = -\frac{1}{dv/du} = \text{constant} \tag{13.6}$$

Thus the lines of constant values of u and v are straight lines and the undetermined function $f(r)$ in Eq. (13.5) must be a constant. Without loss of generality, we can set $f(r) = 0$ and have, for constant values of u and v, the

$$r = -\frac{x}{dv/du} \tag{13.7}$$

surface Eq.(13.7) shows that the lines of constant u and v are radial lines from the origin for the lost solution in the axially symmetrical case. This is much more restrictive than the Prandtl-Meyer flow in which the straight lines of constant velocity need not pass through the same point. This lost solution may be used to describe the flow over a cone in a supersonic uniform flow at zero angle of attack (Fig. 13.1). Let θ_c be the semivertex angle of the cone, and θ_s be the semivertex angle of the conical shock. The boundary conditions are

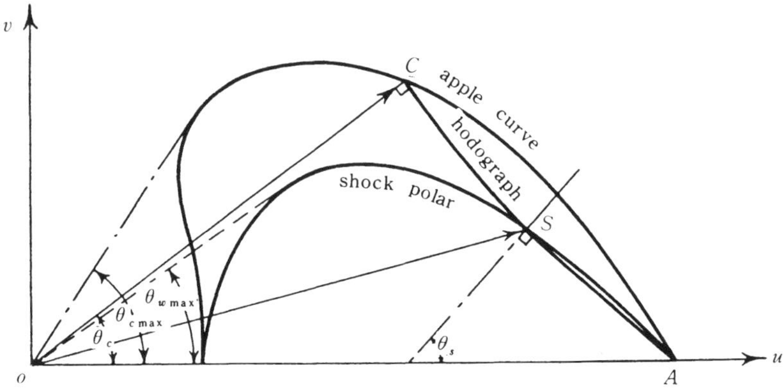

Fig. 13.2 Hodograph, apple curve and shock polar for a supersonic flow a cone at zero angle of attack

such that, on the surface ($\theta = \theta_c$), the velocity vector is tangent to the surface, and on the surface immediately behind the shock ($\theta = \theta_s$), the velocity vector has a constant magnitude which satisfies the Rankine-Hugoniot relations. The lost solution satisfies these conditions. In fact, along any radial line from the origin and between the angles θ_c, and θ_s, the velocity vector is constant. Of course, in general (except at $\theta = \theta_c$), the velocity vector need not be radial.

Let us study the flow field over the cone in the hodograph plane shown in Fig. 13. 2. $OA = U$ is the velocity in front of the shock. This velocity is a constant for a given problem. From OA we may draw the shock polar in accordance with the Rankine-Hugoniot relation (4.34). The point s represents the velocity vector Os immediately behind the shock of shock angle θ_s as shown in the figure. The low between the shock and the shock the surface of the cone is governed by the equation

$$v\frac{d^2v}{du^2} - \left(1 - \frac{v^2}{a^2}\right)\left(\frac{dv}{du}\right)^2 + \frac{2uv}{a^2}\frac{dv}{du} - \left(1 - \frac{u^2}{a^2}\right) = 0 \quad (13.8)$$

obtained form Eqs. (13.1) and (13.4). The boundary conditions for Eq. (13.8) are, (i) at $\theta = \theta_c$, $v/u = \tan\theta_c$, and (ii) at $\theta = \theta_s$, the Rankine-Hugoniot relations are satisfied. In order to solve this equation, numerical or graphical integration must be used. These calculations had been carried out by Taylor[1], Maccoll[2], Busemann[3], and Kopal[4].

One interesting result is that after the shock the flow is continuously compressed until the surface is reached. The hodograph for the flow behind the shock is shown in Fig. 13.2 as the curve sc. The locus of the point s is on the shock polar for a given U, whereas the corresponding locus of the point c will be on another curve first found by Busemann[3] known as the apple curve. The apple curve serves the same function in the flow over a cone as does the shock polar in the flow over a wedge. The maximum cone angle for an attached shock is much larger than the maximum deflection for the case of the wedge

Since the flow is continuously compressed after the shock, there may be transonic flow behind the shock. If the resultant velocity immediately behid the shock is q_s satisfying the relation

$$\frac{q_s^2}{q_m^2} < \frac{\gamma - 1}{\gamma + 1},$$

where q_m is the maximum possible velocity of the flow considered, the flow behind the shock and in front of the cone will be subsonic. If the velocity on the surface of the cone satisfies the relation

$$\frac{q_c^2}{q_m^2} > \frac{\gamma - 1}{\gamma + 1},$$

the flow behind the shock and in front of the cone surface will be supersonic. If at certain angle θ_I, where $\theta_s > \theta_I > \theta_c$, the resultant velocity satisfies the relation

$$\frac{q^2(\theta_1)}{q_m^2} = \frac{\gamma - 1}{\gamma + 1},$$

the flow is supersonic immediately behind the shock. It becomes subsonic when $\theta < \theta_l$.

In Ref. 4, the flow properties of a cone in a uniform stream have been tabulated.

3. Method of characteristics for axially symmetrical irrotational steady flow[5,6]

The general theory of the characteristics of equations of two variables discussed in Chapter XI is applicable to the case of axially symmetric steady flow. The difference between the axially symmetric flow and two-dimensional flow lies in the fact that the equations for axially symmetric flow are irreducible. As a result the characteristics in the hodograph plane depend on the equations as well as on the initial values. Thus in practical calculations, the method of characteristics for axially symmetric flow is much more tedious than the corresponding two-dimensional case. We now derive the characteristic equations for the axially symmetrical steady irrotational flow. The fundamental equations for this case are

$$(a^2-u^2)\frac{\partial u}{\partial x} - uv\frac{\partial u}{\partial r} - uv\frac{\partial v}{\partial x} + (a^2 - v^2)\frac{\partial v}{\partial r} + \frac{a^2 v}{r} = 0 \qquad (13.9)$$

$$\frac{\partial u}{\partial r} - \frac{\partial v}{\partial x} = 0 \qquad (13.10)$$

Comparing Eqs. (13.9) and (13.10) with Eqs. (11.8) shows

$$A_1 = a^2-u^2,\quad B_1 = -uv,\quad C_1 = -uv,\quad D_1 = a^2-v^2,\quad E_1 = \frac{a^2 v}{r} \qquad (13.11)$$

$$A_2 = 0,\quad B_2 = 1,\quad C_2 = -1,\quad D_2 = 0,\quad E_2 = 0$$

From Eqs. (11.22) and (13.11) we have the physical characteristics for the axially symmetrical steady flow as

$$\left(\frac{dr}{dx}\right)_{\pm} = \frac{uv \pm a\sqrt{u^2+v^2-a^2}}{u^2-a^2} = \tan(\theta \mp \alpha) \qquad (13.12)$$

and $C' = \left(\frac{dr}{dx}\right)_{+} = \tan(\theta - \alpha)$, $C'' = \left(\frac{dr}{dx}\right)_{-} = \tan(\theta \mp \alpha)$.

Hence the physical characteristics are the same as those of two-dimensional flow (11.27) because the coefficients of the first derivatives A_1 etc., are the same.

The hodograph characteristics for the axially symmetrical flow are

$$\begin{vmatrix} B_1 & C_1 & D_1 & E_1 \\ B_2 & C_2 & D_2 & E_2 \\ r_\sigma & 0 & 0 & -u_\sigma \\ 0 & x_\sigma & r_\sigma & -v_\sigma \end{vmatrix} = 0 \qquad (13.13)$$

or

$$2uvdudr+(a^2-v^2)dudx-(a^2-v^2)dvdr-\frac{a^2v}{r}(dr)^2=0 \quad (13.13a)$$

Eq. (13.13a) may be written in the form

$$(a^2-u^2)\left(\frac{dr}{dx}\right)_{\pm}du+(a^2-v^2)dv+\frac{a^2v}{r}dr=0 \quad (13.14)$$

where $(dr/dx)_{\pm}$ are given by Eq. (13.12).

Using the polar coordinates q, θ, where q is the magnitude of the velocity and θ is the flow deflection with respect to x-axis, Eq. (13.14) becomes

$$\left(\cos\theta \mp \frac{\sqrt{q^2-a^2}}{a}\sin\theta\right)\left(d\theta \pm \frac{\sqrt{q^2-a^2}}{a}\frac{dq}{q}\right)+\sin\theta\frac{dr}{r}=0 \quad (13.15)$$

If we assume that

$$\cos\theta \neq \pm\frac{\sqrt{q^2-a^2}}{a}=\pm\cot\alpha \quad (13.16)$$

where $\alpha=\sin^{-1}(1/M)$ = local Mach angle, Eq. (13.15) becomes

$$d\theta \pm \cot\alpha\frac{dq}{q}+\frac{\sin\theta\sin\alpha}{\sin(\alpha\mp\theta)}\frac{dr}{r}=0 \quad (13.17)$$

These are the equations for the hodograph characteristics Γ' and Γ'' corresponding to the physical characteristics C' and C'' respectively. They are identical to those of the two-dimensional case, except for the last term. The hodograph characteristics depend on radial distance r as well as q and θ. Thus the hodograph characteristics vary with the initial values. As a result, the actual computation is much more complicated than that for the two-dimensional case.

In the event that

$$\cot\theta = \pm\cot\alpha, \text{ or } \theta = \pm\alpha$$

Eq. (13.17) breaks down. Actually this means that

$$\theta=\pm\alpha,\ dr=0, \text{ and } \frac{dr}{\sin(\theta\pm\alpha)}=\frac{0}{0}$$

This indeterminacy can be eliminated if we use the curvilinear coordinates corresponding to the physical characteristics instead of x and r.
We define

$$\begin{aligned}\xi(x,r) &= \text{constant along characteristics } C''\\ \eta(x,r) &= \text{constant along characteristics } C'\end{aligned} \quad (13.18)$$

Then Eq. (13.17) becomes

$$d\theta \pm \cot\alpha \frac{dq}{q} \mp \frac{\sin\theta \sin\alpha}{r} d\begin{Bmatrix} \xi \\ \eta \end{Bmatrix} = 0 \qquad (13.19)$$

In Ref. 5, Meyer derives the characteristic equations based on the curvilinear coordinates.

Another difficulty in using Eq. (13.19) occurs when $r=0$. For this situation

$$\frac{\sin\theta}{r} = \frac{0}{0}$$

To avoid this difficulty, we use the limiting expression for $\sin\theta/r$ near $r=0$. This expression may be found in the following manner:

$$\lim_{r\to 0} \frac{\sin\theta \sin\alpha}{r} = (\sin\alpha) \lim_{0\, r\to 0} \frac{\Delta\theta}{\Delta r} = \frac{a_0}{q_0^2} \lim_{r\to 0} \frac{\partial v}{\partial r} \qquad (13.20)$$

where the subscript 0 refers to value at $r=0$.

From the equation of continuity,

$$\lim_{r\to 0} \frac{\partial v}{\partial r} = -\frac{1}{2} \lim_{r\to 0} \left(\frac{\partial u}{\partial x} + \frac{u}{\rho} \frac{\partial \rho}{\partial x} \right) \qquad (13.21)$$

From the x-componet of the equation of motion

$$\lim_{r\to 0} \frac{u}{\rho} \frac{\partial \rho}{\partial x} = \lim_{r\to 0} \left(-\frac{u^2}{a^2} \frac{\partial u}{\partial x} \right) = \lim_{x\to 0} \left(-M^2 \frac{\partial u}{\partial x} \right) \qquad (13.22)$$

Furthermore $u_0 = q_0$, $\left(\frac{\partial u}{\partial x}\right)_0 = \left(\frac{\partial q}{\partial x}\right)_0$. Then

$$\lim_{r\to 0} \frac{\sin\theta \sin\alpha}{r} = \left(\frac{\sin\alpha \cot^2\alpha}{2q} \frac{\partial q}{\partial x} \right)_0 \qquad (13.23)$$

Finally on the axis ($r=0$), Eq. (13.19) becomes

$$d\theta \pm \cot\alpha \frac{dq}{q} \mp \frac{1}{2} \frac{\cos^2\alpha}{q \sin\alpha} \frac{\partial q}{\partial x} d\begin{Bmatrix} \xi \\ \eta \end{Bmatrix} = 0 \qquad (13.24)$$

It is sometimes convenient to use the local Mach angle in place of the magnitude of velocity q. With this Eq. (13.17) becomes

$$d\theta \mp \frac{2\cos^2\alpha}{\gamma - 1 + 2\sin^2\alpha} \mp \frac{\sin\theta \sin\alpha}{\sin(\theta \pm \alpha) r} \frac{dr}{r} = 0 \qquad (13.25)$$

For numerical calculation, the characteristic equations are replaced by finite defference equation and solved using a stepwise procedure in exactly the same manner as in the two-dimensional case .

4. Transonic similarity law[7]

In a manner analogous to the two-dimensional case, we may find similarity laws for transonic flow in the axially symmetric case. For irrotational steady flow, the exact differential equation of φ for the axially symmetrical case is

$$\left(1-\frac{u^2}{a^2}\right)\frac{\partial^2\varphi}{\partial x^2}-\frac{2uv}{a^2}\frac{\partial^2\varphi}{\partial x\partial r}+\left(1-\frac{v^2}{a^2}\right)\frac{\partial^2\varphi}{\partial r^2}+\frac{1}{r}\frac{\partial\varphi}{\partial r}=0 \quad (13.26)$$

If the velocity is very close to the local sonic velocity in the whole field, we may write

$$u=a^*+\frac{\partial\varphi}{\partial x}, \quad v=\frac{\partial\varphi}{\partial r} \quad (13.27)$$

where φ is the perturbed velocity potential, a is the local sound velocity, and a^* is the critical sound velocity given by

$$a^2+\frac{\gamma-1}{2}(u^2+v^2)=\frac{\gamma+1}{2}a^{*2} \quad (13.28)$$

For a slender body with thickness ratio, δ, we make the transformation

$$x=L\xi, \quad r=L(\delta\sqrt{\Gamma})^{-n}\eta$$
$$\varphi=La^*\frac{1-M_\infty}{\Gamma}F(\xi,\eta) \quad (13.29)$$

where L is the length of the body and $\Gamma=(\gamma+1)/2$. M_∞=Mach number for the free stream.

Substituting Eq. (13.29) into (13.26), the effective first-order terms are

$$\frac{\partial^2F}{\partial\eta^2}+\frac{1}{\eta}\frac{\partial F}{\partial\eta}=2\frac{1-M_\infty}{(\delta\sqrt{\Gamma})^{2n}}\frac{\delta F}{\partial\xi}\frac{\partial^2F}{\partial\xi^2} \quad (13.30)$$

The boundary conditions are

(a) At infinite

$$\frac{\partial F}{\partial\xi}=-1, \quad \frac{\partial F}{\partial\eta}=0 \quad (13.31)$$

(b) On the surface of the body of revolution

$$\left(r\frac{\partial\varphi}{\partial r}\right)_{r\to 0}=U\bar{r}\frac{d\bar{r}}{dx} \quad (13.32)$$

where $\bar{r}$ is the radius of the body, and U is the velocity of the free stream. In nondimensional form, we have

$$\frac{1-M_\infty}{\Gamma\delta^2}\left(\eta\frac{\partial F}{\partial\eta}\right)_{\eta\to 0}=g(\xi)g'(\xi) \quad (13.33)$$

where $g(\xi)=\bar{r}/L$ is the nondimensional radius of the body. The boundary condition (13.33) is different from the corresponding formula for two-dimensional case. Hence a different transonic similarity parameter will be obtained. This is

$$H=\frac{1-M_{\infty}}{\Gamma\delta^{2}} \tag{13.34}$$

Finally the differential equaiton becomes

$$\frac{\partial^{2}F}{\partial\eta^{2}}+\frac{1}{\eta}\frac{\partial F}{\partial\eta}=2H\frac{\partial F}{\partial\xi}\frac{\partial^{2}F}{\partial\xi^{2}} \tag{13.35}$$

with the boundary conditions:

At infinity, $\frac{\partial F}{\partial\xi}=-1,\ \frac{\partial F}{\partial\eta}=0$

$$\text{On the body, } \left(\eta\frac{\partial F}{\partial\eta}\right)_{\eta\to 0}=\frac{1}{H}g(\xi)g'(\xi) \tag{13.36}$$

Hence if the transonic parameter H is the same, the flow patterns are the same for affinely related bodies, and the flow fields are determined by a single function F.

We may obtain the similarity laws for this case in exactly the same manner as for the two-dimensional case.

For the pressure coefficient C_p, we have

$$C_p=-2\frac{\Delta q}{U}=-\frac{2(1-M_{\infty})}{\Gamma}\left(1+\frac{\partial F}{\partial\xi}\right)-\delta^{2}[g'(\xi)]^{2} \tag{13.37}$$

or

$$C_p=-\delta^{2}\left\{2H\left(1+\frac{\partial F}{\partial\xi}\right)+[g'(\xi)]^{2}\right\}=\delta^{2}P(\xi,H) \tag{13.37a}$$

which is the desired similarity law for pressure coefficient C_p.

Similarly the drag coefficient based on the cross-sectional area of the body of revolution is

$$C_D=\delta^{2}D(H) \tag{13.38}$$

which is the desired similarity law for the drag coefficient.

5. Hypersonic similarity law[8,9]

In a related manner, we may derive the similarity laws for the flow over an axially symmetric slender body . We consider again irrotational flow though , as we shall see in Chapter XIV , the results apply to the rotational flow too .

With the transformation

$$x=L\xi,\ r=L\delta\eta,\ \varphi=a_{\infty}L\frac{1}{M_{\infty}}f(\xi,\eta) \tag{13.39}$$

where $M_{\infty}=\frac{U}{a_{\infty}}$

the nondimensional effective first-order differential equation for velocity potential φ becomes

$$\left[1-(\gamma-1)\frac{\partial f}{\partial \xi}-\frac{\gamma+1}{2}\frac{1}{K^2}\left(\frac{\partial f}{\partial \eta}\right)^2\right]\frac{\partial^2 f}{\partial \eta^2}$$

$$+\left[1-(\gamma-1)\frac{\partial f}{\partial \xi}-\frac{\gamma-1}{2}\frac{1}{K^2}\left(\frac{\partial f}{\partial \eta}\right)^2\right]\frac{1}{\eta}\frac{\partial f}{\partial \eta}=K^2\frac{\partial^2 f}{\partial \xi^2}+2\frac{\partial f}{\partial \eta}\frac{\partial^2 f}{\partial \xi \partial \eta} \tag{13.40}$$

with boundary conditions:

At infinity,

$$\frac{\partial f}{\partial \xi}=-1,\quad \frac{\partial f}{\partial \eta}=0$$

On the surface of the body,

$$\left[\eta\left(\frac{\partial f}{\partial \eta}\right)\right]_{\eta\to 0}=K^2 h(\xi) \tag{13.41}$$

where

$$K=\delta M_\infty=\text{hypersonic similarity parameter} \tag{13.42}$$

The similarity parameter is the same as that for the two-dimensional case.

In a similar way we obtain the similarity laws:

$$C_p=\delta^2 P(k,\gamma,\xi)$$
$$C_D=\delta^2 D(K,\gamma) \tag{13.43}$$

6. Supersonic similarity law[10]

In general, similarity laws for bodies moving in a compressible fluid may be found for bodies having at least one of the two body dimensions perpendicular to the main flow direction small in comparison to that in the dimension parallel to the main flow. In linearized theory, well known similarity laws have been obtained by Glauert and Prandtl for subsonic flow, and by Ackeret for supersonic flow (Chapter VI). However it will be shown later that the similarity laws for linearized theory are arbitrary. Unique similarity laws can be determined only when nonlinear terms are considered.

With the consideration of the nonlinear terms, we have two different types of similarity laws: transonic similarity laws and hyphersonic similarity laws. In this section we shall first discuss these two types of similarity laws for three-dimensional irrotational steady supersonic flow over wings of finite span (i.e., bodies longer axially symmetrical or slender as discussed in § 4 and § 5) and then generalize these laws to large Mach number range.

The differential equation for the velocity potential Φ of a three-dimensional steady irrotational flow of compressible fluid is

$$(a^2-\Phi_x^2)\Phi_{xx}+(a^2-\Phi_y^2)\Phi_{yy}+(a^2-\Phi_z^2)\Phi_{zz}$$

$$-2\Phi_x\Phi_y\Phi_{xy}-2\Phi_y\Phi_z\Phi_{yz}-2\Phi_z\Phi_x\Phi_{zx}=0 \tag{13.44}$$

where x, y, z are the Cartesian coordinates, the subscripts denote partial differentiation, i. e., $\Phi_x=\partial\Phi/\partial x$, etc., and a is the local sound speed, determined by the equation

$$a^2+\frac{\gamma-1}{2}(\Phi_x^2+\Phi_y^2+\Phi_z^2)=a_0^2 \tag{13.45}$$

Here γ is the ratio of the specific heats, and a_0 is the speed of sound in the gas at rest.

Now if a thin wing of finite span is placed in an otherwise uniform stream of velocity V in the x-direction, we may introduce a perturbed velocity potential φ such that

$$\Phi=V(x+\varphi) \tag{13.46}$$

with $\dfrac{\partial\varphi}{\partial x},\ \dfrac{\partial\varphi}{\partial y},\ \dfrac{\partial\varphi}{\partial z}\ll 1.$

Eq. (13.45) may be written as follows:

$$\frac{a^2}{a_1^2}=1-\frac{\gamma-1}{2}M^2(2\varphi_x+\varphi_x^2+\varphi_y^2+\varphi_z^2) \tag{13.47}$$

where a_1 is the speed of sound corresponding to the free stream velocity V, and $M=V/a_1$, the Mach number in the undisturbed stream.

Substituting Eqs. (13.46) and (13.47) into Eq. (13.44) and retaining terms up to the second order, one has, with $\lambda^2=(\gamma-1)/(\gamma+1)$,

$$\left[1-M^2-(\gamma+1)M^2\varphi_x-\frac{\gamma-1}{2}M^2(\varphi_y^2+\varphi_z^2)\right]\varphi_{xx}$$

$$+\left[1-(\gamma-1)M^2\varphi_x-\frac{\gamma+1}{2}M^2(\varphi_y^2+\lambda^2\varphi_z^2)\right]\varphi_{yy} \tag{13.48}$$

$$+\left[1-(\gamma-1)M^2\varphi_x-\frac{\gamma+1}{2}M^2(\lambda^2\varphi_y^2+\varphi_z^2)\right]\varphi_{zz}-2M^2\varphi_y\varphi_{xy}-2M^2\varphi_z\varphi_{xz}=0$$

This is the fundamental differential equation for φ for the consideration of similarity laws. In Eq. (13.48) we have retained terms which while not important everywhere, may be so in certain regions. For example, in hypersonic flow φ_y^2, φ_z^2, φ_x are all of the same order of magnitude. Furthermore while we retain in Eq. (13.48) terms which are usually second order, we shall seek and use those terms which give an effective first-order equation.

The first-order boundary conditions for φ are

(1) At infinity,

$$\varphi_x=\varphi_y=\varphi_z=0 \tag{13.49}$$

(2) On the surface of the wing, which is represented by

$$z=h(x,y) \tag{13.50}$$

the flow must follow the wing surface, i.e.,

$$\left(\frac{\partial\varphi}{\partial z}\right)_{z=0}=\frac{\partial h(x,y)}{\partial x} \tag{13.51}$$

We introduce the transformation

$$x=c\xi,\ y=b\eta,\ z=l\zeta,\ \varphi=\varphi' m \tag{13.52}$$

where c is the mean chord of the wing, b, its span ($A=b/c=$aspect ratio), and l and m are conversion factors which are to be determined.

Substituting Eq. (13.52) into (13.48) gives

$$\begin{aligned}
&\left\{(M^2-1)\left(\frac{l}{c}\right)^2+(\gamma+1)M^2\left(\frac{m}{c}\right)\left(\frac{l}{c}\right)^2\varphi'_\xi\right.\\
&\qquad\left.+\frac{\gamma-1}{2}M^2\left(\frac{m}{c}\right)^2\left(\frac{l}{c}\right)\left[\frac{\varphi_\eta'^2}{A^2}+\varphi_\xi'^2\left(\frac{c}{l}\right)^2\right]\right\}\varphi'_{\xi\xi}\\
&+\left\{-1+(\gamma-1)M^2\frac{m}{c}\varphi'_\xi\right.\\
&\qquad\left.+\frac{\gamma+1}{2}M^2\left(\frac{m}{c}\right)^2\left[\frac{\varphi_\eta'^2}{A^2}+\lambda^2\left(\frac{c}{l}\right)^2\varphi_\zeta'^2\right]\right\}\varphi'_{\eta\eta}\left(\frac{l}{cA}\right)^2\\
&+\left\{-1+(\gamma-1)M^2\frac{m}{c}\varphi'_\xi\right.\\
&\qquad\left.+\frac{\gamma+1}{2}M^2\left(\frac{m}{c}\right)^2\left[\lambda^2\frac{\varphi_\eta}{A^2}+\left(\frac{c}{l}\right)^2\varphi_\zeta'^2\right]\right\}\varphi'_{\zeta\zeta}\\
&+2M^2\frac{m}{c}\left(\frac{l}{cA}\right)^2\varphi'_\eta\varphi'_{\xi\eta}+2M^2\frac{m}{c}\varphi'_\zeta\varphi'_{\zeta\xi}=0
\end{aligned} \tag{13.53}$$

and the nondimensional boundary conditions are

$$\varphi'_\xi=\varphi'_\eta=\varphi'_\zeta=0 \quad \text{(at infinity)} \tag{13.54}$$

$$\left(\frac{\partial \varphi'}{\partial \zeta}\right)_{\zeta=0} = \frac{l\tau}{m} f_{\xi}(\xi, \eta) \text{ (on the wing surface)} \tag{13.55}$$

where $Tf(\xi, \eta) = h(x, y)$ and $T = c\tau$ is the maximum thickness of the wing; it will be supposed that $\tau \ll 1$.

To get the parameters in the similarity laws, we write

$$l = c\tau^{-n}, \quad m = c\tau^{n'} \tag{13.56}$$

where n and n' are exponents to be determined.

If we consider only the linear terms in Eq. (13.53), the necessary conditions for the existence of similarity laws from Eq. (13.53) are

$$\frac{M^2 - 1}{\tau^{2n}} = K_1^2 \tag{13.57}$$

and

$$A\tau^n = K_2' \tag{13.58}$$

where K_1 and K_2' are constants. From the boundary condition (13.55), the necessary condition is

$$\tau^{1-n-n'} = \text{constant}$$

or

$$n + n' = 1 \tag{13.59}$$

From Eqs. (13.57) and (13.58), we have

$$A(M^2 - 1)^{1/2} = \text{constant} = K_2 \tag{13.60}$$

Eq. (13.60) is a unique similarity condition for the aspect ratio of wings of affinely related shapes. The index n in the parameter K_1 is arbitrary since the choice of n' is arbitrary. Hence the similarity laws for linearized theory is arbitrary. To have unique similarity laws it is necessary to study the nonlinear terms. From Eq. (13.53), we see that the nonlinear terms fall into two groups, one group being the more important for transonic-supersonic flow ($n > 0$), the other for supersonic-hypersonic flow ($n < 0$).

For the transonic-supersonic flow, the important nonlinear term is $(\gamma+1)M^2(m/c)(l/c)^2\varphi_{\xi}'$ The similarity condition from this term is

$$M^2\tau^{n'-2n} = \text{constant} = K_3^2 \tag{13.61}$$

In the immediate transonic region ($M = 1$),

$$n' - 2n = 0 \tag{13.62}$$

From Eqs. (13.59) and (13.62),

$$n = 1/3, \quad n' = 2/3 \tag{13.63}$$

which is the well known transonic similarity law due to von Kármán [7]

In the case that the Mach number M is not restricted to the immediate neighborhood of unity, we may eliminate τ and n from Eqs. (13.57), (13.59), and (13.61) to get as a relation between n and M

$$n\{4\log(K_3/M)+3\log[(M^2-1)/K_1^2]\}=\log[(M^2-1)/K_1^2] \tag{13.64}$$

This shows clearly that $n\to 1/3$ as $M\to 1$. Furthermore, by differentiating and proceeding to the limit, it may be shown that,

$$\lim_{M\to 1+0}\frac{dn}{dM}=\begin{cases}0, & K_3=1\\ \infty, & K_3>1\\ -\infty, & K_3<1\end{cases}$$

It is plausible to reject infinite values for dn/dM at $M=1$, in view of the established position of the transonic similarity law which requires $n=1/3$ to be approximately valid in the neighborhood of $M=1$. Then $K_3=1$ and the law of variation of n with M is

$$n=\frac{1}{3}\,\frac{\log\left(\dfrac{M^2-1}{K_1^2}\right)}{\log\left(\dfrac{M^2-1}{K_1^2M^{4/3}}\right)} \tag{13.65}$$

Examples of the variation of n and M, for different values of the parameter K_1 are shown in Fig. 13.3.

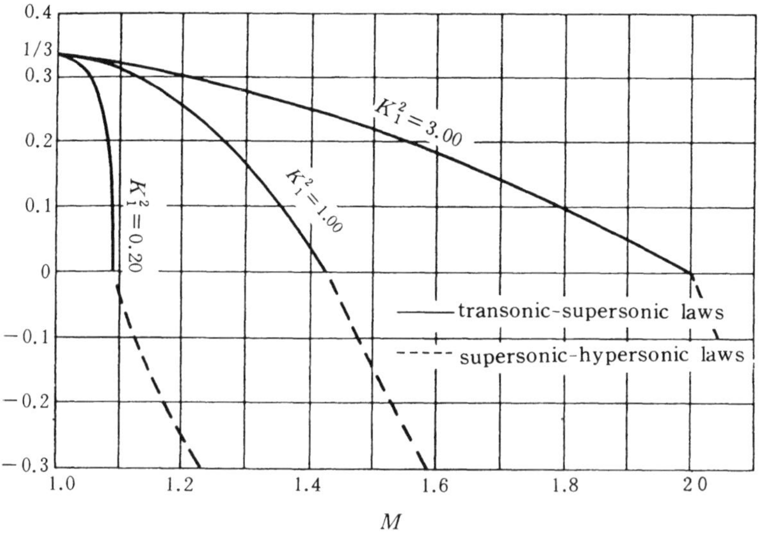

Fig. 13.3 Transonic-supersonic similarity laws

We are now able to deduce similarity laws in the sense that a solution φ (ξ, η, ζ) of the differential equation of motion may be used to give an infinity of flows in the transonic-supersonic region. Given values of the parameters K_1 , K_2 , if we choose a value of M, the corresponding values of τ, A and n are found.

It will be observed that the solid curves in Fig. 13.3 have been stopped at $n=0$. As $n \rightarrow 0$, the relative importance of the nonlinear terms in the second square bracket in Eq. (13.53) increases, and at $n = 0$ these terms are of the same order of magnitude as the "transonic" term.

Another interesting point is that for larger values of K_1, the range of validity of the ordinary transonic similarity law (13.63) is larger.

For the supersonic-hypersonic flow region, the important nonlinear terms are of the same order of magnitude as $(\gamma-1)M^2(m/c)\varphi_\xi'$ and the similarity condition is then

$$M^2\tau^{n'}=\text{constant}=K_4^2 \tag{13.66}$$

As $M \rightarrow \infty$, from Eqs. (13.5 7) and (13.66),

$$n' = -2n \tag{13.67}$$

From Eqs. (13.59) and (13.67)

$$n=-1, \quad n'=2 \tag{13.68}$$

This is the well known hypersonic similarity law due to Tsien.

Since both K_1 and K_4 are constant, and as $M \rightarrow \infty$, $n \rightarrow -1$ and $K_4 \rightarrow K_1$, $K_4=K_1$. Elimination of τ and n' from Eqs. (13.57), (13.59), and (13.66) with $K_4=K_1$ gives

$$n=-\frac{\log\ [(M^2-1)/K_1^2]}{\log\ [M^4/K_1^2(M^2-1)]} \tag{13.69}$$

The variation of n with M for three values of K_1 is shown in Fig.13.4. As before, the reliability of the approximation made falls off as $n \rightarrow 0$. Here for smaller values of K_1, the larger will be the range of validity of the ordinary hypersonic similarity law (13.68) .

In practical cases, K_1 is usually larger than one for the transonic-supersonic region and smaller than one for supersonic-hypersonic region when τ is small.

7. Method of characteristics with three independent variables[11-14]

We have discussed the method of characteristics for two independent variables. This method may be extended to problems with three independent variables, i. e., steady supersonic compressible flow in three-dimensional space, or unsteady compressible flow in two-dimensional space. We may discuss the method of characteristics from a general quasi-linear second-order partial differential equation in three independent variables or a system of three quasi-linear first-order partial differential equations. In the following we shall consider only a general quasi-linear second-order partial differential equation which may be written as:

$$a_{11}\varphi_{x_1x_1}+a_{22}\varphi_{x_2x_2}+a_{33}\varphi_{x_3x_3}+2a_{23}\varphi_{x_2x_3}$$
$$+2a_{31}\varphi_{x_3x_1} \quad 2a_{12}\varphi_{x_1x_2}-b=0 \tag{13.70}$$

where the subscripts denote partial differentiation, and φ is a function of the independent variables x_1, x_2, and x_3. The coefficients a_{ij} and b are functions of φ, x_i and φ_{x_i} only, i and $j=1$, 2, or 3.

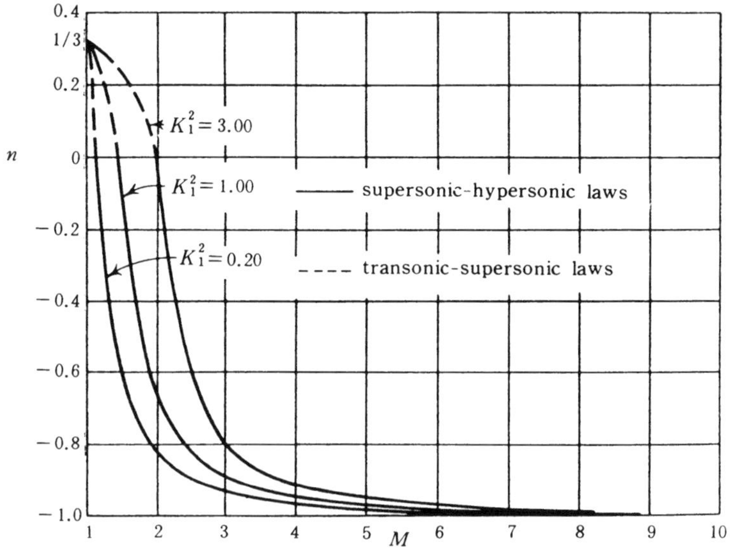

Fig. 13.4 Supersonic-hypersonic similarity laws

Consider a surface such that

$$x_i=x_i(\beta_1,\beta_2) \tag{13.71}$$

where β_1 and β_2 are two parameters defining the surface. The values of φ, φ_{x_1}, φ_{x_2} and φ_{x_3} are suposed to be known at all points on this surface.

At all points on this surface, we have the six relations

$$\frac{\partial\varphi_{x_i x_j}}{\partial\beta_1}=\varphi_{x_i x_j}\frac{\partial x_j}{\partial\beta_1} \tag{13.72}$$

$$\frac{\partial\varphi_{x_i}}{\partial\beta_2}=\varphi_{x_i x_j}=\frac{\partial x_i}{\partial\beta_2} \tag{13.73}$$

Only five of these six relations are independent. Thus any five of these equations and the original equation (13.70) may be solved to give the six second-order partial derivatives $\varphi_{x_i x_j}$ at all points on this surface (13.71) in terms of the initial given data.

If we define the quantities

$$\left.\begin{aligned}
L_1 &= \frac{\partial x_2}{\partial \beta_1}\frac{\partial x_3}{\partial \beta_2} - \frac{\partial x_3}{\partial \beta_1}\frac{\partial x_2}{\partial \beta_2} \\
L_2 &= \frac{\partial x_3}{\partial \beta_1}\frac{\partial x_1}{\partial \beta_2} - \frac{\partial x_1}{\partial \beta_1}\frac{\partial x_3}{\partial \beta_2} \\
L_3 &= \frac{\partial x_1}{\partial \beta_1}\frac{\partial x_2}{\partial \beta_2} - \frac{\partial x_2}{\partial \beta_1}\frac{\partial x_1}{\partial \beta_2}
\end{aligned}\right\} \qquad (13.74)$$

L_1, L_2, and L_3 are proportional to the direction cosines of the normal at (x_1, x_2, x_3) to the surface (13.71). Eqs. (13.72) and (13.73) may be written as

$$\left.\begin{aligned}
\varphi_{x_1x_3}L_2-\varphi_{x_1x_2}L_3 &= \frac{\partial \varphi_{x_1}}{\partial \beta_1}\frac{\partial x_1}{\partial \beta_2} - \frac{\partial \varphi_{x_1}}{\partial \beta_2}\frac{\partial x_1}{\partial \beta_1} = X_{11} \\
\varphi_{x_1x_1}L_3-\varphi_{x_1x_3}L_1 &= \frac{\partial \varphi_{x_1}}{\partial \beta_1}\frac{\partial x_2}{\partial \beta_2} - \frac{\partial \varphi_{x_1}}{\partial \beta_2}\frac{\partial x_2}{\partial \beta_1} = X_{12} \\
\varphi_{x_1x_2}L_1-\varphi_{x_1x_1}L_2 &= \frac{\partial \varphi_{x_1}}{\partial \beta_1}\frac{\partial x_3}{\partial \beta_2} - \frac{\partial \varphi_{x_1}}{\partial \beta_2}\frac{\partial x_3}{\partial \beta_1} = X_{13} \\
\varphi_{x_2x_3}L_2-\varphi_{x_2x_2}L_3 &= \frac{\partial \varphi_{x_2}}{\partial \beta_1}\frac{\partial x_1}{\partial \beta_2} - \frac{\partial \varphi_{x_1}}{\partial \beta_2}\frac{\partial x_2}{\partial \beta_1} = X_{21} \\
\varphi_{x_2x_1}L_3-\varphi_{x_2x_3}L_1 &= \frac{\partial \varphi_{x_2}}{\partial \beta_1}\frac{\partial x_2}{\partial \beta_2} - \frac{\partial \varphi_{x_2}}{\partial \beta_2}\frac{\partial x_2}{\partial \beta_1} = X_{22} \\
\varphi_{x_2x_2}L_1-\varphi_{x_2x_1}L_2 &= \frac{\partial \varphi_{x_2}}{\partial \beta_1}\frac{\partial x_3}{\partial \beta_2} - \frac{\partial \varphi_{x_2}}{\partial \beta_2}\frac{\partial x_3}{\partial \beta_1} = X_{23} \\
\varphi_{x_3x_3}L_2-\varphi_{x_3x_2}L_3 &= \frac{\partial \varphi_{x_3}}{\partial \beta_1}\frac{\partial x_1}{\partial \beta_2} - \frac{\partial \varphi_{x_3}}{\partial \beta_2}\frac{\partial x_1}{\partial \beta_1} = X_{31} \\
\varphi_{x_3x_1}L_3-\varphi_{x_3x_3}L_1 &= \frac{\partial \varphi_{x_3}}{\partial \beta_1}\frac{\partial x_2}{\partial \beta_2} - \frac{\partial \varphi_{x_3}}{\partial \beta_2}\frac{\partial x_2}{\partial \beta_1} = X_{32} \\
\varphi_{x_3x_2}L_1-\varphi_{x_3x_1}L_2 &= \frac{\partial \varphi_{x_3}}{\partial \beta_1}\frac{\partial x_3}{\partial \beta_2} - \frac{\partial \varphi_{x_3}}{\partial \beta_2}\frac{\partial x_3}{\partial \beta_1} = X_{33}
\end{aligned}\right\} \qquad (13.75)$$

Because of the following four identities, only five relations in Eq. (13.75) are independent.

$$\left.\begin{aligned} L_1X_{11}+L_2X_{12}+L_3X_{13}=0 \\ L_1X_{21}+L_2X_{22}+L_3X_{23}=0 \\ L_1X_{31}+L_2X_{23}+L_3X_{33}=0 \\ X_{11}+X_{22}+X_{33}=0 \end{aligned}\right\} \tag{13.76}$$

The solution for $\varphi_{x_ix_j}$ is obtained in the form

$$\varphi_{x_ix_j}=\frac{Y_{ij}}{Z} \tag{13.77}$$

where the six Y_{ij}' s and Z are the seven determinants of the sixth order of matrix with six rows and seven columns. For example, the relations in X_{11}, X_{12}, X_{21}, X_{23}, X_{32} gives the matrix

$$\left\|\begin{matrix} 0 & 0 & 0 & 0 & L_2 & -L_3 & X_{11} \\ L_3 & 0 & 0 & 0 & -L_1 & 0 & X_{12} \\ 0 & -L_3 & 0 & L_2 & 0 & 0 & X_{21} \\ 0 & L_1 & 0 & 0 & 0 & -L_2 & X_{23} \\ 0 & 0 & -L_1 & 0 & L_3 & 0 & X_{32} \\ a_{11} & a_{22} & a_{33} & 2a_{23} & 2a_{31} & 2a_{12} & b \end{matrix}\right\| \tag{13.78}$$

If Z does not vanish, the second derivatives φ_{xixj} at all points of the surface (13.71) are determined. The third and higher derivatives of φ are also determinable. Thus the Cauchy's initial value problem is solved and φ is determined, if it is holomorphic.

If at all points of the surface (13.71) the quantity Z vanishes, all the Y_{ij}' s must also vanish if a nontrivial solution exists. This introduces just one more condition to make the matrix (13.78) of rank 5.

The condition $Z = 0$ gives

$$Z=a_{ij}L_iL_j=0 \tag{13.79}$$

where repeated indices denote summation.

Eq. (13.79) means that the normal (L_1, L_2, L_3) to the surface (13.71) should lie on a cone of the second degree. If the cone (13.79) is imaginary, i.e., the quadratic form (13.79) is either positive definite or negative definite, the differential equation (13.70) is said to be elliptic at the point; if the quadratic from (13.79) is a perfect square at any point, the cone (13.79) degenerates to two coincident planes and Eq. (13.70) is said to be parabolic at the point; if Eq. (13.79) gives a real cone of possible normals, Eq. (13.70) is said to be hyperbolic at the point in question. We shall consider only the hyperbolic case.

In the hyperbolic case, the normal (L_1, L_2, L_3) at any point (x_{10}, x_{20}, x_{30}) to the surface (13.71) lies on this cone. The tangent plane at this point (x_{10}, x_{20}, x_{30}), i.e.,

$$L_1(x_1-x_{10})+L_2(x_2-x_{20})+L_3(x_3-x_{30})=0 \tag{13.80}$$

has an envelope as follows:

$$A_{ij}(x_i - x_{i0})(x_j - x_{j0}) = 0 \tag{13.81}$$

where i, j = 1, 2, or 3, and A_{ij} is the cofactor of a_{ij} in the determinant

$$\Delta = \begin{vmatrix} a_{11} & a_{12} & a_{13} \\ a_{21} & a_{22} & a_{23} \\ a_{31} & a_{32} & a_{33} \end{vmatrix}$$

The real quadratic cone (13.81) is called the characteristic cone at the point (x_{10}, x_{20}, x_{30}).

The displacement (dx_1, dx_2, dx_3) along the surface (13.71) forms an envelope which is given by

$$A_{ij}dx_i dx_j = 0 \tag{13.82}$$

The curvilinear cone (13.82) is called the characteristic conoid at the point.

The general characteristic surface in the case of three independent variables has the property that at any point its normal (L_1, L_2, L_3) lies on the local cone (13.81). The general characteristic surfaces which pass through any point all touch an envelope, and the local characteristic conoid (13.82) is thus a special type of characteristic surface associated with a point. Thus there is an important difference between the case of two independent variables and the case of three independent variables. In the two independent variables case there are two families of characteristic curves, but in the three independent variables case, there are infinitely many families of characteristic surfaces, depending on arbitrary functions. If an arbitrary surface that does not cut the characteristic cones is given, and if, also, a family of curves k upon this surface is given, then there are two families of characteristic surfaces passing through the curves k (bi-characteristic curves); these characteristic surfaces envelop the characteristic conoids emanating from the curve k.

The condition that all Y_{ij}'s should vanish simultaneously with Z is given by equating any Y_{ij} to zero, e.g.,

$$\begin{aligned} Y_{12} = a_{11}L_1X_{12} - a_{22}L_2X_{23} + a_{33}(L_2X_{32} - L_3X_{11}) \\ + 2a_{23}L_2X_{22} - 2a_{31}L_1X_{11} + L_1L_2b = 0 \end{aligned} \tag{13.83}$$

This differential relation holds along any characteristic surface satisfying Eq. (13.79) at every point.

For the case of steady supersonic irrotational flow of compressible fluid in three-dimensional space, we may take φ as the velocity potential. The equation for φ is

$$(u_1{}^2-a^2)\varphi_{x_1x_1}+(u_2{}^2-a^2)\varphi_{x_2x_2}+(u_3{}^2-a^2)\varphi_{x_3x_3}+2u_1u_2\varphi_{x_1x_2}$$
$$+2u_2u_3\varphi_{x_2x_3}+2u_3u_1\varphi_{x_3x_1}=0 \qquad (13.84)$$

where $u_i=\dfrac{\partial\varphi}{\partial x_i}$ =ith velocity component, and i = 1, 2, or 3. The normal (L_1, L_2, L_3) to a characteristic surface satisfies

$$(u_1{}^2-a^2)L_1{}^2+(u_2{}^2-a^2)L_2{}^2+(u_3{}^2-a^2)L_3{}^2+2u_1u_2L_1L_2$$
$$+2u_2u_3L_2L_3+2u_3u_1L_3L_1=0 \qquad (13.85)$$

The condition for Eq. (13.85) to be a definite quadratic form is

$$u_1{}^2+u_2{}^2+u_3{}^2<a^2 \qquad (13.86)$$

where a is the local sound speed. Thus Eq. (13.84) is elliptic where the flow is subsonic, parabolic where the flow is sonic, and hyperbolic where the flow is supersonic.

The equation of the characteristic cone at the point (x_{10}, x_{20}, x_{30}) is

$$(u_2{}^2+u_3{}^2-a^2)(x_1-x_{10})^2+(u_3{}^2+u_1{}^2-a^2)(x_2-x_{20})^2$$
$$+(u_1{}^2+u_2{}^2-a^2)(x_3-x_{30})^2-2u_2u_3(x_2-x_{20})(x_3-x_{30})$$
$$-2u_3u_1(x_3-x_{10})(x_1-x_{20})-2u_1u_2(x_1-x_{10})(x_2-x_{20})=0 \qquad (13.87)$$

This is a circular cone of vertex (x_{10}, x_{20}, x_{30}), axis (u_1, u_2, u_3) and semi-angle to the Mach angle $\alpha=\sin^{-1}(a/\sqrt{u_1{}^2+u_2{}^2+u_3{}^2})$.

The equation of the characteristic conoid is

$$(u_2{}^2+u_3{}^2-a^2)(dx_1)^2+(u_3{}^2+u_1{}^2-a^2)(dx_2)^2+(u_1{}^2+u_2{}^2-a^2)(dx_3)^2$$
$$-2u_2u_3dx_2dx_3-2u_3u_1dx_3dx_1-2u_1u_2dx_1dx_2=0 \qquad (13.88)$$

The differential relation $Y_{12}=0$ is

$$Y_{12}=(u_1{}^2-a^2)L_1X_{13}-(u_2{}^2-a^2)L_2X_{23}$$
$$+(u_3{}^2-a^2)(L_2X_{22}-L_3X_{11}) \qquad (13.89)$$
$$+2u_2u_3L_2X_{22}-2u_3u_1L_1X_{11}=0$$

If (l_1, l_2, l_3) are the actual direction cosines of the normal chosen to be positive in the direction which makes an angle $(\pi/2+\alpha)$ with the direction of the flow, then Eq. (13.85) becomes

$$u_1l_1+u_2l_2+u_3l_3=-a \qquad (13.90)$$

we write

$$\left.\begin{aligned} u_1 &= V \sin\theta \sin\psi \\ u_2 &= V \sin\theta \cos\psi \\ u_3 &= V \cos\theta \end{aligned}\right\} \tag{13.91}$$

where V is the magnitude of the velocity, and θ and ψ are the polar angles of the flow direction. Eq. (13.90) becomes

$$l_1 \sin\theta \cos\psi + l_2 \sin\theta \sin\psi + l_3 \cos\theta = -\sin\alpha \tag{13.92}$$

The bicharacteristic direction corresponding to (l_1, l_2, l_3) is

$$(al_1 + u_1,\quad al_2 + u_2,\quad al_3 + u_3) \tag{13.93}$$

Let the angle δ be the angle between the diametral plane of the characteristic cone passing through the bicharacteristic, and the diametral plane through the x-axis. Then

$$\cos\delta = \frac{l_1 \cos\theta \cos\psi + l_2 \cos\theta \sin\psi - l_3 \sin\theta}{\cos\alpha} \tag{13.94}$$

$$\sin\delta = \frac{l_2 \cos\psi - l_1 \sin\psi}{\cos\alpha} \tag{13.95}$$

We may then express l_1, l_2, l_3 in terms of the single parameter δ. If the bicharacteristic curves are chosen as the parametric curves, β_2=constant and their orthogonal trajectories as the parametric curves β_1 =constant, Eq. (13.89) may be reduced to

$$\tan\alpha\left(\frac{\partial\theta}{\partial\beta_1} + \sin\alpha \sin\theta \frac{\partial\psi}{\partial\beta_2}\right) - \left(\frac{\cos\delta}{V}\frac{\partial V}{\partial\beta_1} - \frac{\sin\alpha \sin\delta}{V}\frac{\partial V}{\partial\beta_2}\right) = 0 \tag{13.96}$$

$$\tan\alpha\left(\sin\alpha \frac{\partial\theta}{\partial\beta_2} - \sin\theta \frac{\partial\psi}{\partial\beta_1}\right) + \left(\frac{\sin\alpha \cos\delta}{V}\frac{\partial V}{\partial\beta_2} + \frac{\sin\delta}{V}\frac{\partial V}{\partial\beta_1}\right) = 0 \tag{13.97}$$

For $\delta = 0$, $\psi = 0$, these reduce to the two-dimensional relations

$$\frac{dV}{V} = \pm \tan\alpha \, d\theta \tag{13.98}$$

along the characteristics

$$\frac{dx_1}{dx_2} = \tan(\theta \mp \alpha) \tag{13.99}$$

The relations (13.98) and (13.99) form the basis of numerical computation suggested by Thornhill along a hexahedral grid.

The general three-dimensional characteristic method is too lengthy and cumbersome. For practical use some reasonable simplifications should be introduced. This method may also be used to solve the linear problem with a complicated boundary.

8. Nonsteady compressible flow [15, 16]

For unsteady two-dimensional or axially symmetric compressible flow, we may use the method of characteristics of three independent variables, x, y, and t. For the case of irrotational flow, we may define a velocity potential φ such that

$$u_1 = \frac{\partial\varphi}{\partial x}, \; u_2 = \frac{\partial\varphi}{\partial y}, \; u_3 = \frac{\partial\varphi}{\partial t} = -\left(\frac{u_1^2 + u_2^2}{2} + \frac{a^2}{\gamma - 1}\right) \tag{13.100}$$

where $x = x_1, y = x_2$, and $t = x_3$. The last identity is obtained by the help of Bernoulli's equation.

The differential equation for φ is

$$(u_1^2 - a^2)\varphi_{xx} + (u_2^2 - a^2)\varphi_{yy} + \varphi_{tt} + 2u_2\varphi_{ty} + 2u_1\varphi_{tx} + 2u_1u_2\varphi_{xy} = 0 \tag{13.101}$$

The normal to a characteristic surface satisfies

$$(u_1^2 - a^2)L_1^2 + (u_2^2 - a^2)L_2^2 + L_3^2 + 2u_2L_2L_3 + 2u_1L_3L_1 + 2u_1u_2L_1L_2 = 0 \tag{13.102}$$

If $a \neq 0$, it is easy to show that Eq. (13.102) is always hyperbolic at every point.

The characteristic cone at the point (x_0, y_0, t_0) is

$$(x - x_0)^2 + (y - y_0)^2 + (u_1^2 + u_2^2 - a^2)(t - t_0)^2 - 2u_1(x - x_0)(t - t_0) - 2u_2(y - y_0)(t - t_0) = 0 \tag{13.103}$$

The equation of the characteristic conoid is

$$(dx - u_1dt)^2 + (dy - u_2dt)^2 = (adt)^2 \tag{13.104}$$

The differential relation $Y_{12} = 0$ gives

$$Y_{12} = (u_1^2 - a^2)L_1X_{13} - (u_2^2 - a^2)L_2X_{23} + L_2X_{32} - L_3X_{11} + 2u_2L_2X_{22} - 2u_1L_1X_{11} = 0 \tag{13.105}$$

If we choose the parameter δ such that

$$\frac{L_1}{\sin\delta} = \frac{L_2}{\cos\delta} = -\frac{L_2}{u_1\sin\delta + u_2\cos\delta + a} \tag{13.106}$$

the bicharacteristic direction is

$$(u_1 + a\sin\delta,\; u_2 + a\cos\delta,\; 1) \tag{13.107}$$

If the bicharacteristic curves are chosen as β_2 = constant and the curves of intersection with the plane t = constant are chosen as β_1 = constant, Eq. (13.105) reduces to

$$\left(\frac{\partial u_1}{\partial \beta_1} - \frac{\partial u_2}{\partial \beta_2}\right) + \frac{2}{\gamma - 1}\left(\sin\delta \frac{\partial a}{\partial \beta_1} + \cos\delta \frac{\partial a}{\partial \beta_2}\right) = 0 \quad (13.108)$$

$$\left(\frac{\partial u_1}{\partial \beta_2} + \frac{\partial u_2}{\partial \beta_1}\right) + \frac{2}{\gamma - 1}\left(\cos\delta \frac{\partial a}{\partial \beta_1} - \sin\delta \frac{\partial a}{\partial \beta_2}\right) = 0 \quad (13.109)$$

These relations are the basis for numerical computations along a hexahedral grid.

For $u_1 = 0$, $\delta = 0$, Eqs. (13.108) and (13.109) reduce to the Riemann invariant of one-dimensional unsteady flow, i.e.,

$$u_2 \pm \frac{2a}{\gamma - 1} = \text{constant along the characteristic } dx_2/dt = (u_2 \pm a) \quad (13.110)$$

Since the method of characteristics for three independent variables is very cumbersome, in ordinary conditions it is advisable to use the method of linearization. It is known that in the steady flow cases there are ranges of Mach numbers in which the fundamental equation cannot be linearized even though the disturbance is small. Such a situation exists in the transonic and the hypersonic regions. It is of interest to find the conditions for linearization of the equation governing the nonsteady motion. This problem was first investigated by Lin, Reissner, and Tsien and further investigated by Miles. We now discuss these conditions.

The differential equation of the velocity potential φ for irrotational nonsteady flow of inviscid compressible fluid in three-dimensional space may be written in vector notation as

$$a^2\nabla^2\varphi = \varphi_{tt} + \left[\frac{\partial}{\partial t} + \frac{1}{2}(\nabla\varphi \cdot \nabla)\right](\nabla\varphi)^2 \quad (13.111)$$

where the local sound speed a is related to the velocity potential by the formula

$$a_0^2 - a^2 = (\gamma - 1)\left[\varphi_t + \frac{1}{2}(\nabla\varphi)^2\right] = \frac{\gamma p_0}{\rho_0}\left[1 - \left(\frac{p}{p_0}\right)^{\frac{\gamma-1}{\gamma}}\right] \quad (13.112)$$

∇ is the gradient operator and $\nabla^2 = \nabla \cdot \nabla$.

Now we consider the case of a uniform stream U (in the x-axis direction) passing over a thin body. The velocity potential φ in this case may be written in the form

$$\varphi(x, y, z, t) = U_x + \varepsilon U l f(\xi, \eta, \zeta, \tau) \tag{13.113}$$

where f is the nondimensional perturbation velocity potential, l is the characteristic length of the body in the direction of flight, i. e., x-axis. The other nondimensional quantities are defined as

$$\xi = x/l, \ \eta = \lambda y/l, \ \zeta = \mu z/l, \ k\tau = k(UT/l) \tag{13.114}$$

The nondimensional parameters λ, μ, k, and ε are to be defined in a given neighborhood such that f and all its derivatives will be of the order of unity. The parameter λ will be taken as the reciprocal of the aspect ratio of the thin body, and k is a measure of the time rate of change.

The velocity vector is given by

$$\boldsymbol{q} = \nabla\varphi = U\left[\boldsymbol{i}(1+\varepsilon f_\xi) + \boldsymbol{j}\varepsilon\lambda f_\eta + \boldsymbol{k}\varepsilon\mu f_\zeta\right] \tag{13.115}$$

If the perturbation velocity is small, we have the conditions

$$\varepsilon \ll 1, \quad \varepsilon\lambda \ll 1, \quad \varepsilon\mu \ll 1 \tag{13.116}$$

The condition that the pressure coefficient

$$C_p = \frac{p-p_0}{\frac{1}{2}\rho_0 U^2} = \frac{2}{\gamma M^2}\left[\left(\frac{a}{a_0}\right)^{\frac{2\gamma}{\gamma-1}} - 1\right] \tag{13.117}$$

is small gives

$$\varepsilon M^2 \ll 1, \quad \varepsilon k M^2 \ll 1, \quad \varepsilon\lambda M \ll 1, \ \varepsilon\mu M \ll 1 \tag{13.118}$$

With the conditions (13.116) and (13.118), we obtain the effective first-order equation for f as

$$\frac{\lambda^2}{M^2} f_{\eta\eta} + \frac{\mu^2}{M^2} f_{\xi\xi} = k^2 f_{\tau\tau} + 2k f_{\xi\tau} + [(1-M^{-2}) + \varepsilon(\gamma+1) f_\xi] f_{\xi\xi} \tag{13.119}$$

while the corresponding pressure coefficient is given by

$$C_p = -2\varepsilon\left[f_\xi + k f_\tau + \frac{1}{2}\varepsilon(\lambda^2 f_\eta^{\,2} + \mu^2 f_\zeta^{\,2})\right] \tag{13.120}$$

The surface of a thin body may be written as

$$z = h(x, y, t) = \delta l g(\xi, \eta, \tau) \tag{13.121}$$

where $g = 0(1)$ and

$$\delta \ll 1, \quad \lambda\delta \ll 1 \tag{13.122}$$

The nondimensional boundary condition at the surface is

$$(f_\xi)_{\xi=\mu\delta g}=\frac{\delta}{\mu\varepsilon}\left[kg_\tau+(1+\varepsilon f_\xi)g_\xi+\varepsilon\lambda^2 f_\eta g_\eta\right] \tag{13.123}$$

The values of μ and ε may be determined from the boundary condition (13.123) and the differential equation (13.119).

Miles found the sufficient conditions for linearization of the velocity potential equation in the neighborhood of a thin body to be

$$\delta \ll 1,\quad k\delta \ll 1,\quad M\delta \ll 1,\quad kM\delta \ll 1 \tag{13.124}$$

and any one of the following:

$$|M-1| \gg \delta^{2/3} \tag{13.125a}$$

$$k \gg \delta^{2/3} \tag{13.125b}$$

$$(AR)^{-1} \gg \delta^{1/3} \tag{13.125c}$$

If these conditions are satisfied, the linearized equation is

$$a_0^2\nabla^2\varphi=\left(\frac{\partial}{\partial t}+U\frac{\partial}{\partial x}\right)^2\varphi \tag{13.126}$$

and the pressure equation becomes

$$p-p_0=-\rho_0\left(\frac{\partial}{\partial t}+U\frac{\partial}{\partial x}\right)\varphi \tag{13.127}$$

If these conditions are not satisfied, the differential equation for the perturbation velocity potential will be nonlinear. The differential equations used in transonic and hypersonic similarity laws are special cases of these perturbation velocity potential.

9. Zonal method for three-dimensional potential flow

The type-dependent difference schemes for the stady two-dimensional potential equations presented in Chapter X, § 10 and 15 can be readily extended to the steady three-dimensional potential equations. For the steady three-dimensional transonic small perturbation flow, the transverse derivatives are all represented by the central differences while the longitudinal derivatives by the type-dependent differences.

The computational work for a three-dimensional flow is usually greater by one order of magnitude than that for the two-dimensional flow. Various efforts to reduce the computational work particularly for three-dimensional flow about complex configurations are under way. An apparent way to economize on the solution of a complex flow field is to use a zonal-equation approach. In this approach, simplified equation sets are used in regions of the flow where it is appropriate. To illustrate the idea of the zonal method, we will combine the linear analytic method of slender body theory with the nonlinear transonic small perturbation potential equation to simulate transonic flow past a body of revolution at an angle of attack. In this technique, the linear theory is used to enforce the body surface boundary condition on a rectilineal grid boundary of cylindrical coordinates. Of course, the zonal methods may combine other equations or methods for compressible flow. For example, one might combine a linear panel method with nonlinear full-potential equations to simulate transonic flow about complex configurations. And for a uniform incoming flow, a zonal code might use boundary-layer equations near the wall, Euler equations away from the wall, Navier-Stokes equations in separated regions, and potential equations in the far field.

Now let us consider the transonic flow about a slender body of revolution at a small angle of attack α. The transonic small perturbation equation in the cylindrical coordinates (x, r, θ) (Fig. 13.15) is

$$\left(1-M_\infty^2-\frac{\gamma+1}{U}M_\infty^2\varphi_x\right)\varphi_{xx}+\varphi_{rr}+\frac{\varphi_r}{r}+\frac{\varphi_{\theta\theta}}{r^2}=0 \tag{13.128}$$

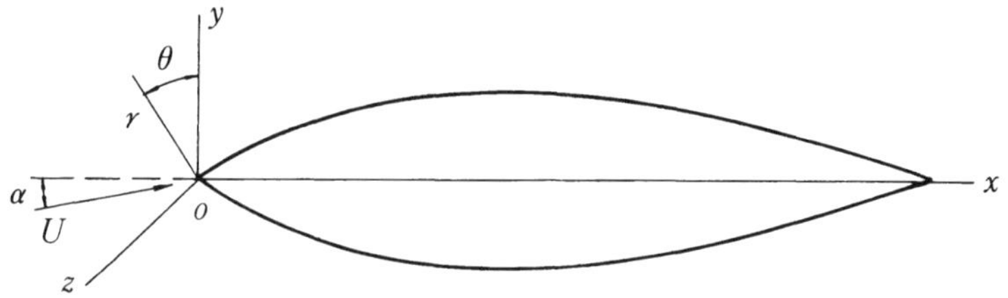

Fig. 13.5 Body of revolution

The boundary condition on the body surface whose equaiton is $r=R(x)$ is given by

$$\varphi_r(x, R, \theta) = UR'(x)-U\alpha\sin\theta \tag{13.129}$$

where the prime denotes the derivative with respect to the argument.

Unlike the thin wing, the boundary condition on the slender body of revolution cannot be satisfied at the x axis because of the singularity behavior of the radial component of the perturbation velocity, $\varphi_r(x, r, \theta)$ at $r = 0$. However, the flow variables in the neighborhood of the body can be related by the slender body theory. In the slender body theory, the transonic potential equation (13.128) reduces to

$$\varphi_{rr}+\frac{\varphi_r}{r}+\frac{\varphi_{\theta\theta}}{r^2}=0 \tag{13.130}$$

and the solution is expressed as

$$\varphi(x, r, \theta)=UR(x)R'(x)\ln r+U\alpha R^2(x)\cos\theta/r+f(x) \tag{13.131}$$

where $f(x)$ has to be determined by the full transonic small perturbation equation (13.128) with the boundary conditions.

Take a cylindrical surface $r = r_0$ in the neighborhood of the body surface. From the solution of the slender body theory (13.131), we have

$$\varphi_r(x, r_0, \theta) = UR'(x)\frac{R(x)}{r_0} - U\alpha\frac{R^2(x)}{r_0^2}\cos\theta \qquad (13.132)$$

This relation will be used as the boundary condition instead of Eq. (13.129) for the finite-difference computations. Let $r = r_0$ be the inner boundary of the computational region and the grid line $j = 0$. The boundary condition (13.132) is embedded in the governing equation (13.128) at $j = 0$ and i between both ends of the body with the following finite-difference representation for φ_{rr}

$$(\varphi_{rr})_{i,o,k} = \frac{2}{\Delta r}\left[\frac{\varphi_{i,j,k} - \varphi_{i,o,k}}{\Delta r} - (\varphi_r)_{i,o,k}\right] + O(\Delta r) \qquad (13.133)$$

The resulting algebraic equation system may be solved by the similar iteration methhods to the two-dimensional problems. After the numerical solution for $\varphi_{i,j,k}$ is obtained in the computational region $r \geqslant r_0$, the flow parameters on the body surface can be evaluated by the use of the slender body theory once more. From the solution by the slender body theory, we have

$$\varphi_x(x, R, \theta) = U[R(x)R'(x)]'\ln R + \frac{U\alpha[R^2(x)]'}{R(x)}\cos\theta + f'(x) \qquad (13.134)$$

and

$$\varphi_x(x, r_0, \theta) = U[R(x)R'(x)]'\ln r_0 + \frac{U\alpha[R^2(x)]'}{r_0}\cos\theta + f'(x) \qquad (13.135)$$

Substracting (13.135) from (13.134) yields

$$\varphi_x(x, R, \theta) - \varphi_x(x, r_0, \theta) = U[R(x)R'(x)]'\ln\frac{R}{r_0}$$

$$+ U\alpha[R^2(x)]'\left(\frac{1}{R(x)} - \frac{1}{r_0}\right)\cos\theta \qquad (13.136)$$

The expression $U\alpha[R_2(x)]'\cos\theta$ in the second term on the right-hand side of Eq. (13.136) obtained from the slender body theory can be replaced by an expression involving the finite-difference solution. Integrating Eq. (13.135) with respect to θ from 0 to 2π and dividing by 2π, we obtain

$$\frac{U\alpha[R^2(x)]'}{r_0}\cos\theta = \varphi_x(x, r_0, \theta) - \frac{1}{2\pi}\int_0^{2\pi}\varphi_x(x, r_0, \theta)\,d\theta \qquad (13.137)$$

Substituting (13.137) into (13.136) gives the formula for the x component of the perturbation velocity on the body surface

$$\varphi_x(x, R, \theta) = \varphi_x(x, r_0, \theta)\, \frac{r_0}{R(x)} + U[R(x)R'(x)]' \ln \frac{R(x)}{r_0}$$

$$- \frac{1}{2\pi} \int_0^{2\pi} \varphi_x(x, r_0, \theta)\, d\theta \left(\frac{r_0}{R(x)} - 1 \right) \tag{13.138}$$

The circumferential component of the perturbation velocity on the body surface can be readily derived from the solution (13.131),

$$\frac{\varphi_\theta(X, R, \theta)}{R} = \frac{\varphi_\theta(x, r_0, \theta)'}{r_0} \cdot \frac{r_0^2}{R^2} \tag{13.139}$$

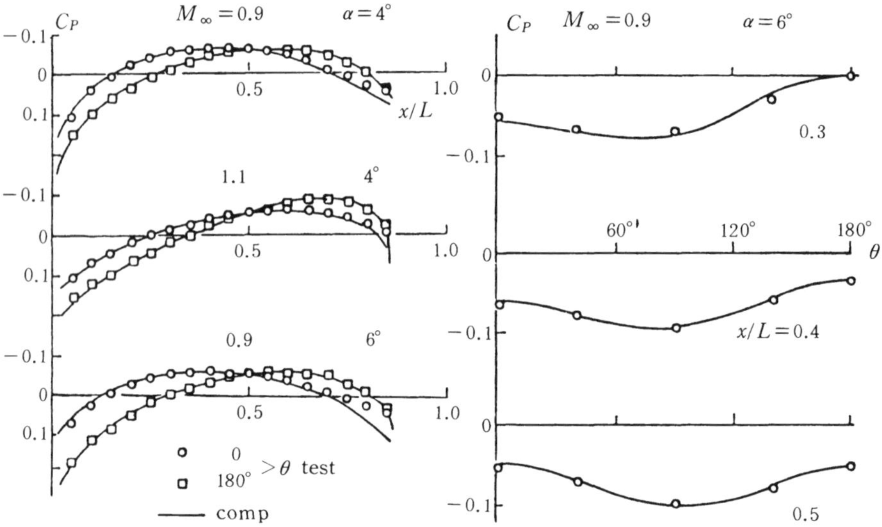

Fig. 13.6 Pressure distribution on paraboloid of revolution, fineness ratio 12

Fig. 13.6[17] compares the computational pressure coefficient results with the wind tunnel data over a paraboloid of revolution, fineness ratio = 12, at M_∞ = 0.9 and 1.1, α = 4° and 6°. In this calculation, the derivatives φ_θ and $\varphi_{\theta\theta}$ are represented by the modified central difference representations

$$\varphi_\theta = (\varphi_{i,j,k+1} - \varphi_{i,j,k-1}) / (2\sin\Delta\theta) \tag{13.140}$$

$$\varphi_{\theta\theta} = (\varphi_{i,j,k+1} - 2\varphi_{i,j,k} + \varphi_{i,j,k-1}) / [2(1 - \cos\Delta\theta)] \tag{13.141}$$

Eqs. (13.140) and (13.141) are exact if $\varphi(x,r,\theta)$ is a linear function of $\sin\theta$ and conθ. By the use of these representations, $\Delta\theta$ as large as $\pi/4$ can be employed to yield sufficiently convergent results. The mesh used for the half space,

$0° \leqslant \theta \leqslant 180°$ in this computation is $65 \times 18 \times 5$ and $r_0 = R_{max}/2$, where R_{max} is the maximum radius of the body. The agreement between the computation and the experiment is quite good.

10. Numerical solution of unsteady transonic flows

The unsteady full potential equation is given by (13.103). Simple treatment of arbitrary airfoils and airfoil motions is the principal advantage of small perturbation formulations. The airfoil boundary condition can be imposed on a flat, mean-surface approximation to the airfoil. In less approximate formulations, coordinate mappings are usually required, and it may be necessary to retransform for each time-step. Of course, the assumptions that permit this simplification place some restrictions on the airfoil thickness-to-chord ratios and motion amplitudes that can be treated.

The unsteady, transonic small perturbation equation can be obtained from Eq. (13.103)

$$\frac{M_\infty^2}{U^2}\varphi_{tt} + \frac{2M_\infty^2}{U}\varphi_{xt} = \left(1 - M_\infty^2 - \frac{\gamma+1}{U}M_\infty^2\varphi_x\right)\varphi_{xx} + \varphi_{yy} \quad (13.142)$$

where φ is the perturbation velocity potential, and U and M_∞ are the free stream velocity and Mach number in the x direction. The righthand side of Eq. (13.142) is the familiar two-dimensional, transonic small perturbation equation for steady flows. Eq. (13.142) is still quasi-linear and hyperbolic. However, the pressure equation can be linearized as Eq. (13.127). The boundary condition on the moving airfoil surface $y = f(x, t)$ is simplified

$$\varphi_y(x, 0, t) = Uf_x + f_t \quad (13.143)$$

The Kutta conditions are along the x axis and downstream of the trailing edge of the airfoil, across which φ is discontinuous,

$$\varphi_y(x, +0, t) = \varphi_y(x, -0, t) \quad (13.144)$$

and

$$\varphi_t(x, +0, t) + U\varphi_x(x, +0, t) = \varphi_t(x, -0, t) + U\varphi_x(x, -0, t) \quad (13.145)$$

In the far field, the perturbation velocity components φ_x and φ_y approach zero as the distance from the airfoil approaches infinity.

For an airfoil of chord length c, traveling with speed U, and executing some unsteady oscillatory motion of angular frequency ω, the reduced frequency is defined as

$$k = \frac{\omega c}{U} \quad (13.146)$$

The reduced frequency demonstrates in terms of radians of oscillatory motion per chord length of airfoil travel. Emphasis may be placed on the treatment of low-frequency oscillatory flows, i. e. $k \ll 1$. These flows are of interest because shock excursion amplitudes, and hence unsteady aerodynamic force amplitudes, usually increase with decreasing frequency for a fixed airfoil oscillatory motion amplitude.

An approximation to Eq. (13.142), valid for low reduced frequencies is the equation

$$\frac{2M_\infty^2}{U}\varphi_{xt}=\left(1-M_\infty^2-\frac{\gamma+1}{U}M_\infty^2\varphi_x\right)\varphi_{xx}+\varphi_{yy} \tag{13.147}$$

In Eq. (13.142), the term of higher order of k is neglected. An alternating-direction implicit (ADI) algorithm is designed for solving the low-frequency transonic equation. In the design of the algorithm, the following features are considered. The type-dependent differencing (cf. Chapter X) has been successfully applied to steady transonic equations and thus may be employed for the right-hand side of the unsteady transonic equation (13.147). To construct a conservative scheme, the differential equation (13.147) should be written in divergence form

$$B(\varphi_x)_t=F_x+(\varphi_y)_y \tag{13.148}$$

where

$$F=(1-M_\infty^2)\varphi_x-\frac{\gamma+1}{2U}M_\infty^2\varphi_x^2 \tag{13.149}$$

and $B=2M_\infty^2/U$.

The partial derivative with respect to t can be represented by a second-order accurate, central difference and thus we have

$$B\frac{(\varphi_x)_{i,j}^{n+1}-(\varphi_x)_{i,j}^{n}}{\Delta t}=(F_x)_{i,j}^{n+\frac{1}{2}}+[(\varphi_y)_y]_{i,j}^{n+\frac{1}{2}} \tag{13.150}$$

where the superscript n is the time level and the subscripts i and j are grid-point indices in the x and y directions, respectively. The conservative, type-dependent difference representations for F_x can be written according to Eq. (10.88)

$$(F_x)_{i,j}^{n+\frac{1}{2}}=(1-\mu_{i,j})\left(F_{i+\frac{1}{2},j}^{n+\frac{1}{2}}-F_{i-\frac{1}{2},j}^{n+\frac{1}{2}}\right)+\mu_{i-1,j}\left(F_{i-\frac{1}{2},j}^{n+\frac{1}{2}}-F_{i-\frac{3}{2},j}^{n+\frac{1}{2}}\right) \tag{13.151}$$

and

$$F_{i,j}^{n+\frac{1}{2}}=(1-M_\infty^2)\frac{(\varphi_x)_{i,j}^{n}+(\varphi_x)_{i,j}^{n+1}}{2}-\frac{\gamma+1}{2U}M_\infty^2(\varphi_x)_{i,j}^{n}(\varphi_x)_{i,j}^{n+1} \tag{13.152}$$

where

$$(\varphi_x)_{i+\frac{1}{2},j}^{n}=\frac{\varphi_{i+1,j}^{n}-\varphi_{i,j}^{n}}{\Delta x} \tag{13.153}$$

$$\mu_{i,j}=\begin{cases}1, & \\ & \text{if } A^n_{i+\frac{1}{2},j}+A^n_{i-\frac{1}{2},j}\begin{cases}\geqslant 0\\ \\ <0\end{cases}\\ 0, & \end{cases} \tag{13.154}$$

and

$$A^n_{i,j}=1-M^2_\infty-\frac{\gamma+1}{U}M^2_\infty(\varphi_x)^n_{i,j} \tag{13.155}$$

The difference representations (13.151) are linearized and second-order accurate at subsonic points and first-order accurate at supersonic points, in terms of Δx.

For the linear term $(\varphi_y)_y$, we use the second-order accurate central differences.

$$[(\varphi_y)_y]^{n+\frac{1}{2}}_{i,j}=\frac{1}{2}\{[(\varphi_y)_y]^n_{i,j}+[(\varphi_y)_y]^{n+1}_{i,j}\} \tag{13.156}$$

and

$$[(\varphi_y)_y]^n_{i,j}=[(\varphi_y)^n_{i,j+\frac{1}{2}}-(\varphi_y)^n_{i,j-1}]/\Delta y \tag{13.157}$$

where

$$(\varphi_y)^n_{i,j+\frac{1}{2}}=(\varphi^n_{i,j+1}-\varphi^n_{i,j})/\Delta y \tag{13.158}$$

Thus, the finite differencings for the right-hand side of Eq. (13.150) are com pleted. Now we come to treat the left-hand side of (13.150). The left-hand side may be written as

$$B\{[(\varphi_x)^{n+1}_{i,j}-(\varphi_x)^{\overline{n+1}}_{i,j}]+[(\varphi_x)^{\overline{n+1}}_{i,j}-(\varphi_x)^n_{i,j}]\}/\Delta t \tag{13.159}$$

The solution of Eq. (13.150) can be advanced from time level n to level $n+1$ by the following two-step procedure:
x-sweep:

$$B[(\varphi_x)^{\overline{n+1}}_{i,j}-(\varphi_x)^n_{i,j}]/\Delta t=(F_x)^{n+\frac{1}{2}}_{i,j}+[(\varphi_y)_y]^n_{i,j} \tag{13.160}$$

y-sweep:

$$B[(\varphi_x)^{n+1}_{i,j}-(\varphi_x)^{\overline{n+1}}_{i,j}]/\Delta t=\{[(\varphi_y)_y]^{n+1}_{i,j}-[(\varphi_y)_y]^n_{i,j}\}/2 \tag{13.161}$$

This splitting may be verified by simple addition of Eqs. (13.160), and (13.161) results in an laternating-direction algorithm. The partial derivative with respect to x on the left-hand sides of Eqs. (13.160) and (13.161) is differenced implicitly and thus the solution algorithm is noniterative. That is, the solution for the $n+1$ time level is obtained directly after two sweeps through the grid.

On the x-sweep, no matter whether the differencing for φ_x on the left-hand side of the equations is central or biased, a matrix is generated that is tridiagonal for subsonic points ($\mu_{i-1,j}=\mu_{i,j}=0$) and lower tridiagonal for supersonic points ($\mu_{i-1,j}=\mu_{i,j}=1$). A quadradiagonal solver is used that solves the matrix equation for $\varphi_{i,j}^{n+1}$ like the Thomas algorithm for tridiagonal matrix equations. On the y-sweep, if the φ_x derivative on the left-hand side of the equations is backward differenced and the sweep is marched from upstream to downstream, a tridiagonal matrix equation for each x = constant line of y grid points is generated and solved by the Thomas algorithm. Therefore, on the left-hand side of Eqs. (13.160) and (13.161), we may use

$$(\varphi_x)_{i,j}^n=(\varphi_{i,j}^n-\varphi_{i-1,j}^n)/\Delta x+ O(\Delta x) \tag{13.162}$$

or

$$(\varphi_x)_{i,j}^n=(3\varphi_{i,j}^n-4\varphi_{i-1,j}^n+\varphi_{i-2,j}^n)/(2\Delta x)+O(\Delta x^2) \tag{13.163}$$

The ADI scheme has no Δt restriction based on a linear stability analysis. However, an instability can be generated by the motion of shocks across which the differencing switches from backward to central. To prevent this instability from occurring, Δt must be chosen small enough that such shock waves do not move a distance greater than one spatial grid point per time step. This restriction is also necessary to maintain accuracy.

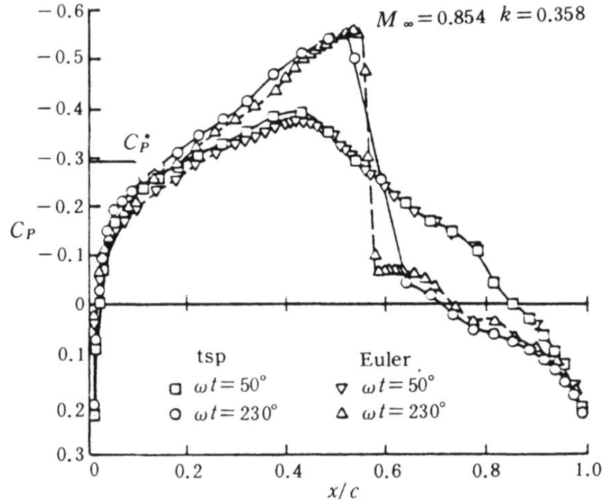

Fig. 13.7 Unsteady upper surface pressure coefficients

Fig. 13.7[18] compares the numerical solutions of the unsteady transonic small perturbation (TSP) equation with those of the unsteady Euler equations for a NACA 64A006 airfoil with sinusoidal-oscillating trailing-edge flap at M_∞ = 0.854, k =0.358 and ωt = 50° and 230°. The computed shock is strongest at ωt = 230°. and weakest at ωt = 50°. In the unsteady TSP computations, the first-order accurate difference representation (13.162) was used. These computations compare favorably with those of the unsteady Euler equations and are obtained in substantially less computer time. The well known Tijdeman's type A, B, and C shock motions observed for the NACA 64A006 airfoil with oscillating flap were demonstrated by the numerical solutions of the unsteady TSP equation.

11. Problems

1. From the hodograph curve, show that the flow after the attached shock of a cone at zero angle of attack is further compressed before the surface of the cone is reached.

2. Plot the apple curve for a cone in a supersonic uniform speed at Mach number of 3.

3. Use the method of characteristics to find the flow pattern in an axially symmetrical jet discharged from a nozzle of uniform velocity at Mach number of 2. The pressure p_2 of the surrounding stream is one half that at the exit of the nozzle p_1.

4. Use experimental data published in any journal to check the three-dimensional transonic similarity laws.

5. Use experimental data published in any journal to check the three-dimensional hypersonic similarity laws.

6. Use the Prandtl-Meyer solution to check the supersonic-similarity laws, and the variation of index n with Mach number M discussed in § 6.

7. Show the detailed procedure for transforming the basic equations (13.96) and (13.97) of the method of characteristics for three independent variables into difference equations for numerical computation along a hexahedral grid.

8. Show the detailed procedure for transforming the basic equations (13.108) and (13.109) of the method of characteristics for nonsteady flow into difference equations for numerical computation along a hexahedral grid.

9. Use the method of characteristics for a three-dimensional linearized equation to calculate the flow pattern over an elliptic cone at zero angle of attack.

10. Show the detailed derivation of the sufficient conditions (13.124) and (13.125) for the linearized velocity potential equation.

11. How can you find a second-order accurate finite-difference representation for φ_{rr} of the transonic small perturbation equation at the boundary grid $j=0$ (or $r = r_0$), in which the boundary condition of $\varphi_r(x, r_0, \theta)$ is embedded?

12. Verify that Eqs. (13.140) and (13.141) are exact if $\varphi(x, r, \theta)$ is a linear function of $\sin\theta$ and $\cos\theta$.

13. Verify Eqs. (13.142) and (13.147).

References

1. Taylor, G. I., and Maccoll, J. W., The air pressure on a cone moving at high speed, *Proc. Roy. Soc. London* **A-139,** p. 278, 1933.
2. Maccoll, J. W., The conical shock wave formed by a cone moving at high speed, *Proc. Roy. Soc. London* **A-159,** pp. 451 –472. 1937.
3. Busemann, A., Die achsensymmetrische kegelige Überschallstroemung, *Luftfahrtforschung* **19,** 137, 1942.
4. Kopal, Z., Tables of supersonic flow around cones, M. I. T. Center of analysis , *Tech. Report* No. 1, U. S. Government Printing Office, 19 47.
5. Meyer, R. E., The method of characteristics for problems of compressible flow involving two independent variables, Part I, Theory, *Quart. J. Mech. Appl. Math.* **1.** pt 2, June 1948.
6. Tollmien, W., Steady two dimensional and axially symmetrical supersonic flows, *Brown University Translation* No. A9-T-1, 1948.
7. von K ámán, Th., The similarity law of transonic flow, *Jour. Math. Phys.* **26,** p. 182, 1947.
8. Tsien, H. S., Similarity laws of hypersonic flows, *Jour. Math. Phys.* **25,** p. 247 1946.
9. Hayes, W. D., On hypersonic similitude, *Quart. Appl.. Math.* **5,** p. 105, 1947.
10. Pack, D. C., and Pai, S. I., Similarity laws for supersonic flows, *Quart. Appl. Math.* **9,** pp. 377-384, 1954.
11. Thornhill, C. K., The numerical method of characteristics for hyperbolic problems in three independent variables, *Armament Research Establishment* Report 29/48, British Ministry of Supply, 1948; also *British ARCR and M* No. 2615, 1952.
12. Coburn, N., and Dolph, C. L., The method of characteristics in three-dimensional stationary supersonic flow of a compressible gas, *Proc. Sym. Appl. Math.* I, pp. 55 –66, Amer. Math. Soc., New York, 1949.
13. Moeckel, W. E., Use of characteristics for unsymmetrical supersonic flow problems, *NACA TN* No. 1849. 1949 .
14. Sauer, R., Recent advances in the theory of supersonic flow, *U. S. Naval Ordnance Report* N o. 1993, 1951.
15. Lin C. C., Reissner, E., and Tsien, H. S., On Two-dimensional nonsteady motion of a slender body in a compressible fluid, *Jour. Math. Phys.* **27,** No. 3, Oct. 1948, pp. 220 –231.
16. Miles, J. W., Linearization of the equations of non-steady flow in a compressible Flu id, *Jour. Math. Phys.* **33,** No. 2, July 1954, pp. 135 – 143.
17. Luo, S.J. and Bao, Y., Computation oftransonic aerodynamicallycompensatingpitot tube,*AIAA*-87-2613-CP, August 1987. *Also Jour. Aircraft* **25,** No. 6, June 1988, pp. 544 –547.
18. Ballhaus, W. F. and Goorjian, P. M., Implicit finite-difference computations of unsteady transonic flows about airfoils, *AIAA Journal* **15,** No. 12, December 1977, pp. 1728 –1735.

Chapter XIV

ANISENTROPIC (ROTATIONAL) FLOW OF INVISCID COMPRESSIBLE FLUID

1. Introduction

In the previous chapters, we considered for the most part the irrotational flow of a compressible fluid. For the adiabatic flow of inviscid fluids, the condition of irrotationality in accordance with Kelvin's theorem is usually satisfied for subsonic flow (Chapter V), but the condition of irrotationality may not be satisfied in supersonic flow if a curved shock occurs. Behind a curved shock, the entropy distribution will not be uniform even if the entropy is uniform in front of the shock. For a nonuniform distribution of entropy, the flow will be rotational and the vorticity will not be zero in the flow field. However, if the strength of the shock is weak, the perturbed entropy is of the third order of the perturbed velocity. In this case, for the first- or second-order theory, the vorticity is negligible and the flow may be still considered to be irrotational. If the strength of the shock is not small, the perturbed vorticity and the perturbed velocity are of the same order of magnitude and we may not neglect the vorticity even in first-order perturbation theory. In this case, the flow should be considered as rotational. It is desirable to find the definite relation between the variation of entropy and the strength of the shock in order to determine under what conditions the nonuniform distribution of entropy is not negligible. We shall discuss this in §2.

If the vorticity in the flow is not negligible, we should consider the flow to be rotational. In §3, we shall discuss the fundamental equations for rotational flow, particularly in the isoenergetic case. Based on these equations and the Rankine-Hugoniot relations, we shall study the flow conditions behind a curved shock in §4 and §5, especially observing attached shocks and the "hedgehog" or "porcupine" diagram.

In §6, we shall discuss briefly the flow field with detached shocks. In § 7, the problem of linearized theory of rotational flow will be discussed. If the steady rotational flow is supersonic, the method of characteristics may be used to find the solution. This will be discussed in §8.

In §9, we shall discuss the solution of the rotational flow equation based on the stream function, particularly when a detached shock is involved. In §10 and §11, unsteady anisentropic flow and the propagation of nonuniform shocks will be treated. Finally in §12, diabatic flow, i.e., flow with heat addition, will be briefly analyzed.

For diabatic flow, the energy release is due to chemical reaction. We shall study the chemical reaction in details in Chapter XVII. However, in this chapter, we shall consider a few simple inviscid flows with heat addition. In §13, two types of surface of discontinuity for diabatic flow will be discussed, i.e., the detonation waves and the flame front while in §14 to §16, inviscid flows of simple dissociating gas will be treated. Finally, in §17 to §23, we will present the most commonly used finite-difference methods for the unsteady Euler equations.

2. Variation of entropy behind a curved shock

For isoenergetic flow, the vorticity may be expressed in terms of the entropy variation as [cf. Eq.(5.45)]

$$\text{vorticity} = \omega = \frac{p}{R}\, y^{\delta} \frac{dS}{d\psi} \tag{14.1}$$

where p is the pressure, R is the gas constant, y is the radial distance, $\delta = 0$ for two-dimensional flow, $\delta = 1$ for axially symmetrical flow, S is entropy, and ψ is stream function.

The increase of entropy across a shock is [cf. Eq.(4.16)]

$$\frac{S_2 - S_1}{c_V} = \log\left[\left(\frac{2\gamma}{\gamma+1} M_1^2 \sin^2\alpha - \frac{\gamma-1}{\gamma+1}\right)\left(\frac{\gamma-1}{\gamma+1} + \frac{2}{(\gamma+1)M_1^2 \sin^2\alpha}\right)^{\gamma}\right] \tag{14.2}$$

where M_1 is the Mach number, α is shock angle with respect to velocity in front of the shock, c_V is specific heat at constant volume, and γ is the ratio of specific heats (c_p/c_V). The subscript 1 refers to the value in front of the shock, whereas the subscript 2 refers to value behind the shock.

To show the order of magnitude of vorticity caused by the shock, we consider a small perturbation from an irrotational flow involving a shock wave as, for example, uniform supersonic flow over a two-dimensional wedge or over an unyawed cone. There are two possible kinds of perturbation from the original irrotational flow :

(a) Disturbances from the front of the shock which affect both M_1 and α.

(b) Disturbances from the rear of the shock which affect only the shock angle.

The change of entropy behind the shock with respect to stream function is

$$\left(\frac{dS_2}{d\psi}\right)_\sigma = \frac{1}{\rho_2 q_2 y^\delta \sin\beta}\left(\frac{\partial S_2}{\partial\alpha}\frac{\partial\alpha}{\partial\sigma} + \frac{\partial S_2}{\partial M_1}\frac{\partial M_1}{\partial\sigma}\right) \tag{14.3}$$

where ρ is the density, q is the resultant velocity, β is the shock angle with respect to q_2, and σ is the arc length along the curved shock. S_1 is here assumed to be constant.

The first-order value of vorticity behind the shock caused by perturbation is , from Eqs.(14.1) and (14.3),

$$\omega = \frac{p_2}{\rho_2 q_2 R \sin\beta}\left(\frac{\partial S_2}{\partial\alpha}\frac{\partial\alpha'}{\partial\sigma} + \frac{\partial S_2}{\partial M_1}\frac{\partial M'}{\partial\sigma}\right) \tag{14.4}$$

where the prime refers to the values of the perturbation quantity.

$$\frac{1}{c_V}\frac{\partial S_2}{\partial\alpha}$$

$$= \frac{2\gamma M_1^2 \sin 2\alpha\,[(\gamma-1)M_1^2\sin^2\alpha+2] - 4\gamma\cot\alpha\,[2\gamma M_1^2\sin^2\alpha-(\gamma-1)]}{[(\gamma-1)M_1^2\sin^2\alpha+2]\,[2\gamma M_1^2\sin^2\alpha-(\gamma-1)]} \tag{14.5}$$

and

$$\frac{1}{c_V}\frac{\partial S_2}{\partial M_1} = \frac{4\gamma(\gamma-1)(M_1^2\sin^2\alpha-1)^2}{M_1[(\gamma-1)M_1^2\sin^2\alpha+2]\,[2\gamma M_1^2\sin^2\alpha-(\gamma-1)]} \tag{14.6}$$

Typical curves of $\frac{1}{c_V}\frac{\partial S_2}{\partial\alpha}$ and $\frac{1}{c_V}\frac{\partial S_2}{\partial M_1}$ as functions of the shock angle α are plotted in Fig. 14.1 for $M_1=3$. This figure shows that these quantities are of the order of magnitude of unity over practically the entire range of shock strengths, except for very weak shocks near the Mach angle. If the strength of the shock is finite, one should consider the nonuniform distribution introduced by the curved shock; in this case, the flow is rotational.

3. Fundamental equations of steady rotational isoenergetic flows

We consider in this section rotational steady flow with constant energy in the whole field, i.e.,

$$\frac{\gamma p}{(\gamma-1)\rho} + \frac{q^2}{2} = \text{constant} = C \tag{14.7}$$

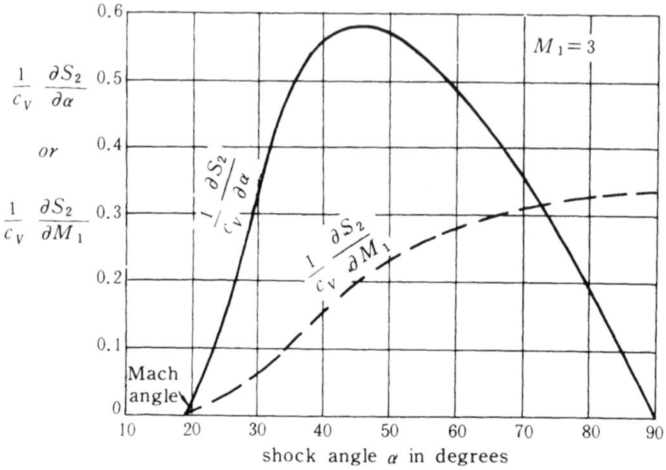

Fig. 14.1 Entropy variation behind a shock wave

We shall consider the two-dimensional or axially symmetric case only. The fundamental equations are :

(a) Equations of motion

$$u\frac{\partial u}{\partial x}+v\frac{\partial u}{\partial y}=-\frac{1}{\rho}\frac{\partial p}{\partial x} \tag{14.8}$$

$$u\frac{\partial v}{\partial x}+v\frac{\partial v}{\partial y}=-\frac{1}{\rho}\frac{\partial p}{\partial y} \tag{14.9}$$

(b) Equation of continuity

$$\frac{\partial \rho u}{\partial x}+\frac{\partial \rho v}{\partial y}=-\delta\frac{\rho v}{y} \tag{14.10}$$

where $\delta=0$ for two-dimensional flow, and $\delta=1$ for axial symmetry.

(c) Equation of state

$$\frac{p}{p_0}=e^{\frac{S-S_0}{c_V}}\left(\frac{\rho}{\rho_0}\right)^{\gamma} \tag{14.11}$$

(d) Equation of energy in an integrated form

$$S=c_V f(\psi) \tag{14.12}$$

where the stream function ψ is defined by the relations

$$\frac{\partial \psi}{\partial y}=\rho y^{\delta}u,\quad \frac{\partial \psi}{\partial x}=-\rho y^{\delta}v \tag{14.13}$$

It is sometimes convenient to express the fundamental equations in intrinsic form with θ denoting the direction of the streamline, and s and n denoting the arc length along and perpendicular to the streamline, respectively. The fundamental equations for rotational steady isoenergetic flow become

$$q\frac{\partial q}{\partial s} = -\frac{1}{\rho}\frac{\partial p}{\partial s} \tag{14.14}$$

$$\rho q^2\frac{\partial \theta}{\partial s} = -\frac{\partial p}{\partial n} \tag{14.15}$$

$$\frac{1-M^2}{q}\frac{\partial q}{\partial s} + \frac{\partial \theta}{\partial n} = -\delta\frac{\sin\theta}{y} \tag{14.16}$$

$$\omega = \frac{\partial q}{\partial n} - q\frac{\partial \theta}{\partial s} = \frac{p}{\rho q R}\frac{\partial S}{\partial n} \tag{14.17}$$

$$\frac{\partial \psi}{\partial n} = \rho q y^{\delta} \tag{14.18}$$

Eqs.(14.11) and (14.12) do not change in form.

4. Conditions across a curved shock [3–7]

As shown in Chapter IV, § 6, if the shock angle α is known, the pressure, density, and velocity behind a shock may be obtained from those in front of the shock according to the Rankine-Hugoniot relations.

$$\rho_1 q_1 \sin\alpha = \rho_2 q_2 \sin(\alpha-\theta) \tag{14.19}$$

$$p_1 + \rho_1 q_1^2 \sin^2\alpha = p_2 + \rho_2 q_2^2 \sin^2(\alpha-\theta) \tag{14.20}$$

$$q_1 \cos\alpha = q_2 \cos(\alpha-\theta) \tag{14.21}$$

$$q_1 q_2 \sin\alpha \sin(\alpha-\theta) = a^{*2} - \frac{\gamma-1}{\gamma+1} q_1^2 \cos^2\alpha \tag{14.22}$$

where

$$\theta = \theta_2 - \theta_1, \quad \alpha = \alpha_1 - \theta_1, \quad \frac{1}{2}\frac{\gamma+1}{\gamma-1} a^{*2} = C \tag{14.23}$$

The angles are defined in Fig. 14.2 , and a^* is the critical sound speed.

If the flow in front of the shock is completely known, the conditions of the flow will be known when the shape of the shock front is known. If we know the angle of shock, the pressure, etc., behind the shock may be obtained from Eqs. (14.19) to (14.22) in terms of α and quantities in front of shock, e.g., p_1 , etc. If the curvature of the shock $\partial\alpha/\partial\sigma$ is known, the first spatial derivatives of the gasdynamic quantities, such as $\partial p_2/\partial s_2$, and $\partial p_2/\partial n_2$, may be obtained from Eqs. (14.14) to (14.17) and Eqs. (14.19) to (14.22) in terms of α and the first spatial derivative of these quantities in front of the shock (such as $\partial p_1/\partial s_1$, and $\partial p_1/\partial n_1$,). We shall derive these relations for the two-dimensional case.

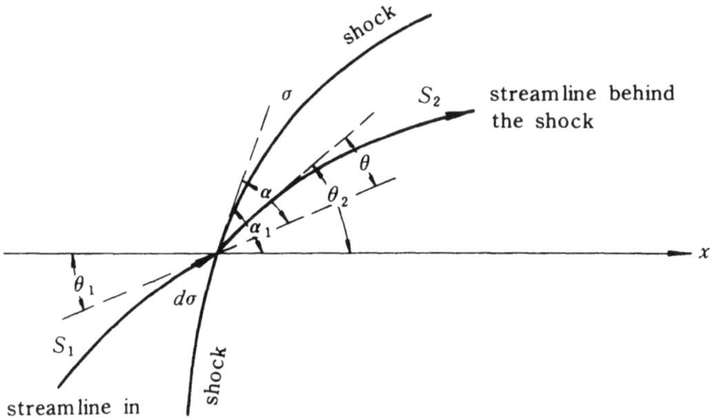

Fig. 14.2 Flow across a curved shock

We use the intrinsic coordinates. The two spatial derivatives for entropy S are

$$\frac{\partial S}{\partial s} = 0 \tag{14.24}$$

$$\frac{\partial S}{\partial n} = \frac{\partial S}{\partial p}\frac{\partial p}{\partial n} + \frac{\partial S}{\partial \rho}\frac{\partial \rho}{\partial n} \tag{14.25}$$

where $\partial S/\partial p$ and $\partial S/\partial \rho$ may be obtained from Eq. (14.11).

We need to consider eight spatial derivatives behind the shock for the variables p, ρ, q, and θ. These derivatives may be obtained from the eight equations (14.14) to (14.17) and (14.19) to (14.22). For simplicity, we shall study the case in which the flow in front of the shock is irrotational. For irrotational flow, all the flow quantities may be expressed in terms of the pressure p and the flow direction θ. All flow quantities behind the shock may be expressed in terms of p_1 and θ_1 and the shock angle α (see Fig. 14.2). In general we may obtain the relations

$$p_2 = p_2(p_1, \theta_1, \alpha_1)$$

$$\theta_2 = \theta_2(p_1, \theta_1, \alpha_1) \tag{14.26}$$

From the shock relations (14.19) to (14.22), we have, in terms of the quantities in front of the shock and the shock angle α, the expressions for the pressure p

$$p_2 = p_1 + \rho_1 \left[q_1^2 \left(1 - \frac{2}{\gamma+1} \cos^2 \alpha \right) - a^{*2} \right] \tag{14.27}$$

$$p_2 = p_1 + \rho_1 q_1^2 \ [\sin^2\alpha - \sin\alpha \cos\alpha \tan(\alpha - \theta)] \tag{14.28}$$

Differentiation of Eq.(14.27) along the shock gives

$$\frac{\partial p_2}{\partial \sigma} = \frac{\partial p_1}{\partial \sigma} + \frac{\partial \rho_1}{\partial \sigma} \left\{ q_1^2 \left(1 - \frac{2}{\gamma+1} \cos^2\alpha \right) - a^{*2} \right\}$$
$$+ 2\rho_1 q_1 \left(1 - \frac{2}{\gamma+1} \cos^2\alpha \right) \frac{\partial q_1}{\partial \sigma}$$
$$+ \rho_1 q_1^2 \frac{4}{\gamma+1} \cos\alpha \sin\alpha \left(\frac{\partial \alpha_1}{\partial \sigma} - \frac{\partial \theta_1}{\partial \sigma} \right) \tag{14.29}$$

The derivative $\partial p_2 / \partial \sigma$ may be expressed in terms of $\partial p_2 / \partial s_2$ and $\partial p_2 / \partial n_2$ as

$$\frac{\partial p_2}{\partial \sigma} = \frac{\partial p_2}{\partial n_2} \sin(\alpha - \theta) + \frac{\partial p_2}{\partial s_2} \cos(\alpha - \theta) \tag{14.30}$$

Furthermore in front of shock we may express all the quantities in terms of p_1 and θ_1. Eq.(14.29) becomes

$$A_1 (\rho_1 q_1^2)^{-1} \left(-\frac{\partial p_1}{\partial s_1} \right) + B_1 \left(\frac{\partial \theta_1}{\partial s_1} \right)$$
$$= A_2 (\rho_1 q_1^2)^{-1} \left(-\frac{\partial p_2}{\partial s_2} \right) + B_2 \left(\frac{\partial \theta_2}{\partial s_2} \right) + C_1 \left(\frac{\partial \alpha_1}{\partial \sigma} \right) \tag{14.31}$$

where

$$A_1 = \frac{2}{\gamma+1} \cos\alpha \left[(3M_1^2 - 4) \sin^2\alpha + \frac{1-\gamma}{2} \right]$$

$$B_1 = \frac{2}{\gamma+1} \sin\alpha \left[(M_1^2 - 4) \sin^2\alpha + \frac{5-\gamma}{2} \right]$$

$$A_2 = \cos(\alpha - \theta), \quad B_2 = \frac{\sin 2\alpha}{2\cos(\alpha - \theta)}, \quad C_1 = \frac{2}{\gamma+1} \sin 2\alpha$$

Similarly, from Eq.(14.28), we have relations

$$A_1' (\rho_1 q_1^2)^{-1} \left(-\frac{\partial p_1}{\partial s_1} \right) + B_1' \left(\frac{\partial \theta_1}{\partial s_1} \right)$$

$$=A_2'(\rho_1 q_1^2)^{-1}\left(-\frac{\partial p_2}{\partial s_2}\right)+B_2'\left(\frac{\partial \theta_2}{\partial s_2}\right)+C_1'\left(\frac{\partial \alpha_1}{\partial \sigma}\right) \quad (14.32)$$

where

$$A_1' = M_1^2 \cos^2\alpha \cos\theta - (M_1^2 - 1)\cos(2\alpha+\theta)$$

$$B_1' = M_1^2 \sin^2\alpha \sin\theta + \sin(2\alpha+\theta)$$

$$A_2' = 1 + (M_2^2 - 2)\sin 2(\alpha-\theta)$$

$$B_2' = \sin 2\alpha,\ C_1' = \sin 2\theta / 2\cos(\alpha-\theta)$$

From Eqs.(14. 31) and (14. 32), we may calculate $\partial p_2/\partial s_2$ and $\partial\theta_2/\partial s_2$ in terms of p_1, θ_1 and α. From Eqs. (14.14) to (14.17), all other first spatial derivatives may be obtained.

5. Shock attached to a solid boundary[5]

We shall apply relations (14.31) and (14.32) to the problem of an attached shock to a solid boundary where the streamline behind the shock is known, this streamline being surface of the solid boundary.

(a) Shock attached to the sharp nose of a symmetrical body in a uniform supersonic stream. If the body is not blunt and if the Mach number is high enough, there is usually an attached shock at the nose. In the present case, all of flow quantities are constants in front of the shock and Eqs. (14.31) to (14.32) become, respectively,

$$\frac{1}{r_2}\sin 2\alpha\left\{\frac{2}{\gamma+1}\sin 2\alpha - \frac{\sin 2\theta}{4\cos^2(\alpha-\theta)}\right\}$$

$$= -\frac{1}{\rho_1 q_1^2}\frac{\partial p_2}{\partial s_2}\left\{\frac{2}{\gamma+1}\sin 2\alpha\,[1+\sin^2(\alpha-\theta)(M_2^2-2)] - \frac{\sin 2\theta}{2}\right\}$$

$$(14.33)$$

$$\frac{1}{r_2}\sin 2\alpha\left\{\cos(\alpha-\theta) - \frac{1+(M_2^2-2)\sin^2(\alpha-\theta)}{2\cos(\alpha-\theta)}\right\}$$

$$= -\frac{\partial \alpha_1}{\partial \sigma}\left\{\frac{2}{\gamma+1}[1+(M_2^2-2)\sin^2(\alpha-\theta)] - \frac{\sin 2\theta}{2}\right\} \quad (14.34)$$

where $r_2 = -\partial s_2/\partial\theta_2$. It is interesting to note that both the curvature of the shock and the pressure gradient behind the shock are proportional to the curvature of the body. For a given M_1, there is a M_2 such that both $\partial p_2/\partial s_2$ and $\partial\alpha_1/\partial\sigma$ are infinite even when the radius of curvature of the body r_2 is finite. This point, known as the Crocco point, is close to the detached shock point.

(b) Shocks attached to solid boundary with continuous slope-weak family. In this case we have

$$p_1 \cong p_2,\ \rho_1 \cong \rho_2,\ q_1 \cong q_2,\ \theta \to 0$$

Eqs.(14.31) and (14.32) give

$$\frac{\partial p_2}{\partial s_2} - \frac{\partial p_1}{\partial s_1} = \rho_1 q_1^2 \left(\frac{1}{r_1} - \frac{1}{r_2} \right) \tan\alpha \tag{14.35}$$

$$\frac{\partial \alpha_1}{\partial \sigma} = \frac{4+(\gamma-3)M_1^2}{2\sqrt{M_1^2-1}} \left(\frac{\sin\alpha}{r_1} + \frac{1}{\rho_1 q_1^2} \frac{\partial p_1}{\partial s_1} \cos\alpha \right) \tag{14.36}$$

Here we have the result that the curvature of the weak shock (Mach line) depends on the pressure gradient $\partial p_1 / \partial s_1$, but not on the radius of curvature r_2 of the body, and the multiplicative factor in Eq.(14. 36) changes sign at $M_1^2 = 4/(3-\gamma)$.

(c) Shocks attached to solid boundary with continuous slope-strong family. In this case we have

$$\theta \to 0 \text{ and } \alpha \to \frac{\pi}{2}$$

If all the quantities in Eqs. (14.31) and (14.32) are finite, we have

$$r_2 = r_1 \frac{\dfrac{1}{M_1^2} + \dfrac{\gamma-1}{2}}{M_1^2 - \dfrac{\gamma+3}{2}} \tag{14.37}$$

If the radii of curvature of the body in front of and behind the shock are equal, i.e., $r_1 = r_2$, M_1 must have the critical value

$$M_1 = M_{1_c} = \left\{ \frac{1}{2} \left[\sqrt{(\gamma+1)^2+4} + (\gamma+1) \right] \right\}^{-1/2} \tag{14.38}$$

if $\gamma = 1.4$, $M_{1_c} \cong 1.67$.

If the curvature of the shock is allowed to become infinite, Eqs. (14.31) and (14.32) give the relation

$$\frac{2}{\gamma+1}\left(M_1^2 - \frac{\gamma+3}{2} \right)\left(-\frac{1}{r_1} \right) - \frac{1}{\gamma+1}\left[\frac{2}{M_1^2} + (\gamma-1) \right]\left(-\frac{1}{r_2} \right) \leqslant 0$$

(14.39)

It should be observed that the results in this section will also give valuable information for detached shock problems. For a detached shock (Fig.14.4), the shock angle varies continuously from the normal shock angle to the Mach angle according to a shock polar corresponding to the Mach number in front of the shock. At each point of the shock polar we may calculate the change in $\partial p_2/\partial s_2$ and that in direction $\partial \theta_2/\partial s_2$ along the streamline from Eqs.(14.33) and (14.34). The initial directions of the streamlines at every point of the shock polar are independent of the body shape in the detached shock case. If we plot the initial directions of the streamlines on the shock polar, we have a diagram known as "hedgehog" or "porcupine." This diagram gives valuable information in solving the whole problem in the hodograph plane. A typical "hedgehog" or "porcupine" diagram is shown in Fig. 14.3. There are two interesting points in this diagram.

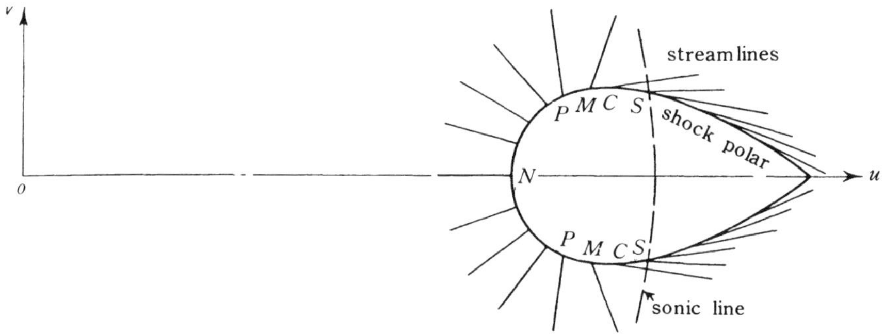

Fig. 14.3 Shock polar with streamlines — hedgehog or porcupine diagram N = normal shock, M = maximum deflection, S = sonic point, C = Crocco point — radial streamline, P = constant pressure point — tangential streamline

(a) Crocco point at which $\partial \theta_2/\partial s_2 = 0$ and the streamline is radial in the hodograph plane. This point is important in the study of detached shock of a wedge.[8]

(b) Constant pressure point at which $\partial p_2/\partial s_2 = 0$ and $\partial q_2/\partial s_2 = 0$ and the streamline is perpendicular to the velocity vector in the hodograph plane. This point is important in the study of shock waves in a supersonic jet in a manner analogous to the Crocco point for the case of detached shock over a wedge.[9]

6. Detached shock[10]

If the Mach number of the flow is relatively low or if the body is blunt, the shock detaches from the body. Behind the shock, there is always a subsonic region, and finally the flow becomes supersonic again (Fig. 14.4). There is no simple method of solution for this problem yet. For the subsonic region, a numerical procedure may be developed for the problem of Fig. 14.4b, treating it as a boundary value problem. For the supersonic region, the method of characteristics may be used (§8). For problems such as that shown in Fig. 14.4a, the numerical procedure has not been developed because the approximate location of the sonic lines is not known. Another method of solution of the detached shock problem is the method of power series expansion, this being applicable to both kinds of problems of Fig. 14.4. It will be briefly described below.

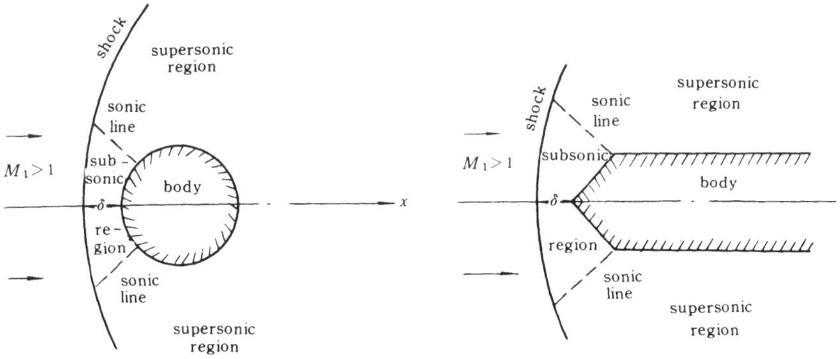

(a) Body without sharp shoulder (b) Body with sharp shoulder

Fig. 14.4 Detached shocks over a bluff body

Our object is to find the shape of the detached shock, the distance of the shock from the body, and the whole flow pattern. For simplicity, we shall consider only the two-dimensional symmetric body. The method may be easily extended into the axially symmetric case or for asymmetric body. This method is particularly good when the Mach number of the uniform stream M_1 is high and the detached shock distance from the body is small. To a first approximation, one would expect that the radius of curvature of the shock and the distance of the shock from the body are proportional to the radius of curvature of the body. To improve on this first-order theory, one may develop the power series expansions in front of the nose of the body and behind the nose of the shock.

First let us develop the flow quantities in power series near the nose of the shock. Along the axis of symmetry and behind the shock, the pressure may be expressed by the power series

$$p_2 = p_{20} + p_2' x_2 + p_2'' x_2^2 + \cdots \tag{14.40}$$

where x_2 is the x distance measured from the nose of the shock. From Eqs. (14.31) and (14.32), if we apply the conditions at the nose of the shock, i.e.,

$$\alpha = \frac{\pi}{2}, \ \theta = 0, \ \frac{\partial p_1}{\partial s_1} = \frac{\partial \theta_1}{\partial s_1} = \frac{\partial \theta_2}{\partial s_2} = 0, \ s_2 = x_2$$

we have

$$(M_2^2-1)(-\rho_1 q_1^2)^{-1}\frac{\partial p_2}{\partial s_2}+\left(\frac{q_2}{q_1}-1\right)\frac{\partial \alpha}{\partial \sigma}=0 \qquad (14.41)$$

Eq.(14.41) may be written as

$$\frac{p'_2}{p_0}=C_1(M_1)\frac{\partial \alpha}{\partial \sigma} \qquad (14.42)$$

where p_0 is the stagnation pressure, $p'_2=\left(\frac{\partial p_2}{\partial s_2}\right)_{x_2=0}$, and $C_1(M_1)$ is the function of M_1 only.

Similarly the coefficients p_2'', etc., of Eq. (14. 40) may be written as

$$\frac{p_2''}{p_0}=C_2(M_1)\left(\frac{\partial \alpha}{\partial \sigma}\right)^2$$

$$\frac{p_2'''}{p_0}=C_3(M_1)\left(\frac{\partial \alpha}{\partial \sigma}\right)^3+C_4(M_1)\frac{\partial^3 \alpha}{\partial \sigma^3} \qquad (14.43)$$

$$\frac{p_2^{\text{iv}}}{p_0}=C_5(M_1)\left(\frac{\partial \alpha}{\partial \sigma}\right)^4+C_6(M_1)\frac{\partial \alpha}{\partial \sigma}\frac{\partial^3 \alpha}{\partial \sigma^3}$$

where $\partial^2\alpha/\partial\sigma^2=\partial^4\alpha/\partial\sigma^4=\cdots=0$ because of the symmetry of the shape of the shock and asymmetry of the shock angle with respect to σ, and $C_i(M_1)$ are functions of M_1 only.

If we take the series (14.40) up to x^4 and apply the condition at the nose of the body, i.e.,

$$x_2=\delta,\ p_2=p_0,\ \frac{\partial p_2}{\partial s_2}=p_2'=0$$

we have

$$\frac{p(\delta)}{p_0}=F\left(\delta\frac{\partial \alpha}{\partial \sigma},\ \delta^3\frac{\partial^3 \alpha}{\partial \sigma^3},\ M_1\right)=1$$

$$\frac{p'(\delta)}{p_0}=F_1\left(\delta\frac{\partial \alpha}{\partial \sigma},\ \delta^3\frac{\partial^2 \alpha}{\partial \sigma^3},\ M_1\right)=0 \qquad (14.44)$$

From Eq.(14.44), we have

$$\delta\frac{\partial \alpha}{\partial \sigma}=f(M_1),\ \delta^3\frac{\partial^3 \alpha}{\partial \sigma^3}=g(M_1) \qquad (14.45)$$

It is interesting to note that to first approximation the relations (14.45) are independent of the shape of the body. Although the first relation of (14.45) seems to agree reasonably well with experimental data, the second is not very accurate.

We may apply the series expansion method to study the flow field near the nose of the body. Taking into account the symmetric nature of the flow field, we write

$$
\begin{aligned}
p &= 1 + p_1 x + (p_{11}x^2 + p_{22}y^2) + (p_{111}x^3 + p_{122}y^2 x) + \cdots \\
\frac{1}{\rho} &= \tau = 1 + \tau_1 x + (\tau_{11}x^2 + \tau_{22}y^2) + (\tau_{111}x^3 + \tau_{122}y^2 x) + \cdots \\
u &= u_1 x + (u_{11}x^2 + u_{22}y^2) + \cdots \\
v &= v_2 y + v_{12}xy + \cdots
\end{aligned}
\tag{14.46}
$$

where the origin is at the nose, and $x=0$, $y=0$, $u=v=0$, and $p=\tau=1$.

If Eq.(14.46) is a solution of the flow problem, it must satisfy the fundamental equations (14.7) to (14.12) for all values of x and y. Substituting Eq.(14.46) into these equations, one observes that the coefficients of the same powers of x and y must be zero. The following relations between the coefficients can be obtained:

From Eq.(14.10), we have

$$
\begin{aligned}
&u_1 + v_2 = 0 \\
&2u_{11} + v_{12} = 0 \\
&\cdot \quad \cdot \quad \cdot \quad \cdot \quad \cdot
\end{aligned}
\tag{14.47}
$$

From Eq.(14.8), we have

$$
\begin{aligned}
&p_1 = 0 \\
&-2p_{11} = u_1^2 = a^2 \text{ (say)} \\
&-p_{111} = u_1 u_{11} \\
&-p_{122} = (u_1 + 2v_2)u_{22} \\
&\cdot \quad \cdot \quad \cdot \quad \cdot \quad \cdot \quad \cdot \quad \cdot
\end{aligned}
\tag{14.48}
$$

From Eq.(14.9), we have

$$
\begin{aligned}
&-2p_{22} = v_2^2 \\
&-2p_{122} = (u_1 + 2v_2)v_{12} \\
&\cdot \quad \cdot \quad \cdot \quad \cdot \quad \cdot \quad \cdot \quad \cdot
\end{aligned}
\tag{14.49}
$$

From Eq.(14.7), we have

$$\tau_1 = 0$$

$$p_{11}+\tau_{11}=-\frac{\gamma-1}{2\gamma}u_1^2$$

$$p_{22}+\tau_{22}=-\frac{\gamma-1}{2\gamma}v_2^2$$

$$p_{111}+\tau_{111}=-\frac{\gamma-1}{\gamma}u_1u_{11}$$

$$p_{122}+\tau_{122}=-\frac{\gamma-1}{\gamma}(u_1u_{22}+v_2v_{12})$$

$$\cdots\cdots\cdots\cdots\cdots \qquad (14.50)$$

On the boundary of the body, we must have

$$v/u=dy/dx \qquad (14.51)$$

For a symmetric body, we may write the shape of the body as

$$y^2=(2x/\kappa_0)+\cdots \qquad (14.52)$$

where κ_0 is the curvature at the nose. Substituting Eqs.(14.46) and (14.52) into (14.51), we have

$$(u_1-2v_2)\frac{\kappa_0}{2}+u_{22}=0 \qquad (14.53)$$

$$\cdots\cdots\cdots$$

Since the flow depends not only on the shape of the body but also on the entropy distribution (which in turn is determined by the curvature of shock) we should also expand the entropy in power series. We write the function $f(\psi)$ of Eq.(14.12) as

$$f=s_1x+(s_{11}x^2+s_{22}y^2)+\cdots \qquad (14.54)$$

Substituting Eqs.(14.46) and (14.54) into Eq.(14.11), it is easy to show that

$$s_1=s_{11}=s_{22}=s_{111}=s_{122}=0$$

and we have

$$S/c_V=f=s_{1111}x^4+s_{1122}x^2y^2+s_{2222}y^4+\cdots \qquad (14.55)$$

For the two-dimensional flow, we have

$$\psi=u_1xy+\cdots$$

Hence we have

$$S\sim\psi^2\sim x^2y^2 \qquad (14.56)$$

and then

$$(df/d\psi)_{\psi=0}=0 \qquad (14.57)$$

for the two-dimensional case. This is not so for the axially symmetirc case.

Summarizing the above results, we have for two-dimensional flow

$$p=1-\frac{a^2}{2}(x^2+y^2)-\frac{3a^2}{2}\kappa_0(x^3+xy^2)+\cdots$$

$$\frac{1}{\rho}=\tau=+\frac{1}{2\gamma}a^2(x^2+y^2)+\frac{3a^2}{2\gamma}\kappa_0\left(x^2+\frac{xy^2}{2}\right)+\cdots$$

$$u=a\left\{x+\frac{3\kappa_0}{2}(x^2-y^2)+\cdots\right\}$$

$$v=-a(y+2\kappa_0 xy+\cdots) \tag{14.58}$$

which depend on a single parameter a [cf. Eq.(14.48)].

Some interesting results may be obtained from the above analysis. Over a straight boundary, we have $\partial\theta/\partial s=0$. From Eq.(14.14),

$$\frac{\partial p}{\partial n}=0 \tag{14.59}$$

From Eqs.(14.17) and (14.57),

$$\omega=\frac{\partial q}{\partial n}=\frac{p}{R}\frac{dS}{d\psi}=0 \tag{14.60}$$

Finally, from Eqs.(14.7), (14.50), and (14.60),

$$\frac{\partial \rho}{\partial n}=0 \tag{14.61}$$

We see that the normal to the straight boundary is locally a line of constant p, q, and ρ. In particular the sonic line is perpendicular to the straight boundary in the two-dimensional case. This is not so in the axially symmetric case because $(dS/d\psi)_{\psi=0}\neq 0$.

Dugundgi[10] applied the above method to the case of a body of revolution and made a comparison with experimental results. Since he used only the first two terms of the series, the agreement of the theory with experimental data is not too good. Improvement is to be expected if more terms are used.

7. Linearized theory of rotational flow[11, 12]

We consider a uniform supersonic flow passing a plane ogive (Fig. 14.5). If the leading-edge angle of the ogive is smaller than the critical angle, a curved shock front is attached to the vertex of the ogive. Behind the curved shock, the flow is rotational with nonuniform entropy. If the surface of the ogive differs only slightly from the corresponding wedge, the linearized theory of rotational flow is applicable. The differential equation for the perturbation stream function ψ_1 is Eq.(6.23), i.e.,

$$(1-M_2^2)\frac{\partial^2\psi_1}{\partial x^2}+\frac{\partial^2\psi_1}{\partial y^2}=-\rho_2\omega_1\,[1+(\gamma-1)M_2^2] \qquad (14.62)$$

Since the flow in front of the shock is supersonic and the curved shock is attached to the ogive, the flow over the upper half of the ogive is independent of that over the lower half. One may thus consider the half plane only. We choose the axis along the surface of the wedge which closely approximates the ogive (Fig.14.5) and the y-axis perpendicular to the x-axis. The surface of the ogive is

$$y=d_1Y(x) \qquad (14.63)$$

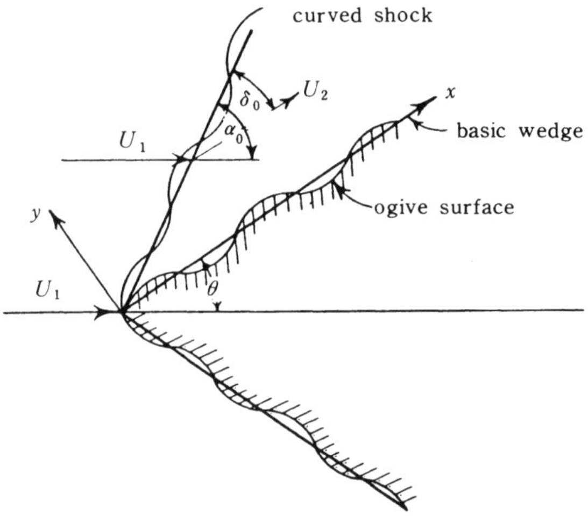

Fig. 14.5 Uniform supersonic flow over a plane ogive

where d_1 is a small quantity of same order of magnitude as ψ_1. $Y(x)$ gives the geometrical shape of the ogive.

The location of the curved shock will be approximately on the line

$$y=x\tan\delta_0 \qquad (14.64)$$

where δ_0 is the shock angle with respect to U_2 (Fig. 14.5).

The basic flow is now a uniform flow of velocity U_2 and density ρ_2. We may define the nondimensional stream function ψ as

$$\psi=y+\psi_1 \qquad (14.65)$$

The relations between the stream function ψ and the resultant velocity components U and V are

$$\frac{\partial\psi}{\partial y} = \frac{\rho U}{\rho_2 U_2} = \left(1 - \frac{\Delta\rho}{\rho_2}\right)\left(1 + \frac{u}{U_2}\right), \quad \frac{\partial\psi}{\partial x} = -\frac{\rho}{\rho_2}\frac{V}{U_2} \tag{14.66}$$

where u and v are the perturbed velocity components such that $U = U_2 + u$, $V = v$, and $u, v \ll U_2$. $\rho = \rho_2 + \Delta\rho$.

The relations between the perturbed stream function and the perturbed velocity components u and v are

$$\frac{\partial\psi_1}{\partial y} = \frac{\Delta\rho}{\rho_2} + \frac{u}{U_2}, \quad \frac{\partial\psi_1}{\partial x} = -\frac{v}{U_2} \tag{14.67}$$

where

$$\frac{\Delta\rho}{\rho} = -M_2^2\frac{u}{U_2} - \frac{\Delta S}{R} \tag{14.68}$$

$\Delta S = S - S_1$ and $M_2 = \dfrac{U_2}{a_2}$.

The boundary conditions of the present problems are:

(1) On the surface of the ogive,

$$\psi[x, d_1 Y(x)] = \psi_0 + \psi_1 + \cdots = y + \psi_1(x, 0) + y\frac{\partial\psi_1(x, 0)}{\partial y} + \cdots = 0$$

and $y = d_1 Y(x)$. Hence to a first approximation,

$$\psi_1(x, 0) = -d_1 Y(x) \tag{14.69}$$

(2) At the shock which is approximately on the line $y + x \tan\delta_0$, the shock angle is $\alpha_0 + \alpha$; we also have the relations

$$u_s = U_\alpha \alpha, \quad v_s = V_\alpha \alpha, \quad \left(\frac{\Delta\rho}{\rho_2}\right)_s = K_\rho U_2 u_s, \quad U_s = U_2 + u_s \tag{14.70}$$

where α is the perturbed shock angle and the subscript s refers to the value immediately behind the shock. We have also

$$\frac{U_1}{U_\alpha} = \frac{\left(\dfrac{\rho_1}{\rho_2} - 1\right)\sin(\alpha_0 + \delta_0)}{1 + K_\rho U_2^2 \sin^2\delta_0},$$

$$\frac{V_\alpha}{U_1} = \frac{\left(\dfrac{\rho_1}{\rho_2} - 1\right)[K_\rho U_2^2 \sin\delta_0 \sin\alpha_0 - \cos(\alpha_0 + \delta_0)]}{1 + K_\rho U_2^2 \sin^2\delta_0}$$

$$K_\rho U_2^2 = -(\gamma-1)M_2^2 \frac{\left(\frac{\gamma+1}{\gamma-1} - \frac{\rho_1}{\rho_2}\right)\left(\frac{\gamma+1}{\gamma-1} - \frac{\rho_2}{\rho_1}\right)}{\frac{\gamma+1}{\gamma-1}\left(\frac{\rho_1}{\rho_2} + \frac{\rho_2}{\rho_1}\right) - 2} \tag{14.71}$$

From Eqs. (14.70) and (14.71), the boundary condition at the shock line is

$$\frac{\partial \psi_1(x,\ x\tan\delta_0)}{\partial y} = -\frac{U_\alpha}{V_\alpha}(1+K_\rho U_{2^2})\frac{\partial \psi_1(x,\ x\tan\delta_0)}{\partial x} \tag{14.72}$$

We shall solve the inhomogeneous partial differential equation (14.62) with the boundary conditions (14.69) and (14.72). In solving this equation we have two different cases: a weak shock case, $M_2 > 1$, and a strong shock case, $M_2 < 1$. These two cases should be treated separately as follows:

(a) Weak shock case $M_2 > 1$. The differential equation (14.62) is of hyperbolic type. We may assume the perturbed stream function to be of the form

$$\psi_1 = \psi_{10} + \psi_{11} \tag{14.73}$$

where ψ_{10} satisfies the homogeneous equation obtained from Eq. (14.62) and boundary condition $\psi_{10}(x,0) = -d_1 Y(x)$.

ψ_{11} satisfies the inhomogeneous equation (14.62) with ω_1 estimated from ψ_{10} and $\psi_{11}(x,0) = 0$.

The general solution of ψ_{10} is

$$\psi_{10} = -d_1[f(x-\mu y) + f_r(x+\mu y)] \tag{14.74}$$

where f and f_r are arbitrary functions still to be determined and $\mu^2 = M_2^2 - 1$. From the boundary condition (14.69),

$$f(x) + f_r(x) = Y(x) \tag{14.75}$$

From the boundary condition (14.72)

$$f_r(x) = (R_f/b)f(bx) \tag{14.76}$$

where

$$R_f = \frac{\mu V_\alpha - U_\alpha(1+K_\rho U_2^2)}{\mu V_\alpha + U_\alpha(1+K_\rho U_2^2)},\quad b = \frac{1-\mu\tan\delta_0}{1+\mu\tan\delta_0}$$

Substituting Eq. (14.76) into Eq. (14.75), we obtain the the first approximate functional relation for $f(x)$ as

$$f(x) + f_r(x) = f(x) + (R_f/b)f(bx) = Y(x) \tag{14.77}$$

After ψ_{10} is obtained, we may derive the expression for ω_1 as

$$\omega_1 = \frac{U_2}{\gamma(\gamma-1)M_2^2}\left(\frac{1}{c_V}\frac{dS}{d\alpha}\right)\frac{d\alpha}{dy}$$

$$= -K_\omega d_{1\mu} \frac{df'\,[(\cot\delta_0-\mu)y]}{dy} = \omega_{10}(y) \qquad (14.78)$$

where

$$K_\omega = -\frac{1-R_f}{\gamma(\gamma-1)M_2^2}\frac{[1+(\gamma-1)M_2^2]}{(1+K_p U_2{}^2)U_{\alpha 1}}\left(\frac{U_2}{U_1}\right)\left(\frac{1}{c_V}\frac{dS}{d\alpha}\right)$$

$$U_\alpha = U_1 U_{\alpha 1}, f' = \frac{df}{dx}$$

The solution of ψ_{11} is then

$$\psi_{11}(x,y) = yW(0) - \int_0^y W(\eta)\,d\eta \qquad (14.79)$$

where
$$\frac{dW(\eta)}{d\eta} = \omega_1(\eta).$$

With the functions ψ_{10} given by Eqs.(14.74), and ψ_{11} by Eq.(14.79), the boundary condition (14.69) is satisfied if the functions f and f_r are related by Eq.(14.75). From the boundary condition (14.72), we obtain the final functional relation for f as

$$f(x) + \left(\frac{R_f}{b} + \frac{1+R_f}{b}K_\omega\right) f(bx) = Y(x) + \frac{1+R_f}{2}K_\omega f'(0)x \qquad (14.80)$$

When f is obtained, the complete solution of this problem is obtained. The general solution may be considered to consist of three parts: the basic disturbance represented by the function f, the reflected disturbance represented by f_r, and the disturbance due to vorticity. In reference 9 some examples are given where it has been found that the effect of rotationality may cause a 20% variation from the corresponding irrotational solution.

(b) Strong shock cases $M_2 < 1$. In this case the differential equation (14.62) is of the elliptic type. We may also put ψ_1 in the form of Eq. (14.74). For the solution of ψ_{10}, it is convenient to use the new coordinates.

$$r = \sqrt{(x^2+\mu^2y^2)}\ , \theta = \tan^{-1}(\mu y/x) \qquad (14.81)$$

The solution of ψ_{10} is then

$$\psi_{10} = -d_1 \sum_{n=0}^{\infty} r^n (A_n \cos n\theta + B_n \sin n\theta) \qquad (14.82)$$

where A_n and B_n are arbitrary constants still to be determined.

If the shape of the surface of the ogive may be expressed as

$$Y(r) = \sum_{n=0}^{\infty} a_n r^n \qquad (14.83)$$

we have

$$A_n = a_n \qquad (14.84)$$

If ψ_{10} alone satisfies Eq.(14.72),

$$B_n = a_n \frac{K_\theta \sin n\theta_{\delta_0} + K_r \cos n\theta_{\delta_0}}{K_\theta \cos n\theta_{\delta_0} - K_r \sin n\theta_{\delta_0}} = b_n \text{ (say)} \qquad (14.85)$$

where

$$K_r = \left[\mu \sin \theta_{\delta_0} + \left(\frac{U_\infty}{V_\infty} \right)(1 + K\rho U_2^2)\cos \theta_{\delta_0} \right]$$

$$K_\theta = \left[\frac{U_\infty}{V_\infty} (1 + K_\rho U_2^2) \sin \theta_{\delta_0} - \mu \cos \theta_{\delta_0} \right]$$

$$\theta_{\delta_0} = \tan^{-1} (\mu \tan \delta_0)$$

From ψ_{11} we have for first approximation ω_1

$$\omega_{10} = \frac{K\omega}{1 - R_f} d_1 \mu k_r \sum_{n=2}^{\infty} n(n-1) r_{\delta_0}^{n-2} [-a_n \sin(n-1)\theta_{\delta_0} + b_n \cos (n-1)\theta_{\delta_0}] = \omega_{10}(y) \tag{14.86}$$

where $r_{\delta_0} = y\sqrt{\mu^2 + \cot^2\delta_0} = yk_r$.

The solution of ψ_{11} is then

$$\psi_{11}(x, y) = \frac{1}{4\pi\mu} \int_0^\infty \omega_{10} d\eta \int_{\eta \cot \delta_0}^\infty \log \frac{(x-\xi)^2 + (y+\eta)^2}{(x-\xi)^2 + (y-\eta)^2} d\eta \tag{14.87}$$

If $\psi_1 = \psi_{10} + \psi_{11}$ satisfies the boundary condition (14.72), we have

$$B_n = b_n - \frac{C_{n-1}}{nk_r^{n-1}} = b_{n1} \tag{14.88}$$

where $C_n = -\left[k_{1n} + \frac{U_\infty}{V_\infty} (1 + K_\rho U_2^2) k_{2n} \right]$

$$\frac{\partial \psi_{11}(y \cot \delta_0, y)}{\partial y} = g_1(y) = \sum_{n=0}^{\infty} k_{1n} y^n$$

$$\frac{\partial \psi_{11}(y \cot \delta_0, y)}{\partial x} = g_2(y) = \sum_{n=0}^{\infty} k_{2n} y^n$$

Eq.(14.84) still holds. The complete solution is then

$$\psi_1 = -d_1 \sum_{n=0}^{\infty} r^n (a_n \cos n\theta + b_{n1} \sin n\theta) + \psi_{11} \tag{14.89}$$

where ψ_{11} is given by Eq.(14.87).

The above solution fails at two singular points: (a) $R_f = \infty$ and (b) $K\rho U_2^2 = -1$. Both points occur at $M_2 > 1$. Near these points, the method of characteristics may be used to find the solution.

The above method may be extended to the case of axial symmetry[12].

8. Method of characteristics[13, 14]

For steady two-dimensional or axially symmetric supersonic rotational flow, the method of characteristics may be used. In these cases, the fundamental equation is Eq.(5.50), i.e.,

$$\left(1-\frac{u^2}{a^2}\right)\frac{\partial^2\psi}{\partial x^2}-\frac{2uv}{a^2}\frac{\partial^2\psi}{\partial x\partial y}+\left(1-\frac{v^2}{a^2}\right)\frac{\partial^2\psi}{\partial y^2}-\frac{\delta}{y}\frac{\partial\psi}{\partial y}$$

$$=-\frac{\rho y^{\delta}}{\gamma R}\frac{a}{M}\frac{dS}{dn}[1+(\gamma-1)M^2] \quad (14.90)$$

The characteristic condition of Eq.(14.90) is

$$\left(1-\frac{u^2}{a^2}\right)\left(\frac{dy}{dx}\right)^2+\frac{2uv}{a^2}\frac{dy}{dx}+\left(1\ \frac{v^2}{a^2}\right)=0 \quad (14.91)$$

or

$$\left(\frac{dy}{dx}\right)_{\pm}=\tan\,(\theta\mp\alpha) \quad (14.92)$$

The physical characteristics are the same as those of irrotational flow. The characteristic curves are Mach curves.

The hodograph characteristics are

$$d\theta\pm\cot\alpha\frac{dq}{q}+\delta\frac{\sin\theta\sin\alpha}{\sin(\theta\mp\alpha)}\frac{dr}{r}\pm\frac{\sin\alpha\cos\alpha}{\gamma R}dS=0 \quad (14.93)$$

The only difference from the corresponding irrotational flow lies in the last term, the variation of entropy. The entropy here is constant along a given stream line except across a shock. In the actual calculation, we must consider three sets of characteristic curves, the two characteristics given by Eq. (14.92) and the streamlines. The computation will be much more complicated than that of the corresponding irrotational flow. Across a shock, there is a jump in entropy S.

9. Stream function for rotational flow

In the last section, we have indicated how to solve Eq. (14. 90) for the supersonic case. There are many interesting cases where both a subsonic and a supersonic flow field exist, such as, for example, the flow behind the detached shock shown in Fig.14.4. In this case, for the supersonic part of the flow field we may still use the method of characteristics. The difficulty lies in the determination of the subsonic flow field and the exact location of the sonic lines. As far as the subsonic flow part is concerned, we are required to solve the boundary value problem for the elliptic differential equation (14.90). The four boundaries are the curved shock, the surface of the body which is a streamline, and the sonic lines on both sides. For these four boundaries, only the surface of the body is given; the other three boundaries depend on the solution of the problem. This is the main difficulty in present problem. Furthermore, Eq.(14.90) itself depends also on the shape of the shock because the variation of entropy, dS/dn, behind the shock depends on the shape of the body. One has to match the solution of the stream function from Eq.(14.90) to the shock relations. There is no known simple method of solution for Eq.(14.90) in the general case, particularly for transonic flow problems. For a simple solid boundary configuration, one may roughly estimate the curved shock and the sonic lines and use relaxation method to calculate the subsonic flow in this region[13]. Once the subsonic flow region is known, from the sonic line on, the method of characteristics may be used for the supersonic region.

From a mathematical point of view, if the shape of the shock wave is given, it is possible to determine the stream function ψ from Eq.(14.90) as an initial value problem for a given shape of body. The approximate analytic solution in this case has been worked out by Shu[15].

10. Nonuniform propagation of shock wave[16]

One of the interesting cases of unsteady anisentropic flow is the decay of a strong shock. To a first approximation, we may consider such a problem as depending on time t and on a single spatical coordinate x i.e., we may consider either a plane, cylindrical, or spherical shock wave. For simplicity we shall consider only plane waves in this section. Since the states of the gas on the two sides of the shock are not constant, the entropy change through the shock also varies. In the case of a strong shock, the entropy change and its variation are not negligible and the differential equation of nonisentropic flow must be used for the analysis.

For one-dimensional anisentropic flow of inviscid fluid, the fundamental equations are [cf. Chapter V]:

$$\frac{\partial u}{\partial t}+u\frac{\partial u}{\partial x}=-\frac{1}{\rho}\frac{\partial p}{\partial x} \tag{14.94}$$

$$\frac{1}{\rho}\frac{\partial \rho}{\partial t}+\frac{u}{\rho}\frac{\partial \rho}{\partial x}=-\frac{\partial u}{\partial x} \tag{14.95}$$

$$\frac{c_V}{RT}\left(\frac{\partial T}{\partial t}+u\frac{\partial T}{\partial x}\right)=-\frac{\partial u}{\partial x} \tag{14.96}$$

$$S=\frac{R}{\gamma-1}\log\theta+\text{constant} \tag{14.97}$$

where S is the entropy and $\theta=p/\rho^{\gamma}$.

From Eqs.(14.95) to (14.97),

$$\left(\frac{\partial}{\partial t}+u\frac{\partial}{\partial x}\right)\log\theta=0 \tag{14.98}$$

This shows that $\theta=$constant, i.e., there is no change in entropy along any line of flow.

If we define the local sound velocity as

$$\text{sound velocity} = a = \sqrt{\frac{\gamma p}{\rho}} = \sqrt{\gamma\theta\rho^{\gamma-1}} \tag{14.99}$$

from Eq.(14.94), we have

$$\frac{\partial u}{\partial t} + u\frac{\partial u}{\partial x} = -\frac{2a}{\gamma-1}\frac{\partial a}{\partial x} + \frac{a^2}{\gamma(\gamma-1)}\frac{\partial \log\theta}{\partial x} \tag{14.100}$$

From Eqs.(14.95) and (14.98)

$$\frac{2}{\gamma-1}\left(\frac{\partial a}{\partial t} + u\frac{\partial a}{\partial x}\right) = -a\frac{\partial u}{\partial x} \tag{14.101}$$

Adding Eqs.(14.100) and (14.101), one obtains

$$\left[\frac{\partial}{\partial t} + (u+a)\frac{\partial}{\partial x}\right]\left(u + \frac{2a}{\gamma-1}\right) = \frac{a^2}{\gamma(\gamma-1)}\frac{\partial \log\theta}{\partial x}$$

$$= \frac{a}{\gamma(\gamma-1)}\left[\frac{\partial}{\partial t} + (u+a)\frac{\partial}{\partial x}\right]\log\theta \tag{14.102}$$

Subtracting Eq.(14.101) from Eq.(14.100) gives

$$\left[\frac{\partial}{\partial t} + (u-a)\frac{\partial}{\partial x}\right]\left(u - \frac{2a}{\gamma-1}\right) = -\frac{a}{\gamma(\gamma-1)}\left[\frac{\partial}{\partial t} + (u-a)\frac{\partial}{\partial x}\right]\log\theta \tag{14.103}$$

From Eqs.(14.98), (14.102), and (14.103) we have three sets of lines of propagation in the (x, t) plane for the present problem. These sets are:

(I) $dx/dt = u$

along which entropy is constant;

(II) $dx/dt = u + a$

which concerns the behavior of $u + \dfrac{2a}{\gamma-1}$; and

(III) $dx/dt = u - a$

which concerns the behavior of $u - \dfrac{2a}{\gamma-1}$.

The last two sets of lines are the characteristic curves of the flow. In the case of isentropic flow, or in the case of a weak shock in which the change in entropy is negligible, the characteristic curves can be used for the flow in the (x, t) plane, i. e., the ordinary method of characteristics (cf. Chapter XI). As soon as θ is not constant, we must use all three sets of lines to find the flow, i. e., the method of characteristics in anisentropic flow.

Another way to analyze this problem is to give primary attention to the lines of flow $dx/dt=u$, along which the entropy is constant, instead of using the characteristic curves $dx/dt=u\pm a$. It is more convenient to apply the Lagrangian description of the motion[16]. In this analysis, we write

$$x=\varphi(s, t) \tag{14.104}$$

where s is a parameter, characterizing a particular element of volume, and thus its path in the x-t plane. In general, the choice of s is immaterial, but in certain problems a special choice of s will be more convenient than others. We will choose s such that $x=s$ when the gas is at rest.

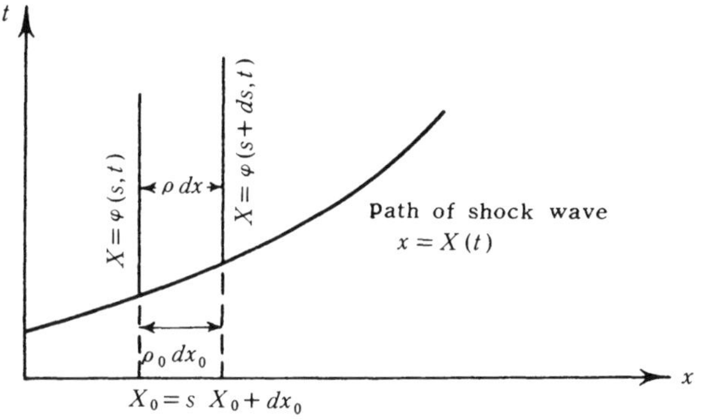

Fig. 14.6 x-t diagram of shock-wave propagation

By definition

$$u=\frac{dx}{dt}=\frac{\partial\varphi}{\partial t} \tag{14.105}$$

The equation of motion is then

$$\frac{du}{dt}=\frac{\partial^2\varphi}{\partial t^2}=-\frac{1}{\rho}\frac{\partial p}{\partial x} \tag{14.106}$$

We write

$$\Phi=\frac{\partial\varphi}{\partial s} \tag{14.107}$$

Eq. (14.107) becomes

$$\Phi_{\rho}\frac{\partial^2\varphi}{\partial t^2}=-\Phi\frac{\partial p}{\partial x}=-\frac{\partial p}{\partial s} \tag{14.108}$$

In the (x, t) plane, we shall draw the path of the shock waves as $x=X(t)$, Fig. 14.6.

Let ρ_0 be the density of the gas at rest and ρ be the density of the gas after crossing the shock. From Fig. 14.6, it is easily seen that $\rho dx = \rho_0 dx_0$.

Now $dx = \dfrac{\partial \varphi}{\partial s} ds = \Phi ds$, and since $ds = dx_0$, we obtain

$$\rho\Phi = \rho_0 \tag{14.109}$$

where ρ_0 may or may not be a constant. In general, one must assume ρ_0 to be a function of s.

If we consider only the case where the flow behind the shock is continuous, so that the entropy is constant along a line of flow, Poisson's equation can be applied, giving

$$p = p_1 \frac{\rho^\gamma}{\rho_1^\gamma} = p_1 \frac{\Phi_1^\gamma}{\Phi^\gamma} \tag{14.110}$$

where the subscript 1 refers to the state immediately behind the shock front. The equation of motion becomes

$$\rho_0 \frac{\partial^2 \varphi}{\partial t^2} = -\frac{\partial}{\partial s}\left(p_1 \frac{\Phi_1^\gamma}{\Phi^\gamma}\right) \tag{14.111}$$

and the velocity of the shock wave is

$$\xi = \frac{dX}{dt} \tag{14.112}$$

At the path of the shock wave, we have

$$\varphi(s, t) = s$$

Hence, taking the time derivative along this path

$$\left(\frac{\partial \varphi}{\partial t}\right)_1 + \left(\frac{\partial \varphi}{\partial s}\right)_1 \frac{ds}{dt} = \frac{ds}{dt}$$

with $ds/dt = \xi$, we have

$$u_1 + \Phi_1 \xi = \xi,$$

It follows that

$$\Phi_1 = \frac{\xi - u_1}{\xi} \tag{14.113}$$

The boundary conditions for this problem are as follows:

When we consider the problem of a shock wave produced by the motion of a body (e.g., by a piston acting on a column of gas in a tube), we can take $s = 0$ for the element of volume of the gas immediately adjacent to the body. If the motion of the body or piston is fully given, $\varphi(0, t)$ will be completely known (Fig. 14.7). In other cases a condition or an equation may be prescribed, determining the motion of this body. In either case, this fixes a boundary condition for the function $\varphi(s,t)$ referring to $s=0$.

A secondary boundary condition refers to the path of the shock wave. It must be obtained from the condition

$$\varphi(s, t) = s$$

together with Eq.(14.113) and the Rankine-Hugoniot condition

$$u_1 = \frac{2}{\gamma+1}\left(\xi - \frac{a_0^2}{\xi}\right) \tag{14.114}$$

(a_0 being the sound velocity if gas is at rest). Both give the relation between $u_1 = \left(\dfrac{\partial \varphi}{\partial t}\right)_1$ and ξ.

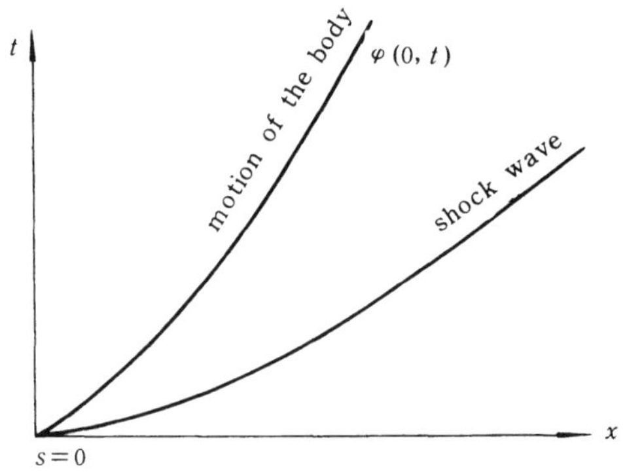

Fig. 14.7 Boundary conditions in x-t diagram for one-dimensional flow with shock wave

For small s and t, a series development for $\varphi(s, t)$ can be used successfully to find the solution $\varphi(s, t)$ which satisfies the boundary conditions. For large values of s and t, the numerical method should be used to continue the solution of series development[16].

11. Spherical and cylindrical shock waves produced by instantaneous energy release

A shock wave may be produced by an instantaneous energy release, such as the explosion of an atomic bomb. Sir Geoffrey Taylor[17] was the first one to analyze the case of a spherical shock wave. Lin[18] has extended Taylor's problem to the eylindrical case. In these problems, we try to find similar solutions for the gasdynamic equations. These solutions also must satisfy the proper intial conditions at the shock. The shock is assumed to be very strong so that some approximations may be used. We assume that a finite amount of energy is suddenly released, in an infinitely concentrated form in the spherical case, and as energy per unit length E in the cylindrical case. If we assume the values of energy release E are the same, these two problems may be treated simultaneously.

We assume that the air is an ideal gas so that the specific heats are constant. The fundamental equations in our problems are

$$\frac{\partial u}{\partial t}+u\frac{\partial u}{\partial r}=-\frac{1}{\rho}\frac{\partial p}{\partial r} \tag{14.115}$$

$$\frac{\partial \rho}{\partial t}+u\frac{\partial \rho}{\partial r}+\rho\left(\frac{\partial u}{\partial r}+\delta\frac{u}{r}\right)=0 \tag{14.116}$$

$$\left(\frac{\partial}{\partial t}+u\frac{\partial}{\partial r}\right)(p/\rho^{\gamma})=0 \tag{14.117}$$

where $\delta=1$ for cylindrical case, $\delta=2$ for spherical case, and u is the radial velocity.

For the similarity solutions of Eqs.(14.115) to (14.117) for an expanding blast wave of constant total energy E, we assume the expressions:

$$\frac{p}{p_0}=\frac{f_1(\eta)}{R^{1+\delta}}$$

$$\frac{\rho}{\rho_0}=\psi(\eta)$$

$$u=\varphi_1(\eta)/R^{\frac{1+\delta}{2}} \tag{14.118}$$

where R is the radius of the shock wave and the subscript 0 represents the values of the undisturbed atmosphere. If r is the radial coordinate, $\eta=r/R$ and f_1, φ_1 are functions of η only.

Substituting Eq.(14.118) into Eqs.(14.115) to (14.117), we have

$$\left(\frac{1+\delta}{2}\varphi_1+\eta\varphi_1'\right)\frac{1}{R^{(3+\delta)/2}}\frac{dR}{dt}-\frac{\varphi_1\varphi_1'}{R^{2+\delta}}=\frac{p_0}{\rho_0}\frac{f_1'}{\psi R^{2+\delta}}$$

$$\eta\psi' R^{\frac{1+\delta}{2}}\frac{dR}{dt}-\varphi_1\psi'-\psi\left(\varphi_1'+\frac{\delta\varphi_1}{\eta}\right)=0 \tag{14.119}$$

$$\left(\gamma\eta f_1\frac{\psi'}{\psi}-\eta f_1'-(1+\delta)f_1\right)R^{\frac{1+\delta}{2}}\frac{dR}{dt}+\varphi f_1'-\frac{\gamma f_1\varphi_1\psi'}{\psi}=0$$

where the prime refers to derivatives with respect to η. The similarity conditions are satisfied if the following conditions are fulfilled:

$$R^{\frac{1+\delta}{2}} \frac{dR}{dt} = A = \text{constant} \tag{14.120}$$

$$\varphi_1 \varphi_1' + \frac{p_0}{\rho_0} \frac{f_1'}{\psi} = A\left(\frac{1+\delta}{2} \varphi_1 + \eta \varphi_1' \right) \tag{14.121}$$

$$\varphi_1 + \left(\varphi_1' + \frac{\delta \varphi_1}{\eta} \right) \frac{\psi}{\psi'} = A\eta \tag{14.122}$$

$$\gamma f_1 \varphi_1 \frac{\psi'}{\psi} - \varphi f_1' = A\left[\gamma \eta f_1 \frac{\psi'}{\psi} - \eta f_1' - (1+\delta) f_1 \right] \tag{14.123}$$

Eqs.(14.121) to (14.123) may be transformed into nondimensional form by the transformations.

$$f_1(\eta) = \frac{A^2}{a^2} f(\eta)$$

$$\varphi_1(\eta) = A\varphi(\eta) \tag{14.124}$$

where $a^2 = \gamma p_0 / \rho_0$. Substituting Eq.(14.124) into Eqs. (14.121) to (14.123), we have

$$f' = \frac{(1+\delta)\eta(\eta-\varphi)\psi + \frac{1-\delta}{2} \gamma f \eta \psi \varphi + \gamma f \psi \delta \varphi^2}{\eta [f - \psi(\eta-\varphi)^2]}$$

$$\psi' = \frac{\psi[(\eta-\varphi) f' + (1+\delta) f]}{\gamma f (\eta-\varphi)} \tag{14.125}$$

$$\varphi' = (\eta - \varphi) \frac{\psi'}{\psi} - \frac{\delta\varphi}{\eta}$$

for a very strong shock, and the Rankine-Hugoniot relations become

$$\frac{p_1}{p_0} = \frac{2\gamma}{\gamma+1} \frac{U^2}{a^2}$$

$$\frac{\rho_1}{\rho_0} = \frac{\gamma+1}{\gamma-1}$$

$$\frac{u_1}{U} = \frac{2}{\gamma+1} \tag{14.126}$$

where $U = \dfrac{dR}{dt}$ and subscript 1 refers to the value immediately behind the shock. Hence the boundary conditions at $\eta = 1$ are

$$f(1) \cong \frac{2\gamma}{\gamma+1},\ \psi(1) \cong \frac{\gamma+1}{\gamma-1},\ \varphi(1) \cong \frac{2}{\gamma+1} \tag{14.127}$$

Eq.(14.125) may be integrated numerically from the boundary conditions (14.127). For $\gamma = 1.4$, Taylor integrated the case of $\delta = 2$ while Lin integrated the case of $\delta = 1$.

We may express the constant A in terms of energy release E by integrating the total energy, kinetic energy and thermal energy, of the disturbance in the whole space, i.e.

$$E = 2\delta\pi \int_0^R \left(\frac{p}{\gamma-1} + \frac{\rho u^2}{2} \right) r^\delta dr = S(\gamma)\rho_0 A^2 \tag{14.128}$$

where $S(\gamma)$ is a function of γ. For $\gamma = 1.4$, we have $S = 5.36$ when $\delta = 2$, and $S = 3.85$ when $\delta = 1$. Hence for $\gamma = 1.4$ and $\delta = 2$,

$$R = 1.025 \left(\frac{E}{\rho_0} \right)^{1/5} t^{2/5} \tag{14.129}$$

while for $\gamma = 1.4$, $\delta = 1$

$$R = 1.009 (E/\rho_0)^{1/4} t^{1/2} \tag{14.130}$$

Eqs.(14.129) and (14.130), respectively, show the propagation of spherical and cylindrical shock due to explosion.

12. Diabatic flow

In most of the previous discussions we considered only adiabatic flow where no heat is added or subtracted from the flow and the total energy of the flow field is constant. In many practical problems, the heat addition to the flow is not zero. Mention may be made of combustion aerodynamics and some meteorological problems. Flow with heat addition is sometimes called "diabatic flow" [19, 20, 21]. In this section, we shall consider the essential influence of heat release on the flow field and neglect those effects due to heat conduction, viscosity, and diffusion. These will be discussed in Chapter XVII. The only explicit difference in the fundamental equations of diabatic flow from those of adiabatic flow lies in the energy equation. The equations of motion, of continuity, and of state are of exactly the same form in both cases. For diabatic flow, the energy equation may be written as [cf. Eq.(5.20)]

$$\frac{DH_0}{Dt} = \frac{\partial H_0}{\partial t} + \boldsymbol{q} \cdot \nabla H_0 = \frac{1}{\rho} \frac{\partial p}{\partial t} + Q_t \tag{14.131}$$

or

$$T\frac{DS}{Dt} = T\left(\frac{\partial S}{\partial t} + \boldsymbol{q} \cdot \nabla S\right) = Q_t \tag{14.132}$$

where Q_t is the rate of heat added to the fluid per unit mass.

For steady flow, we have

$$\boldsymbol{q} \cdot \nabla H_0 = Q_t \quad \text{or } T\boldsymbol{q} \cdot \nabla S = Q_t \tag{14.133}$$

In diabatic flow, both the stagnation enthalpy and entropy are not constant along a streamline, whereas in adiabatic flow they are constant along a streamline. Thus in diabatic flow both the stagnation enthalpy and entropy are nonuniform in the flow field. From Eq.(5.43), we have

$$\boldsymbol{q} \times \boldsymbol{\omega} = T\nabla S - \nabla H_0 \tag{14.134}$$

and the flow in general will be rotational.

Hicks found that it is nore convenient to study diabatic flow in terms of the Crocco vector $\boldsymbol{W}$ and stagnation pressure p_0. The Crocco vector is defined by

$$\boldsymbol{W} = \boldsymbol{q}/q_m \tag{14.135}$$

where q_m is the maximum possible velocity for a given stagnation temperature T_0, i.e.,

$$q_m = (2c_p T_0)^{1/2} \tag{14.136}$$

We then have

$$p = p_t(1 - W^2)^{\frac{\gamma}{\gamma-1}} \tag{14.137}$$

$$T = T_0(1 - W^2) \tag{14.138}$$

In terms of W, the equations of motion, continuity, and energy for steady diabatic flow become, respectively,

$$\nabla \log p_t = \frac{2\gamma}{\gamma - 1}\left[(1 - W^2)^{-1}\boldsymbol{W} \times (\nabla \times \boldsymbol{W}) - q_W \boldsymbol{W}\right] \tag{14.139}$$

$$\nabla \cdot (1 - W^2)^{\frac{1}{\gamma-1}} \boldsymbol{W} = q_W (1 - W^2)^{\frac{1}{\gamma-1}}\left(1 + \frac{\gamma+1}{\gamma-1} W^2\right) \tag{14.140}$$

$$\boldsymbol{W} \cdot \nabla \log q_m = (1 - W^2) q_W \tag{14.141}$$

where

$$q_W = \frac{Q_t}{q_m^3(1 - W^2)} \tag{14.142}$$

We may use the various methods described in the previous chapters to study the diabatic flow equations (14.139) to (14.141). Here we shall discuss only one of the simplest flows, namely, radial flow (cf. Chapter V, § 11). We assume that only the radial velocity component is different from zero and all the quantities are functions of the radial coordinate r only. If we adopt the δ-convention of § 11, we may discuss the cylindrical (two-dimensional) and spherical (three-dimensional) source simultaneously. We write

$$\boldsymbol{W}=W(r)\boldsymbol{i}_r \tag{14.143}$$

where $\boldsymbol{i}_r$ is the unit radial vector. Eqs.(14.139) to (14.141) reduce, respectively, to the simple equations

$$\frac{d}{dr}\log p_t=-\frac{2\gamma}{\gamma-1}Wq_W \tag{14.144}$$

$$r^{-\delta}\frac{d}{dr}\left[r^{\delta}W(1-W^2)^{\frac{1}{\gamma-1}}\right]=q_W(1-W^2)^{\frac{1}{\gamma-1}}\left(1+\frac{\gamma+1}{\gamma-1}W^2\right) \tag{14.145}$$

$$\frac{d}{dr}\log q_m=\frac{1-W^2}{W}q_W \tag{14.146}$$

In Eqs.(14.144) to (14.146), we have four quantities p_t, W, q_W, and q_m which are all functions of r. If we specify any one of these quantities, we may calculate the rest from these three equations. Following Ref. 19, we shall specify $W(r)$ and compute p_t, q_W, and q_m. We consider the cylindrical flow $\delta=1$ with the $W(r)$ given by

$$W=r(1+r^2)^{-1/2} \tag{14.147}$$

or

$$M=\frac{q_r}{a}=\left(\frac{2}{\gamma-1}\right)^{1/2}r=\text{Local Mach number}$$

The heat parameter q_W corresponding to W in Eq.(14.137) is then

$$q_W=2\left[1-\frac{3-\gamma}{2(\gamma-1)}r^2\right]\left(1+\frac{2\gamma}{\gamma-1}r^2\right)^{-1}(1+r^2)^{-1/2} \tag{14.148}$$

and the local maximum velocity q_m is

$$\frac{q_m}{q_{m\,\max}}=\frac{\gamma+1}{2}\left[27(\gamma-1)\right]^{1/2}r^2(1+r^2)(\gamma-1+2\gamma r^2)^{-3/2} \tag{14.149}$$

in which q_m reaches its maximum value $q_{m\,\max}$ when $q_W=0$. Other quantities may easily be obtained. Here we see that, for diabatic radial flow, there is no limiting circle or sphere at $M=1$ and the flow field extends from $r=0$

to $r=\infty$.

13. Flame front and detonation wave

For a diabatic flow of an inviscid fluid, it is possible to have a surface of discontinuity in a manner similar to that for adiabatic flow. However, instead of one, there are types of the surface of discontinuity in the normal velocity component, i.e., the detonation wave and the flame front (cf. Chapter IV, §8 and §10). Across such a surface of discontinuity, one may apply the conservation of energy, mass and momentum in the same manner as that of shock wave analysis in the adiabatic case. However, because of energy release in the process, for the same conditions in front of the surface of discontinuity, there are two solutions for the flow after the surface of discontinuity. For a flame front or a detonation wave, we may use a one-dimensional model which is shown in Fig. 4.14 or Fig. 14.8. In Fig. 4.14, we assume that the flame front or the detonation wave, i.e., the reaction zone, is moving at a speed u_s into a medium at rest; while in Fig. 14.8, we assume that the reaction zone is stationary. If we know the value of the speed of the reaction zone, u_s, the relation between the variables in front of the reaction zone and those behind it may be easily found as shown in Chapter IV, §8. However, in general, the velocity of the reaction zone, u_s, is not known from purely theoretical analysis. It may be obtained by experiment or in the following manner.

For a given amount of total heat release in the reaction zone, we may plot the pressure p against the specific volume $1/\rho$ curve such as the curve $BNGFK$ in Fig. 4.15. The initial condition A is in general not on this Hugoniot curve $BNGFK$. Without heat release, the initial condition A is always on the Hugoniot curve. With heat release, the Hugoniot curve is in fact one of the one-parameter family of curves depending on the amount of heat release. In other words, we have a series of curves similar to $BNGFK$ and each of these curves corresponds to a given amount of heat release. The final state of the flame front or the detonation wave must be on the Hugoniot curve corresponding to the total amount of heat release. The point G represents the final state of combustion at constant volume, i.e., $\rho_1=\rho_2$ where the point F represents the final state of combustion at constant pressure $p_1=p_2$. When the final state lies on the portion FK, we have a flame front. The dotted portion GF does not correspond to any possible physical process. For detonation wave, the Chapman-Jouquet condition is usually assumed for the determination of the detonation velocity u_s. Under this condition, the detonation velocity is given by the point N which is the tangent point on the Hugoniot curve (Fig. 4.15) drawn from the initial state A. The point N is the most stable condition for the detonation wave and the velocity of detonation is then:

$$u_s = \frac{1}{\rho_1} (\tan \alpha)^{1/2}$$

After u_s is known, we may calculate the variables after the reaction zone from the variable given in front of the reaction zone. With the initial and final states of variables in front of and behind the reaction zone, we may calculate the variations of the variables in the reaction zone. In the calculation of the variables in the reaction zone, we have to condider the transport phenomena which will be discussed in Chapter XVII.

In the stationary reaction zone, if the velocity in front of the stationary surface of discontinuity is very great, we have a detonation wave, whereas this velocity is very small, we have a flame wave. For a flame front, it is convenient to write the conservation laws as follows[22]:

$$\rho_1 u_1 = \rho_2 u_2$$

$$\rho_1 u_1^2 + p_1 = \rho_2 u_2^2 + p_2 \qquad (14.150)$$

$$\lambda\left(\frac{u_1{}^2}{2} + \frac{\gamma}{\gamma - 1} \frac{p_1}{\rho_1} \right) = \frac{u_2^2}{2} + \frac{\gamma}{\gamma - 1} \frac{p_2}{\rho_2}$$

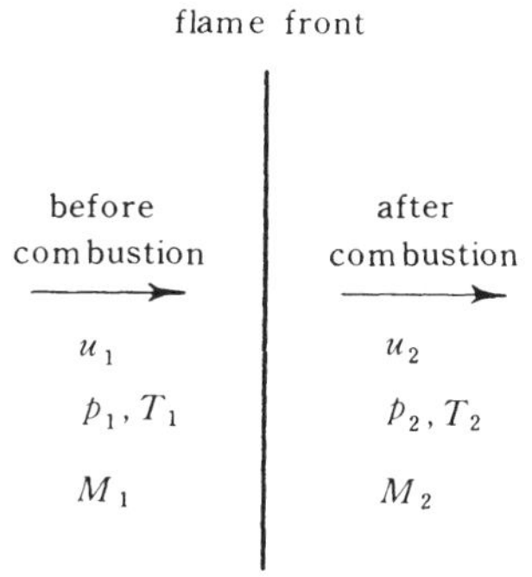

Fig. 14.8 Flow across a flame front

where λ is the ratio of the stagnation temperature after combustion to that before combustion (see Fig. 14.8). From Eq.(14.140),

$$\frac{u_2}{u_1} = \frac{\rho_2}{\rho_2} = \lambda + \frac{\gamma + 1}{2} \lambda(\lambda - 1) M_1^2$$
$$+ \frac{\gamma + 1}{2} \lambda(\lambda - 1)[1 + (\gamma + 1)(\lambda - 1) M_1^4 + \cdots$$

$$\frac{p_2}{p_1} = 1 - \gamma(\lambda - 1) M_1^2 - \frac{\gamma + 1}{2} \gamma\lambda(\lambda - 1) M_1^4$$
$$- \frac{\gamma + 1}{2} \gamma\lambda(\lambda - 1)[1 + (\gamma + 1)(\lambda - 1)] M_1^6 + \cdots$$

$$\frac{T_2}{T_1} = \lambda - \frac{\gamma - 1}{2}\lambda(\lambda - 1)M_1^2 - \frac{\gamma^2 - 1}{2}\lambda^2(\lambda - 1)M_1^4 + \cdots \qquad (14.151)$$

Usually the Mach number M_1 is very small and only the first terms in the expressions of Eq.(14. 151) are retained. We may treat the flow as incompressible both in front and behind the flame front with a jump of density across the flame front.[23] The pressure across the flame front is almost constant.

14. Frozen and equilibrium flows of a simple dissociating gas[24–26]

For a detailed study of diabatic flow, we have to consider the chemical reaction, change of composition of the fluid and the diffusion phenomenon. We shall study these problems in Chapter XVII. In this section, we consider two limiting cases of chemical reaction: frozen and equilibrium flows which give the upper and lower bounds of the flow with chemical reaction. For simplicity, we consider the dissociation of a pure diatomic gas A_2 which may dissociate into two atoms 2A, i.e., $A_2 \rightleftharpoons 2A$. We also assume that the translational, rotational and vibrational energies are in equilibrium with respect to the local temperature T. Furthermore, we assume that the temperature is not high enough so that electronic excitation and ionization are negligible. The enthalpy for this simple dissociating gas is then:

$$H = \left[\frac{7}{2} + \frac{3}{2}\alpha + (1 - \alpha)f(\theta) \right] \frac{p/\rho}{1+\alpha} + \alpha d \qquad (14.152)$$

where

$f(\theta) = \theta / (e^{\theta} - 1)$

$\theta = h^{\nu} / (kT)$

$\alpha =$ degree of dissociation

$d =$ dissociation energy per unit mass

$h =$ Planck constant

$k =$ Boltzmann constant

$\nu =$vibrational frequency

The equation of state for this simple dissociating gas is

$$p = (1+\alpha)\rho R_2 T \qquad (14.153)$$

where R_2 is the gas constant for A_2.

The equation of continuity for the atom A in terms of the degree of dissociation with chemical reaction may be written as follows[24] (cf. Chapter XVII, §4):

$$\frac{D\alpha}{Dt} = \frac{1}{\tau}\,[K(1-\alpha)-\alpha^2] \tag{14.154}$$

where

$$K = [M_2/(4\rho)]\,(k_f/k_b) \tag{14.155}$$

$$\tau = M_2^2/\,[4\rho^2 k_b(1+\alpha)] \tag{14.156}$$

M_2 is the molecular weight of A_2, k_f is the forward specific reaction rate from A_2 to 2A and k_b is the backward specific reaction rate from 2A to A_2.

For the case of steady inviscid flows, the equation of continuity, that of motion and that of energy gives

$$(u^2-a_f^2)\frac{\partial u}{\partial x} + (v^2-a_f^2)\frac{\partial v}{\partial y} + (w^2-a_f^2)\frac{\partial w}{\partial z} + uv\left(\frac{\partial u}{\partial y}+\frac{\partial v}{\partial x}\right)+$$

$$uw\left(\frac{\partial u}{\partial z}+\frac{\partial w}{\partial x}\right)+vw\left(\frac{\partial v}{\partial z}+\frac{\partial w}{\partial y}\right)-a_f^2 G\left(u\frac{\partial \alpha}{\partial x}+v\frac{\partial \alpha}{\partial y}+w\frac{\partial \alpha}{\partial z}\right)$$

$$=0 \tag{14.157}$$

where a_f is known as the frozen speed of sound of the mixture of A_2 and A and is given by the formula:

$$a_f^2 = \left(\frac{\partial p}{\partial \rho}\right)_{S,\alpha} = -\frac{\left(\dfrac{\partial H}{\partial \rho}\right)_{p,\alpha}}{\left(\dfrac{\partial H}{\partial p}\right)_{\rho,\alpha} - \dfrac{1}{\rho}} \tag{14.158}$$

and G is a factor of thermodynamic variables defined by the equation:

$$-G = \frac{\left(\dfrac{\partial H}{\partial \alpha}\right)_{p,\rho}}{\left(\dfrac{\partial H}{\partial \rho}\right)_{p,\alpha}} = -\frac{1}{\rho}\left(\frac{\partial \rho}{\partial \alpha}\right)_{p,T} - \rho\beta_f\frac{1}{c_{pf}}\left(\frac{\partial H}{\partial \rho}\right)_{p,T} \tag{14.159}$$

where $c_{pf} = \left(\dfrac{\partial H}{\partial T}\right)_{p,\alpha}$, is specific heat at constant pressure and frozen composition, and $\beta_f = -\dfrac{1}{\rho^2}\left(\dfrac{\partial \rho}{\partial T}\right)_{p,\alpha}$, volume expansion coefficient at constant pressure and frozen composition.

Eq.(14.157) should be solved simultaneously with Eq.(14.154).Eqs. (14.154) and (14.157) are much more complicated than the corresponding equations of ordinary gasdynamics, because of the interaction between the flow field and the chemical reaction. However, this interaction disappears in the two limiting cases: frozen flow and equilibrium flow.

In the frozen flow, the degree of dissociation is constant and then

$$\alpha = \alpha_0 = \text{constant} \tag{14.160}$$

In the equilibrium flow, the degree of dissociation is a given function of the local thermodynamic state and we have

$$K_e = \frac{\alpha_e^2}{1-\alpha_e^2} \tag{14.161}$$

The equilibrium constant K_e is a function of temperature and density. For an ideal dissociating gas, Lighthill[25] suggested the following simple formula for K_e

$$K_e = \frac{\rho_d}{\rho} \exp\left(- \frac{\rho_d}{R_2 T} \right) \tag{14.162}$$

where ρ_d is a constant reference density of dissociation.

Eq.(14.157) reduces to the simple form for the frozen flow and the equilibrium flow as follows:

$$(u^2 - a_i^2)\frac{\partial u}{\partial x} + (v^2 - a_i^2)\frac{\partial v}{\partial y} + (w^2 - a_i^2)\frac{\partial w}{\partial z} + uv\left(\frac{\partial u}{\partial y} + \frac{\partial v}{\partial x} \right)$$

$$+ vw\left(\frac{\partial v}{\partial z} + \frac{\partial w}{\partial y} \right) + uw\left(\frac{\partial u}{\partial z} + \frac{\partial w}{\partial x} \right) = 0 \tag{14.163}$$

where a_i is equal to a_i for the frozen flow and a_i is equal to a_e for equilibrium flow. The sound speed for the equilibrium flow a_e is defined by the following formula:

$$\frac{1}{a_e^2} = \frac{1}{a_f^2} + \rho G\left(\frac{\partial \alpha_e}{\partial p} \right)_s \tag{14.164}$$

We use the equilibrium degree of dissociation α_e to calculate the sound speed. Eq.(14.163) is of the same form for that for classical gasdynamics whose methods of solution are well known which have been discussed before. The most interesting thing is to extend the method of classical gasdynamics for the non-equilibrium flow as we are going to discuss below.

15. Sound waves in a dissociating gas[26]

In order to bring out the essential features of chemical reaction, particularly in the non-equilibrium flow, we consider first the propagation of waves of small amplitude in the x-direction in a simple dissociating gas. Since the amplitude of the wave is small, we may linearize the fundamental equation; and since we consider waves propagated in the x-direction only, all the unknowns are functions of time t and the x-coordinate only. The fundamental equatin for this one-dimensional acoustic wave is

$$\left(\frac{1}{a_{eo}^2}\frac{\partial^2 u'}{\partial t^2}-\frac{\partial^2 u'}{\partial x^2}\right)+\tau_0'\frac{\partial}{\partial t}\left(\frac{1}{a_{fo}^2}\frac{\partial^2 u'}{\partial t^2}-\frac{\partial^2 u'}{\partial x^2}\right)=0 \qquad (14.165)$$

where

$$\tau_0'=\tau_0\frac{1-\alpha_0^2}{2\alpha_0^2}=\text{characteristic reaction time}$$

and subscript 0 refers to the value in the undisturbed state.

It is evident that if $\tau_0'=0$, the sound wave propagates at the equilibrium sound speed a_{eo} and if $\tau_0'=\infty$, the sound wave propagates at the frozen sound speed a_{fo}. For the non-equilibrium flow with finite τ_0', the small disturbance u' will be propagated at a speed different from both a_{eo} and a_{fo}. Furthermore, even in an inviscid fluid, despersion and absorption phenomena occur in the non-equilibrium flow case.

Let us investigate how a disturbance of a given frequency ω will be propagated in a simple dissociating gas. We assume that the disturbance is of the following type:

$$t>0,\ x=0:\ u'=U\exp(i\omega t)$$

$$x\to\infty,\ u'=0 \qquad (14.166)$$

where U and ω are positive real constants and ω is the frequency of the initial disturbance. The solution of Eq.(14.165) which satisfies the boundary conditions (14.166) is

$$u'=U\exp\left[-\frac{\omega m}{a_{fo}}(\sin A)x\right]\cdot\exp\left\{i\omega\left[t-\frac{m}{a_{fo}}(\cos A)x\right]\right\} \qquad (14.167)$$

where

$$m=\frac{(a_{fo}/a_{eo})^4+(\omega\tau_0')^2}{1+(\omega\tau_0')^4} \qquad (14.168)$$

$$\tan 2A=\frac{[(a_{fo}/a_{eo})^2-1]\omega\tau_0'}{(a_{fo}/a_{eo})^2+\omega^2\tau_0'^2} \qquad (14.169)$$

Eq.(14.167) shows that the disturbances are propagated at a speed

$$a=\frac{a_{fo}}{m\cos A} \qquad (14.170)$$

The interesting points are that the sound speed a in the non-equilibrium flow depends on not only the characteristic reaction time but also the frequency ω of the disturbance. Thus we have the dispersion phenomenon; the amplitude of the disturbance is decaying with the increase of x according to the first exponential term in Eq.(14.167) even though there is no transport phenomenon; all the transport coefficients are assumed to be zero and the gas is considered as an inviscid fluid. The damping factor λ_i may be defined by the following formula:

$$\exp(-\lambda_r x) = \exp\left(-\frac{\omega \cdot m \cdot \sin A}{a_{fo}} x\right)$$

or

$$\lambda_i = \frac{\omega \cdot m \cdot \sin A}{a_{fo}} \tag{14.171}$$

The damping factor will be zero in both the frozen and equilibrium flows and the sound speed reduces to the corresponding sound speed in these two limiting cases.

16. Linear theory of a steady flow in a simple dissociating gas[26]

It is possible to reexamine all the flow problems in classical gasdynamics as we study previously in this book for the case including the chemical reaction. We shall only discuss a few typical flow problems so that the essential features of chemical reaction may be brought out.

First we consider the steady flow of a uniform velocity U passing over a thin body. Under this condition, the fundamental equations may be linearized and we have the following equation for the perturbation velocity components u', v' and w':

$$\tau \frac{\partial}{\partial \xi}\left[(1-M_{fo}^2)\frac{\partial u'}{\partial \xi} + \frac{\partial v'}{\partial \eta} + \frac{\partial w'}{\partial \zeta}\right] + (1-M_{eo}^2)\frac{\partial u'}{\partial \xi}$$

$$+ \frac{\partial v'}{\partial \eta} + \frac{\partial w'}{\partial \zeta} = 0 \tag{14.172}$$

where $M_{fo} = U/a_{fo}$, and $M_{eo} = U/a_{eo}$ are respectively the frozen and the equilibrium Mach number of the free stream; ξ, η, and ζ are respectively the non-dimensional x, y and z coordinates based on the reference length L and $\tau = \tau' U/L$ is the non-dimensional reaction time for the free stream. Eq.(14.172) reduced to the same form as that of ordinary gasdynamics for the frozen or the equilibrium flow when the non-dimensional reaction time τ is zero or infinite. For non-equilibrium flow, we have to use Eq.(14.172). For linearized theory, we may neglect the variation of entropy in the flow field and introduce a velocity potential φ such that

$$\boldsymbol{q} = \nabla \varphi \tag{14.173}$$

Eq.(14.172) in terms of φ becomes:

$$\tau \frac{\partial}{\partial \xi}\left[(1-M_{fo}^2)\frac{\partial^2 \varphi}{\partial \xi^2} + \frac{\partial^2 \varphi}{\partial \eta^2} + \frac{\partial^2 \varphi}{\partial \zeta^2}\right] + (1-M_{eo}^2)\frac{\partial^2 \varphi}{\partial \xi^2}$$

$$+ \frac{\partial^2 \varphi}{\partial \eta^2} + \frac{\partial^2 \varphi}{\partial \zeta^2} = 0 \tag{14.174}$$

Eq.(14.174) has been studied extensively for the two-dimensional cases such as wavy wall in subsonic and supersonic flow field (see problem section 24). It was found that wave drag may occur at subsonic speeds which is contrast to the results of classical gasdynamics.

Let us consider the flow past a sharp corner in a supersonic flow of an ideal dissociating gas based on the linearized equation (14.174). This is the well-known Prandtl-Meyer flow in ordinary gasdynamics (cf. Chapter IX, §2). In ordinary gasdynamics, it is known that linearized equations similar to Eq.(14.174) are not good for studying the Prandtl-Meyer flow and one should use the perturbation theory of hyperbolic characteristics to study such a flow. However, in the present case, the flow Mach number changes during the expansion zone and some approximate results may be obtained from Eq.(14.174). Fig. 14.9 shows a sketch of this flow field which is bounded upstream by the frozen Mach line with a Mach angle

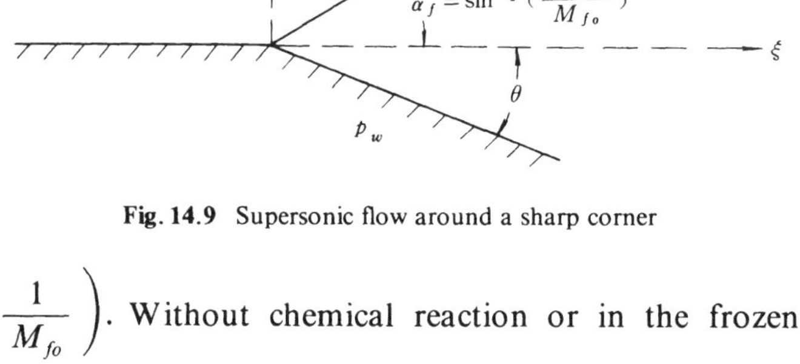

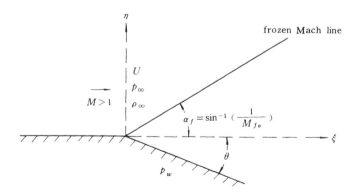

Fig. 14.9 Supersonic flow around a sharp corner

$\alpha_f = \sin^{-1}\left(\frac{1}{M_{fo}}\right)$. Without chemical reaction or in the frozen flow or equilibrium flow cases, the pressure on the wall p_w downstream of the corner is a constant depending on the freestream Mach number and the turning angle θ. However, for non-equilibrium flow, the pressure on the wall downstream of the corner is no longer constant but varies continuously in the relaxation zone from the value of the frozen flow immediately near the corner to the equilibrium flow value far away from the corner. In general, we may write

$$p_w = p_\infty - \frac{U^2\theta}{(M_{fo}^2 - 1)^{1/2}} K(\xi, \tau) \tag{14.175}$$

where $K(\xi, \tau)$ is a function of the distance from the corner ξ and the chemical reaction time τ. If either ξ or τ is zero, $K(\xi, \tau)$ will be unity and we have the frozen flow result. If either ξ or τ is infinite,

$$K(\xi, \tau) = \frac{(M_{fo}^2 - 1)^{1/2}}{(M_{eo}^2 - 1)^{1/2}} \tag{14.176}$$

and we have the equilibrium flow result. The formula (14.175) is applicable for both the expansion ($+\theta$) corresponding to the Prandtl-Meyer flow and compression ($-\theta$) to the wedge flow case, i.e., weak shock case. With the help of such a formula, we may calculate the lift and drag force on an airfoil in a similar manner as those in ordinary gasdynamics.

17. First-order one-dimensional wave equation

In the following sections, we will present the most commonly used finite-difference methods for solving the unsteady Euler equations cf. Chapter V, § 1). The unsteady Euler equations are quasi-linear and hyperbolic. It is not possible to perform the stability and convergence analyses that should be carried out before actual computations are attempted. Therefore the linear wave equation is used to model the behavior of the more complicated Euler equation in a rudimentary fashion. We examine various finite-difference schemes which can be used to solve the simple wave equation. The wave equation has exact solutions for certain initial conditions. We can use this knowledge to evaluate and compare finite-difference methods quickly which we wish to apply to the Euler equations. This simple and well understood linear wave equation has proven to be an amazingly fruitful source of inspiration.

The first-order, one-dimensional wave equation is

$$\frac{\partial u}{\partial t} + a\frac{\partial u}{\partial x} = 0 \tag{14.177}$$

where $u(x, t)$ is a scalar function and a is a positive constant. This linear hyperbolic equation describes a wave propagating in the x direction with a wave speed a in a uniform medium.

The exact solution of the wave equation for the pure initial value problem with initial data

$$u(x,0)=F(x) \quad (-\infty < x < \infty) \tag{14.178}$$

is given by

$$u(x,t)=F(x-at) \tag{14.179}$$

If $F(x)$ is not differentiable, (14.179) is a weak solution (cf. Chapter V, §4). A weak solution that simulates a traveling shock wave arises if F is taken as the Heaviside function.

The characteristic form of Eq.(14.177) is

$$du=0 \quad \text{or} \quad u=\text{constant} \tag{14.180}$$

along the characteristic curve

$$\frac{dx}{dt}=a \tag{14.181}$$

The numerical solution of Eq.(14.177) by the characteristics method is by taking $\Delta x=a\Delta t$ (Fig. 14.10)

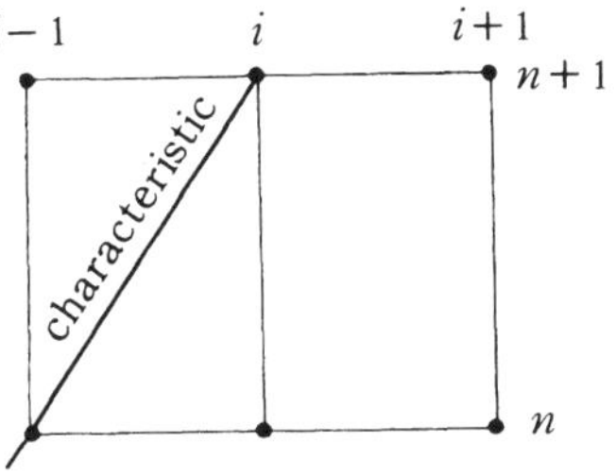

Fig. 14.10 Characteristics method

$$u_i^{n+1}=u_{i-1}^n \tag{14.182}$$

This is an exact difference solution.

Now we consider approximate difference solutions. For approximate difference equation, the truncation error appears. The terms in the truncation error have certain distinctive physical interpretations. We examine the interpretations of truncation error in general. Any difference equation for Eq.(14.177) may be written into the modified differential equation of the form

$$\frac{\partial u}{\partial t}+a\frac{\partial u}{\partial x}+\left[a_2\Delta x\frac{\partial^2 u}{\partial x^2}+a_3\Delta x^2\frac{\partial^3 u}{\partial x^3}+a_4\Delta x^3\frac{\partial^4 u}{\partial x^4}+\cdots\right]=0 \tag{14.183}$$

The quantity in brackets represents the truncation error of the approximate difference equation. Here the time derivative terms appearing in the truncation error have been replaced by spatial derivatives using the modified differential equation itself (see next section). As we have seen before, the lowest-order term of the truncation error contains the partial derivative $\frac{\partial^2 u}{\partial x^2}$ which makes the term similar to the viscous term in one-dimensional fluid flow equation if $a_2 < 0$. We now interpret all terms of the truncation error[27].

The effects of the various terms of the truncation error can be illustrated by a single Fourier component of the numerical solution

$$u(x,t) = e^{\lambda t + ibx} \tag{14.184}$$

where b is any real value, λ is a function of b and may be complex, and $i = \sqrt{-1}$.

Rewrite Eq.(14.183) as

$$\frac{\partial u}{\partial t} + a\frac{\partial u}{\partial x} + \sum_{n=1}^{\infty}\left(b_{2n}\frac{\partial^{2n} u}{\partial x^{2n}} + b_{2n+1}\frac{\partial^{2n+1} u}{\partial x^{2n+1}}\right) = 0 \tag{14.185}$$

Substituting (14.184) into (14.185) and solving for λ

$$\lambda = \sum_{n=1}^{\infty}(-1)^{n-1} b^{2n} b_{2n} - ib\,[a + (-1)^n b^{2n} b_{2n+1}] \tag{14.186}$$

we obtain the solution of the modified equation in terms of the Fourier component

$$u(x,t) = \exp\left\{\sum_{n=1}^{\infty}(-1)^{n-1} b^{2n} b_{2n} t + ib\,[x - at - \sum_{n=1}^{\infty}(-1)^n b^{2n} b_{2n+1} t]\right\} \tag{14.187}$$

If all b_n' s are zero, (14.187) reduces to the exact solution

$$u(x,t) = \exp\{ib\,[x - at]\} \tag{14.188}$$

Comparing the difference solution (14. 187) with the exact solution (14. 188), we find that the even derivative terms which appear in the truncation error tend to reduce the amplitude of the Fourier component if

$$(-1)^n b_{2n+1} > 0 \tag{14.189}$$

This effect is called dissipation. As a result of dissipation, all gradients in the solution are reduced. Another quasi-physical effect of numerical schemes is called dispersion. This is the direct result of the odd derivative terms in the truncation error. If

$$(-1)^n b_{2n+1} > 0 \tag{14.190}$$

the wave speed of the Fourier component tends to increase and the phase shift is leading. Otherwise, the wave speed tends to decrease and the phase snift is lagging. As a result of dispersion, phase relations between various waves are distorted. In general, if the lowest order term in the truncation error contains an even derivative, the resulting solution will predominately exhibit dissipative errors. On the other hand, if the leading term is an odd derivative, the resulting solution will predominately exhibit dispersive errors. Fig. 14.11 illustrates the effects of dissipation and dispersion on the computation of a discontinuity.

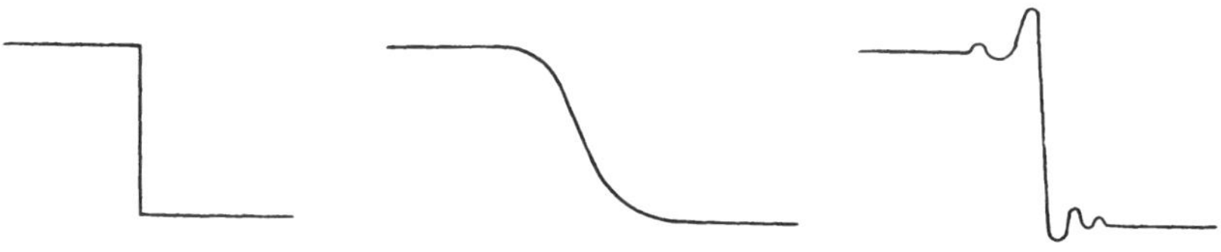

(a) (b) (c)

(a) Exact solution

(b) Numerical solution distorted primarily by dissipation errors

(c) Numerical solution distorted primarily by dispersion errors

Fig. 14.11 Effects of dissipation and dispersion

It is noted that the modified equation approach is challenged by a recent paper[28].

In the following sections, we will discuss several typical finite-difference schemes, each of which exhibits certain basic and distinctive properties of a class of methods. Dissipation will be emphasized, while discussion on dispersion will be omitted due to the space limitations.

18. Explicit difference schemes

The simple explicit difference scheme for the wave equation (14.177), known as Euler method, is

$$\frac{u_i^{n+1}-u_i^n}{\Delta t}+a\frac{u_{i+1}^n-u_{i-1}^n}{2\Delta x}=0 \tag{14.191}$$

We refer it as being first-order accurate since the lowest-order term in the truncation error is first order in Δt. This scheme is explicit since only one unknown u_i^{n+1} appears in the equation. Unfortunately, when the von Neumann stability analysis is applied to this scheme, we find that this scheme is unconditionally unstable. The growth factor of error in this algorithm is (cf. Chapter VII, §8)

$$G=e^{\lambda\Delta t}=1-i\nu\sin\beta \tag{14.192}$$

where

$$\nu=\frac{a\Delta t}{\Delta x}\quad (\nu \text{ is called Courant number}) \tag{14.193}$$

and

$$\beta = b\Delta x \ (b \text{ is the wave number}) \tag{14.194}$$

$| G | \geqslant 1$ for any value of ν. Therefore, this scheme proves by itself to be worthless in solving the wave equation.

Let us examine the truncation error of the Euler method. Substituting Taylor-series expansions into Eq.(14.191) for u_i^{n+1}, u_{i+1}^n and u_{i-1}^n, we have the following equation

$$\frac{1}{\Delta t}\left[\left(u_i^n + \Delta t u_t + \frac{\Delta t^2}{2} u_{tt} + \frac{\Delta t^3}{6} u_{ttt} + \cdots\right) - u_i^n\right] + \frac{a}{2\Delta x}\left[\left(u_i^n + \Delta x u_x + \frac{\Delta x^2}{2} u_{xx} + \frac{\Delta x^3}{6} u_{xxx} + \cdots\right) - \left(u_i^n - \Delta x u_x + \frac{\Delta x^2}{2} u_{xx} - \frac{\Delta x^3}{6} u_{xxx} + \cdots\right)\right] = 0 \tag{14.195}$$

Here, partial derivatives are indicated by subscripts for brevity. Eq.(14.195) simplifies to

$$u_t + a u_x + \frac{\Delta t}{2} u_{tt} + \frac{\Delta t^2}{6} u_{ttt} + \frac{a\Delta x^2}{6} u_{ttt} + \cdots = 0 \tag{14.196}$$

In order to replace u_{tt} by a spatial derivative term, we take the partial derivative of Eq.(14.196) with respect to time, to obtain

$$u_{tt} + a u_{xt} + \frac{\Delta t}{2} u_{ttt} + \cdots = 0 \tag{14.197}$$

and take the partial derivative of Eq.(14.196) with respect to x and multiply by $-a$

$$-a u_{tx} - a^2 u_{xx} - \frac{a\Delta t}{2} u_{ttx} + \cdots = 0 \tag{14.198}$$

Adding Eq.(14.197) and (14.198) gives

$$u_{tt} - a^2 u_{xx} + \frac{\Delta t}{2} u_{ttt} - \frac{a\Delta t}{2} u_{ttx} + \cdots = 0 \tag{14.199}$$

In a similar manner we can obtain the following expressions for u_{ttt} and u_{ttx}

$$\begin{aligned} u_{ttt} &= -a^3 u_{xxx} + \cdots \\ u_{ttx} &= a^2 u_{xxx} + \cdots \end{aligned} \tag{14.200}$$

Combining Eqs.(14.196), (14.199) and (14.200) gives

$$u_t + a u_x + \left[\frac{a^2\Delta t}{2} u_{xx} + \frac{a\Delta x^2}{3}\left(\frac{1}{2} + \nu^2\right) u_{xxx} + \cdots\right] = 0 \tag{14.201}$$

The lowest-order term in the truncation error contains a second-order derivative with a positive coefficient. From the dissipation condition (14. 189), the effect of this term is anti-dissipat ive and tends to increase the amplitudes of the Fourier components in the numerical solution. This is consistent with the fact that the method is unconditionally unstable.

It is important to emphasize that the equation obtained after the substitution of the Taylor-series expansions, i.e. Eq.(14.195), must be used to eliminate the higher-order time derivatives rather than the original equation (14. 177). This is due to the fact that a solution of the original differential equation does not in general satisfy the difference equation. It is the modified differential equation which represents the difference equation. Therefore, the original differential equation should not be used to eliminate the time derivatives.

The Euler method (14.177) can be made stable by replacing u_i^n with the averaged term $(u_{i+1}^n + u_{i-1}^n)/2$. The resulting algorithm is the well known Lax method[29]

$$\frac{u_i^{n+1} - (u_{i+1}^n + u_{i-1}^n)/2}{\Delta t} + a\frac{u_{i+1}^n - u_{i-1}^n}{2\Delta x} = 0 \tag{14.202}$$

This explicit scheme is first-order accurate. The growth factor is

$$G = \cos\beta - i\nu\sin\beta \tag{14.203}$$

$|G| \leqslant 1$, if

$$\nu \leqslant 1 \tag{14.204}$$

This is the Courant-Friedrichs-Lewy (CFL) stability condition (7. 56) which has important physical interpretation as stated in Chapter VII, §8(c). When $\nu = 1$, (14.202) reduces to the exact solution (14.182).

The modified differential equation of (14.202) is

$$u_t + au_x - \frac{a\Delta x}{2}\left(\frac{1}{\nu} - \nu\right)u_{xx} - \frac{a\Delta x^2}{3}(1-\nu^2)u_{xxx} + \cdots = 0 \tag{14.205}$$

The truncation error is predominately dissipative if $\nu \leqslant 1$. Thus, we again demonstrate that the growth factor and the modified equation are related to each other.

In most cases, first-order schemes are not used to solve partial differential equations because of their inherent inaccuracy. A second order accurate method can be derived from a Taylor-series expansion

$$u_i^{n+1} = u_i^n + \Delta t u_t + \frac{\Delta t^2}{2}u_{tt} + O(\Delta t^3) \tag{14.206}$$

where the derivative u_t and u_{tt} are evaluated at the grid point (i, n). Using the wave equation (14.177), we have

$$u_t = -au_x \qquad u_{tt} = a^2 u_{xx} \tag{14.207}$$

Substituting Eq.(14.207) into (14.206), we obtain

$$u_i^{n+1} = u_i^n - a\Delta t u_x + \frac{a^2\Delta t^2}{2} u_{xx} + O(\Delta t^3) \tag{14.208}$$

If u_x and u_{xx} are replaced by second-order accurate central-difference expressions at (i, n), the well known Lax-Wendroff scheme[30] is obtained

$$u_i^{n+1} = u_i^n - \frac{\nu}{2}(u_{i+1}^n - u_{i-1}^n) + \frac{\nu^2}{2}(u_{i+1}^n - 2u_i^n + u_{i-1}^n) \tag{14.209}$$

This second-order accurate explicit scheme is stable, if

$$\nu \leqslant 1 \tag{14.210}$$

since the growth factor is

$$G = 1 - \nu^2(1 - \cos\beta) - i\nu \sin\beta \tag{14.211}$$

If $\nu = 1$, Eq.(14.209) yields the exact solution (14.182).

The modified differential equation for this method (14.209) is

$$u_t + au_x + \frac{a\Delta x^2}{6}(1 - \nu^2)u_{xxx} + \frac{a\Delta x^3}{8}\nu(1 - \nu^2)u_{xxxx} + \cdots = 0 \tag{14.212}$$

The truncation error is predominately dispersive. There is a minor dissipative error if $\nu < 1$, according to the dissipation condition (14.189). The second-order central-difference method when applied to the nonlinear gasdynamic equations was found to be prone to local instability near shocks. Such instability can be cured by adding artificial dissipative terms to the equation. The usual artificial dissipative terms to be added to the right-hand side of Eq.(14.212) are the second derivative term[31] to control shock oscillations

$$\frac{C\Delta x^4}{4|p|}\left|\frac{\partial^2 p}{\partial x^2}\right|\frac{\partial^2 u}{\partial x^2} \tag{14.213}$$

and the fourth derivative term[31] to provide dissipation in smooth regions

$$-K\Delta x^4 \frac{\partial^4 u}{\partial x^4} \tag{14.214}$$

where p is the static pressure of the flow and C, K are positive constants to be determined or tuned by numerical experience.

These artificial dissipative terms are fourth-order of Δx and thus the formal accuracy of the numerical method is unaltered. Both terms produce positive damping according to Eq.(14.189).

For convenience of implementing the second-order accurate algorithm, a two-step variation of the original method can be used. The MacCormack finite-difference method[31] is a two-step variation of the original Lax-Wendroff method and is widely used for solving fluid flow equations. When applied to the linear wave equation, this explicit, predictor-corrector method becomes

$$\text{Predictor:} \quad \overline{u_i^{n+1}} = u_i^n - \nu(u_i^n - u_{i-1}^n) \tag{14.215}$$

$$\text{Corrector:} \quad u_i^{n+1} = \frac{1}{2}\left[u_i^n + \overline{u_i^{n+1}} - \nu(\overline{u_{i+1}^{n+1}} - \overline{u_i^{n+1}})\right] \tag{14.216}$$

Substituting the predictor (14.215) into the corrector (14.216), we obtain the Lax-Wendroff equation (14.209). Therefore, the MacCormack scheme is equivalent to the original Lax-Wendroff scheme. Hence, the growth factor and the modified equation are identical with those of the Lax-Wendroff scheme.

In the two-step algorithm, the term $\overline{u_i^{n+1}}$ is a temporary predicted value of u at the time level $n+1$ and is first-order accurate. The corrector provides the final second-order accurate value of u at the time level $n+1$. Note that in the predictor a backward difference is used for $\partial u/\partial x$, while in the corrector a forward difference is used. This differencing can be reversed to obtain the same solution or can be alternated to obtain other variations to the basic MacCormack scheme.

The second-order accurate explicit schemes such as the MacCormack two-step scheme with an addition of the artificial dissipative terms usually give excellent results for the unsteady Euler equations with a minimum of computational effort.

19. Implicit difference schemes

Hyperbolic systems can be solved by using explicit methods. However, the time step size for most explicit schemes is limited by the CFL condition. This can lead to unreasonably long computation time for some problems. The advantage of implicit methods lies in the unrestricted stablity limit at least linearly speaking. We will present some typical implicit schemes to illustrate the basic properties of this class of methods.

The simple implicit difference scheme for the wave equation (14.177), known as the Euler implicit method, is

$$\frac{u_i^{n+1} - u_i^n}{\Delta t} + a\frac{u_{i+1}^{n+1} - u_{i-1}^{n+1}}{2\Delta x} = 0 \tag{14.217}$$

This scheme is first-order accurate in time and second order accurate in space. The growth factor is

$$G = 1/(1 + i\nu \sin\beta) \tag{14.218}$$

Therefore the scheme (14.217) is unconditionally stable and thus places no constraint on the size of the time step relative to the size of the spatial mesh spacing. The modified differential equation of this method is

$$u_t + au_x - \frac{a^2\Delta t}{2} u_{xx} + \frac{a\Delta x^2}{3}\left(\frac{1}{2} + \nu^2\right)u_{xxx} + \cdots = 0 \tag{14.219}$$

The truncation error is predominately and unconditionally dissipative according to the dissipation condition (14.189). This is consistent with the conclusion drawn from the growth factor (14.218).

In implicit algorithm, a system of algebraic equations for example, a tridiagonal matrix equation, must be solved at each new time level. Thus, more computational effort is required per time step, but the overall time required to obtain a solution may be less since a larger time step is permitted. Of course, for unsteady flow problems, the solution may become meaningless if too large a time step is taken. However, for steady flow problems, we may take the advantage of large time step.

In practical computations, second-order accurate difference schemes are required. An implicit scheme can be obtained from the two Taylor-series expansions

$$u_i^{n+1} = u_i^n + \Delta t(u_t)_i^n + \frac{\Delta t^2}{2}(u_{tt})_i^n + \frac{\Delta t^3}{6}(u_{ttt})_i^n + \cdots \tag{14.220}$$

$$u_i^n = u_i^{n+1} - \Delta t(u_t)_i^{n+1} + \frac{\Delta t^2}{2}(u_{tt})_i^{n+1} - \frac{\Delta t^3}{6}(u_{ttt})_i^{n+1} + \cdots \tag{14.221}$$

Subtracting Eq.(14.221) from Eq.(14.220), we obtain

$$u_i^{n+1} - u_i^n = u_i^n - u_i^{n+1} + \Delta t\,[(u_t)_i^n + (u_t)_i^{n+1}] + \frac{\Delta t^2}{2}[(u_{tt})_i^n - (u_{tt})_i^{n+1}] + O(\Delta t^3) \tag{14.222}$$

Replacing $(u_{tt})_i^{n+1}$ with

$$(u_{tt})_i^{n+1} = (u_{tt})_i^n + O(\Delta t)$$

the resulting expression becomes

$$u_i^{n+1} = u_i^n + \frac{\Delta t}{2}\,[(u_t)_i^n + (u_t)_i^{n+1}] + O(\Delta t^3) \tag{14.223}$$

The time differencing in this equation is known as Crank-Nicolson differencing[32]. Upon substituting the linear wave equation $u_t = -au_x$ we obtain

$$u_i^{n+1} = u_i^n - \frac{a\Delta t}{2}\,[(u_x)_i^n + (u_x)_i^{n+1}] + O(\Delta t^3) \tag{14.224}$$

And finally, if the u_x terms are replaced by second-order accurate central differences, the time-centered, second-order accurate implicit method will result

$$u_i^{n+1} = u_i^n - \frac{\nu}{4}\,(u_{i+1}^n + u_{i+1}^{n+1} - u_{i-1}^n - u_{i-1}^{n+1}) \tag{14.225}$$

The growth factor for Eq.(14.225) is

$$G = \frac{1 - i\dfrac{\nu}{2}\sin\beta}{1 + i\dfrac{\nu}{2}\sin\beta} \tag{14.226}$$

$|G| \equiv 1$ for any value of ν. Hence the numerical stability of the scheme (14.225) is unconditionally neutral. This can also be seen from the modified differential equation for (14.25)

$$u_t + au_x + \frac{a\Delta x^2}{6}\,(1 + \frac{\nu^2}{2})u_{xxx} + \frac{a\Delta x^4}{8}\,(\frac{1}{15} + \frac{\nu^2}{3} + \frac{\nu^4}{10})u_{xxxxx} + \cdots = 0 \tag{14.227}$$

Note that the truncation error contains no even-order derivative terms so that the scheme has no implicit artificial dissipation.

When this scheme is applied to the nonlinear fluid dynamic equations it often becomes necessary to add explicit artificial dissipation to prevent the solution from blowing up. The usual artificial dissipative terms are given by Eqs.(14.213) and (14.214).

There are third-order accurate methods as well as higher-order methods in the literature. The improved accuracy of third-order methods is at the expense of added computer time and additional complexity. These factors must be considered carefully when choosing a finite difference scheme to solve a partial differential equation. In general, second-order accurate methods provide enough accuracy for most practical problems.

20. Characteristic-based differencing

The second-order accurate, explicit and implicit methods presented thus far have required the explicit addition of artificial dissipation to the schemes to prevent the spurious oscillations in the shock region. The addition of artificial dissipation is an empirically based method and may yield a transition region of rather shallow slope in the region where shock wave should have been seen. This feature is a stumbling block in the development of numerical methods for hyperbolic problems. One strategy for getting around it is to combine the physical and mathematical arguments and incorporate ideas drawn from the theory of characteristics into the difference methods. The development and use of the characteristic-based differencing is a relatively recent phenomenon. Publications in this area are still rapidly emerging and tend to unify and consolidate the subject as well as extend its applications to more hyperbolic problems. To illustrate the basic properties of characteristic-based difference algorithm we will discuss the first- and second-order accurate, explicit and implicit methods for the linear wave equation (14.177).

The explicit Euler method (14. 191) can be made stable by replacing

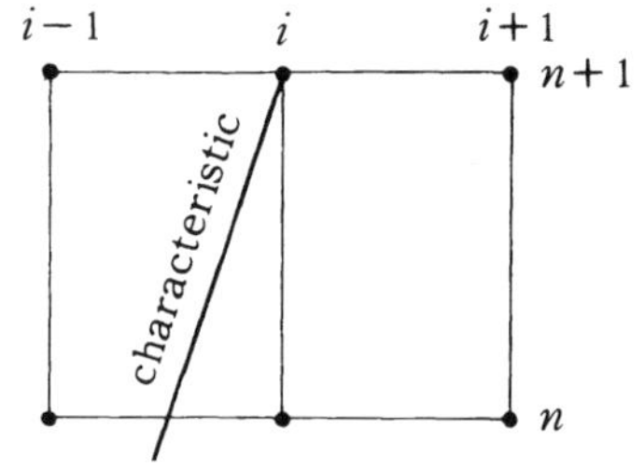

Fig. 14.12 Characteristic-based differencing

the central space difference by a backward space difference provided that the wave speed a is positive. Fig. 14. 12 shows that the characteristic passing through the grid point $(i, n+1)$ intersects the time level n at a backward point. If a backward difference is used, the following algorithm will result:

$$\frac{u_i^{n+1} - u_i^n}{\Delta t} + a\frac{u_i^n - u_{i-1}^n}{\Delta x} = 0 \tag{14.228}$$

This is an upwind differencing method. It is first-order accurate and explicit. The growth factor for this scheme is

$$G = 1 - \nu + \nu \sin\beta - i\nu \sin\beta \tag{14.229}$$

Thus, the stability condition is $\nu \leqslant 1$. When $\nu = 1$, Eq. (14. 228) reduces to exact solution (14.182). The same conclusion can be drawn from the modified differential equation

$$u_t + au_x - \frac{a\Delta x}{2}(1-\nu)u_{xx} - \frac{a\Delta x^2}{6}(1 - 3\nu + 2\nu^2)u_{xxx} + \cdots = 0 \tag{14.230}$$

The truncation error is predominately dissipative if $\nu < 1$.

Eq.(14.228) can be rewritten in the form

$$u_i^{n+1} = \nu u_{i-1}^n + (1-\nu) u_i^n \tag{14.231}$$

This is simply interpreted as the result of tracing back the characteristic from the grid point $(i, n+1)$ to find its intersection at the nth level in the interval $[i-1, i]$ (see Fig. 14.12) and performing linear interpolation. If the characteristic falls outside the assumed interval, so that the method is based on extrapolation, the computations are again unstable, so we must restrict Δt in such a way that $\nu \leqslant 1$.

A second-order accurate, explicit, characteristic-based method can be obtained by using the upwind differences for the spatial derivatives in equation (14.208). Using the second-order accurate upwind difference for u_x and the first-order accurate upwind difference for u_{xx}, we obtain

$$u_i^{n+1} = u_i^n - \frac{\nu}{2}(3u_i^n - 4u_{i-1}^n + u_{i-2}^n) + \frac{\nu^2}{2}(u_i^n - 2u_{i-1}^n + u_{i-2}^n) \tag{14.232}$$

This upwind method is equivalent to the following two steps, known as Warming-Beam method[33]

Predictor: $$\overline{u_i^{n+1}} = u_i^n - \nu(u_i^n - u_{i-1}^n) \tag{14.233}$$

Corrector: $$u_i^{n+1} = \frac{1}{2}[u_i^n + \overline{u_i^{n+1}} - \nu(\overline{u_i^{n+1}} - \overline{u_{i-1}^{n+1}}) - \nu(u_i^n - 2u_{i-1}^n + u_{i-2}^n)] \tag{14.234}$$

The equivalence can be verified by direct substitution. The addition of the second backward difference in the corrector (14.234) makes this scheme second-order accurate.

The growth factor of the scheme (14.232) is

$$G = 1 - 2\nu[\nu + 2(1-\nu)\sin^2\frac{\beta}{2}]\sin^2\frac{\beta}{2} - i\nu\sin\beta[1 + 2(1-\nu)\sin^2\frac{\beta}{2}] \tag{14.235}$$

and the resulting stability condition becomes

$$\nu \leqslant 2 \tag{14.236}$$

The same result can be obtained from the modified differential equation

$$u_t + au_x - \frac{a\Delta x^2}{6}(1-\nu)(2-\nu)u_{xxx} + \frac{a\Delta x^3}{8}(1-\nu)^2(2-\nu)u_{xxxx} + \cdots = 0 \tag{14.237}$$

While the truncation error is predominately dispersive, there exists a dissipative error if $\nu < 2$.

Note that the second-order accurate upwind difference method (14.232) permits a larger time step than the corresponding central difference method (14.209) and yields the exact solution when $\nu=1$ and 2. These are of benifit to the numerical solutions.

Upon incorporation into the implicit methods of the characteristic-based differencing, a first-order accurate upwind difference scheme is obtained

$$\frac{u_i^{n+1}-u_i^n}{\Delta t}+a\frac{u_i^{n+1}-u_{i-1}^{n+1}}{\Delta x}=0 \qquad (14.238)$$

The growth factor of this scheme is

$$G=1/\left[1+\nu\ (1-\cos\beta+i\sin\beta)\right] \qquad (14.239)$$

and thus the scheme is unconditionally stable. This can also be seen from the modified differential equation

$$u_t+au_x-\frac{a\Delta x}{2}(1+\nu)u_{xx}+\frac{a\Delta x^2}{6}(1+\nu)(1+2\nu)u_{xxx}+\cdots=0 \qquad (14.240)$$

The truncation error is predominately and unconditionally dissipative according to the dissipation condition (14.189).

A second-order accurate implicit method can be obtained from Eq. (14.224) if the u_x terms are replaced by second-order accurate upwind differences.

$$u_i^{n+1}=u_i^n-\frac{\nu}{4}(3u_i^n-4u_{i-1}^n+u_{i-2}^n+3u_i^{n+1}-4u_{i-1}^{n+1}+u_{i-2}^{n+1}) \qquad (14.241)$$

The growth factor of this scheme is

$$G=\left[1-\frac{\nu}{2}(1-\cos\beta)^2-i\frac{\nu}{2}\sin\beta(2-\cos\beta)\right]\Big/$$
$$\left[1+\frac{\nu}{2}(1-\mathrm{ocs}\beta)^2+i\frac{\nu}{2}\sin\beta(2-\cos\beta)\right] \qquad (14.242)$$

$|G|\leqslant 1$ for any value of ν. Hence this scheme is unconditionally stable.

Note that the second-order accurate central-differencing implicit scheme (14.225) is neutrally stable and thus when applied to the Euler equations is prone to local instability near shocks. In comparison with this, the present scheme (14.241) is stable and thus when applied to the Euler equations may capture shock waves without inducing spurious oscillations.

21. Numerical solution of unsteady Euler equations

Shock-capturing methods are the most widely used techniques for computing inviscid flows with shocks. In this approach, any shock waves or other discontinuities are predicated as part of the solution with no special treatment required. To this end, the Euler equations are cast in conservation or divergence form (cf. Chapter V, § 2). For simplicity, one-dimensional unsteady Euler equations are considered.

$$\frac{\partial U}{\partial t} + \frac{\partial F}{\partial x} = 0 \tag{14.243}$$

where U is the vector function of conservative variables and F is the vector function of U representing the flux of conservative variables. They are given by

$$U = \begin{bmatrix} \rho \\ \rho u \\ \rho e \end{bmatrix} \qquad F = \begin{bmatrix} \rho u \\ \rho u^2 + p \\ (\rho e + p) u \end{bmatrix} \tag{14.244}$$

This system is hyperbolic and has three characteristics (cf. Chapter V, § 1).

$$\frac{dx}{dt} = u, \ u \pm a \tag{14.245}$$

where a is the speed of sound.

Applying the Lax method (14.202) to the Euler equation (14.243), we obtain

$$U_i^{n+1} = \frac{1}{2} (U_{i+1}^n + U_{i-1}^n) - \frac{\Delta t}{2\Delta x} (F_{i+1}^n - F_{i-1}^n) \tag{14.246}$$

This is a first-order accurate explicit scheme. Upon local linearization, the stability condition is

$$\frac{\Delta t}{\Delta x} \leqslant \min \left\{ \frac{1}{|u|}, \frac{1}{|u+a|}, \frac{1}{|u-a|} \right\}$$

or

$$\frac{\Delta t}{\Delta x} \leqslant \frac{1}{|u| + a} \tag{14.247}$$

Applying the MacCormack two-step method, we obtain a second-order accurate explicit scheme for the Euler equations (14.243)

Predictor: $$\overline{U_i^{n+1}} = U_i^n - \frac{\Delta t}{\Delta x} (F_i^n - F_{i-1}^n) \tag{14.248}$$

Corrector: $$U_i^{n+1} = \frac{1}{2} \left[U_i^n + \overline{U_i^{n+1}} - \frac{\Delta t}{\Delta x} (\overline{F_{i+1}^{n+1}} - \overline{F_i^{n+1}}) \right] \tag{14.249}$$

The same linearized stability condition (14.247) holds.

A second-order accurate explicit method may be formulated in another way. Consider the two-dimensional unsteady Euler equations

$$\frac{\partial U}{\partial t}+\frac{\partial F}{\partial x}+\frac{\partial G}{\partial y}=0 \qquad (14.250)$$

Replacing the spatial derivatives by the central differences and keeping the time derivative, we obtain a system of ordinary differential equations

$$\frac{dU_{i,j}}{dt}+\frac{F_{i+1,j}-F_{i-1,j}}{2\Delta x}+\frac{G_{i,j+1}-G_{i,j-1}}{2\Delta y}=0 \qquad (14.251)$$

This system of ordinary differential equations can be solved by the well known Runge-Kutta method for second- or higher-order accuracy in t[34]. If a four-step scheme is used, the accuracy of the resulting algorithm will be fourth-order accurate in time t but the overall accuracy is still second-order. Applying this scheme to the model equation (14.177), we obtain the stability condition $\nu \leqslant 2\sqrt{2}$.

In multi-dimensional flow computations, the difference equations are usually reinterpreted as expressions of the integral conservation laws (cf. Chapter V, §2)

$$\frac{\partial}{\partial t}\int_V U\, dV+\int_S \boldsymbol{W}\cdot \boldsymbol{n} dS=0 \qquad (14.252)$$

rather than the differential laws. Here $\boldsymbol{W}$ is the flux vector of the conservative variables. There are advantages in this interpretation. Since the integral equations (14.252) hold for flows containing discontinuities in conservative variables, the shock-capturing method can attain mathematical respectability. Moreover through this interpretation, the difference schemes can be applied to any curvilineal computational meshes without alteration. In practical computations, curvilineal meshes conforming to the body surfaces are generally used. Methods based on interpretation (14. 252) are called finite-volume schemes and are widely used for shock-capturing codes today.

Numerical solutions for the Euler equations for flow past a circular cylinder without circulation at $M_\infty=0.2$ and 0.5 given in Figures 7.4 and 10.22 are computed by the finite volume method using Runge-Kutta time stepping schemes[35]. A blend of second and fourth differences, Eqs.(14.213) and (14. 214), is used to construct dissipative terms. Most surprisingly, for $M_\infty= 0.50$ starting from $2x/D \sim 0.78$ the static pressure from the Euler solution is constant and almost equal to $C_p=0$. Fig. 14.13 clarifies the results by presenting the direction of the local velocity vectors, streamlines and isobars. The flow separates from the smooth surface and forms a recirculating "dead air" region with very small velocities ($q \leqslant 0.01\ q_\infty$).

The reason for this inviscid separation is the total pressure loss and the vorticity due to the shock, rather than due to the boundary layer. However, the consequences are similar since the flow due to the total pressure loss at the wall streamline does not have enough kinetic energy to stagnate at the rear stagnation point. It should be noted that this inviscid separation point can be found to be always behind the one known from viscous flow analysis. However, the inviscid prediction may be improved by coupling with a boundary layer computation for flows without massive separation.

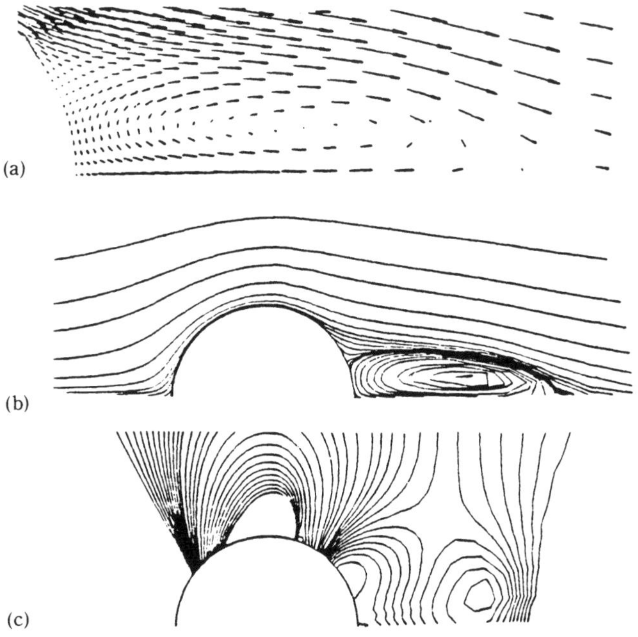

a) Velocity vectors, b) streamlines, and c) isobars

Fig. 14.13 Velocity vector plot and isobar plot for the Euler solution of the circular cylinder flow at $M_\infty = 0.5$

A second-order accurate implicit method can be constructed by the use of the Crank-Nicolson differencing (14.223)

$$U_i^{n+1} = U_i^n - \frac{\Delta t}{2}\left[(F_x)_i^n + (F_x)_i^{n+1}\right] + O(\Delta t^3) \tag{14.253}$$

where $(F_x)_i^n$ and $(F_x)_i^{n+1}$ should be evaluated to second-order accuracy. A Taylor-series expansion of F^{n+1} with respect to U is

$$F^{n+1} = F^n + \left(\frac{\partial F}{\partial U}\right)^n (U^{n+1} - U^n) + O(\Delta t^2) \tag{14.254}$$

where $\dfrac{\partial F}{\partial U}$ is the Jacobian matrix $[A]$.

Putting Eq.(14.254) into (14.253), we obtain[36]

$$\{ I+ \frac{\Delta t}{2} \frac{\partial}{\partial x} [A]_i^n\}(U_i^{n+1} - U_i^n) = -\Delta t(F_x)_i^n \qquad (14.255)$$

where I is the identity matrix and the spatial derivatives are represented by central differences. The scheme is neutrally stable by the model equation analysis and thus the addition of artificial viscosity is required.

The resulting algebraic equations are a linear system for the unknowns U_i^{n+1}. The algorithm requires the solution of a block tridiagonal system of equations. Each block is $m \times m$ if there are m elements in the unknown U vector. For multi-dimensional flow problems, direct solution of the difference equations is usually avoided due to the large operation count. The path usually chosen is to reduce the multi-dimensional problem into a sequence of one-dimensional inversions. This may be done by using the method of approximate factorization (cf. Chapter X, § 14).

The central-differencing second-order accurate schemes with artificial viscosity are now commonly used to solve steady transonic flow problem in two and three dimensions. However, these methods still appear to be unsatisfactory if they are required either to achieve the best possible resolution of discontinuities or to follow rapid transient motion.

22. Flux-vector splitting method

We now turn to the characteristic-based methods. To illustrate the concept of these methods, consider the one-dimensional system of hyperbolic partial differential equations (14.243). This system has three characteristics (14.245) which are shown in Fig. 14.14 for the case $0< u< a$. The upwind direction depends on the sign of the slope $\frac{dx}{dt}$ of the characteristics. If the slope is positive, the upwind direction for spatial differencing is

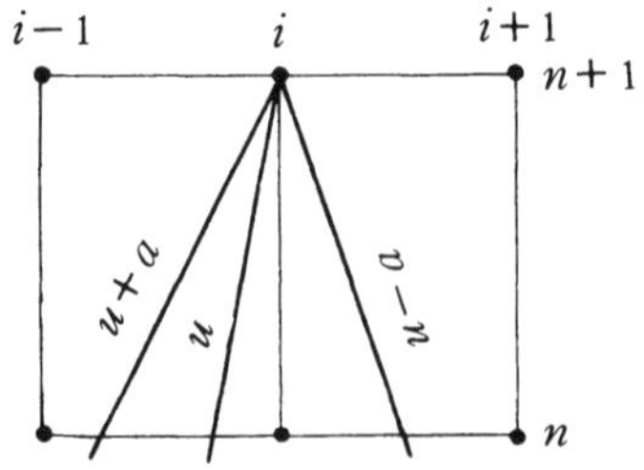

Fig. 14.14 Characteristics, $0 < u < a$

backward. If the characteristic slope is negative, the upwind direction is forward. The simplest way of introducing upwinding into a conservative scheme seems to be based upon the following mathematical trick[37]. We try to split the flux vector F in the governing equations into F^+ and F^-

$$\frac{\partial U}{\partial t} + \frac{\partial F^+}{\partial x} + \frac{\partial F^-}{\partial x} = 0 \tag{14.256}$$

such that the two spatial derivatives $\frac{\partial F^+}{\partial x}$ and $\frac{\partial F^-}{\partial x}$ can be stably computed by the backward- and forward-differencing schemes respectively. While the precise splitting can be accomplished in a number of ways, the usual way is according to the signs of the eigenvalues of the system. The eigenvalues of the system are the characteristic slopes $\frac{dx}{dt}$ of the system.

Let $[\lambda]$ be the diagonal matrix of eigenvalues of the Jacobian matrix $[A]$ of the system. The matrix of eigenvalues is divided into two matrices, one with only positive elements and the other with negative elements

$$[\lambda] = [\lambda^+] + [\lambda^-] \tag{14.257}$$

For hyperbolic systems, we have

$$[A] = [T]\,[\lambda]\,[T]^{-1} \tag{14.258}$$

where $[T]$ is the matrix whose columns are the eigenvectors of $[A]$ taken in order. Now the flux vectors of the Euler equations have the interesting property that (cf. Chapter V, §2).

$$F = [A]\,U \tag{14.259}$$

We may define

$$F = F^+ + F^- \tag{14.260}$$

where

$$F^+ = [A]^+ U, \; F^- = [A]^- U \tag{14.261}$$

and

$$[A]^+ = [T]\,[\lambda^+]\,[T]^{-1}, \; [A]^- = [T]\,[\lambda^-]\,[T]^{-1} \tag{14.262}$$

This split flux idea can be used either for explicit or implicit algorithm. A second-order accurate upwind explicit scheme may be written with the aid of Eqs. (14.233) and (14.234)

$$U_i^{\overline{n+1}} = U_i^n - \frac{\Delta t}{\Delta x}\left(F_i^+ - F_{i-1}^+\right) - \frac{\Delta t}{\Delta x}\left(F_{i+1}^- - F_i^-\right) \tag{14.263}$$

$$\begin{aligned} U_i^{n+1} = \frac{1}{2}\Big[U_i^n + U_i^{\overline{n+1}} &- \frac{\Delta t}{\Delta x}\left(F_i^{+\,\overline{n+1}} - F_{i-1}^{+\,\overline{n+1}}\right) - \frac{\Delta t}{\Delta x}\left(F_i^{+\,n} - 2F_{i-1}^{+\,n} + F_{i-2}^{+\,n}\right) \\ &- \frac{\Delta t}{\Delta x}\left(F_{i+1}^{-\,\overline{n+1}} - F_i^{-\,\overline{n+1}}\right) - \frac{\Delta t}{\Delta x}\left(F_{i+2}^{-\,n} - 2F_{i+1}^{-\,n} + F_i^{-\,n}\right)\Big] \end{aligned} \tag{14.264}$$

The stability condition of the model equation $\nu \leqslant 2$ (14.236) may be applied to this scheme. A second-order accurate upwind implicit scheme may be written with the aid of Eq.(14.255)

$$\left\{ I + \frac{\Delta t}{2}\left(\frac{\partial^+}{\partial x}[A]_i^{+n} + \frac{\partial^-}{\partial x}[A]_i^{-n}\right)\right\}(U_i^{n+1} - U_i^n) = -\Delta t[(F_x^+)_i^n + (F_x^-)_i^n] \tag{14.265}$$

where the plus indicates that a backward second-order accurate difference should be used, and a minus indicates that a forward difference is required. This scheme is unconditionally stable in a linear sense.

The use of split flux techniques can capture shock waves without explicit addition of artificial dissipation and thus produce somewhat better representation of shock waves than the standard central difference schemes. A typical example[38] is shown in Fig. 14. 15 for the pressure distribution around a NACA 0012 airfoil at an angle of attack of 1.25° in a uniform flow with Mach number 0.80. The illustration on the left was obtained by a characteristic-based method using flux-vector splitting. The right-hand side shows results obtained on the same computational mesh by the central-differencing artificial-viscosity scheme . The left-hand side picture defines the shock waves much more crisply, especially the weak one on the lower surface and the transition zone containing only one grid point. Whereas on the right-hand side the shock waves are smeared over several mesh intervals. Therefore, the shocks are resolved more accurately with the flux splitting approach. However, the two results agree closely in overall features.

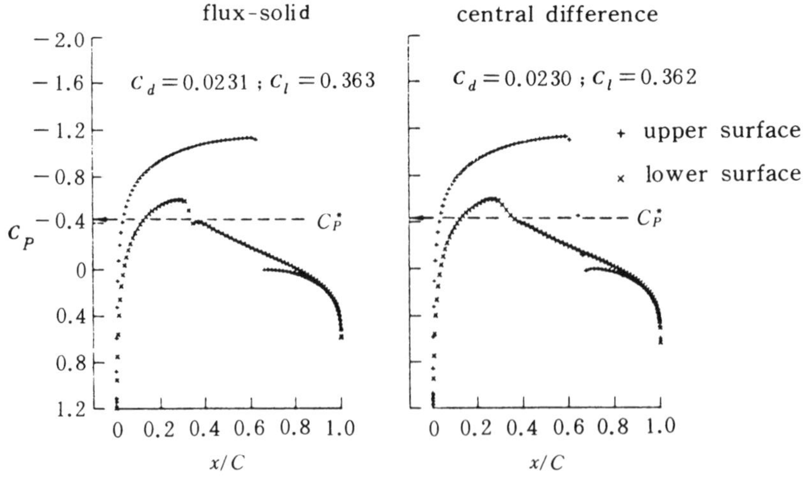

Fig. 14. 15 Two different calculations for the pressure distribution over a NACA 0012 airfoil. C_P^* denotes the pressure coefficient at sonic conditions

Flux-vector splitting is not entirely clear in its physical interpretation. Other ways of introducing upwinding into a conservative scheme have been presented, for example, exact and approximate Riemann solvers. Godunov[39] used a highly ingenious device that advances to the next time level by solving exactly a set of Riemann problems. Space does not permit a discussion of all these methods. For related materials the reader is referred to Roe[40].

In conclusion, the essential idea of the characteristic-based schemes is to better match the numerics to the physics of the flow by identifying the characteristic directions of signal propagation and upwind differencing along these directions. Since the upwind discretization is naturally dissipative, no added artificial viscosity is necessary. The natural dissipation due to the upwind differencing combined with proper treatment of signal propagation enables strong shocks to be captured sharply without oscillation. It is noted that the upwind methods do require more operations to implement and more logic. This inhibits their efficiency on a per iteration per grid point basis. Nevertheless, upwind schemes do seem to be where current research directions are leading.

As to capturing vortical flow by the Euler conservation laws, different methods may give different answers. In the case of delta wings, with subsonic, rounded leading edges, spurious separation was predicated by conventional central difference schemes and shown to occur due to artificial dissipation. In contrast, upwind-difference schemes appear to be much less susceptible to spurious inviscid separation and thus provide a better description of inviscid flow about rounded leading-edge wings. For subsonic, sharp leading edges, central difference methods consistently predict leading-edge separation while upwind difference methods, in some instances, predict attached flow. It is now generally recognized that the Euler equations are not appropriate for high alpha solutions about rounded-edge delta wings especially when secondary separation is considered to be important. For this problem, the Navier-Stokes equations are required.

In practical computations of steady flows, it is often advantageous to replace the differential energy equation in the system of the unsteady Euler equations by the algebraic energy equation for steady adiabatic flow,

$$\frac{\gamma}{\gamma-1}\,\frac{p}{\rho}+\frac{q^2}{2}=\frac{\gamma}{\gamma-1}\,\frac{p_\infty}{\rho_\infty}+\frac{q_\infty^2}{2}$$

if only the steady-state solution is desired[35].

23. Numerical solution of diabatic flow[41]

Consider the process occurring in the closed-end shock tube, which is used to study the thermodynamic and radiative properties of shocked gases. An incident plane shock is reflected from a rigid wall. The Lighthill ideal, dissociating, diatomic gas is selected as the model for the fluid. The governing equations for this one-dimensional flow written in divergence form are

$$\left.\begin{aligned} &\rho_t + (\rho u)_x = 0 \\ &(\rho u)_t + (\rho u^2 + p)_x = 0 \\ &(\frac{1}{2}\rho u^2 + \rho H - p)_t + [\rho u(\frac{1}{2}u^2 + H)]_x = 0 \\ &(\rho c)_t + (\rho c u)_x = \sigma \end{aligned}\right\} \tag{14.266}$$

where

$$H = [(4+c)/(1+c)](p/\rho) + cd$$

$$\sigma = K\rho^2[(1-c)e^{-d/kT} - (\rho/\rho_d)c^2]$$

$$T = p/[R\rho(1+c)]$$

c is the degree of dissociation, σ is the rate of dissociation, d is the energy of dissociation, K is the Freeman rate constant, k is the Boltzmann constant, and ρ_d is the characteristic density of dissociation.

Combine Eqs. (14.266) into a compact vector form

$$U_t + F_x = S \tag{14.267}$$

Applying the Lax method (14.246) to (14.267), we obtain

$$U_i^{n+1} = \frac{1}{2}(U_{i+1}^n + U_{i-1}^n) - \frac{\Delta t}{2\Delta x}(F_{i+1}^n - F_{i-1}^n) + \Delta t S_i^n \tag{14.268}$$

where Δt should be consistent with the CFL stability condition (14.247). Here

$$a^2 = [(4+c)/3](p/\rho) \tag{14.269}$$

At the mesh point $i=1$, a rigid wall is located with the boundary condition $u=0$ for all t. Numerically, this is achieved in the following way. The grid is extended one point beyond the wall, no variable is computed there, but it is given the same value it has at one point in front of the wall, except for u to which the negative of its value in front of the wall is assigned. Such boundary logic is called reflection.

Fig. 14.16 shows the temperature profiles representing the propagation of the incident and reflected shocks in nitrogen. The illustration on the left is without dissociation. The right-hand side shows results with dissocation .The solid curves on Fig . 14.16 indicate the spread of the shock, which should be expected using the Lax method, and the dotted lines show how this spread can be reduced using the more sophisticated differencing method of the Godunov characteristic-based scheme. For the present application of this method to nonequilibrium flows, the gas was assumed to be frozen locally during the resolution of each discontinuity.

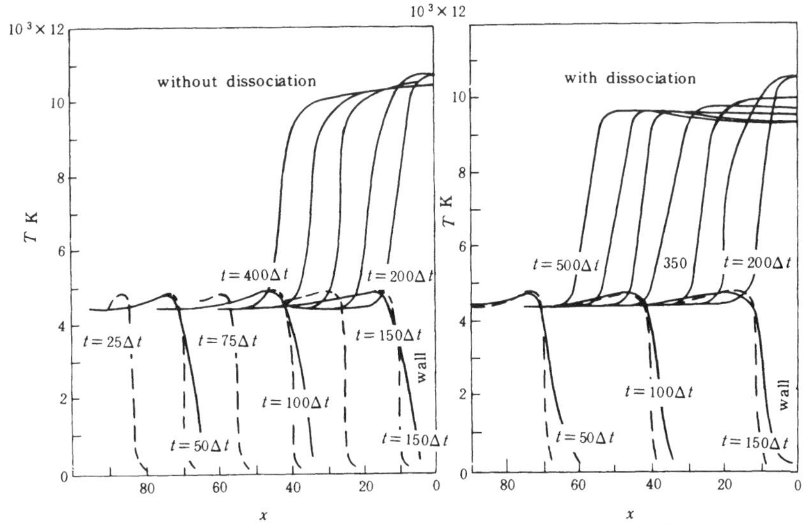

Fig. 14. 16 Temperature profile representing one-dimensional shock reflection

24. Problems

1. Derive the fundamental equations for two-dimensional and axially symmetrical rotational flow in intrinsic form, i. e., Eqs.(14. 14) to (14.18).

2. Derive for axially symmetrical flow curved shock relations similar to Eqs.(14.31) and (14.32).

3. Find the Crocco point when the Mach number in front of a curved shock is 3.

4. Show that Eq.(14.39) is true.

5. Determine the relation between the distance of detached shock over a circular cylinder in the two-dimensional case by the method described in § 6 using the power series up to x^4. Also find the corresponding relation for the curvature of the shock near the axis of symmetry.

6. Using the method described in § 6, find the relation of the detached shock distance and the curvature of the shock at the axis of symmetry in terms of the Mach number in front of the shock over a sphere. Use the velocity profile in power series of x up to x^4.

7. Calculate the flow over an ogive by the linearized theory of §7 with the following data:

$$Y'(x) = \sin x,\ M_1 = 3.00,\ M_2 = 2.00,\ \gamma = 1.4.$$

8. Repeat problem 7 with the following data

$$Y'(x) = e^{-x},\ M_1 = 3.00,\ M_2 = 2.00,\ \gamma = 1.4.$$

9. Repeat problem (7) by the method of characteristics.

10. Derive the equation for the stream function for two-dimensional steady diabatic flow.

11. If the detached shock shape is known, show how to solve the differential equation of the stream function for two-dimensional steady flow (cf. Ref. 12).

12. Consider a vertical column of gas of constant temperature and take the spatial coordinate x as the vertical distance, positive upward. We assume that a shock is traveling through the gas. Determine the flow field behind the shock wave in the initial period by the method of series expansion (§ 10). How does one continue the solution by the series expansion?

13. Calculate the velocity and pressure distribution of a uniform flow of a simple dissociating gas over an infinitely wavy wall.

14. Calculate the lift and drag force on a flat plate airfoil at a small angle of attack in a uniform supersonic inviscid flow of a simple dissociating gas under the condition of (a) equilibrium flow and (b) frozen flow.

15. Compute the lift, drag and moment coefficients for an airfoil or rhomboidal cross section in a supersonic two-dimensional steady flow of an inviscid simple dissociating gas at a small angle of attack.

16. Apply the method of characteristics to the two-dimensional steady supersonic flow of a simple dissociating gas. Find the equations of characteristics and the compatibility conditions along these characteristics curves.

17. Write down the fundamental equations of the one-dimensional steady flow in a nozzle of variable cross section for a simple dissocating gas. Discuss in detail the solution of these equations for a de Laval nozzle.

18. Prove that the system of characteristic equations for the one-dimensional unsteady Euler equations are

$$dp - \rho a\, du = 0 \qquad \text{along} \qquad dx = (u - a)\, dt$$

$$dp - a^2 d\rho = 0 \qquad \text{along} \qquad dx = u\, dt$$

$$dp + \rho a\, du = 0 \qquad \text{along} \qquad dx = (u + a)\, dt$$

19. Let the general difference representation of the linear wave equation (14.177) be

$$u_i^{n+1} = \sum_k c_k u_{i+k}^n$$

where c_k' s are real and $k=0, \pm 1, \pm 2, \cdots$

Show that the scheme is:

(a) first-order accurate if

$$\sum_k c_k = 1 \quad \text{and} \quad \sum_k k c_k = -\nu$$

(b) second-order accurate if

$$\sum_k c_k = 1, \ \sum_k k c_k = -\nu, \ \text{and} \ \sum_k k^2 c_k = \nu^2$$

(c) sth-order accurate (s = 1, 2, ⋯) if

$$\sum_k k^m c_k = (-\nu)^m, \ m = 0, 1, 2, \cdots, s$$

where

$$\nu = a\Delta t / \Delta x.$$

20. Define the scheme to be monotonicity preserving by the requirement that if u_i^o were a monotonic function of i, then u_i^n ($n=1, 2, \cdots$, etc.) should also be a monotonic function of i. Show that the necessary and sufficient condition for the scheme to be monotonicity preserving is that none of c_k (see problem 19) are negative.

21. Show that no second-order accurate scheme which is monotonicity preserving exists. (Godunov's theorem).

22. The leap frog method is given by

$$\frac{U_i^{n+1} - U_i^{n-1}}{2\Delta t} + a\frac{U_{i+1}^n - U_{i-1}^n}{2\Delta x} = 0$$

Evaluate the truncation error to the term of order Δx^4 and the growth factor. Discuss the properties (dissipation or dispersion) of the truncation error and the stability behavior.

23. Write down the MacCormack two-step, second-order explicit scheme for the two-dimensional unsteady Euler equations (14.250). Verify the accuracy of the scheme and give the stability condition.

24. Consider steady supersonic flow of a perfect inviscid gas over a two-dimensional surface. Assume the x axis forms the body surface. Write down the governing equations of steady flow. Show that the equation system is hyperbolic and give the MacCormack two-step, second-order accurate, explicit scheme for the equation system.

25. Compute the flow field produced by a two-dimensional wedge moving at a Mach number of 2.0 if the wedge half angle is 15°. Assume inviscid flow of a perfect gas. Use both the steady and unsteady Euler equations.

26. For steady adiabatic flow, the stagnation enthalpy is constant. If this is used as an equation of state, thus eliminating the need to use the energy equation, show that the one-dimensional unsteady Euler equations remain hyperbolic.

27. Consider the model nonlinear wave equation known as the inviscid Burgers equation

$$\partial u/\partial t+\partial(u^2/2)/\partial x=0, \quad 0< x< 1$$

subject to the periodic boundary conditions

$$u(0, t)=u(1, t), \quad t>0$$

and the initial condition

$$u(x, 0)=u_0+u_1 \sin(2\pi x)$$

Solve this problem up to t = 2.0 for the initial condition corresponding to

$$u_0=0 \quad \text{and} \quad u_1=0.2$$

using any two of the following methods:

(a) Lax method;

(b) A Crank-Nicolson method;

(c) Any Lax-Wendroff method.

You may discover that you need to add additional dissipation in the form of artificial viscosity to stabilize the latter classes of shemes. Discuss the effect of varying Courant numbers for each scheme, as well as comparing the results of the various schemes (including the effect of any artificial viscosity you may find necessary).

28. Repeat the calculations of problem 27 for the initial condition corresponding to

$$u_0=1.0 \quad \text{and} \quad u_1=0.2$$

Make similar observations for the various schemes in this case. Also, describe how this solution is related to that of problem 27.

29. If the influence of boundary conditions on stability is desired, we must use the matrix method (cf. Chapter VII, reference 15). Use the Lax method to solve the linear wave equation (14.177). A four point mesh is used. The boundary conditions are

$$u_1^{n+1}=u_1^n \quad \text{and} \quad u_4^{n+1}=u_3^n$$

Show that the stability condition is

$$\nu\leqslant 2\sqrt{2}-1$$

30. If the boundary conditions of problem 29 are periodic, find the stability condition for the Lax method using both the matrix method and the Von Neumann method. Compare the results obtained by these two methods and discuss the influence of boundary conditions on stability.

31. Stability for systems of equations must be evaluated by the matrix method. Determine the stability requirement necessary in solving the system of first-order linear equations

$$u_t + av_x = 0$$

$$v_t + au_x = 0$$

using the Lax method.

32. Discuss the consistency of the Lax method (14.202) for the first order one-dimensional wave equation.

References

1. Pai, S. I., On the vorticity of the flow behind curved shock, *Jour. Aero. Sci.*, **17**, No. 3, March 1950, pp. 188 — 189.
2. Vazsonyi, A., On rotational gas flow, *Quart. Appl. Math.* **3**, No. 1, April, 1945, pp. 29 — 37.
3. Lin, C. C., and Rubinov, S. I., On the flow behind curved shocks, *Jour. Math. Phys.* **27**, No. 2, July, 1948, pp. 105 — 129.
4. Adams, Mac C., On shock waves in inhomogeneous flow, *Jour. Aero. Sci.* **16**, No. 11, Nov. 1949, pp. 685 — 690.
5. Pai, S. I., On the flow behind an attached curved shock, *Jour, Aero. Sci.* **19**, No. 11, Nov. 1952, pp. 734 — 742.
6. Crocco, L., *L'Aerotechnica* **17**, 519 (1937).
7. Thomas, T. Y., On curved shock waves, *Jour. Math. Phys.* **26**, 1947, pp. 62 — 68.
8. Busemann, A., A review of analytic methods for the treatment of flows with detached shocks, *NACA, TN* No. 1858, April 1949.
9. Oguchi, H., Supersonic jet with the ambient pressure corresponding to its constant pressure point, *Jour. Phys. Soc. Japan* **11**, No. 2, Feb. 1956, pp. 155 — 159.
10. Dugundgi, J., An investigation of the detached shock in front of a body of revolution, *Jour. Aero. Sci.* **15**. No. 12, Dec. 1948, pp. 699 — 705.
11. Pai, S. I., Supersonic rotational flow over two-dimensional ogives, *Proc. Third Midwestern Conf. on Fluid Mech.*, University of Minnesota, 1953, pp. 303 — 317.
12. Pai, S. I., On the flow behind an axially symmetrical attached curved shock, *Jour. Franklin Inst.* **257**, No. 5, May 1954, pp. 383 — 398.
13. Tollmien, W., Steady two dimensional and axially symmetrical supersonic flows, *Brown University Trans.* No. A9 – T – 1, 1948.
14. Holt, M., The method of characteristics for steady supersonic rotational flow in three dimensions, *Jour. Fluid Mech.* **1**, pt **4**, Oct. 1956, pp. 409 — 423.
15. Shu, S. S., On two-dimensional flow after a curved stationary shock, *NACA, TN* No. 2364, May 1951.
16. Burgers, J. M., Non-uniform propagation of shock waves, *Lecture Series No. 10*, Inst. for Fluid Dynamics and Appl. Math., University of Maryland, 1951, prepared by S.I. Pai.
17. Taylor, Sir Geoffrey, The formation of a blast wave by a very intense explosion, *Proc. Roy. Soc.*

London **A −201**, pp. 159 — 186. 1950.

18. Lin, S. C., Cylindrical shock waves produced by instantaneous energy release, *Jour. Appl. Phys.* **25**, No. 1, pp 54 — 57, Jan. 1954.
19. Hicks, B. L., Diabatic flow of a compressible fluid, *Quart. Appl. Math.* **6**, No. 3, pp. 221 — 237, Oct. 1948.
20. Hicks, B. L., On the characterization of fields of diabatic flow, *Quart. Appl, Math.* **6**, No. 4, pp. 405 — 416. Jan. 1949.
21. Hicks, B. L., Aerodynamical effects of heat released by combustion of steadily flowing gases, *Third Sym. of Combustion and Flame and Explosion Phenomena*, The Williams and Wilkins Company, Baltimore, Md., 1949, pp. 212 — 222.
22. Tsien, H. S., Influence of flame front on the flow field, *Jour. Appl. Mech.* **18**, No. 2, June 1951, pp. 188 — 194.
23. Gross, R. A., and Esch, R., Low-speed combusition aerodynamics, *Jet Propulsion*, March-April 1954, pp. 95 — 101.
24. Pai, S. I., *Modern Fluid Mechanics*, Science Press, Beijing, distributed by Van Nostrand Reinhold Co. New York, 1981.
25. Lighthill, M. J., Dynamics of a dissociating gas, part I Equilibrium flow, *Jour. Fluid Mech.* **2**, Pt. 1, 1957, pp. 1 — 32.
26. Li. T. Y., Recent advances in nonequilibrium dissociating gasdynamics, *ARS Jour.* **31**, No. 2, 1961, pp. 170 — 178.
27. Warming, R. F., and Hyett, B. J., The modified equation approach to the stability and accuracy analysis of finite-difference methods, *J. Comp. Phy.* **14**, 1974, pp. 159 — 179.
28. Chang, S. C., On the validity of the modified equation approach to the stability analysis of finite-difference methods, *AIAA Paper*, pp. 87 — 1120.
29. Lax, P. D., Weak solutions of nonlinear hyperbolic equations and their numerical computation, *Comm. Pure Appl. Math.* **7**, 1954, pp. 159 — 193.
30. Lax, P. D., and Wendroff, B., Systems of conservation laws, *Comm. Pure Appl. Math.* **13**, 1960, pp. 217 — 237.
31. MacCormack, R. W., The effect of viscosity in hypervelocity impact cratering, *AIAA Paper*, 69 — 354, 1969.
32. Crank, J., and Nicolson, P., A practical method for numerical evaluation of solutions of partial differential equations of the heat-conduction type, *Proc. Cambridge Philos, Soc.*, **43**, 1947, pp. 50 — 67.
33. Warming, R. F., and Beam, R. M., Upwind second-order difference schemes and applications in unsteady aerodynamic flows, *Proc. AIAA 2nd Computational Fluid Dynamics Coference*, Hartford, Connecticut, 1975, pp. 17 — 28.
34. Jameson, A., Schmidt, W., and Turkel, E., Numerical solutions of the Euler equations by finite volume methods using Runge Kutta time stepping schemes, *AIAA Paper*, pp. 81 — 1259, June, 1981.
35. Schmidt, W., Jameson, A., and Whitfield, D., Finite volume solutions to the Euler equations intransonic flow, *J. Aircraft* **20**, No. 2, February 1983, pp. 127 — 133.
36. Beam, R. M., and Warming, R. F., An implicit finite-difference algorithm for hyperbolic systems in conservation law form, *J. Comp. Phys.* **22**, 1976, pp. 87 — 110.
37. Steger, J. L., and Warming, R. F., Flux vector splitting of the inviscid gasdynamic equations with application to finite-difference methods, *J. Comp. Phys.* **40**, 1981, pp. 263 — 293.
38. Anderson, W. K., Thomas, J. L., and Van Leer, B., Comparison of finite volume flux vector splittings for the Euler equations, *AIAA Journal*, **24**, No. 9, Sept. 1986, pp. 1453 — 1460.
39. Godunov, S. K., A finite difference method for the numerical computation of discontinuous solutions of the equations of fluid dynamics, *Mat. Sb.* **47**, 1959, pp. 357 — 393.
40. Roe, P. L., Characteristic-based schemes for the Euler equations, *Ann. Rev. Fluid Mech.* **18**, 1986, pp. 337 — 365.

41. Bohachevsky, I. O., and Rubin, E. L., A direct method for computation of nonequilibrium flows with detached shock waves, *AIAA Journal*, **4**, No. 4, April 1966, pp. 600 — 607.

Chapter XV

ELECTROMAGNETOGASDYNAMICS

1. Plasma dynamics and electromagnetofluid dynamics [1–7]

At high temperature, above 10,000 K, gas will be ionized. We may use the term *plasma* for ionized gas. The properties of ionized gas or a plasma differ considerably from those of a neutral gas. Hence, we may consider plasma as the fourth state of matter. The most important difference between plasma and ordinary gas is that in plasma the electromagnetic force plays a major role. We should consider simultaneously the electromagnetic forces and the gasdynamic forces in our analysis of the flow problems of plasma. Many new phenomena occur due to the interaction of the gasdynamical and the electromagnetic forces.

The science which deals with the flow problems of plasma is called Plasma Dynamics. The scope of plasma dynamics is very broad. It contains problems from electric discharge in rarefied gas, propagation of electromagnetic waves in ionized medium, to the so-called electromagnetofluid dynamics. In this chapter we shall limit ourselves to the problems of electromagnetogasdynamics.

The plasma may be considered as a mixture of N species which consists of ions, electrons and neutral particles. In the flow field, ionization and recombination of ions and electrons may occur. Hence, in plasma dynamics, we have also the effects of chemical reactions if we consider the ionization process as a chemical reaction. However, since the mass of electrons is much smaller than that of ions or neutral particles and the diffusion velocity of electrons is very large, the treatment of diffusion coefficient approximation will not be satisfactory. We shall discuss this problem in more detail in Chapters XVI[4] and XVII[5]. A better treatment is known as multifluid theory of plasma dynamics which will be discussed in Chapter XVI where we will study the temperature, pressure, density and velocity vector of every species in plasma. But for a first approximation, if the variation of the composition of the plasma in a flow field is small, it is sufficient to consider the plasma as a single fluid and we can use the temperature, pressure, density and velocity vector of the plasma as a whole to describe the flow field. We may employ this point of view in this chapter. We may call it the classical theory of electromagnetofluid dynamics. In Chapter XVI, we shall show that under certain conditions, we may reduce the multifluid theory results to that of the classical theory studied in this chapter.

The most popular name for the branch of fluid dynamics dealing with the flow problems of electrically conducting fluids is known as Magneto-hydrodynamics (MHD)[8]. In the early days of investigations, we could not produce in the laboratory the flow phenomena of ionized gas, where the electromagnetic forces are of the same order of magnitude as the gasdynamic forces. Most of the investigations are theoretical. We could check these theoretical results with astronautical observations.[6, 8] However, it was found that the interaction phenomena could be obtained in the laboratory by investigating the electromagnetic phenomena of the flow of electrically conducting liquid such as mercury.[9] Hence, the term Magnetohydrodynamics was found and extensively investigated. In magnetohydrodynamics, the compressibility effects of the medium are negligible. Magnteohydrodynamics is derived from the term hydrodynamics which is concerned mainly with water and the other liquids.

After 1950, engineers and applied physicists began in earnest the investigation of the flow of an electrically conducting fluid because the importance of controlled fusion research and that of space technology became evident. It is possible to produce the flow of plasma in laboratory in which the electromagnetic forces are of the same order of magnitude as those of gasdynamic forces by means of shock tube and electrical discharge or a combination of them.[1] In such cases, we should consider the effects of compressibility as well as the effects of electromagnetic force. Hence, we use the term electromagnetogasdynamics which is derived from the term gasdynamics and in general both the electric field and the magnetic field effects should be considered. One of the special cases of electromagnetogasdynamics is known as magnetogasdynamics which will be discussed in details later. However, the term Magnetohydrodynamics has been used in such a very loose manner in current practice that it covers all the branches of plasma dynamics. This is not appropriate. The best name should be electromagnetofluid dynamics which is derived from the general term of fluid dynamics including the efffects of both electric and the magnetic fields. Magnetogasdynamics is a very popular branch of the general electromagnetofluid dynamics. For compressible fluid, we may use the term electromagnetogasdynamics which is the title we use for this chapter.

We are going to derive the fundametal equations of electromagnetofluid dynamics in § 2. Since these equations are very complicated, we are going to simplify these equations to obtain several special cases in § 4. Among these special cases, the most popular one is the magnetogasdynamics which will be extensively discussed in the following sections. Some other cases, such as electrogasdynamics, ferrohydrodynamics and electrohydrodynamics will be briefly discussed.

2. Fundamental equations of electromagnetofluid dynamics [1, 3, 5]

We consider the plasma as a single fluid of definite composition. In order to describe the complete flow field for this case, we have to know the following 16 variables :

(i) the velocity vector of the plasma $\boldsymbol{q}$ which has three components in general, i. e., u^i, $i=1$, 2 or 3;

(ii) the temperature of the plasma T;

(iii) the pressure of the plasma p;

(iv) the density of the plasma ρ;

(v) the electric field strength $\boldsymbol{E}$ which has three components E^i;

(vi) the magnetic field strength $\boldsymbol{H}$ which has three components H^i;

(vii) the excess electrical charge ρ_e and

(viii) the electrical current density $\boldsymbol{J}$ which has three components J^i.

In order to find these 16 variables, we have to use 16 relations associated with them as our fundamental equations which are as follows:

(1) The equation of state of the plasma. Plasma may be considered as an ideal gas; hence the simple equation of perfect gas may be used, i. e.,

$$p=\rho RT \tag{15.1}$$

where R is the gas constant of the plasma which depends on the composition of the plasma.

(2) Equation of continuity. The conservation of mass of the plasma gives

$$\frac{\partial \rho}{\partial t}+\nabla \cdot (\rho \boldsymbol{q})=0 \tag{15.2}$$

This is the same as that of ordinary gas [cf Eq. (5.6a)].

(3) Equation of motion. The equation of motion of the plasma is in vector form as follows:

$$\rho \frac{D\boldsymbol{q}}{Dt}=-\nabla p+\nabla \cdot \tau' + \boldsymbol{F}_e \tag{15.3}$$

The viscous stress tensor τ' may be neglected for inviscid flow problem and the only body force $\boldsymbol{F}_e$ which is considered here is the electromagnetic force as follows :

$$\boldsymbol{F}_e=\rho_e \boldsymbol{E}+\boldsymbol{J}\times \boldsymbol{B} \tag{15.4}$$

where $\boldsymbol{B}=\mu_e \boldsymbol{H}$ is the magnetic induction.

(4) Energy equation. The conservation of energy gives

$$\rho \frac{DU_m}{Dt}=-p(\nabla \cdot \mathbf{q})+\Phi+\nabla \cdot (\kappa \nabla T)+\frac{I^2}{\sigma_e} \tag{15.5}$$

where Φ is the viscous dissipation which may be neglected for inviscid flow and the Joue heat is I^2/σ_e where I is the electrical conduction current and σ_e is the electrical conductivity of the gas.

(5) Maxwell's equation of the electromagnetic fields are

$$\nabla\times \boldsymbol{H}=\boldsymbol{J}+\frac{\partial \boldsymbol{D}}{\partial t}=+\boldsymbol{J}+\frac{\partial \varepsilon \boldsymbol{E}}{\partial t} \tag{15.6}$$

$$\nabla\times \boldsymbol{E}=-\frac{\partial \boldsymbol{B}}{\partial t}=-\frac{\partial \mu_e \boldsymbol{H}}{\partial t} \tag{15.7}$$

where $\boldsymbol{D}=\varepsilon\ \boldsymbol{E}$ is the dielectric displacement and $\boldsymbol{B}=\mu_e \boldsymbol{H}$ is the magnetic flux density or magnetic induction, ε is the inductive capacity and μ_e is the magnetic permeability. For Eqs. (15. 6) and (15. 7), the MKS system of units is used so that in free space, we have

$$\mu_e=4\pi\times 10^{-7}\ \frac{\text{kq}\cdot\text{m}}{(\text{coulomb})^2} \tag{15.8}$$

$$\varepsilon=8.854\times 10^{-12}\ \frac{(\text{coulomb})^2\text{sec}^2}{\text{kg}\cdot\text{m}^3} \tag{15.9}$$

The relation between the electric current density $\boldsymbol{J}$ and the electric conduction current $\boldsymbol{I}$ is

$$\boldsymbol{J}=\boldsymbol{I}+\rho_e \boldsymbol{q} \tag{15.10}$$

where $\rho_e \boldsymbol{q}$ is known as the electric convection current and the electric conduction current is due to the relative motion of the charged particles which is a very complicated molecular motion and which will be discussed in Chapter XVI.

(6) Equation of electric current density. The exact equation of the electrical current density $\boldsymbol{J}$ is very complicated because the electric current is due to the complicated motion of all the charged particles, and thus the electric current must depend on the electromagnetic fields as well as all the gasdynamic variables. The exact differential equation of the electric current desnity will be discussed in the multifluid theory section of Chapter XVI. However, if the strength of the magnetic field is not very large and the density of the plasma is not too low, the major influence of the electric current is due to the electromagnetic fields and the so-called generalized Ohm's law may be used as the electric current equation as a first approximation. The generalized Ohm's law is

$$\boldsymbol{I}=\sigma_e(\boldsymbol{E}+\boldsymbol{q}\times \boldsymbol{B})=\sigma_e \boldsymbol{E}_u \tag{15.11}$$

where σ_e is the electrical conductivity of the plasma. In the first approximation, we may assume that σ_e is a scalar quantity. However, if the magnetic field strength is large and the density of the plasma is low, we should consider the electric conductivity as a tensor quantity. In this chapter, we assume that σ_e is a scalar quantity. Another definition of electrical conductivity is that the appearance of conduction current leads to the transformation of electric energy into heat to the amount

$$\text{Joule heat} = \frac{I^2}{\sigma_e} \tag{15.12}$$

(7) Equation of conservation of electric charge

$$\frac{\partial \rho_e}{\partial t} + \nabla \cdot \boldsymbol{J} = 0 \tag{15.13}$$

The fundamental equations (15.1), (15.2), (15.3), (15.5), (15.6), (15.7), (15.11) and (15.13) are known as the classical equations for electro-magneto-gasdynamics. From these equations, we may obtain several simplified equations such as magnetogasdynamic equation in § 4. We are going to solve these fundamental equations with proper boundary conditions of electromagnetogasdynamics.

3. Equations of electromagnetic fields and the boundary conditions of the electromagnetogasdynamics [10]

Since the electromagnetic equations (15. 6), (15. 7), (15. 11) and (15. 13) are new to fluid dynamists, we are going to discuss these equations a little more in details as follows:

In the electromagnetic theory, besides the three fundamental dimensions of mechanics, i. e., length, mass and time, we have another fundamental dimension, an electromagnetic unit. There are a number of different systems of units in electromagnetic theory in common use. This fact complicates the electromagnetic theory. It is often found that the same equations, e. g., the Maxwell's equations (15.6) and (15.7), have different forms in different papers or books. Recently the most common unit system is the MKS unit system which is used in this book and in which the meter is used for unit length, kilogram for unit mass, second for unit time and coulomb for unit of electric charge. For other unit systems, we shall mention in § 10.

In electromagnetic theory, sometimes the vector and scalar potentials are used instead of the electromagnetic fields. These potentials may be defined as follows.

The divergence of Eq. (15.7) gives

$$\frac{\partial}{\partial t}(\nabla \cdot \boldsymbol{B}) = 0 \tag{15.14}$$

or $\nabla \cdot \boldsymbol{B}$=constant at every point in the field. This constant must be zero if ever in its past or future history, the magnetic induction $\boldsymbol{B}$ may vanish. Hence we have

$$\nabla \cdot \boldsymbol{B} = \nabla \cdot (\mu_e \boldsymbol{H}) = 0 \tag{15.15}$$

Because of Eq. (15.15), we may introduce a vector potential $\boldsymbol{A}$ such that

$$\boldsymbol{B} = \nabla \times \boldsymbol{A} \tag{15.16}$$

or

$$\boldsymbol{H} = \frac{1}{\mu_e} (\nabla \times \boldsymbol{A}) \tag{15.16a}$$

with the condition $$\nabla \cdot \boldsymbol{A} = 0 \tag{15.17}$$

Similarly, the divergence of Eq. (15.6) with the help of Eq. (15.13) gives

$$\frac{\partial}{\partial t} (\nabla \cdot \boldsymbol{D} - \rho_e) = 0 \tag{15.18}$$

If in the past or future history, the electric field $\boldsymbol{E}$ and the excess electric charge ρ_e may vanish simultaneously, we have

$$\nabla \cdot \boldsymbol{D} = \nabla \cdot (\varepsilon \boldsymbol{E}) = \rho_e \tag{15.19}$$

Eq. (15.19) is known as the Poisson's equation.

Substituting Eq. (15.16) into (15.7), we have

$$\nabla \times \left(\boldsymbol{E} + \frac{\partial \boldsymbol{A}}{\partial t} \right) = \boldsymbol{0} \tag{15.20}$$

Because of Eq. (15.20), we may introduce a scalar potential φ such that

$$-\nabla \varphi = \boldsymbol{E} + \frac{\partial \boldsymbol{A}}{\partial t} \tag{15.21}$$

Substituting the above relations of these vector and scalar potentials into Eq. (15.6), we have

$$\frac{1}{c^2} \frac{\partial^2 \boldsymbol{A}}{\partial t^2} - \nabla^2 \boldsymbol{A} + \frac{1}{\nu_H} \left[\frac{\partial \boldsymbol{A}}{\partial t} - \boldsymbol{q} \times (\nabla \times \boldsymbol{A}) \right] = -\frac{1}{c^2} \times \left[\frac{\partial \nabla \varphi}{\partial t} + \boldsymbol{q} \nabla^2 \varphi \right] - \frac{1}{\nu_H} \nabla \varphi \tag{15.22}$$

where the generalized Ohm's law (15.11) has been used and

$$\nu_H = \frac{1}{\mu_e \sigma_e} = \text{magnetic diffusivity} \tag{15.23}$$

The Maxwell's equations (15.6), (15.7) and their equivalent equations are valid for those points in whose neighborhood the physical properties of the medium vary continuously. On the boundary of the flow field, the physical properties of the medium may exhibit discontinuities. For instance, at the solid boundary, the electromagnetic properties of the plasma will change to those of a solid. Across such a surface of discontinuity of electromagnetic properties, the following four boundary conditions of electromagnetic fields hold :

(1) The transition of the normal component of the magnetic induction $\boldsymbol{B}=\mu_e \boldsymbol{H}$ is continuous, i.e.,

$$(\boldsymbol{B}_2-\boldsymbol{B}_1) \cdot \boldsymbol{n}=0 \tag{15.24}$$

where $\boldsymbol{n}$ is the unit normal of the surface of discontinuity. Subscripts 1 and 2 refer to the value immediately on each side of the surface of discontinuity.

(2) The behavior of the magnetic field $\boldsymbol{H}$ at the boundary is

$$\boldsymbol{n} \times (\boldsymbol{H}_2 - \boldsymbol{H}_1) = \boldsymbol{J}_s \tag{15.25}$$

where $\boldsymbol{J}_s$ is the surface current density. For finite electrical conductivity $\sigma_e \neq \infty$, $\boldsymbol{J}_s$ is zero; while for infinite electric conductivity $\sigma_e=\infty$, $\boldsymbol{J}_s$ may be different from zero.

(3) The transition of the tangential component of the electric field E is continuous, i. e.,

$$\boldsymbol{n} \times (\boldsymbol{E}_2 - \boldsymbol{E}_1) = 0 \tag{15.26}$$

(4) The bahavior of the dielectric displacement $\boldsymbol{D}=\varepsilon \boldsymbol{E}$ at this boundary is

$$\boldsymbol{n} \cdot (\boldsymbol{D}_2-\boldsymbol{D}_1) = \rho_{es} \tag{15.27}$$

where ρ_{es} is the surface free electric charge density.

For most of our problems of electromagnetogasdynamics, we may neglect the surface current density $\boldsymbol{J}_s$ and the surface free charge density ρ_{es}. Hence our boundary conditions of the electromagnetic fields are that the tangential components of $\boldsymbol{H}$ and $\boldsymbol{E}$ and the normal components of $\boldsymbol{B}$ and $\boldsymbol{D}$ are all continuous across a surface of discontinuity which separates a solid and a fluid or two different fluids. The distinctions between $\boldsymbol{H}$ and $\boldsymbol{B}$ and between $\boldsymbol{D}$ and $\boldsymbol{E}$ should be noticed because the values of μ_e and ε may be different on both sides of the boundary.

Because of the electromagnetic boundary conditions, the flow field for a given configuration depends on the magnetic properties of the body, e. g. whether it is a conducting body or an insulated body.

In electromagnetogasdynamics, besides the boundary conditions of the electromagnetic field, we have also considered the boundary conditions of the gasdynamic variables. The boundary conditions of gasdynamical variables are exactly the same as those of ordinary gasdynamics as we have discussed in Chapter I, §4.

4. Classifications of flow problems of electromagnetofluid dynamics [4, 5]

The electromagnetofluid dynamical equations are too complicated to be used to study any practical flow problems. However, for many practical cases, we may simplify the complete systems of equations of electromagneto-fluid dynamics so that many essential features of the effects of the electromagnetic fields on the flow field may be easily and clearly brought out. According to the importance of the electromagnetic variables, we have four special cases of electromagnetofluid dynamics which will be discussed as follows:

(i) Magnetogas dynamics[1,3]

The most popular subject of the electromagnetofluid dynamics is the magnetofluid dynamics which includes the magneto-hydrodynamics and the magnetogasdynamics. In magnetofluid dynamics, the magnetic field is the most important electromagnetic variable. For such flow problems, the following magneto-fluid dynamic approximations are satisfied:

(1) The time scale t_0 for this problem is of the same order of magnitude as L/U where L is the characteristic length and U is the characteristic velocity of the flow field. Hence, the time parameter

$$R_t = \frac{t_0}{L/U} \tag{15.28}$$

is of the order of unity. It means that we shall not consider the phenomena of very high frequency with the time scale t_0 being very small.

(2) The electric field strength characterized by a value $\boldsymbol{E}_0$ is of the same order of magnitude as the induced electric field $\boldsymbol{q} \times \boldsymbol{B}$ or UB_0. Hence, the electric field parameter

$$R_E = \frac{E_0}{B_0 U} \tag{15.29}$$

is of the order of unity, where B_0 is a characteristic magnetic induction. This is a good approximation for very large electric conductivity σ_e, because as σ_e approaches infinity we expect that

$$\boldsymbol{E} = -\boldsymbol{q} \times \boldsymbol{B} \tag{15.30}$$

Otherwise, the electric conduction current $\boldsymbol{I}$ would become very large for a slight motion of the electrically conducting fluid.

(3) The velocity of the flow of the fluid is much small than the velocity of light $c = 1/(\mu_e \varepsilon)^{1/2}$, i. e., the relativistic parameter

$$R_C = U^2/c^2 \tag{15.31}$$

is much smaller than unity.

(4) Both the electric and magnetic polarization, i. e., $\boldsymbol{P}$ and $\boldsymbol{M}$ are negligible. The electric polarization $\boldsymbol{P}$ is defined by the relation :

$$\boldsymbol{D} = \varepsilon \boldsymbol{E} = \varepsilon_0 \boldsymbol{E} + \boldsymbol{P} \tag{15.32}$$

where ε_0 is the value of ε in the free space.

The magnetic polariztation $\boldsymbol{M}$ is defined by the relation

$$\boldsymbol{B}=\mu_e \boldsymbol{H}=\mu_{e0}(\boldsymbol{H}+\boldsymbol{M}) \tag{15.33}$$

where μ_{e0} is the value of μ_e in free space.

In this approximation, we use the values of μ_e and ε in free space in our fundamental equations of electromagnetofluid dynamics.

Under the above four approximations, the displacement current $\partial D/\partial t$ and the excess electric charge ρ_e are negligible in comparison with terms with magnetic field $\boldsymbol{H}$ in the fundamental equations of electromagneto-fluid dynamics. For instance, the electromagnetic force terms have the relation :

$$\rho_e \boldsymbol{E}=R_c \boldsymbol{J}\times\boldsymbol{B} \tag{15.34}$$

Hence $\rho_e \boldsymbol{E}$ is negligible in comparison with $\boldsymbol{J}\times\boldsymbol{B}$. In fact, we may neglect all the terms with ρ_e and the displacement current in the fundamental equations. The electrical current density and the electric field strength may be expressed in terms of the magnetic field as follows:

$$\rho_e=0,\ \partial D/\partial t=0,\quad \boldsymbol{J}=\boldsymbol{I}=\nabla\times\boldsymbol{H} \tag{15.35}$$

$$\boldsymbol{E}=\frac{\boldsymbol{J}}{\sigma_e}-\boldsymbol{q}\times\boldsymbol{B}=\frac{1}{\sigma_e}\nabla\times\boldsymbol{H}-\boldsymbol{q}\times\boldsymbol{B} \tag{15.36}$$

As a result, we need to consider the interaction between the magnetic field strength $\boldsymbol{H}$ and the fluid dynamic variables only. It is the reason why we call this subject as magnetofluid dynamics, or magnetohydrodynamics, or magnetogasdynamics. The ten electromagnetic equations of § 2 may be replaced by the vector equation for the magnetic field $\boldsymbol{H}$ as follows :

$$\frac{\partial \boldsymbol{H}}{\partial t}=\nabla\times(\boldsymbol{q}\times\boldsymbol{H})-\nabla\times[\nu_{H}(\nabla\times\boldsymbol{H})] \tag{15.37}$$

Furthermore, the electromagnetic variables may be expressed in terms of the magnetic field by the help of Eqs. (15.35) and (15.36). Thus the equation of motion in magnetofluid dynamics becomes

$$\rho\frac{D\boldsymbol{q}}{Dt}-(\boldsymbol{B}\cdot\nabla)\boldsymbol{H}=-\nabla(p+\frac{1}{2}\boldsymbol{B}\cdot\boldsymbol{H})\nabla\tau' \tag{15.38}$$

and similarly the energy equation in the magnetofluid dynamics becomes

$$\begin{aligned}\rho\frac{Dh_0}{Dt}=\frac{\partial p}{\partial t}+\nabla\cdot(\boldsymbol{q}\tau')+\nabla\cdot(K\nabla T)+\\(\nabla\times\boldsymbol{H})\cdot(\nu_{H}\nabla\times\boldsymbol{B}-\boldsymbol{q}\times\boldsymbol{B})\end{aligned} \tag{15.39}$$

where $h_0=U_m+RT+\frac{1}{2}q^2$ is the stagnation enthalpy of the gas. Eqs. (15.37), (15.38) and (15.39) together with Eqs. (15.1) and (15.2) are the fundamental equations of magnetogasdynamics in which we have nine variables: $\boldsymbol{H}$, p, ρ, T and $\boldsymbol{q}$. The new important parameters of magnetogasdynamics are :

The magnetic Reynolds number R_σ defined as

$$R_\sigma = \frac{UL}{\nu_H} \tag{15.40}$$

For a large magnetic Reynolds number, the magnetic field would be influenced greatly by the flow field velocity while for small magnetic Reynolds number it would not be affected significantly by the flow velocity.

The magnetic pressure number R_H defined as

$$R_H = \frac{\text{magnetic pressure}}{\text{dynamic pressure}} = \frac{B_0 H_0}{\rho_0 U^2} = \frac{V_H^2}{U^2} = \frac{1}{M_m^2} \tag{15.41}$$

where

$$V_H = \left(\frac{B_0 H_0}{\rho_0} \right)^{1/2} = \text{speed of Alfven's wave.} \tag{15.42}$$

$M_m = U/V_H$ is known as the magnetic Mach number, $B_0 = \mu_e H_0$ is the reference magnetic induction and ρ_0 is the reference density of the gas. The Alfven's wave speed V_H is a speed pf propagation of small disturbance in magnetogasdynamics due to the effect of magnetic field and the inertial force which was first found by Alfven. When R_H is large, the velocity field would be influenced significantly by the magnetic field, and for very small R_H, the velocity field would not be influenced significantly by the magnetic field.

It is interesting to notice that there is an analogy between the magnetic field of magnetogasdynamics and the vorticity in ordinary gasdynamics. If we assume that the electrical conductivity is a constant, the equations which govern the magnetic field $\boldsymbol{H}$ in magnetogasdynamics are as follows:

$$\nabla \cdot \boldsymbol{H} = 0 \tag{15.43a}$$

$$\frac{\partial \boldsymbol{H}}{\partial t} + (\boldsymbol{q} \cdot \nabla)\boldsymbol{H} - (\boldsymbol{H} \cdot \nabla)\boldsymbol{q} + \boldsymbol{H}(\nabla \cdot \boldsymbol{q}) = \nu_H \nabla^2 \boldsymbol{H} \tag{15.43b}$$

Eqs. (15. 43) are formally identical to the equations which govern the vorticity $\boldsymbol{\omega} = \nabla \times \boldsymbol{q}$ in ordinary gasdynamics if we assume that the coefficient of viscosity μ is a constant. The equations which govern the vorticity $\boldsymbol{\omega}$ are

$$\nabla \cdot \boldsymbol{\omega} = 0 \tag{15.44a}$$

$$\frac{\partial \boldsymbol{\omega}}{\partial t} + (\boldsymbol{q} \cdot \nabla)\boldsymbol{\omega} - (\boldsymbol{\omega} \cdot \nabla)\boldsymbol{q} + \boldsymbol{\omega}(\nabla \cdot \boldsymbol{q}) = \nu \nabla^2 \boldsymbol{\omega} \tag{15.44b}$$

where $\nu = \mu/\rho$. In Eqs. (15.44), we assume that the fluid is baratropic, i.e., the pressure is a function of density only. [cf. Eq. (5.47)].

Since Eqs. (15.43) and (15.44) are identical in form, wẹ may apply all the well known theorems of vorticity in a baratropic gas of ordinary gas dynamics to the magnetic field strength in magnetogasdynamics. Particularly, by Helmholtz's theorem, we know that vortex lines move with the fluid if $v=0$ (Chapter V, § 8). Hence, we may also show that the lines of magnetic force move with the fluid if $v_H=0$, i. e, when $\sigma_e=\infty$.

We are going to discuss (i) one-dimensional steady flow (ii) waves of small amplitude and the shock waves in Magnetogasdynamics in sections 5 to 7.

(ii) Electrogasdynamics[11, 12]

In magnetogasdynamics, the electric conduction current is much larger than the electric convection current, and the number of ions is about the same as that of electrons so that the excess electric charge is approximately zero. There is another limiting case in which the electrically conducting fluid consists of essentially one kind of charges particles, e. g. electrons alone. The study of the motion of electron gas has been made by many authors from the point of view of electronics, but only recently by authors from a gasdynamic point of view. Such a study would give us many interesting new phenomena about the motion of charged particles. In this case, the electric field strength E is more important than the magnetic field. Hence we use the name of electrogasdynamics in which we study the interaction between the gasdynamic variables and the electric field and excess electric charge. The fundamental equations of electrogasdynamics may be derived from the fundamental equations of electromagneto-fluid dynamics by the following electrogasdynamic approximations:

In the present case, the electric field strength is very large and the excess electric charge is far from zero. For a gas of one kind of charged particles only, the excess electric charge is simply the electric charge density of the gas, i. e., $\rho_e=Zen$, where n is the number density of the charged particles and Ze is the electric charge of each charged particles in the gas. For electron gas $Z=-1$ and for gas consisting of singly positive charged ions only, $Z=1$. The electric current density is then $\boldsymbol{J}=\rho_e\boldsymbol{q}$. From Eq. (15.19) with constant ε, the order of magnitude of the electric field strength E is

$$E\cong\rho_e L/\varepsilon \tag{15.45}$$

The order of magnitude of the electric current density is

$$J\cong\rho_e U \tag{15.46}$$

If we assume that the time parameter R_t of Eq. (15.28) is of the order of unity, by Eq. (15.6), the magnetic field strength has the following order of magnitude :

$$H\cong\rho_e UL \tag{15.47}$$

From the relations (15.45) to (15.47), it is easy to show that those terms with a magnetic field H or B in the fundamental equations of electromagneto-fluid dynamics are negligibly small in comparison with those with excess electric charge ρ_e. For instance:

$$\rho_e \boldsymbol{E} = \left(\frac{c^2}{U^2} \right) \boldsymbol{J} \times \boldsymbol{B} \tag{15.48}$$

because c^2/U^2 is much large than unity, $\boldsymbol{J} \times \boldsymbol{B}$ is negligibly small in the equations of motion of electrogasdynamics. The fundamental equations of electrogasdynamics which contain the ten unknowns p, ρ, T, $\boldsymbol{q}$, E and ρ_e are as follows:

$$P = \rho RT \tag{15.49a}$$

$$\frac{\partial \rho}{\partial t} + \nabla \cdot (\rho \boldsymbol{q}) = 0 \tag{15.49b}$$

$$\rho \frac{D\boldsymbol{q}}{Dt} = -\nabla p + \nabla \cdot \tau' + \rho_e \boldsymbol{E} \tag{15.49c}$$

$$\rho \frac{DU_m}{Dt} = p(\nabla \cdot \boldsymbol{q}) + \Phi + \nabla \cdot (K \nabla T) \tag{15.49d}$$

$$\nabla \times \boldsymbol{E} = \boldsymbol{0} \tag{15.49e}$$

$$\nabla \cdot \boldsymbol{E} = \rho_e / \varepsilon \tag{15.49f}$$

There are some interesting practical applications of electrogasdynamics such as electrogasdynamic power generation and propulsion device. We are going to discuss in details the one dimensional steady flow of electrogasdynamics in Chapter XVI, §6 which is useful in the study of electrogasdynamic power generation.

(iii) Ferrohydrodynamics[13, 14]

In sections (i) and (ii), we neglect the electric and the magnetic polarizations. In order words, we assume that the inductive capacity ε and the magnetic permeability μ_e are constants equal to their values in the free space. There are many interesting practical problems in which the electric or the magnetic polarization is not negligible. In this section, we consider the case in which the magnetic polarization is important, while in the next section, we consider where the electric polarization is important.

Magnetically responsive fluids are composed of very small particles dispersed in a liquid carrier. These particles which are magnetics are dispersed homogeneously in the liquid under all the flow conditions. We thus have a two-phase flow of solid magnetic particles and liquid. Under a magnetic field gradient, a magnetic force is generated by the particles. Even though the liquid may be considered as an incompressible fluid, the present problem is greatly influenced by the heat transfer and we have to study simultaneously the fluid dynamics with heat transfer which may be considered as a special case of compressible flow. We are going to discuss this problem in details as follows:

For simplicity, we assume that the small magnetic particles are mixed homogeneously with a gas. Furthermore, we assume that the slip between the gas and the magnetic particles is negligibly small and the temperature of the particles is equal to that of the gas. Thus we may treat the flow problem by the single-fluid theory instead of the more general two-phase flow problem as we shall discuss in Chapter XVI. The only new point in this problem is the body force due to magnetization which is

$$\boldsymbol{F}_m = \mu_{eo}(\boldsymbol{M} \cdot \nabla)\boldsymbol{H} \tag{15.50}$$

First we study the thermodynamics of the mixture of the gas and the magnetic particles as a whole. The new parameter is that the thermodynamic state as well as other thermodynamic variables such as temperature T, depends on the magnetic field strength H. For instance, we may consider the magnetization M as a function of both temperature T and the magnetic field H, i.e.,

$$M = M(T, H) \tag{15.51}$$

One of the simple expressions for the magnetization M of Eq. (15.51) is the following linear state:

$$M = M^* - K(T - T^*) \tag{15.52}$$

where M^* and T^* are reference values of M and T respectively. The factor $K = \partial M/\partial T$ is known as the pyromagnetic coefficient which may be assumed as a constant in the present case. The first law of thermodynamics, i.e., the conservation of energy of this magnetic mixture may be written as follows :

$$dQ = dU_m + dW_R + dW_{ir} \tag{15.53}$$

where U_m is the internal energy of the magnetized mixture, dW_R is the reversible work on the mixture which may be given by the formula

$$dW_R = -\mu_{eo}\ H\ dM \tag{15.54}$$

and dQ is the heat addition to the mixture which consists of two parts, the reversible part dQ_r and the irreversible part dQ_{ir}, i. e.,

$$dQ = dQ_r + dQ_{ir} \tag{15.55}$$

For the reversible part of heat addition, we may write

$$dQ_r = c(T, H)dT + g(T, H)dH \tag{15.56}$$

The irreversible work done is dW_{ir} which is equal to the irreversible heat addition dQ_{ir}, i. e.,

$$dQ_{ir}=dW_{ir} \tag{15.57}$$

The coeffieients c and g may be determined from the thermodynamic relations. The internal energy dU_m is an exact differential so that

$$dU_m=dQ_r-dW_R=\left(c+\mu_{eo}H\frac{\partial M}{\partial T}\right)dT+\left(g+\mu_{eo}H\frac{\partial M}{\partial H}\right)dH \tag{15.58}$$

is an exact differential and we have

$$\left(\frac{\partial c}{\partial H}\right)_T=-\mu_{eo}\left(\frac{\partial M}{\partial T}\right)_H+\left(\frac{\partial g}{\partial T}\right)_H \tag{15.59}$$

According to the second law of thermodynamics, the entropy dS is given by the following relation :

$$\frac{dS}{T}=\frac{dQ_r}{T}=\frac{c}{T}dT+\frac{g}{T}dH \tag{15.60}$$

is an exact differential, and we have

$$\left(\frac{\partial c}{\partial H}\right)_T=\left(\frac{\partial g}{\partial T}\right)_H-\frac{g}{T} \tag{15.61}$$

From Eqs. (15.59) and (15.61), we have

$$g=\mu_{eo}T\left(\frac{\partial M}{\partial T}\right)_H \tag{15.62}$$

and

$$\left(\frac{\partial c}{\partial H}\right)_T=\mu_{eo}T\left(\frac{\partial^2 M}{\partial T^2}\right)_H \tag{15.63}$$

For the linear state (15.52), we have simply $c=c(T)$ and $g=-\mu_{eo}TK$.

Now we are going to derive the fundamental equations of ferrohydrodynamics. In general, for ferro-fluid dynamics, we should have consider nine variables; the flow velocity vector $\boldsymbol{q}$, the density ρ, the pressure p, the temperature T of the mixture and the magnetization M; we have to find nine fundamental equations. They are similar to those of magnetogasdynamics but the magnetization M replaces the magnetic field H. If we derive the fundamental equations of magnetogasdynamics including the magnetic polarization, those equations are applicable for ferrohydrodynamics too. In order to show the essential features of ferrohydrodynamics, we consider the case of incompressible fluid with ρ=constant. We have eight variables and the fundamental equations are as follows :

(a) Equation of continuity

$$\nabla \cdot \boldsymbol{q}=0 \tag{15.64}$$

(b) The equation of motion

$$\rho \frac{D\boldsymbol{q}}{Dt} = -\nabla p+ \mu_{eo} (\boldsymbol{M} \cdot \nabla) \boldsymbol{H}+ \boldsymbol{F}_b+ \mu\nabla^2\boldsymbol{q} \tag{15.65}$$

where $\boldsymbol{F}_b$ is the body force other than the magnetic force and μ is the coefficient of viscosity of the fluid. For our present case, the magnetic force is given by Eq. (15. 50) which was neglected in our fundamental equations of magnetogasdynamics of section (i).

(c) Equation of energy

$$c(T) \frac{DT}{Dt} = \mu_{eo} KT \frac{DH}{Dt} + \nabla \cdot (\kappa\nabla T) + \Phi \tag{15.66}$$

where κ is the coefficient of heat conductivity of the fluid and is the viscous dissipation. The new term is the internal energy due to the magnetization which is the first term on the right of Eq. (15.66).

(d) Equation of magnetization

Similar to the magnetogasdynamics, we may neglect the displacement current, and the Maxwell's equations of electromagnetic fields may be written as follows :

$$\nabla \cdot \boldsymbol{B}=0 \tag{15.67}$$

and

$$\nabla \times \boldsymbol{H}=\boldsymbol{J} \tag{15.68}$$

we assume that the mixture is non-electrically conducting so that $\boldsymbol{J}=\boldsymbol{0}$. The magnetic field $\boldsymbol{H}$ consists of two parts : one is the applied magnetic field $\boldsymbol{H}_0$ and the orher is the induced magnetic field $\boldsymbol{H}_i$, i. e.,

$$\boldsymbol{H}=\boldsymbol{H}_0+ \boldsymbol{H}_i \tag{15.69}$$

Since $\boldsymbol{J}=\boldsymbol{0}$, we have $\nabla \times \boldsymbol{H}_i=\boldsymbol{0}$ and we may introduce a scalar potential φ_m such that

$$\boldsymbol{H}_i= -\nabla\varphi_m \tag{15.70}$$

From Eq. (15.67) with $\nabla \times \boldsymbol{H}_0=\boldsymbol{0}$, we have then

$$\nabla^2\varphi_m=\nabla \cdot \boldsymbol{M} \tag{15.71}$$

Hence, our fundamental equations are Eqs. (15.64), (15.65), (15.66) and (15.71) for the unknowns p, $\boldsymbol{q}$, T and φ_m with a given $M(T, H)$. This set of equaiton is similar to that of ordinary gasdynamics. We may derive similar expressions from this set of equations for ferrohydrodynamics as those of ordinary hydrodynamics. For instance, we have the Bernoulli's equation for ferrohydrodynamics as follows :

$$\frac{\partial \varphi_V}{\partial t} + 1/2\, q^2 + \frac{p}{\rho} + \frac{\psi}{\rho} - \frac{\psi_m}{\rho} = f\,(t) \qquad (15.72)$$

where φ_V is the velocity potential, i. e., $\boldsymbol{q} = \nabla\varphi_V$, ψ is the potential energy for those body forces such as gravitational force and ψ_m is the magnetic scalar potential so that

$$\mu_{eo}(M \cdot \nabla) H = \nabla\psi_m \qquad (15.73)$$

The vorticity $\boldsymbol{\omega}$ equation is

$$\frac{D\boldsymbol{\omega}}{Dt} = (\boldsymbol{\omega} \cdot \nabla)\boldsymbol{q} + \frac{\mu_{eo}}{\rho} \frac{\partial M}{\partial T} \nabla T \times \nabla H + \frac{\mu}{\rho} \nabla^2 \boldsymbol{\omega} \qquad (15.74)$$

Hence we have the extra-term $\frac{\mu_{eo}}{\rho} \frac{\partial M}{\partial T} \nabla T \times \nabla H$ which may produce vorticity in the flow field and which is a new term for ferrohydrodynamics.

In Ref. 13, some examples are given for flow of ferrohydrodynamics such as equilibrium of a free surface due to the effects of magnetization and energy conversion and two-dimensional source flow of a saturated ferrofluid with heat addition. In Ref. 14, it has been discussed the power produced by cycling a ferromagnetic material thermally througha range of temperature such that its magnetization changes appreciably and of utilizing the change of magnetization and its interaction with a magnetic field.

(iv) Electrohydrodynamics (Electric polarization)[15]

Even though for most fluids the inductive capacity ε may be considered as a scalar quantity or a constant, when we study the two-phase flow (Chapter XVI) between two different kinds of fluid, the value of inductive capacity changes abruptly from one value to another. Hence near the interface, the electric polarization may not be neglected. For such flow problems, the basic equations are the same as those given for electrogasdynamics of section (ii) but the diselectrical displacement $\boldsymbol{D}$ is not equal to $\varepsilon_0 \boldsymbol{E}$($\varepsilon_0$ is the inductive capacity in the free space). Hence, we should consider the electric polarization of Eq. (15.32). We are going to discuss a few problems of two-phase flows with electric polarization in Chapter XVI. There is no special name for this type of two-phase flow. Since it associates with a liquid and a gas or two liquids, we use the conventional name Electrohydrodynamics for such problems, even though the problems discussed in section (ii) for incompressible fluid may also be called electrohydrodynamics. Probably, the name of flow with electrical polarization may be used for the present flow problem.

5. One-dimensional steady flow of electromagnetogasdynamics (EMGD)[16,17,18]

We have studied the one-dimensional flow of an inviscid, compressible but non-electrically conducting fluid in Chapter III. We may extend those analysis into one-dimensional flow of electromagnetogasdynamics. However, there are some interesting complication occurred in the case of electromagnetogasdynamics. We have two distinct types of analyses of one-dimensional flow problem of EMGD. One is refered to as approximately one-dimensional flow and the other as strictly one-dimensional flow. The main difference in these analyses is in the expression of the electrical current density. In order to show these differences we derive exactly the one-dimensional flow equations by the quasi-one-dimensional analysis as follows[16].

In the quasi-one-dimensional flow analysis, we use the average value of EMGD variables over the slowly varying cross-sectional area of a pipe or a channel in which the electrically conducting fluid flows under transverse electromagnetic fields and an axial pressure gradient. Let the cross-sectional area $A(x)$ be

$$A(x) = \iint dy\, dz \tag{15.75}$$

at a fixed distance x along the axis of the pipe or a channel. The average value of any EMGD variables $Q(x, y, z, t)$ is defined by the formula

$$\bar{Q}(x, t) = \frac{1}{A(x)} \iint_{A(x)} Q(x, y, z, t)\, dy\, dz \tag{15.76}$$

The EMGD variable $Q(x, y, z, t)$, such as the x-wise velocity u, the pressure p, etc., is in general a function of all three spacial coordinates x, y and z and time t. But the average value of $\bar{Q}(x, t)$ is in general a function of the axial coordinate x and time t only.

The average value of the product of two quantities f and g is in general not equal to the product of the average values of the two quantities. The difference between these two averages is known as the covariance of f and g, i.e.,

$$\overline{fg} - \bar{f}\,\bar{g} = \operatorname{cov}(f, g) \tag{15.77}$$

where bar refers to the average value of a quantity defined by Eq.(15.76).

Now we will derive the basic equations of the quasi-one-dimensional flow by integrating the fundamental equations of EMGD over a cross-section $A(x) = b_z b(x)$ at a fixed value of x where the width of the channel in the z-direction is assumed to be a constant b_z for simplicity and the width of the channel in the y-direction is assumed as a function of x, $b(x)$.

The equation of continuity (15.2) in Cartesian coordinates is

$$\frac{\partial \rho}{\partial t} + \frac{\partial \rho u}{\partial x} + \frac{\partial \rho v}{\partial y} + \frac{\partial \rho w}{\partial z} = 0 \tag{15.2a}$$

If we integrate Eq. (15.2a) over the cross section $A(x)$ or Eq. (15.75) and use the definition of the average value (15.76) and the boundary conditions that $v=0$ and $w=0$ on the surface of the channel, we have

$$\frac{\partial A\bar{\rho}}{\partial t} + \frac{\partial A\overline{\rho u}}{\partial x} = 0 \tag{15.78}$$

or

$$\frac{\partial \bar{\rho}}{\partial t} + \bar{u}\frac{\partial \bar{\rho}}{\partial x} + \bar{\rho}\frac{\partial \bar{u}}{\partial x} + \frac{\partial}{\partial x}[\text{cov}(\rho, u)] + [\bar{\rho}\,\bar{u} + \text{cov}(\rho, u)]\frac{1}{A}\frac{dA}{dx} = 0 \tag{15.79}$$

In ordinary quasi-one-dimensional analysis, the cov(ρ, u) is usually neglected.

For steady flow , Eq. (15.79) gives

$$A\overline{\rho u} = A\,\bar{\rho}\,\bar{u} + A\,\text{cov}(\rho, u) = \text{constant} = m \tag{15.80}$$

Since it is inconvenient to use Eq, (15.79) or (15.80) because we do not know the functional variation of cov (ρ, u), a non-dimensional covariance coefficient k_ρ may be introduced in the expression $\overline{\rho u}$ such that

$$\overline{\rho u} = k_\rho\,\bar{\rho}\,\bar{u} \tag{15.81}$$

The factor k_ρ depends on the shape of the channel and its operating conditions and should be determined experimentally. However, for a given channel at a given operating condition, we may assume that k_ρ is a constant. If one of the variable ρ or u is a constant, k_ρ is unity.

Fig. 15.1 shows a rough estimate of k_ρ and cov(ρ, u) by assuming $u^* = 1 - y^{*2}$ and $\rho^* = [1 - 0.2M^2(1 - y^{*2})^2]^{2.5}$ where u^* and ρ^* are respectively the non-dimensional velocity and density with respect to their values at the center of the channel and y^* is the non dimensional y-distance and M is the Mach number of the average flow velocity. Fig. 15.1 shows that k_ρ may differ significantly from unity and that the value k_ρ depends not only on the velocity profile but also on the Mach number M which represents the operating conditions.

Similarly we may apply the average process to the equation of state, equation of motion, equation of energy and electromagnetic equations with the proper definitions of various covariance factors to obtain the accurate quasi-one-dimensional equations of EMGD as follows:[2]

$$\frac{\partial \rho A}{\partial t} + \frac{\partial \rho u k_\rho A}{\partial x} = 0 \tag{Ia}$$

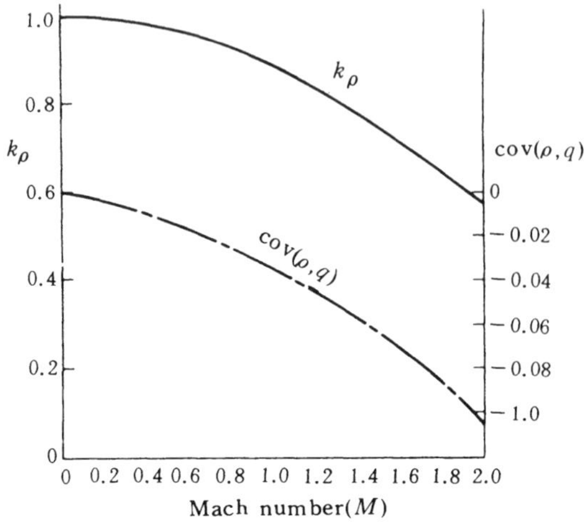

Fig. 15.1 Covariance of density and velocity cov(ρ, u) and its corresponding factor k_ρ

$$p = k_p \, \rho \, RT \tag{Ib}$$

$$\rho \frac{\partial k_\rho u}{\partial t} + \rho u \frac{\partial k_u u}{\partial x} = -\frac{\partial p}{\partial x} - k_\sigma \sigma_e B_y \, (k_E E_Z + k_B u B_y) + \frac{C_f}{2b} \rho u^2 \tag{Ic}$$

$$\rho \frac{\partial k_e U_m}{\partial t} + \rho u \frac{\partial k_e U_m}{\partial x} + \frac{p}{A} \frac{\partial k_{pu} A u}{\partial x} = \frac{1}{b} Q_c - k_\tau \frac{C_f}{2b} \rho u^3$$
$$+ k_{uB} \sigma_e u B_y \, (k_E E_Z + k_B u B_y) + k_{EJ} k_{\sigma e} \, (E_z + k_j \, u B_y) \tag{Id}$$

$$\frac{1}{A} \frac{\partial A H_y}{\partial x} = k_o \sigma_e \, (E_z + k_J \, u \, B_y) \tag{Ie}$$

$$A \frac{\partial B_y}{\partial t} = \frac{\partial}{\partial x} (A E_z) \tag{If}$$

where all the variables refer to their mean values and we drop the bar for the mean value from now on. The factors k_ρ, k_o etc. are the corresponding covariance factors which are given in details in Reference 3. Since we know very little about these covariance factors, we shall from now then assume that all the covariance factors are unity except k_0 which is the covariance factor for electrical conductivity which determines whether the flow is a strictly one-dimensional flow or approximate one dimensional flow. If $k_0 = 0$, we have the approximately one-dimensional flow which corresponds to the case of very small electrically conductivity; and if $k_0 = 1$, we have the strictly one-dimensional flow which corresponds to very large electrical conductivity. We are going to discuss two limiting cases of these quasi-one dimensional flows as follows :

(i) Strictly one dimensional steady magnetogasdynamic flow. $k_0 = 1$. We shall consider only the inviscid and non-heat conducting flow so that the friction coefficient $c_f = 0$ and the heat flux $Q_c = 0$. For simplicity, we consider a constant area channel. $A =$ constant. If $\sigma_e = \infty$, we have the frozen flow relation $H_y/\rho=$constant and an effective sound speed a_e :

$$a_e=\sqrt{a^2+\frac{\mu_e H_y^2}{\rho}} \tag{15.82}$$

is the characteristic speed of the flow and "a" is the ordinary sound speed, i.e, $a^2=\gamma RT$.

(ii) Gasdynamic approximation

For many practical applications, the electrical conductivity σ_e is small, i.e. $\sigma_e \rightarrow 0$, which corresponds to take $k_0=0$, we have

$$AH_y=\text{constant};\ AE_z=\text{constant} \tag{15.83}$$

Since both the eletric field E_z and the magnetic field H_y are independent of the flow field, we may assume that both E_z and H_y are given functions of x. Hence we need to solve the gasdynamic varialbes u, ρ, p and T with given functions of $A(x)$, $E_z(x)$ and $H_y(x)$ from the following basic equations of approximately quasi-one dimensional inviscid and non-heat-conducting fluid but the fluid is electrically conducting as follows :

We consider now the steady flow case and have

$$A\rho u=\text{constant} \tag{15.84}$$

$$p=\rho RT \tag{15.85}$$

$$u\frac{du}{dx}=-\frac{dp}{dx}-\sigma_e B_y(E_z+uB_y) \tag{15.86}$$

$$u\frac{dU_m}{dx}+\frac{p}{A}\frac{duA}{dx}=\sigma_e(E_z+uB_y)^2 \tag{15.87}$$

The system of Eqs. (15.84) to (15.87) may be reduced to two firstorder, non-linear differential equations of the variable u and Mach number M as follows :

$$\frac{du}{dx}=\frac{1}{M^2-1}\left\{\frac{u}{A}\frac{dA}{dx}-\frac{\sigma_e B_y^2}{p}\left(u+\frac{E_z}{B_y}\right)\left(u+\frac{\gamma-1}{\gamma}\frac{E_z}{B_y}\right)\right\}$$

$$=\frac{1}{M^2-1}F_1(u, M, x) \tag{15.88}$$

$$\frac{dM}{dx}=\frac{1}{M^2-1}\left\{\left(1+\frac{\gamma-1}{2}M^2\right)\frac{M}{A}\frac{dA}{dx}-\left(1+\frac{\gamma-1}{2}M^2\right)\frac{\sigma_e B_y^2 M}{up}\right.$$

$$\left.\times\left(u+\frac{E_z}{B_y}\right)\left[u+\frac{(1+\gamma M^2)(\gamma-1)E_z}{[2+(\gamma-1)M^2]\gamma B_y}\right]\right\}=\frac{1}{M^2-1}F_2(u, M, x) \tag{15.89}$$

where the functions $F_1(u, M, x)$ and $F_2(u, M, x)$ are the abbreviations of the corresponding expressions in the curl brackets. Eqs. (15.88) and (15.89) illustrate that there are three characteristic velocities which show the effects of electromagnetic fields and which are

$$u_1 = -\frac{\gamma-1}{\gamma}\frac{E_z}{B_y}\ ;\ u_2 = \frac{1+\gamma M^2}{2+(\gamma-1)M^2}\,u_1;\ u_3 = \frac{\gamma}{\gamma-1}\,u_1 \qquad (15.90)$$

It is not easy to integrate Eqs. (15.88) and (15.89) from a given initial condition for given values of $A(x)$, $E_z(x)$ and $B_y(x)$. In general, numerical method are required. However, for special cases, simplifications may be made and these equations are easily integrated. Now we are going to consider two special cases as follows :

(a) Constant area channel. A =constant.

For this case, Eqs. (15.88) and (15.89) become respectively:

$$\frac{du}{dx} = -\frac{\sigma_e B_y^2}{(M^2-1)p}(u-u_1)(u-u_3) \qquad (15.91)$$

and

$$\frac{dM}{dx} = -\frac{1+[1/2(\gamma-1)]M^2}{M^2-1}\frac{\sigma_e B_y^2 M}{up}(u-u_2)(u-u_3) \qquad (15.92)$$

These equations show that the characteristics of the flow depend on the initial value of the velocity u_0 relative to the characteristic velocities u_1, u_2 and u_3. For instance, if we take $M_0 < 1$, $u_0 < u_2 < u_1 < u_3$, both du/dx and dM/dx are positive initially. This is the case of magnetogasdynamic acceleration. If the initial Mach number M_0 is sufficiantly close to unity, the acceleration will soon become infinite and the channel is choked. For other initial conditions, the velocity u may increase monotonically toward u_1 while the Mach number M first increases, then decreases ($u > u_2$) and finally approaches an asymptotic value $M_a < 1$. The third possibility is the case of a particular set of initial conditions that M reaches unity and u reaches u_1 simultaneously. We have a smooth acceleration from a subsonic to a supersonic flow in a nozzle of constant cross-sectional area which is not possible in ordinary gasdynamics.

A summary of various possibilities for constant area channel flow provided by Sears and Resler, [19] is presented in Fig. 15.2, where the signs of du/dx and dM/dx in a $u-M$ plane are indicated. There are eight distinct regions in the u-M plane in which different behaviors of $u(x)$ and $M(x)$ can be predicted from Eqs. (15.91) and (15.92). These are identified in Fig. 15.2 as follows:

(I) $M > 1$
- (A) $u_3 < u$
- (B) $u_2 < u < u_3$
- (C) $u_1 < u < u_2$
- (D) $u < u_1$

(II) $M < 1$
- (A) $u_3 < u$
- (B) $u_1 < u < u_3$
- (C) $u_2 < u < u_1$
- (D) $u < u_2$

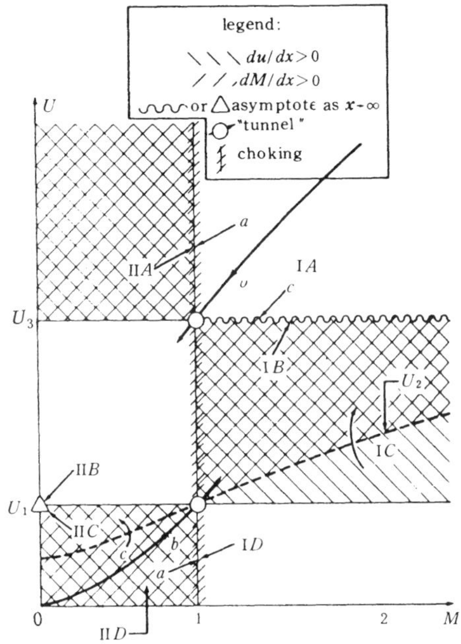

Fig. 15.2 Magnetogas dynamic acceleration of one-dimensional flow in a constant area channel (E_z/B_y =constant)

In general, a "barrier" exists at $M=1$, where the phenomenon of choking occurs when the denominators of both Eqs. (15.91) and (15.92) disappear. This represents choking since the curve of u versus x turn back upon itself and becomes double valued after reaching the barrier. Thus the sonic speed can never actually be attained in such a channel, and the phenomenon would have to occur, if at all at the open end of a channel.

There are two exceptional cases, denoted "tunnels" in Fig. 15.2. These are two one-way tunnels through the barrier at sonic speed. For these special combinations of parameters (IAb) and (IIDb) the numerators and denominators of both Eqs. (15.91) and (15.92) vanish simultaneously, and finite and continuous deceleration or acceleration, respectively occur. The curves that lead to these tunnels have been plotted in Fig. 15.2 as they separate choking from non-choking channels and represent the special combinations that permit smooth passage through the sonic speed.

(b) Channel of variable cross-section $A(x)$ but with zero electric fiels. $E_z=0$.

In this case, we have

$$\frac{dM}{du}=\frac{F_2(u,M,x)}{F_1(u,M,x)}=\frac{M}{u}\left(1+\frac{\gamma-1}{2}M^2\right) \tag{15.93}$$

Integration of Eq. (15.93) gives

$$1+\frac{\gamma-1}{2}\frac{u^2}{a^2}=\frac{a_0^2}{a^2} \tag{15.94}$$

where a_0 is the stagnation sound speed which is a constant. These relations are the same as those in ordinary gas dynamics as we discuss in Chapter III, Eq.(3.12). Thus, we need to consider Eq.(15.91) alone which becomes[16,17]

$$\frac{du}{dx}=\frac{u}{M^2-1}\left(\frac{1}{A}\frac{dA}{dx}-\frac{\sigma_e B_y^2 u}{p}\right) \tag{15.95}$$

Eq. (15.95) shows that the ponderomotive force in the present case retards the velocity of the flow in the channel if the flow is supersonic $(M>1)$ and accelerates the flow if it is subsonic $(M<1)$. We consider a channel with one minimum section, de Laval nozzle as shown in Fig. 15. 3 with B_y=constant. For $\sigma_e=0$ or $B_y=0$, the flow field for a given entrance pressure and various exit pressures is identical to that of ordinary gasdynamics, i. e., Fig. 15. 3 whose velocity distributions are shown in the dotted lines in

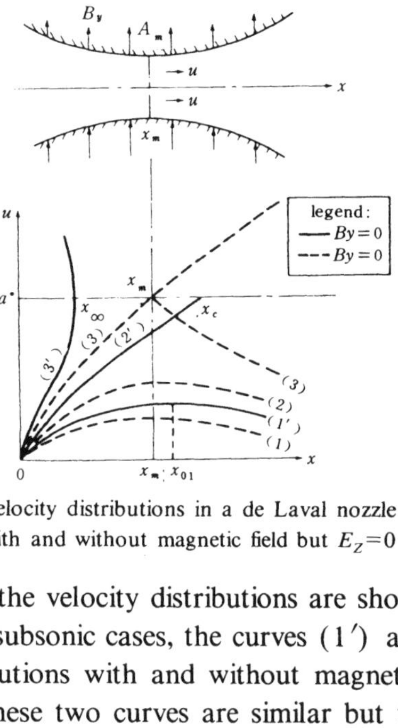

Fig. 3 Velocity distributions in a de Laval nozzle with and without magnetic field but $E_Z=0$

Fig.15.3. When $B_y\neq 0$, the velocity distributions are shown in solid curves in Fig.15.3. For the pure subsonic cases, the curves (1′) and (1) are the corresponding velocity distributions with and without magnetic field respectively. The general shapes of these two curves are similar but the velocity on curve (1′) is always larger than that on curve (1) at the same x and the maximum velocity on curve (1) occurs at $x=x_m$ where $dA/dx=0$ while the maximum velocity on curve (1′) occurs at $x=x_{01}$ where

$$\frac{L}{A}\frac{dA}{dx}=\gamma R_\sigma \frac{M^2}{M_m^2} \tag{15.96}$$

In the above equation, L is the reference length, $R_\sigma=\sigma_e\mu_e$ uL is the local magnetic Reynolds number, $M=u/a$ is the local Mach number, and $M_m=u/V_H$ is the local magnetic Mach number where $V_H=H_Y(\mu_e/\rho)^{1/2}$ is the local Alfven's wave speed. For given values of A and B_y, the exact location x_{01} depends on the velocity or the axial pressure gradient in the channel .

For the subsonic-supersonic cases, i. e., curves (3) and (3′), the velocity on curve (3′) with B_y is always larger than that on curve (3) without B_y on the subsonic side . Hence the curve (3′) reaches the local Mach number unity ($u=a^*$) before the minimum section i .e ., at $x=x_\infty < x_m$. At $x=x_\infty$, the slope du/dx is infinite and our assumption of inviscid fluid breaks down .

There is a critical case (2′) where the conditions $M=1$ and Eq. (15.96) occur simultaneously. Let us call this location $x=x_c$. Hence $x=x_c$ is a singular point of (15 .95). The nature of this singular point is one of the essential features of MGD nozzle flow . Depending on the magnetic field and the shape of the nozzle , this singular point may be a saddle point , nodal point, vortex point or a spiral point type. By comparison, in ordinary gasdynamics of nozzle flow, only the saddle point type occurs. The nature of the flow in the region near x_c, will be determined by the type of singularity. However, this feature of MGD nozzle flow has not been studied in details. In reference 16, Pai discussed the integral curves for various singularities.

6. Waves of small amplitude in electromagnetogasdynamics[1, 5]

We consider the wave motion of infinitesimal amplitude in an ionized and radiating gas but without transport phenomena. We shall conisder the effects of transport phenomena in Chapter XVII. Now we assume that the gas is of sufficient opacity that the radiation can be considered as being trapped in it and as in equilibrium with the plasma.

Now we assume that originally the plasma is a rest with a pressure p_0, temperature T_0 and density ρ_0 and that it is subjected to an externally applied uniform magnetic field $\boldsymbol{H}_0=\boldsymbol{i}\ H_x+\boldsymbol{j}\ H_y+\boldsymbol{k}\ 0$ where $\boldsymbol{i}$, $\boldsymbol{j}$, and $\boldsymbol{k}$ are respectively the x-, y-, and z-wise unit vectors; and H_x and H_y are constants. There is no electric current, nor excess electric charge, nor externally applied electric field. The plasma is perturbed by a small disturbance so that in the resultant disturbed motion, we have

$$\left.\begin{aligned} u=u\,(x,t),\quad v=v\,(x,t),\quad w=w\,(x,t) \\ p=p_0+p'(x,t),\quad T=T_0+T'(x,t) \\ \rho=\rho_0+\rho'(x,t),\quad \boldsymbol{E}=\boldsymbol{E}(x,t) \\ \boldsymbol{H}=\boldsymbol{H}_0+\boldsymbol{h}\,(x,t),\quad \boldsymbol{J}=\boldsymbol{J}\,(x,t),\quad \rho_e=\rho_e(x,t) \end{aligned}\right\} \tag{15.97}$$

where u, v, and w are respectively the perturbed x-, y-, and z- velocity components; p', ρ', T' are respectively the perturbed pressure, density and temperature; and $\boldsymbol{E}$, $\boldsymbol{h}$, $\boldsymbol{J}$, and ρ_e are respectively electric field strength, magnetic field strength, electrical current density and the excess electric charge. We assume that all the perturbed quantities are small so that the second or higher order terms in these quantities are negligible. For simplicity, we assume that the perturbed quantities are functions of one space coordinate x and time t only. Thus we discuss only the wave propagation in the direction of x-axis.

Substituting Eq. (15.97) into the fundamental equations of radiation electromagnetogasdynamics, i.e., the fundamental equations of §2 with the addition of radiation pressure and radiation energy density for an inviscid and non-heat-conducting and non-radiative transfer gas[2], we have the following linear equations for the wave motion of electromagnetogasdynamics of an opical, inviscid and non-heat-conducting gas.

(i) Maxwell equations for the electromagnetic field (15.6) and (15.7) become

$$\frac{\partial \mu_e h_x}{\partial t} = 0 \tag{15.98a}$$

$$\frac{\partial \mu_e h_y}{\partial t} = \frac{\partial E_z}{\partial x} \tag{15.98b}$$

$$\frac{\partial \mu_e h_z}{\partial t} = -\frac{\partial E_v}{\partial x} \tag{15.98c}$$

$$J_x = \frac{\partial \varepsilon E_x}{\partial t} = 0 \tag{15.98d}$$

$$J_x + \frac{\partial \varepsilon E_y}{\partial t} = -\frac{\partial h_z}{\partial x} \tag{15.98e}$$

$$J_z + \frac{\partial \varepsilon E_z}{\partial t} = \frac{\partial h_y}{\partial x} \tag{15.98f}$$

where the subscripts x, y, or z refer to the corresponding component of a vector.

(ii) The generalized Ohm's law (15.11)

$$J_x = \sigma_e\,(E_x - \mu_e w H_y) \tag{15.98g}$$

$$J_y = \sigma_e\,(E_y + \mu_e w H_x) \tag{15.98k}$$

$$J_z = \sigma_e\,[E_z + \mu_e\,(uH_y - vH_x)] \tag{15.98i}$$

(iii) The equation of conservation of electric charge (15.13)

$$\frac{\partial p_e}{\partial t} + \frac{\partial J_x}{\partial x} = 0 \tag{15.98j}$$

(iv) The equation of state (15.1)

$$\frac{p'}{p_0} \doteq \frac{\rho'}{\rho_0} + \frac{T'}{T_0} \tag{15.98k}$$

(v) The equation of continuity (15.2)

$$\frac{\partial \rho'}{\partial t} + \rho_0 \frac{\partial u}{\partial x} = 0 \tag{15.98l}$$

(vi) The equation of motion (15.3)

$$\rho_0 \frac{\partial u}{\partial t} = -\frac{\partial p'}{\partial x} + 4RR_p\rho_0 \frac{\partial T'}{\partial x} - \mu_e J_z H_y \tag{15.98m}$$

$$\rho_0 \frac{\partial v}{\partial t} = \mu_e J_z H_x \tag{15.98n}$$

$$\rho_0 \frac{\partial w}{\partial t} = \mu_e (J_x H_y - J_y H_x) \tag{15.98o}$$

where we neglect the viscous terms but add the radiation pressure term (3.61) and $R_p = a_R T_0^3/(3R\rho_0)$ is the radiation pressure number based on the undisturbed state variables.

(vii) The energy equation (15.5)

$$C_p^* \rho_0 c \frac{\partial T'}{\partial t} - 4RR_p T_0 \frac{\partial \rho'}{\partial t} = \frac{\partial p'}{\partial t} \tag{15.98p}$$

where $C_p^* = C_p + 12\ RR_p$ = effective specific heat at constant pressure including the radiation effect. Here we neglect the effects of viscosity and heat conductivity but add the radiation energy density per unit mass E_R/ρ.

Examining the linearized equations (15.98), we see that the quantity h_x is independent of all the other variables and is given by Eq. (15.98a) alone. Since the divergence of h is zero, h_x must be a constant which may be put equal to zero.

The rest of the perturbed quantities may be divided into two groups : One group consists of the variables w, h_z, J_x, J_y, E_x, E_y and ρ_e which are governed by equations (15.98c), (15.98d), (15.98e), (15.98g), (15.98h), (15.98j) and (15.98o). This group characterizes a transverse wave, since it deals with the velocity component w and the magnetic field h_z which are perpendicular to the externally applied magnetic field H_0.

The second group consists of the variables u, v, p', T', ρ', h_y, J_z, and E_z which are governed by the rest equations of (15.98). This group characterizes a longitudinal wave in which ordinary sound wave is a special case.

We are looking for periodic solutions in which all the perturbed quantities are proportional to

$$\exp\ [i\ (\omega t - \lambda x)] = \exp\ [-i\lambda\ (x - Vt)] \tag{15.99}$$

where ω is a given angular frequency, λ is the wave number which is equal to 2π (wave length), V is the velocity of wave propagation and $i = (-1)^{1/2}$. Substituting these variables into equations (15.98), we obtain one determinantal equation for each group of the above quantities. The eigen values of these determinantal equations give the different modes of wave propagation through the plasma.

(i) Transverse waves

The determinantal equation for the transverse waves gives the following relation :

$$\left(i\omega - \nu_H \frac{\omega^2}{c^2}\right)\left[i\omega\left(i\omega + \nu_H\lambda^2 - \nu_H \frac{\omega^2}{c^2}\right) + V_x^2\left(\lambda^2 - \frac{\omega^2}{c^2}\right)\right]$$

$$-\frac{\omega^2}{c^2} V_y^2 \left(i\omega + \nu_H\lambda^2 - \nu_H \frac{\omega^2}{c^2}\right) = 0 \qquad (15.100)$$

where c is the speed of light such that $c^2 = 1/(\varepsilon\mu_e)$,

$$\left.\begin{aligned} V_x &= H_x\ (\mu_e/\rho_o)^{1/2} = x\text{-component of Alfven's wave speed} \\ V_y &= H_y\ (\mu_e/\rho_o)^{1/2} = y\text{-component of Alfven's wave speed} \\ \nu_H &= \frac{1}{\mu_e\sigma_e} = \text{magnetic diffusivity} \end{aligned}\right\} \qquad (15.101)$$

Eq. (15.100) does not contain the radiation parameters. Hence the transverse wave is independent of the radiation field and it is simply the regular transverse wave of magnetogasdynamics which is independent of the compressibility properties of the plasma. Since we neglect the effect of viscosity here, there is only one transverse electromagnetic wave. If there is no external applied magnetic field, this transverse electromagnetic wave reduces to the simple form :

$$\lambda^2 = \frac{\omega^2}{c^2} - i\frac{\omega}{\nu_H} \qquad (15.102)$$

Eq. (15.102) gives a damped electromagnetic wave in a conducting medium. For a non-conducting medium, $\nu_H = 0$, this wave propagates at the speed of light c without damping.

For an ideal plasma, inviscid, non-heat conducting and infinitely electrical conducting. $\mu = 0$, $\nu_H = 0$, Eq. (15.100) gives

$$V = \frac{\omega}{\lambda} = V_x\left(1 + \frac{V_x^2 + V_y^2}{c^2}\right)^{-1/2} \qquad (15.103)$$

This may be called the modified Alfven's wave. For magnetogasdynamic waves, $(V_x^2 + V_y^2)$ is much smaller than c^2, Eq. (15.103) becomes

$$V = \frac{\omega}{\lambda} = V_x \qquad (15.104)$$

This is the well-known Alfven's wave which shows that the disturbance in an inviscid and infinitely electrically conducting fluid of density ρ_0 propagates as a wave in the direction of H_x with the Alfven's wave speed V_x.

(ii) Longitudinal waves [20]

The dispersion relation of the longitudinal waves in an inviscid, non-heat-conducting but electrically conducting fluid has the general form as follows:

$$\left[(C_p + 20\,R\,R_p + 16R\,R_p^2)\lambda^2 - \frac{\omega^2}{T_0(\gamma-1)}(1 + 12(\gamma-1)R_p\right]\cdot$$

$$\left[\left(\nu_H\lambda^2 + i\omega - \nu_H\frac{\omega^2}{c^2}\right)i\omega + V_x^2\left(\lambda^2 - \frac{\omega^2}{c^2}\right)\right] - \left(\lambda^2 - \frac{\omega^2}{c^2}\right)\cdot$$

$$\left[V_y^2 \frac{\omega^2}{T_0(\gamma-1)} \quad (1+12\,(\gamma-1)R_p)\right]=0 \qquad (15.105)$$

The first square bracket characterizes the sound wave in a radiating gas. If there is no trasverse magnetic field, i. e., $V_y=0$, this sound wave will not interact with the magnetic fields. If $R_p=0$, this sound wave reduces to ordinary sound wave. The speed of radiating sound wave is C_R which is given in Eq. (3.67).

For an ideal plasma, i. e., inviscid, non-heat-conducting and infinitely electrical conducting gas, $\nu_H=0$, we have two electromagnetogasdynamic longitudinal waves given by the following relation :

$$\left(C_R^2-\frac{\omega^2}{c^2}\right)\left[V_x^2-\frac{\omega^2}{\lambda^2}\left(1+\frac{V_x^2}{c^2}\right)\right]-V_y^2\frac{\omega^2}{\lambda^2}\left(1+\frac{\omega^2}{c^2\lambda^2}\right)=0 \qquad (15.106)$$

The two solutions of Eq. (15. 106) give the fast and slow waves of electromagnetogasdynamics. In magnetogasdynamics, both V_x and V_y are negligibly small in comparison with the speed of light c, and Eq. (15. 106) gives two solutions as follows :

$$V=\left[\frac{(a^2+V_H^2)\pm\sqrt{(a^2+V_H^2)^2-4a^2V_H^2\cos^2\theta}}{2}\right]^{1/2} \qquad (15.107)$$

In Eq.(15.107), we assume that R_p is zero. If R_p is different from zero, we simply replace a by c_R in Eq. (15.107); the angle θ is the angle between the magnetic field strength H and the velocity vector at any local point. $V_H=H\,(\mu_e/\rho_0)^{1/2}$ is the resultant speed of Alfven's wave. The plus sign is for the fast wave V_{fast} and the minus sign is fow the slow wave V_{slow}. We have the inequalities;

$$V_{fast}\geqslant a\geqslant V_{slow} \qquad (15.108a)$$

$$V_{fast}\geqslant V_x\geqslant V_{slow} \qquad (15.108b)$$

For an ideal plasma, we have three characteristic velocities of magnetogasdynamic waves, i. e., V_x, V_{fast} and V_{slow}. The wave pattern in magnetogasdynamics is then considerably different from that in ordinary gasdynamics. For instance, in ordinary gas dynamics, we have only one set of downstream inclined waves when the steady flow is supersonic (cf. Fig. 4.6). In magnetogasdynamics, we may have zero, one or two sets of wave patterns over a body in a steady flow. The standing wave may incline downstream or upstream. These wave patterns can be easily studied by the help of Friedrichs' diagram, Fig. 15.4. Similar to Fig. 4.6, the Friedrichs diagram for the shape of the disturbance that propagates from a point disturbance in two-dimensional flow at the origin is shown in Fig. 15.4. The shape of the propagation of disturbance is not a circle as in the ordinary gas dynamics. The shape of fast wave differs. In Fig. 15.4, the abscissa is in the direction of the

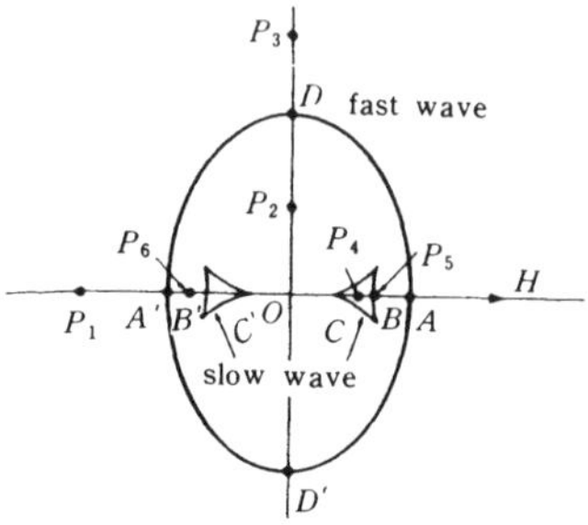

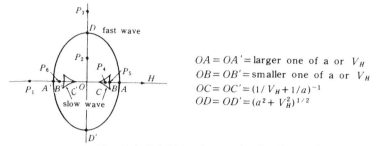

$OA = OA' =$ larger one of a or V_H
$OB = OB' =$ smaller one of a or V_H
$OC = OC' = (1/V_H + 1/a)^{-1}$
$OD = OD' = (a^2 + V_H^2)^{1/2}$

Fig. 15.4 Friedrichs diagram showing shapes of wave front from a point disturbance at origin 0

magnetic field H and the direction of the flow velocity OP may be in arbitrary direction. The angle θ is the angle between the direction of the magnetic field H and the velocity of the point source $OP = U$. The velocity of the point source may be in any direction such as OP_1, OP_2 and so forth. The wave pattern depends on the location of the point P, i.e. it depends on both the direction and the amplitude of the velocity U. Fig. 15. 5 shows typical cases for the wave patterns. When the velocity U is OP_1, i.e., the velocity U is in the same direction as H but its magnitude is larger than both a and V_H, we may call it super-Alfven flow and we have a similar wave pattern as that in ordinary gas dynamics, i.e., we have a set of backward inclined waves (Fig.15.5 a). If the magnitude of U is between a and V_H, there is no standing wave which corresponds to the subsonic case of ordinary gas dynamics. This is the case $U - OP_2$. However, if the velocity vector $U = OP_4$ is within the curved triangle of the slow wave, we have a set of forward inclined wave patterns which is not possible in ordinary gas dynamics (Fig. 15.5b). When the veloci-

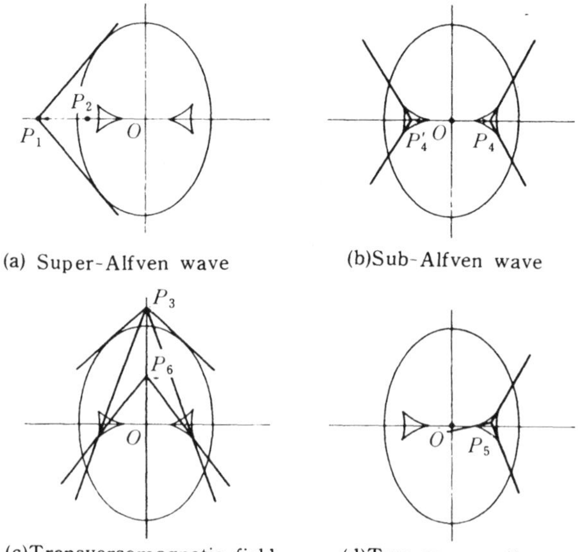

(a) Super-Alfven wave (b)Sub-Alfven wave

(c)Transversemagnetic field (d)Two wave patterns

Fig. 15.5 Wave patterns in magnetogas dynamics

ty vector U is perpendicular to the magnetic field H, we have the wave patterns shown in Fig. 15. 5c. When $U=OP_3>(a^2+V_H^2)^{1/2}$, we have two sets of wave patterns and when $OP_6=U<(a^2+V_H^2)^{1/2}$, we have only one set of wave pattern. The case of two sets of waves is also not possible in ordinary gas dynamics. Finally if $U=OP_5$ where P_5 is inside the curved triangle but OP_5 is not parallel to H, we have again two sets of wave patterns: one set is inclined forward while the other is inclind backward (15. 5d). This is the case which cannot occur in ordinary gas dynamics. The forward inclined waves may be obtained by drawing the envelope of the disturbance shape of the slow wave. (Fig. 15.6)

These wave patterns are mathematical facts but they have not been-verified experimentally. Experimental obsevation of these wave phenomena will be most interesting and enlightening. Not only are there fast and slow waves of infinitesimal amplitude, there are waves of finite amplitude and shock waves of both types, fast and slow, in magnetogasdynamics of an ideal plasma.

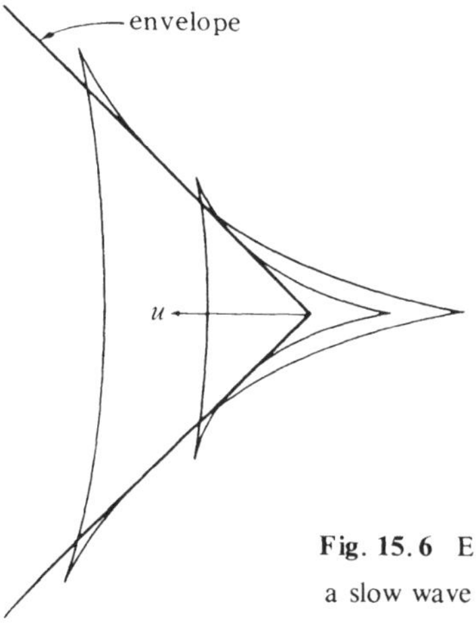

Fig. 15.6 Envelope of disturbance of a slow wave - forward inclined wave

7. Shock waves in magnetogasdynamics [21, 22]

Since the shock wave is developed from the compression wave of magnetogasdynamics, one would expect that there will be shock waves corresponding to the fast wave, slow wave and transverse wave. In studying the shock waves, two things are of interest: one is the Rankine-Hugoniot relations across a shock wave and the other is the shock structure. We shall study the shock structure in Chapter XVII because we have to consider the transport phenomena in such a study. In this section we first consider the normal magnetogasdynamic shock under the influence of a transverse magnetic field. The Rankine-Hugoniot relation of this shock gives the ratio of velocity behind the shock u_2 to that in front of the shock u_1 (cf. Chapter XVII, § 16) and is

$$\frac{u_2}{u_1} = 1/2 \left\{ \frac{\gamma-1}{\gamma+1} + \frac{2}{(\gamma+1)M_1^2} + \frac{2\gamma h_1^2}{\gamma+1} \right\}$$

$$+ 1/2 \left\{ \left[\left(\frac{\gamma-1}{\gamma+1} \right)^2 + \frac{2}{(\gamma+1)M_1^2} \right]^2 + \frac{8(2-\gamma)h_1^2}{\gamma+1} \right\}^{1/2} \qquad (15.109)$$

where $M_1 = u_1/a_1$ is the Mach number in front of the shock and $h_1 = H_1/(2\rho_1 u_1^2/\mu_e)^{1/2}$ is the non-dimensional magnetic field in front of the shock. When there is no magnetic field, $h_1 = 0$, Eq. (15.109) reduces to the relation for a normal shock in ordinary gasdynamics, i. e., (Eq. 4.12). For a given value of M_1, the ratio u_2/u_1 increases or the strength of the shock decreases with the increase of h_1. This occurs since effective Mach number $M_{e1} = u_1/a_{e1}$, where a_e is given by Eq. (15. 82), decreases as h_1 increases. When $M_{e1} = 1$, $u_2/u_1 = 1$. For a given Mach number, if $h_1^2 > h_c^2 = 1/2\ (1-(1/M_1^2))$, $M_{e1} < 1$ and no shock will exist.

We shall discuss the flow in the transition region from u_1 to u_2 in Chapter XVII, § 16 when we discuss the structure of the shock.

For a magnetogasdynamic oblique shock, we have to consider both the direction of the magnetic field and that of the flow velocity with respect to the shock front. In ordinary gas dynamics, by proper choice of the coordinate system, it is always possible to reduce the oblique shock to the corresponding normal shock case. (cf. Chapter IV, § 5). However, it is, in general, not possible to do so in magnetogasdynamics. Since the shock wave is developed from the ordinary waves, and there are fast, slow and transverse waves in magnetogasdynamics, we would expect that several different types of shock waves exist in magnetogasdynamics.

We shall assume that both the velocity vector $\boldsymbol{q}$ and the magnetic field vector $\boldsymbol{H}$ have arbitrary direction with respect to the shock front whose normal vector is $\boldsymbol{n}$. The jump of any physical quantity Q across the shock front may be denoted by

$$[Q] = Q_1 - Q_0 \qquad (15.110)$$

where subscript 1 refers to the value on the side to which the normal n points and subscript 0 refers to the value on the other side of the shock. The conservation laws across the surface of discontinuity are :

$$[H_n] = 0 \qquad (15.111a)$$

$$[(q_n - U)\boldsymbol{B} - B_n \boldsymbol{q}] = \boldsymbol{0} \qquad (15.111b)$$

$$\left[(q_n - U)\rho\, \boldsymbol{q} + \left(p + \frac{\boldsymbol{B} \cdot \boldsymbol{H}}{2} \right) \boldsymbol{n} - B_n \boldsymbol{H} \right] = 0 \qquad (15.111c)$$

$$[(q_n - U)] = 0 \tag{15.111d}$$

$$\left[(q_n - U)\left(\frac{\rho q^2}{2} + \rho U_m + \frac{\boldsymbol{B}\cdot\boldsymbol{H}}{2}\right) + q_n\left(p + \frac{\boldsymbol{B}\cdot\boldsymbol{H}}{2}\right) - H_n(\boldsymbol{B}\cdot\boldsymbol{q})\right] = 0 \tag{15.111e}$$

where subscript n refers to the component of a vector in the direction of $\boldsymbol{n}$ and U is the normal velocity of the surface of discontinuity. In order that there exists such a surface of discontinuity, there must be a surface current along this surface of discontinuity with a value of

$$\boldsymbol{J}^* = \boldsymbol{n} \times [\boldsymbol{H}] \tag{15.112}$$

From Eq. (15.111b) we see that if we choose the coordinate system such that the flow velocity $\boldsymbol{q}$ is parallel to the magnetic field $\boldsymbol{H}$ on one side of the shock wave front and $(q_{n1} - U)B_1 - B_n q_1 = 0$, these two vectors $\boldsymbol{q}$ and $\boldsymbol{H}$ are automatically parallel on the other side of the shock front and $(q_{n0} - U)$ $B_0 - B_n q_0 = 0$.

Let us introduce the following notations:

$$\tau = \rho^{-1} \tag{15.113}$$

$$\widetilde{Q} = 1/2(Q_0 + Q_1) \tag{15.114}$$

$$m = \rho\ (q_n - U) \tag{15.115}$$

where the normal n is so chosen that across a shock wave

$$q_n - U > 0 \tag{15.116}$$

Eq. (15.111a) gives

$$H_{n0} = H_{n1} = H_n = \text{constant} \tag{15.117}$$

From Eqs. (15.111b) to (15.111d), we have

$$\tilde{\tau}^2 m(\tilde{\tau}m^2 - B_n H_n)\{\tilde{\tau}m^4 + (\tilde{\tau}\,[\tau]^{-1}[p] - \widetilde{\boldsymbol{B}}\cdot\widetilde{\boldsymbol{H}})m^2 - [\tau]^{-1}[p]B_n H_n\} = 0 \tag{15.118}$$

Eq. (15. 118) shows that there are four different types of surface of discontinuity in magnetogasdynamics according to the four roots of m in this equation.

$m = 0$. This is a contact discontinuity. If $H_n \neq 0$, the only jump for this case is $[\rho] = 0$, because $[s] \neq 0$. If $H_n = 0$, we can have $[p + 1/2\ \boldsymbol{B}\cdot\boldsymbol{H}] = 0$.

Transverse shock or Alfven's shock A. This corresponds to the root:

$$m = m_{tr} = (\tilde{\tau}^{-1} B_n H_n)^{1/2} \tag{15.119}$$

In this case, both the density and entropy are continuous across the surface of discontinuity but the tangential projection of the magnetic field undergoes a rotation through an arbitrary angle on crossing the surface of discontinuity.

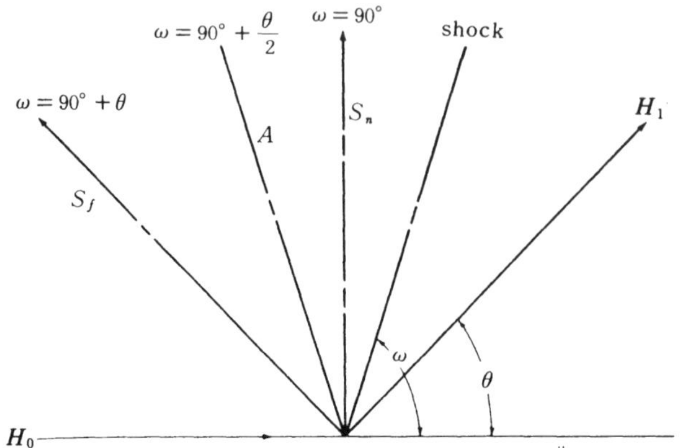

Fig. 15.7 Magnetogasdynamic shock when $\boldsymbol{q} \| \boldsymbol{H}$

The most interesting magnetogasdynamic shocks are the fast and the slow shocks which correspond to the two roots given by the curl bracket of Eq.(15. 118). The oblique shock relations have been calculated by many authors. We shall give here some descriptions of the general features of such magnetogasdynamic shocks.

Let us follow Ref. 23 by considering the case where the velocity and the magnetic field ane parallel on both sides of the shock front. Let θ be the turning angle of the magnetic field and ω be the shock angle. Fig. 15.7 shows the orientations of fast and slow shocks :

1. When $90° > \omega$, we have the fast shock.

2. When $90° < \omega < 180°$, we have the slow shock. The inclination of a slow shock is upstream which is not possible in ordinary gasdynamics.

3. When the shock angle is 90°, the tangential component of the magnetic field is zero in front of the shock but different from zero behind the shock. It is known as switch-on shock *Sn*.

4. When $\omega = 90° + \theta$, the tangential magnetic field is different from zero in front of the shock but vanishes behind the shock. It is known as switch-off shock Sf which is a special slow shock.

5. $\omega = 90° + \frac{1}{2}\theta$, it is the Alfven's shock *A*.

8. One-dimensional steady flow of electrogasdynamics [11]

We consider the one-dimensional steady flow of an inviscid and non-conducting charged fluid in a nozzle of cross-sectional area $A(x)$. Similar to the MGD case, we shall assume that all the covariance factors are unity. In the present case, we have four unknowns; p n, u and $E_x = E$ where the temperature T may be expressed in terms of p and n, the number density of the gas. The equations governing these four unknowns are :

$$n A u = \text{constant} = A_0 N \qquad (15.120a)$$

$$\frac{dp}{dx} + mnu\frac{du}{dx} = ZeEn \qquad (15.120b)$$

$$\frac{dE}{dx} = Zen \qquad (15.120c)$$

$$p = p\,(n) \qquad (15.120d)$$

where m is the mass of a charged particle of the gas and Ze is the charge of a particle in the gas. We replace the energy equation by the barotropic relation (15.120d), i.e., the pressure p is a function of the number density n only. Since the energy equation for the adiabatic case can be reduced to the following equation:

$$\frac{dS}{dx} = 0 \qquad (15.121)$$

we have $p=(\text{constant})\,n$. The other simple case is the isothermal process $T=\text{constant}$ and $p=(\text{constant})\,n$.

Eqs. (15.120) are a set of nonlinear differential equation and the well-known method of nonlinear mechanics may be used to solve them. For simplicity, let us consider the case of isothermal flow $T=\text{constant}$ in a channel of constant cross-sectional area $A(x) = A_0 = \text{constant}$. For this case, Eqs. (15.120) give

$$p + \frac{mN^2}{n} - 1/2\,\varepsilon E^2 = \text{constant} \qquad (15.122)$$

Since $p=p(n)$, Eq.(15.122) gives the phase relation between n and E. Clauser[11] discussed extensively the case of isothermal flow, i.e, $p=knT$. Clauser introduced the following non-dimensional quantities for the isothermal case:

$$n^* = \frac{n}{n_0}\ ; \ E^* = \frac{E}{\left(\dfrac{2n_a kT}{\varepsilon}\right)^{1/2}} \qquad (15.123)$$

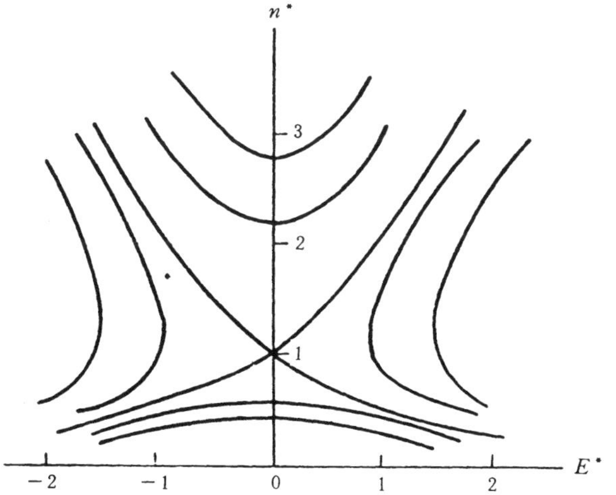

Fig. 15.8 One-dimensional flow in an isothermal constant area channel in electrogasdynamics

where $n_a = [mN^2/(kT)]^{1/2}$ is the value of n at $u=a=(kT/m)^{1/2}$. Eq.(15.122) becomes

$$n^* + \frac{1}{n^*} - E^{*2} = \text{constant} \qquad (15.124)$$

Eq. (15. 124) has been plotted in Fig. 15. 8. There is a singular point at $E^*=0$ and $n^*=1$ and it is a saddle point. If we substitute the relation $E^*=E^*(n^*)$ obtained from Eq. (15. 124) into Eq. (15. 120c), we have a first order differential equation in E (x) which can be numerically integrated. Some typical results are shown in Fig. 15.9. The most interesting point is that the integral curve of E^* has cusp at the sonic line $n^*=1$ or $M=u/a=1$. This means that they cannot represent the real flow condition at the sonic point. The only curves which are reasonable are those integral curves whcih pass through the saddle point.

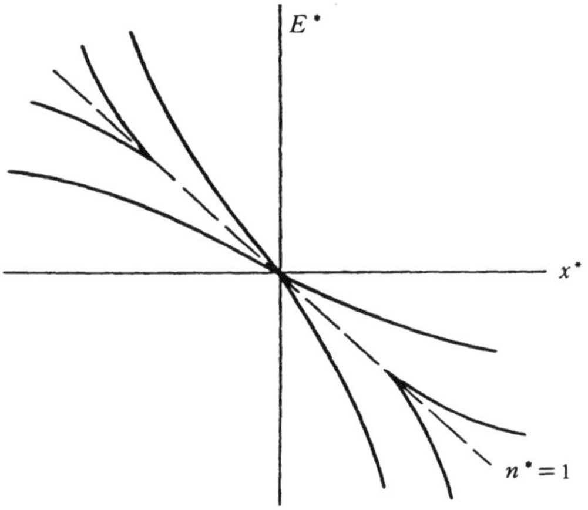

Fig. 15.9 Electric field along the x-direction in a constant area channel of an isothermal flow in electrogasdynamics

9. Units in electromagnetic theory[10]

In the electromagnetic theory, besides the three fundamental dimensions of mechanics, e. g., length, mass, and time, we have another fundamental dimension——an electromagnetic unit. There are a number of different systems of units in electromagnetic theory in common use. This fact complicates the electromagnetic theory. It is often found that the same equation, e. g., Maxwell's equation, has different forms in different papers or books. In order that the reader may transform our equations into a form with which he is familiar, we summarize briefly here those systems of unit commonly used in electromagnetic theory.

The most common systems of units used in electromagnetic theory are :

(i) The absolute cgs system of electrostatic units in which the basic unit is usually abbreviated as esu.

(ii) The absolute cgs system of electromagnetic units in which the basic unit is usually abbreviated as emu.

(iii) Gaussian or mixed units which are a mixture of esu and emu.

(iv) Practical units which are largely based on emu but differ from emu by a power of ten in order to get convenient size.

(v) MKS or Giorgi system which is the one used in this book.

Furthermore in all these systems, we have a choice of "rationalized" and "unrationalized" units differing by factors of 4π.

The essential points of these systems are as follows :

(i) Unrationalized electrostatic units (esu)

In this system we take the centimeter as the unit of length, the gram as the unit of mass, and the second as the unit of time. The fundamental electrostatic unit is the unit charge which is defined by Coulomb's law:

$$F = \frac{QQ'}{r^2} \text{ dynes} \tag{15.125}$$

where F is the force between two charges Q and Q' and is expressed in dynes; r is the distance between these two charges in centimeters. The charges Q and Q' are expressed in unit charges. If Q and Q' are unit charges and r is one centimeter, the force F will be one dyne.

The corresponding rationalized electrostatic unit charge would be defined by

$$F = \frac{Q_r Q_r'}{4\pi r^2} \text{ dynes} \tag{15.126}$$

Hence the unit charge in rationalized unit is $1/\sqrt{4\pi}$ times that in unrationalized units. We shall no longer consider the rationalized esu system.

The unit current is equal to unit charge passing any point of circuit per second.

The unit electric field $\boldsymbol{E}$ is defined by the equation :

$$\boldsymbol{F} = Q\boldsymbol{E} \text{ dynes} \tag{15.127}$$

In a unit $\boldsymbol{E}$, the unit charge is acted on by a force of one dyne.

The unit of electric displacement $\boldsymbol{D}$ in the unrationalized esu system is determined by the arbitrary assumption that $\boldsymbol{D} = \boldsymbol{E}$ in free space, i.e.,

$$\boldsymbol{D} = \varepsilon \boldsymbol{E} = \boldsymbol{E} \tag{15.128}$$

Hence the induced capacity ε is taken as unity in free space.

The electrical conductivity σ is defined by Ohm's law:

$$\boldsymbol{J} = \sigma \boldsymbol{E} \tag{15.129}$$

where $\boldsymbol{J}$ is the current density in esu.

(ii) Unrationalized electromagnetic units (emu)

In this system, we still take cgs units for length, mass, and time, but the fundamental electric unit is the unit current in the electromagnetic system which is defined by the formula

$$dF = I_1 I_2 \frac{ds_1 \times (ds_2 \times r)}{|r|^3} \text{ dynes} \tag{15.130}$$

where Ids is the vector current element, I is the current in emu, and r is the spatial vector between the two current elements 1 and 2. The unit current in emu is that current which, flowing in a one-centimeter length of wire, acts on a similar current one centimeter away, appropriately oriented, with a force of one dyne.

The unit charge in emu is defined as the charge crossing a given point per second, when unit current is flowing.

The unit of electric field is still defined by Eq. (15.127) except that here both E and Q are in emu.

From experiments, it was found that the following relations between esu and emu of charge, current, and electric field hold:

$$\begin{gathered} 1 \text{ emu of charge} = c \times 1 \text{ esu of charge} \\ 1 \text{ emu of current} = c \times 1 \text{ esu of current} \\ c \times 1 \text{ emu of electric field} = 1 \text{ esu of electric field} \end{gathered} \tag{15.131}$$

where

$$\begin{gathered} \text{velocity of light in free space} = 1/\sqrt{\varepsilon\mu_e} = \\ 2.99790 \times 10^{10} \text{cm/sec} \cong 3 \times 10^{10} \text{cm/sec} \end{gathered} \tag{15.132}$$

in which ε = inductive capacity in free space, and μ_e = magnetic permeability in free space.

In emu, we define the magnetic induction B by

$$F = Q(q \times B) \text{ dynes} \tag{15.133}$$

or

$$F/\text{volume} = J \times B \text{ dynes/cm} \tag{15.134}$$

where J is the current density, q is the velocity vector. Thus in the emu system the unit magnetic induction is that induction by which a unit current flowing in a wire one centimeter long at right angle to the magnetic field is acted on by a force of one dyne. This emu of magnetic induction is known as a gauss.

In emu, the magnetic field H in free space is defined as

$$H = B = \mu_e H \tag{15.135}$$

μ_e in free space is arbitrarily chosen as unity.

The electric displacement in free space is then

$$D = \varepsilon E = E/c^2 \tag{15.136}$$

In emu, Maxwell equations in free space may be written as follows:

$$\nabla \times \boldsymbol{E} = -\frac{\partial \boldsymbol{H}}{\partial t}, \nabla \cdot \boldsymbol{H} = 0$$

$$\nabla \times \boldsymbol{H} = \frac{1}{c^2}\frac{\partial \boldsymbol{E}}{\partial t} + 4\pi \boldsymbol{J}, \nabla \cdot \boldsymbol{E} = 4\pi c^2 \rho_e \tag{15.137}$$

(iii) Gaussian or mixed units

In this system we use emu for B and H and esu for all other quantities including current and charge. Maxwell equations in free space in Gaussian units are then

$$\nabla \times \boldsymbol{E} = -\frac{1}{c}\frac{\partial \boldsymbol{H}}{\partial t}, \nabla \cdot \boldsymbol{H} = 0.$$

$$\nabla \times \boldsymbol{H} = \frac{1}{c^2}\frac{\partial \boldsymbol{E}}{\partial t} + 4\pi \frac{\boldsymbol{J}}{c}, \nabla \cdot \boldsymbol{E} = 4\pi c^3 \rho_e \tag{15.138}$$

(iv) Practical units

They are based on the emu but differ from emu by some arbitrary power of 10. Some of the important relations or practical units and emu are given below:

Quantity	Practical Unit	Emu
current	1 ampere	$=10^{-1}$ emu of current
potential	1 volt	$=10^8$ emu of potential
work/sec.	1 ampere × 1 volt	$=1$ watt $=10^7$ erg/sec
resistance	1 ohm	$=10^9$ emu of resistance
magnetic induction	1 weber/cm	$=10^8$ gausses

The weber is unit of flux, not of magnetic induction. Other practical units are :

$$1 \text{ ampere} = 1 \text{ coulomb/sec}.$$

A coulomb is a unit charge in practical units. The unit electric field in practical units is 1 volt/cm.

(v) MKS or Giorgi system

Probably this is the most common unit used in recent years. It is a rationalized MKS unit system in which the meter is used as the unit of length; the kilogram, as the unit of mass; the second, as the unit of time and any electric quantity belonging to the practical unit system such as the coulomb, ampere, or ohm as the fourth unit. We shall use the coulomb as the fourth unit.

In this system, the magnetic permeability in free space has the magnitude

$$\mu_e = 4\pi \times 10^{-7} \frac{\text{kg} \cdot m}{(\text{coulomb})^2} \tag{15.139}$$

The value of ε in free space may be determined by Eqs. (15.132) and (15.139), i.e.,

$$\varepsilon = 8.854 \times 10^{-12} \frac{\text{coulomb}^2 \ \text{sec}^2}{\text{kg} \cdot m} \tag{15.140}$$

Maxwell's equations in this system are then

$$\begin{aligned} \nabla \times \boldsymbol{E} &= -\frac{\partial \mu_e \boldsymbol{H}}{\partial t}, \quad \nabla \cdot \boldsymbol{H} = 0 \\ \nabla \times \boldsymbol{H} &= \boldsymbol{J} + \frac{\partial \varepsilon \boldsymbol{E}}{\partial t}, \quad \nabla \cdot \varepsilon \boldsymbol{E} = \rho_e \end{aligned} \tag{15.141}$$

10. Problems

1. Discuss the one-dimensional steady flow between two parallel plates of incompressible fluids under the influence of an external magnetic field perpendicular to the plates. These plates may be at rest or moving. (This is a generalization of plane Poiseulle flow or plane Couette flow of ordinary hydrodynamics.)

2. Calculate the pressure jumps across the shock wave of § 7 in terms of M_1 and h_1.

3. If $M_1 = 2.00$, $h_1^2 = 0.1$, $\gamma = 1,4$, calculate the upper and lower critical magnetic field h_c and h_b for the shock wave discussed in § 7.

4. If $M_1 = 2.00$, $h_1^2 = 0.1$, $P_r = \frac{3}{4}$, and $\gamma = 1.4$, $\sigma = \infty$, calculate the velocity distribution and magnetic field distribution in the shock of § 7.

5. Use the method of characteristics of Chapter XI to find the characteristic curve in the x-t plane for one-dimensional unsteady flow of magnetogasdynamics when $\nu = \nu_H = \kappa = V_x = 0$ but $V_y \neq 0$.

6. Give the detailed derivation of the quasi-one dimensional equation of motion, i.e., Eq. (Ic).

7. Give the detailed derivation of the quasi-one dimensional equation of energy, i.e., Eq. (Id).

8. Give the detailed derivation of the quasi-one-dimensional equation of the magnetic field strength, i.e., Eq. (Ie).

9. Choose an area distribution for an MGD nozzle and calculate the velocity variation for a subsonic - supersonic flow from Eqs. (15.84) to (15.87) with constant values of E_z and B_y.

10. Discuss the possible singularities for the flow in a nozzle of magnetogasdynamics as given by Eqs. (15.88) and (15.89).

11. Find the velocity distributions for the case of a uniform flow over a magnetized sphere in an inviscid and electrically conducting gas. Assume that both the magnetic field strength and the electrical conductivity of the gas are small.

12. Derive the fundamental equations of the linearized theory of magnetogasdynamics for the case of a uniform velocity over a thin body and an arbitrary.orientated externally applied magnetic field. Discuss the general solution of the resultant equation.

13. Discuss the system of characteristics of magnetogasdynamics of an ideal plasma for three-dimensional unsteady flow.

14. Calculate some typical cases of oblique shocks in magnetogasdynamics of an ideal plasma and show that the shock angle may be greater than 90°.

15. Find the Rankine-Hugoniot relations across a normal shock in radiation magnetogasdynamics of an ideal plasma including the effects of radiation pressure and radiation energy density.

References

1. Pai, S. I., *Magnetogasdynamics and Plasma Dynamics*. Springer Verlag, Vienna, Austria and Prentice Hall, New Jersey, 1962.
2. Pai, S. I., *Radiation Gasdynamics*. Springer Verlag, Vicnna, Austria, 1966.
3. Pai, S. I. and Cramer K. R., *Magnetofluid Dynamics for Engineers and Applied Physicists*. McGraw-Hill Co. 1973.
4. Pai, S. I., *Two-Phase Flows*. Vieweg Verlag, Braunschweig, West Germany, 1977.
5. Pai, S. I., *Modern Fluid Mechanics*. Science Press, Beijing, China and Van Nostrand-Reinhold, N. Y. 1981.
6. Burgers, J. M. and Van de Hulst, H. V. (editors), *Gas Dynamics of Cosmic Clouds*. Interscience Publishers, Inc. New York, 1955.
7. de Hoffmann F. and Teller E., Magnetogasdynamic shock. *Phys. Rev*. **80**, No. 4, pp. 692—703, Nov. 15, 1950.
8. Alfven, H., *Cosmical Electrodynamics*. Oxford University Press, 1950.
9. Cowling, T. G., *Magnetohydrodynamics*, Interscience Publishers, Inc. New York, 1957.
10. Stratton, J. A., *Electromagnetic Theory*, McGraw-Hill Book Co. Inc., New York, 1941.
11. Clauser, F. H., Plasma dynamics. *Aero. & Astro. Int. Ser. of Aero. Sci. & Space Flight*, Div. IX, **4**, pp. 305— 343, Pergamon Press, New York, 1961.
12. Kahn, B. and Gourdine, M. C., Electrogasdynamic power generation. *AIAA Jour*. **2**, No. 8, pp. 1423— 1427, Aug. 1964.
13. Neuringer, J. L. and Rosenweig, R. E., Ferrohydrodynamics, *Phys. of Fluids*, **7**, No. 12, pp. 1927 — 1937, 1964.
14. Resler, E. L. jr. and Rosenweig, R. E., Magnetocalorie Power, *AIAA Jour*. **2**, No. 8 pp. 1418— 1422 , 1964.
15. Melcher, J. R. and Taylor, G. I., Electrohydrodynamics, A review of the role of interfacial shear stresses. *Annual Rev. of Fluid Mech*. **1**, pp. 111— 146, 1969.
16. Pai, S. I., Quasi-one-dimensional analysis of magnetogasdynamics Proc. *Electricity from MHD*, **1**, Symp. MHD Power Gen. Salzburg, International Atomic Energy Agency, pp. 283— 294. July 1966.

17. Dahlberg, E., Quasi-one-dimensional analysis of magnetohydrodynamic Channel flows. *Trans. Roy. Inst. Tech.* (Stockholm) No. 209, pp. 1— 16, 1963.
18. Rosa, R. J., *Magnetohydrodynamic Energy Conversion*. McGraw-Hill Book Co. New York, 1968.
19. Reslet, E. L. jr. and Sears, W. R., The prospect for Magnetoaerodynamics. *Jout. Aero. Sci.* **25**, No. 4, pp. 235— 245, April, 1957.
20. Sears, W. R. and Resler, E. L. Jr., *Sub- and Super-Alfvenic Flows past bodies*.
21. Pai, S. I, On exact solutions of one-dimensional flow equations of magnatogasdynamics, *Proc. 9th International congress of Appl. Mech.*.
22. Marshall, W., The structure of magneto-hydrodynamic shock waves, *Proc. Roy. Soc.*, London A-233, pp. 367— 376, 1955.
23. Bazer, J. and Ericson, W. B., Oblique Shock Waves in a Steady Two Dimensional Hydromagnetic Flow. *Proc. Symp. on Electromagnetics and Fluid Dynamics of Gaseous Plasma*, **IX**. Interscience Publishers, Inc. New York, 1961.

Chapter XVI

MULTIPHASE FLOWS

1. Introduction[1]

In the previous chapters, we consider the compressible flow of a simple fluid which is in one state or phase only, i.e., mainly in gas or plasma state. In such problems, the solid bodies in the flow field are usually assumed to be rigid bodies so that the solids may be considered as given boundary conditions of the fluid flow problem. As we discuss in Chapter II, liquid in most cases may be considered as an incompressible fluid. Thus in most studies in previous chapters, we did not deal with liquid flow. However, in many engineering problems as well as fluid flow in nature, we have to treat the flow problems of a mixture of substances in different states and the solid bodies may not be considered as rigid bodies and the liquid may be changed into gas or vapor. In these cases, the corresponding flow may be called multi-phase flow[1, 2]. The most common types of multi-phase flows consist of two-phases of some substances only, i.e., any combination of two of the four phases of matter (cf. Chapter II) of the same or different substances. In this chapter, we discuss various types of two phase flows only and the results my be generalized into multiphase cases.

As we discussed in Chapter II, matter may be divided into four phases or four states: solid, liquid, gas and plasma (ionized gas). The term fluid has been used as the general name for the last three states: liquid, gas and/or plasma because they may be deformed without applying any force, provided that the change of shape occurs very slowly. Furthermore, when a large number of small solid particles flow in a fluid, and the velocity of the fluid is sufficiently high, the behavior of such solid particles is similar to those of ordinary fluid. We may consider these solid particles as a pseudofluid. Under proper conditions, we may treat fluid flow problems for a mixture of solid (pseudo-fluid), liquid, gas and/or plasma.

It should be noticed that the concept of phases is a microscopic description of the properties of matter (cf. Chapter II) and it is an ancient concept, which is known as the theory of elements. Aristotle had this idea over two thousand years ago. He thought that all matters consists of four elements: Earth, water, air and fire. Of course, the meaning of these four words: earth, water, air and fire are not the same in their modern sense. In modern language, we probably should translate them as solid (earth), liquid (water), gas (air) and plasma (fire). A better description of matter is the atomic theory of matter[2] as we discuss in Chapter II. However, because of many physical and mathematical difficulties, the atomic or kinetic theory of matter in general, and that of the liquid in particular is still far from well developed. For engineering problems, the macroscopic theory is always used in the analysis of flow problems. Hence, in this book, only the macroscopic treatment is considered.

2. Classification of two-phase flows[1]

Before we study the two-phase flows, we should give a definition of two-phase flow. By two-phase flow, we mean a special flow problem in which we consider the mechanics of two phases of matter simultaneously. In general, two-phase flows may be divided into two groups: The first group consists of the flow problems of the mixture of two phases of matter of the four states: solid (pseudo-fluid), liquid, gas and/or plasma. These two phases may be of the same or different substances and may be mixed homogeneously or inhomogeneously. Customarily, the two-phase flows refer to these problems in this first group only. The second group consists of the study of the interaction between two phases of matter through their interface. In each phase, the matter is a homogeneous medium of single phase, but we have to solve the mechanics of these two phases simultaneously because they are coupled through their interface.

Since the properties of a substance in different states are greatly different (cf. Chapter II), the two-phase flows should be classified according to the states in the flow field[1]. Each class should be treated independently from the other even though there may be similarity between some of the two-phase flows.

We may classify the two-phase flows as follows:

(i) Liquid-gas flow

Here we may again deal with two kinds of liquid-gas flow: One is the case in which the substances of these two phase are the same. In the flow, there is a phase change of this substance. We shall discuss this case in §3. The other is the case in which the gas and the liquid in the mixture are two different substances. We are going to discuss some of these two-phase liquid-gas flows in §4.

(ii) Gas-solid flow[3]

We are going to discuss some of these two-phase flows of a gas and solid particles in §5.

(iii) Liquid-solid flows

In many cases, there are some similarities between the liquid-solid flows and the gas-solid flows. Hence many of the analysis in § 5 may be applied to the solid-liquid flow.

(iv) Plasma-solid flow

This two-phase flow includes the case of a mixture of charged particles in a neutral gas which will be discussed in §6.

(v) Liquid-plasma flow

This case includes the case of the mixture of electrically conducting liquid and a gas which will be discussed in §7.

(vi) Gas-plasma flow

This two-phase flows include the cases of a mixture of different gases. We have so-called multi-fluid theory of a plasma which will be discussed in §8. We study the wave motion in a plasma by two-fluid theory in § 9. We apply this theory of three-fluids to study the general electrical conductivity of a plasma in § 10.

There is a very interesting two-phase flow of gas-liquid for the compressible flow which is the ablation in hypersonic flow which will be discussed in Chapter XVII, §11 in which the transport phenomena should be included.

Since two-phase flow is a very complicated flow phenomenon, special treatise should be referred to for the detailed knowledge (cf. Ref. 1). In this chapter, we shall discuss only some essential points of two -phase flow and their main new features which are different from the single phase flows which have been studied in the previous chapters of this book.

3. Two-phase flow including the effect of change of phase in the flow field[4]

In many practical fluid flow problems, there are phase changes in the flow field, i.e., the effects of condensation and/or evaporation are important. One of such problems is that associated with meteorology and the another is the well-known condensation shock in a supersonic wind tunnel (cf. Chapter IV, §9). In this section, we are going to consider the flow of a moist air in a de Laval nozzle with arbitrary degree of humidity. Three possible types of flow for this case have been observed:[4,5,6]

(a) The flow exhibits a steady and continuous behavior;

(b) A steady shock occurs at the start of condensation region;

(c) The flow becomes unsteady even though the boundary conditions are steady.

There is no satisfactory theory to explain the possibility of these three different flow patterns, particularly the prediction of the starting point of condensation and the conditions when the unsteady periodic processes set in. Most of the theoretical analyses, starting with the classical work of Oswatitsch[7] and those of Wegener and his associates[8], are concerned with both microscopic (formation of droplets) and macroscopic (continuous flow) points of view. It seems that such analysis is rather inconsistent and difficult to deal with. In this section, we are going to study this problem by continuum theory only. By means of a relaxation time, we formulate the problem by two-fluid theory because the problem is quite similar to the well-known analysis of flow with chemical reactions (cf. Chapter XVII, §10). In our approach, the effects of both condensation and evaporation will be considered depending whether the dryness factor λ is increasing (evaporation) or decreasing (condensation). We are going to discuss first the essential features of our two-fluid theory as follows:

We consider the flow of a mixture of two fluids, i.e., fluid 1 and fluid 2. The fluid 1 is always in the gaseous state in the whole flow field and subscript 1 or a is used for the value of any flow variable of fluid 1. The fluid 2 may be in the vapor, liquid or coexistence of liquid and vapor states and subscript 2 or w is used for the value of any flow variables of fluid 2. For each of these two fluids, we have in general, six flow variables: 3 velocity components and three state variables, i.e.,

For fluid 1, we have $\boldsymbol{q}_1\ (u_1, v_1, w_1),\ p_1, \rho_1, T_1$ and

For fluid 2, we have $\boldsymbol{q}_2\ (u_2, v_2, w_2),\ p_2, \rho_2, T_2$ (or λ)

where $\boldsymbol{q}$ is the velocity vector with u, v and w as respectively the x-, y- and z-component of the velocity vector; p is the pressure, ρ is the density and T is the temperature. For fluid 2, because of the phase change in the flow field, it is convenient to introduce a dryness factor λ such that

$$\frac{1}{\rho_2} = V_2 = V_L + \lambda(V_V - V_L) = V_2(p, T, \lambda) \qquad (16.1)$$

where $V = 1/\rho$ is the specific volume of the fluid and the value of λ lies between 0 and 1 in the following manner:

$$\text{when} \quad \left.\begin{array}{ll} T > T_S(p), & \lambda = 1 \\ T < T_S(p), & \lambda = 0 \\ T = T_S(p), & 0 \leqslant \lambda \leqslant 1 \end{array}\right\} \qquad (16.2)$$

The saturated temperature of the fluid 2, T_S, is a given function of the pressure p. Even though we use seven variables for fluid 2, at any given point in the flow field, only six variables of the fluid 2 are independent variables.

For two-fluid theory, we have to distinguish the species density from the partial density[1]. Let us consider an element of the mixture of the fluid 1 and the fluid 2 with total mass $\overline{M} = \overline{M}_1 + \overline{M}_2$ and total volume $\overline{V} = \overline{V}_1 + \overline{V}_2$. The value for the mixture as a whole is that without subscript.

The species density of fluid 1 is

$$\rho_1 = \frac{\overline{M}_1}{\overline{V}_1} = \rho_a \qquad (16.3)$$

and the species density of fluid 2 is

$$\rho_2 = \frac{\overline{M}_2}{\overline{V}_2} = \rho_w \tag{16.4}$$

The partial density of fluid 1 is

$$\overline{\rho}_1 = \frac{\overline{M}_1}{\overline{V}} = (1-Z)\,\rho_a \tag{16.5}$$

and the partial density of fluid 2 is

$$\overline{\rho}_2 = \frac{\overline{M}_2}{\overline{V}} = Z\,\rho_w \tag{16.6}$$

where

$$Z = \frac{\overline{V}_2}{\overline{V}} = \text{volume fraction of fluid 2 in the mixture} \tag{16.7}$$

In the fundamental equations of two-fluid theory, the partial density is always used. The total density of the mixture is

$$\rho = \frac{\overline{M}}{\overline{V}} = \overline{\rho}_1 + \overline{\rho}_2 = (1-Z)\,\rho_a + Z\,\rho_w \tag{16.8}$$

The total pressure of the mixture p is

$$p = p_1 + p_2 \tag{16.9}$$

Since fluid 1 is always in the gaseous state, we assume that the perfect gas law holds for fluid 1, i.e.,

$$p_1 = \overline{\rho}_1 R_1 T_1 = (1-Z)\,\rho_1 R_1 T_1 = (1-Z)\,p \tag{16.10}$$

where R_1 is the gas constant of fluid 1.

From Eqs.(16.9) and (16.10) we have

$$p_2 = Z\,p \tag{16.11}$$

In our two-fluid theory, it is convenient to use the following 13 variables:

$$\boldsymbol{q}_1(u_1, v_1, w_1),\ T_1, \rho_1, p, Z;\ \boldsymbol{q}_2(u_2, v_2, w_2),\ T_2, \rho_2, \lambda \tag{16.12}$$

We should have 13 fundamental equations for these 13 variables. The fundamental equations are given in Ref.4.

Now we are going to discuss in details the one-dimensional unsteady inviscid flow through a de Laval nozzle with condensation and evaporation effect. Let the cross-sectional area of the nozzle be $A(x)$. The moist air is from a reservior of constant pressure p_r at constant temperature T_r with a specific humidity

$$k_{20} = \frac{\overline{\rho}_{20}}{\overline{\rho}_{a0} + \overline{\rho}_{20}} = \frac{\overline{\rho}_{20}}{\rho_0} = \frac{Z_0\,\rho_{20}}{\rho_0} \tag{16.13}$$

The x is the distance along the axis of the nozzle. We shall consider only the main effect of heat release or absorption due to the condensation or the evaporation of the water in the mixture in the flow field. Hence we may neglect the effects of viscosity, heat conduction and diffusion phenomena.
Our variables are:

(1) p = the pressure of the mixture;
(2) $T = T_a = T_w$ = the temperature of the mixture;
(3) ρ = the density of the mixture;
(4) $U = U_a = U_w$ = the velocity of the mixture in the direction along the axis of the nozzle;
(5) Z = volume fraction of the water in the mixture;
(6) λ = dryness fraction of the water in the mixture.

Here we assume that the flow is in equilibrium as far as the fluid dynamic variables are concerned, i. e., $U = U_a = U_w$ and $T = T_a = T_w$, where subscript a refers to the value of the air and subscript w refers to the value of the water which may be in vapor, liquid or coexistence of vapor and liquid. Without the diffusion effect, $k_2 = k_{2o}$ is a constant in the whole flow field. Hence the volume fraction Z is related to the other variables by the following relation:

$$k_2 = \frac{\overline{\rho_2}}{\rho} = \frac{Z\rho_2}{\rho} \tag{16.14}$$

Let us introduce the following non-dimensional variables:

$$\left.\begin{aligned} &\overline{x} = \frac{x}{b},\ \overline{T} = \frac{T}{T_0},\ \overline{p} = \frac{p}{p_0},\ \overline{\rho} = \frac{\rho}{\rho_0},\ \overline{t} = \frac{t}{b/U_0},\ \overline{U} = \frac{U}{U_0} \\ &\overline{T_S} = \frac{T_S}{T_0},\ \overline{U}_m = \frac{U_m}{c_{va}T_0},\ \overline{\rho}_w = \frac{\rho_w}{\rho_0},\ \overline{L} = \frac{L}{R_w T_0} \end{aligned}\right\} \tag{16.15}$$

where bar refers to the non-dimensional quantities, b is the length of the nozzle subscript o refers to the value at the entrance of the nozzle $x = 0$, and L is the latent heat of evaporation of fluid 2 which is a function of the saturated temperature T_S.

The non-dimensional fundamental equations of our problems are as follows:

$$\overline{p} = \overline{\rho}\,\overline{T}\,\frac{(1 - Z_0)}{(1 - Z)} \tag{16.16a}$$

$$\frac{\partial\overline{\rho}}{\partial\overline{t}} = -\overline{U}\frac{\partial\overline{\rho}}{\partial\overline{x}} - \overline{\rho}\frac{\partial\overline{U}}{\partial\overline{x}} - \frac{\overline{\rho U}}{A}\frac{dA}{dx} \tag{16.16b}$$

$$\frac{\partial \overline{U}}{\partial \overline{t}} = -\overline{U}\frac{\partial \overline{U}}{\partial \overline{x}} - \frac{1}{\gamma_a M_0^2}\frac{1}{\overline{\rho}}\frac{\partial \overline{p}}{\partial \overline{x}} \tag{16.16c}$$

$$\frac{\partial \overline{U}_m}{\partial \overline{t}} = -\overline{U}\frac{\partial \overline{U}_m}{\partial \overline{x}} - \frac{(1-k_{20})(\gamma_a - 1)}{(1-Z_0)}\frac{\overline{p}}{\overline{\rho}}\left(\frac{\partial \overline{U}}{\partial \overline{x}} + \frac{1}{A}\frac{dA}{d\overline{x}}\right) \tag{16.16d}$$

$$\frac{\partial \lambda}{\partial \overline{t}} = -\overline{U}\frac{\partial \lambda}{\partial \overline{x}} + \frac{1}{R_{t\lambda}}(\lambda_e - \lambda) \tag{16.16e}$$

$$Z = \frac{k_{20}\overline{\rho}}{\overline{\rho}_w} \tag{16.16f}$$

$$\frac{d\overline{p}}{d\overline{T}_S} = \frac{\overline{L}\,(1-Z_0)\,\dfrac{R_v}{R_a}}{\left(\dfrac{1}{\overline{\rho}_{wvs}} - \dfrac{1}{\overline{\rho}_{wls}}\right)\overline{T}_S\,(1-k_{20})} \tag{16.16g}$$

where

$$M_0 = \frac{U_0}{a_0} = \text{initial Mach number at } x=0 \tag{16.17}$$

and $a_0 = \sqrt{\gamma_0 p_0/\rho_0}$ is the sound speed of the mixture when both Z and k_{20} are negligibly small. In general, the sound speed of the mixture is different from a_0 (cf. § 4).

$$R_t = \frac{t_\lambda U_0}{b} = \text{non-dimensional relaxation time} \tag{16.18}$$

and t_λ is the relaxation time for evaporation or condensation and

$$\lambda_e = \frac{\dfrac{1}{\rho_2} - \dfrac{1}{\rho_{2s}}}{\dfrac{1}{\rho_{vs}} - \dfrac{1}{\rho_{1s}}} = \lambda_e(\rho_2, T_S) \tag{16.19}$$

λ_e is the equilibrium value of λ which is a function of the density ρ_2 and the saturated temperature T_S.

In this problem, we have four non-dimensional parameters, i.e., M_0, R_{t_λ}, k_{20} and $\overline{L}$ for given initial and boundary conditions. For the boundary conditions, we have given values of the variables at the entrance section $x=0$ such as λ_0, p_0, T_0 and ρ_0. At the exit section $x=b$, there might be a steady uniform state or not. Hence the values of the variables at the exit section may be a function of time. We have to study the case whether there is a steady uniform state at a section far downstream for a given uniform state at the entrance section.

To investigate the possibility of a steady uniform state far downstream, it is convenient to solve this problem by comparing the analysis of our present two-fluid theory with the conventional single fluid theory as we discussed in Chapter III, § 8. To make this comparison, we first reduce Eqs. (16.16) into those forms in single fluid theory with heat addition. For simplicity, we consider the case of a nozzle with constant cross-sectional area, i.e., $A(x)=\text{constant}$.

In ordinary single fluid theory, the following approximations are usually made[5]:

(1) The volume fraction of the water is assumed to be negligibly small, i.e., $Z=0$ and $Z_0=0$.

(2) The density of the mixture ρ is approximately equal to that of the air, i.e., $\rho=\rho_a$, $R_M=R_a$.

(3) We do not calculate how the dryness fraction λ varies in the flow field, i.e., we do not calculate the relaxation zone of condensation and evaporation but consider the limiting equilibrium cases before and after the relaxation zone. Hence we take $\lambda=\lambda_0=1$ before the relaxation zone and $\lambda=\lambda_\infty=0$ after the relaxation zone.

With the above assumptions, for the case of steady flow, Eqs. (16.16) are reduced to the following forms[5]:

$$p=\rho_a R_a T \tag{16.20a}$$

$$\rho_a U=\rho_{a0} U_0=\text{constant} \tag{16.20b}$$

$$\rho_a U^2+p=\rho_{a0} U_0^2+p_0=\text{constant} \tag{16.20c}$$

$$H_a+\frac{1}{2}U^2=H_{a0}+\frac{1}{2}U_0^2+k_{20}L \tag{16.20d}$$

where

$$L=H_{wvo}-H_{w1}=\text{latent heat of condensation of the water in the mixture} \tag{16.21}$$

and H is the enthalpy. Eqs.(16.20) are the equations of single-fluid theory (cf Chapter III, § 8) with heat addition due to condensation of water vapor, $dQ=Q_w=k_{20}L$. As we showed in Chapter III, § 8, if the initial Mach number M_0 is supersonic, for a given Q_w below a critical value Q_{wc}, the final Mach number M_∞ decreases with the increase of Q_w and when $Q_w=Q_{wc}$, $M_\infty=1$. When $Q_w>Q_{wc}$, there will be no steady state solution for the flow in this nozzle.

The corresponding steady state equation of our two-fluid theory, i.e., Eqs.(16.16) give the following equations:

$$p=\frac{\rho R_M T}{1-Z} \tag{16.22a}$$

$$\rho\ U=\rho_0\ U_0=\text{constant} \tag{16.22b}$$

$$\rho\ U^2+p=\rho_0 U_0^2+p_0=\text{constant} \tag{16.22c}$$

$$k_{20}\ U_{mw}+(1-k_{20})c_{va}\ T+\frac{p}{\rho}+\frac{1}{2}\ U^2=\text{constant} \tag{16.22d}$$

$$=k_{20}\ U_{mwo}+(1-k_{20})c_{va}T_0+\frac{p_0}{\rho_0}+\frac{1}{2}\ U_0^2$$

$$U\ \frac{\partial\lambda}{\partial x}=\frac{1}{t_\lambda}\ (\lambda_e-\lambda) \tag{16.22e}$$

$$Z=\frac{k_{20}}{\rho_w(p,\ T,\ \lambda)} \tag{16.22f}$$

The main difference between Eqs.(16.20) and (16.22) is that by the two-fluid theory, Eqs.(16.22), we may calculate the flow field in the relaxation zone. Furthermore, the assumptions (1) and (2) for the single fluid theory, Eq.(16.20), will introduce some errors in the analysis if k_{20}, Z or Z_0 is not negligibly small. Since the heat addition Q_w is proportional to k_{20}, the interesting case that the flow will be unsteady for steady uniform entrance condition occurs when k_{20} and then Z_0 is sufficiently large. For large value of k_{20}, Eqs.(16.22) would be more accurate than those of Eqs.(16. 20). At the final equilibrium condition far downstream, the dryness fraction may be a constant but different from zero. From Eqs.(16.22a) to (16.22d), we may calculate a critical value of $\lambda_{\infty c}$ such that $M_\infty=1$ for a given large value of k_{20}. Hence for $\lambda_\infty>\lambda_{\infty c}$, steady state uniform solution will always exist far downstream. We will not be able to determine this $\lambda_{\infty c}$ in the conventional single-fluid theory of Eqs.(16.22). When $Q_w>Q_{wc}$ or $\lambda_\infty<\lambda_{\infty c}$, we have to use Eqs.(16.16) to study the unsteady flow field. The results of this calculation of the unsteady flow field will be given as problem 3 in §11.

It should be noticed that the sound speed of fluid 2 which may be a mixture of the liquid and its own vapor behaves quite different from the ordinary gas in certain domain. It can be shown by the following analysis:

The sound speed "a" of a medium is defined as

$$a=\left(\frac{dp}{d\rho}\right)_S^{\frac{1}{2}}=V\left[-\left(\frac{dp}{dV}\right)_S\right]^{\frac{1}{2}} \tag{16.23}$$

where $V=1/\rho$ is the specific volume and subscript S refers to the isentropic process and S is the entropy.

In general, we may consider the pressure p as a function of the specific volume V and temperature T, i.e., $p=p(V,T)$. Then

$$\left(\frac{dp}{dV}\right)_S=\left(\frac{\partial p}{\partial V}\right)_T+\left(\frac{\partial p}{\partial T}\right)_V\left(\frac{dT}{dV}\right)_S \tag{16.24}$$

The first law of thermodynamics gives for isentropic process

$$T\,dS=\left(\frac{\partial U_m}{\partial T}\right)_V dT+\left[\left(\frac{\partial U_m}{\partial V}\right)_T+p\right]dv$$

$$=C_v\,dT+\left[\left(\frac{dU_m}{dV}\right)_T+p\right]dV=0 \tag{16.25}$$

From Eq.(16.25), we have

$$\left(\frac{dT}{dV}\right)_S=-\frac{1}{C_v}\left[\left(\frac{dU_m}{dV}\right)_T+p\right] \tag{16.26}$$

Since $\dfrac{\partial^2 S}{\partial V\partial T}=\dfrac{\partial^2 S}{\partial T\partial V}$, we have from Eq.(16.25)

$$\left(\frac{\partial U_m}{\partial V}\right)_T+p=T\left(\frac{\partial p}{\partial T}\right)_V \tag{16.27}$$

From Eqs. (16.26) and (16.27), we have

$$\left(\frac{dT}{dV}\right)_S=-\left(\frac{\partial p}{\partial T}\right)_V\frac{T}{C_v} \tag{16.28}$$

From Eqs.(16.23), (16.24) and (16.28), we have

$$a^2=-V^2\left[\left(\frac{\partial p}{\partial V}\right)_T-\left(\frac{\partial p}{\partial T}\right)_V^2\frac{T}{C_v}\right]=\left(\frac{\partial p}{\partial \rho}\right)_T+\frac{1}{\rho^2}\left(\frac{\partial p}{\partial T}\right)^2\frac{T}{C_v} \tag{16.29}$$

For a perfect gas with the equation of state

$$p=\rho\,RT \tag{16.30}$$

Eq.(16.29) becomes

$$a^2=RT\left(1+\frac{R}{C_v}\right)=\gamma\,RT \tag{16.31}$$

Eq.(16.31) is the wellknown results of the sound speed of a perfect gas. [cf. Eq.(3.10)].

In the coexistence region of the liquid and its own vapor, the pressure p is a function of the temperature T only and Eq.(16.29) becomes

$$a=\frac{1}{\rho}\left(\frac{dp}{dT}\right)\left(\frac{T}{C_v}\right)^{\frac{1}{2}} \tag{16.32}$$

In this coexistence region $p=p(T)$ is given by the Clausius-Clapeyron equation (2.58). Hence from Eqs.(2.58) and (16.32), we have

$$a=\left(\frac{\rho_v}{\rho}\right)\frac{L}{\left(1-\frac{\rho_v}{\rho_L}\right)}\left(\frac{1}{C_v T}\right)^{\frac{1}{2}} \tag{16.33}$$

If we consider the case near the saturated vapor line, $\rho_v \simeq \rho$ and $\rho_v \ll \rho_L$ and L = constant, Eq.(16.33) becomes

$$a=L\left(\frac{1}{C_v T}\right)^{\frac{1}{2}} \tag{16.34}$$

Now the sound speed of the medium is inversely proportional to $(T)^{\frac{1}{2}}$ rather than proportional to $(T)^{\frac{1}{2}}$ in the case of a perfect gas.

4. Two-phase flows of gas and liquid

In §3, we consider a special case of the two-phase flow of gas (or vapor) and liquid. In general, the two-phase flows of gas and liquid are much more complicated than what we discuss in §3. The phenomenon of a small amount of gas in a large amount of liquid is different from the case of a small amount of liquid in a large amount of gas. Furthermore, the relative position of the liquid and gas also introduces complications in the treatment of liquid-gas flows. Hence it is convenient to divide the liquid-gas flows into several special classes as follows:

(i) Bubble flow and cavitation[9, 10]

These are phenomena of a small amount of gas in a large amount of liquid. When we mix a small amount of gas with liquid, there will be many bubbles in the liquid. There is a special subject known as bubble dynamics[10] which studies the formation of bubbles in liquid, stability problems of bubbles and statistical representation of the bubbles. We do not study these problems in this book and special treatises should be referred to for this subjects. (Ref. 1, 11, 12.)

A special case of bubbles in liquid is known as cavitation[11]. In a liquid whose stagnation or undisturbed pressure p_0 is larger than its vapor pressure p_v, the motion of this liquid may reduce the pressure of the liquid p to a value less than its vapor pressure. Vapor is then suddenly formed and the pressure of the liquid rises to its vapor pressure p_v. Because of the formation of vapor in the liquid, the streamline of the flow field will be different from the case without such a cavitation. When the pressure of the liquid rises above the vapor pressure, the cavitation may collapse suddenly with a large sound. If there is no gas in the liquid, a cavity may not form even if the liquid pressure is below the vapor pressure. However, if there are some gases in the liquid, cavitation begins when the vapor pressure is reached. Cavitation frequently begins when the flow of liquid separates from the wall. Cavitation is very important in the high speed flow of water such as high speed water turbines, ship propeller, and hydrofoils. Special treatise of cavitation should be referred to for the study of cavitation (cf. Ref. 11).

Another similar phenomenon to ordinary cavitation is the boiling effects (Ref. 12, 13 and 14). If the temperature of the body is high enough, either due to heat addition[12, 13] or due to high speed viscous dissipation[14], the liquid begins to boil at the surface of the body. Bubbles and cavitation may occur in such cases. Such phenomena have been studied from the heat transfer point of view[12, 13] and we shall discuss some of these problem in Chapter XVII, § 11.

(ii) Atomization of liquid and spray flow[1, 9, 15]

These are phenomena of a small amount of liquid in a large amount of gas. There are many engineering problems in which we inject liquid in a stream of gas. When the jet of liquid issues into a gaseous medium, the slender jet will break up into drops due to instability of the jet. The atomization of a liquid jet has been extensively studied in connection with combustion problems. In nature, the most common example of this case is the rain, i.e., water drops in the air. If the size of the water drops is very small, the shape of the drop is practically spherical. We may consider the water drops as small rigid spheres in the air. This problem may be treated in the same way as the mixture of solid particles and gas as we shall discuss in §5.

(iii) Froth flow and spray flow[16−19]

Another case which differs from the above two cases is one in which the volume of the gas and that of the liquid are of the same order of magnitude and the gas and the liquid mix homogeneously or inhomogeneously

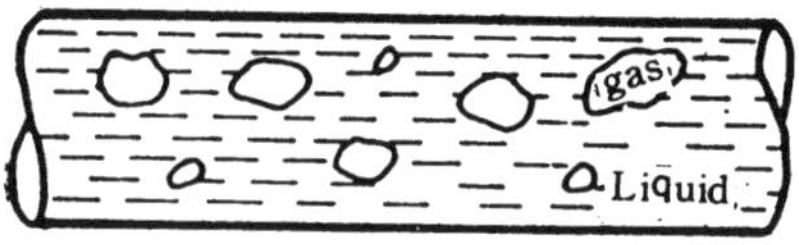

(a) Bubble flow

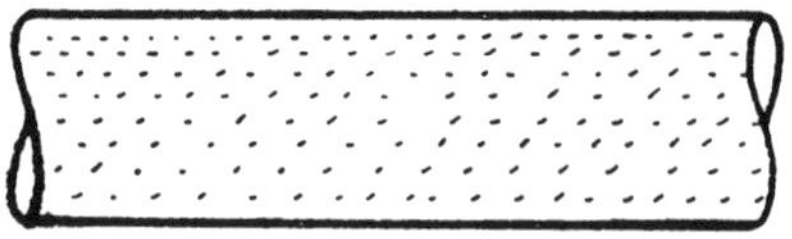

(b) Froth flow or spary flow

Fig. 16.1 Bubble flow and froth or spray flow in a pipe (First group of two-phase flow of liquid and gas)

(Fig. 16.1). If we spray the liquid drops into a stream of gas, we call it spray flow and if we introduce bubbles into a liquid, we call it bubble flow in which the bubbles of gas do not mix homogeneously. However, if the bubbles are very small, and the gas and the liquid mix homogeneously, we call it froth flow or homogeneous bubble flow. The treatment of froth flow may be used the continuum theory as we did in §3 and we will further discuss the treatment of froth flow in later part of this section. One of the new features of the froth flow is that we have to consider the change of density of the mixture due to the change of relative volume of the gas and the liquid. Hence the new variable, relative volume of the gas to that of the liquid plays an important role in the flow of the mixture. This relative volume may be expressed by the volume fraction Z of §3.

(iv) Plug flow, stratified flow, wavy flow and annular flow[1, 9, 15]

These are flow problems of the second group of two-phase flows of liquid and gas. They are classified according to the relative positions between the liquid and gas. Because these flows were first studied in connection with pipe flows, various name were introduced accordingly. In each of the flow regions, we should consider a single phase. Only on the boundary of two regions, we should consider the two phases. One of the difficulties of such problems is that the boundary between the two phases is not known a priori and depends on the solution of the problem and that the boundary may be irregular in shape. Because of these difficulties, engineers classify the flow empirically as shown in Fig. 16.2. We are going to describe these flows briefly as follows:

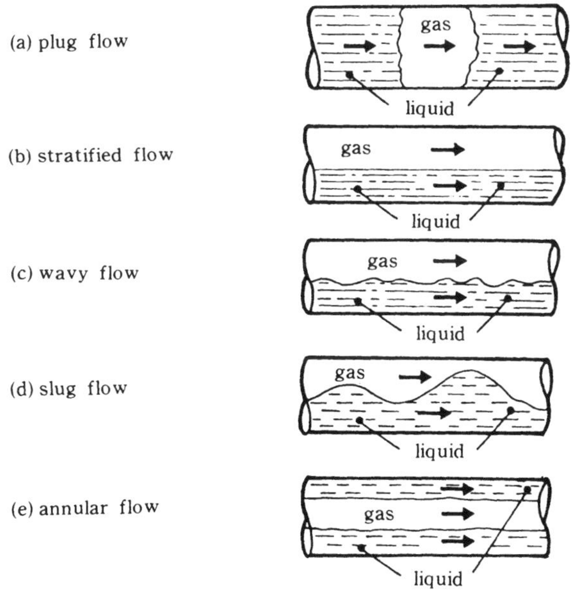

Fig. 16.2 Second group of two-phase flow of liquid and gas. Various types of two-phase flow of liquid and gas in a pipe or a channel

(a) Plug flow (Fig.16.2a)

In the plug flow in a pipe, the flow of liquid and that of gas are in tandem position. In a way, we may consider the plug flow as a special case of flow with cavitation.

(b) Stratified flow (Fig. 16. 2b), wavy flow (Fig. 16. 2c) and slug flow (Fig.16.2d)

In these three types of two-phase flow, the gas flows on one side of the channel while the liquid flows on the other side. If the pipe is horizontal and the flow is steady and laminar, the interface between the gas and the liquid will be straight horizontal surface. Such a flow is known as stratified flow (Fig.16.2b). If the interface is not a straight surface but a wavy surface (Fig.16.2c), it is called wavy flow. If the interface of this stratified flow has large variation of shape as shown in Fig.16.2d, the flow is called slug flow. The only difference between stratified flow, wavy flow and slug flow is the shape of the interface. Hence they are really the same type of flows which may be simply called stratified flow in a general sense. The main feature of these three flows is the interaction on the flow of the gas and that of liquid through its interface which has been extensively studied[20]. One of the best known examples of stratified flow is the interaction of atmosphere with the ocean. In the case of calm weather, the ocean surface is a flat surface which is that shown in Fig. 16.2(b). When there are waves on the surface of the ocean, we have the wavy flow as shown in Fig.16.2(c). When the weather is very rough, we have slug flow as shown in Fig.16.2(d).

(c) Annular flow (Fig.16.2e)

In Fig.16.2(e)., we consider a stream of gas submerged in a stream of liquid. Since we may consider the flow of Fig.16.2(e)as a pipe flow, we have a ring of liquid surrounding a stream of gas. Thus we call it annular flow. In a similar manner, we may have an annular flow in which a stream of liquid is submerged in a stream of gas.

In engineering literature, it should be noted that the above classifications are empirical in nature and have been widely used as a description of two-phase flow of liquid and gas.[15] In most engineering practices, we do not solve the flow problem but some empirical formulas to estimate the pressure drop and flow rate in pipe are given[15]. It is advisable to solve these problems analytically so that a better understanding of the problem may be obtained.

Even though there is no simple way to predict various types of two-phase flows of liquid and gas shown in Figs.16.1 and 16.2, the type of flow depends mainly on the relative rate of flow of liquid and gas. For instance, we consider a horizontal pipe with concurrent flow of a liquid and a gas. If the flow rate of the gas is very small, we have the bubble flow of Fig.16.1(a). As the rate of the gas increases the bubble increases in size and finally, we have very large bubble and the plug flow of Fig.16.2(a). Further increase of the flow rate of the gas will produce the stratified flow so that the liquid flows along the bottom of the pipe when the gas flows on the upper half with a smooth gas-liquid interface (Fig.16.2b). When the rate of flow of the gas increases further, the interface is no longer smooth and we have first wavy flow (Fig.16.2c) and then slug flow (Fig.16.2d), which are due to the instability of the interface. When the gas flow rate is very high, the gas becomes a jet in the liquid and we have the annular flow of Fig.16.2(e). At very high rate of the flow of the gas, the gas and the liquid will mix homogeneously and we have the spray flow or froth flow of Fig.16.1(b).

(v) Liquid-vapor flow; superspeed hydrodynamics[1, 14]

In the above four types of liquid-gas flow, we implicitly assume that the gas and liquid are not the same substance. Hence at the same temperature and pressure, one substance is always in the gaseous state while the other is in the liquid state. But at very high or very low temperature, there may be phase change of the substance as we study in §3. New phenomena may occur for the flow with phase change. The classical example of this type of flow is the flow in a boiler[1, 12]. A modern example of this type flow is superspeed hydrodynamics[14]. We may image the case in which a very slender body or a flat plate moves at a very high speed in the ocean. If the speed of the body is high enough, the surface temperature due to skin friction will be so high that the water begins to boil. But at a large distance from the body, the medium is still in its liquid state, i.e., ordinary water. If the speed of the body is very high, we would expect there would be a water vapor boundary layer surrounding this thin body which would significantly affect the skin friction and heat transfer of the body in comparison with the case that the medium remains in its liquid state. This problem has been treated in reference 14, and some numerical results are given.

(vi) Ablation[1, 21]

Another interesting liquid-gas flow problem is ablation which is an important method in protecting the surface of a space vehicle from overheating. This is a liquid boundary layer in a stream of gas. We shall also discuss this two-phase problem in Chapter XVII, § 11.

Among various types of two-phase flow of liquid-gas, the froth flow has been well developed from the analytical treatment. We are going to discuss the froth flow in the detailed treatment as follows:

In froth flow we have 12 variables:

$$\boldsymbol{q}_L, \boldsymbol{q}_g, p_L, p_g, \overline{\rho}_L, \overline{\rho}_g, T_L, T_g \tag{16.35}$$

where subscript L refers to the value of the liquid species and subscript g refers to the value of the gaseous species. In liquid-gas two-phase flows, void fraction f is used instead of the volume fraction Z of Eq.(16.5). The relation between f and Z is as follows:

$$f = \frac{\text{volume of the gas in an element of the mixture}}{\text{volume of the element of the mixture}}$$

$$= \frac{V_L}{V} = (1 - Z) \tag{16.36}$$

Furthermore the relations (16.10) and (16.11) still hold in the general two-phase flows of liquid-gas so that we may replace p_L and p_g by p, the total pressure of the mixture (cf Eq.16.12).

We have 12 equations for these 12 variables as follows:

(1) Equation of state of the gas, i.e., Eq.(16.10).

(2) Equation of state of the liquid, i.e.,

$$\rho_L = \rho_L\ (T_L) \tag{16.37}$$

In many practical cases, we may assume that ρ_L is a constant except when condensation and evaporation effects become important. Then we should introduce dryness fraction λ as we did in §3.

(3) Equations of continuity.

The conservation of mass of the gas and that of the liquid give respectively the corresponding equation of continuity as follows:

For the gas, we have

$$\frac{\partial f \rho_g}{\partial t} + \frac{\partial}{\partial x^i} (f \rho_g u_g^i) = \sigma_g \tag{16.38}$$

and for the liquid, we have

$$\frac{\partial (1-f) \rho_L}{\partial t} + \frac{\partial}{\partial x^i} \left[(1-f) \rho_L u_L^i \right] = \sigma_L \tag{16.39}$$

where u^i is the ith component of the velocity vector q and σ_g and σ_L are respectively the source terms of the gas and the liquid. Ordinarily, we usually assume that both σ_g and σ_L are zero, i.e., no mass transfer between the gas and the liquid. In general, $\sigma_g = -\sigma_L$, adding Eqs.(16.38) and (16.39), we have the equation of continuity for the mixture as follows:

$$\frac{\partial \rho}{\partial t} + \frac{\partial \rho u^i}{\partial x^i} = 0 \tag{16.40}$$

where ρ is the density of the mixture defined in Eq.(16.8) and u^i is the ith component of the velocity vector of the mixture defined as follows:

$$\rho\ u^i = (\bar{\rho}_g u_g^i + \bar{\rho}_L u_g^i) \tag{16.41}$$

(4) Equations of motion.

The equations of motion for the gas and those for the liquid are as follows:

$$\bar{\rho}_g\left(\frac{\partial u_g^i}{\partial t}+u_g^j\frac{\partial u_g^i}{\partial x_i^j}\right)=-\frac{\partial p_g}{\partial x^i}+\frac{\partial \tau_g^{ij}}{\partial x^j}+K_{gL}(u_L^i-u_g^i) \quad (16.42)$$

$$\bar{\rho}_L\left(\frac{\partial u_L^i}{\partial t}+u_L^j\frac{\partial u_L^i}{\partial x^j}\right)=-\frac{\partial p_L}{\partial x^i}+\frac{\partial \tau_L^{ij}}{\partial x^j}+K_{Lg}(u_g^i-u_L^i) \quad (16.43)$$

where $K_{gL}=K_{Lg}$ is the friction coefficient between the gas and the liquid which gives the interaction forces between them. The viscous stresses of the gas and those of the liquid are respectively τ_g^{ij} and τ_L^{ij}.

If we add Eqs. (16.42) and (16.43), we have the equations of motion of the mixture in the following form:

$$\rho\left(\frac{\partial u^i}{\partial t}+u^j\frac{\partial u^i}{\partial x^j}\right)=-\frac{\partial p}{\partial x^i}+\frac{\partial \tau^{ij}}{\partial x^j} \quad (16.44)$$

The viscous stresses of the mixture include not only the viscous stresses of each species and also the effects due to diffusion velocity which will be discussed in Chapter XVII.

(5) Equations of energy.

The energy equations for gas and that for the liquid are respectively:

$$\frac{\partial\bar{\rho}_g(U_{mg}+\frac{1}{2}u_g^2+\varphi_g)}{\partial t}+\frac{\partial}{\partial x^j}\left[\bar{\rho}_g u_g^j(U_{mg}+\frac{1}{2}u_g^2+\varphi_g)-u_g^i\tau_g^{ij}+\right.$$

$$\left.+\delta^{ij}u_g^i p_g-Q_{cg}^j\right]=K_T(T_L-T_g) \quad (16.45)$$

$$\frac{\partial\bar{\rho}_L(U_{mL}+\frac{1}{2}u_L^i+\varphi_L)}{\partial t}+\frac{\partial}{\partial x^j}\left[\bar{\rho}_L u_L^j(U_{mL}+\frac{1}{2}u_L^2+\varphi_L)-u_g^i\tau_g^{ij}+\right.$$

$$\left.+\delta^{ij}u_g^i p_g-Q_{cL}^j\right]=K_T(T_g-T_L) \quad (16.46)$$

where U_{ms} is the internal energy of species s per unit mass, $\delta^{ij}=0$ if $i\neq j$ and $\delta^{ij}=1$ if $i=j$; Q_{cs}^j is the jth component of the heat conductive flux of sth species; and K_T is the thermal friction coefficient between the gas and the liquid and φ_s is the potential energy of sth species. If we add Eqs.(16.45) and (16.46), we may have the energy equation of the mixture in the same form as that of a single fluid but in the viscous stresses of the mixture and the heat flux of the mixture, we have additional terms due to diffusion velocities which will be discussed in Chapter XVII.

In conclusion, the fundamental equations of the froth flow, i.e., for the homogeneous mixture of a gas and a liquid are Eqs.(16.10), (16.37), (16.38), (16.39), (16.42), (16.43), (16.45) and (16.46).

Since it is very complicated to consider all the 12 variables, we may simplify this problem by various approximations:

The simplest model of the froth flow has the following assumptions:

(a) We assume the temperature of the gas and that of the liquid are the same, i.e.,

$$T_g = T_L = T \tag{16.47}$$

(b) We assume that the velocity of the gas and that of the liquid are the same, i.e.,

$$\boldsymbol{q}_g = \boldsymbol{q}_L = \boldsymbol{q} \tag{16.48}$$

With the assumption of Eq.(16.48) and with $\sigma_g = 0$, Eq.(16.38) becomes

$$\frac{\partial f \rho_g}{\partial t} + \frac{\partial f \rho_g u^i}{\partial x^i} = 0 \tag{16.49}$$

From Eqs.(16.40) and (16.49), we have the formal relation that

$$\frac{f \rho_g}{\rho} = k = \text{mass concentration of the gas in the mixture}$$
$$= \text{constant} \tag{16.50}$$

With this simplest model, we study the sound wave in a froth flow as follows:

Initially the mixture of gas and liquid is at rest with a temperature T_0, a pressure p_0, void fraction f_0 and species densities ρ_{go} and ρ_{Lo}. The mixture is perturbed by a small disturbance so that in the perturbed flow we have the following perturbed variables:

$$u(x,t);\; p = p_0\,[1 + p'(x,t)]\,;\; \rho = \rho_0\,[1 + \rho'(x,t)]\,\text{dn}$$

$$T = T_0\,[1 + T'(x,t)] \tag{16.51}$$

The perturbed variables are assumed to be function of x and t only. Hence we study the one dimensional motion with u, the x-component of velocity only. The linearized equations for u, T', p' and ρ' are obtained from the fundamental equations of froth flow as follows:

$$p' = \rho_g' + T' = \frac{1}{f_0}\rho' + T' \tag{16.52a}$$

$$\frac{\partial \rho'}{\partial t} = -\frac{\partial u}{\partial x} \tag{16.52b}$$

$$\frac{\partial u}{\partial t} = -\frac{p_0}{\rho_0}\frac{\partial p'}{\partial x} \tag{16.52c}$$

$$\rho_0 C T_0 \frac{\partial T'}{\partial t} = -p_0 \frac{\partial u}{\partial x} \tag{16.52d}$$

where we consider only the inviscid and non-heat-conducting fluids. From Eqs. (16.52), we have the wave equation of the perturbed velocity u as follows:

$$\frac{\partial^2 u}{\partial t^2} - a^2 \frac{\partial^2 u}{\partial x^2} = 0 \tag{16.53}$$

where "a" is the sound speed of the mixture of gas and liquid. If we drop the subscript o for the values of the mixture, we have the sound speed of the mixture as follows:

$$a^2 = \frac{p}{\rho}\left(\frac{1}{f} + \frac{p}{\rho CT}\right) \tag{16.54}$$

The specific heat C of the mixture at constant volume is given by the relation

$$C = (1-k)\, C_{vL} + k\, C_{vg} \tag{16.55}$$

It is evident that for the case of gas alone, $f=1$ and $C=C_{vg}$ and Eq. (16.54) becomes $a^2 = \gamma\, p/\rho$.

When the void fraction f is finite which is not close to either 0 not 1, the sound speed of the mixture is less than the sound speed of both species. In our approximation of incompressible liquid, the sound speed of the liquid is infinite while that of the gas is the value of an ideal gas. For instance, for a mixture of air and water with $f_0 = 10$, the sound speed of the mixture is about 1/9 of the value of the sound speed of the air.

We may reexamine many problems of single fluid discussed previously in this book for the cases of two-phase flows of a mixture of gas and liquid. (cf. problems, § 11).

5. Two-phase flows of gas and solid particles[3]

The study of the fluid flow containing solid particles has been the subject of scientific and engineering research for a long time. In the early days, the sediment transport in open channel flow was one of the most interesting research problems since J. Boussinesq's time of the 19th century[22]. The sediment transport by water and by air are important in problems which are current interesting problems. The physical situation of the dynamics of the fluid-particle systems is very complicated[1]. It is advisable to look in details how the fluid-particle system behaves before we try some analytical treatment. One of the examples of the fluid-particle system is the fluidized bed and the other is the burning of well-packed gunpowder. There are five different phases of the flow of a mixture of fluid and solid particles which are dependent on the rate of the fluid flow[1].

(i) Porous medium phase-fixed bed stage

Let us consider a fluid which may be either a gas or a liquid flowing through a well-packed solid particles by means of pressure gradient. If the rate of the fluid flow is very small, the well-packed solid particles will not be disturbed and the fluid motion is the same as that through a porous medium[1]. We may consider the solid particles as fixed in space. This phase is usually called the fixed bed stage.

(ii) Sedimentation stage

When the flow rate increases, some of the small particles may first move with the fluid flow. As the flow rate further increases, the amount of the solid particles moved with the fluid flow increases. From now on, we may call it the sedimentation stage in which some solid particles are transported by the fluid flow. At first the solid particles may not be considered as a pseudo-fluid and the individual properties of these solid particles in the fluid flow plays an important role in the fluid flow.

(iii) Fluidized stage

As the rate of the fluid flow reaches a critical value, it is called the flow of incipient fluidization at which the character of the solid particles changes abruptly to a pseudo-fluid and waves can be set in the solid particles bed. The pseudo-fluid of the solid particles has similar behavior as ordinary fluid such as to form a level surface. It is usually called the dense phase of the fluidized bed. Ordinarily the overall density of the mixture of the solid particles and the fluid (gas) decreases only fractionally -10% to 50% as compared to the fixed bed. In other words, the volume occupied by the solid particles in the mixture would be 10% to 50% of the total volume of the mixture.

(iv) Slugging phase

Further increase of the fluid flow rate would cause the flow of the mixture irregular. For instance, if the fluid is a gas, bubbles of gas rise through the packed solid particles and burst and more and more particles will be carried out by the fluid.

(v) Two-phase flow of a mixture of solid particles and fluid

For still further increase of the flow rate; the solid particles occupy less than 5% of the total volume of the mixture and mix well with the fluid in the flow field. This is known as the dilute phase of the two-phase flow of a mixture of solid particles and a fluid in a narrow sense. Many two-phase flow problems[1, 23] of gas-solid mixture discuss this dilute phase.

In this section, we study only the last three phases (iii) to (v). In principle, the fundamental equations for the mixture of the gas and the gas for these three phases are the same because as long as the solid particles may be considered as a pseudo-fluid, we may apply the general theory of the mixture of two fluids for such problems. Hence the present fundamental equations of the mixture of solid particles and a fluid are similar to those of froth flow discussed in § 4, but the calculation of the friction coefficients etc. may be different.

The fundamental equations of a mixture of a gas and solid particles (pseudo-fluid) are as follows:

For simplicity, we assume that the particles are spheres with a radius of r_p and a volume of each particle is $\bar{\tau}_p = 4\pi\, r_p^3/3$. The number density of the solid particle in the mixture is n_p.

The species density of the solid particles is

$$\rho_{sp} = \frac{m_p}{\bar{\tau}_p} \tag{16.56}$$

where m_p is the mass of a solid particle. The partial density of the solid particles is then

$$\bar{\rho}_{sp} = m_p\, n_p = Z\, \rho_{sp} = \rho_{sp}\, \bar{\tau}_p\, n_p \tag{16.57}$$

The volume fraction of the solid particles in the mixture is

$$Z = n_p\, \bar{\tau}_p \tag{16.58}$$

The partial density of the gas is then

$$\bar{\rho}_g = (1 - Z)\, \rho_g \tag{16.59}$$

where ρ_g is the species density of the gas.

The fundamental equations of the mixture of solid particles and a gas are as follows:

(i) Equation of state

For each species in the mixture of a gas and a pseudo-fluid of solid particles, we have one equation of state. For the gas, we have

$$p_g = R\, \bar{\rho}_g\, T_g = R(1 - Z)\, \rho_g\, T_g = (1 - Z)\, p \tag{16.60}$$

Eq.(16.60) is identical to Eq.(16.10). Hence we have also

$$p_p = Z\, p \tag{16.61}$$

where p is the total pressure of the mixture.

The equation of state for the pseudo-fluid of solid particles is simply

$$\rho_{sp} = \text{constant} \tag{16.62}$$

(ii) Equation of continuity

Similar to Eqs. (16.38) and (16.39), we have the equation of continuity for the pseudo-fluid of solid particles as follows:

$$\frac{\partial Z\rho_{sp}}{\partial t} + \frac{\partial}{\partial x^i}(Z\,\rho_{sp}\,u_p^i) = -\sigma_p \tag{16.63}$$

where subscript p refers to the value for pseudo-fluid of solid particles.

For the gas, we have the equation of continuity as follows:

$$\frac{\partial}{\partial t}\left[(1-Z)\,\rho_g\right] + \frac{\partial}{\partial x^i}\left[(1-Z)\,\rho_g\,u_g^i\right] = \sigma_p \tag{16.64}$$

If we add Eqs. (16.63) and (16.64), we have the equation of continuity of the mixture (16.40).

(iii) Equations of motion

For each species, the conservation of momentum gives the corresponding equations of motion.

For the pseudo-fluid of solid particles, we have the equations of motion as follows:

$$Z\,\rho_{sp}\left(\frac{\partial}{\partial t} + u_p^i\frac{\partial}{\partial x^i}\right)\boldsymbol{q}_p = Z\,\rho_{sp}\frac{D_p\boldsymbol{q}_p}{Dt} = -\nabla p_p + \nabla\cdot\tau_p + \boldsymbol{F}_{bp} + \boldsymbol{F}_p - \sigma_p\boldsymbol{Z}_p \tag{16.65}$$

and for the gas, the equation of motion is

$$(1-Z)\rho_g\left(\frac{\partial}{\partial t} + u_g^i\frac{\partial}{\partial x^i}\right)\boldsymbol{q}_g = (1-Z)\rho_g\frac{D_g\boldsymbol{q}_g}{Dt}$$
$$= -\nabla p_g + \nabla\cdot\tau_g + \boldsymbol{F}_{bg} + \boldsymbol{F}_g + \sigma_p\,\boldsymbol{Z}_p \tag{16.66}$$

where τ_p and τ_g are respectively the viscous stress tensor of the pseudo-fluid of solid particles and that of the gas. We shall neglect these viscous stress in this chapter.

The body forces of the species are $\boldsymbol{F}_{bp}$ and $\boldsymbol{F}_{bg}$ which consist of the gravitational forces and the electromagnetic forces and other body forces. For instance, for the gravitational force, we have

$$\boldsymbol{F}_{bpg} = Z\,\rho_{sp}\,\boldsymbol{g}\ ;\ \boldsymbol{F}_{bgg} = (1-Z)\,\rho_g\,\boldsymbol{g} \tag{16.67}$$

where $\boldsymbol{g}$ is the gravitational acceleration vector.

The most difficult forces in the present analysis are the interaction forces between the gas and the pseudo-fluid of solid particles shown in

Eqs.(16.55) and (16.66) as F_p and F_g. For first approximation, we may use the Stokes relation for these interaction forces:

$$F_p = 6\pi\, r_p\, \mu_g\, n_p\, (q_g - q_p) = -F_g \tag{16.68}$$

where μ_g is the coefficient of viscosity of the gas. If Z is very small, say $Z < 0.1$, Eq.(16.68) may be considered as a good relation. In general, the interaction force F_p depends on Z as well as the Reynolds number Re_p of the solid particles and the Mach number M of the flow. There is no theoretical formula for F_p which includes the effects of Z, Re_p and M. Some empirical formulas have been suggested for the effects of Z, Re_p and M. For very small values of Re_p and M, an empirical equation which fits the experimental data well was found for the pressure gradient due to viscous loss (see Fig.16.3) for Z from 0 to 0.5 as follows[23]:

$$\frac{dp}{dy} = -\frac{9\mu}{2r_p^2}\frac{Z}{(1-Z)^2}\left[1 + \frac{0.68}{(1.35Z)^{-1/3} - 1}\right] \cdot (u_g - u_p) \tag{16.69}$$

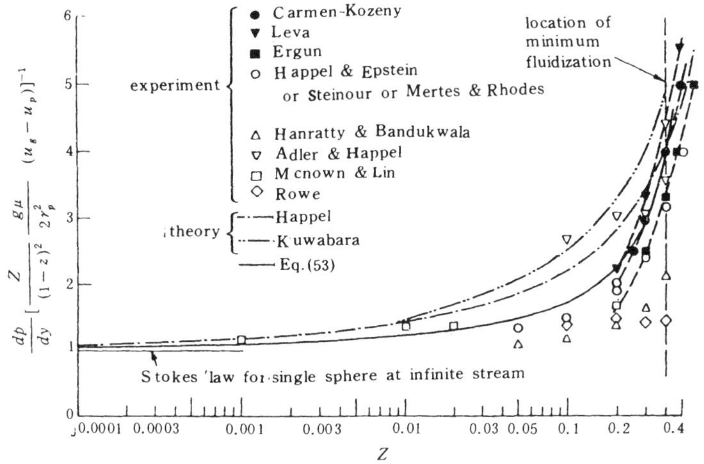

Fig. 16.3 Pressure gradient as a function of volume fraction Z of solid particles in gas-solid two-phase flow

The drag coefficient at small Re_p and M may then be calculated from the pressure gradient by assuming that the force is evenly distributed over each particle, we have

$$C_D = \frac{F_p}{\pi r_p^2 \, \frac{1}{2}\rho_g\, |q_g - q_p|(q_g - q_p)} = \frac{24}{Re_p}\cdot\frac{1}{(1-Z)^2}\left[1 + \frac{0.168}{(1.35Z)^{-1/3} - 1}\right]$$

$$= \frac{24}{Re_p} F_1(Z) \tag{16.70}$$

where the direction of $\boldsymbol{F}_p$ is the same as that of $(\boldsymbol{q}_g - \boldsymbol{q}_p)$ and

$$Re_p = 2\, r_p \mid \boldsymbol{q}_g - \boldsymbol{q}_p \mid \rho_g / \mu \tag{16.71}$$

The function $F_1(Z)$ is the correction of the interaction force for finite Z. If Z is much less than unity, $F_1(Z)$ is approximately zero to unity. Eq. (16.70) is good for $Re_p < 1$.

If Re_p is large than 1 but smaller than 100, it was found that a good formula for the drag coefficient is as follows:

$$C_D = \frac{24}{Re_p} \cdot F_1(Z)\,(1 + 0.15\, Re_p^{0.687}) = \frac{24}{Re_p} F_1(Z)\, F_2(Re_p) \tag{16.72}$$

where $F_2(Re_p)$ is the correction of the interaction force for large Re_p. When Re_p is very small, $F_2(Re_p) \simeq 1$.

Similarly if the Mach number of the flow M is not small, we have the empirical correction factor as follows:

$$C_D = \frac{24}{Re_p} \cdot F_1(Z) \cdot F_2(Re_p) \cdot F_3(M) \tag{16.73}$$

where the correction function for Mach number is given as follows:

$$\begin{aligned} F_3(M) &= 1 \qquad \text{for } 0 < M < 0.5 \\ &= \frac{1}{0.192}\left[0.3198 + 0.298(M-1) - 0.0809(M-1)^2 - 0.3606(M-1)^3\right] \\ &\qquad \text{for } 0.4 < M < 1.5 \\ &= \frac{1}{0.192}\left[0.3812 + 0.1199\left(\frac{1}{M} - \frac{1}{2.75}\right) - 0.2375\left(\frac{1}{M} - \frac{1}{2.75}\right)^2\right] \\ &\qquad \text{for } 1.5 < M < 4.5 \\ &= 1.88 \qquad \text{for } 4.5 < M < 10 \end{aligned} \tag{16.74}$$

It should be noted that $C_D = \frac{24}{Re_p}$ corresponds to Eq.(16.68), i.e., the Stokes formula for the drag of a sphere at very small Reynolds number and very small Mach number.

If we consider the gravitational force as a body force, we should include the bouyancy force as an additional interaction force to the interaction force due to viscous drag as given in Eq.(16.73). The bouyancy force is simply

$$\boldsymbol{F}_{p(b)} = -Z\, \rho_g\, \boldsymbol{g} \tag{16.75}$$

The term $\sigma_p\, \boldsymbol{Z}_p$ is the force associated with the momentum due to the source term σ_p.

(iv) Equation of energy

For each species, the conservation of energy gives the corresponding equation of energy for that species.

For the pseudo-fluid of solid particles, we have the energy equation as follows:

$$\frac{\partial\left[Z\rho_{sp}\left(U_{mp}+\frac{1}{2}q_p^2+\varphi_p\right)\right]}{\partial t}+\frac{\partial}{\partial x^j}\left[Z\rho_{sp}\,u_p^j\left(U_{mp}+\frac{1}{2}q_p^2+\varphi_p\right)\right.$$

$$\left.-u_p^i\,\tau_p^{ij}+\delta^{ij}\,u_p^i\,p_p-Q_{cp}^j\right]=K_T\,(T_g-T_p)+\varepsilon_p \tag{16.76}$$

$$\frac{\partial\left[(1-Z)\rho_g\left(U_{mg}+\frac{1}{2}q_g^2+\varphi_g\right)\right]}{\partial t}+\frac{\partial}{\partial x^j}\left[(1-Z)\rho_g\,u_g^j\left(U_{mg}+\frac{1}{2}q_g^2+\varphi_g\right)\right.$$

$$\left.-u_g^i\,\tau_g^{ij}+\delta^{ij}\,u_g^i\,p_g-Q_{cg}^j\right]=K_T\,(T_p-T_g)+\varepsilon_g \tag{16.77}$$

where U_{mr} is the internal energy per unit mass of the rth species, τ_r^{ij} is the ijth components of the viscous stress tensor of the rth species; Q_{cr}^j is the jth component of the heat conduction flux of the rth species and φ_r is the potential energy of the rth species; q_r is the magnitude of the velocity vector of rth species and ε_r is the energy source due to chemical reaction and/or electromagnetic forces and other heat addition terms of the rth species. The K_T is the thermal friction coefficient between the solid particles and the gas. Our knowledge of the expression of K_T is still very meager. In our approximation with Stokes law for the interaction force, we may take

$$K_T=4\pi\,r_p\,\kappa_p\,n_p \tag{16.78}$$

where κ_p is the coefficient of heat conductivity of the solid particles.

Since the density of the solid particles ρ_{sp} is considered as a constant, we have eleven variables: $\boldsymbol{q}_p$, $\boldsymbol{q}_g$, p, Z, ρ_g, T_p and T_g of the study of flow problems of a mixture of a gas and small solid particles which are governed by the fundamental equations (16.60), (16.63), (16.64), (16.65), (16.66), (16.76) and (16.77).

First we are going to derive some simple relations for a mixture of a gas and a pseudo-fluid of solid particles under thermodynamic equilibrium conditions such that $T_g=T_p=T$. The mass concentration of the pseudo-fluid of solid particles is defined as

$$k_p = \frac{\overline{\rho}_p}{\rho} = \frac{Z\,\rho_{sp}}{\rho} \tag{16.79}$$

The total pressure of the mixture is

$$p = p_p + p_g \tag{16.80}$$

The equation of state of the mixture under equilibrium condition is

$$p = \frac{1-k_p}{1-Z}\,\rho\,RT = \frac{\rho\,R_M\,T}{1-Z} \tag{16.81}$$

where ρ is the density of the mixture, i.e., $\rho = \overline{\rho}_p + \overline{\rho}_g$, and

$$R_M = (1-k_p)\,R \tag{16.82}$$

where R is the gas constant of the gas.

The internal energy of the mixture per unit mass is related to the internal energies of the two species by the following relation:

$$U_{mM} = Z\,\rho_{sp}\,c_{sp}\,T_p + (1-Z)\,\rho_g\,c_v\,T_g \tag{16.83}$$

where c_{sp} is the specific heat of the solid particles. Eq.(16. 83) may be written as

$$U_{mM} = k_p\,c_{sp}\,T_p + (1-k_p)\,c_v\,T_g \tag{16.83a}$$

For thermodynamic equilibrium condition, the specific heat of the mixture at constant volume c_{vM} is as follows:

$$c_{vM} = k_p\,c_{sp} + (1-k_p)\,c_v \tag{16.84}$$

where c_v is the specific heat of the gas at constant volume.

The enthalpy of the mixture per unit mass is

$$H_M = U_{mM} + \frac{p}{\rho} = k_p\left(c_{sp}\,T_p + \frac{p}{\rho_{sp}}\right) + (1-k_p)\,c_p\,T_g \tag{16.85}$$

For thermodynamic equilibrium condition, the specific heat of the mixture at constant pressure is then

$$c_{pM} = k_p\,c_{sp} + (1-k_p)\,c_p \tag{16.86}$$

where c_p is the specific heat of the gas at constant pressure.

The specific heats of the mixture are independent of the volume fraction Z but depend on the mass concentration k_p of the solid particles.

The ratio of the specific heats of the mixture is

$$\Gamma = \frac{c_{pM}}{c_{vM}} = \frac{(1-k_p)\,c_p + k_p\,c_{sp}}{(1-k_p)\,c_v + k_p\,c_{sp}} = \gamma \cdot \frac{\left(1+\eta\dfrac{\delta}{\gamma}\right)}{1+\eta\delta} \tag{16.87}$$

where $\gamma = c_p/c_v$, $\delta = c_{sp}/c_v$, and $\eta = k_p/(1-k_p)$. The ratio Γ is always smaller than γ of the gas if k_p is different from zero. As $k_p = 0$, $\Gamma = \gamma$.

If we consider the mixture as a homogeneous medium, the first law of thermodynamics for the mixture gives

$$dQ = dU_{mM} - \frac{1}{\rho^2}\, p\, d\rho \tag{16.88}$$

where dQ is the heat addition to the mixture. Eq.(16.88) is the energy equation of the mixture as a whole.

For isentropic change of state of the gas-solid particle mixture, we have $dQ = 0$, and Eq.(16.88) gives

$$\frac{1}{\Gamma - 1}\frac{dT}{T} = \frac{1}{1-Z}\frac{d\rho}{\rho} \tag{16.89}$$

Since $Z = k_p\, \rho/\rho_{sp}$, for constant k_p and $T_p = T_g = T$, the integration of Eq.(16.89) gives

$$T\left(\frac{\rho}{1-Z}\right)^{-(\Gamma-1)} = \text{constant} \tag{16.90}$$

If $Z \ll 1$, the isentropic change of state of the mixture has a similar relation as that for a pure gas with an effective ratio of specific heats Γ. In general, the volume fraction Z has some influence on the isentropic change of the mixture.

Similarly, for a given k_p and under thermodynamic equilibrium condition, we have

$$\frac{dp}{p} = \frac{dT}{T} + \frac{1}{1-Z}\frac{d\rho}{\rho} \tag{16.91}$$

From Eqs.(16.89) and (16.91), we have

$$\frac{dp}{p} = \frac{\Gamma}{1-Z}\frac{d\rho}{\rho} \tag{16.92}$$

or

$$p\left(\frac{\rho}{1-Z}\right)^{-\Gamma} = \text{constant} \tag{16.93}$$

Again, if $Z \ll 1$, Eq.(16.93) is identical in form from the corresponding relation of a pure gas but with an effective ratio of specific heats.

We may calculate the so-called equilibrium speed of sound of the mixture a_M from Eq.(16.92) as follows:

$$a_M^2 = \frac{dp}{d\rho} = \frac{\Gamma(1-k_p)RT}{(1-Z)^2} = \frac{\Gamma R_M T}{(1-Z)^2} \tag{16.94}$$

If $Z \ll 1$, Eq.(16.94) is identical in form as that for a pure gas, but the effective ratio of specific heats and effective gas constant R_M are used. The ratio of the equilibrium sound speed of the mixture a_M to that of the gas $a = (\gamma RT)^{\frac{1}{2}}$ is

$$\frac{a_M}{a} = \frac{1}{1-Z}\left[\frac{\Gamma}{\gamma}(1-k_p)\right]^{\frac{1}{2}} = \frac{1}{1-Z}\left[\frac{(1+\frac{\eta\delta}{\gamma})(1-k_p)}{1+\eta\delta}\right]^{\frac{1}{2}} \tag{16.95}$$

We may reexamine all the ordinary gasdynamic problems for the mixture of a gas and solid particles[1]. In order to show some interesting effects of the solid particles on the compressible flow, we consider a normal shock wave through a gas-solid particle mixture[23]. Since we assume that solid particles are large with respect to molecular dimensions, the thickness of a gasdynamic shock is negligible in comparison with the region for appreciable changes of the momentum and the temperature of the solid particles. Thus, the shock wave structure of the gas-solid particle flow may be divided into three distinct regions: the frozen flow region, the relaxation region and the equilibrium flow region as shown in Fig.16.4. In front of the

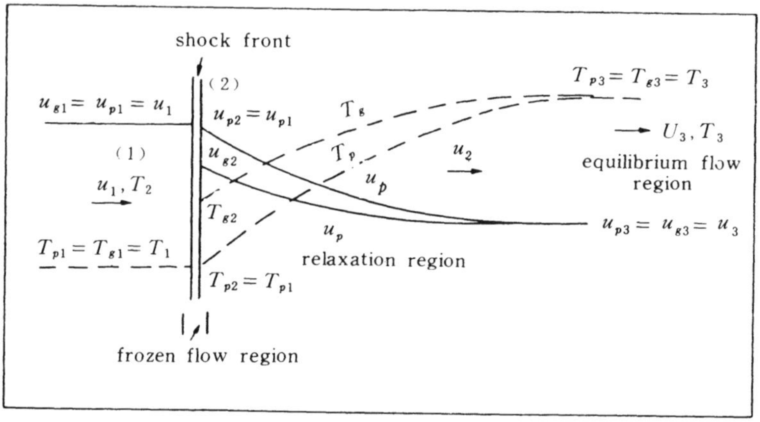

Fig.16.4 A sketch for the flow regions in a normal shock wave in a gas-solid particles mixture

shock, the temperature of the gas and that of the solid particles are equal and the velocities of the gas and the solid particles are also equal. The frozen flow region includes the sharp shock front and the immediate region behind the shock front in which the temperature of the solid particles and the velocity of the solid particles change little from their values in front of the shock. The gas is under a big change in both temperature and velocity according to the normal shock relations in a pure gas. The relaxation region follows immediately from the frozen flow region in which the temperatures of the gas and the solid particles as well as the velocities of the gas and the solid particles vary with the downstream distance toward the final equilibrium values. Far downstream, we reach the equilibrium flow region in which the velocities of the gas and the solid particles are equal and the temperatures of the gas and the solid particles are also equal but their values are different from those in front of the shock wave. The distance in which the temperatures T_g and T_p reach their equilibrium values is in general different from the distance in which the velocities q_g and q_p reach their equilibrium values. We are going to study the flow field in these three regions. Let us denote the values in front of the shock wave by subscript 1, those immediately behind the shock by subscript 2, and far downstream in the equilibrium flow region by subscript 3.

First we consider the frozen and the equilibrium flow regions by considering the conservation laws of the mass, momentum and energy across the shock for each of the species as follows:

$$N_p = Z\,\rho_{sp}\,u_p = Z_1\,\rho_{sp}\,u_1 \tag{16.96}$$

$$N_g = (1-Z)\,\rho_g\,u_g = (1-Z_1)\,\rho_{g1}\,u_1 \tag{16.97}$$

$$N_g\,u_g + N_p\,u_p + p = (N_p + N_g)\,u_1 + p_1 \tag{16.98}$$

$$N_g\left(\frac{1}{2}u_g^2 + c_p\,T_g\right) + N_p\left(\frac{1}{2}u_p^2 + c_{sp}\,T_p + \frac{p}{\rho_{sp}}\right)$$

$$= (N_g + N_p)\frac{1}{2}u_1^2 + (N_g\,c_p + N_p\,c_{sp})T_1 + N_p\frac{p_1}{\rho_{sp}} \tag{16.99}$$

where we use u for the velocity because it is a one-dimensional flow.

Furthermore, if we use the Stokes' law for the interaction force, we have the following relations for the temperatures and velocities of the gas and the solid particles:

$$(1-Z)u_p\frac{du_p}{dx} = C_D\frac{3}{2r_p}\,|\,u_g - u_p\,|\,(u_g - u_p) \tag{16.100}$$

$$u_p\frac{dT_p}{dx} = \frac{3Nu\,k_p}{\rho_{sp}\,c_{sp}\,r_p^2}(T_g - T_p) \tag{16.101}$$

where Nu is the Nusselt number, k_p is the coefficient of heat conductivity of the solid particles and C_D is the drag coefficient of the solid particles with an average radius r_p.

Now we have seven variables u_g, u_p, T_g, T_p, p, ρ_g and Z from the seven equations (16.60), (16.69) to (16.101).

Let us consider the three regions of the normal shock (Fig.16.4) in the gas-solid particle flow separately.

(i) Frozen flow region

Across the shock front, from state 1 to state 2, the effects of the shock on the properties of the solid particles are small. One of the reasonable approximations is to assume that the temperature of the solid particles remains unchanged through the shock front i.e.,

$$T_{p1} = T_{p2} = T_1 \tag{16.102}$$

If the volume fraction of the solid particles Z is negligible in comparison to unity, the conditions in the frozen flow region are exactly the same as those in ordinary gasdynamic shock (cf. Chapter IV, § 2) and we have the following Rankine-Hugoniot relations for the gas at state 2:

$$\frac{u_{g2}}{u_{g1}} = \frac{2}{(\gamma+1)M_1^2}\left(1 + \frac{\gamma-1}{2} M_1^2\right) = \frac{\rho_{g1}}{\rho_{g2}} \tag{16.103}$$

$$\frac{p_2}{p_1} = 1 + \frac{2\gamma}{1+\gamma}(M_1^2 - 1) \tag{16.104}$$

$$\frac{T_{g2}}{T_{g1}} = 1 + \frac{2(\gamma-1)}{(\gamma+1)^2}\,\frac{1+\gamma M_1^2}{M_1^2}(M_1^2 - 1) \tag{16.105}$$

and the velocity of the solid particles at state 2 is $u_{p2} = u_1$ if the volume fraction Z is negligibly small. The Mach number $M_1 = u_1/a_1$. For finite value of Z, the solid particle velocity at state 2 is a little smaller than its value in front of the shock and is given by the relation[24]

$$u_{p2}^2 = u_1^2 - 2\frac{\rho_{g1}}{\rho_{sp}}\,\frac{p_2 - p_1}{\rho_{g1}} \tag{16.106}$$

Eq. (16.106) shows that u_{p2} depends on the density ratio ρ_{g1}/ρ_{sp}. For small but not negligible value of Z, we have

$$u_{p2} = u_1 - \frac{Z_1\,(p_2 - p_1)}{N_p} \tag{16.107}$$

and $Z_2 = Z_1$.

(ii) Equilibrium flow region

Far downstream, the velocities of the gas and the solid particles will be equal and also are the temperatures of the gas and the solid particles, and we have

$$u_{p3} = u_{g3} = u_3\,; \quad T_{p3} = T_{g3} = T_3 \tag{16.108}$$

Hence we call the region 3 the equilibrium flow region. Since both the velocity and the temperature of the solid particles are different from their corresponding values in front of the shock, u_{g3} is different from u_{g2} and T_{g3} is different from T_{g2}. We should calculate the values of u_{g3} and T_{g3} from Eqs. (16.96) to (16.99) with the help of the relations (16.108). We have

$$\frac{u_{q3}}{u_1}=\frac{\rho_{g1}}{\rho_{g3}}=\frac{(\Gamma-1)M_{el}^2+2}{(\Gamma+1)M_{el}^2} \quad (16.109)$$

$$\frac{p_3}{p_1}=1+\frac{2\Gamma}{\Gamma+1}(M_{el}^2-1) \quad (16.110)$$

$$\frac{T_3}{T_1}=\frac{p_3}{p_1}\frac{g_1}{g_3} \quad (16.111)$$

where Γ is the effective ratio of the specific heats of the mixture given in Eq.(16.87) and the effective Mach number of the mixture M_e is

$$M_e=\frac{u_e}{a_M} \quad (16.112)$$

where u_e is the velocity of the gas or the particles in the equilibrium flow condition, $u_e=u_p=u_g$ and a_M is the sound speed of the equilibrium flow given by Eq.(16.94).

(iii) Relaxation region[23]

The relaxation flow region is the region from state 2 to state 3. We may calculate the velocities u_p and u_g and the temperatures T_p and T_g as a function of x from Eqs.(16.100) and (16.101). Fig. 16.4 shows some sketch of the variations of u_g, u_p, T_p and T_g in the relaxation region. The detailed calculations and results are given in reference 24.

6. One-dimensional flow of electrogasdynamics[24, 25]

Now we are going to study the one-dimensional steady flow of a mixture of charged particles and a neutral gas in a channel of cross-sectional area $A(x)$. Such a flow is useful in the investigation of EGD generator. For simplicity, we assume that the volume fraction Z of the particles is negligibly small so that the total pressure of the mixture is equal to the gas pressure. Our unknowns are the pressure p, the density of the gas ρ_g, the temperature of the gas T_g, the x-wise velocity of the gas u_g, the number density of the particles n_p, the temperature of the particles T_p, the x-wise velocity of the particles u_p and the radial velocity of the particles v_p, the x-wise component of the electric field E_x and the radial component of the electric field E_r. Hence in general, we have ten unknowns in our problem. We have to find ten fundamental equations to govern these ten unknowns as follows:

(i) The equation of state of the gas is the perfect gas law

$$p=\rho_g R T_g \quad (16.113a)$$

(ii) Equation of continuity of the gas

$$\rho_g u_g A=\text{constant}=M_g \quad (16.113b)$$

(iii) Equation of continuity of the charged particles

$$m_p\, n_p\, u_p\, A = \text{constant} = M_p \tag{16.113c}$$

(iv) The x-wise equation of the charged particels

$$u_p \frac{du_p}{dx} = \frac{KeE_x}{m_p} + \left(\frac{C_D\, \rho_g\, \pi\, d_p^2}{8m_p} \right) |\, u_g - u_p| \,(u_g - u_p) \tag{16.113d}$$

where m_p is the mass of a charged particle, d_p is the diameter of the charged particle, Ke is the electric charge on a particle, and C_D is the drag coefficient of a charged particle in the stream of the gas. Hence the first term of the right-hand side of Eq. (16. 113d) is the body force due to the electric field E_x and the second term is the drag force on the charged particles. We assume that the particales are spheres of diameter d_p whose C_D has been discussed in §5.

(v) The x-wise equation of motion of the mixture as a whole is

$$\frac{d}{dx} \left(\frac{m_g\, u_g}{A} + \frac{M_p\, u_p}{A} \right) = - \frac{dp}{dx} + n_p\, KeE_x + 2f\, \rho_g\, u_g^2 / D_H \tag{16.113e}$$

where f is the non-dimensional friction coefficient and D_H is the hydraulic diameter of the channel. Hence, the last term in Eq.(16.113e) is the friction force in the channel which is the overall effect due to the boundary layer of the channel (cf. Chapter XVII). We neglect the friction force in Eq. (16.113d) because Z is very small.

(vi) The energy equation for the charged particles as a pseudofluid is

$$u_p \frac{dT_p}{dx} = \frac{K_e\, u_p\, E_x}{m_p\, c_s} + Nu\, \kappa_p\, \pi\, d_p\, (T_g - T_p)/(4m_p\, c_s) \tag{16.113f}$$

where c_s is the specific heat of the charged particles, Nu is the Nusselt number based on the particle diameter (cf. Chapter XVII) and κ_p is the coefficient of heat conductivity of the gas.

(vii) The energy equation of the mixture as a whole is

$$\frac{d}{dx} \left[\rho_g\, u_g\, (c_p\, T_g + \frac{1}{2} u_g^2) + m_p\, u_p\, u_p\, (c_s\, T_p + \frac{1}{2} u_p^2) \right] = n_p\, Keu_p\, E_x \tag{16.113g}$$

(viii) The Maxwell's equations for electric field may be written as follows

$$\nabla \cdot \boldsymbol{J} = 0 \tag{16.113h}$$

and

$$\nabla \cdot \boldsymbol{E} = n_p Ke/\varepsilon \tag{16.113i}$$

(ix) The equation of motion for v_p

For the radial velocity component v_p of the charged particles, we should use the radial equation of motion of the charged particles. However, since the essential force in the radial direction is the electrostatic force, we may use an approximation for the radial velocity in terms of the mobility of the charged particles μ_m which is assumed to be a constant. In other words, we have

$$v_p = \mu_m E_r \tag{16.113j}$$

It should be noticed that the electric current density $\boldsymbol{J}$ is

$$\boldsymbol{J} = n_p K_e \boldsymbol{q}_p \tag{16.114}$$

Since we consider a one-dimensional flow problem with a two-dimensional electrostatic field such that the effect of the space charge may be included in the analysis, we have to make some approximations in Eqs. (16.113h) and (16.113i). In fact we consider only the slender channel so that the axial distance along the channel is much larger than the radial distance across the channel. Hence we may assume that the gradient of the electric field in the x-direction is much smaller than that in the radial direction. As a result, Eq. (16.113i) becomes

$$\frac{1}{r}\frac{d}{dr}(E_r) = n_p Ke/\varepsilon_0 \tag{16.115}$$

or

$$E_r = n_p Ker^2/(2\varepsilon_0) \tag{16.116}$$

From Eqs. (16.113h), (16.113j), (16.114), (16.116), we have

$$\frac{dJ_x}{dx} = -\mu_m J_x^2 / u_p^2 \varepsilon_0 \tag{16.117}$$

The total x-wise electric current is $I_x = J_x A$.

The total electric field E depends on the applied electric potential Φ_L and the space charge field. We assume that Φ_L is a constant and has only the axial component but the space charge electric field E_{sc} has both axial E_{scx} and radial E_{scr} components. The radial space charge electric field E_{scr} is the total radial electric field E_r given by Eq. (16.116) and which is expressed in terms of the number density of the charged particles n_p. The total axial electric field may be written as folows:

$$E_x = \frac{\Phi_L}{L} + E_{sox} \tag{16.118}$$

where L is the length of the conversion section of the EGD generator across which the electric potential Φ_L is applied. The axial space charge electric field E_{scx} depends on the total current flowing through the channel I_x and the geometry of the channel. In reference 26, the following formula was found for a slender circular channel of diameter d_c with the total length of sections of conversion and collector of the EGD generator L_1 as follows:

$$E_{scx}=I_c L_1$$

$$\times \frac{\ln [\frac{(1+gx/L)^2}{1-g}]+2\{\tanh^{-1}b+\tanh^{-1}b_1\}+\frac{2}{s}\{\tanh^{-1}(\frac{G-2}{s})+\tanh^{-1}[-\frac{2b_1+G}{s}]\}}{g\cdot 2\, u_p A\, \varepsilon_0} \tag{16.119}$$

where I_c is I_x at $x=0$ and I_L is the load current, $g=(I_c/I_L)-1$,

$b=\tan[\tan^{-1}\frac{1}{2}(2x/d_c)]$; $G=-gd_c/(1+gx)$, $b=\tan[\tan^{-1}(L-x)/d_c]$

and $s=(G^2+r^2)^{\frac{1}{2}}$.

Since we assume that the mobility of the charged particles is a constant, v_p may be expressed in terms of E_r by Eq.(16.113j) and E_r may be expressed in terms of n_p by Eq.(16.116). Furthermore, the axial electric current density J_x may be expressed in terms of n_p and u_p and the gas density ρ_g is expressed in terms of the gas pressure p and the gas temperature T_g and ρ_g may also be expressed in terms of u_g by Eq.(16.113b). With the help of these algebraic relations, we have only six variables left which are u_g, T_g, u_p, T_p, E_x and J_x while E_x is given by Eq.(16.118). Hence, we have five unknowns u_g, T_g, u_p, T_p and J_x which may be solved from the five first order differential equations (16.113d), (16.113e), (16.113f), (16.113g) and (16.117) from some initial values at an initial point $x=0$. Numrical examples are given in reference 26. When we start with similar initial conditions, the flow conditions depend on the mass ratio M_p/M_g. For large M_p/M_g, the temperature of the gas drops faster than that for small mass ratio downstream. The influence of the mass ratio on the temperature of the charged particless is small. Similarly, the decrease of the gas pressure with axial distance is larger for large mass ratio. The rate of increase of axial velocity of the gas is larger for larger mass ratio.

For rapid estimation of the performance of a EGD generator, we may simplify the above calculations by neglecting the space charge electric field and assume that $u_g=u_p$ and $T_g=T_p$. We need to solve three differential equations for three variables u_g, T_g and J_x as a first approximation[24].

7. Two-phase flow of electrohydrodynamics[26]

In previous sections, we assume that the inductive capacity ε is a constant and equal to the value in free space. If the value of the inductive capacity ε is not a constant throughout the flow field, additional electric force will be introduced in comparison with those given in the case of constant inductive capacity. The electric force in general is

$$\boldsymbol{F}_e = K e n_p \boldsymbol{E} - \frac{1}{2} E^2 \nabla \varepsilon \tag{16.120}$$

where the last term on the right-hand side is due to the non-uniformity of the inductive capacity. Eq.(16.120) may also be written as follows:

$$F_e^i = \frac{\partial \tau_e^{ij}}{\partial x^j} \tag{16.121}$$

where τ_e^{ij} is the ijth component of the electric stress tensor which is given by the formula:

$$\tau_e^{ij} = \varepsilon E^i E^j - \frac{1}{2} \varepsilon \delta^{ij} E^k E^k \tag{16.122}$$

In general, the fundamental equations of electrohydrodynamics with electric polarization are the same as those discussed in Chapter XV, § 4(ii) except that the inductive capacity is no longer constant. In reference 26, a special class of electrohydrodynamic problems has been reviewed in which the interfacial coupling between the flow fields of two media has been studied. Many interesting results are obtained. Here we are going to give a simple example of such problems as shown in Fig.16.5. We consider a two-dimensional electrohydrodynamic flow. The electrically conducting liquid 2 is filled in a rectangular tank of length L and height h_2. The interface between liquid 2 and liquid 1 which may be a gas or air is on the x-axis, i.e.,

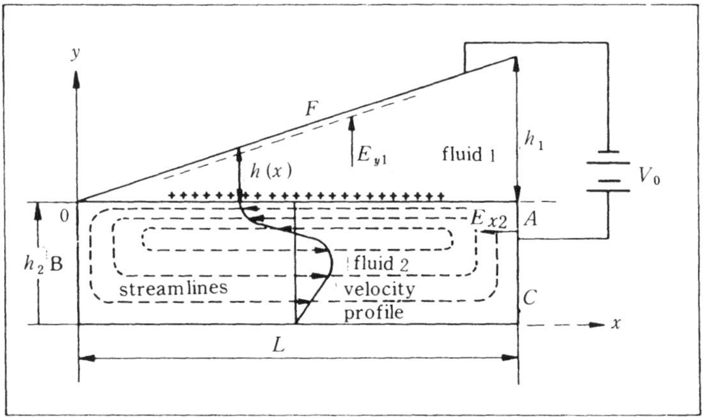

Fig. 16.5 An example of the two-phase flow of electrohydrodynamics with cellular convection

the surface OA. We have three electrodes B, C and F with a battery of electric potential V_0. The position of the straight electrode F is such that the gap $h(x)$ is essentially linear and the height h_1 is much smaller than the length of the tank L. On the interface, the x-component of the electric field is

$$E_{x2} = -\frac{V_0}{L} = \text{constant} \tag{16.123}$$

where subscript 2 refers to the value in fluid 2. We also have $E_{y2}=0$. On the side of fluid 1, we have both the x-component of the electric field $E_{x1}=E_{x2}$ and the y-component of the electric feld E_{y1} which is approximately

$$E_{y1} = \frac{V_0}{h_1} \tag{16.124}$$

If we assume that the length L is much larger than the height of the tank h_2, in the central portion of the tank, i.e., far away from the end of the electrodes B and C, the velocity induced by the voltage V_0 has only x-component $u(y)$ which is a function of y only. The flow induced by the electric force which has an electric stress component on the interface

$$\tau_e^{xy} = \varepsilon_0 E_x E_y = -\frac{\varepsilon_0 V_0^2}{Lh_1} \tag{16.125}$$

Hence the velocity distribution $u(y)$ is given by the Navier-Stokes equation with the boundary condition that:

$$\text{at } y=0: u=0; \quad y=h_2: \mu \frac{\partial u}{\partial y} = -\frac{\varepsilon_0 V_0^2}{Lh_1} \tag{16.126}$$

Hence, we have the velocity distribution:

$$u(y) = -\frac{\varepsilon_0 V_0^2 h_2}{\mu 2Lh_1}\left[\frac{3}{2}\left(\frac{y}{h_2}\right)^2 - \left(\frac{y}{h_2}\right)\right] \tag{16.127}$$

In Fig.16.5, the dotted lines are roughly the streamlines in liquid 2 which is counter-clockwise cellular.

With different arrangement of the electrodes, we may have different flow patterns. For instance, if we apply sinusoidal potential to the liquid, periodic convection may be obtained[26]. Another interesting problem is the convection of electrified drops and bubbles which moves in the electric field of a surrounding fluid[26]. There are many engineering applications of electrohydrodynamics. For instance, the condensation heat transfer coefficient increases significantly with applied electric field as found by Choi[27].

8. Multifluid theory of a plasma[1, 2]

In Chapter XV , we consider only the classical electromagnetogasdynamics of single fluid theory in which we assume that (i) the gas has a scalar electrical conductivity, (ii) a simple generalized Ohm's law may be used for the electric current density equations, (iii) the composition of the mixture of gases does not change so that the mixture may be considered as a single fluid and (iv) the temperatures of all species are the same. Such a system of equations gives good results for an electrically conducting liquid as well as an ionized gas if the strength of the magnetic field is not too large and the density of the gas is not too small. However, in current practice, the strength of the magnetic field gradually increases and the density of the gas decreases. Hence a better and more logic approach should be used .One of the improvement of the classical theory of electromagnetogasdynamics is the multifluid theory of a plasma which will be discussed in this section.

We consider the plasma as a mixture of N species which consist of ions, electrons, and neutral particles. From the macroscopic point of view, the following variables should be used to describe the flow field of a plasma:

The temperature of the sth species T_s,
The pressure of the sth species p_s,
The density of the sth species $\rho_s = m_s n_s$,
The velocity vector of the sth species $\boldsymbol{q}_s$ with components u_s^i,
The electric field strength $\boldsymbol{E}$ with components E^i and
The magnetic field strength $\boldsymbol{H}$ with components H^i.

where $s = 1, 2, \cdots$, or N, $i = 1, 2$ or 3 which represents one of the three spatial directions. The number density of sth species is n_s, the mass of a particle of sth species is m_s and the electric charge on a particle in sth species is e_s. There are $6N + 6$ variables in the multifluid theory instead of 16 variables in the classical theory of EMGD.

The variables T_s etc., are known as partial variables in the multifluid theory. The relations between these partial variables and the gross variables of the mixture as a whole are as follows:

Pressure of the mixture $p = \sum_{S=1}^{N} p_s$ (16.128)

number density of the mixture $n = \sum_{S=1}^{N} n_s$ (16.129)

density of the mixture $\rho = mn = \sum_{S=1}^{N} \rho_s = \sum_{S=1}^{N} m_s n_s$ (16.130) where m is the mass of a particle in the mixture which is a function of the composition of the mixture.

Temperature of the mixture $T = \frac{1}{n} \sum_{S=1}^{N} n_s T_s$ (16.131)

ith velocity component of the mixture $u^i = \frac{1}{\rho} \sum_{S=1}^{N} \rho_s u_s^i$ (16.132)

ith component of diffusion velocity of sth species $w_s^i = u_s^i - u^i$ (16.133)

From the definition of the flow velocity u^i and the diffusion velocity w_s^i, we have

$$\sum_{S=1}^{n} \rho_s w_s^i = 0 \quad (16.134)$$

Excess electric charge $\rho_e = \sum_{S=1}^{N} \rho_{es} = \sum_{S=1}^{N} e_s n_s$ (16.135)

ith component of the electric current density $j^i = \sum_{S=1}^{N} j_s^i$

$$= \sum_{s=1}^{N} \rho_{es} u_s^i = \sum_{S=1}^{N} \rho_{es} w_s^i + u^i \sum_{S=1}^{N} \rho_{es} = I^i + \rho_e u^i \quad (16.135a)$$

where I^i is the ith component of the electric conduction current density and $\rho_e u^i$ is the ith component of the electric convection current density.

Now we are going to derive the fundamental equations of multifluid theory of a plasma and their relations with the fundamental equations of the classical theory of EMGD as follows:

Since the electromagnetic equation for E^i and H^i are the same in the multifluid theory as those in the single fluid theory, we need to consider the gasdynamic equations only:

(i) Equation of state. The ideal gas law may be used. For sth species, we have

$$p_s = k n_s T_s \quad (16.136)$$

where k is the universal gas constant. The sum of N equations of Eq. (16.136) gives the equation of state of the mixture:

$$p = k n T = \rho RT \quad (16.137)$$

where $R = k/m$ is the gas constant of the mixture which is a function of the composition of the mixture and which is identical in form as Eq. (15.1).

(ii) Equation of continuity. The conservation of mass of each species gives

$$\frac{\partial \rho_s}{\partial t} + \nabla \cdot (\rho_s \boldsymbol{q}_s) = \sigma_s \tag{16.138}$$

where σ_s is the mass source per unit volume of sth species. The sum of the source functions must be zero because of the conservation of total mass of the mixture, i.e.,

$$\sum_{S=1}^{N} \sigma_s = 0 \tag{16.139}$$

The sum of N equations of the type of Eq.(16. 138) with the help of Eq. (16.139), gives

$$\frac{\partial \rho}{\partial t} + \nabla \cdot (\rho \boldsymbol{q}) = 0 \tag{16.140}$$

It is interesting to notice that Eq.(16.138) is the diffusion equation used in ordinary gasdynamics (cf. Chapter XVII). Eq.(16. 140) is the same form as that of the single fluid theory Eq.(15.2).

If we multiply $\gamma_s = e_s/m_s$ to Eq.(16.138), we have

$$\frac{\partial \rho_{es}}{\partial t} + \nabla \cdot (\rho_{es} \boldsymbol{q}_s) = \gamma_s \sigma_s \tag{16.141}$$

By conservation of total electric charge, we have

$$\sum_{S=1}^{N} \gamma_s \sigma_s = 0 \tag{16.142}$$

The sum of N equations of the type of Eq.(16. 141) with the help of Eq. (16.142) gives

$$\frac{\partial \rho_e}{\partial t} + \nabla \cdot \boldsymbol{J} = 0 \tag{16.143}$$

Eq.(16. 143) is the equation of conservation of electric charge (15. 13) which is one of the fundamental equations of the single fluid theory of EMGD but in the multifluid theory, it is simply another form of the equation of continuity and may be used to replace one of the equations of continuity (16.138).

(iii) Equations of motion. The conservation of momentum of sth species gives the equation of motion of the sth species as follows:

$$\frac{\partial \rho_s u_s^i}{\partial t} + \frac{\partial}{\partial x^{ij}} (\rho_s u_s^i u_s^j - \tau_s^{ij}) = X_s^i + \sigma_s Z_s^i \tag{16.144}$$

where the summation convention is used for the repeated tensorial indices i or j but not for the indices distinguishing the species s. Eq.(16.144) is similar to that of two-phase flow equation discussed before. The term $\sigma_s Z_s^i$ is the

ith component of the momentum source of sth species. In general, we have

$$\sum_{S=1}^{N} \sigma_s Z_s^i = 0 \tag{16.145}$$

The term τ_s^{ij} is the ijth component of the stress tensor of the sth species. For a first approximation, the Navier-Stokes relations may be used with a coefficient of viscosity μ_s for sth species.

The body force X_s^i consists of the electromagnetic force F_{es}^i, a nonelectric body force such as gravitational force F_{gs}^i and the interaction force between species F_{os}^i which are given below:

$$F_{es}^i = \rho_{es} \left[E^i + (\boldsymbol{q}_s \times \boldsymbol{B})^i\right] \tag{16.146}$$

$$F_{gs}^i = \rho_s g^i \tag{16.147}$$

$$F_{os}^i = \sum_{S=1}^{N} K_{st} (u_t^i - u_s^i) \tag{16.148}$$

where K_{st} is known as the friction coefficient between species t and s and

$$\sum_{S=1}^{N} F_{os}^i = 0 \tag{16.149}$$

The ith component of the gravitational acceleration is g^i.

If we add all the N-equations of the type of Eq.(16.144), we have the equation of motion of the mixture as a whole:

$$\frac{\partial \rho u^i}{\partial t} + \frac{\partial \rho u^i u^j}{\partial x^j} = \rho \left(\frac{\partial u^i}{\partial t} + u^j \frac{\partial u^i}{\partial x^j} \right) = \rho \frac{Du^i}{Dt}$$

$$= -\frac{\partial p}{\partial x^i} + \frac{\partial \tau_v^{ij}}{\partial x^j} + F_e^i + F_g^i \tag{16.150}$$

The form of Eq.(16.150) is exactly the same as that of the single fluid theory (15.3) except in Eq.(15.3), we neglect the gravitational force. However, the definition of various terms is different, particularly the stress tensor terms.

The nonelectric body force such as gravitational force is

$$F_g^i = \sum_{S=1}^{N} F_{gs}^i = \rho g^i \tag{16.151}$$

The electromagnetic force is

$$F_e^i = \sum_{S=1}^{N} F_{es}^i = \rho_e E^i + (\boldsymbol{J} \times \boldsymbol{B})^i \tag{16.152}$$

The stress tensor of sth species may be written as below:

$$\tau_s^{ij} = -p_s\,\delta^{ij} + \tau_{os}^{ij} \tag{16.153}$$

and

$$\tau_{os}^{ij} = \mu_s\left(\frac{\partial u_s^i}{\partial x^j} + \frac{\partial u_s^j}{\partial x^i}\right) - \frac{2}{3}\,\mu_s\,\frac{\partial u_s^k}{\partial x^k}\,\delta^{ij} \tag{16.154}$$

The viscous stress tensor of the plasma as a whole is

$$\tau_v^{ij} = \sum_{s=1}^{N} \tau_{os}^{ij} - \sum_{s=1}^{N} \rho_s\, w_s^i\, w_s^j \tag{16.155}$$

The interesting point is that the viscous stress tensor τ_v^{ij} depends on the viscous stress of each species and the diffusion velocities between the species. Since the diffusion velocities are complicated, only when the diffusion velocities are small, simple expressions may be used for τ_v^{ij}. In general, the equation of diffusion velocity $w_s^i = u_s^i - u^i$ has the following form:

$$\frac{\partial w_s^i}{\partial t} + u_s^j\,\frac{\partial u_s^i}{\partial x^j} - u^j\,\frac{\partial u^i}{\partial x^j} = \frac{1}{\rho}\,\frac{\partial p}{\partial x^i} + \frac{1}{\rho_s}\,\frac{\partial \tau_s^{ij}}{\partial x^j} - \frac{1}{\rho}\,\frac{\partial \tau_v^{ij}}{\partial x^j}$$

$$+ \frac{x_s^i}{\rho_s} - \frac{F_e^i + F_q^i}{\rho} + \frac{\sigma_s}{\rho_s}\,(Z_s^i + u_s^i) \tag{16.156}$$

(iv) Equation of energy . The conservation of energy of sth species gives the energy equation of the sth species:

$$\frac{\partial \bar{e}_s}{\partial t} + \frac{\partial}{\partial x^j}\,(\bar{e}_s\, u_s^i - u_s^i\, \tau_s^{ij} - Q_s^j) = \bar{\varepsilon}_s \tag{16.157}$$

where $\bar{e}_s = \rho_s\, \bar{e}_{ms}$ is the total energy of sth species of the mixture per unit volume which consists of the internal energy U_{ms} of sth species, kinetic energy $\frac{1}{2}\, u_s^i\, u_s^i$ of sth species and potential energy of sth species. The heat conduction flux of sth species is Q_s^j. The energy source $\bar{\varepsilon}_s$ consists of the terms due to electromagnetic field $\bar{\varepsilon}_{es}$, that due to chemical reaction $\bar{\varepsilon}_{cs}$ and that due to elastic collision between species $\bar{\varepsilon}_{os}$.

If we add all the N equations of the type of (16.157), we have the energy equation of the mixture as a whole as follows:

$$\frac{\partial \rho \bar{e}_m}{\partial t} + \frac{\partial \rho u^j \bar{e}_m}{\partial x^j} = -\frac{\partial\, p u^j}{\partial x^j} + \frac{\partial u^i\, \tau_v^{ij}}{\partial x^j} - \frac{\partial Q^j}{\partial x^j} + \bar{\varepsilon}_T \tag{16.158}$$

The form of Eq.(16.158) is the same as that of the single fluid theory, Eq. (15.5), but the meaning of various terms is different.

The total energy $\bar{e}_m$ consists of the internal energy, kinetic energy and the potential energy of the mixture as a whole but the definition of the internal energy per unit mass of the mixture U_m consists of the sum of the ordinary molecular internal energy of all species and the diffusion kinetic energy of all species, i.e.,

$$U_m = \frac{1}{\rho} \sum_{s=1}^{N} (\rho_s U_{ms} + \frac{1}{2} \rho_s w_s^2) \tag{16.159}$$

The energy source $\bar{\varepsilon}_T$ consists of the resultant energy source $\bar{\varepsilon}_c$ due to chemical reaction and that due to electromagnetic field, $\bar{\varepsilon}_e = E^i J^i$.

The electromagnetic field equations for the multifluid theory are the same as those for the single fluid theory, i.e., Eqs. (15.6) and (15.7).

There are two ways to describe the flow of a plasma by the multifluid theory.

(a) We may use the partial variables (T_s, p_s, n_s, u_s^i, E^i, H^i) to describe the flow field of a plasma where $s = 1, 2, \cdots$, N. The fundamental equations are Eqs. (16.136), (16.138), (16.144), (16.157), (15.6) and (15.7); or

(b) We may use the gross variables (T, p, n, u^i, E^i, H^i) and some of the partial variables (T_r, p_r, n_r, u_r^i) where $r = 1, 2. \cdots, (N-1)$. The fundamental equations are Eqs. (16.137), (16.140), (16.150), (16.158), (15.6), (15.7), together with the corresponding equations of the partial variables (16.136), (16.138), (16.144) and (16.157). We may also replace the partial velocity u_r^i by the diffusion velocity w_r^i and the partial density n_r or ρ_r by the concentration $c_r = n_r/n$ or by the mass concentration $k_r = \rho_r/\rho$.

In principle, the above two approaches are identical. But from the easy application of the multifluid theory to the flow problem point of view, approach (a) is preferable because we do not need to consider the complicated function U_m etc., which includes many diffusion phenomena. We use approach (a) to discuss some simple flow problems in the next two sections.

It seems to be easier to find the relation between the multifluid theory and the classical single fluid theory from approach (b). To show this relation, we consider a case of partially ionized monatomic gas which consists of atoms, singly charged ions and electrons, i.e., $N = 3$. According to the approach (b), we should use the following 24 variables to describe the flow field of such a plasma:

$$T, p, \rho, u^i, E^i, H^i, \rho_e, J^i, T_1, p_1, \rho_1, w_2^i, T_2, p_2$$

where subscript 1 refers to the value of electrons and subscript 2 refers to the value of ions.

For the classical single fluid theory of a plasma, discussed in Chapter XV, we use only the following 16 variables:

$$T, p, \rho, u^i, E^i, H^i, \rho_e, J^i$$

The reduction of the 24 variables to the 16 variables is based on the following approximations:

(i) We assume that the temperature of all species are equal, i.e.,

$$T = T_1 = T_2 \tag{16.160}$$

Hence we do not have to consider the variables T_1 and T_2. This assumption is reasonably good if the masses of all species in the mixture of gases are approximately equal. Thus it is a good approximation for neutral air as we shall use in Chapter XVII for the problems of flow with chemical reactions. But for a partially ionized air, the electrons are much lighter than the ions and the neutral particles. The temperature of electrons T_1 may be different considerably from that of ions T_2 or that of atoms T_a. At high temperature range, it is advisable to consider at least two temperatures in the flow field of a plasma: one is the electron temperature T_1, and the other is the temperature of heavy particles $T_2 = T_a$.

(ii) Usually we are not interested in the partial pressures p_1 and p_2 in most engineering probelms. Furthermore, if we know the temperature and the density of a species, we can easily calculate the partial pressure from the equation of state (16.136). Hence we do not need to analyze the partial pressures in our calculation of the flow variables.

(iii) If the mass of all species in the mixture are approximately equal, the diffusion velocities w_r^i are usually small. We may either neglect the diffusion velocities completely or use some simple formula for the diffusion velocities as we shall discuss in Chapter XVII. If the diffusion velocities may be completely neglected, the concentrations of all species in the mixture of gases are constant. As a result, we need to consider the gross variables p, ρ, T and u^i only. This is the reason why the single fluid theory gives good results for a mixture of neutral gases such as ordinary air. However, in a plasma, the electrons are much lighter than all the other heavy particles and we can not neglect the diffusion velocity of electrons w_1^i. If the diffusion velocity of the ions w_2^i is still negligible, the electric current density J^i is proportional to the electron diffusion velocity w_1^i. The diffusion phenomena of electrons may be expressed in terms of the two variables J_i and ρ_o.

With the assumptions (i) to (iii), the mixture of neutral particles, ions and electrons can be treated by the single fluid theory of classical electromagnetogasdynamics with the variables p, ρ, T, u^i, E^i, H^i, J^i and ρ_e as we did in Chapter XV. Strictly speaking, J^i and/or w_1^i should be governed by a differential equation such as Eq.(16.156) or similar equation. Since Eq.(16.156) is too complicated to be useful in the single fluid theory, we use some simple formula instead of Eq.(16.156). We shall discuss this point in §10.

When the degree of ionization of the plasma is not small and the temperature range of the flow field is large, the assumption (i) and (iii) may not be good. The temperature of electrons may differ significantly from that of the heavy particles and the gasdynamic forces may have influence on the electrical current density. Thus the classical theory of single fluid will not be sufficient to predict the flow field of a plasma. It is then advisable to use the multifluid theory to analyze the flow problems of a plasma, particularly by approach (a). Furthermore, the multifluid theory will give much more informations than those by single fluid theory as we shall see in the next section.

9. Waves of small amplitude in a plasma obtained from the multifluid theory[28, 29]

Now we use the multifluid theory to study the waves of infinitesimal amplitude in a plasma similar to the case of single fluid theory studied in Chapter XV, § 6. For simplicity, we consider the case of a fully ionized plasma in which $N=2$. We consider the plasma as a mixture of singly charged ions (subscript 2) and electrons (subscript 1). Originally the plasma is at rest under an external magnetic uniform field H_0, i.e.,

$$\boldsymbol{H}_0 = \boldsymbol{i}H_x + \boldsymbol{j}H_y + \boldsymbol{k}0 \tag{16.161}$$

where H_x and H_y are constants which may or may not be zero. The unit vectors in the x-, y- and z-directions are respectively $\boldsymbol{i}$, $\boldsymbol{j}$ and $\boldsymbol{k}$. In the undisturbed state of the field, there is neither electric current nor excess electric charge, nor electric field. In the disturbed state, we have

$$\left.\begin{aligned} \boldsymbol{q}_s &= \boldsymbol{i}u_s(x,t) + \boldsymbol{j}v_s(x,t) + \boldsymbol{k}w_s(x,t) \\ p_s = p_o + p_s'(x,t),\quad & n_s = n_o + n_s'(x,t),\quad T_s = T_o + T'(x,t) \\ \boldsymbol{E} &= \boldsymbol{i}E_x(x,t) + \boldsymbol{j}E_y(x,t) + \boldsymbol{k}E_z(x,t) \\ \boldsymbol{H} = \boldsymbol{H}_o + \boldsymbol{h}(x,t) &= \boldsymbol{H}_o + \boldsymbol{i}h_x(x,t) + \boldsymbol{j}h_y(x,t) + \boldsymbol{k}h_z(x,t) \end{aligned}\right\} \tag{16.162}$$

where $s=1$ or 2. All the perturbed quantities are functions of x and t and subscript o refers to the value in the undisturbed state.

We make the following assumptions in our fundamental equations:

(i) Both the ions and the electrons may be considered as inviscid and non-heat-conducting perfect gas.

(ii) The interaction forces between the ions and the electrons are proportional to the difference between their mean velocities:

$$\boldsymbol{F}_{12} = K_{ie}(\boldsymbol{q}_1 - \boldsymbol{q}_2) = -\boldsymbol{F}_{21} \tag{16.163}$$

The linearized fundamental equations for the two-fluid theory of a plasma under the above two assumptions are as follows:

(a) Maxwell's equations for the electromagnetic field become

$$\frac{\partial \mu_e h_x}{\partial t} = 0 \tag{16.164a}$$

$$\frac{\partial \mu_e h_y}{\partial t} = \frac{\partial E_z}{\partial x} \tag{16.164b}$$

$$\frac{\partial \mu_e h_z}{\partial t} = -\frac{\partial E_y}{\partial x} \tag{16.164c}$$

$$\frac{\partial \varepsilon E_x}{\partial t} + en_o (u_2 - u_1) = 0 \tag{16.164d}$$

$$\frac{\partial \varepsilon E_y}{\partial t} + en_o (v_2 - v_1) = -\frac{\partial h_z}{\partial x} \tag{16.164e}$$

$$\frac{\partial \varepsilon E_z}{\partial t} + en_o (w_2 - w_1) = \frac{\partial h_y}{\partial x} \tag{16.164f}$$

(b) Equation of state for sth species

$$\frac{p_s'}{p_o} = \frac{n_s'}{n_o} + \frac{T_s'}{T_o} \tag{16.164g}$$

(c) Equation of continuity for sth species

$$\frac{\partial n_s'}{\partial t} + n_o \frac{\partial u_s}{\partial x} = 0 \tag{16.164h}$$

(d) Equations of motion for sth species

$$m_s n_o \frac{\partial u_s}{\partial t} = -\frac{\partial p_s'}{\partial x} + e_s n_o E_x - e_s n_o B_y w_s + K_{st} (u_s - u_t) \tag{16.164i}$$

$$m_s n_o \frac{\partial v_s}{\partial t} = e_s n_o E_y + e_s n_o B_x w_s + K_{st} (v_s - v_t) \tag{16.164j}$$

$$m_s n_o \frac{\partial w_s}{\partial t} = e_s n_o E_z + e_s n_o (u_s B_y - v_s B_x) + K_{st} (w_s - w_t) \tag{16.164k}$$

where $e_1 = -e$, $e_2 = e$, $B_o = \mu_e H_o$, $s, t = 1$ or 2.

(e) Energy equation for sth species

$$m_s n_o c_{ps} \frac{\partial T_s'}{\partial t} = \frac{\partial P_s'}{\partial t} \tag{16.164l}$$

where $m_s c_{ps} = 5k/2$.

Among the 18 perturbed quantities, only h_x is independent of all the other 17 quantities, and is given by Eq. (16.164a) alone. This is similar to the result of the single fluid theory and we may take $h_x=0$. In contrast to the single fluid theory, all the other 17 perturbed quantities are interrelated by Eqs.(16.164b) to (16.164l). However, we may consider these 17 perturbed quantities as four basic modes of waves which are interacted by the external magnetic field. Without the external magnetic field, the 17 perturbed quantities may be divided into three independent groups:

(i) The first group consists of w_1, w_2, E_z and h_y which represents a transverse wave.

(ii) The second group consists of v_1, v_2, E_y and h_z which represents another transverse wave.

(iii) The third group consists of u_1, u_2, p_1', p_2', n_1', n_2', T_1', T_2' and E_x which represents two longitudinal (sound) waves.

Now we are going to look for the periodic solutions similar to the single fluid theory by using the form of Eq.(15.99) and we obtain the dispersion relations for these basic waves.

(a) Transverse waves

The dispersion relations for the two basic transverse waves are the same, because, without the external magnetic field, we cannot distinguish y and z. The dispersion relation for the basic transverse wave is then:

$$c^2\lambda^2=\omega^2-\omega_e^2\left[(1-iK_{ie}^*)/(1+K_{ie}^*)\right] \tag{16.165}$$

where c is the velocity of light, $i=(-1)^{\frac{1}{2}}$,

$$\omega_e=e\left(\frac{n_o}{\varepsilon m_1}\right)^{\frac{1}{2}}=\text{electron plasma frequency} \tag{16.166}$$

and

$$K_{ie}^*=\frac{K_{ie}}{m_1\omega n_o} \tag{16.167}$$

For an ideal plasma, $K_{ie}=0$. Eq.(16.165) gives the speed of propogation of this transverse wave as

$$V=\frac{\omega}{\lambda}=\frac{c}{\left(1-\frac{\omega_e^2}{\omega^2}\right)^{\frac{1}{2}}} \tag{16.168}$$

Eq.(16.168) is the well-known result of elementary theory of plasma oscillation. It shows that there is a large damping for the low frequency ($\omega<\omega_e$) which is the essential mechanism of blackout of radio communication during the reentry of a space vehicle. This result can not be obtained in the single fluid theory of Chapter XV.

In vacuum, $n_o=0$ and $\omega_e=0$. Eq.(16.168) shows that $V=c$ and the electromagnetic wave is propagated in vacuum at a speed of light c.

The effect of electrical conductivity σ_e or K_{ie} is to introduce a damping on this transverse wave.

(b) Longitudinal waves.

The dispersion relation of the longitudinal wave gives two roots of λ^2 for the longitudinal wave of group (c) as follows:

$$\lambda^2 = \frac{\omega^2}{2a_i^2}\left\{\left[(1-2\frac{\omega_i^2}{\omega^2})+2iK^*_{ie}\frac{m_e}{m_i}\right]\pm\left\{\left[(1-2\frac{\omega_i^2}{\omega^2})+2iK^*_{ie}\frac{m_e}{m_i}\right]^2\right.\right.$$

$$\left.\left.-4\frac{a_i^2}{a_e^2}(1-\frac{\omega_e^2}{\omega^2})+iK^*_{ie}\right\}^{\frac{1}{2}}\right\} \tag{16.169}$$

where

$$a_s = [\gamma\, p_o/(m_s\, n_o)]^{\frac{1}{2}} = \text{sound speed of } s\text{th species} \tag{16.170}$$

$$\omega_s = e\,[n_o/(m_s\, n_o)]^{\frac{1}{2}} = \text{plasma frequency of } s\text{th species} \tag{16.171}$$

There are two basic longitudinal waves from Eq.(16.169). One is the ion sound speed (corresponding to + sign and shown in Fig.16.6 which is always an undamped wave). As the frequency ω of this wave approaches to zero, the velocity of propagation of this wave is $V_i = \sqrt{2}\, a_i = a_p$ which is the sound speed of the plasma as a whole. The value of V_i decreases continuously as ω increases and as $\omega \to \infty$, $V_i = a_i$, the sound speed of ions alone. Hence this mode is closely associated with ions and we may call it the ion sound speed.

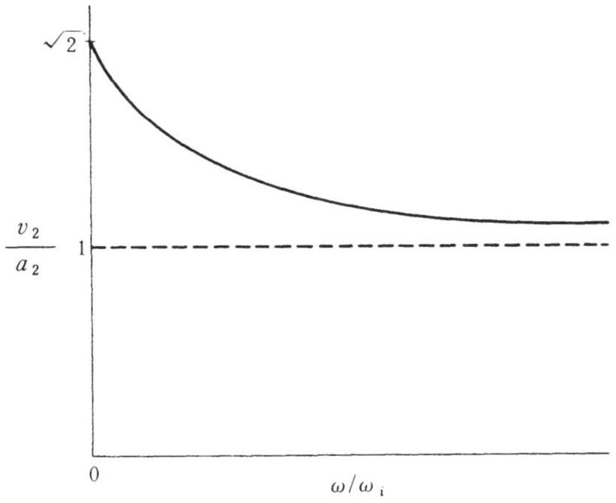

Fig. 16.6 Speed of propagation of ion sound wave in an ideal plasma

The other longitudinal wave is the electron sound wave (corresponding to the – sign and shown in Fig.16.7). This wave is a damped wave when $\omega < \omega_e$ and an undamped wave when $\omega > \omega_e$. As $\omega \to \infty$, the spoed of propagation of this sound wave is $V_e = a_e$, the speed of sound wave for electron alone. Hence we call this mode the electron sound wave. The single fluid theory of Chapter XV gives the results corresponding to $\omega \to 0$ only.

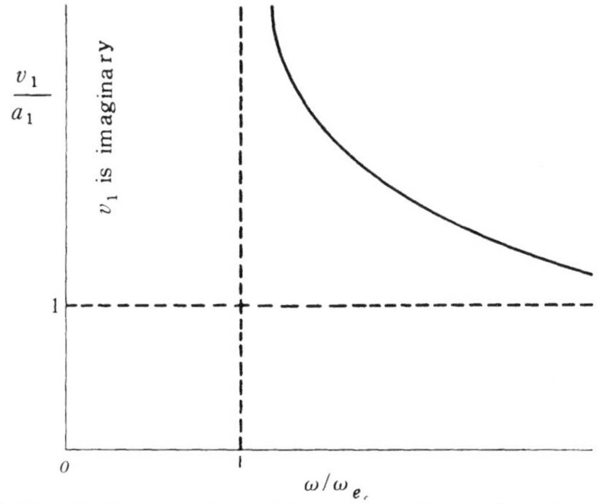

Fig.16.7 Speed of propagation of electron sound wave in an ideal plasma

(c) Waves under longitudinal external magnetic field

If $H_x \neq 0$ and $H_y = 0$, the basic longitudinal waves are not affected by the external magnetic field and the two basic transverse waves are interacted. The resultant dispersion relation of these interacted transverse waves is

$$A_o \lambda^4 + 2 A_2 \lambda^2 + A_4 = 0 \tag{16.172}$$

The coefficients A_0, A_2 and A_4 are functions of plasma frequencies, cyclotron frequencies of the species, frequency ω and the friction coefficient K^*_{ie}. For an ideal plasma, we have the following expressions for these coefficients:

$$A_0 = (1 - \frac{\omega_{xi}^2}{\omega^2})(1 - \frac{\omega_{xe}^2}{\omega^2})$$

$$A_2 = \frac{\omega_e^2}{c^2}\left[(1 - \frac{\omega^2}{\omega_e^2}) - \frac{\omega_{xi}\,\omega_{xe}}{\omega^2}(1 - \frac{\omega^2}{\omega_i^2} + \frac{\omega_{xi}\,\omega_{xe}}{\omega_e^2})\right] \tag{16.173}$$

$$A_4 = \frac{\omega_e^4}{c^4}\left[(1 - \frac{\omega^2}{\omega_e^2} + \frac{\omega_{xi}\,\omega_{xe}}{\omega^2})^2 - \frac{\omega_{xi}\,\omega_{xe}\,\omega^2}{\omega_i^2\,\omega_e^2}\right]$$

and

$$\omega_{xs} = \frac{e B_x}{m_s} = x\text{-wise cyclotron frequency of } s\text{th species} \tag{16.174}$$

The two solutions of Eq.(16.172) are:

$$\lambda_1^2 = \frac{-A_2 - (A_2^2 - A_0 A_4)^{\frac{1}{2}}}{A_0} = \text{ordinary wave (electrons)} \qquad (16.175)$$

$$\lambda_2^2 = \frac{-A_2 + (A_2^2 - A_0 A_4)^{\frac{1}{2}}}{A_0} = \text{extraordinary wave (ions)} \qquad (16.176)$$

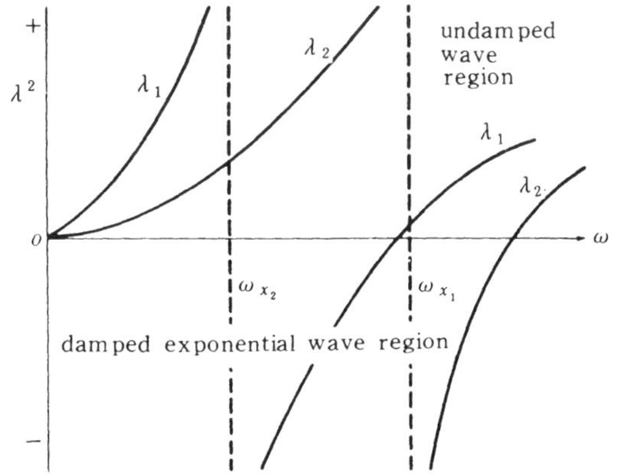

Fig. 16.8 Interacted transverse waves in an ideal plasma under the longitudinal applied magnetic field H_x

The variations of λ_1^2 and λ_2^2 with frequency ω are shown in Fig.16.8. One interesting result is that as $\omega \to 0$, the speed of wave propagation of these two interacted transverse waves is given by the same formula:

$$V = \frac{\omega}{\lambda} = \frac{V_x}{(1 + \frac{V_x^2}{c^2})^{\frac{1}{2}}} \qquad (16.177)$$

where $V_x = H_x(\mu_e/\rho_o)^{\frac{1}{2}}$ is the speed of Alfven's wave based on H_x. As $V_x \ll c$, Eq.(16.177) reduces to Eq.(15.104). Here again, the single fluid theory gives only the results as $\omega \to 0$. With the applied longitudinal magnetic field, we have undamped transverse waves at low frequencies, instead of damped waves for the case without applied magnetic field. Thus the applied magnetic field improves the transverse wave propagation at low frequencies. In the intermediate frequencies, one or both of the transverse waves may change into damped waves ($\lambda^2 < 0$) but at high frequencies, they become undamped waves again. As $\omega \to \infty$, these two basic waves are not interacted.

(d) Waves under transverse applied magnetic field

If $H_x = 0$ and $H_y \neq 0$, the second basic transverse wave (v_1, v_2, E_y and h_z)

is not affected by the applied magnetic field and the first transverse wave and the two longitudinal waves are interacted and thus three new transverse-longitudinal waves can be obtained. The dispersion relation of these three interacted waves is

$$\lambda^6 + S_4 \lambda^4 + S_2 \lambda^2 + S_0 = 0 \tag{16.178}$$

where the coefficients S_4, S_2 and S_0 are functions of the sound speeds, plasma frequencies, cyclotron frequencies of the two species and the friction coefficients.[28] For an ideal plasma, the solutions of Eq.(16.178) are shown in Fig.16.9. There are three interacted waves. The first wave λ_1^2 represents the interaction between the ion sound wave and the magnetic field H_y and it is always an undamped wave. As $\omega \to 0$, the velocity of propagation of this wave when $V_y = H_y\,(\mu_e/\rho_o)^{\frac{1}{2}} \ll c$, is

$$V_1 = (a_p^2 + V_y^2)^{\frac{1}{2}} \tag{16.179}$$

Eq.(16.179) is the same as Eq.(15.82). Hence the single fluid theory gives again the results of multifluid theory at $\omega \to 0$.

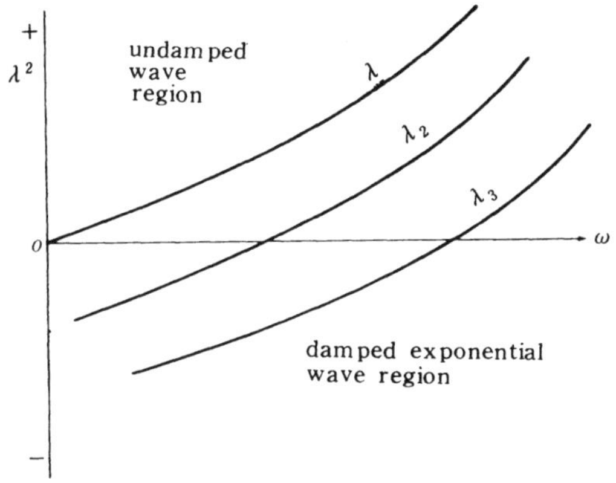

Fig. 16.9 Dispersion relations of transverse-longitudinal waves in an ideal plasma under transverse magnetic field H_y

At low frequencies, the other two interacted waves are damped waves. At high frequencies, these two damped waves turn into undamped waves. The points at which these damped waves turn into undamped waves depend on the relative magnitude of the plasma frequencies of the species to the cyclotron frequencies of the species.[29] At very high frequencies, all the three basic waves do not interact.[29]

(e) Waves under arbitrarily orientated applied magnetic field

If both H_x and H_y are different from zero, all the four basic waves interact and we have four new modes of waves in a plasma whose dispersion relation is:

$$\lambda_0 \lambda^8 + C_1 \lambda^6 + C_2 \lambda^4 + C_3 \lambda^2 + C_4 = 0 \qquad (16.180)$$

The coefficient C_0, C_1 etc., are functions of plasma frequencies, cyclotron frequencies, sound speeds of these two species and friction coefficient.[28] In Fig. 16.10, we sketch the solutions of Eq.(16.180). At low frequencies,

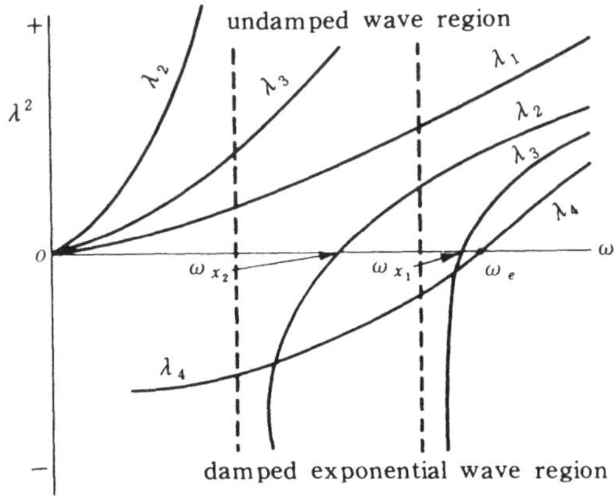

Fig. 16.10 Dispersion relations of magnetogasdynamic waves in an ideal plasma under arbitrarily orientated applied magnetic field H_x and H_y

there are three undamped waves and one damped wave. These three undamped waves are the fast wave, the slow wave and the transverse wave given in the single fluid theory of Chapter XV, § 6 when $\omega \to 0$. Hence the single fluid theory gives only the results at very low frequencies. As the frequency of the wave increases, only one of the undamped waves remains undamped all the time; the other two waves may change first into damped waves at the ion and electron x-wise cyclotron frequencies respectively. But at high frequencies, they will return to the state of undamped wave. The fourth wave which is associated with the electron sound wave becomes undamped when $\omega > \omega_e$.

From the above results, we see that the multifluid theory gives us much more informations than that from the classical single fluid theory of Chapter XV.

10. Generalized Ohm's law and Hall current and ion slip[1, 2]

Let us examine the equation of the electric current density from the multifluid theory point of view. The electric current density J^i consists of an electric conduction current density I^i and an electric convection current density $\rho_e u^i$. Since the electric convection current density $\rho_e u^i$ is already expressed in terms of ρ_e and u^i, we need to find an equation for the electric conduction current density I^i only. Eq.(16.135) shows that the electric current density I^i depends on the diffusion velocities w_s^i. Hence we may find a formula for I^i by studying the diffusion velocities w_s^i.

We may calculate the diffusion velocities w_s^i from Eq.(16.156). A generalized Ohm's law may be derived from Eq.(16.156) under the following assumptions:

(a) The equation is explicitly independent of time and spatial coordinates.

(b) The electromagnetic force is the only dominant force, and

(c) There is no source terms, i.e., $\sigma_s=0$.

Under these three conditions, Eq.(16.156) reduces to the form

$$\frac{F_{er}^i+F_{or}^i}{\rho_r}=\frac{F_e^i}{\rho} \tag{16.181}$$

or

$$(\rho\,\rho_{er}-\rho_r\rho_e)\boldsymbol{E}_u+(\rho\,\rho_{er}\,\boldsymbol{w}_r-\rho_r\,\boldsymbol{I})\times\boldsymbol{B}=\rho\sum_{s=1}^{N}K_{sr}\,(\boldsymbol{w}_r-\boldsymbol{w}_s) \tag{16.182}$$

where $\boldsymbol{E}_u=\boldsymbol{E}+(\boldsymbol{q}\times\boldsymbol{B})$. If we consider that ρ_r, ρ_{er}, $\boldsymbol{E}_u$, $\boldsymbol{B}$ and K_{sr} are given, Eq.(16.182) gives a set of linear algebraic equations for $\boldsymbol{w}_r$ because $\boldsymbol{I}$ is a linear function of $\boldsymbol{w}_r$. We may solve for $\boldsymbol{w}_r$ from Eq.(16.182).

To illustrate the solution of Eq.(16.182), we consider the case of a partially ionized monatomic gas which consists of (i) electrons, (subscript e), (ii) singly charged ions (subscript i) and (iii) atoms (subscript a). The number densities of electrons, ions, and atoms are respectively n_e, n_i and n_a. We consider a slightly ionized gas so that $n_e \simeq n_i \ll n_a$. The density of the gas as a whole is

$$\rho=m_e\,n_e+m_i\,n_i+m_a\,n_a=m_a\,(n_e+n_a) \tag{16.183}$$

where $m_e \ll m_i \cong m_a$.

From Eq.(16.134), we have

$$m_e\,n_e\,w_e^i+m_i\,n_i\,w_i^i+m_a\,n_a\,w_a^i=0 \tag{16.184}$$

Since $m_e \ll m_i=m_a$ and $n_e \cong n_i \ll n_a$ and the three terms of Eq.(16.184) should be of the same order of magnitude, we conclude that

$$w_e^i \gg w_i^i \gg w_a^i \tag{16.185}$$

From Eq.(16.182) we have

(a) For electrons

$$-en_e\,\boldsymbol{E}_u-en_e\,\boldsymbol{w}_e\times\boldsymbol{B}=K_{ie}\,(\boldsymbol{w}_e-\boldsymbol{w}_i)+K_{ea}\,(\boldsymbol{w}_e-\boldsymbol{w}_a) \tag{16.186}$$

(b) For ions

$$+en_e\,\boldsymbol{E}_u+en_e\,\boldsymbol{w}_i\times\boldsymbol{B}=K_{ie}\,(\boldsymbol{w}_i-\boldsymbol{w}_e)+K_{ia}\,(\boldsymbol{w}_i-\boldsymbol{w}_a) \tag{16.187}$$

where we neglect the term of the order of m_e/m_a. Eqs.(16.186) and (16.187) have three unknowns $\boldsymbol{w}_e$, $\boldsymbol{w}_i$ and $\boldsymbol{w}_a$. By the relations (16.185), we see that (i) for first approximation, we may neglect both $\boldsymbol{w}_i$ and $\boldsymbol{w}_a$ and Eq.(16.186) gives $\boldsymbol{w}_e$ and then the electric conduction current density I^i; (ii) for second approximation, we may neglect $\boldsymbol{w}_a$ and solve Eqs.(16.186) and (16.187) for $\boldsymbol{w}_e$ and $\boldsymbol{w}_i$ and then I^i and (iii) for the third approximation, we may estimate $\boldsymbol{w}_a$ from Eq.(16.184), i.e.,

$$\boldsymbol{w}_a \cong -\frac{n_e}{n_a}\boldsymbol{w}_i \tag{16.188}$$

and solve Eqs.(16.186), (16.187) and (16.188) for $\boldsymbol{w}_e$, $\boldsymbol{w}_i$ and $\boldsymbol{w}_a$ and then I^i.

After $\boldsymbol{w}_e$ and $\boldsymbol{w}_i$ are obtained, the total electric conduction current density is

$$I^i = en_e\,(w_i^i - w_e^i) = I_i^i + I_e^i \tag{16.189}$$

and the electron current density is

$$I_e^i = -en_e\,w_e^i \tag{16.190}$$

From Eqs. (16.186) to (16.188), we find the relation between $\boldsymbol{I}$ and $\boldsymbol{I}_e$ as follows:

$$\boldsymbol{I}_e = \frac{1}{1+\varepsilon_0}\left(\boldsymbol{I} + \beta_i\,\frac{\boldsymbol{B}}{B}\times\boldsymbol{I}\right) \tag{16.191}$$

where

$$\varepsilon_0 = \frac{K_{ea}}{\dfrac{n_e}{n_a}K_{ea} + \left(1+\dfrac{n_e}{n_a}\right)K_{ai}} \tag{16.192}$$

$$\beta_i = \frac{en_e\,B}{\dfrac{n_e}{n_a}K_{ea} + \left(1+\dfrac{n_e}{n_a}\right)K_{ai}} = \text{ion slip factor} \tag{16.193}$$

From Eqs.(16.186) to (16.191), we obtain the generalized Ohm's law as follows:

$$\sigma_e\,\boldsymbol{E}_u = A_1\,\boldsymbol{I} + A_2\,\boldsymbol{I}\times\boldsymbol{B} + A_3\,(\boldsymbol{I}\times\boldsymbol{B})\times\boldsymbol{B} \tag{16.194}$$

where

$$A_1 = 1 + \beta_a \tag{16.195a}$$

$$A_2 = \frac{\sigma_e}{en_e(1+\varepsilon_0)}\left[1-\varepsilon_0\left(1+\frac{n_e}{n_a}\right)\right] \tag{16.195b}$$

$$A_3 = -\frac{1}{1+\varepsilon_0}\,\frac{\beta_i\beta_e}{B^2} \tag{16.195c}$$

$$\beta_a = \frac{K_{ea}}{K_{ei}}\,\frac{1-\dfrac{n_e}{n_a}\varepsilon_0}{1+\varepsilon_0} = \text{atom collision factor} \tag{16.195d}$$

$$\beta_e = \frac{en_e B}{K_{ei}} = \frac{\omega_e}{f} = \text{hall current factor} \tag{16.195e}$$

$$\sigma_e = \frac{e^2 n_e^2}{K_{ei}} = \text{scalar electric conductivity} \tag{16.195f}$$

where ω_c is the electron cyclotron frequency and f is the collision frequency. If we neglect K_{ea}, the collision between electrons and atoms, Eq. (16.194) may be written in the following form:

$$\boldsymbol{I} = \frac{1+\beta_i\beta_e+\beta_e^2}{(1+\beta_i\beta_e)^2+\beta_e^2}(\sigma_e \boldsymbol{E}_u'') + \frac{(1+\beta_i\beta_e)\sigma_e \boldsymbol{E}_u^{\perp}}{(1+\beta_i\beta_e)^2+\beta_e^2} + \frac{\beta_e}{(1+\beta_i\beta_e)^2+\beta_e^2}\left(\frac{\boldsymbol{B}}{B}\times \boldsymbol{E}_u\right) = \sigma_T \boldsymbol{E}_u \tag{16.196}$$

where $\boldsymbol{E}_u = \boldsymbol{E}_u'' + \boldsymbol{E}_u^{\perp}$ and

$\boldsymbol{E}_u'' = (\boldsymbol{E}_u \cdot \boldsymbol{B})\boldsymbol{B}/B^2$ = component of $\boldsymbol{E}_u$ parallel to $\boldsymbol{B}$ and B is the magnitude of $\boldsymbol{B}$,

$\boldsymbol{E}_u^{\perp} = [\boldsymbol{B}\times(\boldsymbol{E}_u\times \boldsymbol{B})]/B^2$ = component of $\boldsymbol{E}_u$ perpendicular to $\boldsymbol{B}$.

σ_T is the tensor electrical conductivity defined in Eq.(16.196).

In general, the electrical conductivity of a plasma is a tensor so that the directions of $\boldsymbol{I}$ and $\boldsymbol{E}_u$ are not the same. Only when both β_e and β_i are negligible, Eq. (16.196) reduces to Eq.(15.11), i.e.,

$$\boldsymbol{I} = \sigma_e \boldsymbol{E}_u \tag{16.197}$$

we have a scalar electrical conductivity.

Let us consider the case where the ion slip factor β_i is negligible but the Hall current factor β_e is not. Eq.(16.196) becomes

$$\boldsymbol{I} = \sigma_e \boldsymbol{E}_u'' + \sigma^{\perp} \boldsymbol{E}_u^{\perp} + \sigma_h(\boldsymbol{E}_u \times \boldsymbol{B}_1) \tag{16.198}$$

where $\boldsymbol{B}_1$ is the unit vector in the direction of $\boldsymbol{B}$. Eq.(16.198) shows that in the direction of the magnetic field ***H*** or ***B***, the electrical conductivity of the plasma has the value of its scalar electrical conductivity σ_e while the electrical conductivity in the other two directions perpendicular to B has a value less than the scalar electrical conductivity σ_e. In the direction of $E_u^{\perp}$, the electrical conductivity is $\sigma^{\perp}$ which may be written as

$$\sigma^{\perp} = \frac{\sigma_e}{[1+(\omega_c/f)^2]} \tag{16.199}$$

where $\omega_c = en_e B/m_e$ = cyclotron frequency of electron and $f = K_{ei}/m_e$ = collision frequency between ions and electrons with m_e as the mass of the electron. It is evident that $\sigma^{\perp}$ decreases with increase of magnetic induction B. Only when the ratio ω_c/f is negligibly small, $\sigma^{\perp}$ will be equal to σ_e.

The last term of Eq.(16.198) is the Hall current which is in the direction of $(\boldsymbol{E}_u \times \boldsymbol{B}_1)$, i.e., perpendicular to both $\boldsymbol{E}_u$ and $\boldsymbol{B}$. The electrical conductivity of the Hall component of the electrical conduction current may be written as

$$\sigma_h = \frac{\omega_c}{f}\sigma^{\perp} \tag{16.200}$$

Hence when ω_c/f is negligibly small, we may neglect the Hall current.

11. Problems

1. Discuss the thermodynamic relations of a mixture of a liquid and its own vapor.

2. Study the one dimensional steady flow in a nozzle of the mixture of a liquid and its own vapor with condensation and evaporation effects.

3. Calculate the non-steady one-dimensional flow in a nozzle of a mixture of air and steam including the condensation and evaporation effects.

4. Discuss briefly the supercavitating flow (see Ref. 30).

5. Study the one-dimensional steady froth flow in a nozzle.

6. Find the Rankine-Hugoniot relations of a normal shock in a froth flow.

7. Discuss the various possible flow regions in a high pressure and high temperature steam boiler from the two-phase flow point of view (see Refs. 9 and 10).

8. By dimensional analysis, discuss the (a) relaxation time and length in velocity of a gas-solid particle two phase flow and (b) the relaxation time and length in temperature of a gas-solid particle two-phase flow.

9. Calculate the sound speed in a mixture of gas-solid particle two-phase flow including the viscosity and heat-conduction effects.

10. Study the one-dimensional flow of a mixture of a gas and small solid particles in a nozzle.

11. Calculate one example of the transition flow region behind a normal shock of a mixture of a gas and small solid particles by using the Stokes formula for the interaction forces.

12. Calculate the one-dimensional steady flow in a nozzle of electrogasdynamics.

13. Calculate the flow pattern of Fig.16.5.

14. Calculate the unsteady flow of a mixture of a gas and solid particles in a nozzle.

15. Calculate the coefficients S_0, S_2 and S_4 of Eq.(16.178) and the curves in Fig.16.9.

16. Calculate the coefficients C_0, C_1, C_2 and C_3 of Eq.(16.180) and the curves in Fig.16.10.

17. Study the waves of small amplitude in an ideal plasma with three species: ions, electrons and neutral atoms.

18. Study one dimensional steady flow of an ideal plasma in a nozzle by two-fluid theory such that it is a fully ionized gas which consists of only ions and electrons.

19. Derive the linearized flow equations for a magnetogasdynamics with tensor electrical conductivity.

20. Using the fundamental equations derived from problem 19, study the flow field of a uniform stream over a thin airfoil at zero angle of attack.

References

1. Pai, S.I., *Two-Phase Flows,* Vieweg Verlag, Braunschweig, West Germany, 1977.
2. Pai, S.I., *Modern Fluid Mechanics*. Science Press, Beijing, Van Nostrand-Reinhold, New York, 1981.
3. Pai, S.I., Two-Phase Flow and Lunar Ash Flows. *Meca.Appl.* **20**, No.1, pp. 3 — 25, 1975.
4. Two-Fluid Theory of the Condensation and Evaporation Effects in Fluid Flow, *Archives of Machanics,* **34**, No.5 — 6, pp. 661 — 674, 1982.
5. Zierep, J., Theory of Flows in Compressible Media with heat addition *AGARDograph,* No. 191, 1974.
6. Frank, W., Stationaere und unstationaere Kondensationavornaenge bei einer Prandtl-Meyer Expansion, *Special report* 25/78 *of the Institute fuer Stroemungs lehre und Stroemungsmaschinen,* Universitaet Karlsruhe, 1978.
7. Oswatitsch, K., Kondensationserscheinungen in Oberschallduesen, *ZAMM,* **22**, pp. 1 — 14, 1942.
8. Wegener, P.P. and Mach L.M., Condensation in supersonic and hypersonic Wind Tunnels, *Adv. Appl. Mech.* **5**, pp.307 — 442, 1958.
9. Brodkey, R.S., *The Phenomena of Fluid Motions,* Addison Wesley Publishing Co., Reading, Mass., 1967.
10. Plesset, M.S. and Hsieh, D.Y., Theory of Bubble Dynamics in oscillating pressure filed. *Phys. of Fluids,* **3**, pp. 882 — 892, 1960.
11. Eisenberg, P. and Tulin, M. P., Cavitation, Sec. 12, *Handbook of Fluid Dynamics.,* Ed. V. L. Streeter, McGraw-Hill Book Co., New York, 1961.

12. Hsu, Y. Y. and Graham, R. W., *Transport Processes in Boiling and Two-Phase Systems*, Hemisphere Publishing Corp., Washington and McGraw-Hill Book Co., New York, 1976.
13. Tong, L.S., Heat Transfer in Two-Phase, *Lecture Series*, No.31, von Karman Institute for Fluid Dynamics, Rhode-Saint-Genese, Belgium, 1971.
14. Van Driest, E.R., Problems of high speed hydrodynamics, *Jour. Eng. for Industry*, ASME Trans. **91**, series B, No.1, pp. 1 — 12, 1969.
15. Tek, M. R., Two-Phase Flow, Sec. 17, *Handbook of Fluid Dynamics*, ed. V. L. Streeter, McGraw-Hill Book Co., New York, 1961.
16. Campbell, I.J. and Pitcher, A.S., Shock waves in a liquid containing gas bubbles, *Proc. Roy. Soc.*, London, A. 243, pp. 534 — 545, 1958.
17. Eddington, R.B., Investigation of supersonic shock phenomena in a two-phase (liquid-gas) tunnel, *Tech. Note*, No.32 — 1096, Jet Prop. Lab., California Institute of Tech., March 15, 1967.
18. Muir, J. F. and Eichhorn, R., Compressible flow of air water mixture through a vertical two-dimensional convergent-divergent nozzle, *Proc. Heat Transfer and Fluid Mech.*, Inst. Stanford Univ. Press, pp. 183 — 204, 1963.
19. Tangren, R.F., Dodge, C.H. and Seifert, H.S., Compressibility effects in two-phase flow, *Jour. Appl. Phys.* **20**, No.7, pp. 637 — 645, 1945.
20. Yih, C.S., *Dynamics of Non-homogeneous Fluids*, Macmillan Co., New York, 1965.
21. Lees, L., Ablation in Hypersonic flow, *Proc. 7th Anglo-Amer. Aero. Conf.*, Inst. of Aerosci., pp. 344 — 362, 1959.
22. Boussinesq, J., Essay on the theory of flowing water, *Mem. Acad. Sci.* Paris, **23**, pp. 1 — 680, 1877.
23. Pai, S.I. and Hsieh, T., Shock wave relations in liuar Ash flow, *ZAMM*. **55**, pp. 243 — 258, 1975.
24. Kahn, B. and Gourdine, M. C., Electrogasdynamic power generation, *AIAA Jour.*, **2**, No. 8, pp. 1423 — 1427, Aug. 1964.
25. Trezek, G. J. and France, D. M., One-dimensional particulate electrogasdynamics., *AIAA Jour.*, **8**, No.8, pp. 1386 — 1391, 1970.
26. Melcher, J.R. and Taylor, G.I., Electrohydrodynamics: A review of the role of interfacial shear stresses, *Annual Rev. of Fluid Mech.*, **1**, pp. 111 — 146, 1969.
27. Choi, H.Y., Electrohydrodynamic condensation heat transfer, *Trans. ASME, Jour. Heat Transfer*, **90**, pp. 98 — 102, 1968.
28. Pai, S.I., Wave motion of small amplitude in a fully ionized plasma under applied magnetic field, *Phys. of Fluids*, **5**, No.2, pp. 243 — 240, 1962.
29. Tanenbaum, S.B. and Mintzer, D., Wave propagation in a partially ionized gas, *Phys. of Fluids*, **5**, No.10, pp. 1226 — 1237, 1962.

Chapter XVII

FLOWS OF A COMPRESSIBLE FLUID WITH TRANSPORT PHENOMENA

1. Introduction

So far we have considered mainly the theory of the inviscid fluid flow. In spite of its idealization, the inviscid theory gives satisfactory explanations for many phenomena of fluid flow. In the inviscid theory, we have to introduce, on many occasions , surfaces of discontinuity, such as shock waves, vortex sheets and flame fronts. These surfaces of discontinuity are allowable in the flow of inviscid fluid but not in a real fluid. In the real fluid, these surfaces of discontinuity should be replaced by transition regions. If the thickness of the transition region is very thin, to a first approximation, we may replace it by a surface of discontinuity. The flow pattern outside such a transition region will not be influenced greatly by such a replacement. This is why the theory of inviscid fluid flow gives satisfactory explanations for a great many phenomena of fluid flow outside such a transition region. If we are interested in the flow inside the transition region or in flow phenomena closely related to the transition region, we must consider the effect of viscosity . For instance, the theory of inviscid flow fails to explain such phenomena as skin friction, form drag of a body, no slippage on the surface of a solid body, etc. To explain these phenomena, we must take into consideration the effect of viscosity. In §2, we shall discuss the viscosity of a fluid and its influences on the fluid flow.

Besides the viscous effects, heat conduction is also important in those transition regions mentioned above. We shall discuss heat conduction and its effects on fluid flow in §3.

In many problems of fluid flow, the fluid is actually a mixture of several gases, e.g., air is a mixture of oxygen, nitrogen, and other gases. If the concentrations of the gases in the mixture remain unchanged during the flow field, we may consider the mixture as a single fluid. Then in our analysis of flow field we may use the average fluid properties of the mixture as those of the corresponding single fluid. However if the concentrations of the gases in the mixture change within the flow field, such as in a chemical reaction or in the mixing of other fluids, it becomes necessary to consider the diffusion phenomena simultaneously with the other governing processes. In § 4, we shall briefly discuss the diffusion and chemical reactions and their influence on the fluid flow.

It is interesting to notice that the effects of viscosity, heat conduction and diffusion are due to molecular motions and they may be referred to as the transport phenomena. The effect of viscosity is the transport of momentum of the fluid flow, the effect of heat conduction is the transport of heat energy in the fluid flow and the effect of diffusion is the transport of mass in the fluid flow. In a more accurate descriptions of these transport phenomena, we have to use the molecular theory of matter[1–4]. But for many practical problems, it is sufficient to use some simple formulas to describe these transport phenomena. We shall use these simple formulas in this book. For more exact theory, Refs. 1 to 4 should be referred to.

For high temperature flows of a gas, two additional transport phenomena may be important. One is the electrical conductivity of an ionized gas which will be discussed in §5. The other is the thermal radiative transfer which will be discussed in §6.

In §7, we discuss the fundamental equations of the flow of a compressible fluid including the effects of viscosity, heat-conduction, diffusion, chemical reaction , electrical conductivity and radiative heat transfer based on the continuum theory in which simple formulas are used to describe the transport phenomena. The general discussion of the properties of these fundamental equations is very difficult. We shall discuss some properties of these fundamental equations under certain simplified conditions so that we may know some essential features of these equations. From these essential properties, we may simplify these equations so that practical problems of fluid flow may be solved.

One of the most useful and important approximations to simplify the fundamental equations of §7 is the boundary layer concept which was first introduced by Prandtl[5, 6] in 1904. The boundary layer concept has been one of the most powerful approaches to solve the flow problems with transport phenomena which will be discussed in §8. In §9 to §13, we shall discuss some special types of boundary layer flows with various transport phenomena.

The transport phenomena have significant influences on the wave motion in a compressible fluid. In §14, we study the wave of small amplitude with the effect of transport phenomena. In §15 and §16, we study the shock wave structures by including the transport phenomena. Finally, some improvements of boundary layer theory will be given in §17.

In §18, we shall briefly discuss the most difficult problems of transport phenomena, i.e., the turbulent transport phenomena which are still not quite understood.

In §19 to §21, numerical solutions of flow problems with transport phenomena will be discussed, which included (i) heat equation, (ii) steady boundary layer equations and (iii) numerical solutions of unsteady Navier-Stokes equations.

2. Viscosity

Viscosity represents that property of an actual fluid which exhibits a certain resistance to alteration of form. Although this resistance is comparatively small for many practically important fluids, such as water or gases, it is not negligible. For other fluids, such as oil, glycerine, etc., this resistance is quite large. In a viscous fluid, both tangential and normal forces exist. Some of the kinetic energy of flow will be dissipated as heat through the viscous forces.

We shall consider only the so-called Newtonian fluids, these representing most of the fluids encountered in ordinary engineering problems. The following discussion is applicable to such fluids.

Let the fluid be between two parallel plates separated by a distance y_0 from each other (Fig. 17.1). Let the lower plate be fixed, while the upper

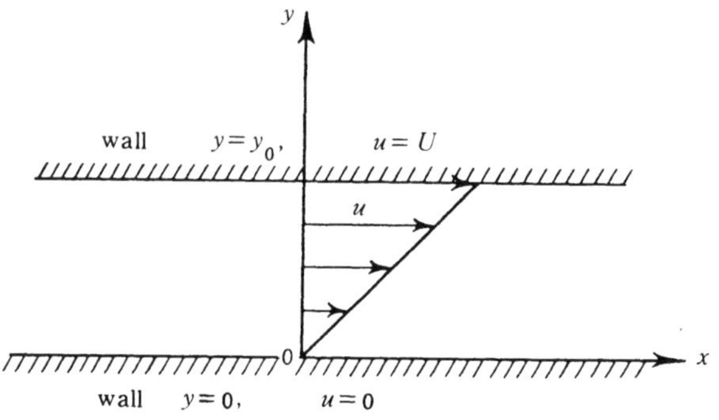

Fig. 17. 1 Viscous fluid between two parallel plates (plane Couette flow)

plate is moving uniformly with a velocity U and in a direction parallel to the lower one. A resistance D is experienced which is given by the formula

$$D = A_0 \mu \frac{U}{y_0} \tag{17.1}$$

where A_0 is the area of the upper plate and μ is a constant of proportionality called the coefficient of viscosity.

It is an experimental fact that for an ordinary fluid the relative velocity at the solid surface is zero, i.e., there is no slip at the wall. The fluid in Fig. 17.1 is displaced in such a manner that the various layers of the fluid slide uniformly over one another, and the velocity u of a layer of the fluid at a distance y from the lower plate is then

$$u = Uy/y_0 \tag{17.2}$$

Experimental results show that the tangential force per unit area, or the shearing stress τ, is proportional to the slope of the velocity, i.e.,

$$\tau = \mu \partial u / \partial y \tag{17.3}$$

This linear relation is found to be very closely correct. The factor μ depends on the temperature T, but is independent of the pressure p for gases at ordinary temperature. Discrepancies from the above law are observed only at very high velocities.

The dimensions of the coefficient of viscosity μ are easily determined from Eq. (17.3)

$$\mu = \frac{\text{shearing stress}}{\text{velocity gradient}} = \frac{mL/t^2L^2}{L/tL} = \frac{m}{tL} \tag{17.4}$$

where m is the mass, t is the time, and L is the length.

For air at $T = 20°C$, $\mu = 3.77 \times 10^{-7}$ slug/ft-sec.

From the simple kinetic theory of gases, one may show that the coefficient of viscosity μ is proportional to the square root of the absolute temperature T. In actual fact the viscosity of a gas does rise with a rise of temperature, but the square-root variation of the simple kinetic theory of gases is only qualitatively correct. In practice we usually assume that the coefficient of viscosity is proportional to a power of the absolute temperature, i. e.,

$$\frac{\mu}{\mu_0} = \left(\frac{T}{T_0}\right)^n \tag{17.5}$$

where the subscript 0 refers to a certain reference value, and n is a factor between $\frac{1}{2}$ and 1. For air at ordinary temperature, n is usually taken as 0.76. As the temperature increases, n decreases toward $\frac{1}{2}$.

A more accurate equation for the variation of viscosity with temperature, known as Sutherland's formula,

$$\frac{\mu}{\mu_0} = \frac{a_1 (T/T_0)^{1/2}}{1 + \dfrac{c_1}{T/T_0}} \tag{17.6}$$

where a_1 and c_1 are constants for a given gas. The reference temperature T_0 may be arbitrarily chosen. For air, Eq. (17.6) may be written as

$$\frac{\mu}{\mu_0} = \left(\frac{T}{T_0}\right)^{3/2} \frac{1 + b}{\dfrac{T}{T_0} + b} \tag{17.7}$$

where if we take $T_0 = 400°R$, $b = 0.522$.

The shearing stress of Eq. (17.3) is only one component of the stress tensor in the more general case. In a three-dimensional flow, the stress tensor has the nine components

$$\begin{matrix} \sigma_x & \tau_{xy} & \tau_{xz} \\ \tau_{yx} & \sigma_y & \tau_{yz} \\ \tau_{zx} & \tau_{zy} & \sigma_z \end{matrix} \tag{17.8}$$

where σ denotes the normal stress on the surface, i.e., the stress perpendicular to the surface considered. Hence σ_x is the normal stress on the surface perpendicular to the x-axis. The shearing stress is denoted by τ which is the stress in the surface considered. The first subscript refers to the direction of the axis perpendicular to the surface considered, and the second subscript refers to the direction of the force in the surface. Thus τ_{xy} denotes the component of the shearing stress in the surface perpendicular to the x-axis in the direction of the y-axis. It can be shown that, for the six tangential stresses, those which have the same suffixes but in reversed order are equal. This result follows from the condition of equilibrium of moments on an element in the continuum, i.e., $\tau_{xy}=\tau_{yx}$, etc.

For an ideal fluid, the tangential stresses are zero. By definition, the pressure is taken as the negative value of the normal stress. In an ideal fluid at any point, we have

$$\left.\begin{aligned} \sigma_x=\sigma_y=\sigma_z=-p \\ \tau_{xy}=\tau_{yz}=\tau_{zx}=0 \end{aligned}\right\} \tag{17.9}$$

In the previous chapters dealing with inviscid fluid flow, the equations of motion are based on Eqs. (17.9). For the general case, the stresses should be given by (17.8). It is easy to show that the x-, y-, and z-components of force per unit volume due to the nonhomogeneous state of stresses are, respectively:

$$\left.\begin{aligned} X &= \frac{\partial \sigma_x}{\partial x} + \frac{\partial \tau_{yx}}{\partial y} + \frac{\partial \tau_{zx}}{\partial z} \\ Y &= \frac{\partial \tau_{xy}}{\partial x} + \frac{\partial \sigma_y}{\partial y} + \frac{\partial \tau_{zy}}{\partial z} \\ Z &= \frac{\partial \tau_{xz}}{\partial x} + \frac{\partial \tau_{yz}}{\partial y} + \frac{\partial \sigma_z}{\partial z} \end{aligned}\right\} \tag{17.10}$$

For an inviscid fluid, Eqs. (17.10) reduces simply to $X=-\partial p/\partial x$, $Y=-\partial p/\partial y$ and $Z=-\partial p/\partial z$.

For a viscous fluid, we must express the viscous stresses in terms of the rate of change of velocity. In a fluid, there is a resistance to the time rate of change of shape, i. e., the deformation of velocity which may be called the strain in fluid flow. In three-dimensional flow, there are six quantities for the strain.

$$\varepsilon_x = \frac{\partial u}{\partial x}, \; \varepsilon_y = \frac{\partial v}{\partial y}, \; \varepsilon_z = \frac{\partial w}{\partial z}$$

$$\gamma_{xy} = \frac{\partial u}{\partial y} + \frac{\partial v}{\partial x}, \; \gamma_{yz} = \frac{\partial v}{\partial z} + \frac{\partial w}{\partial y}, \; \gamma_{zx} = \frac{\partial w}{\partial x} + \frac{\partial u}{\partial z} \tag{17.11}$$

where ε is the normal strain, γ is the shearing strain, and u, v, and w are the x-, y- and z-components of velocity, respectively. The strains also form a tensor of second order, namely,

$$\begin{matrix} \varepsilon_x & \frac{\gamma_{xy}}{2} & \frac{\gamma_{xz}}{2} \\ \frac{\gamma_{xy}}{2} & \varepsilon_y & \frac{\gamma_{yz}}{2} \\ \frac{\gamma_{xz}}{2} & \frac{\gamma_{yz}}{2} & \varepsilon_z \end{matrix} \tag{17.12}$$

The sum of two of the quantities $\partial u/\partial y$, $\partial v/\partial x$, $\cdots$, etc., gives the shearing strain, whereas the difference of two of these quantities gives the angular rotation of the fluid element. The components of rotation of a fluid element are

$$\omega_z = \frac{1}{2}\left(\frac{\partial v}{\partial x} - \frac{\partial u}{\partial y}\right), \; \omega_x = \frac{1}{2}\left(\frac{\partial w}{\partial y} - \frac{\partial v}{\partial z}\right), \; \omega_y = \frac{1}{2}\left(\frac{\partial u}{\partial z} - \frac{\partial w}{\partial x}\right) \tag{17.13}$$

where ω_x, ω_y, and ω_z are the average rate of rotation of the fluid element about the x-, y-, and z-axes, respectively. These angular rotations do not give internal stresses but the strains do. The vorticity is defined as twice the rate of rotation.

To a first approximation, the relations between stress and strain in a viscous fluid are

$$\sigma_x = -p + \lambda\left(\frac{\partial u}{\partial x} + \frac{\partial v}{\partial y} + \frac{\partial w}{\partial z}\right) + 2\mu\frac{\partial u}{\partial x}$$

$$\sigma_y = -p + \lambda\left(\frac{\partial u}{\partial x} + \frac{\partial v}{\partial y} + \frac{\partial w}{\partial z}\right) + 2\mu\frac{\partial v}{\partial y} \tag{17.14}$$

$$\sigma_z = -p + \lambda\left(\frac{\partial u}{\partial x} + \frac{\partial v}{\partial y} + \frac{\partial w}{\partial z}\right) + 2\mu\frac{\partial w}{\partial z}$$

$$\tau_{xy} = \mu\gamma_{xy}, \; \tau_{yz} = \mu\gamma_{yz}, \; \tau_{zx} = \mu\gamma_{zx}$$

where p is the hydrostatic pressure in a frictionless fluid, μ is the ordinary coefficient of viscosity, and λ is the second coefficient of viscosity. The only restrictions on the existence of two independent coefficients of viscosity are

$$\mu \geqslant 0 \tag{17.15}$$

$$2\mu + 3\lambda \geqslant 0 \tag{17.16}$$

so that the general dissipation function

$$\Phi = 2\mu\left[\left(\frac{\partial u}{\partial x}\right)^2 + \left(\frac{\partial v}{\partial y}\right)^2 + \left(\frac{\partial w}{\partial z}\right)^2 + \frac{\gamma_{xy}^{\ 2}}{2} + \frac{\gamma_{yz}^{\ 2}}{2} + \frac{\gamma_{zx}^{\ 2}}{2}\right] + \lambda\left(\frac{\partial u}{\partial x} + \frac{\partial v}{\partial y} + \frac{\partial w}{\partial z}\right)^2 \tag{17.17}$$

may never be negative.

The relation $3\lambda + 2\mu = 0$ is true only for a monatomic gas. If we follow the analysis of the simple kinetic theory of gases, we obtain $\lambda = -2\mu/3$. For ordinary fluid dynamic problems, the relation $3\lambda + 2\mu = 0$ is adequate. We adopt this relation in this book. Under this assumption, we see that the hydrostatic pressure p in Eqs. (17.14) is the average normal stress in the viscous fluid. The second coefficient of viscosity plays an important role in the explanation of the results of sound dissipation measurements.

Relations (17.14) are confirmed by the kinetic theory of nonuniform gases developed by Maxwell, Boltzmann, Chapman[3], Enskog and others[4]. Eqs. (17.14) are only the first approximation of the exact relation. According to the kinetic theory of gases, the second-order approximation will introduce additional highly complicated terms connecting the stresses with the heat flux and vice versa. These additional terms are negligible for gases under ordinary pressure but are important for rarefied gases.

Even though some of the most popular fluids such as water and air under normal conditions are Newtonian fluids, there are many fluids which do not satisfy the simple Navier-Stokes relations of Eqs. (17.14). These fluids may be called non-Newtonian fluids. Special treaties such as Refs. 1, 7 and 8 should be referred to for the study of the flow of non-Newtonian fluid flow.

3. Heat Conduction

Practical experience shows that compression and expansion of the fluid are accompanied by a temperature change. In the study of the flow of compressible fluids we must simultaneously study thermodynamics and the mechanics of continuous media. Whenever there is a temperature gradient in the flow field, there is a flow of heat through the mechanisms of conduction, convection, and radiation. These phenomena of heat transfer should be considered together with momentum transfer in the study of the flow of viscous compressible fluids.

The heat transfer consists of heat convection, heat conduction and thermal radiation. Thermal radiation will be discussed in §6.

Heat convection, which depends on the velocity field will be discussed in §7 in connection with the equation of energy. Heat conduction, which depends on the physical property of the fluid, will be discussed in this section.

The mechanism of heat conduction in a fluid is analogous to that producing viscous stresses; hence, heat conduction should not be omitted from the consideration if viscous stresses are retained.

If two parallel layers of fluid separated by a distance y_0 are kept at different temperatures T_1 and T_2, respectively (one of the layers may be a solid surface), a flow of heat is set up through the layer. The quantity of heat Q transferred through unit area per unit time is

$$Q=\kappa\frac{T_1-T_2}{y_0} \tag{17.18}$$

where κ is the coefficient of thermal conductivity or heat conductivity of the fluid, a physical property of the fluid. Its dimensions are

$$\kappa=\left(\frac{mL^2}{t^2}\ \frac{L}{\text{deg}}\ \frac{1}{tL^2}\right)=\left(\frac{mL}{t^3}\ \frac{1}{\text{deg}}\right) \tag{17.19}$$

where the degree is the unit of temperature. For air at 32°F, $\kappa=0.0708$ B.t.u. /ft.hr.°F.

If the distance y_0 between two layers of fluid is infinitesimal, we have

$$Q=-\kappa\frac{\partial T}{\partial y} \tag{17.20}$$

where the negative sign has been inserted because the heat flows in the direction of decreasing temperature.

From the simple kinetic theory of gases, it is easy to show that

$$\mu=\frac{\kappa}{c_p}$$

or

$$Pr=\frac{\mu c_p}{\kappa}=1 \tag{17.21}$$

where Pr is known as the Prandtl number of the gas. When the Prandtl number is equal to unity, the mechanism of heat conduction and viscous stress are exactly the same, because

$$Pr=\frac{\mu/\rho}{\kappa/(c_p\rho)}=\frac{\text{kinematic viscosity}}{\text{thermal diffusivity}}=\frac{\text{momentum diffusivity}}{\text{thermal diffusivity}}$$

Actually the Prandtl number of a gas is different from unity because of the rotation and vibration of the molecules transferring energy. A more refined theory gives

$$Pr = \frac{\mu c_p}{\kappa} = \frac{4\gamma}{9\gamma - 5} \tag{17.22}$$

For air, $\gamma=1.4$, $Pr=0.74$. For most of the practical problems, we may assume that the Prandtl number of a gas is a constant, independent of temperature.

In three-dimensional flow, Eq. (17.20) may be generalized and written in the vector form

$$\boldsymbol{Q} = -\kappa \nabla T \tag{17.23}$$

where ∇T is the gradient of temperature and $\boldsymbol{Q}$ is the vector rate of heat conduction.

The total amount of heat flowing out a parallelepiped per unit volume is equal to the divergence of the vector $\boldsymbol{Q}$, i.e.,

$$Q_e = \frac{\partial}{\partial x}\left(\kappa \frac{\partial T}{\partial x}\right) + \frac{\partial}{\partial y}\left(\kappa \frac{\partial T}{\partial y}\right) + \frac{\partial}{\partial z}\left(\kappa \frac{\partial T}{\partial z}\right) \tag{17.24}$$

4. Diffusion and chemical reactions[1, 4]

In many problems of fluid flow, the fluid actually is a mixture of several gases, e. g., air is a mixture of oxygen, nitrogen, and other gases. For the flow of a mixture of gases, new variables, in the form of the concentrations of any one gas in the mixture, enter our problems along with the ordinary fluid dynamic variables. If the concentrations remain unchanged during the flow field, we may consider the mixture as a single fluid in our fluid dynamic analysis. If the concentrations change in the flow field, we must introduce new relations to determine these concentrations. These new relations are expressions of the conservation of mass of each of the constituents in the mixture. The new equation is known as the equation of diffusion. The diffusion equation consists of three parts: diffusion due to convection, a term due to molecular transfer, and a term due to chemical reaction. Diffusion due to convection, which will be discussed in §7, depends on the velocity field. The other two terms depend on the physical properties of the mixture. These will be discussed in the present section.

Let n be the total number of molecules in the mixture of gases per unit volume under the pressure p and temperature T; let n_i be the corresponding number of molecules of ith gas in the mixture. We have

$$n = \sum_i n_i \tag{17.25}$$

We define the concentration of ith gas in the mixture as

$$c_i = \frac{n_i}{n} \tag{17.26}$$

By definition

$$\sum_i c_i = 1 \tag{17.27}$$

Let m_i be the mass of a molecule of ith gas. The mass density of the mixture under the pressure p and temperature T is

$$\rho = \sum_i m_i n_i = mn \tag{17.28}$$

where m is the average mass of a molecule of the mixture which is a function of the concentrations c_i and mass m_i. The mass concentration of ith gas in the mixture is defined as

$$k_i = \frac{m_i n_i}{mn} = \frac{m_i n_i}{\rho} \tag{17.29}$$

The diffusion due to molecular transfer, or simply molecular diffusion, depends on diffusion velocity. The diffusion velocity $\boldsymbol{V}_i$ for the ith gas in the mixture is the average velocity with which the molecules of the ith gas diffuse relative to the average velocity $\boldsymbol{q}$ of the mixture. From the kinetic theory of gases, the diffusion velocity may be calculated in terms of the properties of the molecules of the ith gas, the concentration c_i, the pressure, and the temperature of the mixture. From a macroscopic point of view, we may write the results in terms of coefficients of diffusion in the following manner:

Molecular diffusion term in diffusion equation

$$= \text{grad}\,(\rho D_i \,\text{grad}\, c_i) + D_{pi}\,\text{grad}\,(\log p) - D_{Ti}\,\text{grad}\,(\log T) \tag{17.30}$$

where we assume that there is no body force on the molecules of the gas. D_i is the ordinary coefficient of diffusion of the ith gas, D_{pi} and D_{Ti} are the coefficients of diffusion due to gradients of pressure and temperature, respectively. These coefficients are given in standard textbooks on the kinetic theory of gases[3, 4]. For a mixture of more than two constituents, the expressions for these coefficients of diffusion are rather complicated. In most textbooks, only the values for binary mixture are given. For the analysis of fluid dynamics, we assume that the necessary coefficients are given. For a binary mixture, we have only one coefficient of diffusion D_i which is proportional to $T^{3/2}p^{-1}$, the proportionality constant depending on the nature of the molecules present.

In a mixture of gases, chemical reactions may take place. The molecules of the ith gas may be formed or destroyed by the chemical reaction. Heat may be released or absorbed during the chemical reaction. The study of the mechanism of chemical reaction is a special science known as chemical kinetics, a science of considerable complexity[9, 10, 11]. Here we shall mention briefly those points which are useful in the analysis of problems of fluid dynamics.

The simplest chemical reaction is the first-order process. Consider a binary mixture of gases A and B. In the first-order chemical reaction process, we have

$$A \rightarrow bB \tag{17.31}$$

where A and B are the molar concentrations of the two components and b is the number of moles of B formed by decomposition of one mole of A. In this process, the rate that gas A is transformed into gas B is proportional to the molar concentration of A or n_A. If this rate of transformation is denoted by K_A, we have

$$K_A = -k_f n_A \tag{17.32}$$

Because K_A is proportional to n_A we say that it is a first-order process. For all practical purposes, if a first-order reaction occurs in a mixture of more than two constituents, the rate of transformation of ith gas may be written as

$$K_i = -k_f n_i \tag{17.33}$$

where k_f is called the specific reaction rate coefficient.

Many reactions are bimolecular and proceed as the result of reactions following binary collisions. For second-order bimolecular reaction, we have

$$b_1 A + b_2 B \rightarrow b_3 C + b_4 D \tag{17.34}$$

where b_1, b_2, b_3 and b_4 are the number of moles of the gases A, B, C and D, respectively. The rate of transformation of gas A into C and D is proportional to the product of the molar concentrations of both A and B, i.e.,

$$K_A = -k_f n_A n_B \tag{17.35}$$

Now the rate of reaction is proportional to the product of molar concentration of two gases. For this reason we call it a second-order process.

There is also a third-order process. In general, chemical reactions can proceed both in the forward direction [such as shown in Eqs. (17.31) and (17.34)] and in the reverse direction [i.e., from B to A in Eq. (17.31)] with a rate constant k_b. At thermodynamic equilibrium there is no net change in composition. Hence the rate constants k_f and k_b must be related.

The rate coefficient k_f is a function of temperature. The dependence of k_f on temperature was first found experimentally by Arrhenius and has been confirmed by theoretical studies. We have

$$k_f = B \exp\left(-E_A/R'T\right) \tag{17.36}$$

where B is called the frequency factor. In some literature one writes $B=1/\tau$, τ being the characteristic time of reaction, E_A is the activation energy which is a measure of the energy that a molecule must possess in order to react successfully.

The heat release per unit time during the chemical process is

$$Q_{ch}=-Q_i k_i m_i \tag{17.37}$$

where Q_i is the heat release per unit mass due to chemical reaction as a result of the change of internal energy of ith gas.

5. Electrical conductivity

The scalar electrical conductivity of a medium is one of the most important physical properties of an electrically conducting medium. As shown in Chapter XVI, § 10, the electrical conductivity represents a very complicated relative motion between charged particles. In the simplest case, the scalar electrical conductivity σ_e is defined by the simple generalized Ohm's law (16.197). It is interesting to notice that the Joule heat, i.e., the energy dissipation due to the conduction electrical current I is

$$\text{Joule heat}=I^2/\sigma_e \tag{17.38}$$

Eq. (17.38) holds whether the Hall current is negligible or not.

The electrical resistivity R is the inverse of electrical conductivity, i. e., $R=1/\sigma_e$. We know that the electrical conductivity is essentially a transport property which may be calculated by the kinetic theory of the electrically conducting medium, particularly that of a plasma.[2] The collision processes[2, 12] have great influence on the value of the scalar electrical conductivity. For a slightly ionized gas, the binary collision dictates the electron mobility[12]. Chapman and Cowling[3] gave the formula for the electrical conductivity for a slightly ionized gas with the assumption of rigid spherical molecules:

$$\sigma_e=0.532\ \frac{e^2\alpha}{(m_e kT)^{1/2}}\ \frac{1}{Q}=\sigma_c \tag{17.39}$$

where α is the degree of ionization and Q is the electron-atom collision cross section which is about 10^{-15} cm^2 for air at 5,000 K.

For a fully ionized plasma, the distance encounters between ions dominate the situation. Spitzer and Harem[13] gave the following formula for electrical conductivity of a fully ionized plasma:

$$\sigma_e=\frac{01591\,(kT)^{3/2}}{m_e^{1/2}e^2\ln(L_d/b_o)}=\sigma_d \tag{17.40}$$

where $L_d=[(kT)/(8\pi n_e e^2)]^{1/2}$ is the Debye shielding distance and $b_o=e^2/3kT$

is the impact parameter.

For intermediate degree of ionization, no simple formula for the electrical conductivity has been found. Kantrowitz, Lin and Resler[14] suggested the following formula for the approximate value of the electrical conductivity for a plasma with medium degree of ionization:

$$\frac{1}{\sigma_e} = \frac{1}{\sigma_c} + \frac{1}{\sigma_d} \tag{17.41}$$

The electrical conductivity of an ionized gas may be increased by seeding with additive of low ionization potential, e.g. cesium or potassium. Fig. 17.2 shows the tensor electrical conductivity of argon seeded with 1% potassium as a function of density and temperature where $\beta_e = WB\rho_o en_e$.

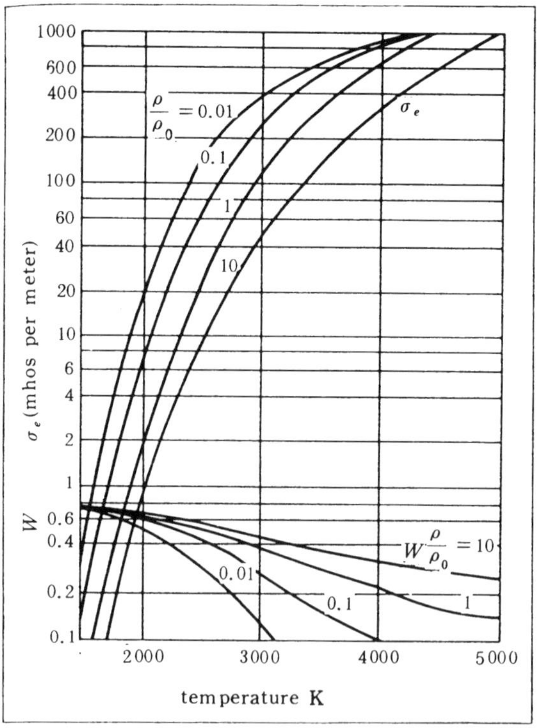

Fig. 17. 2 Tensor electrical conductivity of Argon seeded with 1% potassium as a function of density and temperature (Fig. 2 of Ref. 15, courtesy of AVCO)

The kinetic theory of plasma gives us much interesting information. For instance, from one simple case of electric current without the influence of pressure gradient nor gravitational force, the cross section area in the calculation of electrical conductivity was found to decrease with increasing the relative velocity between ions and electrons. Thus the electric resistance decreases with high electric current strength and tends to be zero at very high electrical current. This is known as the runaway effect[2].

Since the kinetic theory of liquid has not been well developed, we do not know the electric conductivity of a liquid as well as we know that of an ionized gas. Some typical values of electrical conductivity σ_e of various conductors are listed below:

Copper	600,000 mho/cm
Mercury	10,800 mho/cm
Air (seeded with 0.1K) (shock Mach No. =16)	4 mho/cm
Air (shock Mach No. =14)	0.5 mho/cm
Salt water (saturated at 25°C)	0.25 mho/cm
Pure water	0.000,000,2 mho/cm

For liquid and for solid, we may consider the electrical conductivity as a scalar quantity. For ionized gas under a strong magnetic field, the electrical conductivity should be considered as a tensor quantity, as we have shown in Chapter XVI, § 10.

6. Thermal radiation[1, 16]

At very high temperature, $T > 10^4$ K, and low density, the thermal radiation effects may not be negligible. There are three main thermal radiation effects on the flow field of a high temperature gas which are the radiation energy density E_R, the radiation stresses (the radiation pressure p_R is a special case of these stresses) and the heat flux of radiation. In previous chapters, we have considered the effects of radiation energy density and radiation pressure on the compressible flow. Now we are going to consider the thermal radiation effects in general, particularly the heat flux of radiation which may be considered as a special transport phenomenon.

For the radiation heat flux, we may estimate its value by the simple equilibrium formula:

$$q_R = \frac{\sigma T^4}{\rho} \tag{17.42}$$

where $\sigma = ca_R/4 = 5.75 \times 10^{-5}$ erg • cm^{-2} sec^{-1} K^{-4} is the Stefan-Boltzmann constant for radiative heat transfer and c is the velocity of light in vacuum ($c = 3 \times 10^8$m/sec). The heat flux by convection may be written as

$$q_v = U\ c_v T \tag{17.43}$$

where U is a typical flow velocity. If we take $U = 6.5 \times 10^4$ m/sec which is an average speed of a satellite, we find that at $T = 10^4$ K, the heat flux of radiation q_R and the heat flux by convection q_v are of the same order of magnitude. At present time, many reentry problems in space flight have a temperature of the order of 10^4 K. It is one of the reasons that aerospace engineers study the gasdynamics including the radiative heat flux. However, the radiation energy density and radiation pressure as shown in Chapter III, § 9 (cf. Fig. 3.7) become important at much higher temperature, say $T > 10^5$ K. Hence in many aerospace problems, we may still neglect the effects of radiation energy density and radiation pressure. Of course for the flow of very high temperature and very low density, all these three radiation effects should be considered.

We may consider the radiation either as a stream of photons or as electromagnetic waves. For continuum theory, it is more convenient to consider the thermal radiation as electromagnetic waves which will be used in this book. Reference 16 gives both the approaches of stream of photons which should be studied by relativistic theory and that of electromagnetic waves which is studied by the method of geometric optics.

When we consider the thermal radiation as electromagnetic waves, the thermal radiation may be expressed in terms of a specific intensity I_ν which is defined as follows:

$$I_\nu = \lim_{d\sigma_0,\, d\nu,\, dt,\, d\omega \to 0} \left(\frac{dE_\nu}{d\sigma_0 \cdot \cos\theta \cdot d\nu \cdot dt \cdot d\omega} \right) \tag{17.44}$$

where I_ν is a function of time t, spatial coordinates, direction θ which is the angle between the direction of the ray L and the normal of the elementary area $d\sigma_0$ (Fig. 17.3), and the frequency of the wave ν. The amount of radiative energy flowing through the area $d\sigma_0$ in the frequency range ν and $\nu + d\nu$, in the direction of the ray L which makes an angle θ with the normal $\boldsymbol{n}$ of the area $d\sigma_0$ within the solid angle $d\omega$ in the time interval dt is dE_ν. The

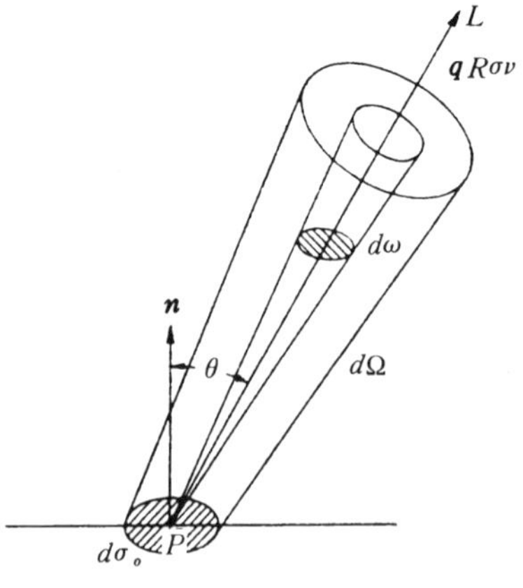

Fig. 17.3 A heat ray in a radiation field

total amount of energy radiated over the whole spectrum is

$$dE = \int_0^\infty \frac{dE_\nu}{d\nu}\, d\nu = I \cos\theta \cdot d\sigma_0 \cdot d\omega \cdot dt \tag{17.45}$$

where

$$I = \int_0^\infty I_\nu\, d\nu = \text{integrated intensity of thermal radiation} \tag{17.46}$$

The main interesting point for the specific intensity I_ν is that it depends on the frequency ν. In general, the specific intensity I_ν tends to be zero for both very high and very low frequencies as shown in Fig. 17.4. There is a maximum in the intermediate frequency. Hence strictly speaking the radiation effects would depend on the frequency too. However, for the study of engineering flow problems, we may use some average value so that we do not study the detailed effects due to the variation of frequency as we shall show later.

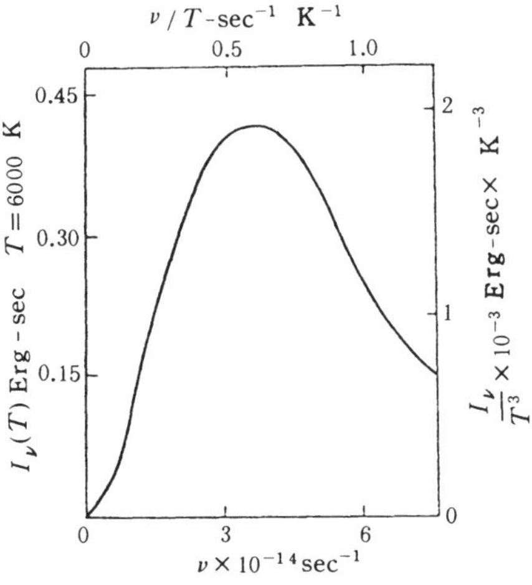

Fig. 17.4 The specific intensity of a black body

The radiative properties of a medium may be expressed by the absorption coefficient k_ν of radiation. The absorption coefficient k_ν may be considered as a transport coefficient of radiation energy. The loss of specific intensity along the ray of radiation over a distance ds in the medium of density ρ is given by the relation:

$$dI_\nu = -\rho k_\nu I_\nu\, ds \tag{17.47}$$

where k_ν is knwon as the mass absorption coefficient or simply absorption coefficient of the medium and ρk_ν is known as linear absorption coefficient of the medium because its dimension is $1/L$ where L is the dimension of a length. Some times, we introduce the term of mean free path of radiation as

$$L_{R\nu} = \frac{1}{\rho k_\nu} \tag{17.48}$$

The mean free path of radiation $L_{R\nu}$ represents the average distance between collision of a photon and a molecule and plays a similar role in radiative transfer as ordinary mean free path in gasdynamics. The integration of Eq. (17.47) gives

$$I_\nu(s) = I_\nu(s_0)\exp\left(-\int_{s_0}^{s} \rho k_\nu \, ds\right) = I_\nu(s_0)\exp(-\tau_\nu) \tag{17.49}$$

where τ_ν defined in Eq. (17.49) is known as optical thickness of radiation of the layer $(s-s_0)$ and s_0 is a reference point where the specific intensity is $I_\nu(s_0)$. The optical thickness τ_ν is a non-dimensional distance which shows the effective length in absorption of radiation. For a given physical length $L=s-s_0$, if τ_ν is large, the medium is said to be optically thick, while if τ_ν is small, the medium is said to be optically thin.

The absorption coefficient k_ν is a function of temperature and the pressure of the medium as well as the frequency ν. We shall discuss these variations later.

The specific intensity I_ν is governed by the equation of radiative transfer which is the conservation of radiative energy. In deriving the equation of radiative transfer, we have to consider both the abosrption coefficient as well as the emission coefficient[1, 16]. Special treatises should be referred to for the general discussion of the equation of radiative transfer[1, 16]. For the gasdynamic problems, we may assume that the gas is in local thermodynamic equilibrium and the equation of radiative transfer is given as follows:

$$\frac{1}{c}\frac{\partial I_\nu}{\partial t} + \frac{\partial I_\nu}{\partial s} = \rho k_\nu'(B_\nu - I_\nu) \tag{17.50}$$

where s is the distance along the radiation ray and

$$k_\nu' = k_\nu\left[1-\exp\left(-\frac{h}{kT}\right)\right] = \text{reduced aborption coefficient} \tag{17.51}$$

and

$$B_\nu = \frac{2h\nu^3}{c^2}\,\frac{1}{\exp[h\nu/(kT)]-1} = \text{Planck radiation function} \tag{17.52}$$

where $h=6.62\times10^{-27}$ erg. sec is the Planck constant and $k=1.379\times10^{-16}$ erg/K is the Boltzmann constant.

If we obtain the specific intensity I_ν, we may calculate the radiation terms in gasdynamics as follows:

(i) Radiation energy and radiation stresses

The energy density of radiation U_ν is the amount of radiant energy per unit volume in the state of frequency interval ν and $\nu + d\nu$ which is on course of transit in the neighborhood of the point considered. It is easy to show that[16]

$$U_\nu = \frac{1}{c} \int_{4\pi} I_\nu \, d\omega \tag{17.53}$$

The total energy density of radiation for the whole spectrum at any point in space at any time is then

$$E_R = \int_0^\infty U_\nu \, d_\nu = \frac{1}{c} \int_{4\pi} I \, d\omega \tag{17.54}$$

The energy density of radiation per unit mass is then E_R/ρ which should be added to the internal energy due to molecular motion so that we may have the total or effective internal energy of a radiating gas (cf. Chapter III, §9).

A quantum of energy $h\nu$ is associated with a momentum $h\nu/c$. Hence we may calculate the radiation stress tensor from the rate of change of momentum associated with the energy of photons. The ijth component of the radiation stress tensor τ_R is

$$\tau_R^{ij} = -\frac{1}{c} \int_{4\pi} I \, n^i \, n^j \, d\omega \tag{17.55}$$

We may define a radiation pressure p_R as:

$$p_R = -\frac{1}{3} \left(\tau_R^{11} + \tau_R^{22} + \tau_R^{33} \right) = \frac{1}{3c} \int_{4\pi} I \, d\omega = \frac{E_R}{3} \tag{17.56}$$

Hence we see that the radiation pressure and the radiation energy density are always of the same order of magnitude.

For isotropic radiation, we have

$$p_R = -\tau_R^{11} = -\tau_R^{22} = -\tau_R^{33} = \frac{1}{3} E_R, \; \tau_R^{ij} = 0, \text{ if } i \neq j \tag{17.57}$$

(ii) Radiative heat flux

The flux of heat energy by thermal radiation is q_R^i which is given by the following formula:

$$q_R^i = \int_{4\pi} I \, n^i \, d\omega \tag{17.58}$$

We should add the divergence of this radiation heat flux in the energy equation of gasdynamics when thermal radiation is included, i.e.,

$$Q_R = \nabla \cdot \boldsymbol{q}_R \tag{17.59}$$

In spherical coordinates, Eq. (17.58) may be written as

$$q_{RL}(\theta, \varphi, r, t) = \int_0^{\infty}\int_0^{2\pi}\int_0^{\pi} I_\nu(\theta, \varphi, r, t)\sin\theta\cos\theta d\theta d\varphi d\nu \tag{17.60}$$

where the ray of radiation L is specified by the angular variables θ $(0 \leqslant \theta \leqslant \pi)$ and the azimuth angle φ $(0 \leqslant \varphi \leqslant 2\pi)$. In many engineering problems and astrophysical problems, we consider the radiative transfer with respect to a straight plane (Fig. 17.5). It is convenient to split the net radiative flux q_{RL} into two parts: One part q_{RL}^{+} represents the contribution coming from the side of the unit normal vector $\boldsymbol{n}$, the incoming rays, and the other part $\boldsymbol{q}_{RL}^{-}$ represents the contribution from the opposite side, the outgoing rays, i.e.,

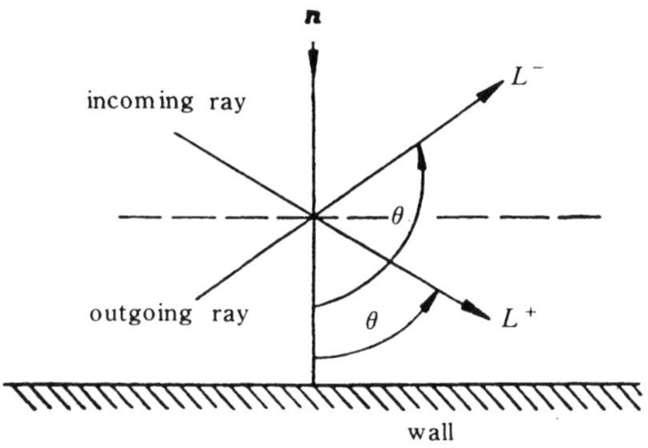

Fig. 17.5 Incoming and outgoing rays of radiation with respect to a straight wall

$$q_{RL}^{+} = \int_0^{\infty}\int_0^{2\pi}\int_0^{\pi/2} I_\nu(\theta, \varphi, r, t)\sin\theta\cos\theta\, d\theta d\varphi d\nu \tag{17.61a}$$

$$q_{RL}^{-} = \int_0^{\infty}\int_0^{2\pi}\int_{\pi/2}^{\pi} I_\nu(\theta, \varphi, r, t)\sin\theta\cos\theta d\theta d\varphi d\nu \tag{17.61b}$$

and the net radiative flux through a unit area of the straight wall with normal $\boldsymbol{n}$ is

$$q_{RL} = q_{RL}^{+} - q_{RL}^{-} \tag{17.61c}$$

Customarily we choose the normal $\boldsymbol{n}$ toward the wall and q_{RL} is the net radiative heat flux toward the wall.

(iii) Some approximation on the radiation terms

In general, the radiation terms are integrals which are difficult to deal with for practical problems. Hence we use some approximate formulas to solve practical problems. The following are some of the common approximations which are used to solve engineering problems with thermal radiation effects:

(a) Optically thick medium

When the mean free path of radiation $L_{R\nu}$ is very small in comparison with the typical dimension of the flow field, the solution of Eq. (17.50) by neglecting the unsteady term because $1/c$ is very small, may be written as follows:

$$I_\nu = B_\nu - L_{R\nu}\left(n^i \frac{\partial B_\nu}{\partial x^i}\right) + O(L_{R\nu}^2) \tag{17.62}$$

where the summation convention in i is used. If we neglect the second and higher order terms of $L_R^2\nu$, the radiation terms in gasdynamics can be easily evaluated, i.e.,

$$E_R = a_R T^4 = 3\, p_R \tag{17.63}$$

and all shearing stresses of radiation vanish. The formula (17.63) is the one we used before (cf. Chapter III, §9).

The radiative heat flux (17.58) with the help of Eq. (17.62) gives for first approximation:

$$\boldsymbol{q}_R = D_R \nabla E_R = \kappa_R \nabla T \tag{17.64}$$

where

$$D_R = c/(3\,\rho K_R) = \text{Rosseland diffusion coefficient of radiation} \tag{17.65}$$

and

$$K_R = \left(\int_0^\infty \frac{\partial B_\nu}{\partial T}\, d\nu\right) \bigg/ \left(\int_0^\infty \frac{1}{k_\nu'} \frac{\partial B_\nu}{\partial T}\, d\nu\right)$$

$$= \text{Rosseland mean absorption coefficient} \tag{17.66}$$

and

$$\kappa_R = 4 D_R a_R T^3 = \text{coefficient of heat conductivity by thermal radiation} \tag{17.67}$$

(b) Grey gas approximation

The absorption coefficient k_ν' is in general a function of the frequency ν. As a result, the optical thickness τ_ν is a function of ν. Thus under the assumption of local thermodynamic equilibrium and neglecting the term $1/c$, Eq. (17.50) may be intergated to obtain the specific intensity $I_\nu(s)$. In general, such an integration may be carried out numerically, but seldom analytically because k_ν' is usually a very complicated function of ν (see Fig. 17.6). However, if we assume that the gas is grey so that the absorption coefficient k_ν' is independent of the frequency ν, we can integrate Eq. (17.50) with respect of the frequency ν because

$$\int_0^\infty B_\nu d\nu = B(T) = \frac{\sigma}{\pi} T^4 \tag{17.68}$$

It is easy to show that the radiative heat flux over a half plane, i.e., q_{RL}^+ or q_{RL}^- with the help of Eq. (17.68) gives the result of Eq. (17.42) as we used to estimate the radiative heat flux. By means of Eq. (17.68), we obtain for the grey gas the following formula for the radiative heat flux for a two-dimensional problem over a flat plate (Fig. 17.5) and all variables are independent of the azimuth angle φ:

$$q_{Ry} = 2\left\{\int_0^{\pi/2}\int_\tau^\infty \sigma T^4(t,\theta)\exp[-m(t-\tau)]\,dt\,\sin\theta\,d\theta\right.$$
$$\left. + \int_{\pi/2}^{\pi} B_0(0,\theta)\cos\theta\,\exp(m\tau) - \int_0^\tau \sigma T^4(t,\theta)\exp[-m(t-\tau)\,dt]\sin\theta d\theta\right\} \tag{17.69}$$

where $\tau = \tau_\nu$ because we assume that τ is independent of the frequency ν. Actually, we should use a Planck mean absorption coefficient K_p in Eq. (17.69) which is defined as

$$K_p = \frac{1}{B(T)}\int_0^\infty k_\nu' B_\nu d\nu \tag{17.70}$$

and

$$\tau = \int_0^y K_p dy$$

and

$$B_0(0,\theta) = \int_0^\infty I_{\nu_0}(0,\theta)\,d\nu$$

(c) One-dimensional approximation

Eq. (17.69) may be further simplified if we assume that the temperature T depends only on τ but independent of θ. This is known as the one-dimensional approximation. In other words, we assume that the variation of the temperature with respect to y is much larger than that with respect to x. Such an approximation is good for boundary layer flow and other one-dimensional flow. If we use this approximation, we may carry out the integration with respect to θ in Eq. (17.69) and obtain

$$q_{Ry} = 2\pi\int_\tau^\infty B(t)\varepsilon_2(t-\tau)\,dt - 2\pi\int_0^\tau B(t)\varepsilon_2(\tau - t)\,dt - q_R(0)\varepsilon_3(\tau) \tag{17.71}$$

where the exponential integral ε_n is defined by the formula:

$$\varepsilon_n(t) = \int_1^\infty m^{-n}\exp(-mt)\,dm = \int_0^1 z^{n-2}\exp(-t/z)\,dz \tag{17.72}$$

and $q_R(0) = B_0(0)$.

(d) Differential approximation

Eq. (17.71) is much simpler than Eq. (17.69). But it is still an integral expression and the fundamental equations of radiation gasdynamics with the expression of Eq. (17.71) is still a system of integral-differential equations which is difficult to solve. If we approximate the exponential integrals in Eq. (17.71) by the exponential such that

$$\varepsilon_2(t) = \frac{m^2}{3} \exp(-mt) \tag{17.73}$$

where m is a constant and apply the conditions that the limiting values of optical thick medium and optical thin medium are satisfied. Eq. (17.71) gives the following differential equation for the radiative heat flux q_{Ry}:

$$\frac{d^2 q_{Ry}}{d\tau^2} - 3\, q_{Ry} + 16\sigma T^3 \frac{dT}{d\tau} = 0 \tag{17.74}$$

If we consider q_{Ry} as a new variable in the energy equation, we may solve Eq. (17.74) with the other fundamental equations of radiation gasdynamics simultaneously for the variables T, p, ρ, $\boldsymbol{q}$ and q_{Ry}. In this case, we neglect the radiation energy density and radiation stresses.

For a three-dimensional case of a grey gas, Eq. (17.74) may be replaced by the following equation:

$$\frac{1}{\rho K} \frac{\partial}{\partial x^j} \left(\frac{1}{\rho K} \frac{\partial q_{Rj}}{\partial x^j} \right) - 3\, q_{Rj} + 16\sigma T^3 \frac{1}{\rho K} \frac{\partial T}{\partial x^j} = 0 \tag{17.75}$$

(e) Optically thin approximation

For an optically thin medium, the optical thickness is usually small. It should be noticed that the upper limit ∞ in the first integral of Eq. (17.71) should be replaced by a finite value τ_2 in the optically thin medium, because in the actual flow field the region which we are interested in is small compared with the mean free path of radiation. For the optically thin medium, we may expand the exponential integral as a power series of τ. If we keep only the lowest term, we have

$$\frac{dq_{Ry}}{dt} = \frac{4\sigma}{L_R} \left(-T^4 + 1/2\, T_W^4 \right) \tag{17.76}$$

where $L_R = 1/(\rho K)$ and T_w is the temperature of the wall. In Eq. (17.76) we neglect completely the absorption of radiation. The formula (17.76) was one of the most popular formulae when the radiative heat flux first became important in aerospace problem, and the medium is optically thin. However as we shall show in §13, such an approximation may give very wrong results.

(iv) Absorption coefficient of radiation and mean free path of radiation

In studying the effects of thermal radiation in flow problem, the absorption coefficient of radiation and the mean free path of radiation are the most important physical properties of the gas. These properties represent a new diffusive phenomenon in the medium in addition to the momentum diffusivity by viscosity and thermal diffusivity by heat conductivity. Many new and interesting phenomena may be introduced by this radiative diffusivity. The value of the mean free path of radiation or the absorption coefficient of radiation should be determined either by experiments or from some microscopic theory.

In general the mean free path of radiation $L_{R\nu}$ or absorption coefficient k_ν is a function of frequency ν of the radiation ray and the state variables of the medium. For instance[17], Fig. 17.6 shows the variation of mean free path of radiation with frequency ν or wave length λ and the temperature of hydrogen gas. If we use the actual variation of the mean free path of radiation with frequency and state variables, the gas is said to be nongrey and the radiative terms such as radiative heat flux should be expressed in integral form. The solution of the radiation gasdynamic equations would be very complicated and extensive numerical computations are required. In engineering problems, it is advisable to use some simple expressions of the average value

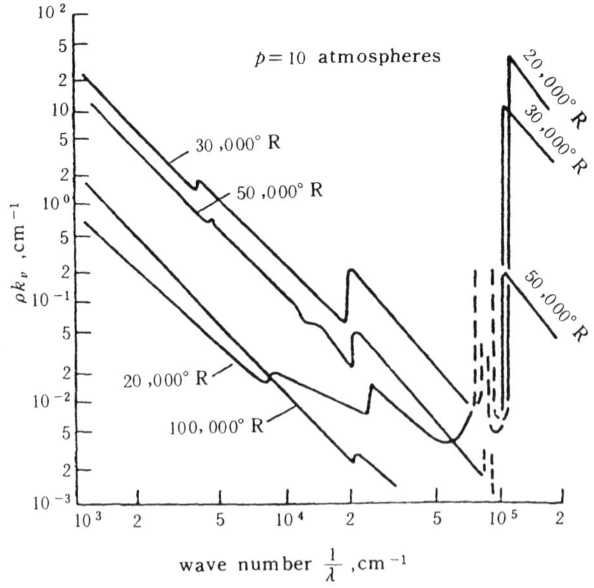

Fig. 17.6 Variation of absorption coefficient of hydrogen gas with wave number at a pressure of 10 atmospheres and various temperatures (Fig. 4 of Ref. 17 by N. L. Krascella, courtesy of NASA)

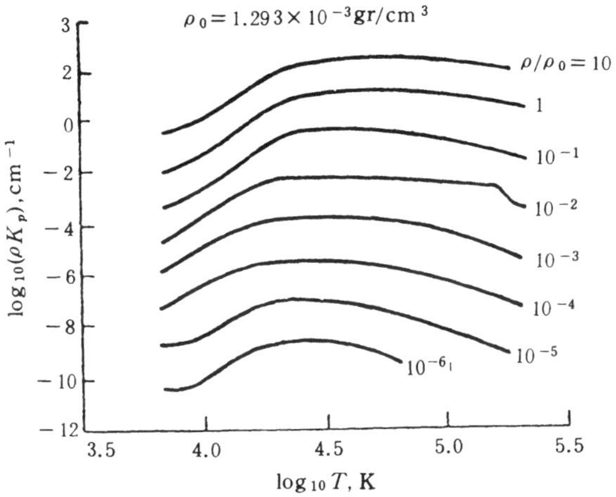

Fig. 17.7 Planck mean absorption coefficient of air (Fig. 11 of Ref. 18, by B. H. Armstrong, J. Sokoloff, R. W. Nicholls, D. H. Holland and R. E. Meyerott, courtesy of Pergamon Press. Ltd)

of the mean free path of radiation over the whole spectrum of frequency. For an optically thick medium, we use the Rosseland mean absorption coefficient given by Eq. (17.66). For an optically thin medium or finite mean free path of radiation case, we may use the Planck mean absorption coefficient given by Eq. (17.70).

It should be noticed that the value of the mean free path of radiation alone does not determine whether the medium is optically thick or thin. We have to compare the mean free path of radiation with the representative length L of the flow field in order to determine whether the medium may be regarded as optically thick or thin in a given problem. In other words, the optical thickness defined by Eq. (17.49) determines the optical properties of the medium. In fact, for a given state of a medium, the Planck mean free path of radiation L_p is usually smaller than the Rosseland mean free path of radiation L_R. For high temperature air up to a temperature of 20,000 K, we may take $L_R = 8.3\ L_p$ as a first approximation.

After we have taken the average of the mean free path of radiation over the whole spectrum, we should find some simple formulas for the variation of the average mean free path of radiation with the temperature and density or pressure of the medium. Since the absorption coefficient increases with density, the mean free path of radiation decreases with the increase of density. The variation of the mean free path with temperature is very complicated. At low temperature, say $T < 10{,}000$ K, the mean free path of radiation decreases with the increase of temperature, while at very high temperature,

$T > 10{,}000$ K, the mean free path of radiation increases with the temperature and there is a minimum in the intermediate temperature range depending on the density and composition of the medium. Fig. 17.7 shows some typical variations of abosrption of air[18]. If the temperature range in the flow field is not too large, a power law for the average values of the mean free path of radiation may be used, i.e.,

$$\frac{L_R}{L_{Ro}} = \frac{L_p}{L_{po}} = \left(\frac{T_o}{T}\right)^{m_1}\left(\frac{\rho_o}{\rho}\right)^{m_2} \tag{17.77}$$

where subscript o refers to the value at some reference conditions. The powers m_1 and m_2 shoud be so chosen that the formula (17.77) gives the best fit with the opacity data over the range of temperature considered. For instance, in the temperature range of 7,000 K to 12,000 K, the following values may be used for air:

$$m_1 = 4.4,\ m_2 = 1,\ \rho_o = 1.23 \times 10^{-3}\ \mathrm{gr/cm^3}$$

$$T_o = 10{,}000\ \mathrm{K},\ L_{po} = 0.5\ \mathrm{meter}$$

For higher temperature in the neighborhood of 20,000 K, we may take $m_1 = 2.5$ and $m_2 = 1$. At very high temperature in some astrophysical problems, we may take $m_1 = -7/2$, and $m_2 = 2$.

If we consider a large range of temperature including the minimum value of the mean free path of radiation, the following formula may be used:

$$\frac{L_{Ro}}{L_R} = \frac{L_{po}}{L_p} = \frac{L_{pol}}{L_{pl}} + \frac{L_{po2}}{L_{p2}} \tag{17.78}$$

where L_{pl} is the mean free path of radiation given by Eq. (17.77) which is good for low temperature range, i.e., m_1 is positive while L_{p2} is the corresponding value for high temperature range, i.e., m_1 is negative.

There are some other formulas for the mean free path of radiation with the state variables of the medium which have been used. For instance, Scala and Sampson used the following formula for the Rosseland mean free path of radiation for air[19]:

(i) For linear representation,

$$K_R = \frac{1}{L_R} = 4.86 \times 10^{-7} p^{1.31} \exp(4.56 \times 10^{-4} T) \tag{17.79}$$

and (ii) for quadratic expression,

$$K_R = \frac{1}{L_R} = 4.52 \times 10^{-7} p^{1.31} \exp(5.18 \times 10^{-4} T - 7.13 \times 10^{-9}\, T^2) \tag{17.80}$$

where the mean free path of radiation L_R is in centimeters, the pressure p of the air is in atmospheres and the temperature T is in K. Eq. (17.79) is good for low temperature range which corresponds to the case of positive m_1 while

Eq. (17.80) is for the case of large variation of temperature including the minimum vale of L_R which corresponds to the formula (17.78).

7. Fundamental equations of a viscous, heat-conducting, electrically conducting and thermal radiating compressible fluid flow

We consider this compressible fluid as a mixture of N reacting gases which are viscous, heat-conducting, electrically conducting and thermal radiating. We shall use the single fluid approach and consider $N+15$ variables, i.e., the velocity vector $\boldsymbol{q}$ with three components u^i, the temperature T, the pressure p, and the density ρ of the mixture as a single fluid and the $(N-1)$ concentration c_s (or the $(N-1)$ mass concentration k_s), the electromagnetic variables E^i, H^i and I^i and the specific intensity of radiation I_ν where we neglect the excess electrical charge ρ_e and $i=1$, 2 or 3 and $s=1$, 2, $\cdots$, $(N-1)$. Hence we have the $N+15$ fundamental equations for our problems as follows:

(i) Equation of state of the mixture

We consider the case that all the species in the mixture are perfect gas and then for the mixture, we have [see Eq. (16.137)]

$$p=knT=\rho RT \tag{17.81a}$$

(ii) The equation of continuity of the mixture as a whole is [see Eq. (16.140)]

$$\frac{\partial \rho}{\partial t}+\nabla \cdot (\rho \boldsymbol{q})=0 \tag{17.81b}$$

(iii) The diffusion equation for the sth species[1]

$$\frac{Dk_s}{Dt}=-\sum_{q\neq s}^{N}\frac{\partial}{\partial x^j}\left(\rho D_{qs}\frac{\partial c_q}{\partial x^j}\right)-\frac{\partial}{\partial x^j}\left(D_p\frac{\partial \ln p}{\partial x^j}\right)+\frac{\partial}{\partial x^j}\left(D_T\frac{\partial \ln T}{\partial x^j}\right)+\sigma_s \tag{17.81c}$$

where D_{qs} is the binary diffusion coefficient between species q and s, D_p is the coefficient of pressure diffusion and D_T is the coefficient of thermal diffusion and σ_s is the source term of sth species which depends on the chemical reactions or ionization processes. Eq. (17.81c) is derived from the equation of diffusion velocity which is assumed to be small and may be expressed in terms of diffusion coefficients. (see Ref. 1).

(iv) Equations of motion

$$\frac{D\boldsymbol{q}}{Dt}=\boldsymbol{X}+\nabla \cdot \tau \tag{17.81d}$$

where $\boldsymbol{X}$ is the body forces which consist of the gravitational force $\boldsymbol{X}_g=\rho \boldsymbol{q}$ and the electromagnetic force $\boldsymbol{F}_e=\boldsymbol{X}_e=\boldsymbol{I}\times \boldsymbol{B}$ where we neglect the terms with ρ_e.

The stresses term is τ which consists of the viscous stress tensor τ_v and the radiation stress tensor τ_R. For the viscous stress tensor, we have the ijth component as follows:

$$\tau_v^{ij}=\mu\left(\frac{\partial u^i}{\partial x^j}+\frac{\partial u^j}{\partial x^i}\right)+\left[-p-\frac{2\mu}{3}\left(\frac{\partial u^k}{\partial x^k}\right)\right]\delta^{ij} \tag{17.82}$$

For the radiation stresses tensor, we may neglect the shearing stress components and use the expression for radiation pressure only, i.e.,

$$\tau_R^{ij}=-p_R\delta^{ij} \tag{17.83}$$

where $\delta^{ij}=0$ if $i\neq j$ and $\delta^{ij}=1$ if $i=j$. The coefficient of viscosity of the mixture is μ.

(v) Equation of energy

The conservation of energy gives the equation of energy as follows:

$$\rho\frac{D\bar{e}_m}{Dt}=\nabla\cdot(\boldsymbol{q}\cdot\tau_v)+\nabla\cdot(\boldsymbol{q}\cdot\tau_R)+\nabla\cdot(\kappa\nabla T)+\nabla\cdot\boldsymbol{q}_R+Q \tag{17.81e}$$

where $\bar{e}_m=U_m+\frac{1}{2}q^2+\varphi+E_R/\rho=$total energy of the gas per unit mass, U_m is the internal energy of the mixture per unit mass, $\frac{1}{2}q^2$ is the kinetic energy of the mixture per unit mass, φ is the potential energy of the mixture per unit mass and E_R is the radiation energy density per unit volume, κ is the coefficient of heat conductivity of the mixture, $\boldsymbol{q}_R$ is the radiative heat flux and Q is the total energy input by the electromagnetic fields, chemical reaction and other heat sources, i.e., $Q=Q_e+Q_c$ where $Q_e=\boldsymbol{E}\cdot\boldsymbol{J}$ and Q_c is energy input due to chemical reaction and other heat source.

(vi) Electromagnetic equation

The electromagnetic equations are the Maxwell's equations (see Eqs. (15.6) and (15.7)):

$$\nabla\times\boldsymbol{H}=\boldsymbol{J}+\frac{\partial\varepsilon\boldsymbol{E}}{\partial t} \tag{17.81f}$$

$$\nabla\times\boldsymbol{E}=-\frac{\partial\mu_e\boldsymbol{H}}{\partial t} \tag{17.81g}$$

In most engineering problems without consideration of very high frequency phenomena, the displacement current $\partial\boldsymbol{D}/\partial t$ may be neglected.

(vii) The generalized Ohm's law (cf. Eq. (15.11))

$$\boldsymbol{I}=\sigma_e\boldsymbol{E}_u \tag{17.81h}$$

or most accurate formula such as Eq. (16.196) may be used instead of Eq. (17.81 h).

(viii) Radiative transfer equation

In general, we may use Eq. (17.50) to solve the specific intensity of radiation I_ν. After I_ν is obtained, we may calculate the radiation stress tensor and radiative heat flux. However, for engineering problems, the following approximations may be used:

(a) we neglect radiation shearing stress and use the simple formula [cf Eq. (17.63)]

$$p_R = E_R/3 = \frac{1}{3}\, a_R T^4 \tag{17.81i}$$

for the radiation pressure and Eq. (17.64) for the radiative heat flux

$$\boldsymbol{q}_R = \kappa_R \nabla T \tag{17.81j}$$

With the approximations (17.81i) and (17.81j), we express the radiation terms in terms of temperature. Hence we do not have to consider the radiation effect as a new variable.

However, for many aerospace reentry problems, a better approximation such as Eq. (17.74) or (17.75) may be used so that we consider $\boldsymbol{q}_R$ as a new variable.

In general our fundamental equations for a viscous, heat-conducting, electrically conducting and thermal radiating are Eqs. (17.81). They are nonlinear differential equations. There is no general way to solve these equations for reasonable complicated boundary conditions. We have to use numerical computations to solve practical problems. However, one of the important simplifications for these fundamental equations is known as boundary layer equations which are very useful to deal with engineering problems. We are going to discuss various boundary layer equations in the following sections.

8. Boundary layer concepts

Most of the important practical problems in the study of the flow of a compressible fluid involve very large Reynolds numbers, i.e., large velocities and body dimensions, and fluid such as air or other gases with vanishing small transport coefficients, viscosity, heat condutivity etc. In dealing with these problems, two different approximations may be used. The most obvious one involves the droping of the transport terms such as viscous stress and heat conducting terms which are small in comparison with the inertial terms and convectional terms.

For simplicity and clarification, let us consider first only the effects of viscosity. If we drop the viscous terms, the Navier-Stokes equations[1, 6, 20] are then reduced to the equation for inviscid fluid as we discussed in Chapters III to XIV. In the inviscid flow equations the order of the differential equations is one lower than the order of those of a viscous fluid. In addition, one of the boundary conditions of the actual fluid cannot be satisfied. Although the inviscid fluid solutions give good approximate results in regions distant from the solid wall, they do not give a good approximation for the region near the wall. Thus, many important phenomena, such as skin friction and heat transfer, cannot be explained by the solution for inviscid fluids. To explain these important phenomena, other approximations must be used to obtain results for the flow in the vicinity of the solid wall.

It was Prandtl[6] who in 1904, first introduced the concept of the boundary layer by stating that the viscous effects are confined to a very thin layer near the boundary. The theory of the boundary layer opened a new era for the fluid dynamics of viscous fluids. Today the theory of boundary layer has become the most important branch of the fluid dynamics of viscous flow, and all the important problems of viscous effects in aerodynamics are treated by this theory.

For high Reynolds number, the viscous effects are confined to a very thin boundary layer whose thickness tends to be zero as the coefficient of kinematic viscosity goes to zero. The Navier-Stokes equations can be simplified by such considerations. For the sake of clarity, let us consider two-dimensional flow over a flat plate. Let x be the direction along the plate and also that of the main flow. Let y be the direction perpendicular to the plate. As far as the region of the boundary layer over the plate is concerned, the y-dimension is much smaller than the x-distance. As the coefficient of kinematic viscosity ν tends to zero, the y-dimension of the boundary layer also tends to zero. To see what is inside the boundary layer we must amplify the y-dimension of the boundary layer as ν decreases. As we shall show later, if we put

$$\lim_{\nu \to 0} \frac{y}{\sqrt{\nu}} = \text{finite distance} = \eta \tag{17.84}$$

into the Navier-Stokes equations, we have the boundary-layer equations (§ 9) which give a good approximation for the flow in the boundary layer. The boundary-layer equation is of the same order as the Navier-Stokes equation; hence, the same number of boundary conditions may be satisfied. In particular, the boundary conditions on the solid wall are satisfied. However, the boundary condition at $y=\infty$ for the boundary-layer equation is not the boundary condition at $y=\infty$ for the actual flow, i.e., for the Navier-Stokes equations. Because we amplify the y-dimension in the boundary layer, the boundary condition at $y=\infty$ will be that in the actual flow just outside the boundary layer. This is usually assumed to be the value on the wall ($y=0$) for the solution of an inviscid fluid. This point can be illustrated by a simple mathematical example discussed in the next paragraph. When the interaction between the boundary layer and the main flow is small, it is current practice to solve first the problem for inviscid flow and then solve the boundary layer problem taking the boundary condition at $y=\infty$ equal to the inviscid flow value at $y=0$. This scheme fails when the interaction between the boundary layer and the outside flow is appreciable, as in flows with interaction of a shock wave and the boundary layer, or in hypersonic viscous flow where the boundary conditions at the outer edge of the boundary layer depend on the boundary layer itself. In the case of boundary-layer shock-wave interaction, the complete Navier-Stokes equation should be used to obtain a solution. In the hypersonic viscous flow case, the boundary-layer equations may be used providing the influence of the boundary layer upon its own edge condition is properly taken into account (§9). The mathematical structure of the boundary-layer equation[21] may be illustrated by the following simple mathematical problem.

Consider the differential equation:

$$c-f_y=\nu f_{yy} \tag{17.85}$$

(where c is a positive constant, ν is a small positive parameter, and subscript y denotes differentiation with respect to y) for the function $f(y)$ defined throughout the interval $0\leqslant y\leqslant 1$ and subject to the boundary conditions

$$f(0)=0, f(1)=1 \tag{17.86}$$

We shall find the solution of Eq. (17.85) with the boundary conditions (17.86) for the case when ν is vanishingly small. If we simply put $\nu=0$ in Eq. (17.85), the order of the differential equation is lowered by one, and hence its solution can satisfy only one boundary condition. The number of boundary conditions in this problem changes discontinuously as ν tends to zero. Such a problem is known as a singular perturbation problem[22]. The theory of singular perturbation problems is rather incomplete when compared with that of regular perturbations problem where the number of boundary conditions is independent of the small parameter ν. It is evident that the boundary-layer theory of Prandtl can be considered as a perturbation procedure for a singular problem.

Friedrichs[21] divided the whole problem into two separate problems: one is called the "interior limit problem" and the other, the "boundary limit problem."

In the interior limit problem, we simply put $\nu=0$ in Eq. (17.85) and the differential equation becomes

$$c-f_y=0 \tag{17.87}$$

with the boundary condition

$$f(1)=1 \tag{17.88}$$

The solution of equation (17.87) with the boundary condition (17.88) is

$$f = 1 + c(y-1) \tag{17.89}$$

and at $y=0$ we have

$$f = 1 - c \tag{17.90}$$

In the boundary-limit problem, we try to magnify the region near $y=0$ by putting $y=\nu\eta$ and $g(\eta) = f(y)$ in Eq. (17.85) and letting ν tend to zero. The differential Eq. (17.85) becomes

$$-g_\eta = g_{\eta\eta}, \quad 0 \leqslant \eta \leqslant \infty \tag{17.91}$$

the subscript η denoting differentiation with respect to η. The boundary conditions become

$$g(0) = 0, \; g(\infty) = F \tag{17.92}$$

where F is to be determined.

The solution of Eq. (17.91) with the boundary condition (17.92) is

$$g(\eta) = F(1 - e^{-\eta}) \tag{17.93}$$

We must now find the value F in the boundary-limit problem. The exact solution of Eq. (17.85) with the boundary condition (17.86) is

$$f(y) = (1-c)\frac{1-e^{-y/\nu}}{1-e^{-1/\nu}} + cy \tag{17.94}$$

We see that, when $\nu \to 0$ with $y \neq 0$, Eq. (17.94) becomes Eq. (17.89). On the other hand, when $\nu \to 0$, $\eta = y/\nu$ tends to be finite, and Eq. (17.94) becomes Eq. (16.10), if we take

$$F = 1 - c \tag{17.95}$$

Thus we find that the solution of the boundary-limit problem at $y=\infty$ is equal to the solution of the interior limit problem at $y=0$.

In the present problem we have found the explicit form of the solution of the equation and shown the relation between the boundary conditions for these two approximate problems. Since in general the explicit solution of the Navier-Stokes equations cannot be found, we cannot prove mathematically that the relation between the boundary conditions for the approximate methods used in solving the Navier-Stokes equations is the same as shown in this simple example. It is, however, quite plausible to assume that such a relation holds for the Navier-Stokes equations. For the Navier-Stokes equations the interior limit problem is the inviscid flow problem, and the boundary-limit problem is the boundary-layer problem. It is emphasized, once again, that the foregoing relation between the boundary conditions is only an assumption, since it has not been proved. Experience shows that the assumption is a good one for a great many practical problems. In some special cases, however, such as hypersonic viscous flow or the interaction between a shock wave and the boundary layer, the assumption is not valid. In these cases modifications must be made.

For the effects of viscosity, we have the boundary layer of velocity whose thickness δ_v/L is inversely proportion to $(Re)^{1/2}$ where $Re = \dfrac{UL}{\nu}$, U is a typical velocity and L is a typical length of the flow field. For the effect of heat conductivity, we have the boundary layer of temperature whose thickness δ_T/L is inversely proportional to $(PrRe)^{1/2}$ where $Pr = \mu c_p/\kappa$ is the Prandtl number. We have $\delta_v/\delta_T \sim (Pr)^{1/2}$. For ordinary gas, the Prandtl number is of the order of unity. Hence δ_v and δ_T are of the same order of magnitude. For some fluids, the value of Prandtl number may differ greatly from unity. Then δ_v and δ_T are not of the same order of magnitude.

In §9, we consider the boundary layer flow of an ordinary gas without the effects of electromagnetic field, thermal radiation nor chemical reaction. But we have to consider the boundary layer of velocity and that of temperature simultaneously because the Prandtl number is of the order of unity. Hence we have to consider the effects of viscosity and heat conductivity simultaneously for such a compressible flow.

We have other boundary layer flows due to the effects of other transport coefficients. In §10, we consider the boundary layer flow with chemical reactions in which the diffusion coefficient plays an important role in the boundary layer of mass transfer. In §11, we consider the boundary layer flow with phase change, i.e., the ablation problem which is a special two-phase flow problem. In §12, we discuss the boundary layer flow in magnetogasdynamics in which the magnetic diffusivity plays an important role. Finally in §13, we consider the boundary layer flow with thermal radiative transfer in which the absorption coefficient of radiation plays an important role.

9. Two-dimensional boundary-layer equations for a compressible fluid

Let us consider the two-dimensional flow of a compressible fluid over a flat plate. Let the plate lie in the x-z plane and the flow of the free stream be in the x-direction. When the Reynolds number based on the free stream is large, the viscous effects are confined to a narrow region near the plate, i.e., the boundary layer. Let δ be a measure of the thickness of this boundary layer. The characteristic length, l, of the plate, is much larger than δ. The fundamental equations of the present problem are

$$\frac{Du}{Dt} = \frac{\partial u}{\partial t} + u\frac{\partial u}{\partial x} + v\frac{\partial u}{\partial y}$$

$$= -\frac{1}{\rho}\frac{\partial p}{\partial x} + \frac{1}{\rho}\frac{\partial}{\partial x}\left\{\mu\left[2\frac{\partial u}{\partial x} - \frac{2}{3}\left(\frac{\partial u}{\partial x} + \frac{\partial v}{\partial y}\right)\right]\right\}$$

$$+ \frac{1}{\rho}\frac{\partial}{\partial y}\left[\mu\left(\frac{\partial u}{\partial y} + \frac{\partial v}{\partial x}\right)\right] \tag{17.96}$$

$$\frac{Dv}{Dt} = \frac{\partial v}{\partial t} + u\frac{\partial v}{\partial x} + v\frac{\partial v}{\partial y}$$

$$= -\frac{1}{\rho}\frac{\partial p}{\partial y} + \frac{1}{\rho}\frac{\partial}{\partial y}\left\{\mu\left[2\frac{\partial v}{\partial y} - \frac{2}{3}\left(\frac{\partial u}{\partial x} + \frac{\partial v}{\partial y}\right)\right]\right\}$$

$$+ \frac{1}{\rho}\frac{\partial}{\partial x}\left[\mu\left(\frac{\partial u}{\partial y} + \frac{\partial v}{\partial x}\right)\right] \tag{17.97}$$

$$\frac{\partial \rho}{\partial t} + \frac{\partial \rho u}{\partial x} + \frac{\partial \rho v}{\partial y} = 0 \tag{17.98}$$

$$\frac{Dc_pT}{Dt} - \frac{Dp}{Dt} = \frac{\partial}{\partial x}\left(\kappa\frac{\partial T}{\partial x}\right) + \frac{\partial}{\partial y}\left(\kappa\frac{\partial T}{\partial y}\right)$$

$$+\mu\left[2\left(\frac{\partial u}{\partial x}\right)^2 + \left(\frac{\partial v}{\partial y}\right)^2 + \left(\frac{\partial v}{\partial x} + \frac{\partial u}{\partial y}\right)^2 - \frac{2}{3}\left(\frac{\partial u}{\partial x} + \frac{\partial v}{\partial y}\right)^2\right] \tag{17.99}$$

$$p = \rho RT \tag{17.100}$$

where the body forces are assumed to be zero. The flow is adiabatic.

We now put the above equations in nondimensional form by setting

$$x = l\xi,\ y = \delta\eta,\ u = \bar{u}U,\ v = \bar{v}V,\ t = \bar{t}\frac{l}{U},\ \rho = \rho_0\bar{\rho}$$
$$T = T_0\bar{T},\ p = p_0\bar{p},\ \mu = \mu_0\bar{\mu},\ \kappa = \kappa_0\bar{\kappa},\ c_p = c_{p0}\bar{c}_p \tag{17.101}$$

where U, V, p_0, T_0, ρ_0, μ_0, κ_0, and c_{p0} are certain reference values of these corresponding quantities such as those at the edge of the boundary layer, while ξ, η, $\bar{u}$, $\bar{v}$, $\bar{t}$, $\bar{\rho}$, $\bar{T}$, $\bar{p}$, $\bar{\mu}$, $\bar{\kappa}$, and $\bar{c}_p$ are the corresponding nondimensional quantities. These nondimensional quantities are all of the order of unity.

First we shall find the order of magnitude of u and v (or U and V). We integrate the equation of continuity (17.98) for the steady case with the boundary conditions: $y=0$, $v=0$; $y=\delta$, $v=V$, $\rho=\rho_0$,

$$\rho_0 V = -\int_0^\delta \frac{\partial \rho u}{\partial x}\, dy$$

or

$$\frac{V}{U} = -\frac{\delta}{l}\left[\int_0^1 \frac{\partial \bar{\rho}\bar{u}}{\partial \xi}\,\partial\eta\right] \tag{17.102}$$

The quantity in the bracket is of the order of unity; hence, V/U is of the order of δ/l, i.e., $V \ll U$. The nondimensional form of Eq. (17.96) is

$$\begin{aligned}\underset{1}{\frac{\partial \bar{u}}{\partial \bar{t}}} + \underset{1}{\bar{u}\frac{\partial \bar{u}}{\partial \xi}} + \bar{v}\frac{\partial \bar{u}}{\partial \eta}\underset{\frac{1}{\delta}}{\frac{l}{\delta}}\,\underset{\delta}{\frac{V}{U}} &= -\frac{p_0}{\rho_0 U^2 \bar{\rho}}\frac{\partial \bar{p}}{\partial \xi} \\ &+ \underset{\delta^2}{\frac{1}{Re}}\,\frac{1}{\bar{\rho}}\left\{\bar{\mu}\left[2\underset{1}{\frac{\partial \bar{u}}{\partial \xi}} - \frac{2}{3}\left(\underset{1}{\frac{\partial \bar{v}}{\partial \eta}\frac{V}{U}\frac{l}{\delta}} + \underset{1}{\frac{\partial \bar{u}}{\partial \xi}}\right)\right]\right. \\ &\left. + \underset{\frac{1}{\delta}}{\frac{l}{\delta}}\frac{\partial}{\partial \eta}\left[\bar{\mu}\left(\underset{\frac{1}{\delta}}{\frac{\partial \bar{u}}{\partial \eta}\frac{l}{\delta}} + \underset{\delta}{\frac{\partial \bar{v}}{\partial \xi}\frac{V}{U}}\right)\right]\right\}\end{aligned} \tag{17.103}$$

The order of magnitude of various terms is also indicated under Eq. (17.103). The same procedure is followed below. Since in the boundary layer, the inertial force and the viscous force are of the same order of magnitude, we conclude that the Reynolds number, $Re = U\rho_0 l/\mu_0$, must be of the order of $1/\delta^2$.

The nondimensional form of Eq. (17.97) is

$$\begin{aligned}\underset{\delta}{\frac{V}{U}}\frac{\partial \bar{v}}{\partial \bar{t}} + \bar{u}\frac{\partial \bar{v}}{\partial \xi}\underset{\delta}{\frac{V}{U}} + \bar{v}\frac{\partial \bar{v}}{\partial \eta}\underset{\frac{1}{\delta}}{\frac{l}{\delta}}\underset{\delta^2}{\left(\frac{V}{U}\right)^2} &= -\frac{p_0}{\rho_0 U^2 \bar{\rho}}\frac{\partial \bar{p}}{\partial \eta}\frac{l}{\delta} \\ &+ \underset{\delta^2}{\frac{1}{Re}}\,\frac{1}{\bar{\rho}}\left\{\underset{\frac{1}{\delta}}{\frac{l}{\delta}}\frac{\partial}{\partial \eta}\left[\bar{\mu}\left(2\frac{\partial \bar{v}}{\partial \eta}\underset{\delta}{\frac{V}{U}}\underset{\frac{1}{\delta}}{\frac{l}{\delta}} - \frac{2}{3}\left(\underset{1}{\frac{\partial \bar{u}}{\partial \xi}} + \underset{1}{\frac{\partial \bar{v}}{\partial \eta}\frac{V}{U}\frac{l}{\delta}}\right)\right)\right]\right. \\ &\left. + \frac{\partial}{\partial \xi}\left[\bar{\mu}\left(\underset{\frac{1}{\delta}}{\frac{\partial \bar{u}}{\partial \eta}\frac{l}{\delta}} + \underset{\delta}{\frac{\partial \bar{v}}{\partial \xi}\frac{V}{U}}\right)\right]\right\}\end{aligned} \tag{17.104}$$

The nondimensional form of Eq. (17.98) is

$$\underset{1}{\frac{\partial \bar{\rho}}{\partial \bar{t}}} + \underset{1}{\frac{\partial \bar{\rho}\bar{u}}{\partial \xi}} + \frac{\partial \bar{\rho}\bar{v}}{\partial \eta}\underset{\frac{1}{\delta}}{\frac{l}{\delta}}\,\underset{\delta}{\frac{V}{U}} = 0 \tag{17.105}$$

The nondimensional form of Eq. (17.99) is

$$\frac{c_{p0}T_0}{U^2}\,\bar{c}_p\bar{\rho}\left(\frac{\partial T}{\partial \bar{t}}+\bar{u}\frac{\partial \overline{T}}{\partial \xi}+\bar{v}\frac{\partial \overline{T}}{\partial \eta}\frac{V}{U}\frac{l}{\delta}\right)$$

$$\underset{1}{}\qquad\underset{1}{}\qquad\underset{1}{}\qquad\qquad\underset{\delta\,\frac{1}{\delta}}{}$$

$$=\frac{p_0}{\rho_0 U^2}\left(\frac{\partial \bar{p}}{\partial \bar{t}}+\bar{u}\frac{\partial \bar{p}}{\partial \xi}+\bar{v}\frac{\partial \bar{p}}{\partial \eta}\frac{V}{U}\frac{l}{\delta}\right)$$

$$\qquad\qquad\underset{1}{}\qquad\underset{1}{}\qquad\qquad\underset{\delta\,\frac{1}{\delta}}{}$$

$$+\frac{1}{PrRe}\frac{T_0 c_{p0}}{U^2}\left[\frac{\partial}{\partial \xi}\left(\bar{\kappa}\frac{\partial \overline{T}}{\partial \xi}\right)+\frac{\partial}{\partial \eta}\left(\kappa\frac{\partial \overline{T}}{\partial \eta}\right)\left(\frac{l}{\delta}\right)^2\right]+\frac{1}{Re}\left\{\bar{\mu}\left[2\left(\frac{\partial \bar{u}}{\partial \xi}\right)^2\right.\right.$$

$$\underset{1\;\delta^2}{}\quad\underset{1}{}\qquad\quad\underset{1}{}\qquad\qquad\underset{\frac{1}{\delta^2}}{}\qquad\underset{\delta^2}{}\qquad\qquad\underset{1}{}$$

$$+2\left(\frac{\partial \bar{v}}{\partial \eta}\right)\left(\frac{V}{U}\right)^2\left(\frac{l}{\delta}\right)^2+\left(\frac{\partial \bar{v}}{\partial \xi}\frac{V}{U}+\frac{\partial \bar{u}}{\partial \eta}\frac{l}{\delta}\right)^2-\frac{2}{3}\left(\frac{\partial \bar{u}}{\partial \xi}+\frac{\partial \bar{v}}{\partial \eta}\frac{V}{U}\frac{l}{\delta}\right)^2\right]\right\}$$

$$\underset{\delta^2}{}\qquad\underset{\frac{1}{\delta^2}}{}\qquad\qquad\underset{\delta^2}{}\qquad\qquad\underset{\frac{1}{\delta^2}}{}\qquad\qquad\underset{1}{}\qquad\qquad\underset{\delta\;\frac{1}{\delta}}{}$$

$$\tag{17.106}$$

where the Prandtl number $Pr=\dfrac{c_{p0}\mu_0}{\kappa_0}$ is of the order of unity and is a charac-

teristic physical constant of the fluid. The quantities $\dfrac{c_{p0}T_0}{U^2}$ and $\dfrac{p_0}{\rho_0 U^2}$ are

of the order of unity. From Eq. (17.104) we see that $\partial p/\partial \eta$ is of the order of δ^2.

The nondimensional form of Eq. (17.100) is

$$\frac{\bar{p}}{\bar{\rho}\overline{T}}=\frac{Re_0 T_0}{p_0}\tag{17.107}$$

$$\underset{1}{}\qquad\quad\underset{1}{}$$

If we neglect the terms of the order of δ and smaller in Eqs. (17.103) to (17.107), we have the following boundary-layer equations for the two-dimensional flow of a compressible fluid over a flat plate:

$$\frac{\partial u}{\partial t}+u\frac{\partial u}{\partial x}+v\frac{\partial u}{\partial y}=-\frac{1}{\rho}\frac{\partial p}{\partial x}+\frac{1}{\rho}\frac{\partial}{\partial y}\left(\mu\frac{\partial u}{\partial y}\right)\tag{17.108a}$$

$$\frac{\partial p}{\partial y}=0\tag{17.108b}$$

$$\frac{\partial \rho}{\partial t}+\frac{\partial \rho u}{\partial x}+\frac{\partial \rho v}{\partial y}=0\tag{17.108c}$$

$$\rho\left(\frac{\partial c_p T}{\partial t}+u\frac{\partial c_p T}{\partial x}+v\frac{\partial c_p T}{\partial y}\right)$$

$$=\frac{\partial p}{\partial t}+u\frac{\partial p}{\partial x}+\frac{\partial}{\partial y}\left(\kappa\frac{\partial T}{\partial y}\right)+\mu\left(\frac{\partial u}{\partial y}\right)^2 \qquad (17.108\text{d})$$

$$p=\rho RT \qquad (17.108\text{e})$$

The above boundary-layer equations are still nonlinear, but they are much simpler than the original equations. Not only is the number of viscous terms reduced, but the pressure p is a known function in the boundary-layer equations instead of an unknown as in the original equations. Eq. (17.108b) shows that the pressure variation across the boundary layer is so small that it can be neglected. The pressure is a function of x only. In the ordinary case, the pressure just outside the boundary layer is presumed to be equal to that on the surface of the same body in an inviscid flow. The pressure may also be determined by experiment.

The above system of boundary-layer equations may be applied to the boundary layer over a curved plate except that Eq. (17.108b) should be replaced by

$$-Ku^2=-\frac{1}{\rho}\frac{\partial p}{\partial y} \qquad (17.109)$$

provided that x is the curvilinear coordinate along the curved surface, y is the perpendicular distance, and the curvature K of the surface and its variation with x, $\partial K/\partial x$, are of the order of unity or smaller.

From Eq. (17.109) we see that the total change of pressure across the boundary layer is still of the order of δ and therefore may be neglected. As a result, the v-component of the fluid just outside the boundary layer is still negligible and we have

$$-\frac{1}{\rho_\infty}\frac{\partial p}{\partial x}=\frac{\partial u_\infty}{\partial t}+u_\infty\frac{\partial u_\infty}{\partial x} \qquad (17.110)$$

where the subscript ∞ refers to the value at $y=\infty$, i.e., the assumed known value just outside of the boundary layer.

In the investigation of the flow in the boundary layer, we must solve the boundary-layer equations with proper boundary conditions. These conditions for the velocity require the fluid velocity components at the surface of the body to be equal to those of the surface. These are usually zero. At the outer edge of the boundary layer, the boundary conditions require that the velocity components tend toward the free stream values and that the derivatives of the velocity components with respect to the normal coordinate be equal to zero in order that the boundary-layer velocity components pass over smoothly into the velocity components of the main stream.

The boundary conditions on the velocity at the outer edge of the boundary layer are

$$y=\infty:\ u=u_\infty,\ \frac{\partial u}{\partial y}=\frac{\partial^2 u}{\partial y^2}=\cdots=0 \tag{17.111}$$

The boundary conditions on the velocity at the surface of a body at rest are

$$y=0:\ u=v=0$$

$$\frac{\partial p}{\partial x}=\frac{\partial}{\partial y}\left(\mu\frac{\partial u}{\partial y}\right)_w \tag{17.112}$$

$$\left[\frac{\partial}{\partial y}\left(\frac{\partial}{\partial y}\mu\frac{\partial u}{\partial y}\right)\right]_w=\left\{\frac{\partial^2}{\partial y^2}\left[\frac{\partial}{\partial y}\left(\mu\frac{\partial u}{\partial y}\right)\right]\right\}_w=\cdots=0$$

where the second condition is obtained by substituting the first into the equation of motion of the boundary layer and the third is obtained by differentiating the second with respect to y subject to the condition that p is independent of y in the boundary layer.

The first derivative of u with respect to y, $\partial u/\partial y$, determines the shearing stress in the boundary layer. The value of $\partial u/\partial y$ at the wall $(\partial u/\partial y)_w$, is what we determine. Ordinarily, $(\partial u/\partial y)_w$ is positive. As the longitudinal distance from the leading edge, x increases, $(\partial u/\partial y)_w$ might vanish and change in sign, particularly if dp/dx is positive. In this case the flow separates from the surface of the body and a backward flow takes place. After separation of the flow, the boundary-layer approximations no longer hold true. Hence the solution of the boundary-layer equations is valid only to the point of separation, i.e., $(\partial u/\partial y)_w=0$. The flow conditions downstream from the point of separation are too complicated to admit of accurate mathematical treatment in the present state of theory.

For the boundary conditions of temperature, at the surface of the body, the temperature of the fluid is equal to the temperature of the wall, and the fluid normal temperature fluid gradient is equal to the body normal temperature gradient. Hence we have

$$y=0:\ T=T_w,\ \frac{\partial T}{\partial y}=-\frac{Q}{\kappa} \tag{17.113a}$$

where Q is rate of heat transferred from the wall to the fluid.

At the outer edge of the boundary layer, the temperature of the fluid must be that of the free stream, and the gradient of temperature mut be zero, i. e.,

$$y=\infty: T=T_\infty, \quad \frac{\partial T}{\partial y}=0 \tag{17.113b}$$

The boundary conditions discussed above are based on the assumption that there is no interaction between the boundary-layer flow and the outside free stream. In certain cases, such interactions exist. Whenever they do exist, either the boundary conditions should be modified to take care of this interaction, or the complete Navier-Stokes equations should be used if the boundary-layer approximation breaks down.

Special treatises such as Refs. 6 and 20 should be referred to for the general discussion of the solutions of the boundary layer Eq. (17.108) with proper initial and boundary conditions. Here we shall give only a few important and interesting results of Eq. (17.108).

For steady and two-dimensional flow, Eq. (108) may give the following relation:

$$\rho u \frac{\partial H+\frac{u^2}{2}}{\partial x}+\rho v \frac{\partial H+\frac{u^2}{2}}{\partial y}$$

$$=\frac{\partial}{\partial y}\left(\mu \frac{\partial H+\frac{u^2}{2}}{\partial y}\right)+\left(\frac{1}{P_r}-1\right)\frac{\partial}{\partial y}\left(\mu \frac{\partial c_p T}{\partial y}\right) \tag{17.114}$$

If $Pr=1$, we see that

$$H+\frac{u^2}{2}=H\left(1+\frac{\gamma-1}{2} M^2\right)=c_p T_0=\text{constant} \tag{17.115}$$

is a particular integral of equation (17.114) no matter what the pressure distribution is. $M=u/\sqrt{\gamma RT}$ is the local Mach number of the flow. This corresponds to the case of an insulated body, since at $y=0$, $u=0$, and then $\partial H/\partial y=0$. This relation was first found by Busemann[6].

If $dp/dx=0$ and $Pr=1$, we see, with the help of the equation of motion in (17.108a), that

$$H+\frac{u^2}{2}=au+b \tag{17.116}$$

is a solution of Eq. (17.114). The constants a and b are determined by the heat-transfer conditions. This relation was first found by Crocco[6]. At $u=0$, we have $b=c_p T_0$, where T_0 is the stagnation temperature. At the wall, $y=0$ and

$$\left(\frac{\partial H}{\partial y}\right)_{y=0}=a\left(\frac{\partial u}{\partial y}\right)_{y=0} \tag{17.117}$$

Hence a is determined by the condition of heat transfer from the wall to the fluid. The constants a and b may be expressed in terms of the temperature at the wall and free stream conditions, i.e.,

$$T^* = T_w^* + u^*\left[\frac{\gamma-1}{2}M_\infty^2 - (T_w^* - 1)\right] - u^{*2}\frac{\gamma-1}{2}M_\infty^2 \tag{17.118}$$

where the non-dimensional quantities are defined as follows:

$$\left.\begin{aligned} u^* &= \frac{u}{U},\ T^* = \frac{T}{T_\infty},\ x^* = \frac{x}{L},\ M_\infty = \frac{U}{\sqrt{\gamma R T_\infty}} \\ T_w^* &= \frac{T_w}{T_\infty},\ \rho^* = \frac{\rho}{\rho_\infty},\ \mu^* = \frac{\mu}{\mu_\infty} \end{aligned}\right\} \tag{17.119}$$

where U is the free stream velocity, L is a characteristic length, subscript ∞ refers to the value in free stream and subcript w refers to the value at the wall.

10. Boundary layer flow with chemical reactions

In this section, we consider the boundary layer flow of a compressible fluid with chemical reactions. Here we have to deal with three transport coefficients, the coefficient of viscosity, the coefficient of heat conductivity and the coefficent of diffusion. For simplicity, we consider a two-dimensional steady flow in which the y-direction is the direction of boundary layer thickness and the x-direction is that of the main flow velocity. We further consider the case of binary mixture with first order reaction and neglect the pressure and thermal diffusion terms. The boundary layer equations for this simple case are as follows:

$$p = knT \tag{17.120a}$$

$$\frac{\partial(\rho u)}{\partial x} + \frac{\partial(\rho v)}{\partial y} = 0 \tag{17.120b}$$

$$\rho u\frac{\partial k_1}{\partial x} + \rho v\frac{\partial k_1}{\partial y} = \frac{\partial}{\partial y}\left(\rho D\frac{\partial k_1}{\partial y}\right) + \sigma_1 \tag{17.120c}$$

$$\rho u\frac{\partial u}{\partial x} + \rho v\frac{\partial u}{\partial y} = -\frac{\partial p}{\partial x} + \frac{\partial}{\partial y}\left(\mu\frac{\partial u}{\partial y}\right) \tag{17.120d}$$

$$\rho u\frac{\partial H}{\partial x} + \rho v\frac{\partial H}{\partial y} = u\frac{\partial p}{\partial x} + \frac{\partial}{\partial y}\left(\frac{\kappa}{c_p}\frac{\partial H}{\partial y}\right) + \mu\left(\frac{\partial u}{\partial y}\right)^2 + Q^*\sigma_1 \tag{17.120e}$$

where σ_1 is the source term of the unburned mixture with concentration k_1. We neglect the thermal radiation and the effects of electromagnetic fields.

Eq. (17.120c) gives the boundary layer of concentration k_1, Eq. (17.120d) gives the boundary layer of velocity u and Eq. (17.120e) gives the boundary layer of temperature T. If we transform these three equations in non-dimensional form, we will obtain Reynolds number, Schmidt number $(\mu/\rho D)$ and Prandtl number as the important non-dimensional parameters in the present problem.

Now we solve Eqs. (17.120) for the case of two uniform streams of two different gases: one is the combustible gas 1 and the other is the combustion product 2 (Fig. 17.8). When the temperature of the combustion product is high enough, combustion may occur in the mixing zone which may be regarded as a laminar jet flame in which sharp flame front occurs.

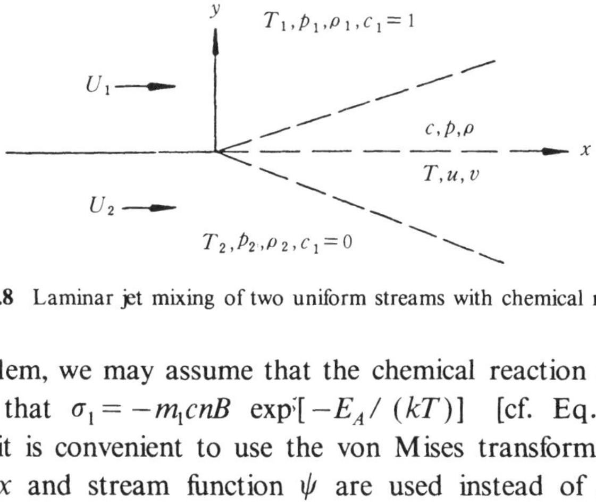

Fig. 17.8 Laminar jet mixing of two uniform streams with chemical reaction

For such problem, we may assume that the chemical reaction is a first order reaction such that $\sigma_1 = -m_1 cnB \exp[-E_A/(kT)]$ [cf. Eq. (17.36)]. In this problem, it is convenient to use the von Mises transformation in which the variables x and stream function ψ are used instead of x and y. The stream function ψ is defined as follows:

$$\frac{\partial\psi}{\partial y} = \rho u, \quad \frac{\partial\psi}{\partial x} = -\rho v \tag{17.121}$$

In terms of x and ψ, Eq. (17.120b) is automatically satisfied. Eqs. (17.120c) to (17.120e) become respectively

$$\frac{\partial c}{\partial x} = \frac{\rho^2}{n^2 m_1 m_2} \frac{\partial}{\partial\psi}\left(\rho^2 u D \frac{\partial c}{\partial\psi}\right) - \frac{c(\beta c + 1)B}{u} \exp\left(-\frac{E_A}{kT}\right) \tag{17.122a}$$

$$\frac{\partial u}{\partial x} = \frac{\partial}{\partial\psi}\left(\rho\mu u \frac{\partial u}{\partial\psi}\right) \tag{17.122b}$$

$$\frac{\partial H}{\partial x} = \frac{\partial}{\partial\psi}\left(\frac{\kappa\rho u}{c_p} \frac{\partial H}{\partial\psi}\right) + \mu\rho u\left(\frac{\partial u}{\partial\psi}\right)^2 + Q^* \frac{m_1 cnB}{u} \exp\left(-\frac{E_A}{kT}\right) \tag{17.122c}$$

where $\beta = (m_1/m_2) - 1$.

Eqs. (17.122) are a set of three generalized heat conduction equations which may be easily integrated by stepwise numerical procedure from the following initial conditions

$$x=0: \; c=\frac{n_1}{n}=c_0(\psi), \quad u=u_0(\psi), \quad T=T_0(\psi) \tag{17.123}$$

where subscript o refers to the values at $x=0$. After $c(x, \psi)$, $u(x, \psi)$ and $T(x, \psi)$ are obtained, the y-coordinate for any given x may be obtained by the following formula:

$$y-y_o=\int_0^{\psi} \frac{d\psi}{\rho u} \tag{17.124}$$

where y_o is a constant corresponding to $\psi=0$ at a given x.

In general, we have to solve the three partial differential equations (17.122) simultaneously. As an example, we consider the simple case of isovel jet mixing in which the velocities of the two streams shown in Fig. 17.8 are the same and the whole flow field has a constant velocity U. We need to solve Eqs. (17.122a) and (17.122c) only for c and T.

We consider a two-dimensional jet with initial conditions at $x=0$.

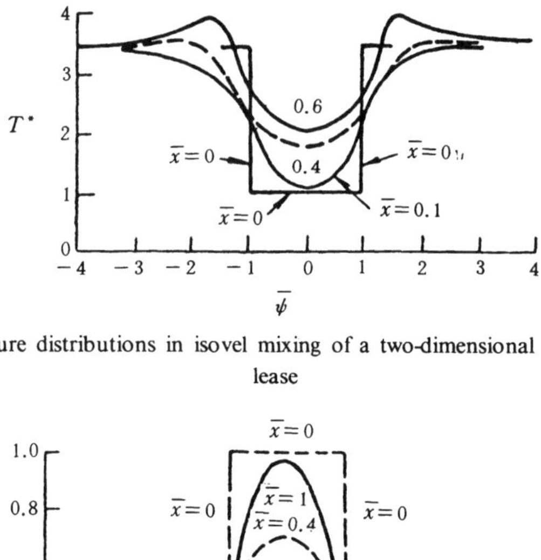

Fig. 17.9 Temperature distributions in isovel mixing of a two-dimensional laminar jet with heat lease

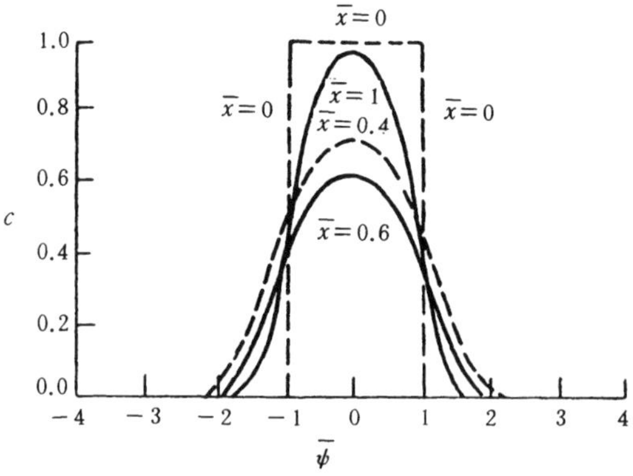

Fig. 17.10 Concentration distributions in isovel mixing of a two-dimensional laminar jet with chemical reaction

$$\psi < 1 \text{ and } \psi > -1 : c=1,\ T^* = \frac{T}{T_1} = 1$$

$$\psi > 1 \text{ and } \psi < -1 : c=0,\ T^* = 3.5$$

The temperature and concentration distributions for the above initial conditions are shown in Fig. 17.9 and Fig. 17.10 respectively. Fig. 17.9 shows that there is a tendency to develop a flame front at the edge of the jet, i. e. a peak in temperature distribution occurs.

Special treatises may be referred to for other cases of the flow problems with chemical reaction (see Ref. 1 and problem section).

11. Boundary layer flow with phase change: ablation problem

One of the most important problems during the reentry of a space vehicle is to protect the surface of the space vehicle from over-heating. One of the effective methods of heat protection for a blunt body is ablation which is a process of absorbing heat energy by removal of surface material, either by melting or by sublimation. For a more accurate calculation, we should include the motion of the melting liquid in our analysis. Let us consider a blunt body

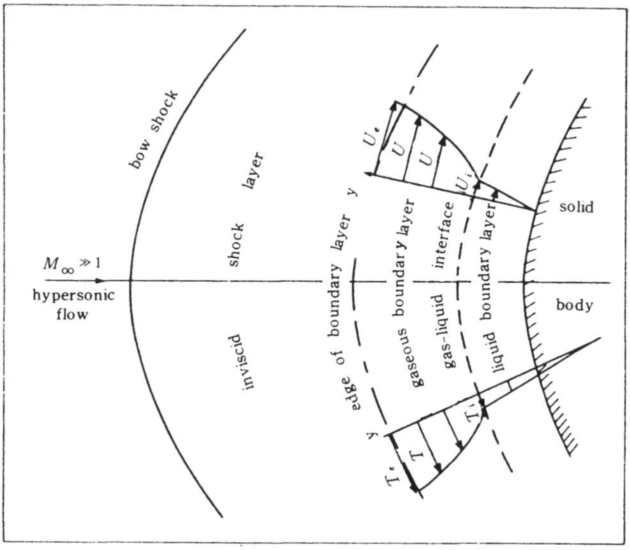

Fig. 17.11 A sketch of two-phase boundary layer over a melting blunt body in a hypersonic flow

of revolution in a hypersonic flow (Fig. 17.11). There is a bow shock wave in front of the body. Behind the bow shock, the gas in the shock layer is very hot and is partly or fully dissociated. On the surface of the body, we have a boundary layer of the hot gas. In ablation, the surface of the body is allowed to melt or vaporize. Hence, there will be a molten liquid layer between the gas boundary layer and the solid body. The melting rate, which is the rate at which the surface of the body receded, depends on the external conditions and the boundary layer flow of the liquid layer. The problem is complicated by the fact that the interface temperature between the gaseous boundary layer and the liquid layer is not known a priori. We have to match the solutions of these two boundary layers. Hence our problem consists of three parts: the first is the gaseous boundary layer problem, the second is the liquid boundary layer problem and the third is the interaction between the gas phase and the liquid phase.

In the gaseous phase laminar boundary layer, we assume that the gas is a mixture consisting of air atoms, air molecules and vaporing molecules. It is assumed that the flow is quasi-steady and vaporizing material will not affect the thermodynamic properties of the air appreciably and has the same properties as the air molecules. Thus the gaseous boundary layer may be treated by the regular method of dissociated gas boundary layer[23]. The unknowns are the velocity components u and v, the temperature T, the density ρ, the mass concentration of atom c_A and the mass conecntration of vaporizing material c_k. The fundamental equations for these unknowns are

$$p=\rho RT \tag{17.125a}$$

where R is the gas constant of the gas mixture which depends on the concentration of various species in the mixture.

Equation of continuity

$$\frac{\partial}{\partial x}(\rho\, u r_0)+\frac{\partial}{\partial y}(\rho\, v r_0)=0 \tag{17.125b}$$

where r_0 is the radius of the cross-section of the blunt body.

Equation of diffusion of the atomic species

$$\rho u\frac{\partial c_A}{\partial x}+\rho v\frac{\partial c_A}{\partial y}=\frac{\partial}{\partial y}\left(\rho\, D_{Aj}\frac{\partial c_A}{\partial y}\right) \tag{17.125c}$$

where we assume that the gas reaction is frozen and the thermal diffusion is negligible. D_{Aj} is the coefficient of diffusion of the atom in the mixture.

Equation of diffusion of the vaporing species

$$\rho u\frac{\partial c_k}{\partial x}+\rho v\frac{\partial c_k}{\partial y}=\frac{\partial}{\partial y}\left(\rho D_{Ak}\frac{\partial c_k}{\partial y}\right) \tag{17.125d}$$

Equation of motion

$$\rho u\frac{\partial u}{\partial x}+\rho v\frac{\partial u}{\partial y}=-\frac{\partial p}{\partial x}+\frac{\partial}{\partial y}\left(\mu\frac{\partial u}{\partial y}\right) \tag{17.125e}$$

Equation of energy of the mixture:

$$\rho c_p\left(u\frac{\partial T}{\partial x}+v\frac{\partial T}{\partial y}\right)=u\frac{\partial p}{\partial x}+\mu\left(\frac{\partial u}{\partial y}\right)^2+\frac{\partial}{\partial y}\left(\kappa\frac{\partial T}{\partial y}\right)$$
$$+\sum_n\left(c_{pn}D_{nj}\frac{\partial c_A}{\partial y}\right)\frac{\partial T}{\partial y} \qquad (17.125f)$$

We are going to study the ablation near a stagnation point based on the above equations (17.125) and the following boundary conditions:

$$\left.\begin{aligned} &y=0: u=u_i,\ T=T_i,\ v=v_i,\ c_A=c_{Ai},\ c_k=c_{ki}\\ &y\to\infty: u\to u_e=x\,u_{eo},\ T\to T_e,\ c_A\to c_{Ae},\ c_k\to 0\end{aligned}\right\} \qquad (17.126)$$

It should be noticed that the values at the interface, i.e., T_i, u_i, etc. are not known a priori. The values at the outer boundary of the gaseous boundary layer are given by the solution of inviscid shock layer.

In the liquid phase laminar boundary layer, the fluid is an incompressible liquid and we assume that the surface is melting at a steady rate such that the derivative with respect to time is much smaller than the spatial derivatives. The aerodynamic pressure gradient and the shear stress cause the molten material to flow. The fundamental equations are then:

Equation of continuity

$$\frac{\partial}{\partial x}(ur_o)+\frac{\partial}{\partial y}(vr_o)=0 \qquad (17.127a)$$

Equation of motion

$$\rho u\frac{\partial u}{\partial x}+\rho v\frac{\partial u}{\partial y}=-\frac{\partial p}{\partial x}+\frac{\partial}{\partial y}\left(\mu\frac{\partial u}{\partial y}\right) \qquad (17.127b)$$

Equation of energy

$$\rho c_p\left(u\frac{\partial T}{\partial x}+v\frac{\partial T}{\partial y}\right)=u\frac{\partial p}{\partial x}+\frac{\partial}{\partial y}\left(\kappa\frac{\partial T}{\partial y}\right)+\mu\left(\frac{\partial u}{\partial y}\right)^2 \qquad (17.127c)$$

We shall take the coordinate system so that the y-coordinate is zero at the interface. Hence, the boundary conditions for the liquid layer at $y=0$ are the same as those given in Eqs. (17.126). If we take y as positive toward the solid body for the liquid layer, the boundary conditions at infinity are:

$$y\to\infty: u\to 0,\ T\to T_b \qquad (17.128)$$

where T_b is the temperature of the solid body.

The next step is to match the solutions of the liquid phase and the gas phase. A unique solution is obtained when the interface mass transfer $\dot{m}_i$, the interface velocity u_i, the interface shear stress τ_i, the interface temperature T_i

and the interface energy transfer Q_i and the mass fraction of the vaporizing species c_{ki} are the same for both the gas and the liquid phases.

It has been found that u_i/u_e is of the order of 10^{-4}. Hence, as far as the velocity distribution of the gaseous boundary layer is concerned, we may consider $u=0$ at $y=0$. The solution of the liquid phase equations at proper match point will yield the value u_i/u_e a posteriori. The quantities to be matched are as follows:

In the gaseous boundary layer, we have

$$\dot{m}_i=\dot{m}_i\,(c_{ki},\, T_i) \tag{17.129a}$$

$$\tau_i=\tau_i(\dot{m}_i,\, T_i) \tag{17.129b}$$

$$Q_i=Q_i(\dot{m}_i,\, T_i) \tag{17.129c}$$

Eq. (17.129a) means that the mass transfer is a function of the gaseous mass concentration of the evaporated species at the interface and the interface temperature. The relation may be obtained by solving the gaseous boundary layer equations for various values of c_{ki} and T_i. Similarly, we may obtain the relations (17.129b) and (17.129c).

The solution to the liquid phase boundary layer equations gives

$$Q_i=Q_i(\tau_i,\, \dot{m}_i,\, T_i) \tag{17.129d}$$

Finally from the assumed dependence of vapor pressure on temperature, we obtain the relation:

$$c_{ki}=c_{ki}(T_i) \tag{17.129e}$$

From the five relations of Eqs. (17.129), we obtain the five unknowns: $\dot{m}_i$, τ_i, Q_i, T_i and c_{ki}. For a given condition of the environment, we may obtain the solution of this two-phase boundary layer problem. Refs. 24 and 25 give some numerical results of this problem. Some of the general conclusions of the ablation in a hypersonic flow in the case of reentry of a space vehicle are as follows:

(i) The most important physical property of a melting ablator is the viscosity of the liquid layer. The larger the viscosity is, the greater the liquid layer thickness will be. So that are the temperature rise across the layer and the magnitude of the heat energy absorbed by the ablator.

(ii) If the ablator melts but not evaporizes, the total heat capacity is about 1,200 B. t. u. 1b for laminar gas flow and about 1,500 B. t. u. for turbulent gas flow. The vaporization or sublimation causes a large increase in heat capacity of the ablator.

12. Boundary layer flows in magnetogasdynamics

In this section, we are going to study the boundary layer flow of a viscous, heat-conducting and electrically conducting fluid. We shall assume that the Prandtl number of the fluid is of the order of unity and the Reynolds number is very large. We have a new transport coefficient, the magnetic diffusivity [cf. Eq. (15.23)]. Here we may define a magnetic Prandtl number as P_m:

$$P_m = \frac{\mu\mu_e\sigma_e}{\rho} = \frac{\nu}{\nu_H} = \frac{R_\sigma}{Re} \tag{17.130}$$

where R_σ is known as the magnetic Reynolds number. It should be noticed that the magnetic Prandtl number may be of the order of unity or very small. In fact, for many engineering problems, the magnetic Prandtl number is very small so that even though the Reynolds number Re is very large, the magnetic Reynolds number is small. As a result, we may have the boundary layer of velocity and temperature but we do not have the magnetic boundary layer. Only when the magnetic Reynolds number is large, shall we have the magnetic boundary layer. We have to treat the cases of small and large magnetic Prandtl number separately as follows:

(i) Case of small magnetic Reynolds number

We first consider the case of large Reynolds number Re but small magnetic Reynolds number R_σ. This is the case for many engineering problems. We have boundary layer flows of velocity and temperature but not for the magnetic field. Hence in the boundary layer flow, we may assume that the applied electromagnetic fields E_o and H_o or B_o are unaffected by the flow field. For simplicity, let us consider the case where there is an externally applied transverse magnetic field H_{oy} in the direction normal to the main flow direction u. The boundary layer direction is y with $u \gg v$. The ponderamotive force has a component in the x-direction only, i.e.,

$$\boldsymbol{F}_e = \boldsymbol{J} \times \boldsymbol{B} = -\boldsymbol{i}\ \sigma_e u B_{oy}^2 \tag{17.131}$$

where $\boldsymbol{J} = \boldsymbol{k}\sigma_e u B_{oy}$ is in the z-direction only and $\boldsymbol{i}$ is the unit vector in the x-direction. The Joule heat in the present case is

$$\frac{J^2}{\sigma_e} = \sigma_e u^2 B_{oy}^2 \tag{17.132}$$

The two-dimensional boundary layer equations for the present case without chemical reaction nor thermal radiation effects are

$$\frac{\partial \rho}{\partial t} + \frac{\partial \rho u}{\partial x} + \frac{\partial \rho v}{\partial y} = 0 \tag{17.133a}$$

$$\rho\left(\frac{\partial u}{\partial t} + u\frac{\partial u}{\partial x} + v\frac{\partial u}{\partial y}\right) = -\frac{\partial p}{\partial x} + \frac{\partial}{\partial y}\left(\mu\frac{\partial u}{\partial y}\right) - \sigma_e u\, B_{oy}^2 \tag{17.133b}$$

$$\rho\left(\frac{\partial h}{\partial t}+u\frac{\partial h}{\partial x}+v\frac{\partial h}{\partial y}\right)=\frac{\partial p}{\partial t}+u\frac{\partial p}{\partial x}+\frac{\partial}{\partial y}\left(\kappa\frac{\partial T}{\partial y}\right)+\mu\left(\frac{\partial u}{\partial y}\right)^2+\sigma_e u^2 B_{oy}^2 \tag{17.133c}$$

$$p=\rho RT \tag{17.133d}$$

where $h=c_pT$ is the enthalpy of the electrically conducting gas and the pressure p is assumed to be a given function of x and t.

For steady flow, Eqs. (17.133) give the relation:

$$\rho\left(u\frac{\partial h_o}{\partial x}+v\frac{\partial h_o}{\partial y}\right)=\frac{\partial}{\partial y}\left(\mu\frac{\partial h_o}{\partial y}\right)+\left(\frac{1}{P_r}-1\right)\frac{\partial}{\partial y}\left(\mu\frac{\partial h}{\partial y}\right) \tag{17.134}$$

If the Prandtl number $Pr=c_p\mu/\kappa=1$, we have

$$h_o=h+\frac{1}{2}u^2=\text{constant}=\text{stagnation enthalpy} \tag{17.135}$$

as a particular integral of Eq. (17.134). This is known as the Busemann relation in ordinary gasdynamics (17.115) which holds here too. But the corresponding Crocco relation (17.116) of ordinary gasdynamics does not hold true here.

Eqs. (17.133) may be solved by the usual method of solution of boundary layer equations[6]. For instance, if we consider the case of boundary layer flow over a flat plate for an incompressible steady fluid flow, we may use the power series method in x for the stream function ψ as follows:

$$\psi=(U\nu x)^{1/2}[f_o(\eta)+mxf_2(\eta)+(mx)^2f_4(\eta)+\cdots] \tag{17.136}$$

where $m=\sigma_e B_{oy}^2/(\rho U)$, $\eta=y[U/(\nu x)]^{1/2}$ and U is the free stream velocity. Substituting Eq. (17.136) into Eq. (17.133b), we have a set of total differential equations for f_n as follows:

$$2f_o'''+f_o''f_o=0 \tag{17.137a}$$

$$2f_2'''=2f_o'f_2'-f_of_2''-3f_2f_o''+2f_o' \tag{17.137b}$$

$$2f_4'''=4f_o'f_4'+2f_2'f_2'-f_of_4''-3f_2f_2''-5f_4f_o''-2f_2' \tag{17.137c}$$

etc., with the boundary conditions:

at $\quad \eta=0:\ f_o=f_2=f_4=\cdots=0$

$$\eta=0:\ f_o'=f_2'=f_4'=\cdots=0 \tag{17.138}$$

$$\eta\to\infty:\ f_o'=1,\ f_2'=-1,\ f_4'=\cdots=0$$

Eq. (17.137a) is the well-known Blasius equation of ordinary boundary layer flow. Eqs. (17.137b), (17.137c) etc. are linear equations which can be numerically integrated from the boundary conditions (17.138). Up to the term f_2, the skin friction coefficient is[26]

$$c_f = \frac{\mu}{1/2\,\rho U^2}\left(\frac{\partial u}{\partial y}\right)_{y=0} = \frac{0.664 - 1.788mx + \cdots}{(Re_x)^{1/2}} \qquad (17.139)$$

where $Re_x = Ux/\nu$.

After $u(x, y)$ is obtained, we may either use the relation (17.135) to find the temperature distribution if the wall is insulated or to solve Eq. (17.133c) by the method of series expansion with the known values of velocity and stream function.

(ii) Case of large magnetic Reynolds number

If the magnetic Reynolds number and the ordinary Reynolds number are large, we have the boundary layer equations for velocity, magnetic field and temperature. We have to solve the magnetic field simultaneously with the velocity and temperature. Hence for two-dimensional flow, we have the following boundary layer equations of an electrically conducting fluid with large Reynolds number and large magnetic Reynolds number as follows:

$$p = \rho RT \qquad (17.140a)$$

$$\frac{\partial \rho}{\partial t} + \frac{\partial \rho u}{\partial x} + \frac{\partial \rho v}{\partial y} = 0 \qquad (17.140b)$$

$$\frac{\partial H_x}{\partial t} + u\frac{\partial H_x}{\partial x} + v\frac{\partial H_x}{\partial y} + H_x\frac{\partial v}{\partial y} - H_y\frac{\partial u}{\partial y} = \frac{\partial}{\partial y}\left(\nu_H \frac{\partial H_x}{\partial y}\right) \qquad (17.140c)$$

$$\frac{\partial H_x}{\partial x} + \frac{\partial H_y}{\partial y} = 0 \qquad (17.140d)$$

$$\rho\left(\frac{\partial u}{\partial t} + u\frac{\partial u}{\partial x} + v\frac{\partial u}{\partial y}\right) - B_y\frac{\partial H_x}{\partial y} = -\frac{\partial p}{\partial x} + \frac{\partial}{\partial y}\left(\mu\frac{\partial u}{\partial y}\right) \qquad (17.140e)$$

$$p + \frac{B_x H_x}{2} = p_o(x) = \text{a given function of } x \qquad (17.140f)$$

$$\rho\left(\frac{\partial h}{\partial t} + u\frac{\partial h}{\partial x} + v\frac{\partial h}{\partial y}\right) = \frac{\partial p}{\partial t} + u\frac{\partial p}{\partial x} + \frac{\partial}{\partial y}\left(\kappa\frac{\partial T}{\partial y}\right) + \mu\left(\frac{\partial \mu}{\partial y}\right)^2 + \nu_H\left(\frac{\partial H_x}{\partial y}\right)^2 \qquad (17.140g)$$

In solving Eqs. (17.140), we have the new parameters of magnetic Reynolds number and the magnetic pressure number or the magnetic Mach number. There are many new phenomena. For instance, in sub-Alfven flow, i. e., $M_m = U/V_x < 1$, we have an upstream influence of the boundary layer flow. Without going to very lengthy calculations, we shall consider the following simple wake problems to illustrate this phenomenon.

(iii) Magnetohydrodynamic wakes behind a two-dimensional body[26, 27]

We consider a two-dimensional flow of a uniform stream U over a body (Fig. 17.12) under a uniform magnetic field $\boldsymbol{H}_0 = \boldsymbol{i}H_{xo} + \boldsymbol{j}H_{yo} + \boldsymbol{k}0$. We assume that the compressibility effect is negligible and the deviation of the velocity and magnetic field from the uniform state are small. Hence we write:

$$u = U + u', \quad v = v', \quad H_x = H_{xo} + h_x, \; H_y = H_{yo} + h_y$$

such that u', $v' \ll U$ and h_x, $h_y \ll H_{xo}$, H_{yo}

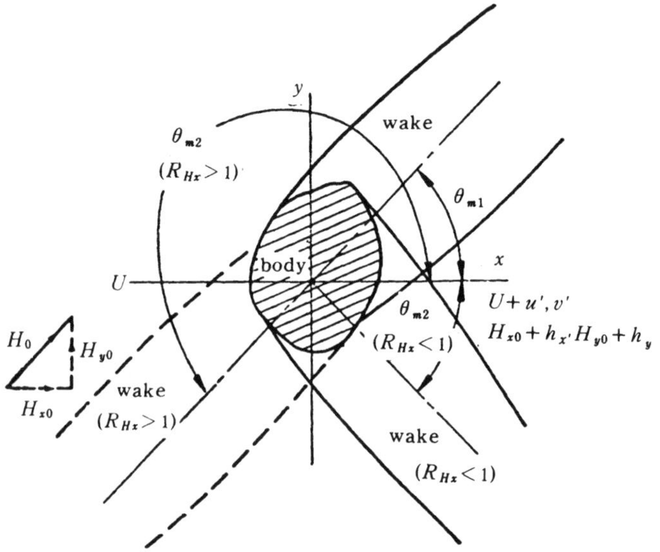

Fig. 17.12 MHD wakes around a body in an electrically conducting liquid

If we neglect the higher order terms of the perturbed quantities: u', v' etc., we may obtain the following non-dimensional linearized equation:

$$\nabla^2\left[\left(\frac{\partial}{\partial x} - \frac{1}{R_e}\nabla^2\right)\left(\frac{\partial}{\partial x} - \frac{1}{R_\sigma}\nabla^2\right) - \left(R_{Hx}\frac{\partial}{\partial x} - R_{Hy}\frac{\partial}{\partial y}\right)^2\right]u^* = 0 \qquad (17.141)$$

where x and y are the non-dimensional coordinates in terms of a reference length L, and

$$u^* = \frac{u'}{U}, \quad \nabla^2 = \frac{\partial^2}{\partial x^2} + \frac{\partial^2}{\partial y^2}, \quad Re = \frac{UL}{\nu}, \quad R_\sigma = \frac{UL}{\nu_H},$$

$$R_{Hx} = \frac{\mu_e H_{xo}^2}{\rho U^2}, \quad R_{Hy} = \frac{\mu_e H_{yo}^2}{\rho U^2}$$

The operator of the square bracket may be considered as a product of two Oseen-type operators. For instance, if we assume that $Re = R$, Eq. (17.141) becomes

$$\nabla^2\left[(1+R_{Hx})\frac{\partial}{\partial x}+R_{Hy}\frac{\partial}{\partial y}-\frac{1}{Re}\nabla^2\right]\cdot\left[(1-R_{Hx})\frac{\partial}{\partial x}-R_{Hy}\frac{\partial}{\partial y}-\frac{1}{Re}\nabla^2\right]u^*=0 \tag{17.142}$$

The two operators in the square bracket of Eq. (17.142) are the Oseen-type operator, which may be written in the general form:

$$\left(A_1\frac{\partial}{\partial x}+B_1\frac{\partial}{\partial y}-\frac{1}{Re}\nabla^2\right)u^*=0 \tag{17.143}$$

The solution of Eq. (17.143) is

$$u^*=\text{constant}\cdot\exp\left[\frac{(A_1x+B_1y)Re}{2}\right]K_o\left[\frac{(A_1^2+B_1^2)Rer}{2}\right] \tag{17.144}$$

where $r^2=x^2+y^2$ and K_0 is the Bessel function of zeroth order.

For a given radial distance from the origin, the maximum value of u^* occurs at

$$\tan\theta=\frac{y}{x}=\frac{B_1}{A_1} \tag{17.145}$$

Thus we have a wake along the direction

$$\theta=\theta_m=\tan^{-1}\left(\frac{B_1}{A_1}\right) \tag{17.146}$$

Since there are two Oseen operators in Eq. (17.142), we have two wakes which are respectively along the lines (Fig. 17.12):

$$\theta_{m1}=\tan^{-1}\left(\frac{R_{Hy}}{1+R_{Hx}}\right);\ \theta_{m2}=\tan^{-1}\left(\frac{-R_{Hy}}{1-R_{Hx}}\right) \tag{17.147}$$

If $R_{Hx}>1$, i. e., $M_m<1$, we have an upstream inclined wake in sub-Alfven flow, which is not possible in ordinary hydrodynamics. There are two wakes instead of one wake in ordinary hydrodynamics.

From Eq. (17.144), we see that the rate of decrease of u^* with r increases with Reynolds number R_e and we have boundary layer phenomena for large value of R_e, i. e., the spread of the wake is within a narrow region from the axis which makes angles θ_{m1} and θ_{m2} with respect to the direction of the main flow U. Thus if $R_{Hx}>1$, we have the upper stream influence of the boundary layer flow.

13. Boundary layer flow with thermal radiation effects

In this section, we are going to consider the additional transport phenomenon due to radiative heat flux in addition to viscosity and heat conductivity, i.e., radiation absorption coefficient or radiation mean free path. In many current engineering problems, the radiation pressure or the radiation stresses are still negligible. Hence in this section, we shall consider only the effect of the radiation heat flux. The important non-dimensional parameter for this case is known as the radiation flux number which is defined as follows:

$$\text{Radiation flux number} = R_F = \frac{\text{radiative heat flux}}{\text{heat conduction flux}} \tag{17.148}$$

Since in §6, we show that the exact expression of the radiative heat flux depends on the optical thickness of the medium, we have different expressions of R_F for the optically thick medium and the optically thin medium. For optically thick medium, the radiative heat flux is given by Eq. (17.64) and then the radiation flux number is

$$R_{Fl} = \frac{\kappa_R}{\kappa} = 4\left(\frac{\gamma-1}{\gamma}\right)PeK_rR_p/R_r \tag{17.149}$$

where $Pe=UL/(\kappa/c_p\rho)=PrRe$ is the Peclet number, $K_r=\dfrac{1}{\rho K_R L}$ is the Knudsen number of radiation, R_p is the radiation pressure number and $R_r=U/c$ is the relativistic parameter where c is the speed of light, and γ is the ratio of the specific heats.

For optically thin medium, the radiative heat flux is proportional to ca_RT^4/K_r and then the radiative heat flux number becomes

$$R_{F2} = 3\left(\frac{\gamma-1}{\gamma}\right)PeR_p/(K_rR_r) \tag{17.150}$$

The main difference between R_{F1} and R_{F2} lies in the parameter Knudsen number of radiation K_r. For small Knudsen number of radiation, R_{F1} is proportional to K_r but for a large Knudsen number of radiation, R_{F2} is inversely proportional to K_r. The radiative heat flux tends to be zero for both a very large and a very small mean free path of radiation.

Now we are going to consider a uniform flow over a semi-infinite plate for a steady boundary layer flow with radiative heat flux effect. If we consider the case of finite mean free path of radiation, we may apply the boundary approximation to the velocity and the heat conduction terms but not the thermal radiative heat flux term in general. We consider the case where the pressure is constant over the whole flow field. Our fundamental equations are as follows:

(i) Equation of state

$$P = \rho RT \tag{17.151a}$$

(ii) Equation of continuity

$$\frac{\partial \rho u}{\partial x} + \frac{\partial \rho v}{\partial y} = 0 \tag{17.151b}$$

(iii) Equation of motion

$$\rho u \frac{\partial u}{\partial x} + \rho v \frac{\partial u}{\partial y} = \frac{\partial}{\partial y}\left(\mu \frac{\partial u}{\partial y}\right) \tag{17.151c}$$

(iv) Equation of energy

$$\rho u \frac{\partial c_p T}{\partial x} + \rho v \frac{\partial c_p T}{\partial y} = \mu\left(\frac{\partial u}{\partial y}\right)^2 + \frac{\partial}{\partial y}\left(\kappa \frac{\partial T}{\partial y}\right) + \frac{\partial q_{Rx}}{\partial x} + \frac{\partial q_{Ry}}{\partial y} \tag{17.151d}$$

If the radiative flux number is not very large, we may apply the boundary layer approximation to the radiative heat flux terms and neglect $\partial q_{Rx}/\partial x$. Then we may use the one-dimensional approximation for the radiative heat flux q_{Ry}. If we further assume that the gas is a grey gas and local thermodynamic equilibrium is applicable, we have

$$\frac{\partial q_{Ry}}{\partial y} = 2\sigma\rho K_p\left[\left[\int_0^{\tau} T^4\varepsilon_1(\tau - t')\,dt' + \int_{\tau}^{\infty} T^4\varepsilon_1(t' - \tau)\,dt'\right] - 2T^4 + T_w^4\varepsilon_2(\tau)\right] \tag{17.152}$$

where K_p is the Planck mean abosrption coefficient and τ is the optical thickness defined by the formula:

$$\tau = \int_0^y \rho K_p dy \tag{17.153}$$

The boundary conditions of our problem are

$$\left.\begin{aligned} &x > 0,\ y = 0:\ u = v = 0,\ T = T_w,\ \rho = \rho_w \\ &y \to \infty:\ u \to U,\ T \to T_\infty,\ \rho \to \rho_\infty \end{aligned}\right\} \tag{17.154}$$

where U is the free stream velocity which is a constant and T_w is the temperature of the plate which is also a constant. Subscript ∞ refers to the values in the free stream.

Since the thermal radiation has a larger influence on the temperature distribution than on the velocity distribution, we consider the similar solution of the velocity profile[6] as follows:

$$\frac{u}{U} = \frac{df(\xi)}{d\xi} = f'(\xi) \tag{17.155}$$

where we assume that the coefficient of viscosity is proportional to the temperature, i.e., $\mu/\mu_\infty = C(T/T_\infty)$ and the variable ξ is defined as follows:

$$\xi = \frac{Y}{x}\left(\frac{\rho_\infty U x}{\mu_\infty}\right)^{1/2} \tag{17.156}$$

and

$$Y = \int_0^y \frac{\rho}{\rho_\infty} dy \tag{17.157}$$

The function $f(\xi)$ is the non-dimensional stream function and is related to the stream function ψ by the following relation

$$w = -\left(\frac{\partial \psi}{\partial x}\right)_Y = 1/2\left(\frac{C\mu_\infty U}{x}\right)^{1/2}(f' - f) \tag{17.158}$$

and the stream function ψ is defined by the relations:

$$\frac{\partial \psi}{\partial y} = \frac{\rho}{\rho_\infty} u \, ; \quad \frac{\partial \psi}{\partial x} = -\frac{\rho}{\rho_\infty} v \tag{17.159}$$

The function $f(\xi)$ is the well-known Blasius function of an incompressible boundary layer flow which has been tabulated in the standard textbook[6].

In terms of Y and x, the energy equation (17.151d) without the x-wise radiative heat flux becomes:

$$c_p u \frac{\partial T}{\partial x} + c_p w \frac{\partial T}{\partial y} = \frac{C\mu_\infty}{\rho_\infty}\left(\frac{\partial u}{\partial Y}\right)^2 + \frac{C\mu_\infty c_p}{\rho_\infty P_r}\frac{\partial^2 T}{\partial Y^2}$$

$$+\left[T_w^4 \varepsilon_2(\tau) - 2T^4\right] 2\rho\sigma K_p + 2\sigma\rho K_p\left[\int_0^\tau T^4 \varepsilon_1(\tau - t')\, dt' + \int_\tau^\infty T^4 \varepsilon_1(t' - \tau)\, dt'\right] \tag{17.160}$$

with the boundary conditions:

$$x > 0 : Y = 0, \; T = T_w; \; Y \rightarrow \infty, \; T \rightarrow T_\infty \tag{17.161}$$

Even though we have similar solution for the velocity $u(\xi)$, there will be no similar solution for temperature T when the absorption mean free path of radiation is finite or large. Only when the absorption mean free path of radiation is very small, and Rosseland approximation (17.64) may be used, may we have a similar solution in temperature $T(\xi)$. In Eq. (17.160), we assume that both c_p and Pr are constant.

We integrate Eq. (17.160) with the expression (17.80) for K_p; $Pr = 0.74$; $T_w = 20{,}000$ K and T_∞ at several temperatures. The results are shown in Figs. 17.13 to 17.16.

Fig. 17.13 is the case of low Mach number $M_\infty = 1.25$. Since there is no similar solution in termperature, the temperature profiles change with x-distance. We use the following definition for the radiative flux number

$$R_f = \frac{\sigma T_\infty^3 x^2}{K_\infty L_p} = \frac{\sigma T_\infty^3 x^2 \rho_\infty K_p}{K_\infty} \tag{17.162}$$

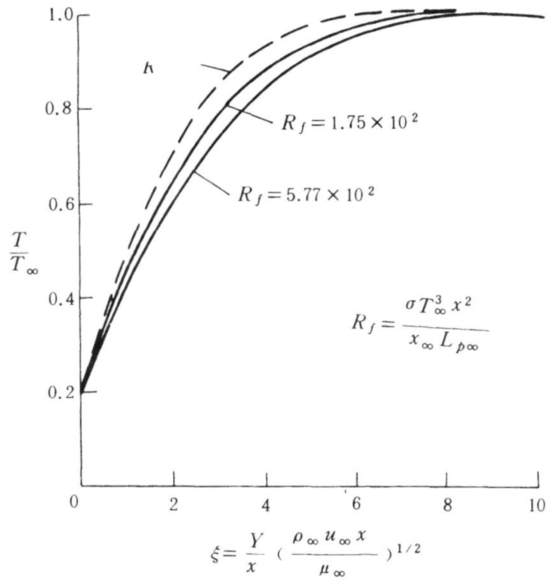

Fig. 17.13 Temperature distribution in a boundary layer flow over a semi-infinite plate at $M_\infty = 1.25$, $T_\infty = 10{,}000$ K and $p_\infty = 5.0$ atmospheres

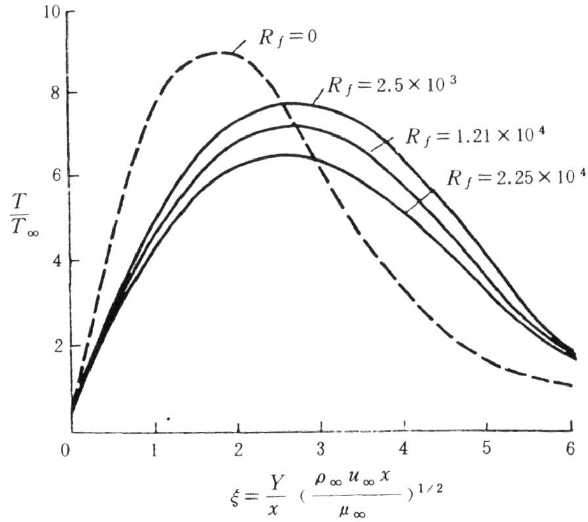

Fig. 17.14 Temperature distributions in a boundary layer flow over a semi-infinite plate at $M_\infty = 15.0$, $T_\infty = 20{,}000$ K, $p_\infty = 1.0$ atmosphere

When $R_f = 0$, i.e., without the radiative heat flux, we have a similar solution for temperature $T(\xi)$ which is shown in the dotted curve in Figs. 17.13 to

17.16.

Fig. 17.14 shows a corresponding case for high Mach number $M_\infty = 15$.

In the numerical calculations of Figs. 17.13 and 17.14, we use the approximate formula (17.73) with $m = 1.562$.

The main effects of the radiative heat flux are (i) to decrease the maximum temperature in the boundary layer at high Mach number case, (ii) to increase the boundary layer thickness of temperature and (iii) to decrease the slope of $[d(T/T_\infty)/d\xi]_w$ at the wall.

It is interesting to compare the results of the integral expression of

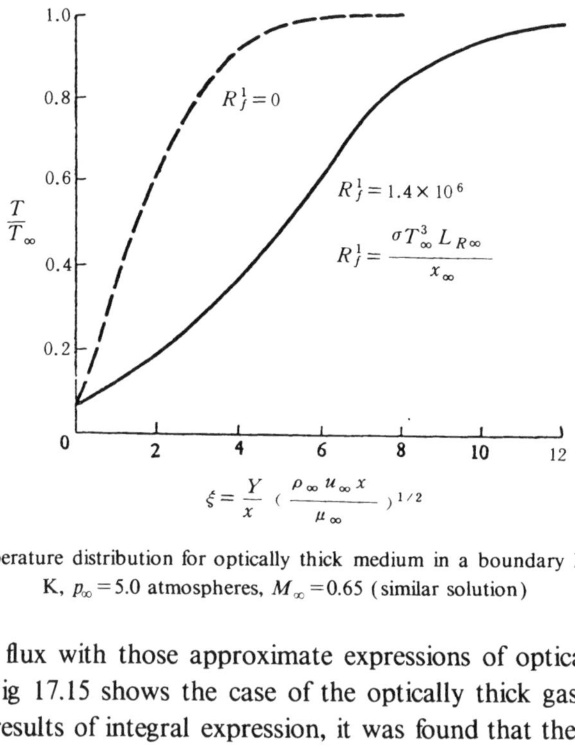

Fig. 17.15 Temperature distribution for optically thick medium in a boundary layer. $T_\infty = 40{,}000$ K, $p_\infty = 5.0$ atmospheres, $M_\infty = 0.65$ (similar solution)

radiative heat flux with those approximate expressions of optically thick and thin media. Fig 17.15 shows the case of the optically thick gas. In comparison with the results of integral expression, it was found that the general trend of the results of the optically thick approximation is the same as that of the integral expression, while the optically thick medium approximation overestimates the effect of thermal radiation.

Fig. 17.16 shows the results of the optically thin approximation (17.76). Since the wall temperature T_w is always much smaller than the local temperature T except in the neighborhood of the wall, Eq. (17.76) shows that the radiative heat flux acts as a heat source in the boundary layer. In our numerical results of Fig. 17.16, the optically thin medium approximation gives an entirely wrong results in comparison with the results of the integral expression. Since the optically thin approximation (17.76) was a popular expression in many literatures, one should be very careful to use such an optically thin approximation when the temperature range is very high.

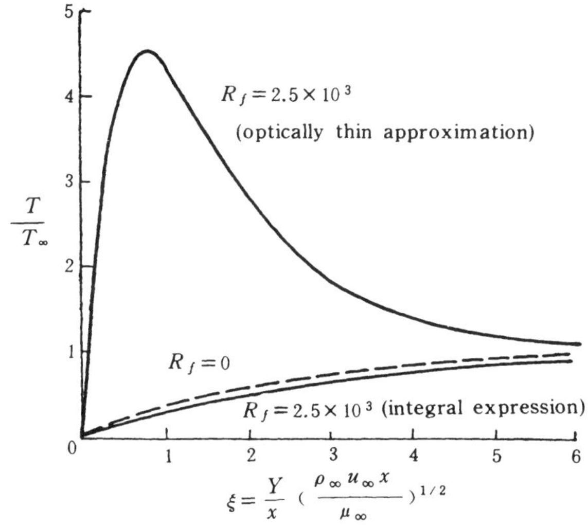

Fig. 17.16 Temperature distributions in a boundary layer flow of an optically thin gas. $T_\infty = 20{,}000$ K, $p_\infty = 1.0$ atmosphere, $M_\infty = 0.9$

When the temperature of the flow field is very high, the thickness of the thermal boundary layer will be very large that the x-wise radiative heat flux will not be negligible. When we include the x-wise radiative heat flux in the energy equation (17.151d), there will be an upstream influence[28]. For instance, when the wall temperature is much smaller than that of the free stream, we have an upstream wake of temperature. At a given x-station upstream, the temperature is lowest at $y=0$ and increases with y. The defect of temperature along the x-axis increases from zero at minus infinity to the value of $T_\infty - T_w$ at $x=0$.

14. Waves of small amplitude in a compressible gas with transport phenomena

In previous chapters, we discuss some wave motions of infinitesimal amplitude for inviscid fluid without transport phenomena. The main results are the sound waves and their modifications by thermal radiation and magnetic field. In this section, we consider the influences of the transport phenomena on these wave motions of infinitesimal amplitude. Many new phenomena occur such as new modes of these wave motions and the modifications on the sound waves for inviscid fluid. We consider in this section the influences of the wave motion of infinitesimal amplitude by (i) viscosity, (ii) heat conductivity, (iii) scalar electrical conductivity, (iv) optical thick medium with both radiation pressure and radiative heat flux and (v) the effects of displacement current and excess electrical charges in the electromagnetic field equations.

Now we assume that originally the plasma or ionized gas is at rest with a pressure p_0, temperature T_0 and density ρ_0 and that is subjected to an externally applied uniform magnetic field $H_0 = \boldsymbol{i}H_x + \boldsymbol{j}H_y + \boldsymbol{k}0$ where $\boldsymbol{i}$, $\boldsymbol{j}$ and $\boldsymbol{k}$ are respectively the x-, y- and z-wise unit vector; and H_x and H_y are constants. There is no electric current, nor excess electric charge, nor externally applied electric field. The ionized gas is perturbed by a small disturbance so that in the resultant disturbed motion we have:

$$\left.\begin{aligned} u=u(x,t),\quad v=v(x,t),\quad w=w(x,t) \\ p=p_0+p'(x,t),\quad T=T_0+T'(x,t) \\ \rho=\rho_0+\rho'(x,t),\quad \boldsymbol{E}=\boldsymbol{E}(x,t) \\ \boldsymbol{H}=\boldsymbol{H}_0+\boldsymbol{h}(x,t),\quad \boldsymbol{J}=\boldsymbol{J}(x,t),\quad \rho_e=\rho_e(x,t) \end{aligned}\right\} \tag{17.163}$$

where u, v, and w are respectively the perturbed x-, y- and z-velocity components; p', ρ' and T' are respectively the perturbed pressure, density and temperature; and $\boldsymbol{E}$, $\boldsymbol{h}$, $\boldsymbol{J}$ and ρ_e are respectively the perturbed electric field strength, magnetic field strength, electric current density and the excess electric charge. We assume that all the perturbed quantities are small so that the second and higher order terms in these quantities are negligible. For simplicity, we assume that the perturbed quantities are functions of one space coordinate x and time t only. Thus we discuss only wave propagation in the direction of x-axis.

Substituting Eq. (17.163) into the fundamental equations of the compressible fluid with transport phenomena, i.e., Eqs. (17.81) without the effects of chemical reaction and diffusion and neglecting the higher order terms of the perturbed quantities, we have the following linear equations for the wave motion of a compressible fluid with transport phenomena without chemical reaction nor diffusion effects as follows:

(i) Maxwell's equations for the electromagnetic fields

$$\frac{\partial \mu_e h_x}{\partial t} = 0 \tag{17.164a}$$

$$\frac{\partial \mu_e h_y}{\partial t} = \frac{\partial E_z}{\partial x} \tag{17.164b}$$

$$\frac{\partial \mu_e h_z}{\partial t} = -\frac{\partial E_y}{\partial x} \tag{17.164c}$$

$$J_x + \frac{\partial \varepsilon E_x}{\partial t} = 0 \tag{17.164d}$$

$$J_y + \frac{\partial \varepsilon E_y}{\partial t} = -\frac{\partial h_z}{\partial x} \tag{17.164e}$$

$$J_z + \frac{\partial \varepsilon E_z}{\partial t} = \frac{\partial h_y}{\partial x} \tag{17.164f}$$

where the subscripts x, y, or z refers to the corresponding component of a vector.

(ii) The generalized Ohm's law with scalar electrical conductivity

$$J_x = \sigma_e(E_x - \mu_e w H_y) \tag{17.164g}$$

$$J_y = \sigma_e(E_y + \mu_e w H_x) \tag{17.164h}$$

$$J_z = \sigma_e[E_z + \mu_e(uH_y - vH_x)] \tag{17.164i}$$

(iii) The equation of conservation of electric charge

$$\frac{\partial \rho_e}{\partial t} + \frac{\partial J_x}{\partial x} = 0 \tag{17.164j}$$

(iv) The equation of state

$$\frac{p'}{p_0} = \frac{\rho'}{\rho_0} + \frac{T'}{T_0} \tag{17.164k}$$

(v) The equation of continuity

$$\frac{\partial \rho'}{\partial t} + \rho_0 \frac{\partial u}{\partial x} = 0 \tag{17.164l}$$

(vi) The equations of motion

$$\rho_0 \frac{\partial u}{\partial t} = -\frac{\partial p'}{\partial x} + 4RR_p\rho_0 \frac{\partial T'}{\partial x} + \frac{4}{3}\mu \frac{\partial^2 u}{\partial x^2} - \mu_e J_x H_y \tag{17.164m}$$

$$\rho_0 \frac{\partial v}{\partial t} = \mu \frac{\partial^2 v}{\partial x^2} + \mu_e J_z H_x \tag{17.164n}$$

$$\rho_0 \frac{\partial w}{\partial t} = \mu \frac{\partial^2 w}{\partial x^2} + \mu_e(J_x H_y - J_y H_x) \tag{17.164o}$$

where R_p is the radiation pressure number based on the undisturbed state variables, i.e., $R_p = a_R T_0^3/(3R\rho_0)$. The coefficient of viscosity corresponds to the temperature T_0.

(vii) The equation of energy

$$C_p^* \rho_0 \frac{\partial T'}{\partial t} - 4RR_p T_o \frac{\partial \rho'}{\partial t} = \frac{\partial p'}{\partial t} + K^* \frac{\partial^2 T'}{\partial x^2} \tag{17.164p}$$

where

$$C_p^* = C_p + 12RR_p = \text{effective specific heat at constant pressure including radiation effect} \tag{17.165}$$

$$K^* = \kappa + 12\, RR_p\rho_o D_R = \kappa + \kappa_R = \text{effective coefficient of heat conductivity including the radiation effect} \tag{17.166}$$

The second term on the left-hand side of Eq. (17.164p) is due to the variation of the energy density of radiation per unit mass E_R/ρ and the work done by the radiation pressure.

Examining the linearized equations (17.164), we see that the quantity h_x is independent of all the other quantities and is given by Eq. (17.164a). Since the divergence of $\boldsymbol{h}$ is zero, h_x must be a constant which may be put equal to zero.

The rest of the perturbed quantities may be divided into two groups: One group consists of the variables w, h_z, J_x, J_y, E_x, E_y and ρ_e which are governed by Eqs. (17.164c), (17.164d), (17.164e), (17.164g), (17.164h), (17.164j) and (17.164o). This group characterizes a transverse wave, since it deals with the velocity component w and the magnetic field h_z which are perpendicular to the externally applied magnetic field $\boldsymbol{H}_o$.

The second group consists of the variables u, v, p', T', ρ', h_y, J_z and E_z which are governed by the rest equations of (17.164). This group characterizes a longitudinal wave in which ordinary sound wave is a special case.

We are looking for periodic solutions in which all the perturbed quantities are proportional to

$$\exp i(\omega t - \lambda x) = \exp\left[-i\lambda(x - Vt)\right] \tag{17.167}$$

where ω is a given angular frequency, λ is the wave number which is equal to $2\pi/(\text{wave length})$, and V is the velocity of wave propagation and $i=(-1)^{1/2}$. Substituting these variables into Eqs. (17.164), we obtain one determinantal equation for each group of the above quantities. The eigen values of these determinantal equations give the different modes of wave propagation through the ionized gas.

(i) Transverse waves

The determinantal equation for the transverse waves gives the following relation:

$$\left(i\omega - \nu_H \frac{\omega^2}{c^2}\right)\left[(i\omega + \nu\lambda^2)\left(i\omega + \nu_H\lambda^2 - \nu_H\frac{\omega^2}{c^2}\right) + V_x^2\left(\lambda^2 - \frac{\omega^2}{c^2}\right)\right]$$
$$- \frac{\omega^2}{c^2} V_y^2\left(i\omega + \nu_H\lambda^2 - \nu_H\frac{\omega^2}{c^2}\right) = 0 \tag{17.168}$$

where $V_x = H_x(\mu_e/\rho_o)^{1/2}$ = x-component of Alfven's wave velocity

$V_y = H_y(\mu_e/\rho_o)^{1/2}$ = y-component of Alfven's wave velocity

$\nu = \mu/\rho_o$; $\nu_H = 1/(\mu_e\sigma_e)$ = magnetic diffusivity

Eq. (17.168) does not contain the radiation parameter. Hence the transverse wave is independent of the radiation field and it is simply the regular transverse wave of magnetogasdynamics, which is independent of the compressibility properties of the gas. Since Eq. (17.168) is a quadratic equation in λ^2, there are two different modes of transverse waves. These transverse waves are the interaction of an electromagnetic wave and a viscous wave through the action of the externally applied magnetic field. These two basic waves can be obtained if the external magnetic field is zero, i.e., $V_x = V_y = 0$. Then Eq. (17.168) gives the following two roots:

(a) A damped electromagnetic wave in a conducting medium

$$\lambda_1^2 = \frac{\omega^2}{c^2} - i\frac{\omega}{\nu_H} \tag{17.169}$$

For non-conducting medium $\nu_H = \infty$, this wave propagates at the speed of light c without damping.

(b) A damped wave in a viscous fluid

$$\lambda_2^2 = -i\frac{\omega}{\nu} \tag{17.170}$$

These two basic waves (17.169) and (17.170) are coupled through the magnetic field H_o. For an ideal plasma, i.e., $\nu = 0$ and $\nu_H = 0$, inviscid and infinitely electrically conducting gas, Eq. (17.168) gives

$$V = \frac{\omega}{\lambda} = V_x\left(1 + \frac{V_x^2 + V_y^2}{c^2}\right)^{-1/2} \tag{17.171}$$

This may be called the modified Alfven's wave. For magnetogasdynamic waves, $(V_x^2 + V_y^2)$ is much smaller than c^2, Eq. (17.171) becomes

$$V = \frac{\omega}{\lambda} = V_x \tag{17.172}$$

This is the well-known Alfven's wave which shows that the disturbance in an inviscid and infinitely electrically conducting gas of density ρ_o propagates as a wave in the direction of H_x with the Alfven's wave speed V_x.

It is interesting to notice that the first bracket of Eq. (17.168) gives a special value of ω, i.e.,

$$\omega = i\frac{c^2}{\nu_H} = i\frac{\sigma_e}{\varepsilon} \tag{17.173}$$

This is the well-known damped electromagnetic wave in ordinary electrodynamics.

(ii) Longitudinal waves

The dispersion relation of the longitudinal waves has the following general form:

$$\left[K^*\left(\frac{1}{\rho_o} + \frac{4}{3}i\omega\frac{\nu}{p_o}\right)\lambda^4 - \left\{\frac{\omega^2 K^*}{p_o} + \frac{4}{3}\frac{\nu\omega^2}{T_o(\gamma-1)}(1 + 12(\gamma-1)R_p)\right.\right.$$

$$-i\omega(C_p+20RR_p+16RR_p^2)\Big\}\lambda^2-\frac{i\omega^3}{T_o(\gamma-1)}(1+12(\gamma-1)R_p)\Big]$$

$$+\left[\left(\lambda^2\nu_H+i\omega-\nu_H\frac{\omega^2}{c^2}\right)(i\omega+\nu\lambda^2)+\nu_x^2\left(\lambda^2-\frac{\omega^2}{c^2}\right)\right]-\left(\lambda^2-\frac{\omega^2}{c^2}\right)(i\omega+\nu\lambda^2)$$

$$+\left[\frac{\omega^2}{T_o(\gamma-1)}(1+12(\gamma-1)R_p)-\frac{i\omega K^*\lambda^2}{p_o}\right]V_y^2=0 \qquad (17.174)$$

Eq. (17.174) is a quadratic equation in λ^2. Hence there are four different modes of longitudinal waves in the present case. These waves are the modified heat wave and modified magnetogasdynamic waves.

The heat wave depends on the effective coefficient of heat conductivity K^*. When K^* vanishes, there will be no heat wave. Even in a nonheat conducting and inviscid gas, the radiative heat transfer will in troduce the heat wave in ordinary non-radiating fluid.

Even though Eq. (17.174) is rather complicated, the meaning of various terms is very clear. The first square bracket characterizes sound waves in a viscous heat-conducting and radiating gas. If there is no transverse magnetic field, i.e., $V_y=0$, these sound waves will not interact with the electromagnetic variables. Hence the first square bracket gives the ordinary sound waves in a viscous, heat-conducting and radiating medium. Since this bracket gives a quadratic equation in λ^2, there are two modes of sound waves: one corresponds to the ordinary sound wave and the other corresponds to the heat wave. There are two types of radiation effects on these sound waves: One is due to the radiation pressure and radiation energy density and the other is due to the radiative heat flux.

In Chapter III, § 9, we have discussed this effect of sound speed due to radiation pressure and radiation energy density, i. e., for inviscid and non-heating conducting fluid without radiative heat flux, the sound speed is C_R of Eq. (3.67) shown in Fig. 3. 8 as a function of R_p and T_o. This result may be obtained by putting $K^*=\nu=0$ and the first square bracket of Eq. (17.174) equal to zero.

If the radiative heat flux is different from zero, K^* will be different from zero. It will introduce two effects on the sound waves: One is to introduce a damping on the above sound wave and modifies its speed and the other is to introduce a heat wave. In order to show the effects of the radiative heat flux, we consider a radiating, inviscid and non-heat conducting gas with a small amount of heat flux of radiation so that K^* is a very small quantity. Now the first bracket of Eq. (17.174), i.e.,

$$\frac{K^*}{\rho_0}\lambda^4 - \left\{\frac{\omega^2 K^*}{p_o} - i\omega(c_p + 20RR_p + 16RR_p^2)\right\}\lambda^2$$

$$-\frac{i\,\omega^3}{T_o(\gamma-1)}\,[1+12(\gamma-1)R_p] = 0 \tag{17.175}$$

gives the following two modes as a first approximation:

(a) Radiation sound wave

$$\lambda_1 = \pm\frac{\omega}{C_R}\,[1 - i\omega f(R_p) D_R] \tag{17.176}$$

(b) Radiation heat wave

$$\lambda_2 = \pm\frac{\omega}{C_R}\left[\frac{g(R_p)}{\omega D_R}\right]^{1/2}(-1+i) \tag{17.177}$$

where

$$\left.\begin{aligned} f(R_p) &= \left(\frac{\gamma}{a_0^2} - \frac{1}{C_R^2}\right)\frac{a_0^2 6(\gamma-1)R_p}{C_R^2\gamma[1+12(\gamma-1)R_p]} \\ g(R_p) &= \frac{C_R^4\gamma[1+12(\gamma-1)R_p]}{24a_0^4(\gamma-1)R_p} \end{aligned}\right\} \tag{17.178}$$

It is evident that the first mode λ_1 of Eq. (17.176) is the radiation sound wave whose speed of propagation is C_R given by Eq. (3.67). The radiative heat flux introduces the damping term in this sound wave which is indicated by the imaginary part of λ_1. The second mode λ_2 of Eq. (17.177) is the heat wave which is propagated at a speed

$$V_2 = \frac{\omega}{\lambda_{2R}} = C_R\left(\frac{\omega D_R}{g}\right)^{1/2} \tag{17.179}$$

For small radiative heat flux, V_2 is very small. V_2 is proportional to C_R and the square root of ω. This heat wave is a damped wave with the damping factor given by the imaginary part of λ_2. The damping factor is inversely proportional to the square root of D_R. As D_R tends to zero, the damping of the sound wave due to the radiative heat flux tends to be zero while the damping of the heat wave tends to be infinity so that the heat wave will disappear. On the other hand for a very large value of D_R, the sound wave will be highly damped and the heat wave will be only slightly damped.

As the radiation pressure number R_p decreases, the radiation sound speed C_R tends to be the value of ordinary sound speed a_o and the damping of the sound wave tends to zero. Thus as R_p tends to zero, the radiation sound speed tends to the ordinary sound speed. As R_p tends to zero, the damping of heat wave tends to be infinity and the heat wave will disappear. As R_p tends to infinity, the damping of the sound wave tends to zero because $f(R_p)$ tends to zero. The sound wave will be undamped at very large value of R_p. As R_p tends to infinity, the damping of the heat wave tends to infinity and the heat wave will again disappear.

Another interesting result is that the damping of the sound wave is proportional to the square of the frequency ω while the damping of the heat wave is proportional to the square root of the frequency ω. Thus at very high frequency, the sound wave will be damped out more than the heat wave.

The second square bracket of Eq. (17.174) is the transverse wave of Eq. (17.168) without the transverse magnetic field, i.e., $V_y=0$ and the frequency is different from that given by Eq. (17.173).

The last term of Eq. (17.174) gives the effect on the longitudinal wave due to the transverse magnetic field. When V_y is different from zero, we have four waves: one modified heat wave and three modified magnetogasdynamic waves. This modified heat wave depends on essentially the effective heat conductivity K^*. In Chapter XV, § 6, we consider the case without effective heat conductivity, i. e., when $K^*=0$, there is no heat wave and we have only three modified magnetogasdynamic waves. If we put $K^*=\nu=\nu_H=0$, Eq. (17.174) reduces to Eq. (15.105). We have already discussed these modified magnetogasdynamic waves, i.e., fast wave, slow wave and Alfven's wave in Chapter XV, § 6. Hence we do not discuss them any more here.

When any one of the transport coefficients is different from zero, i. e., $\nu\neq 0$, $\nu_H\neq 0$ or $K^*\neq 0$, it will introduce damping on the fast and slow waves and will also introduce a new damped wave which will be a modified heat wave if K^* is different from zero and which will be a transverse longitudinal viscous or electromagnetic wave if ν and ν_H or both are different from zero. In order to show this damping effect, we now consider a simple case that $\nu=0$, $\nu_H=0$ and $K^*\neq 0$ but small. Now the damping terms of the waves as well as their speed of propagation depend on the acute angle θ between the planar externally applied magnetic field and the direction of wave propagation. If $\theta=0$, i.e., $V_y=0$ but $V_x\neq 0$, there is no interaction of the radiation gasdynamic wave Eq. (17.175) and the electromagnetic wave. From Eq. (17.174), we see that the radiation gasdynamic wave (17.175) and the magnetogasdynamic transverse wave (17.168) are not coupled. If θ is different from zero, we have three damped waves which are the interaction of the heat wave, fast wave and slow wave. We are of special interest to see the interaction of the heat wave and the magnetogasdynamic wave. To show the essential features of this interaction, we consider the case $\theta=90°$, i.e., $V_y\neq 0$ and $V_x=0$. In this case, the slow wave vanishes and we have only the fast wave. Eq. (17.174) reduces again to a quadratic equation which gives two modes of waves which are the results of the interaction of the heat wave and the fast magnetogasdynamic wave. If we assume K^* or D_R to be small so that the square or higher order terms of K^* may be neglected, the two roots of λ are as follows:

$$\lambda_1 = \pm \frac{\omega}{C_R} \frac{A_2}{A_3} \left[1 - i\,\omega \frac{A_2^2}{A_3^2} f\,(R_p) D_R \right] \tag{17.180a}$$

$$\lambda_2 = \pm \frac{\omega}{C_R} \frac{A_3}{A_1} [g(R_p)/D_R]^{1/2} (-1+i) \tag{17.180b}$$

where

$$A_1^2 = i + \frac{\gamma V_y^2}{a_0^2} \;;\; A_2^2 = 1 + \frac{V_y^2}{c^2} \;;\; A_3^2 = 1 + \frac{V_y^2}{C_R^2}$$

the functions $f\,(R_p)$ and $g(R_p)$ are given in Eqs. (17.178).

When $V_y=0$, Eqs. (17.180a) and (17.180b) reduce respectively to Eqs. (17.176) and (17.177). Hence root λ_1 represents a modified fast magnetogasdynamic wave and the root λ_2 represents a modified heat wave.

The effect of the magnetic field increases the speed of propagation of the radiation sound speed (17.176) by a factor A_3/A_2 and decreases the damping factor by a factor A_2^3/A_3^3. The resultant speed of propagation is the same as that given by Eq. (15.106) with $V_x=0$, i.e.,

$$V_{\text{fast}} = \frac{\omega}{\lambda_{1R}} = (C_R^2 + V_y^2)^{1/2} \left(1 + \frac{V_y^2}{c^2} \right)^{-1/2} \tag{17.181}$$

The variation of the damping terms with ω and D_R for the cases with and without V_y are exactly the same. When D_R tends to zero, the heat wave disappears and only the fast wave survives.

The effects of the magnetic field on the heat wave are that it changes the speed of propagation by a factor A_1/A_3 and the damping term by a factor A_3/A_1. It is interesting to notice that in the factor A_1 the isothermal sound speed occurs instead of the adiabatic spound speed, i.e., $a_T = a_o/(\gamma)^{1/2}$.

As V_y increases from zero to infinity, the speed of propagation for the modified fast wave varies monotonously from the radiation sound C_R to the speed of light c and the corresponding damping factor from $-(\omega^2/C_R) f\,(R_p)\, D_R$ to $-(\omega^2/c)(C_R^2/c^2) f\,(R_p) D_R$. Similarly for the heat wave the speed of propagation varies from $C_R[\omega D_R/g(R_p)]^{1/2}$ to $(C_R^2/a_T)\,[\omega D_R/g(R_p)]^{1/2}$ and the damping factor varies from $-(1/C_R)\,[\omega g(R_p)/D_R]^{1/2}$ to $-(a_T/C_R^2)\,[\omega g\,(R_p)/D_R]^{1/2}$.

15. Shock wave structure with relaxation

In Chapter IV, we consider essentially the Rankine-Hugoniot relations of an ideal gas in which only the translational and rotational internal energies are considered and the shock wave is considered as a surface of discontinuity. In fact, the shock wave is not a surface of discontinuity but a thin finite transition region. In order to study the flow field in this transition region, we should consider the transport phenomena which will be discussed in next section. In this section, we discuss the effects of relaxation of various modes of internal energies only and consider the fluid as an inviscid and non-heat-conducting fluid. At high temperature over 2,000 K, the specific heat of a gas is no longer constant because of the excitation of the vibrational energy, dissociation, and ionization as we have discussed in Chapter II, § 5. Furthermore the time to reach the equilibrium condition for various modes of the internal energies is not the same. Usually, the translational and rotational energies of a gas molecule will reach their equilibrium conditions in a very short time such as a few mean free paths. If we consider only the translational and rotational energies, the thickness of the shock wave is of the order of a few free mean paths. The vibrational energy will reach its equilibrium condition in a longer time and the dissociation will need an even longer time than that for vibrational energy to reach its equilibrium condition. In order to determine these relaxation time effects on the shock wave structure and the shock transition is divided into various subdivisions (Fig. 17.17) as a first approximation. In

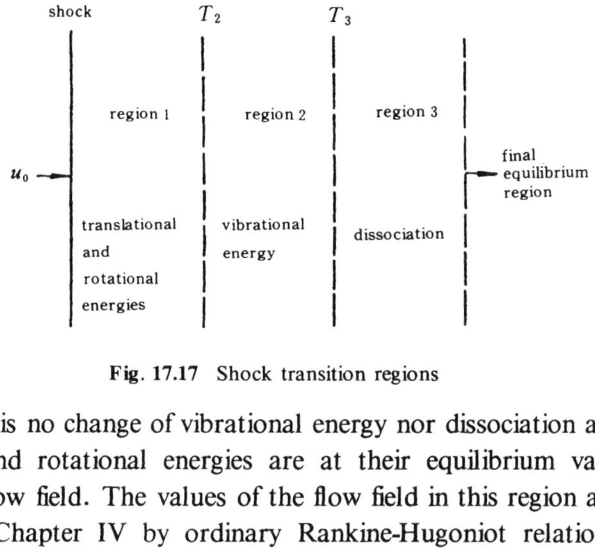

Fig. 17.17 Shock transition regions

region 1, there is no change of vibrational energy nor dissociation and that the translational and rotational energies are at their equilibrium values at all points in the flow field. The values of the flow field in this region are given by the results of Chapter IV by ordinary Rankine-Hugoniot relations. At the end of region 1, the translational temperature of the gas increases from the temperature T_1 to the value of T_2, but the vibrational temperature T_v is still equal to T_1. Hence the local vibrational temperature T_v is not equal to the translational temperature T_2. In region 2, the vibrational temperature T_v of the gas gradually increases from the value T_1 to its final value of equilibrium and at the same time the translational temperature T_t of the gas decreases from its highest value T_2 to its final equilibrium temperature T_e. In the equilibrium condition, we have $T_t = T_r = T_v = T_e = T_3$, where T_3 is the equilibrium temperature at the end of region 2. In region 3, we may assume that translational, rotational, and vibrational temperature of the gas are equal and that dissociation of the gas gradually takes place. At first, the composition of the gas will not be its equilibrium value corresponding to the local translational temperature, but as we go down the stream, the composition of the gas will eventually reach its equilibrium value corresponding to the local translational temperature, i.e., finally all the internal energies have their equilibrium values corresponding to the local temperature. From these results, we see that the Rankine-Hugoniot relations of a shock wave are affected by the modes of various internal energies of the gas.

Of course, in exact theoretical analysis, we do not divide the regions behind the initial shock front arbitrarily, as shown in Fig. 17.17, and sometimes these regions are overlapped. However, such a simple picture gives qualitative results of a shock wave structure with relaxation, i.e., the temperature in the shock wave first increases from T_1 to a high value T_2 and then gradually decreases to its final equilibrium value (Fig. 17.18). Ordinarily, region 1 is controlled by viscosity and heat conductivity, which are proportional to a few mean free paths of the gas as we shall discuss them in §16. Hence the thickness of region 1 is of the order of a few mean free paths of the gas. Region 2 or 3 is governed by the reaction time τ or a relaxation length $L_r = U\tau$. The reaction time τ in the present case may be introduced in the following manner:

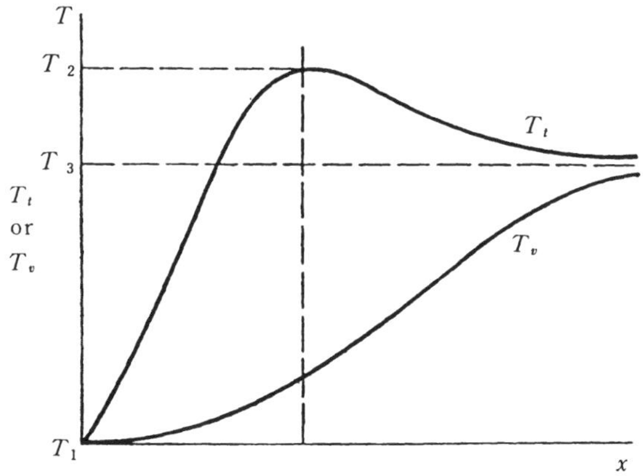

Fig. 17.18 Shock wave structure with vibrational relaxation

For simplicity, we consider the case of shock wave structure where the vibrational relaxation is important and where both the translational and rotational energies are in equilibrium and thier temperature is denoted by a single translational temperature T_t. Since the vibrational energy is not in equilibrium, we may assume that there is a vibrational temperature T_v which is in general different from T_t. Now we assume that there is no dissociation of the gas molecules. Hence the change of internal energy of the gas is

$$dU_m = c_{vo} dT_t + c_{vv} dT_v \tag{17.182}$$

The expression of Eq. (17.182) should be used for the total internal energy of the gas in the energy equation. Since we introduce a new variable, the vibrational temperature T_v, we have to add a reaction equation in our fundamental equations:

$$\frac{DT_v}{Dt} = \frac{1}{\tau}(T_t - T_v) \tag{17.183}$$

where c_{vo} is the specific heat at constant volume of both the translational and rotational energies, which is a constant, and c_{vv} is the specific heat at constant volume of the vibrational energy which is a function temperature T_v. The factor τ is the vibrational relaxation time. For the one-dimensional flow, Eq. (17.183) becomes

$$\frac{dT_v}{dx} = \frac{1}{\tau u}(T_t - T_v) = \frac{1}{L_r}\left(\frac{U}{u}\right)(T_t - T_v) \tag{17.184}$$

where $L_r = \tau U$ and U is a reference velocity. We should solve Eq. (17.184) with other equations of a compressible flow[1]. Some typical results are shown in Fig. 17.18.

The relaxation length L_r is usually much larger than the mean free path of the gas. Thus the length of the thickness of region 2 is much larger than that of region 1. For a first approximation, we may assume that region 1 is a surface of discontinuity, i.e., shock wave in an inviscid flow. We may calculate the temperature T_t and T_v in region 2 by neglecting the effects of viscosity and heat conductivity. The results in region 2 check well with that exact solution including viscosity and heat conductivity. Such an inviscid shock transition region 2 is usually referred to as the partially dispersed shock. If the relaxation length is long enough, it is possible to obtain the solution of both region 1 and region 2 without viscosity and heat conductivity. This case is known as fully dispersed shock. In the fully dispersed shock, the transition region is so wide that the gradients of velocity and temperature are small and the viscosity and heat conductive effects are thus negligible. We shall discuss similar cases in details in §16. It is interesting to notice that our present results are similar to the shock relaxation analysis of two-phase flow of a mixture of a gas and solid particles discussed in Chapter XVI, §5.

16. Shock wave structure including transport phenomena

When we study the shock wave as a transition region from one equilibrium condition to another equilibrium condition, we should include the transport phenomena, i.e., the effects of viscosity, heat conductivity, electrical conductivity, and radiative trasfer. Now we discuss the case of a normal shock in an ionized and radiating gas. We shall choose the coordinate system such that the shock is stationary. (Fig. 17.19).

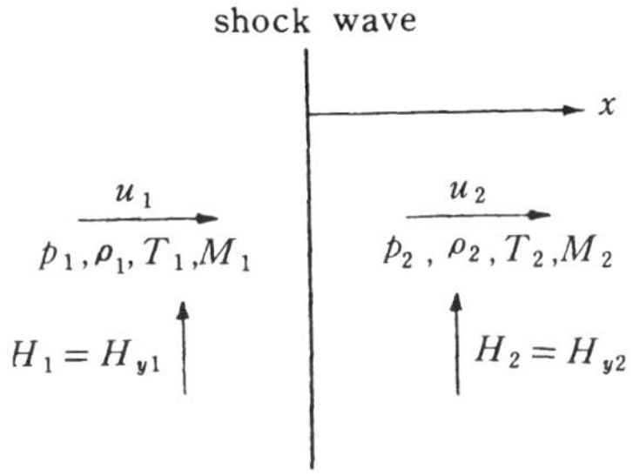

Fig. 17.19 Normal shock in a transverse magnetic field.

In our system, the gas flow is parallel to the *x*-axis and has the *x*-component of velocity u only. We assume that the magnetic field strength H is planar and perpendicular to the velocity field and has the *y*-component only $H=H_y$. Both the velocity and the magnetic field strength are uniform in front of the shock and far behind the shock and there are large variations of all the variables u, H, p, ρ and T in the shock transition region. In general, the shock transition region is very narrow and then for first approximation it may be considered as a surface of discontinuity in many flow problems. The fundamental equations which govern the flow field with shock wave shown in Fig. 17.19 are as follows:

$$\rho u = \text{constant} = m \tag{17.185a}$$

$$mu + p_t + \mu_e \frac{1}{2} H^2 - \frac{4}{3} \mu \frac{du}{dx} = \text{constant} = mC_1 \tag{17.185b}$$

$$mh_R + up_t - \frac{4}{3} \mu u \frac{du}{dx} - K^* \frac{dT}{dx} + E\mu_e H = \text{constant} = mC_2 \tag{17.185c}$$

$$uH - \nu_H \frac{dH}{dx} = \text{constant} = E \tag{17.185d}$$

Here we assume that the gas is optically thick so that Rosseland approximation (17.64) is used for radiative transfer. Furthermore, we use the strictly one-dimensional flow of magnetogasdynamics (Chapter XV, §5) in deriving these fundamental equations. The total pressure p_t is the sum of the gas pressure p and the radiation pressure p_R, H is the *y*-component of the magnetic field; $h_R = \frac{1}{2} u^2 + C_v T + E_R/\rho$ and C_v is the specific heat of the gas at constant volume; $K^* = \kappa + \kappa_R$ is the effective coefficient of heat conductivity with radiation. The rest symbols have their usual meanings.

When we consider the Rankine-Hugoniot relations between the two uniform states separated by the shock transition, we put

$$\frac{du}{dx} = \frac{dH}{dx} = \frac{dT}{dx} = 0 \tag{17.186}$$

into Eqs. (17.185). The resultant equations give the Rankine-Hugoniot relations in radiation magnetogasdynamics which are similar to those given in Chapter IV, § 11 except that we have here additional effect due to magnetic field. For instance, the non-dimensional velocity §2 behind the normal shock is:

$$\xi_2 = \frac{1}{2}\left[\frac{\gamma_e - 1}{\gamma_e + 1} + \frac{2\gamma_e(Pe + h_1^2)}{\gamma_e + 1}\right] + \frac{1}{2}\left\{\left[\frac{\gamma_e - 1}{\gamma_e + 1} + \frac{2\gamma_e(Pe + h_1^2)}{\gamma_e + 1}\right]^2 + 8\frac{2 - \gamma_e}{\gamma_e + 1} h_1^2\right\}^{1/2} \tag{17.187}$$

where $h_1 = H_1/(2mu_1/\mu_e)^{1/2}$ and the other symbols have the same meanings as those given in Chapter IV, § 11 except $Pe = p_t^*$. The non-dimensional variable h_1 shows the effect of the magnetic field H. When $h_1 = 0$, Eq. (17.187) reduces to Eq. (4.57)

In the transition region, we should solve Eqs. (17.185) for the flow field, with the initial conditions 1 of the uniform stream. The general solution of Eqs. (17.185) has not been worked out. However, the results for the case without magnetic field are given as follows. Eqs. (17.185b) and (17.185c) in non-dimensional form are

$$\frac{4}{3}\frac{\mu}{m}\xi\frac{d\xi}{dx} = T^* - T_\infty^*(\xi, R_p) \tag{17.188a}$$

and

$$\frac{K^*}{m\,C_p}\frac{dT^*}{dx} = T^* - T_o^*(\xi, R_p) \tag{17.188b}$$

where $T_\infty^* = -\xi^2 - \frac{1}{3} Q\xi T^{*4} + (1 + T_1^* + \frac{1}{3} QT_1^{*4})\xi$ (17.189a)

$$T_o^* = -(\gamma - 1)\left[-\frac{1}{2}\xi^2 + (1 + T_1^*)\xi + Q\xi T^{*4} + \frac{1}{3} QT_1^{*4}(\gamma - 4) - 1/2 - \frac{\gamma T_1^*}{\gamma - 1}\right] \tag{17.189b}$$

and $Q = a_R u_1^6/(R^4 \rho_1)$

Eliminating x from Eqs. (17.188), we have

$$\frac{dT^*}{d(\frac{1}{2}\xi^2)} = \frac{4}{3}\frac{Pr_R}{\gamma}\frac{T^* - T_o^*}{T^* - T_\infty^*} \tag{17.190}$$

where $Pr_R = \mu C_P/K^*$ = effective Prandtl number with radiation effect.

Eq. (17.190) may be regarded as a differential equation for T^* in terms of ξ. Regardless of whether or not the effective Prandtl number is a constant, if $0 < Pr_R < \infty$, Eq. (17.190) has a singularity whenever the numerator and the denominator of the right-hand side of Eq. (17.190) vanish simultaneously. The conditions $T^* = T_o^*$ and $T^* = T_\infty^*$ lead to the Rankine-Hugoniot relations of the shock wave. We are interested in the integral curve of Eq. (17.190) in the $\xi - T^*$ plane joining the two uniform states $\xi = 1$ and $\xi = \xi_2$ Since $T^* = T_o^*$,

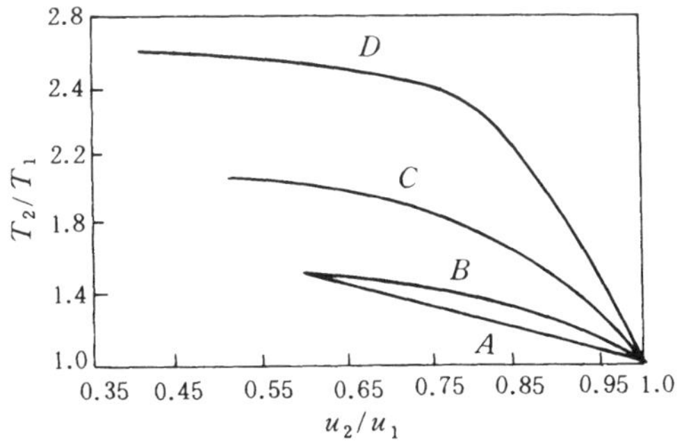

Fig. 17.20 Variation of temperature with velocity in a shock transition: A. $M_1 = 1.5$ without radiation; B. $M_1 = 1.5$ with radiation; C. $M_1 = 2.0$ with radiation; D. $M_1 = 2.5$ with radiation (Fig. 1 of Ref. 29 by H. K. Sen and A. W. Guess, courtesy of American Institute of Physics, Physical Review)

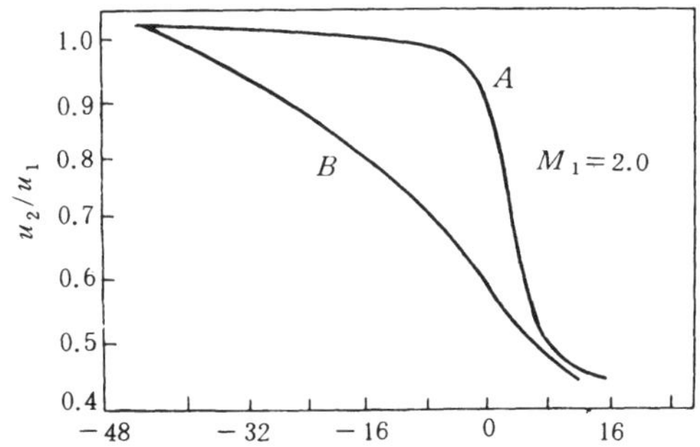

Fig. 17.21 Velocity distributions in a shock transition region: A. without radiation; B. with radiation (Fig. 2 of Ref. 29 by H. K. Sen and A. W. Guess, courtesy of American Institute of Physics, Physical Review)

$dT^*/d(\frac{1}{2}\xi^2) = 0$ and $T^* = T_\infty^*$, $dT^*/d(\frac{1}{2}\xi^2) = \infty$, the integral curve must lie between the two curves $T^* = T_o^*$ and $T^* = T^*$ in the $\xi - T^*$ plane. In general, we have

$$T^* = T^*(\xi) \tag{17.191}$$

from Eq. (17.190) by numerical or graphical integration. The general case for finite R_{pl} has not been calculated. Guess and Sen calculated the case for $R_{pl}=0$ but finite radiative heat transfer and the results are shown in Figs. 17.20 and 17.21. The main result is that the radiation effect broadens the shock transition region. For finite value of R_p, we know from Figs. 4.16 and 4.17, the change of velocity ratio across the shock is small but the ratio of temperatures across the shock reduces. This will further increase the thickness of the shock transition region.

After $T^*=T^*(\xi)$ is obtained, the x-coordinate in the shock transition region may be obtained by simple quadrature, i.e.,

$$\frac{\rho_1 u_1}{\mu_1} x = \frac{4}{3} \int \frac{(\mu/\mu_1)}{T^* - T^*_\infty} d\left(\frac{1}{2}\xi^2\right) \tag{17.192}$$

17. Improvements of boundary layer theory

The boundary-layer theory discussed in previous sections may be considered as a first approximation toward obtaining asymptotic expansions for large Reynolds number. For instance, if for a simple example of two-dimensional flow past a semi-infinite flat plate parallel to the main stream, we substitute

$$\bar{y} = Re^{1/2} y;\ \bar{\psi} = Re^{1/2} \psi \tag{17.193}$$

where $Re = UL/\nu$ in the Navier-Stokes equations, and assume that the derivatives which occur are all $0(1)$ as $Re \to \infty$, we obtain the boundary-layer equations by putting $1/Re = 0$ in the resulting equations. Obviously a method of successive approximation may be used to improve the results of the ordinary boundary-layer theory for moderate Reynolds numbers. For instance, instead of Eq. (17.193), we may write

$$\psi = \varepsilon \psi^{(0)} + \varepsilon^2 \psi^{(1)} + \cdots \tag{17.194}$$

where $\varepsilon = Re^{1/2}$. Substituting Eq. (17.194) into the Navier-Stokes equations and collecting terms of the same powers of ε gives the differential equations for $\psi^{(n)}$. The differential equation for $\psi^{(0)}$ is the boundary-layer equation. However this classical iteration method fails in the present problem because of higher-order singularities in the higher approximations (equations for $\psi^{(n)}$) so that the iteration process diverges. A special technique should be used to render the approximate solution uniformly valid[30].

For ordinary boundary-layer theory, the conditions at infinity for the actual flow will not be satisfied. For instance, in the case of a flat plate in a uniform stream, the v-component of velocity will not be zero at the outer edge of the boundary layer. However, if one uses the proper coordinate system the boundary conditions at infinity may be satisfied[31].

Theoretically, the better results may be obtained if we solve the complete generalized Navier-Stokes equations of § 7, i. e., Eqs. (17.181). Eqs. (17.181) are too difficult to be solved analytically. However, with the help of high speed computer, Eqs. (17.181) may be solved numerically for some cases of simple configurations. We shall discuss some simple numerical solutions of the complete Navier-Stokes equations and compare them with the corresponding numerical solutions of boundary-layer equations.

18. Turbulent transport phenomena

So far we have considered only laminar flow in which the transport phenomena are due to molecular chaotic motion. Actual flow at large Reynolds number are characterized by a peculiar phenomenon known as "turbulence". In such flows, the apparently steady motion of fluids is steady only in so far as the temporal mean values of velocities, pressure and other flow properties are concerned; in reality, velocities, pressure and other flow properties are subject to irregular fluctuations. The essential characteristic of turbulent motion is that turbulent fluctuations are random in nature. Hence the final and logical solution of turbulence problems requires the application of statistical mechanics. It is not possible to discuss even briefly the statistical theory of turbulence in this section. Special treatises such as refernces 20, 32, 33 and 34 should be referred to for the discussion of the statistical theory of turbulence. In this section, we consider briefly the engineering application of turbulent flow which is known as the semi-empirical theory of turbulent flow (cf. Refs. 20, 32, 34, 35 and 36).

It was Osborne Reynolds[20] who first introduced elementary statistical motions into the consideration of turbulence problem. The motion of a fluid in laminar flow is obtained by considering the means of molecular motions according to the kinetic theory of gases. The velocity components, pressure, density, temperature, etc., defined in this sense for a gas in motion approximately satisfy the Navier-Stokes equations. In turbulent motion, however, even if the fluid is regarded as a continuum, we must still deal with turbulent fluctuations superimposed on a mean motion. We may write:

$$u=\overline{u}+u', \quad v=\overline{v}+v'$$
$$p=\overline{p}+p', \quad \rho=\overline{\rho}+\rho' \tag{17.195}$$

etc. where u and v are the instantaneous velocity components in x- and y- direction respectively; the bar refers to the mean values and the prime to the fluctuation values.

Consider a steady two-dimensional flow. If the viscous force can be neglected, the equation of motion in the x-direction is

$$\rho u\frac{\partial u}{\partial x}+\rho v\frac{\partial u}{\partial y}=-\frac{\partial p}{\partial x} \tag{17.196}$$

Substituting Eq. (17.195) into Eq. (17.196) and averaging it, we have, with the approximations of boundary layer,

$$\bar{\rho}\,\bar{u}\frac{\partial \bar{u}}{\partial x}+\bar{\rho}\,\bar{v}\frac{\partial \bar{u}}{\partial y}+\overline{\rho' v'}\frac{\partial \bar{u}}{\partial y}+\bar{\rho}\,\overline{v'\frac{\partial u'}{\partial y}}=-\frac{\partial \bar{p}}{\partial x} \qquad (17.197)$$

Eq. (17.197) may be considered as an equation of motion for the mean flow $\bar{u}$, $\bar{v}$ and $\bar{p}$. The difficult point is that we introduce new unknowns due to the correlation of turbulent fluctuations. We do not have enough equations for all the unknowns. This is known as the closure problem of turbulent flow which has not been solved yet.

For practical applications, we may use some approximate mathematical model for those turbulent fluctuations so that these turbulent fluctuations may be expressed in terms of the values of the mean flow variables. There are many of these mathematical models. Special treatises should be referred to for the detailed discussions of these mathematical models such as references 20, 32, 34, 35 and 36. Here we shall discuss one of the simpliest models and also one of the most successful models for simple practical problems, known as Prandtl's mixing length theory[20], particularly the Taylor's modified vorticity transport theory to find the relation between $\partial u'/\partial y$ and the mean flow variables. Taylor writes:

$$\frac{\partial \bar{\omega}}{\partial y}=\Delta\omega=l\frac{\partial^2 \bar{u}}{\partial y^2} \qquad (17.198)$$

where $\Delta\omega$ is the change of vorticity of the mean flow in y-direction and l is the mixing length. We also write

$$\rho'=l\frac{\partial \bar{\rho}}{\partial y} \qquad (17.199)$$

Strictly speaking, the mixing length for velocity distribution may be different from that for density distribution. For first approximation, we assume that they are equal.

Substituing Eqs. (17.198) and (17.199) into Eq. (17.197) gives

$$\bar{\rho}\bar{u}\frac{\partial \bar{u}}{\partial x}+\overline{\rho v}\frac{\partial \bar{u}}{\partial y}=-\frac{\partial \bar{p}}{\partial x}+\varepsilon\frac{\partial}{\partial y}\left(\bar{\rho}\frac{\partial \bar{u}}{\partial y}\right) \qquad (17.200)$$

where $\varepsilon=-\overline{lv'}$ is the turbulent exchange coefficient. The turbulent exchange coefficient plays the role for the momentum transfer in turbulent flow just as the coefficient of kinematic viscosity of the laminar flow. Since the viscous stresses are much smaller than the turbulent stresses, we may neglect the viscous stresses in most studies of turbulent flow.

Similarly the energy equation for the turbulent flow is

$$\bar{\rho}\bar{u}\frac{\partial \bar{h}}{\partial x}+\overline{\rho v}\frac{\partial \bar{h}}{\partial y}=\bar{u}\frac{\partial \bar{p}}{\partial x}+\frac{\partial}{\partial y}\left(c_p\varepsilon_T\bar{\rho}\frac{\partial \bar{T}}{\partial y}\right)+\varepsilon\bar{\rho}\left(\frac{\partial \bar{u}}{\partial y}\right)^2 \tag{17.201}$$

where $\bar{h}$ is the mean enthalpy and $\varepsilon_T=-\overline{l_T v'}$ is the turbulent thermal exchange coefficient. If we assume that $\varepsilon=\varepsilon_T$ and they are independent of y, for the case of $d\bar{p}/dx=0$, we find that the Crocco's relation holds for turbulent flow too, i.e.,

$$c_p\bar{T}+\frac{1}{2}\bar{u}^2=a\,\bar{u}+b \tag{17.202}$$

One of the difficult problems of the investigation of turbulent flow is that we do not have any theoretical description of the turbulent exchange coefficients. Hence we have to make some semi-empirical assumptions about the variation of ε or the variations of l and v'. In the simple boundary layer flow of nearly parallel flow[20], we may write:

$$l\sim y \tag{17.203}$$

where y is the distance from the wall and

$$v'\sim l\frac{\partial \bar{u}}{\partial y} \tag{17.204}$$

where the proportional constants will be determined experimentally.

For jet mixing problems, with $d\bar{p}/dx=0$, we may assume that ε is a function of x only, x being the distance along the axis of the jet. Prandtl proposed that:

$$\varepsilon=kb(\bar{u}_{\max}-\bar{u}_{\min}) \tag{17.205}$$

where k is a constant and b is the width of the mixing zone. For the mixing of two uniform streams, $\bar{u}_{\max}-\bar{u}_{\min}=$constant, then:

$$\varepsilon=\varepsilon_0(x/L) \tag{17.206}$$

where L is a reference length. In general, we may assume

$$\varepsilon=\varepsilon_0(x/L)^n \tag{17.207}$$

If we neglect the term $\overline{\rho' u'}$ in comparison with $\bar{\rho}\bar{u}$, we may define the stream function ψ for the turbulent flow as

$$\frac{\partial \psi}{\partial y}=\bar{\rho}\bar{u},\quad \frac{\partial \psi}{\partial x}=-\overline{\rho v} \tag{17.208}$$

In terms of x and ψ, Eq. (17.200) becomes

$$\frac{\partial \bar{u}}{\partial x}=\left(\frac{x}{L}\right)^n\frac{\partial}{\partial \psi}\left(\varepsilon_0\bar{\rho}^2\bar{u}\frac{\partial \bar{u}}{\partial \psi}\right) \tag{17.209}$$

We now introduce the new independent variable for

$$\frac{X}{L} = \frac{(x/L)^{n+1}}{n+1} \tag{17.210}$$

In terms of X and ψ, Eq. (17.209) becomes

$$\frac{\partial \bar{u}}{\partial X} = -\left(\varepsilon_0 \bar{\rho}^2 \bar{u} \frac{\partial \bar{u}}{\partial \psi}\right) \tag{17.211}$$

Eq. (17.211) is identical to the corresponding equation for laminar flow with $\varepsilon_0 \bar{\rho}$ for $\bar{\mu}$. Eq. (17.211) is a generalized heat conduction equation.

Because of the development of high speed computer, we may calculate the turbulent flow with more complicated mathematical models[32, 36]. For instance, in the one-equation mathematical model of turbulence, we assume

$$v' \sim \sqrt{k} = \sqrt{u'^2 + v'^2 + w'^2} \tag{17.212}$$

Then we derive the differential equation for the turbulent kinetic energy k which may be solved simultaenously with the Reynolds equations for the turbulent flow, i.e., the fundamental equations of the mean flow variables of turbulent flow. Here we have to make simple assumption for the mixing length.

In the two-equations mathematical model of turbulent flow we derive one equation for k and another equation for l or a variable $z = l^m k^n$ with m and n as constants. We solve these two equations of k and z with the Reynolds equations of the mean variables of turbulent flow. Refs. 32 and 36 should be referred to for the detailed discussions of these mathematical model. In general, we may have the multi-equations mathematical model which has not been used in any practical problems yet.

Similar to the turbulent momentum transfer, we may have the turbulent heat transfer by using the turbulent thermal exchange coefficient

$$\varepsilon_T = -\overline{l_T v'} \tag{17.213}$$

where l_T is the mixing length for the temperature distribution, i.e.,

$$T' = l_T \frac{\partial \bar{T}}{\partial y} \tag{17.214}$$

We also have turbulent mass transfer by using the turbulent diffusion coefficient

$$\varepsilon_C = -\overline{l_c v'} \tag{17.215}$$

where l_c is the mixing length of concentration distance C, i.e.,

$$C' = l_c \frac{\partial \bar{c}}{\partial y} \tag{17.216}$$

In general, we have

$$l_T \sim l_c > l \tag{17.217}$$

But for a first approximation, we may assume that l, l_T and l_c are all equal[20].

19. One-dimensional heat equation

To model in a rudimentary fashion the parabolic boundary-layer equations we study the one-dimensional heat equation

$$u_t = \alpha u_{xx} \tag{17.218}$$

This linear parabolic partial differential equation describes heat conduction or diffusion in the x direction with a conductivity coefficient $\alpha\ (>0)$ in an isotropic medium.

The exact solution of the heat equation for the initial condition

$$u(x, 0) = f(x) \tag{17.219}$$

and boundary conditions

$$u(0, t) = u(1, t) = 0 \tag{17.220}$$

is

$$u(x, t) = \sum_{n=1}^{\infty} A_n e^{-\alpha n^2 \pi^2 t} \sin(n\pi x) \tag{17.221}$$

where

$$A_n = 2\int_0^1 f(x) \sin(n\pi x)\, dx \tag{17.222}$$

Note that in the exact solution (17.221) of the parabolic equation the amplitude of each Fourier component decreases exponentially with time t. In opposition to this, hyperbolic equations are wave-like and do not possess solutions which vary exponentially with time.

The characteristic of the heat equation (14.213) is

$$dt = 0 \text{ or } t = \text{constant} \tag{17.223}$$

The effects of truncation error of the finite-difference method for parabolic equations are the same as those for hyperbolic equations (cf. Chapter XIV, § 17). Let us now examine some important finite-difference algorithms which can be used to solve the heat equation.

A simple explicit scheme is

$$\frac{u_i^{n+1} - u_i^n}{\Delta t} = \alpha \frac{u_{i+1}^n - 2u_i^n + u_{i-1}^n}{\Delta x^2} \tag{17.224}$$

The growth factor for this scheme is

$$G=1-2r(1-\cos\beta) \tag{17.225}$$

where

$$r=\frac{\alpha\Delta t}{\Delta x^2} \tag{17.226}$$

and $\beta=b\Delta x$ (b is the wave number).
$|G|\leqslant 1$ if

$$r\leqslant\frac{1}{2} \tag{17.227}$$

This is the stability condition for the simple explicit scheme.

The modified differential equation for the scheme (17.224) is

$$u_t-\alpha u_{xx}+\left(\frac{\alpha^2\Delta t}{2}-\frac{\alpha\Delta x^2}{12}\right)u_{xxxx}-\left(\frac{\alpha^3\Delta t^2}{3}-\frac{\alpha^2\Delta t\Delta x^2}{12}+\frac{\alpha\Delta x^4}{360}\right)u_{xxxxxx}+\cdots=0 \tag{17.228}$$

The scheme is first-order accurate in time and second-order accurate in space with truncation error of $O(\Delta t,\ \Delta x^2)$. It is noted that when $r<\frac{1}{6}$ the coefficient of u_{xxxx} is negative. According to Eq. (14.189) the truncation error in this case is anti-dissipative, i.e., the truncation error tends to increase the amplitude of the Fourier component in the numerical solution. However, from Eq. (17.227), this scheme is still stable. The reason is that for the heat equation the exact solution decreases exponentially with time.

It is interesting to note that no odd derivative terms appear in the

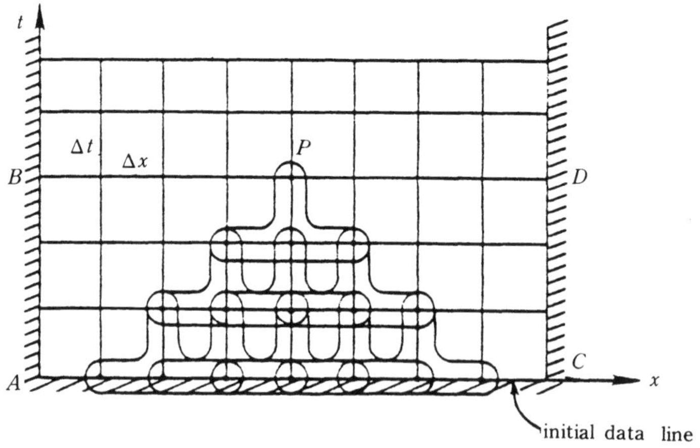

Fig. 17.22 Zone of dependence of simple explicit scheme

truncation error. As a consequence, this scheme has no dispersive error. This fact can also be ascertained by examining the growth factor for this scheme having no imaginary part and hence no phase shift.

Physical insight must be used when the suitability of a finite-difference method is investigated. The present simple explicit scheme matches the solution outward from the initial data line in much the same manner as the explicit schemes for the wave equation. This is illustrated in Fig. 17.22. In this figure we see that the unknown u can be calculated at a point P without any knowledge of the boundary conditions along AB and CD. We know, however, that point P should depend on the boundary conditions along AB and CD since the parabolic heat equation has the characteristic $t=$constant. From this we conclude that the present explicit scheme (with a finite Δt) does not properly model the physical behavior of a parabolic partial differential equation.

By the use of the Crank-Nicolson differencing (17.223), a second-order accurate implicit scheme is obtained

$$u_i^{n+1} = u_i^n + \frac{\alpha \Delta t}{2\Delta x^2}\left(u_{i+1}^n - 2u_i^n + u_{i-1}^n + u_{i+1}^{n+1} - 2u_i^{n+1} + u_{i-1}^{n+1}\right) \tag{17.229}$$

The growth factor is

$$G = \frac{1 - r(1-\cos\beta)}{1 + r(1-\cos\beta)} \tag{17.230}$$

Hence the scheme is unconditionally stable. The modified equation is

$$u_t - \alpha u_{xx} - \frac{\alpha \Delta x^2}{12} u_{xxxx} - \left(\frac{\alpha^3 \Delta t^2}{12} + \frac{\alpha \Delta x^4}{360}\right) u_{xxxxxx} + \cdots = 0 \tag{17.231}$$

The truncation error is of $O(\Delta t^2, \Delta x^2)$. Upon examining Eq. (17.229) it is apparent that a tridiagonal system of linear algebraic equations must be solved at each time level $n+1$.

An explicit finite-difference method which has worked well for boundary layer equations is the DuFort-Frankel scheme (7.52). This scheme is obtained from the well known, unconditionally unstable Richardson method[37]

$$\frac{u_i^{n+1} - u_i^{n-1}}{2\Delta t} = \alpha \frac{u_{i+1}^n - 2u_i^n + u_{i-1}^n}{\Delta x^2} \tag{17.232}$$

by replacing u_i^n with the time-averaged expression $(u_i^{n+1} + u_i^{n-1})/2$. The resulting scheme is Eq. (7.52), i.e.,

$$\frac{u_i^{n+1} - u_i^{n-1}}{2\Delta t} = \alpha \frac{u_{i+1}^n - u_i^{n+1} - u_i^{n-1} + u_{i-1}^n}{\Delta x^2} \tag{17.233}$$

The growth factor is given by

$$G=\frac{2r\cos\beta\pm\sqrt{1-4r^2\sin^2\beta}}{1+2r} \tag{17.234}$$

$|G|\leqslant 1$ for any r. Hence the explicit DuFort-Frankel scheme[38] has the unusual property of being unconditionally stable.

The modified equation for the DuFort-Frankel scheme is

$$u_t-\alpha u_{xx}-\left[\frac{\alpha\Delta x^2}{12}-\alpha^3\left(\frac{\Delta t}{\Delta x}\right)^2\right]u_{xxxx}-\left[\frac{\alpha\Delta x^4}{360}-\frac{\alpha^3\Delta t^2}{3}+2\alpha^5\left(\frac{\Delta t}{\Delta x}\right)^4\right]$$

$$\times u_{xxxxxx}+\cdots=0 \tag{17.235}$$

The truncation error is of $O[\Delta x^2, (\Delta t/\Delta x)^2, \Delta t^2]$. If this scheme is to be consistent, $(\Delta t/\Delta x)^2$ must approach zero as Δt and Δx approach zero. If we let r remain constant as Δt and Δx approach zero, the term $(\Delta t/\Delta x)^2$ becomes formally a first-order term of $O(\Delta t)$. Consequently, the zone of dependence for the explicit DuFort-Frankel scheme does approach the zone of dependence for the parabolic differential equation. Unlike the simple explicit scheme (17.224), the explicit DuFort-Frankel scheme can model the physical behavior of the parabolic partial differential equation in the asymptotic sense.

20. Numerical solutions of steady boundary-layer equations

Consider the steady two-dimensional flow of a compressible fluid over a flat plate. The boundary-layer equations are obtained from Eqs. (17.108) with $\partial/\partial t=0$. The momentum and energy equations are parabolic with x as the marching coordinate. The DuFort-Frankel representation of the x-momentum equation may be written as

$$\frac{\rho_j^n u_j^n(u_j^{n+1}-u_j^{n-1})}{2\Delta x}+\frac{\rho_j^n v_j^n(u_{j+1}^n-u_{j-1}^n)}{2\Delta y}=-\frac{p_j^{n+1}-p_j^{n-1}}{2\Delta x}$$

$$+\frac{1}{\Delta y^2}\left[\mu_{j+\frac{1}{2}}^n(u_{j+1}^n-\bar{u}_j^n)-\mu_{j-\frac{1}{2}}^n(\bar{u}_j^n-u_{j-1}^n)\right] \tag{17.236}$$

where

$$\bar{u}_j^n=(u_j^{n+1}+u_j^{n-1})/2 \tag{17.237}$$

$$\mu_{j+\frac{1}{2}}^n=(\mu_{j+1}^n+\mu_j^n)/2 \tag{17.238}$$

n denotes the marching coordinate x, and j denotes the y coordinate. The formal truncation error for the difference equation (17.236) is $O[\Delta x^2, \Delta y^2, (\Delta x/\Delta y)^2]$. However, the leading term in the truncation error represented by $O(\Delta x/\Delta y)^2$ is actually $(\Delta x/\Delta y)^2(\partial^2u/\partial x^2)$ and $\partial^2u/\partial x^2$ is presumed to be very small for boundary-layer flows. The boundary layer energy equation has a similar form to the momentum equation and thus can be represented by a

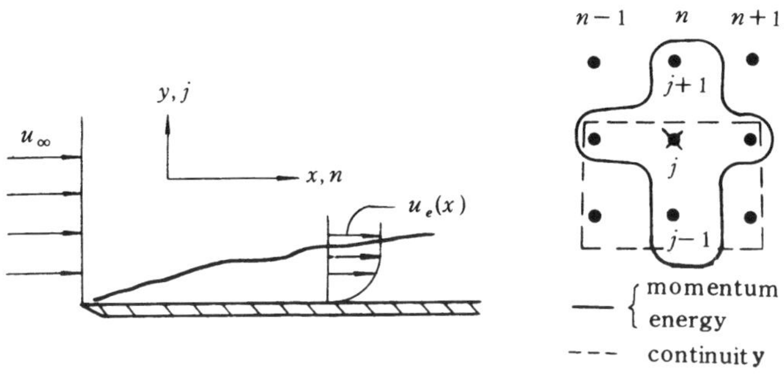

Fig. 17.23 DuFort-Frankel explicit scheme

similar difference equation. A consistent treatment of the continuity equation is given by (Fig. 17.23)

$$\frac{\rho_j^{n+1}u_j^{n+1}-\rho_j^{n-1}u_j^{n-1}+\rho_{j-1}^{n+1}u_{j-1}^{n+1}-\rho_{j-1}^{n-1}u_{j-1}^{n-1}}{4\Delta x}+\frac{\rho_j^{n+1}v_j^{n+1}-\rho_{j-1}^{n+1}v_{j-1}^{n+1}}{\Delta y}=0 \qquad (17.239)$$

with truncation error $O(\Delta x^2, \Delta y^2)$.

For flow over a flat plate, the computation is usually started by assuming the free stream parameters at the leading edge. Since the DuFort-Frankel procedure requires information at two streamwise levels in order to advance the calculation, some other method must be used to obtain a solution for at least one streamwise station before the DuFort-Frankel scheme can be employed. A simple explicit scheme may be used to provide these starting values. A typical calculation would require the solution to the momentum, energy, and continuity equations. The equations can be solved sequentially starting with the momentum equation in an uncoupled manner. The usual procedure is to solve first for the unknown streamwise velocities from the momentum equation starting with the point nearest the wall and working outward to the outer edge of the boundary layer. The outer edge of the boundary layer is located when the velocity from the solution is within a prescribed tolerance of the velocity u_e specified as the outer boundary condition. The energy equation can be solved in a similar manner for the thermal variable. The density at the new station can be evaluated from an equation of state. Finally, the continuity equation is used to obtain the normal component of velocity at the $n+1$ level starting from the point adjacent to the wall and working outward.

For laminar flows in which the boundary-layer approximation is valid, finite difference predictions can easily be made to agree with results of more exact theories to several significant figures. Fig. 17.24[39] compares the velocity profile computed by the DuFort-Frankel finite-difference scheme with ana-

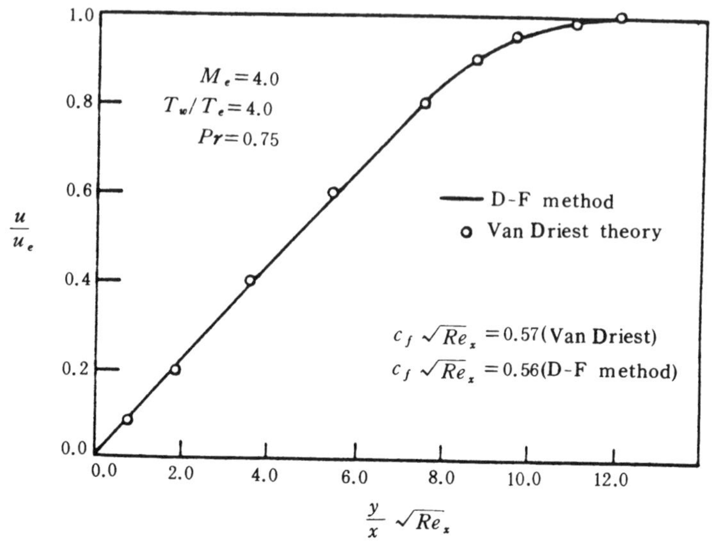

Fig. 17.24 Velocity profile comparisons for a lamilar compressible boundary layer

lytical results for laminar flow at Mach number of 4 and $T_w/T_e=4$. The agreement is quite good and typical of what can be expected for attached laminar boundary-layer flows. The prediction of turbulent boundary-layers is another matter. The issue of turbulence modeling adds complexity and uncertainty to the prediction. However, for turbulent boundary-layer flows in zero or mild pressure gradients a simple algebraic turbulence model can give good predictions over a wide range of Mach numbers.

Let us consider an expansion of the range of usefulness for the boundary-layer approximation. The conventional formulation for boundary-layer problems is to specify the pressure distribution on the outer boundary and is called the direct boundary-layer method. When the pressure gradient is fixed as in the direct methods, the normal component of the velocity v tends toward infinity at the point of separation. This is the well-known singularity of the conventional boundary-layer formulation. This singular behavior is mathematical rather than physical and can be eliminated by an inverse procedure[40]. An inverse method is to replace the outer boundary condition of pressure distribution by the specification of a displacement thickness or wall shear stress which must be satisfied by the solution. The pressure distribution is determined as part of the solution. The inverse formulation is regular at separation and evidence has been accumulating that the boundary-layer equations provide a useful approximation for flows containing small, confined (bubble) separated regions. In support of the validity of the boundary-layer approximation, it is noted that the formulation of a separation bubble normally does not cause the thickness of the viscous region to increase by an order of magnitude; that is, the boundary-layer measure of thickness, $\delta / l \ll 1$ is still met.

With the singularity removed, the remaining problems for calculating separated flows are: 1) the computation of boundary layers with separation, and 2) the coupling of the viscous and inviscid regions.

Finite-difference methods have been developed for computing separated boundary-layer flows. There is a widespread interest in the use of integral boundary-layer procedures in viscous-inviscid interaction analyses due to their rapid computational speed and relative simplicity. Integral method transforms the partial differential equations into one or more ordinary differential equations by integrating out the dependence of one independent variable (usually the normal coordinate) in advance by making assumptions about the general form of the velocity and temperature profiles for both attached and separated flow. A Runge-Kutta method is usually used to solve the ordinary differential equation system.

The interaction calculation proceeds in the following way. First, the velocity distribution over the body is obtained by an inviscid flow solver and the viscous flow is computed up to a point $x=x_1$ upstream of separation by a conventional direct method. Next, an initial displacement thickness distribution $\delta^*(x)$ is chosen over the region $x > x_1$. The initial guess is purely arbitrary but should match the $\delta^*(x)$ of the boundary layer computed by the direct method at $x=x_1$. The boundary-layer solution is then obtained by an inverse procedure using this $\delta^*(x)$ as a boundary condition. An edge velocity distribution $u_{e,\,v}(x)$ is obtained as an output. Now, adding this $\delta^*(x)$ to the ordinate of the body surface and solving the inviscid flow problem, a new distribution for edge (surface) velocity $u_{e,\,inv}(x)$ is obtained. The $u_e(x)$ from the two calculations, boundary layer and inviscid, will not agree until convergence has been achieved. The difference between $u_e(x)$ calculated both ways can be used as a potential to calculate an improved distribution for $\delta^*(x)$.

$$\delta^*_{k+1} = \delta^*_k \left(\frac{u_{e,\,v,\,k}}{u_{e,\,inv,\,k}} \right) \tag{17.240}$$

where k denotes iteration level.

The viscous-inviscid interaction calculation is completed by making successive passes first through the inverse boundary-layer scheme, then through the inviscid flow procedure with δ^* being computed by Eq. (17.240) prior to each boundary-layer calculation. When $|u_{e,\,v} - u_{e,\,inv}|$ is less than a prescribed tolerance, convergence is considered to have been achieved. In some applications of this iteration procedure, over- and under-relaxation of δ^* in Eq. (17.240) have been observed to speed convergence and to avoid divergence respectively.

The inviscid-viscous coupling solution given in Fig. 10.18 is obtained in 350 relaxations of a steady transonic small perturbation potential equation in which an inverse intergal boundary-layer solution is calculated about every 50 relaxations.

Fig. 17.25[41] gives the comparisons of the viscous-inviscid interaction calculations with experimental data, and Reynolds averaged Navier-Stokes calculations for flow over an 18% thick circular-arc airfoil at $M_\infty = 0.7425$, $\alpha = 0^\circ$, and $Re_{\infty,\,c} = 4 \times 10^6$. The inviscid region is computed using an unsteady finite-volume Euler equation code which uses the MacCormack second-order accu-

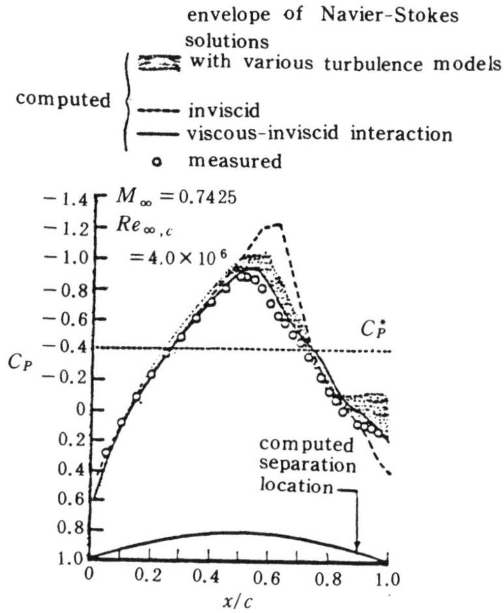

Fig. 17.25 Pressure distributions on 18%-thick circular-arc airfoil

rate, explicit, two-step scheme (Chapter XIV, § 21). The boundary-layer method is an inverse integral technique. For a typical viscous-inviscid interaction solution where convergence is obtained in about 8000 iterations, an inverse boundary-layer solution is obtained about every 400 iterations. The Navier-Stokes solutions are obtained using various turbulence models and in dicated by the shaded area (see §21). The viscous-inviscid interaction calculations gave about the same quality of agreement with experimental pressure data as the Navier-Stokes solutions.

However, for certain viscous flow problems, it is not possible to obtain an accurate solution using the coupled inviscid-viscous flow equations. Such flows involve strong viscous-inviscid interactions in which the classical boundarylayer formulation fails, because of the appearance of strong normal pressure gradients across the boundary layer and large separated flow regions. For these cases, it becomes necessary to solve the complete set of the Navier-Stokes equations.

21. Numerical solutions of unsteady Navier-Stokes equations

The unsteady compressible Navier-Stokes equations are a mixed set of hyperbolic-parabolic equations in time. If the unsteady terms are dropped from these equations, the resulting equations become a mixed set of hyperbolic-elliptic equations which are difficult to solve because of the differences in numerical techniques required for hyperbolic and elliptic type equations. As a consequence, many successful solutions of the compressible Navier-Stokes equations have employed the unsteady form of the equations. The steady-state solution is obtained by marching the solution in time until convergence is achieved.

Consider two-dimensional flow. The unsteady compressible Navier-Stokes equations without body forces and external heat addition are Eqs. (17.96) to (17.99). Written in the conservation or divergence form, they are

$$\frac{\partial U}{\partial t}+\frac{\partial F}{\partial x}+\frac{\partial G}{\partial y}=0 \tag{17.241}$$

where

$$\left.\begin{aligned} U&=\begin{bmatrix}\rho\\ \rho u\\ \rho v\\ \rho e\end{bmatrix},\quad F=\begin{bmatrix}\rho u\\ \rho u^2+p-\tau_{xx}\\ \rho uv-\tau_{xy}\\ (\rho e+p)u-u\tau_{xx}-v\tau_{xy}+q_x\end{bmatrix}\\ G&=\begin{bmatrix}\rho v\\ \rho uv-\tau_{xy}\\ \rho v^2+p-\tau_{yy}\\ (\rho e+p)v-u\tau_{xy}-v\tau_{yy}+q_y\end{bmatrix}\end{aligned}\right\} \tag{17.242}$$

$$\left.\begin{aligned}\tau_{xx}&=\frac{2}{3}\mu\left(2\frac{\partial u}{\partial x}-\frac{\partial v}{\partial y}\right)\\ \tau_{xy}&=\mu\left(\frac{\partial u}{\partial y}+\frac{\partial v}{\partial x}\right)\\ \tau_{yy}&=\frac{2}{3}\mu\left(2\frac{\partial v}{\partial y}-\frac{\partial u}{\partial x}\right)\end{aligned}\right\} \tag{17.243}$$

$$q_x=-\kappa\frac{\partial T}{\partial x},\quad q_y=-\kappa\frac{\partial T}{\partial y} \tag{17.244}$$

In comparison with the unsteady Euler equations which are hyperbolic, the unsteady Navier-Stokes equations have additional terms in the flux vectors, F and G. These terms due to viscosity and heat conduction are linearly expressed into the first derivatives of velocity components and temperature with respect to space variables. Thus the equations become hyperbolic-parabolic. Since both equations have the hyperbolic property, any difference schemes developed for the unsteady Euler equations can be applied to the unsteady Navier-Stokes equations if the viscous stress and heat transfer terms in the flux vectors are represented by the central differences consistent with the scheme.

The evolution of finite-difference methods for the unsteady Navier-Stokes equation has also been from explicit central differencing to implicit upwind differencing. The basic explicit, predictor-corrector, central-difference scheme (cf. Chapter XIV, § 21) has been the most widely used and is given here in the finite-volume representation. Write Eqs. (17.241) in the integral form

$$\frac{\partial}{\partial t}\int_V U\, dV+\int_S \boldsymbol{W}\cdot \boldsymbol{n}\, dS=0 \tag{17.245}$$

where

$$\boldsymbol{W}=F\, \boldsymbol{i}_x+G\, \boldsymbol{i}_y \tag{17.246}$$

and $\boldsymbol{i}_x$, $\boldsymbol{i}_y$ are unit vectors along the Cartesian coordinate axes x, y, and $\boldsymbol{n}$ is a unit outward normal vector to the surface S enclosing an arbitrary volume V. For the two-dimensional case, V is an arbitrary quadrilateral volume element as shown in Fig. 17.26. In Fig. 17.26, (ξ, η) are the local coordinates

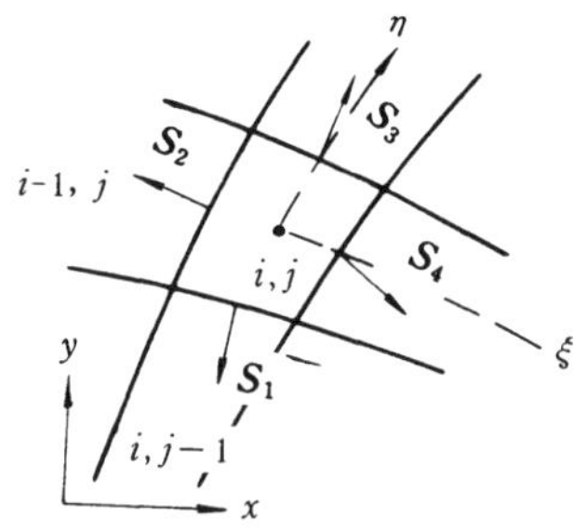

Fig. 17.26 Quadrilateral volume element and nonorthogonal mesh notation

of the nonorthogonal mesh, i and j are the spatial index in the ξ and η directions respectively, and $\boldsymbol{S}_1$, $\boldsymbol{S}_2$, $\boldsymbol{S}_3$, $\boldsymbol{S}_4$ are the four surface area vectors of the quadrilateral volume element $\Delta V_{i,j}$.

The two-dimensional difference scheme can be split [42] into a sequence of one-dimensional operations, L_η and L_ξ. The one-dimensional second-order accurate explicit MacCormack scheme is given by Eqs. (14.248) and

(14.249). Reinterprete these equations as the representations for the integral equations (17.246). The $L_\eta(\Delta t)$ operator is

$$\text{Predictor: } \overline{U_{i,j}^{n+\frac{1}{2}}} = U_{i,j}^{n} - \frac{\Delta t}{\Delta V_{i,j}}\left(\boldsymbol{W}_{i,j}^{n} \cdot \boldsymbol{S}_3 + \boldsymbol{W}_{i,j-1}^{n} \cdot \boldsymbol{S}_1\right) \tag{17.247}$$

$$\text{Corrector: } U_{i,j}^{n+\frac{1}{2}} = \frac{1}{2}\left[U_{i,j}^{n} + \overline{U_{i,j}^{n+\frac{1}{2}}} - \frac{\Delta t}{\Delta V_{i,j}}\left(\overline{\boldsymbol{W}_{i,j+1}^{n+\frac{1}{2}}} \cdot \boldsymbol{S}_3 + \overline{\boldsymbol{W}_{i,j}^{n+\frac{1}{2}}} \cdot \boldsymbol{S}_1\right)\right] \tag{17.248}$$

The $L_\xi(\Delta t)$ operator is

$$\text{Predictor: } \overline{U_{i,j}^{n+1}} = U_{i,j}^{n+\frac{1}{2}} - \frac{\Delta t}{\Delta V_{i,j}}\left(\boldsymbol{W}_{i,j}^{n+\frac{1}{2}} \cdot \boldsymbol{S}_4 + \boldsymbol{W}_{i-1,j}^{n+\frac{1}{2}} \cdot \boldsymbol{S}_2\right) \tag{17.249}$$

$$\text{Corrector: } U_{i,j}^{n+1} = \frac{1}{2}\left[U_{i,j}^{n+\frac{1}{2}} + \overline{U_{i,j}^{n+1}} - \frac{\Delta t}{\Delta V_{i,j}}\left(\overline{\boldsymbol{W}_{i+1,j}^{n+1}} \cdot \boldsymbol{S}_4 + \overline{\boldsymbol{W}_{i,j}^{n+1}} \cdot \boldsymbol{S}_2\right)\right] \tag{17.250}$$

where the superscript n denotes the time step index.

A second-order accurate (formally) scheme can be constructed by applying the L_η and L_ξ operators to $U_{i,j}^{n}$ in the symmetric sequence

$$U_{i,j}^{n+2} = L_\eta(\Delta t)\, L_\xi(\Delta t)\, L_\xi(\Delta t)\, L_\eta(\Delta t)\, U_{i,j}^{n} \tag{17.251}$$

This scheme is actually second-order accurate if the mesh is uniform and orthogonal.

An empirical stability formula for the L_η and L_ξ operators is

$$\Delta t \leqslant \frac{h}{|w| + a + \alpha\mu/\rho h} \tag{17.252}$$

where h is the appropriate mesh spacing, w is the appropriate velocity component, a is the local speed of sound and α is a function of the mesh aspect ratio. It is evident that the stability condition for the one-dimensional operators is less restrictive than that for the multidimensional scheme. Moreover, the splitting makes it possible to advance the solution in each direction with the maximum allowable time step if the sums of the time steps for each of the operators are equal. For example,

$$U_{i,j}^{n+1} = L_\eta\left(\frac{\Delta t}{2}\right) L_\xi(\Delta t)\, L_\eta\left(\frac{\Delta t}{2}\right) U_{i,j}^{n} \tag{17.253}$$

This time-splitting technique is particularly advantageous if the allowable time steps are much different because of differences in the mesh spacings.

Now we evaluate the viscous and heat-conducting derivative with respect to the orthogonal coordinates x and y from the solutions in the nonorthogonal mesh system (ξ, η). Start from the fundamental formulas

$$\frac{\partial \varphi}{\partial \xi} = \frac{\partial \varphi}{\partial x}\cos(\xi, x) + \frac{\partial \varphi}{\partial y}\cos(\xi, y)$$
$$\frac{\partial \varphi}{\partial \eta} = \frac{\partial \varphi}{\partial x}\cos(\eta, x) + \frac{\partial \varphi}{\partial y}\cos(\eta, y) \tag{17.254}$$

where φ is a dummy-dependent variable. In terms of the notation in Fig. 17.26, we may write

$$\frac{\partial \varphi}{\partial \xi} = \frac{\Delta\varphi_\xi}{\Delta\xi}, \quad \frac{\partial \varphi}{\partial \eta} = \frac{\Delta\varphi_\eta}{\Delta\eta}$$
$$\cos(\xi, x) = \frac{\Delta x_\xi}{\Delta x}, \quad \cos(\xi, y) = \frac{\Delta y_\xi}{\Delta y} \tag{17.255}$$
$$\cos(\eta, x) = \frac{\Delta x_\eta}{\Delta x}, \quad \cos(\eta, y) = \frac{\Delta y_\eta}{\Delta y}$$

Substitute Eqs. (17.255) into (17.254) and solve for $\dfrac{\partial \varphi}{\partial x}$ and $\dfrac{\partial \varphi}{\partial y}$.

$$\frac{\partial \varphi}{\partial x} = \frac{\Delta\varphi_\xi \Delta y_\eta - \Delta\varphi_\eta \Delta y_\xi}{\Delta x_\xi \Delta y_\eta - \Delta x_\eta \Delta y_\xi}$$
$$\frac{\partial \varphi}{\partial y} = \frac{\Delta\varphi_\xi \Delta x_\eta - \Delta\varphi_\eta \Delta x_\xi}{\Delta y_\xi \Delta x_\eta - \Delta y_\eta \Delta x_\xi} \tag{17.256}$$

The differences in Eqs. (17.256) should be central and consistent with the operators where they appear. Consider the L_η operator.

$$\Delta\varphi_\xi = \varphi_{i+1,jj} - \varphi_{i-1,jj} \tag{17.257}$$

where

$$jj = \begin{cases} j-1, \text{ on } \boldsymbol{S}_1 \\ j\ \ , \text{ on } \boldsymbol{S}_3 \end{cases} \text{ in predictor of } L_\eta$$
$$ \begin{cases} j\ \ , \text{ on } \boldsymbol{S}_1 \\ j+1, \text{ on } \boldsymbol{S}_3 \end{cases} \text{ in corrector of } L_\eta \tag{17.258}$$

$$\Delta\varphi_\eta = \varphi_{i,\,jj+1} - \varphi_{i,\,jj} \tag{17.259}$$

where

$$jj = \begin{cases} j-1, \text{ on } \boldsymbol{S}_1 \\ j\ \ , \text{ on } \boldsymbol{S}_3 \end{cases} \tag{17.260}$$

For the L_ξ operator, we have

$$\Delta\varphi_\xi = \varphi_{ii+1,j} - \varphi_{ii,\,j} \tag{17.261}$$

where

$$ii = \begin{cases} i-1, \text{ on } \boldsymbol{S}_2 \\ i\ \ , \text{ on } \boldsymbol{S}_4 \end{cases} \tag{17.262}$$

$$\Delta\varphi_\eta = \varphi_{ii,\,j+1} - \varphi_{ii,\,j-1} \tag{17.263}$$

where

$$ii = \begin{cases} \left.\begin{array}{l} i-1, \text{ on } \boldsymbol{S}_2 \\ i \;\;\;\;, \text{ on } \boldsymbol{S}_4 \end{array}\right\} \text{ in predictor of } L_\xi \\ \left.\begin{array}{l} i \;\;\;\;, \text{ on } \boldsymbol{S}_2 \\ i+1, \text{ on } \boldsymbol{S}_4 \end{array}\right\} \text{ in corrector of } L_\xi \end{cases} \tag{17.264}$$

Δx_ξ, Δx_η, Δy_ξ and Δy_η are evaluated similarly. That is, here φ is standing for u, v, T, x and y. This treatment of the viscous and heat-conducting derivatives results in central differences and maintains the second order accuracy for the entire algorithm.

For the high Reynolds number flows of interest, the mesh resolution near and normal to the body surface must be extremely fine to resolve the boundary layer to the sublayer scale. This sublayer scale is nearly proportional to $1/(Re_c)^{1/2}$. As a rule of thumb, a first mesh spacing of $h_{\min} = 2c/3\sqrt{Re_c}$ is adequate.

Computations involving the compressible Navier-Stokes equations sometimes blow up because of numerical oscillations. These oscillations are the result of inadequate mesh refinement in regions of large gradients. In many cases, it is impractical to refine the mesh in these regions, particularly if they are far removed from the region of interest. For such situations, the artificial dissipation terms (14.213) and (14.214) can be added to the right-hand side of Eqs. (17.247) to (17.250).

The Navier-Stokes solutions given in Fig. 17.25[43] are computed by the above method. For many of the viscous flow problems where the boundary-layer equations are not applicable, it is possible to solve a reduced set of equations that fall between the complete Navier-Stokes equations and the boundary-layer equations in terms of complexity. These reduced equations belong to a class of equations which are referred to as the thin-layer or parabolized Navier-Stokes equations. In the thin-layer approximation to the Navier-Stokes equations, the viscous terms containing derivatives in the directions parallel to the body surface are again neglected in the unsteady Navier-Stokes equations, but all other terms in the momentum equations are retained. One of the principal advantages of retaining the terms which are normally neglected in the boundary-layer theory is that separated and reverse flow regions can be computed in a straight-forward manner. Also, flows which contain a large normal pressure gradient can be readily computed. And the required computation time is reduced relative to the complete Navier-Stokes equations.

The mathematical character of the unsteady thin-layer Navier-Stokes equations is identical to that of the original Navier-Stokes equations, and as a result, the two sets of equations are normally solved in the same manner. Another advantage of the thin-layer Navier-Stokes equations is the fact that most of steady equations are a mixed set of hyperbolic-parabolic equations in the streamwise direction. In other words the Navier-Stokes equations are parabolized in the streamwise direction. As a consequence, the steady equations can be solved using a boundary-layer type of marching technique.

Quite recently, upwind methods (cf. Chapter XIV, §22) have matured to the point where general three-dimensional algorithms for viscous flow about arbitrary geometries have been developed. The approximate factorization, finite-volume, implicit, upwind, thin-layer Navier-Stokes code on grids of up to 550,000 points has been applied to the flow about delta wings.

A comparison of the surface pressure coefficient at several streamwise stations is given in Fig. 17.27[44] for a delta wing of aspact ratio $AR=1$ at

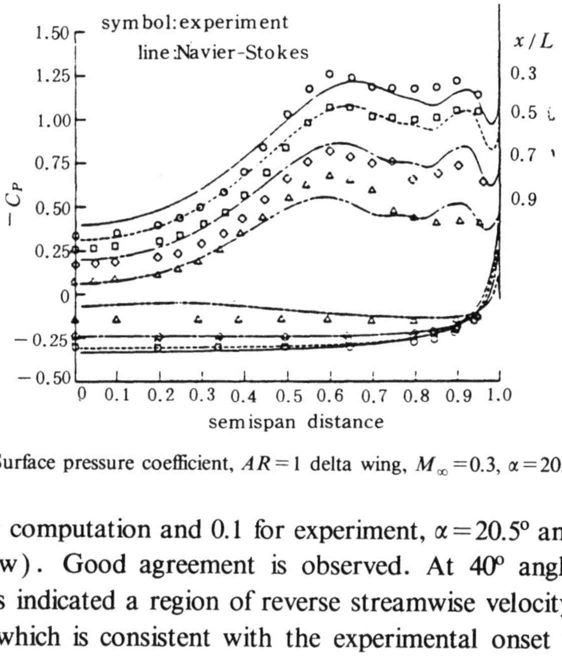

Fig. 17.27 Surface pressure coefficient, $AR=1$ delta wing, $M_\infty=0.3$, $\alpha=20.5°$, $Re=0.95\times10^6$

$M_\infty=0.3$ for computation and 0.1 for experiment, $\alpha=20.5°$ and $Re=0.95\times10^6$ (laminar flow). Good agreement is observed. At 40° angle of attack, the computations indicated a region of reverse streamwise velocity in the primary vortex core which is consistent with the experimental onset of vortex breakdown between 30° and 35° angle of attack.

It is interesting to note that both the periodic aerodynamic oscillations about a rigid airfoil known as the transonic buffet[45] and the one-degree-of-freedom aeroelastic instabiltiy of aileron known as the transonic aileron buzz[46] were demonstrated by the numerical solutions of the unsteady Navier-Stokes equations.

22. Problems

1. What is the Sutherland formula for the coefficient of viscosity of air? Calculate the coefficient of viscosity of air for the temperature range from 0°C

to 1,000°C by (a) Sutherland formula and (b) a power law with $n=0.76$. Eq. (17.5).

2. Calculate the Prandtl number Pr from the values of the coefficient of thermal conductivity κ, the coefficient of viscosity μ and the specific heat c_p at $T=273$ K and 1,500 K for air. Also calculate the Prandtl number for (a) water and (b) mercury at $T=273$ K and $T=350$ K.

3. Derive the Saha relation for singly ionized plasma of monoatomic gas for degree of ionization in terms of the pressure p, temperature T and the ionization potential V_i of the gas.

4. Derive the Planck radiation law, i.e., the relation of the spectral radiation of thermal radiation as a function of frequency ν and the temperature T.

5. Calculate the flow field in a stagnation point region of a two-dimensional blunt body in a hypersonic flow of a simple dissociating gas.

6. Calculate a typical flame structure where $M_1=0.1$ and the temperature across the flame is $T_2/T_1=4.0$.

7. Write down the fundamental equations of one-dimensional steady flow of a viscous, heat-conducting, and simple dissociating gas and discuss the singularities of this system of equations.

8. By von Mises transformation, i.e., using variables x and streamfunction ψ instead of x and y, write down the two-dimensional steady flow equations of the boundary layer flow of a binary mixture with first order chemical reaction.

9. Find the similar solution at the stagnation point of a blunt body in a hypersonic flow in which the binary gas mixture is in non-equilibrium dissociating condition.

10. Calculate the velocity and the temperature distributions; the skin friction and the heat transfer at the wall over a flat plate in a uniform flow of velocity U parallel to the plate and an applied magnetic field perpendicular to the plate but fixed to the fluid. Assume that the magnetic Reynolds number is small, boundary layer approximations are applicable, and the fluid is incompressible. Both the temperature of the free stream T_e and that of the plate T_w are constant.

11. Discuss the turbulent boundary layer flow in magnetohydrodynamics by means of mixing length theory.

12. Calculate the shock structure at shock Mach number of 10 in an optically thick, viscous and heat-conducting gas in which the radiation pressure number is negligibly small but the radiative flux number is not negligible.

13. Calculate the velocity and temperature distributions over an infinite flat plate setting into motion in its own plane impulsively. The fluid is assumed to be viscous, heat conducting and thermally radiating for the cases

(i) $R_p=0$, $R_F\neq 0$, (ii) $R_p\neq 0$ but $R_F=0$. For simplicity, we assume that the transport properties are constant and the gas is optically thick.

14. Repeat the case (i) of the above problem by assuming that the gas is grey and of finite mean free path of radiation.

15. Calculate the heat transfer near the stagnation point of a two-dimensional blunt body in a viscous, heat-conducting and radiating gas in which $R_p=0$ and $R_F\neq 0$.

16. Consider two sets of Cartesian coordinate axes (x, y, z) and (x', y', z'). Find the relation between the stress tensor $\sigma_x{}'$, etc, referred to the (x', y', z') axes and those referred to the (x, y, z) axes. Show that

$$\sigma_x+\sigma_y+\sigma_z=\sigma_x{}'+\sigma_y{}'+\sigma_z{}'=-3p$$

17. Show that the following expressions are invariants of strain with respect to orthogonal coordinate axes.

$$\varepsilon_x+\varepsilon_y+\varepsilon_z=\varepsilon_x{}'+\varepsilon_y{}'+\varepsilon_z{}'=\text{constant}$$

$$\varepsilon_y\varepsilon_z+\varepsilon_z\varepsilon_x+\varepsilon_x\varepsilon_y-\frac{1}{4}(\gamma_{yz}{}^2+\gamma_{zx}{}^2+\gamma_{xy}{}^2)=$$
$$\varepsilon_y{}'\varepsilon_z{}'+\varepsilon_z{}'\varepsilon_x{}'+\varepsilon_x{}'\varepsilon_y{}'-\frac{1}{4}(\gamma_{yz}{}'^2+\gamma_{zx}{}'^2+\gamma_{xy}{}'^2)=\text{constant}$$

18. Derive the fundamental equations of a viscous compressible fluid in cylindrical coordinates.

19. Derive the fundamental equations of a viscous compressible fluid in spherical coordinates.

20. Derive the differential equation for the stream function of a viscous incompressible fluid.

21. Discuss the flow between two parallel plates without pressure gradient but with one plate having a relative velocity U parallel to the other plate, i.e., plane Couette flow, for a viscous compressible fluid.

22. Discuss the simple shearing motion between two rotating cylinders of a viscous compressible fluid.

23. Discuss the flow in the transition region of a flame front if thermal diffusion is neglected.

24. Derive the equations of boundary-layer flow in an axially symmetric flow in which the extent of the radial direction is the order of the thickness of the boundary layer δ and is much smaller than the extent of the region in the axial direction.

25. Derive the boundary-layer equations for the flow over a yawed infinite wing.

26. With the help of Mangler's transformation discuss the boundary-layer flow over a cone.

27. Discuss similarity solutions of the boundary-layer equations for the two-dimensional steady flow of an incompressible fluid. By a similar solution we mean that for any two x-coordinates, say x_1 and x_2, we have

$$\frac{u\left[x_1, \dfrac{y}{g(x_1)}\right]}{u_\infty(x_1)} = \frac{u\left[x_2, \dfrac{y}{g(x_2)}\right]}{u_\infty(x_2)}$$

28. Discuss the similarity solutions for the hypersonic viscous boundary-layer flow in the two-dimensional steady case.

29. Derive the differential equation for $\psi^{(1)}$ of Eq. (17.194) for a two-dimensional steady flow of an incompressible fluid over a flat plate. Discuss the solution of such a differential equation.

30. Derive the boundary layer equation in parabolic coordinates ζ and η where $x+iy=(\zeta+i\eta)^2$ for problem (9). Discuss the solution of this problem. $i=\sqrt{-1}$.

31. Derive the fundamental equations of the jet mixing problem with first order chemical reaction where the pressure is constant over the whole flow field in terms of axial distance x and stream function for a steady two-dimensional flow.

32. Derive the boundary layer equations over a body of revolution where the radius of the body is much larger than the thickness of the boundary layer. Show that by proper transformation known as Mangler transformation, those boundary layer equations over a body of revolution may be reduced into the same form as those of two-dimensional flow.

33. Derive the boundary layer equations over a very slender body such that the radius of the body is of the same order of magnitude as the thickness of the boundary layer.

34. Study the interaction of the shock wave with the boundary layer flow in a hypersonic viscous flow by considering a uniform stream of very high Mach number passing over a flat plate placed along the direction of the main flow.

35. Solve the heat equation with a source term

$$u_t = \alpha u_{xx} + cu \quad (c>0)$$

using the simple explicit finite-difference method. Show that the solution of the difference equation may grow with time and still satisfy the von Neumann stability condition.

36. A solution of the two-dimensional heat equation

$$u_t = \alpha(u_{xx} + u_{yy})$$

is desired using the simple explicit scheme. Find the truncation error and the stability requirement for the method.

37. Solve the two-dimensional heat equation using the Crank-Nicolson scheme. Find the truncation error and the stability requirement if any.

38. Solve the heat equation $u_t = 0.2u_{xx}$ on a digital computer using simple explicit method for the initial condition

$$u(x, 0) = 100 \sin \frac{\pi x}{L}, \quad L = 1$$

and boundary conditions

$$u(0, t) = u(L, t) = 0$$

Compute to $t = 0.5$ using the parameters in Tab. 17.1 if possible and compare graphically with the exact solution.

Table 17.1

Case	Number of grid points	r
1	11	0.25
2	11	0.50
3	16	0.50
4	11	1.00
5	11	2.00

39. Solve the steady-state, two-dimensional heat conduction equation in the unit square, $0 < x < 1$, $0 < y < 1$, by finite-differences using mesh increments $\Delta x = \Delta y = 0.2$ and 1. Compare the center temperatures with the exact solution. Use boundary conditions:

$$T = 0 \text{ at } x = 0 \text{ and } 1$$

$$\frac{\partial T}{\partial y} = 0 \text{ at } y = 0$$

$$T = \sin \pi x \text{ at } y = 1$$

40. The one-dimensional linearized Burgers equation is

$$u_t + a u_x = \mu u_{xx} \quad (a, \mu > 0)$$

Derive stability conditions for the FTCS method (forward-time and centered space difference scheme) applied to this equation.

41. Use the FTCS method of problem 40 to solve the linearized Burgers equation for initial condition

$$u(x, 0) = 0, \qquad 0 \leqslant x \leqslant 1$$

and the boundary conditions

$$u(0, t) = 100, \; u(1, t) = 0$$

on a 21 grid point mesh. Find the steady-state solution for the conditions in

Tab. 17.2 and compare the numerical solutions with the exact solution.

Table 17.2

case	γ	r
1	0.25	0.50
2	1.00	0.50
3	0.40	0.10
4	0.50	0.05

42. Generalize the DuFort-Frankel representations (17.236) and (17.239) to provide second-order accurate difference equations when the mesh increments, Δx and Δy, are not constant.

43. Develop a finite-difference scheme for compressible laminar boundary-layer flow. Solve the energy equation in an uncoupled manner. Use the computer program to predict the skin-friction coefficient and Stanton number distributions for the flow of air over a flat plate at $M_e = 4$ and $T_w/T_e = 2$. Use the Sutherland equation to evaluate the fluid viscosity as a function of temperature. Assume constant values of Pr and C_p, $Pr = 0.75$, $C_p = 1 \times 10^3$ J/Kg • K. Compare the numerical solutions with the analytical results of Van Driest [NACA TN 2597 (1952), heat transfer results can be found in Kays, W. M., et al: Convective Heat and Mass Transfer (1980)].

44. Apply the time-splitting technique to the two-dimensional flow scheme obtained in problem 23 of Chapter XIV. Show that the splitting is second-order accurate if and only if the sequence of operators is symmetric.

45. Write the unsteady Navier-Stokes equations in cylindrical coordinates, apply the explicit MacCormack scheme to these equations and show how all the terms in the r momentum equation are differenced.

References

1. Pai, S. I., *Modern Fluid Mechanics*, Science Press, Beijing, 1981, distributed by Van Nostrand Reinhold Co., New York.
2. Burgers, J. N., *Flow Equations for Composite Gases*, Academic Press, New York, 1969.
3. Chapman, S. and Cowling, T. G., *The Mathematical Theory of Non-Uniform Gases*, Cambridge University Press, 1939.
4. Hirschfelder, J. O., Curtis, C. F. and Bird, R. B., *Molecular Theory of Gases and Liquid*, John Wiley, Inc., N. Y., 1954.
5. Prandtl, L., Ueber Fluessigkeits bewegung bei sehr kleiner Reibung, *Verhanlg. d. III Intern. Math. Kongr.*, Heidelberg, 1904.
6. Pai, S. I., *Viscous Flow Theory. I-Laminar Flow*, D.Van Nostrand Company, Princeton, N. J., 1956.
7. Skelland, A. H., *Non-Newtonian Flow and Heat Transfer*, John Wiley, 1967.
8. Metzner, A. B., Flow of non-Newtonian fluids, Sec. 7 in *Handbook of Fluid Dynamics*, V. L. Streeter (ed), McGraw-Hill, 1961.

9. Penner, S. S., Chemical reactions in flow systems, *AGARD NATO AGARDograph*, No. 7, Butterworths Sci. Pub., London, 1955.
10. Glasstone S., Laidler, K. J. and Eyring, H., *The theory of Rate Processes*, McGraw-Hill Book Co. Inc, New York, 1941.
11. Frost, A. A. and Pearson, R. G., *Kinetics and Mechanism*, John Wiley and Sons, Inc., New York, 1953.
12. Pai, S. I., *Two-Phase Flows*, Viewey, Verlag, Braunschweig, West Germany, 1977.
13. Spitzer, L. jr. and Haerm, R., Transport phenomena in a completely ionized gas, *Phys. Rev.*, **89**, pp. 977 — 981, 1953.
14. Lin, S. C., Resler, E. L. jr. and Kantrowitz, A., Electrical conductivity of highly ionized Argon produced by shock wave, *Jour. Appl. Phys.*, **26**, No. 1, pp. 95 — 109, 1955.
15. Rosa, R. J., Physical principles of MHD power generation. *AVCO research report*, **69**, 1960.
16. Pai, S. I., *Radiation Gasdynamics*, Springer-Verlag, New York Inc., 1966.
17. Krascella, N. L., Tables of composition, opacity and thermodynamic properties of hydrogen at high temperature, *NASA SP* — 3005, 1963.
18. Armstrong, B. H., Sokoloff, J., Nicholis, R. W., Holland, D. H. and Meyerott, R. E., Radiative properties of high temperature air, *Jour. Quant. Spec. Rad. Transfer*, **1**, No. 2, pp. 143 — 162, Pergamon Press, 1961.
19. Scala, S. M. and Sampson, D. H., Heat transfer in hypersonic flow with radiation and chemical reaction in Supersonic Flow, *Chemical Processes and Radiative Transfer*. pp. 319 — 354, Pergamon Press, 1964.
20. Pai, S. I., *Viscous Flow Theory*, II-Turbulent Flow, D. Van Nostrand Company, Inc., Princeton, N. J., 1957.
21. von Mises, R. and Friedrichs, K. O., Fluid Dynamics, *Brown University Notes*. p. 171, 1946.
22. Friedrichs, K. O. and Wasow, W., Singular perturbations of non-linear oscillation. *Dukes Math. Jour.*, **13**, p. 367, 1946.
23. Riddle, F. R. and Fay, A. J., Theory of stagnation point heat transfer in dissociated air, *Jour. Aero Sci*, **25**, No. 2, p. 73, 1958.
24. Lees, L., Ablation in hypersonic flows, *Proc. 7th Anglo-American Conf.*, IAS, New York, pp. 344 — 362, 1959.
25. Scala, S. M. and Sutton, S. W., The two-phase hypersonic laminar boundary layer-A study of surface melting, *Proc.* 1958 *Heat Transfer & Fluid Mech.*, Inst. Stanford University Press, pp. 231 — 240, 1958.
26. Cramer, K. R. and Pai, S. I., *Magnetofluid Dynamics for Engineers and Applied Physicists*, McGraw-Hill Book Co., New York, 1973.
27. Hasimoto, H., Magnetohydrodynamic wakes in a viscous conducting fluid, *Rev. Mod. Phys.*, **32**, No. 2, pp. 860 — 866, 1960.
28. Pai, S. I. and Tsao, C. K., A Uniform flow of a radiating gas over a flat plate, *Proc. 3rd Intern. Conf. of heat transfer*, pp. 129 — 137. ASME, 1966.
29. Sen, H. K. and Guess, A. W., Radiation effects in shock wave structure, *Phys. Rev.*, **108**, No. 3, pp. 560 — 564, Nov. 1, 1957.
30. Goldstein, S., Some developments of boundary layer theory in hydrodynamics, *Lecture Series*, No. 33, Inst. for Fluid Dynamics and Applied Mathematics, University of Maryland, 1956.
31. Kaplun, S., The role of coordinate systems in boundary layer theory, *Zeit. Ang. Math. Phys.*, **5**, pp. 111 — 135, 1954.
32. Launder, B. E. and Spalding, *Mathematical Models of Turbulence*, Academic Press, New York, 1972.
33. Hinze, J. O., *Turbulence*, McGraw-Hill Inc., New York, 1975, Second Edition.
34. Bradshaw, P. Ed., *Turbulence*, Springer Verlag, Berlin and New York, 1978.
35. Compressible Turbulent Boundary Layer, *NASA SP* —216, 1969.

36. Kline, S. J., Morkovin, M. V., Sovran, G., and Cockrell, D. J. (Ed.), Computation of turbulent flow, 1968-*AFOSR-IFP-Stanford Conference Proc.*, **1**, Thermoscience Div., Mech. Eng. Department, Stanford University.
37. Richardson, L. F., The approximate arithmetical solution by finite differences of physical problems involving differential equations, with an application to the stresses in a masonry dam, *Philos. Trans.*, Royal Soc., London, Ser. A, **210**, pp. 307 — 357, 1910.
38. DuFort, E. C. and Frankel, S. P., Stability conditions in the numerical treatment of parabolic differential equations, *Math. Tables and other Aids to Computation*, **7**, pp.135— 152, 1953.
39. Pletcher, R. H., On a calculation method for compressible boundary layers with heat transfer, *AIAA Paper*, No. 71 — 165, 1971.
40. Catherall, D. and Mangler, K. W., The integration of the two-dimensional laminar boundary-layer equations past the point of vanishing skin friction, *Jour. of Fluid Mech.*, **26**, Pt. 1, pp. 163 — 183, 1966.
41. Whitfield, D. L., Swafford, T. W., and Jacocks, J. L., Calculation of turbulent boundary layers with separation, and viscous-inviscid interaction, *AIAA Journal.*, **19**, pp. 1315 — 1322, Oct. 1981.
42. MacCormack, R. W. and Paullay, A. J., Computational efficiency achieved by time splitting of finite difference opperators, *AIAA Paper*, No. 72—154, 1972.
43. Derweit, G. S., Computation of separated transonic turbulent flows, *AIAA Journal*, **14**, pp. 735 — 740, June 1976.
44. Thomas, J. L., Taylor S. L., and Anderson, W. K., Navier-Stokes computations of vortical flows over low aspect wings, *AIAA Paper*, No. 87 — 0207, Jan. 1987.
45. Levy, L. L. Jr., Experimental and computational steady and unsteady transonic flows about a thick airfoil, *AIAA Journal*, **16**, No. 6, pp. 564 — 572, June 1978.
46. Steger, J. L. and Bailey, H. E., Calculation of transonic aileron buzz, *AIAA Journal*, **18**, No. 3, pp. 249 — 255, March 1980.

AUTHOR INDEX

SUBJECT INDEX